Ferrite Control Components

Volume 1
Junction Circulators, YIG Filters and Limiters

MODERN FRONTIERS IN APPLIED SCIENCE

Noise Performance Factors in Communications Systems
by W.W. Mumford and E.H. Scheibe

Laser Applications
by Dr. William V. Smith

Microwave Engineer's Handbook
by Ted Saad

Logarithmic Video Amplifiers
by R.S. Hughes

Parallel Coupled Lines and Directional Couplers
by Dr. Leo Young

Microwave Filters Using Parallel Coupled Lines
by Dr. Leo Young

Microwave Semiconductor Control Components
by Dr. Kenneth E. Mortenson & Jose M. Borrego

Phased-Array Antennas
by Dr. A.A. Oliner & G.H. Knittel

Data Modem Selection and Evaluation Guide
by V.V. Vilips

Infrared to Millimeter Wavelength Detectors
by Dr. Frank R. Arams

Adaptive Electronics
by Dr. Wolfgang W. Gaertner

Gallium Arsenide Microwave Bulk and Transit-Time Devices
by Dr. Lester F. Eastman

Avalanche Transit-Time Devices
by Dr. George I. Haddad

Microwave Filters for Communication Systems
by C.M. Kudsia & M.V. O'Donovan

Radar Detection and Tracking Systems
by Dr. S.A. Hovanessian

Significant Phased Array Papers
by Dr. Robert C. Hansen

Stripline Circuit Design
by Harlan Howe

Spectrum Analyzer Theory and Applications
by Morris Engelson & Fred Telewski

Bubble Domain Memory Devices
by Alan Smith

Advances in Computer Communications
by Wesley W. Chu

Electronic Information Processing
by Dr. William V. Smith

ARTECH HOUSE, INC.
affiliated with Horizon House – Microwave, Inc.
610 Washington Street
Dedham, Massachusetts 02026

Ferrite Control Components

Volume 1
Junction Circulators, YIG Filters and Limiters

Lawrence R. Whicker

ARTECH HOUSE, INC.

Printed and bound in the United States of America.

Standard Book Number: 0-89006-025-8

Library of Congress Catalog Card Number: 73-81247

Ferrite Control Components

TABLE OF CONTENTS

INTRODUCTION

This reprint volume is concerned with three types of ferrite components including junction circulators, YIG filters, and limiters. This reprint volume is the first of a two-volume series on ferrite control components. The other volume (Ferrite Control Components Volume II) is devoted to ferrite phase shifters. The two volumes are largely independent although the first section of this volume, which is composed of historical papers, contains some early information on phase shifters.

The ferrite junction circulator, perhaps, has received more attention in the literature than any other microwave control component. The wide range of system application of circulators as transmit-receive devices, as isolators, and as switches, justifies these efforts. The compactness of junction circulators has contributed to the interest in these components. Readers of this Volume are referred to *Lax and Button* (1) and to the chapter by Bosma (2) in the *ADVANCES IN MICROWAVES* (Edited by L. Young) for discussion of early work in the circulator field. The chapter by Bosma (2), in addition to providing a good technical discussion of circulators provides a list of one hundred and fourteen references.

Although somewhat less technical efforts have been expended in the development of YIG filters and limiters than junction circulators, these components find important systems usage. The YIG filter provides a nonmechanically tuned, narrow band filter which is tunable over a broad frequency range and has found wide application in microwave receivers. Limiters are universally found in receivers to provide receiver protection.

In compiling this reprint volume, selecting specific papers to use has been a formidable job and many good works had to be left out due to space limitations. An effort has been made to include those papers which are considered useful in designing control components. The organization of this volume is as follows:

- General Historical Ferrite Device Papers

- Junction Circulators

  —Historical Papers

  —Theory and Practice for Strip Transmission Line and Waveguide Junction Circulators

—Switching Circulators

—Microstrip Circulators

—Lumped Constant Circulators

—Additional Circulator References

- YIG Filters
- Limiters

The papers reprinted here are grouped together under the four headings listed above. Generally the oldest paper is given first under each heading or sub-heading, followed by the newer papers in chronological order.

ACKNOWLEDGMENTS

I wish to express my appreciation to all the authors and publishers for permission to reprint the papers in this volume. The majority of these papers originated in publications of the IEEE.

REFERENCES

1. B. Lax, K.J. Button, *Microwave Ferrites and Ferrimagnetics,* McGraw-Hill, Inc., New York, 1962

2. H. Bosma, "Junction Circulators," in *Advances in Microwaves* — Vol. 6 (Leo Young, Editor), Academic Press, New York, 1971

HISTORICAL PAPERS

I

In organizing the two reprint volumes, an effort has been made to reprint a few early papers of historic value in each major section. An example of this is found in the next section on Junction Circulators. In this section, a few of the many important, early, general papers are included.

The first paper by C. Lester Hogan (1) provides a good introduction to nonreciprocal components and provides an introduction to these reprint volumes. Another earlier classic paper by Hogan (2) has not been included because of its length. The important application paper by Fox, Miller, and Weiss (3) could not be included also because of its length. The second paper in this section by Sakiotis, Simmons and Chait (4) describes the dual mode reciprocal phaser which is considered in detail in Volume II in this series. Two additional short papers by Chait and Sakiotis are included (5, 6) which treat reciprocal and nonreciprocal effects in ferrite loaded waveguide. A paper by Sharfman (7) is included which describes important phase shifter geometries. A final 1956 paper by Kales (8) described mode propagation in loaded waveguide.

The papers reprinted in this section are intended to be only a sampling of the work in the late 1940, early 1950 era. The reader is referred to Lax and Button (9) for detailed discussions of the early history of various components. A feeling for the state of the art in the mid 1950's may be obtained by reviewing the October 1956 IRE Proceedings which is a special issue on Microwave Ferrites.

References

1. C.L. Hogan, "The Elements of Nonreciprocal Microwave Devices," IRE Proc., Vol. 44, Oct. 1956, pp. 1345-1368.
2. C.L. Hogan, "The Microwave Gyrator," Bell Sys. Tech. J., Vol. 31, Jan. 1952, pp. 1-31.
3. A.G. Fox, S.E. Miller, M.T. Weiss, "Behavior and Applications of Ferrites in the Microwave Region," Bell Sys. Tech. J., Vol. 34, Jan. 1955, pp. 6-103.
4. N.G. Sakiotis, A.J. Simmons, H.N. Chait, "Application of Ferrites to the Microwave Antenna Problem," NRL Report of Progress, Jan. 1952.
5. N.G. Sakotis, H.N. Chait, "Ferrites at Microwaves," Proc. IRE, Vol. 41, Jan. 1953, pp. 87-95.
6. N.G. Sakiotis, H.N. Chait, "Properties of Ferrites in Waveguides," MTT-1, Nov. 1953, pp. 11-16.
7. H. Sharfman, "Three New Ferrite Phase Shifters," IRE Proc., Vol. 44, Oct. 1956, pp. 1456-1459.
8. M.L. Kales, "Topics in Guided-Wave Propagation in Magnetized Ferrites," IRE Proc., Vol. 44, Oct. 1956, pp. 14 3-1410.
9. B. Lax, K.J. Button, *Microwave Ferrites and Ferrimagnetics*, McGraw-Hill, Inc., New York, 1962.

The Elements of Nonreciprocal Microwave Devices*

C. LESTER HOGAN†, SENIOR MEMBER, IRE

Summary—**This paper gives an elementary but nevertheless quite complete introduction to the basic theory of nonreciprocal microwave devices. The concept of reciprocity, itself, is introduced and some of the consequences which result when one violates reciprocity are pointed out. The basic theory of microwave ferrite devices is developed from a classical theory which gives one a strong intuitive feeling for extrapolating the theory to devices not discussed in the paper.**

The analysis is extended to include basic design criterion for the three general types of microwave ferrite devices which exist today. The operation of these devices are compared and advantages and disadvantages of the various devices are pointed out for the benefit of the component engineer.

INTRODUCTION

UNTIL A few years ago, all known linear passive electrical networks obeyed the theorem of reciprocity. Today several different types of passive nonreciprocal microwave networks are in practical use, and it is to be expected that the frequency range in which they are practical will continuously expand until eventually they can be constructed to operate at any frequency which is of interest to the communications engineer. Our recent ability to build these particular nonreciprocal networks now permits the communications engineer to perform system functions which were previously impossible with passive elements. Examples of these particular functions are numerous, and many of the more practical ones will be discussed in further detail in this article. However, to give the reader an idea of the new techniques which have recently been made possible by the advent of these new circuit elements, it is worth while to point out some of the more important system functions which can now be performed by nonreciprocal microwave devices. In the first place, it is now possible to build so-called one-way transmission systems. For instance, one can now construct a microwave system which can propagate a wave in one direction with negligible attenuation, but which almost completely absorbs a wave which is simultaneously traveling in the opposite direction through the system. In addition, it is now possible to perform the operation of duplexing without power loss. In other words, the ability to violate reciprocity allows one, in principle at least, to send and receive simultaneously the same frequency from the same antenna. To be sure, this last operation would require an antenna which is much better matched to the transmission line than is the case for most systems today, but the very fact that this is at least theoretically possible without power loss gives the reader an inkling of the advance which has recently been made possible in microwave techniques by the advent of passive non reciprocal networks.

The theorem of reciprocity can be stated in many equivalent forms, and perhaps the simplest to conceive and understand is the particular form which it takes when it is expressed especially for a two terminal microwave network. Thus if we conceive of the box in Fig. 1

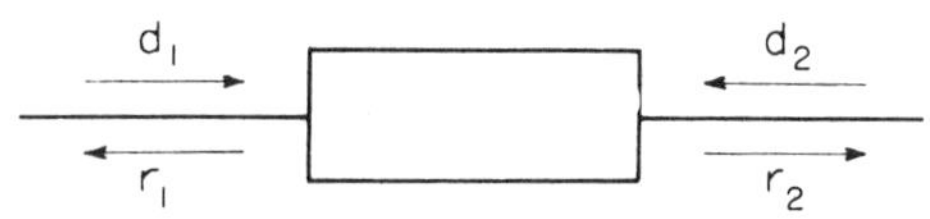

Fig. 1—Scattering matrix representation of a two terminal microwave network.

as representing such a microwave network and let d_1 represent the wave incident on the network from the left, and r_1 the wave propagating out from the network on the same side, and let d_2 and r_2 represent the same quantities on the other side, it is relatively easy to see what the theorem of reciprocity means.

It seems to be intuitively obvious that in a linear passive network, both r_1 and r_2 are made up of two components. Thus r_1 consists of a part which arises from a partial reflection of d_1 at the input terminals of the network and a part consisting of that fraction of d_2 which is transmitted through the network. Thus one can write

$$\begin{aligned} r_1 &= S_{11}d_1 + S_{12}d_2 \\ r_2 &= S_{21}d_1 + S_{22}d_2. \end{aligned} \tag{1}$$

The significance of the above coefficients becomes obvious if one lets d_1 and d_2 be separately equal to zero. Thus when d_2 is zero

$$S_{21} = \left(\frac{r_2}{d_2}\right)_{d_2=0}. \tag{2}$$

In other words, S_{21} gives the amplitude and phase of the wave emerging on the right-hand side of the network when a unit wave (with zero phase angle) is incident on the left-hand side. Thus if the network has a zero insertion loss for propagation from left to right, and if the input and output terminals are chosen to be at such a plane that the phase of r_2 is identical to that of d_1, then the quantity S_{21} would be equal to unity. The significance of all the other coefficients above should be obvious to the reader by similar arguments. Thus it is seen that if the network is perfectly matched when looking in both directions, then S_{11} and S_{22} are zero. These coefficients turn out to be very useful in the analysis and

* Original manuscript received by the IRE, June 13, 1956. This paper was presented at the Symposium on the Microwave Properties and Applications of Ferrites, Harvard Univ., Cambridge, Mass., April 2–4, 1956. The research reported in this document was supported by the U. S. Air Force, Cambridge Res. Center, under contract AF 19(604)-1084.

† Div. of Engrg. and Appl. Phys., Harvard University, Cambridge, Mass.

design of microwave networks and are referred to as the scattering matrix of the network.

Now the theorem of reciprocity demands that

$$S_{12} = S_{21}. \tag{3}$$

In other words, the network has the same transfer characteristic for one direction of propagation as it has for the other. Thus if a two-terminal matched network has a certain insertion loss or causes a certain phase shift for one direction of propagation, then it must have the same insertion loss and phase shift for the other direction of propagation.

For the reader who is more familiar with ordinary four-pole circuit analysis than he is with microwave networks, the reciprocity theorem can be equally well stated. Thus if some general network is cut at any two points and these terminals are considered as the input and output terminals of a four-pole network, then the theorem of reciprocity demands that the open circuit transfer impedance between these two sets of terminals be the same regardless of the direction in which it is measured. In other words (see Fig. 2), if a current is

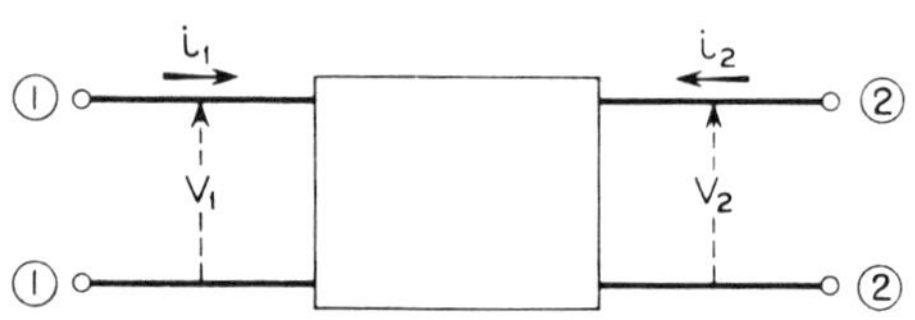

Fig. 2—Impedance matrix representation of a four-pole network.

injected into terminals ①–① and the resulting voltage is measured at terminals ②–②, then the ratio of these two quantities,

$$Z_{21} = \left(\frac{V_2}{i_1}\right)_{i_2=0}$$

is the same as the ratio,

$$Z_{12} = \left(\frac{V_1}{i_2}\right)_{i_1=0}$$

which is determined by measuring voltage which results at terminals ①–① from a current injected at ②–②.

In linear four-pole circuit analysis, it is a familiar fact that the relation between the voltages and currents at the terminals of the four-pole can be written as

$$V = Zi \tag{4}$$

where V, i, and Z are the voltage, current, and impedance matrices, respectively

$$V = \begin{bmatrix} V_1 \\ V_2 \end{bmatrix} \quad i = \begin{bmatrix} i_1 \\ i_2 \end{bmatrix} \quad Z = \begin{bmatrix} Z_{11} & Z_{12} \\ Z_{21} & Z_{22} \end{bmatrix}. \tag{5}$$

Thus (4) can be written in the more familiar form

$$\begin{aligned} V_1 &= Z_{11}i_1 + Z_{12}i_2 \\ V_2 &= Z_{21}i_1 + Z_{22}i_2. \end{aligned} \tag{1a}$$

The theorem of reciprocity can be simply stated for a four-pole by stating the condition

$$Z_{12} = Z_{21}. \tag{6}$$

The theorem of reciprocity has, in the past, been considered so universally valid that present-day textbooks still make the statement that if the condition stated in (6) is valid, the four-pole is passive, and if the condition does not hold for a particular network, the network must of necessity be an active network.

Actually, the theorem of reciprocity is as universally valid in mechanical, acoustical, or optical systems as it is in electrical systems, but it is known that passive systems can be built in all these fields which violate the theorem of reciprocity. These systems are so unique, though, that they have obtained special attention in the technical literature for over fifty years. Thus it has been known for some time that a mechanical system which contains a gyroscopic coupler does not obey the theorem of reciprocity,[1] and as early as the turn of the century, scientists were worried about the apparent lack of reciprocity which arose in the operation of certain electromechanical transducers.[2]

Probably one of the first passive nonreciprocal systems proposed in the literature was an optical one-way transmission system proposed by Lord Rayleigh.[3] Lord Rayleigh's "one-way" system made use of the Faraday rotation of the plane of polarization and consisted of two polarizing Nicol prisms (oriented so that their planes of acceptance made an angle of 45° with each other), with the material causing the Faraday rotation placed between them. Thus, light which was passed by the first crystal and whose plane of polarization was rotated 45° would be passed by the second crystal also. But, in the reverse direction, the rotation would be in such a sense that light which was admitted to the system by the second crystal would not be passed by the first. This device is shown in Fig. 3.

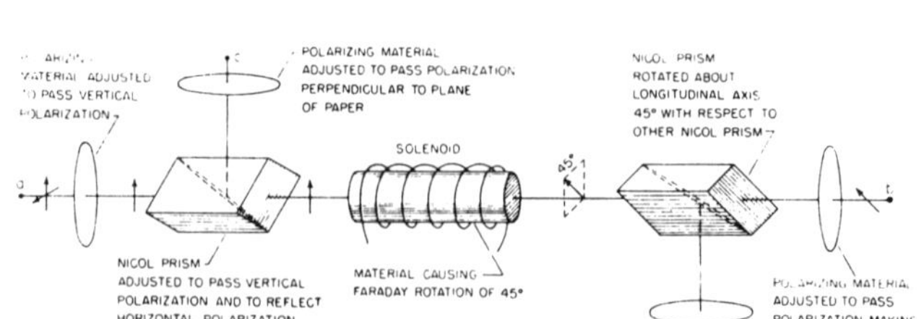

Fig. 3—Lord Rayleigh's optical one-way transmission system.

[1] A Bloch, "A new approach to the dynamics of systems with gyroscopic coupling terms," *Phil. Mag.*, Series 7, vol. 35, p. 315; 1944.

[2] H. Poincare, "Étude du récepteur téléphonique," *L'Éclairage Electrique*, vol. 50., p. 221; 1907.

For an excellent analysis of this subject, see Ph. LeCorbeiller, "Origine des termes gyroscopiques," *Annales des Postes, Télégraphes et Telephones*, vol. 18, pp. 1–22; 1929.

[3] Lord Rayleigh, "On the magnetic rotation of light and the second law of thermodynamics," *Nature*, vol. 64, p. 577; 1901.

In 1946, McMillan[4] published an interesting paper in which he showed that an antireciprocal four-pole electrical network could be obtained by properly coupling electromechanical transducers. This probably constituted the first practical proposal for realizing a passive electrical system which violated the reciprocity theorem. The main results of McMillan's paper can be demonstrated from the following rather simple analysis.

A linear electromechanical transducer can be treated as a four terminal network with voltages and currents applied at one end, and forces and velocities at the other end. In a steady state condition these quantities can be related to each other by

$$\begin{aligned} V &= ai + bv \\ F &= ci + dv \end{aligned} \tag{7}$$

where

V = voltage, F = force, v = velocity, and i = current.

For electrostatic transducers, such as a crystal or condenser transducer, $b=c$ in the above equations. For electromagnetic transducers it can be easily shown[5] that $b=-c$ in the above equations. Hence, it is obvious that electrostatic transducers obey reciprocity, and electromagnetic transducers violate reciprocity, provided that the correlation is made between electrical and mechanical quantities

Force ~ Voltage

velocity ~ current.

It should be pointed out, however, that this is not the only correlation that can be made. It is equally possible to make the correlation

Force ~ current

velocity ~ Voltage.

With this analogy, it is more enlightening to put (7) in the following form

$$\begin{aligned} V &= \frac{ad-bc}{d} i + \frac{b}{d} F \\ v &= \frac{-c}{d} i + \frac{1}{d} F. \end{aligned} \tag{8}$$

It is seen that with this analogy one is led to the opposite conclusion, namely, that the electromagnetic transducer is reciprocal and the electrostatic transducer is antireciprocal. Be that as it may, in any one system containing both kinds of transducers, one must use a consistent analogy, and as McMillan showed, it is possible to build an electrical network which is antireciprocal by mechanically coupling an electromagnetic and an electrostatic transducer. If these are coupled as shown in Fig. 4, then the equation relating the voltages and currents at the input and output of this combination can be written as

$$V = Zi$$

where

$$Z = \begin{vmatrix} \dfrac{a_1a_2 + a_1d_1 - b_1c_1}{a_2+d_1} & \dfrac{b_1b_2}{a_2+d_1} \\ \dfrac{c_1c_2}{a_2+d_1} & \dfrac{d_1d_2 + a_2d_2 - b_2c_2}{a_2+d_1} \end{vmatrix}. \tag{9}$$

If transducer 1) is electrostatic, $b_1=c_1$, and if transducer 2) is electromagnetic, $b_2=-c_2$, and the matrix expressed in (9) becomes

$$Z = \begin{vmatrix} \dfrac{a_1a_2 + a_1d_1 - b_1^2}{a_2+d_1} & \dfrac{b_1b_2}{a_2+d_1} \\ \dfrac{-b_1b_2}{a_2+d_1} & \dfrac{d_1d_2 + a_2d_2 + b_2^2}{a_2+d_1} \end{vmatrix}. \tag{10}$$

Thus the network in Fig. 3 is seen to be antireciprocal.

McMillan also showed that a one-way transmission system could be built by using any such antireciprocal network. For instance, if the network shown in Fig. 4 is connected to an impedance as shown in Fig. 5, and

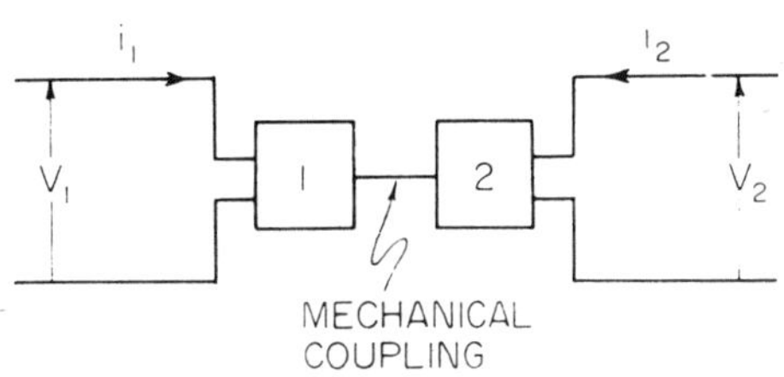

Fig. 4—Gyrator constructed from coupled electromagnetic and electrostatic transducers.

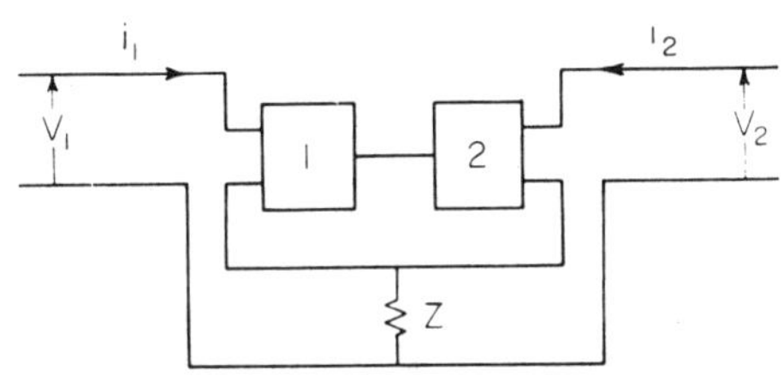

Fig. 5—Network which transmits signals in one direction only.

if we denote the matrix given in (10) by the usual notation, *i.e.*,

$$Z_{11} = \frac{a_1a_2 + a_1d_1 - b_1^2}{a_2+d_1} \tag{11}$$

$$Z_{12} = \frac{b_1b_2}{a_2+d_1} \tag{12}$$

$$Z_{21} = -Z_{12} \tag{13}$$

[4] E. M. McMillan, "Violation of the reciprocity theorem in linear passive electromechanical systems," *J. Acoust. Soc. Amer.*, vol. 18, pp. 344; 1946.

[5] W. P. Mason, "Electromechanical Transducers and Wave Filters," D. Van Nostrand Co., Inc., New York, N. Y.

Also, E. V. Hunt, "Electroacoustics," Harvard University Press, Cambridge, Mass.; 1954.

5

$$Z_{22} = \frac{d_1 d_2 + a_2 d_2 + b_2^2}{a_2 + d_1} \quad (14)$$

then the equations of the network in Fig. 5 are

$$\begin{aligned} V_1 &= (Z_{11} + Z)_{i_1} + (Z_{12} + Z)_{i_2} \\ V_2 &= (Z - Z_{12})_{i_1} + (Z_{22} + Z)_{i_2}. \end{aligned} \quad (15)$$

If the impedance Z is taken equal to Z_{12}, then (15) becomes

$$\begin{aligned} V_1 &= (Z_{11} + Z_{12}) i_1 + 2Z_{12} i_2 \\ V_2 &= (Z_{22} + Z_{12}) i_2. \end{aligned} \quad (16)$$

Thus currents or voltages applied to the first set of terminals produce no effect at the second set, while signals are transmitted in the opposite direction. This system violates reciprocity in an extreme way, even though it is made up of linear passive elements.

There are several other methods of realizing one-way transmission systems from any antireciprocal device such as the coupled transducers. One of these, which is of interest because of its close analogy to some microwave devices, is shown in Fig. 6, where:

$$Z = \frac{Z_{11} Z_{22} + Z_{12}^2}{Z_{12}}. \quad (17)$$

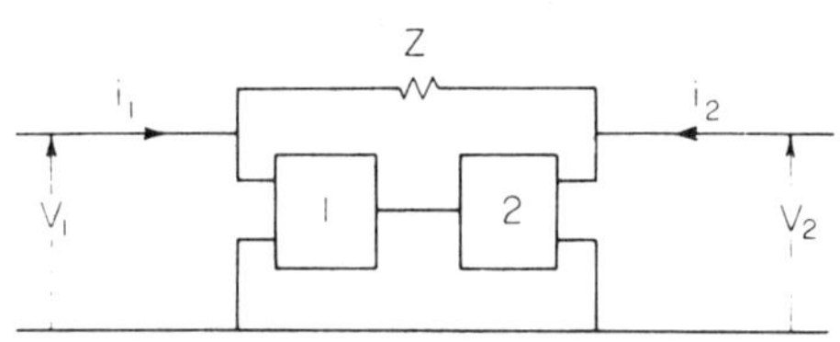

Fig. 6—One-way transmission network with properties similar to network shown in Fig. 5.

The first comprehensive treatment of antireciprocal four-poles was given in 1948 by Tellegen,[6] and he coined the word gyrator to describe such networks. Tellegen restricted his analysis to a so-called ideal gyrator in which the impedance matrix had the following form

$$Z_{11} = Z_{22} = 0 \quad (17a)$$

$$Z_{12} = -Z_{21} = -R. \quad (17b)$$

Now it is easily shown that for such a gyrator to be nondissipative, the transfer impedances must be real as Tellegen has shown them; however, it is not necessary that the diagonal terms of the impedance matrix be zero (*i.e.*, Z_{11} and Z_{22}). In addition, the restriction imposed by (17a) is also not necessary in order to realize two of the most important properties of the gyrator (*i.e.*, construction of one-way transmission systems and phase inversion of signal polarity for one direction of transmission). As a result of the restriction placed on the so-called ideal gyrator by (17a), Tellegen's gyrator has the additional interesting property of impedance inversion. Thus, if Tellegen's gyrator is terminated by an impedance Z_L, the input impedance of the gyrator is

$$Z_{in} = \frac{V_1}{i_1} = \frac{R^2}{Z_L}. \quad (18)$$

This property of impedance inversion is not, however, a distinguishing property of an antireciprocal network; for if we make the same requirement that in a nondissipative reciprocal four-pole, $Z_{11} = Z_{22} = 0$, we find that the input impedance of the terminated network is

$$Z_{in} = \frac{V_1}{i_1} = \frac{X_{12}^2}{Z_L}. \quad (19)$$

Such a reciprocal impedance inverting network can be easily realized and has been used by Doherty[7] in his high efficiency amplifier.

Much has been written about the conditions under which it might be possible to build an antireciprocal element such as the gyrator. Onsager[8] has recently shown that in order for a *completely* electrical passive system to violate reciprocity, a nonchanging magnetic field must be present. Only two types of practical gyrators have been built to date. Needless to say, they both satisfy the requirement of having a dc biasing magnetic field acting on them. One gyrator, proposed by Goldstein and Lampert,[9] arises from the behavior of electric charges moving in a magnetic field, and the other, proposed by the author, arises from the behavior of magnetic dipoles in a magnetic field. The basic operation of both types can be understood from a rather simple physical or mathematical analysis, but since the most useful devices today result from the use of magnetic materials (ferrites) in the presence of a dc magnetic field, it is this type which will be discussed in detail

Basic Theory of Operation of Microwave Ferrite Devices

Modern physical analysis shows that the ferromagnetic property of matter arises from the spin property of the electron. Since the effective quantum number of concern in the ferromagnetic case is extremely large, most of the important aspects of ferromagnetism can be understood from a rather simple classical theory. In this classical theory, the magnetic moment of the electron arises from its spinning motion and hence the electron can be regarded as a spinning magnetic top. Thus every electron has an invariable angular momentum due to spin which is given by

$$|\vec{J}| = 1/2 \frac{h}{2\pi}.$$

[6] B. D. H. Tellegen, "The gyrator, a new electric network element," *Philips Res. Reps.* vol. 3, p. 81; 1948.

[7] W. H. Doherty, "A new high efficiency power amplifier for modulated waves," Proc. IRE, vol. 24, pp. 1163–1182; September, 1936.

[8] H. B. G. Casimir, "On Onsager's principle of microscopic reversibility," *Rev. Mod. Phys.*, vol. 17, pp. 343; 1945.

[9] L. Goldstein and M. A. Lampert, "New linear passive nonreciprocal microwave circuit component," Proc. IRE, vol. 41, p. 295; February, 1953.

$\vec{J}$ = Angular momentum of electron $\frac{(\text{gm cm}^2)}{\text{sec}}$.

h = Planck's constant (erg–sec).

The magnetic moment which arises due to this rotation is

$$\vec{\mu_\beta} = g\frac{e}{2mc}\vec{J} = \gamma\vec{J}.$$

μ_β = magnetic moment of electron (emu).
γ = gyromagnetic ratio of electron.
e = charge on electron (esu).
m = mass of electron (gm).
g = Landé g factor for electrons of concern (approximately 2 for ferromagnetic electrons in ferrites).

If an electron is placed in a magnetic field which is not parallel to its magnetic moment as indicated in Fig. 7,

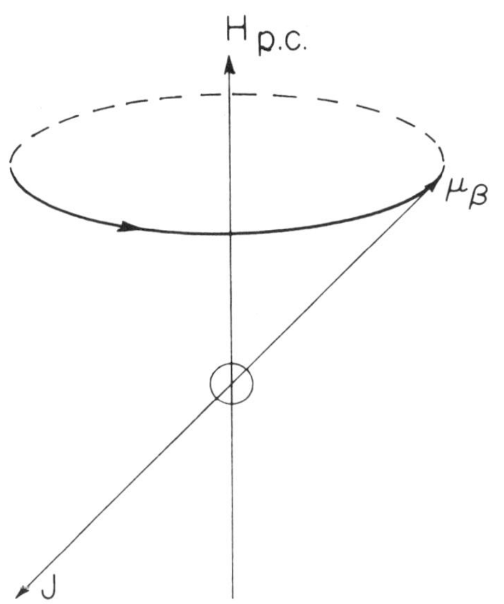

Fig. 7—Classical representation of electron precessing about an applied magnetic field (Larmour precession).

the torque exerted on the electron by the magnetic field causes the electron to precess about the field with an angular velocity given by

$$\omega_0 = \gamma H_{dc}. \tag{20}$$

This precessional motion is highly damped in ferromagnetic solids so that the magnetic moment vector gradually spirals in until it comes into alignment with the magnetic field. If it were not for this damping, it would be impossible to magnetize a ferromagnetic material, since the electrons would just precess indefinitely around the applied field without ever coming into alignment with the field. In most ferromagnetic materials, the effective damping of this precessional motion is large enough so that it requires only about one one-hundredth of a microsecond for the precessional motion to become effectively damped out, and hence this represents a theoretical lower limit to the time required to magnetize a ferromagnetic body.

Now if one makes a simple analysis of an electron placed in a dc magnetic field acting along the Z axis and an alternating magnetic field acting in the XY plane, a rather unusual property of ferromagnetic materials which arises from the gyroscopic property of the electron can be readily understood. To do this the linearly polarized alternating magnetic field is decomposed into two counter-rotating circularly polarized components in the XY plane, and the circular component which is rotating in the direction which the electron precesses about the dc field is called a positive circular component, (h_+), and the component rotating in the opposite direction is labeled the negative circular component. Now it is well known that the torque exerted on the electron by the positive circular component is given by

$$\vec{\tau_+} = (\vec{\mu_\beta} \times \vec{h_+}) \tag{21}$$

and hence the torque is a vector which is perpendicular to the plane of μ_β and h_+. In Fig. 8 these vectors are

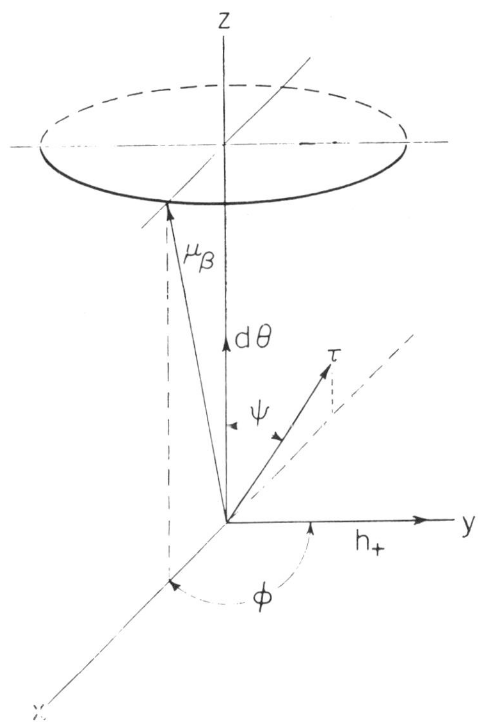

Fig. 8—Classical representation of electron motion when subjected to a dc magnetic field along z axis and a positive circularly polarized magnetic field in xy plane.

shown at a given instant under certain assumed conditions concerning their orientation. The magnetic moment vector is shown at the particular instant to be lying in the XZ plane, and the rotating magnetic field vector is shown lying along the positive Y axis. Under these conditions, the torque vector given by (21) would also be in the XZ plane, perpendicular to the plane formed by μ_β and h_+. An exact analysis of this problem shows that these relative orientations are correct provided the frequency of the alternating field corresponds to the natural precessional frequency of the electron about the dc field H_Z as given by (20). Now during an infinitesimal rotation of all these vectors about the Z axis, work is done on the electron by the positive circular component of the alternating field. This work is given by

$$dW = \vec{\tau} \cdot \vec{d\Theta} = |\vec{\tau}|\,|\vec{d\Theta}| \cos \psi, \tag{22}$$

and in one complete cycle the total work done is

$$W = \int_0^{2\pi} \vec{\tau} \cdot \vec{d\Theta} = 2\pi\tau \cos \psi. \tag{23}$$

Hence there is a net transfer of energy from the positive circularly polarized alternating field to the electron. This causes the electron's magnetic moment vector to *fan out* so that it makes a larger and larger angle with the direction of dc field acting along the Z axis. Thus a positive circular component in the XY plane can act to maintain the precessional motion of the electron at some constant amplitude, even though this motion is heavily damped. The energy fed into the electron by the rotating magnetic field just makes up for the energy dissipated by the damping.

If the same analysis is made for the negative circular component of the alternating field (*i.e.*, the component which is rotating in the opposite direction to the precessional motion), it is seen that there is no net transfer of energy from the rotating field to the electron since the torque reverses twice each cycle. Hence the magnetic moment vector of the electron would not be expected to respond to a negative circular component in the X-Y plane.

In other words, if a block of ferromagnetic material is saturated by a large dc magnetic field acting in the Z direction, and if oppositely rotating circularly polarized magnetic fields acting in the XY plane are applied to it, the material will present a different effective permeability to the two oppositely rotating circularly polarized fields.

An exact analysis of this problem is relatively simple and was first carried out by Polder[10] and extended to include damping by the author.[11] In the general case when the frequency of the alternating field does not necessarily coincide with the natural resonant frequency given in (20), the magnetic moment vector does not lag the magnetic field vector by exactly 90 degrees as shown in Fig. 7 (the angle ϕ) but might lag by anywhere from zero to one hundred and eighty degrees as they both rotate about the Z axis. Hence, the effective permeability which relates the circularly polarized flux in the medium with the circularly polarized field must be complex to indicate the component of flux in phase (or 180° out of phase) with the field and the component which is exactly 90° out of phase with the field. Thus the relation between the circularly polarized flux (in the XY plane) and the circularly polarized magnetic field in the XY plane is

$$\vec{b}_\pm = \mu_\pm \vec{h}_\pm \tag{24}$$

where the plus or minus sign refer to the direction of rotation as previously defined. Under the conditions shown in Fig. 7, μ_+ is imaginary since the component of flux in phase with the rotating magnetic field is zero. However, in general, both μ_+ and μ_- are complex with their real parts indicating the reactive component of the flux and their imaginary parts indicating the lossy or resistive component of the alternating flux.

If now we consider a block of ferromagnetic material subjected to the above dc and circularly polarized fields, it is enlightening to plot the behavior of μ_+ and μ_- at a given frequency as a function of the strength of the dc field acting along the Z axis. This is done in Figs. 9 and 10 opposite. The curves are calculated for a frequency of 9000 mc, and the constants chosen for the medium are typical of many ferrites.

The above discussion has attempted to explain in a rather qualitative way the basic features of the phenomenon of ferromagnetic resonance which one must understand in order to understand the operation of microwave ferrite devices. These relations can be derived in a more quantitative way following the theory developed in the paper by Bloembergen.[12]

As shown in Bloembergen's paper, the equation of motion of magnetization of a saturated ferromagnetic medium can be written as

$$\frac{d\vec{M}}{dt} = \gamma(\vec{M} \times \vec{H}) - \lambda\left[\frac{(\vec{H}\cdot\vec{M})\vec{M}}{M^2} - \vec{H}\right]. \tag{25}$$

To obtain quantitative relations for the phenomenon of ferromagnetic resonance, it is necessary to solve the above equation under the conditions that both a dc and an alternating magnetic field are applied to the medium. It turns out that the proper method of solving this problem is not at first completely obvious. The reason for this is, that in practice, the actual macroscopic magnetic field within a ferromagnetic material is seldom equal to the applied magnetic field. This results because: 1) the sample has a finite size and hence there is a divergence of magnetization on the surface of the sample, which sets up magnetic fields in addition to those applied, or 2) the sample is large compared to the wavelength of the alternating field and in some cases there results a volume divergence of the magnetization, which also sets up magnetic fields in addition to those which

[10] D. Polder, "On the theory of ferromagnetic research," *Phil. Mag.*, vol. 40, pp. 99–115; 1949.

[11] C. L. Hogan, "The microwave gyrator," *Bell Sys. Tech. J.*, vol. 31, pp. 1–31; 1952.
Also, C. L. Hogan, *Rev. Mod. Phys.*, vol. 25, p. 253; 1953.

[12] N. Bloembergen, p. 1259, this issue.

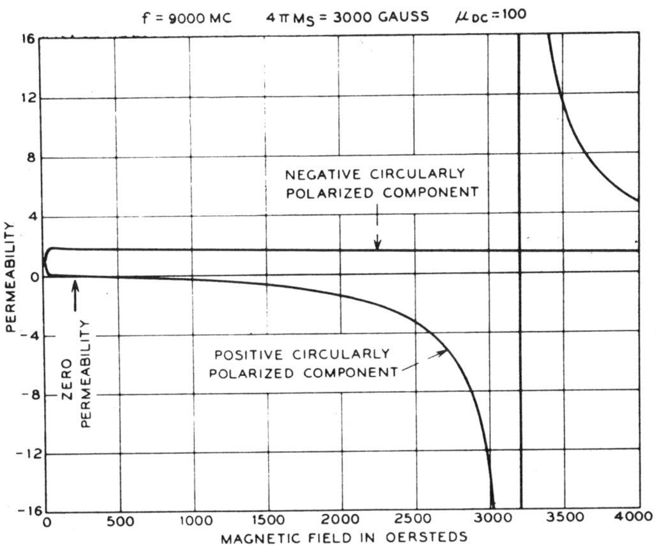

Fig. 9—Real part of the effective permeability of a ferromagnetic medium which is magnetized parallel to the direction of propagation of an infinite plane wave. The curves are computed for a medium whose saturation moment $4\pi M_s = 3000$ gauss and for a wave whose frequency is 9000 $\mu dc = 100$.

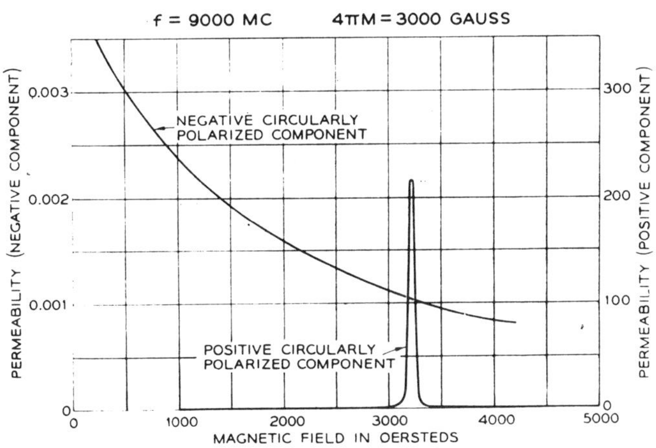

Fig. 10—Imaginary part of the effective permeability of a ferromagnetic medium which is magnetized parallel to the direction of propagation of an infinite plane wave. The constants chosen to compute the curves were the same as those in Fig. 1. In addition the relaxation frequency, λ, was arbitrarily taken as 1.9×10^7 for the curves.

arise from electric currents (applied fields). Kittel[13] first showed how to solve the problem provided the sample was small compared to the wavelength of the ac magnetic field and Polder[14] then solved the problem more generally so that it could be applied to the case which Kittel treated, or to the more general case listed in 2) above. In a paramagnetic material these problems do not arise, because the magnetic fields set up by the poles, which exist where the magnetization has a divergence, are negligible compared to the applied fields.

[13] C. Kittel, "On the theory of ferromagnetic resonance absorption," *Phys. Rev.*, vol. 73, p. 155; 1948.

[14] Polder, *loc. cit.*

These ideas will become clearer after we have solved the problem and applied the solution to certain specific cases.

We shall follow the general method of Polder in obtaining the solution. What we are attempting to find is a relation between the alternating field and the alternating flux, so that we can have some expression which we can call the permeability of the medium. Now, as generally used in conjunction with Maxwell's equations, the permeability gives the relation between the flux density and the total macroscopic field. As such, it is not a property of the size and shape of the sample but is an intrinsic property of the material. Once we know the intrinsic permeability of the material, we can, in principle at least, find the flux density which results from any particular applied field to any particularly shaped sample by simply solving the appropriate boundary value problem, which will automatically take into account the additional magnetic fields set up by a divergence in the magnetization, if it exists.

9

Thus, in order to find the intrinsic permeability of the medium, we can neglect any surface divergence by assuming that the sample is infinite in size,[15] and we can avoid any volume divergence by making the wavelength of the alternating magnetic field infinite. Thus, let

$$H_z = \text{dc magnetic field in } Z \text{ direction}$$

$$\chi_0 = \text{static susceptibility}\left(\frac{M_z}{H_z}\right) \tag{26}$$

$$H_0 = H_z\left[1 + \frac{\lambda^2}{\gamma^2 M^2}\right]^{1/2} \approx H_z.$$

Now if the alternating components of the magnetic field are small compared to the dc field applied along the Z axis, and if the alternating field has a harmonic time dependence, exp $[j\omega t]$, then it can be shown from (25) that the correct relation between the alternating magnetic field and the alternating flux density is given by

$$\vec{b} = |\mu_{ij}| \vec{h} \tag{27}$$

(lower case letters denote alternating quantities; upper case, dc) where the permeability is a tensor of the following form

$$|\mu_{ij}| = \begin{vmatrix} \mu & -j\kappa & 0 \\ +j\kappa & \mu & 0 \\ 0 & 0 & 1 \end{vmatrix} \tag{28}$$

[15] The mathematically-inclined reader will recognize that if we approach infinite size as a limiting process the demagnetizing fields do not disappear and hence we can in practice never build a sample in which the surface can really be neglected. If no volume divergence exists a more correct statement might be to simply state that we are concerned with internal fields.

and

$$\mu = \mu' - j\mu'' \tag{29}$$

$$\kappa = \kappa' - j\kappa'' \tag{30}$$

$$\mu' = 1 + \frac{[\gamma^2 H_0{}^2 - \omega^2]4\pi M_z \gamma^2 H_0 \chi_0{}^2 + 8\pi\omega^2\lambda^2\chi_0}{X_0{}^2[\gamma^2 H_0{}^2 - \omega^2]^2 + 4\omega^2\lambda^2} \cdot \tag{31}$$

$$\kappa' = \frac{4\pi M_z \gamma\omega\chi_0{}^2[\gamma^2 H_0{}^2 - \omega^2]}{\chi_0{}^2[\gamma^2 H_0{}^2 - \omega^2]^2 + 4\omega^2\lambda^2} \tag{32}$$

$$\mu'' = \frac{4\pi\lambda\omega\chi_0{}^2[\gamma^2 H_0{}^2 + \omega^2]}{\chi_0{}^2[\gamma^2 H_0{}^2 - \omega^2]^2 + 4\omega^2\lambda^2} \tag{33}$$

$$\kappa'' = \frac{8\pi\omega^2\gamma\lambda H_z\chi_0{}^2}{\chi_0{}^2[\gamma^2 H_0{}^2 - \omega^2]^2 + 4\omega^2\lambda^2} \cdot \tag{34}$$

The constant λ in the above expressions can be easily related to the ferromagnetic resonance line width which is a more easily determined quantity by the relation

$$\lambda = \frac{\gamma M \Delta H}{2H_r} \tag{35}$$

where

ΔH = resonance line width in oersteds (*i.e.*, distance between half-power points of resonance line shown in Fig. 10).

H_r = magnetic field required for resonance.

As was pointed out previously, these expressions are valid only for the internal fields within the ferrite and then only if the alternating fields are uniform throughout all space. If we are dealing with a finite sized sample it is often more convenient to deal with the so-called external magnetic fields since they are more readily measured. For mathematical reasons it is necessary to confine our interest to ellipsoids. All other shapes have nonuniform magnetization, and hence the macroscopic field varies from point to point within the sample, and there is thus no macroscopic field which can be defined for the sample as a whole. In ellipsoids, which are small compared to the wavelength, the relation between the external field and internal field can be very simply written as

$$\begin{aligned} h_x &= h_x{}^e - N_x m_x \\ h_y &= h_y{}^e - N_y m_y \\ H_z &= H_z{}^e - N_z M_z \end{aligned} \tag{36}$$

where $h_x{}^e$, $h_y{}^e$, and $H_z{}^e$ are the externally applied magnetic fields, and N_x, N_y, and N_z are the so-called demagnetization factors.

Now from (27) and (28) we may write

$$\begin{aligned} b_x &= h_x + 4\pi m_x = \mu h_x - j\kappa h_y \\ b_y &= h_y + 4\pi m_y = j\kappa h_x + \mu h_y \\ b_z &= h_z + 4\pi m_z = h_z. \end{aligned} \tag{37}$$

If the expressions for the internal magnetic fields given in (36) are inserted into the above expressions and then these expressions are solved for the alternating components of magnetization, we have

$$m_x = (\mu - 1)\frac{\left\{4\pi + (\mu-1)N_y - \frac{\kappa^2}{\mu-1}N_y\right\}h_x{}^e}{[4\pi+(\mu-1)N_x][4\pi+(\mu-1)N_y] - \kappa^2 N_x N_y} - \frac{j\kappa 4\pi h_y{}^e}{[4\pi+(\mu-1)N_x][4\pi+(\mu-1)N_y] - \kappa^2 N_x N_y} \tag{38}$$

$$m_y = \frac{j\kappa 4\pi h_x{}^e}{[4\pi+(\mu-1)N_x][4\pi+(\mu-1)N_y] - \kappa^2 N_x N_y} - \frac{(\mu-1)\left\{4\pi + (\mu-1)N_x - \frac{\kappa^2}{\mu-1}N_x\right\}h_y{}^e}{[4\pi+(\mu-1)N_x][4\pi+(\mu-1)N_y] - \kappa^2 N_x N_y} \cdot \tag{39}$$

In order to find the susceptibility tensor in terms of the external fields, it is necessary to insert the values for μ and κ, given in (29) to (34), into the above expressions. The algebraic manipulation required is quite formidable, and so we shall find the expressions for the case in which the damping parameter, λ, is zero. It turns out that the expressions obtained are quantitatively valid everywhere, except in the vicinity of the ferromagnetic resonance absorption line. Under these conditions, the equations for μ and κ are

$$\mu = \mu' = 1 + \frac{4\pi M\gamma^2 H_z}{\gamma^2 H_z{}^2 - \omega^2}, \qquad \mu'' = 0 \tag{40}$$

$$\kappa = \kappa' = \frac{4\pi M\gamma\omega}{\gamma^2 H_z{}^2 - \omega^2}, \qquad \kappa'' = 0. \tag{41}$$

Using these, (38) and (39) become

$$m_x = \frac{\gamma^2 M[H_z{}^e + (N_y - N_z)M]h_x{}^e}{\gamma^2 H_{\text{eff}}{}^2 - \omega^2} - \frac{jM\gamma\omega h_y{}^e}{\gamma^2 H_{\text{eff}}{}^2 - \omega^2} \tag{42}$$

$$m_y = \frac{jM\gamma\omega h_x{}^e}{\gamma^2 H_{\text{eff}}{}^2 - \omega^2} + \frac{\gamma^2 M[H_z{}^e + (N_x - N_z)M]h_y{}^e}{\gamma^2 H_{\text{eff}}{}^2 - \omega^2} \tag{43}$$

where

$$H_{\text{eff}}{}^2 = [H_z{}^e + (N_x - N_z)M][H_z{}^e + (N_y - N_z)M]. \tag{44}$$

This equation was first derived by Kittel and will be referred to as Kittel's equation in the future.

The most important result of this calculation is the fact that the resonant frequency now depends upon the shape of the sample. Thus, in the infinite medium where no divergence of the magnetization exists, the ferromagnetic resonance frequency occurs for

$$\omega_0 = \gamma H_z. \tag{45}$$

For an ellipsoidal specimen which is small compared to the wavelength of the radiation being used, the resonant frequency occurs at

$$\omega_0 = \gamma H_{\text{eff}}. \tag{46}$$

This effect is sometimes confusing to the novice in this field since it is obvious that if a magnetized piece of ferrite is subjected to an applied alternating magnetic field, the frequency of the internal field must be equal to the frequency of the applied field. This paradox can be removed in the following way.

Consider a cylinder of ferrite whose axis lies along the Z axis as shown in Fig. 11. Assume $N_z = 0$. It is exposed to an external rf magnetic field which lies in the xy plane. Now, it is known from the previous discussion that if the magnetization is expressed in terms of the internal fields, a resonance exists at the frequency given by (45). If, however, the magnetization is expressed in terms of the external fields, the resonance occurs at a frequency [given by (46)]

$$\omega = \gamma(H_z + 2\pi M_s). \tag{47}$$

The question logically arises: How is this possible? The frequency of the internal and external fields must be identical.

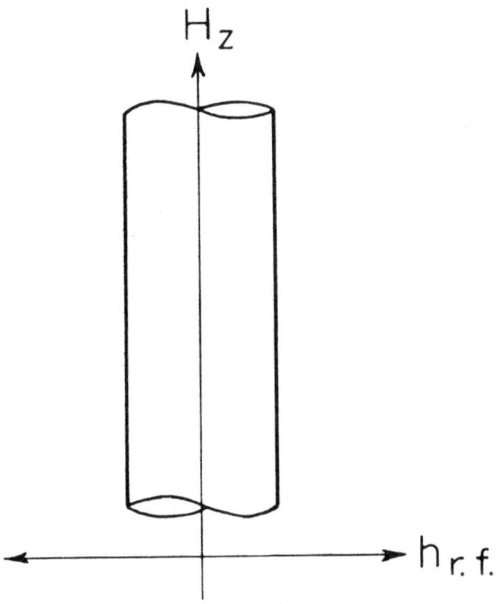

Fig. 11—Infinite cylinder subjected to a axial dc magnetic field and an rf magnetic field perpendicular to axis.

This question can be answered from purely physical reasoning. It is known that the internal fields are given by (36), and if we imagine the rf field in Fig. 11 to start with a zero frequency and to gradually have the frequency increased until it passes through both frequencies given in (45) and (47), the following sequence of events takes place. When the frequency is far below the resonant frequencies, the response of the magnetization vector is small. In other words, m_x and m_y are small, and according to (36), there is little difference between the magnitudes of the internal and external fields. However, as the frequency corresponding to (45) is approached, the magnetization responds more readily to the applied field, and according to (36), the magnitude of the internal field approaches zero. [In the absence of damping, the internal field is exactly zero at the frequency given in (45).] Since the magnitude of the internal field becomes vanishingly small at this frequency, the magnetization vector is not fanned out into a violent precessional motion as would at first be expected. In fact, it fans out just enough to reduce the internal field to just the right magnitude to just make up for losses of energy. As the frequency is increased beyond this point the phase of the alternating magnetization reverses with respect to the applied fields and hence when dealing with magnitudes we can replace the minus signs in (36) by plus signs. Thus the magnitude of the internal field increases and reaches its maximum amplitude at the frequency given in (47).

These results can be easily shown mathematically. However, to do it simply we shall neglect damping. This allows the magnetization to go to infinity and hence reduces the internal field to exactly zero at the frequency given in (45), and allows the magnitude of the internal field to become infinite at the frequency given in (47). In any actual case, it is not quite zero (but almost) at the one point and is not quite infinite at the other.

From (36) and (37), it is possible to write

$$\begin{aligned} h_x^e &= (1 + N_x\chi_{xx})h_x - jN_x\chi_{xy}h_y \\ h_y^e &= (1 + N_y\chi_{xx})h_y + jN_y\chi_{xy}h_x \end{aligned} \tag{48}$$

where the χ's are the diagonal and off-diagonal components of the susceptibility tensor.

At the *infinite medium resonance* of (45), the susceptibilities in the above equation are infinite. Hence, if the external fields are finite, the internal fields are zero.

If (41) is inverted we get

$$h_x = \frac{(1 + N_y\chi_{xx})h_x^e + iN_x\chi_{xy}h_y^e}{(1 + N_x\chi_{xx})(1 + N_y\chi_{xx}) - N_xN_y\chi_{xy}^2} \tag{49}$$

with a similar equation for h_y. The denominator of this expression has a zero at the frequency given in (47), and hence the internal fields become *infinite* at this frequency, and thus the magnetization vector is driven into a large amplitude of precessional motion. The reader should be warned that this analysis is valid only if the diameter of the cylinder is small enough so that transverse propagation in the cylinder can be neglected; and in addition it refers only to that component at the magnetic field for which a resonance occurs (*i.e.*, h_+).

Another problem of some interest deals with the behavior of an infinite saturated ferromagnetic medium when the wavelength of the radiation is finite. This problem was first solved by Polder, who showed that the resonant frequency of the medium depended upon the polarization and direction of propagation of the wave. Actually, the reason for this is quite simple. The magnetic dipoles in the medium are caused to precess around the dc magnetic field by the rf field of the wave. Since the phase of this precessional motion is determined by the strength of the dc field and the relative phase of the rf magnetic field, it follows that the magnetic dipoles will be 180° out of phase with those one-half wavelength away. Thus, at one instant of time the magnetic dipoles in the sample are not all pointing in the same direction. This may or may not cause a divergence (volume divergence) in the magnetization of the medium. If a divergence does result, then additional internal fields not normally present in the wave motion arise, and these cause the resonant frequency to shift in the same way that the demagnetizing fields which arise from a surface divergence shift the resonant frequency.

This can be understood better by looking in detail at a few particular problems of wave propagation in an infinite medium. Thus, if an infinite plane wave is propagated through an infinite medium saturated in the Z direction, it can be easily shown by solving Maxwell's equation in conjunction with (27) that the propagation constant of the wave is given by

$$\Gamma_{\pm} = j\omega(\mu_0\epsilon)^{1/2}\left[\frac{\mu^2 - \mu - \kappa^2)\sin^2\theta + 2\mu \pm (\mu^2 - \mu - \kappa^2)^2\sin^4\theta + 4\kappa^2\cos^2\theta^{1/2}}{2[(\mu - 1)\sin^2\theta + 1]}\right]^{1/2} \tag{50}$$

12

where the angle between the Z axis and the direction of propagation of the wave is θ, and polarization of the wave is determined by the choice of signs above. Now, when the wave is propagated along the Z axis (around which the magnetic dipoles are precessing), there is no divergence of the magnetization. (If the reader will take a few seconds to sketch the orientation of the dipoles as functions of X, Y, and Z, this statement will be obvious.) Hence we should expect the resonant frequency to be given by (45). To check this prediction, we need only set θ equal to zero in (50). When this is done, it is found that the propagation constants are given by

$$\Gamma_{\pm} = j\omega\left[\epsilon\mu_0(\mu \pm \kappa)\right]^{1/2}. \tag{51}$$

This expression has a resonance at the value of ω, given by (45), for a circularly polarized wave that is rotating in the direction which the electrons precess about the dc field. If, however, the wave is propagated along the Y axis ($\theta = 90°$) and is polarized so that the rf magnetic field vector lies along the X axis, the propagation constant as found from (50) is

$$\Gamma = j\omega\left[\epsilon\mu_0\frac{\mu^2 - \kappa^2}{\mu}\right]^{1/2}. \tag{52}$$

If one substitutes into this expression the values of μ and κ given in (40) and (41), the propagation constant given in (52) becomes

$$\Gamma = j\omega\left[\epsilon\mu_0\frac{\gamma^2 B_z^2 - \omega^2}{\gamma^2 H_z B_z - \omega^2}\right]. \tag{53}$$

The resonance occurs at the same frequency as it would if the sample were a thin disc rather than an infinite medium, and for approximately the same reason. In the case of a thin disc, an rf magnetic field is set up in the Y direction, due to the surface divergence of M on the faces of the disc as the dipoles precess about the Z axis. In the case of the infinite medium above, an rf magnetic field is set up in the Y direction, due to a volume divergence in the magnetization which results from the fact that the spins are not all parallel. In both cases, the rf magnetic field set up in the Y direction is

$$h_y = -4\pi m_y. \tag{54}$$

In both cases this additional magnetic field is in the direction in which the wave is being propagated. In other words, it is a longitudinal component. In a sense, then, the infinite plane wave is no longer a purely transverse wave in the ferromagnetic medium. However, it is transverse in terms of the B vector, since the net flux density in the longitudinal direction is always zero. Thus, there is no electric field vector directly associated with this longitudinal magnetic field vector.

A problem of more concern to the engineer than Polder's infinite medium or Kittel's finite medium which is smaller than a wavelength, is the problem of a finite medium in which at least one dimension is large compared to the wavelength of the radiation. This problem cannot be solved exactly, except for particularly shaped samples; but in most cases approximate relations can be deduced from the above arguments, which are often quite valid for quantitative applications. A few examples will make this method clear.

Assume that an infinite cylindrical rod is exposed to an external alternating magnetic field with a finite wavelength, as shown in Fig. 12. The axis of the cylindrical specimen lies along the Z axis, and the wave is being propagated along the Z axis.

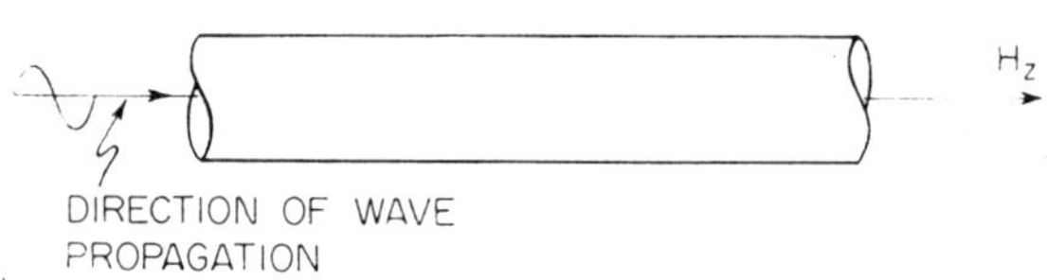

Fig. 12—Infinite plane wave being propagated along axis of an infinitely long magnetized ferrite cylinder.

The exact solution to this problem involves the solution of Maxwell's equations, both inside and outside the medium, matching boundary conditions. However, if the diameter of the rod is small compared to the wavelength (roughly $5d < \lambda$), the resonant frequency can be found approximately, using the above reasoning. The only divergence in the magnetization occurs on the cylindrical surface (assuming infinite length). This results from the precessional motion of the magnetization vector about the Z axis. Hence if the appropriate rf demagnetizing factors for this shape can be deduced, the resonant frequency can be obtained from Kittel's equation. In order to find the correct rf demagnetizing factors, it is necessary to find the magnetic field in the X and Y directions which results from the divergence of magnetization on the cylindrical surface. The pole distribution on the surface is a helical distribution of north and south poles as shown in Fig. 13. Obviously, if the wavelength is very large compared to the diameter of the rod, then the magnetic fields in the X and Y direction are the same as they would be for an infinite wavelength, and hence Kittel's equation should hold exactly where

$$N_x = N_y = 2\pi, \qquad N_z = 0.$$

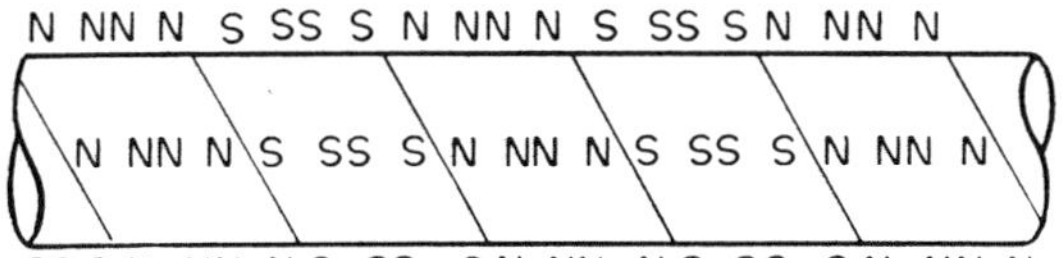

Fig. 13—Resulting pole distribution on surface of cylinder shown in Fig. 12.

If in the other extreme, the diamenter of the rod is very large compared to the wavelength (roughly $d=5\lambda$), no demagnetizing factor can be defined, due to the fact that right under the surface the effective demagnetizing field is just $h_x = -2\pi M_x$. However, in the bulk of the material the north poles on the surface cancel the south poles (due to their proximity when $d \gg \lambda$) and the effective demagnetizing field is zero. This effect tends to broaden the absorption line, since dipoles at different radii see different effective fields. However, in the limit as $d \rightarrow \infty$ one can neglect the small amount of material near the surface, and the correct effective demagnetizing factors are $N_x = N_y = N_z = 0$, and the resonant frequency of a large cylinder is then the same as for an infinite medium regardless of the fact that it has cylindrical symmetry. Where $d \approx \lambda$, there must be a smooth transition from the resonant frequency.

$$\omega = \gamma(H_z + 2\pi M) \tag{55}$$

to

$$\omega = \gamma H_z. \tag{56}$$

The trend from a resonant frequency given by (55) to that given by (56) has actually been observed experimentally as the diameter is increased. However, before the frequency given in (56) is reached the diameter is so large that transverse propagation in the cylinder becomes important and the problem is much more complicated than the above analysis would indicate.

If now the cylinder is finite in length in the Z direction, the same problem arises that no demagnetizing factor can be defined for the entire length of the cylinder. If we assume the cylinder is uniformly magnetized (not a bad approximation if the dc field in the Z direction is very large, *i.e.*, $H_z > 2\pi M_s$), then again, there is a field in the Z direction just inside the butt ends of the cylinder equal to $H_z(\text{demag}) = 2\pi M_z$. However, again if the cylinder is long, the demagnetizing field in the Z direction near the center is zero. Obviously then, a finite cylinder cannot be uniformly magnetized in weak fields, and hence the absorption line is broadened especially if the resonant field for the major portion of the sample is less than $2\pi M_s$. If the ends of the cylinder are tapered in order to approximate an ellipsoid, the broadening of the absorption line due to the finite length can be greatly reduced. In order to obtain sharp absorption lines using the above geometry, it is obvious that either extremely small diameter tapered cylinders should be used or very thin discs with large diameter.

Another configuration of engineering interest is that shown in Fig. 14. In this configuration a wave polarized with its magnetic field vector along the X axis is propagated along the Y axis past a thin slab of material which is uniformly magnetized in the Z direction. In this problem, a true volume divergence of the magnetization exists such that there is a magnetic pole distribution which varies harmonically along the Y axis. If the wavelength is very long compared to the thickness of the slab, this volume divergence can be neglected, since it will lead to a small density of magnetic poles within the material. Thus a close approximation would allow one to set the effective demagnetizing factors equal to the following

$$N_z = N_y = 0, \qquad N_x = 4\pi.$$

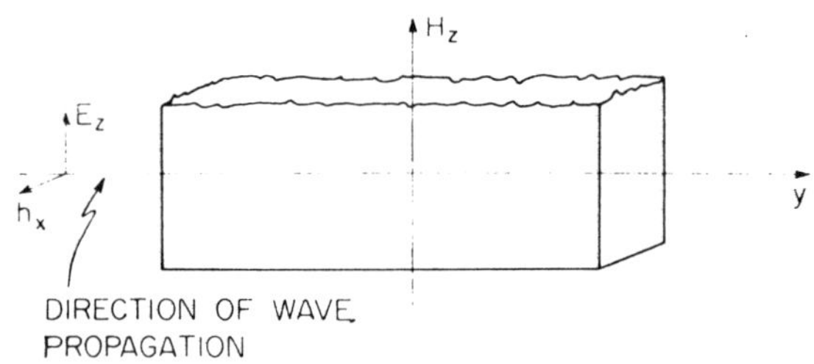

Fig. 14—Infinite plane wave being propagated past a slab of magnetized ferrite.

If, however, the thickness of the slab becomes appreciable compared to the wavelength, the effective demagnetizing factors will change to the extreme values, $N_z = N_x = 0$, $N_y = 4\pi$ (as the wavelength becomes very small compared to the thickness of the slab). In any case, the effective demagnetizing fields in the X and Y direction are not uniform throughout the volume of the sample, and hence there is a broadening of the absorption line in this geometry.

In any actual problem the resonant frequency is not easily obtained, due to the effects of volume and surface divergence of the magnetization. However, close inspection of the problem often leads to the establishment of effective rf demagnetizing factors from which the resonant frequency can be calculated.

Thus even though the problem of wave propagation in magnetized ferrites is extremely complicated the major point of the above analysis is that the permeability which an infinite saturated ferromagnetic material exhibits to superposed alternating magnetic fields is a tensor quantity. Likewise, if the magnetization of a finite ellipsoid is expressed in terms of the externally applied magnetic fields, the permeability or susceptibility which results is also a tensor quantity. If the ellipsoid has cylindrical symmetry ($N_x = N_y$), the tensors are practically identical, except for the value of the resonant frequency. Thus for both the case of an infinite medium and an ellipsoid with cylindrical symmetry, the diagonal component of the tensor can be written as

$$\mu' = 1 + \frac{(4\pi - N_t)\gamma M \omega_0}{\omega_0^2 - \omega^2} \tag{57}$$

and the off-diagonal component becomes

$$\kappa' = \frac{(4\pi - N_t) M \gamma \omega}{\omega_0^2 - \omega^2} \tag{58}$$

where ω_0 is the appropriate resonant frequency for the particular sample, and N_t is the transverse demagnetizing factor which is set equal to zero if one is dealing with internal fields rather than external fields. It should be pointed out, though, that if the finite sized ellipsoid does not have cylindrical symmetry, then the diagonal components of the tensor are all different when the magnetization is expressed in terms of the externally applied fields.

If now we restrict ourselves to circularly polarized alternating magnetic fields, it turns out that the tensor describing an infinite medium or a cylindrically symmetrical ellipsoid can be reduced to a scalar quantity. This has an important engineering advantage, because it is easy to think intuitively in terms of scalar permeabilities but difficult, without more practice than most of us have, to think intuitively in terms of tensor permeabilities. There is really no great restriction imposed by requiring that we limit our discussion to circularly polarized fields, because the most general polarization can be decomposed into counter-rotating circularly polarized fields with arbitrary amplitudes. The only restriction which we shall find annoying is the fact that the ellipsoid must have cylindrical symmetry in order to simply reduce its tensor permeability to a scalar quantity for circularly polarized external fields.

Thus for any material which is described by

$$b_x = \mu h_x - j\kappa j_y,$$
$$b_y = j\kappa h_x + \mu h_y, \tag{59}$$

whether the magnetic fields are internal or external fields, it is possible to write

$$\vec{b}_{cp} = (\mu \pm \kappa)\vec{h}_{cp} \tag{60}$$

for the circularly polarized magnetic fields. The choice of signs is determined by the direction of rotation of the fields.

From (57) and (58) we may write (neglecting the damping of precessional motion)

$$\mu' \pm \kappa' = 1 + \frac{(4\pi - N_t)\gamma M}{\omega_0 \pm \omega}. \tag{61}$$

Since the Kramers Kronig relation must hold for these quantities, we can write the imaginary (or lossy) part of the above effective scalar permeability from (29) to (34) above (neglecting the difference between H_z and H_0).

$$\mu'' \pm \kappa'' = \frac{4\pi\lambda\omega[\omega_0 \pm \omega]^2}{[\omega_0^2 - \omega^2]^2 + \dfrac{4\omega^2\lambda_0^2\omega_0^2}{\gamma^2 M_z^2}}. \tag{62}$$

Everywhere except in the vicinity of the absorption line, (62) reduces to

$$\mu'' \pm \kappa'' = \frac{4\pi\lambda\omega}{(\omega_0 \mp \omega)^2}. \tag{63}$$

The permeabilities given in (61) and (63) are plotted in Figs. 9 and 10.

Thus it is apparent that if an infinite plane wave is propagated in the Z direction through the medium described by (27), it will split into two counter-rotating circularly polarized components which will travel with different velocities, since the medium presents a different permeability to these two components.

Since the two circularly polarized components travel at different velocities there is a continuous rotation of the plane of polarization of the net wave. It is to be expected that the rotation of the plane of polarization will be proportional to the difference between the square roots of the two permeabilities shown in Fig. 9. Fig. 15

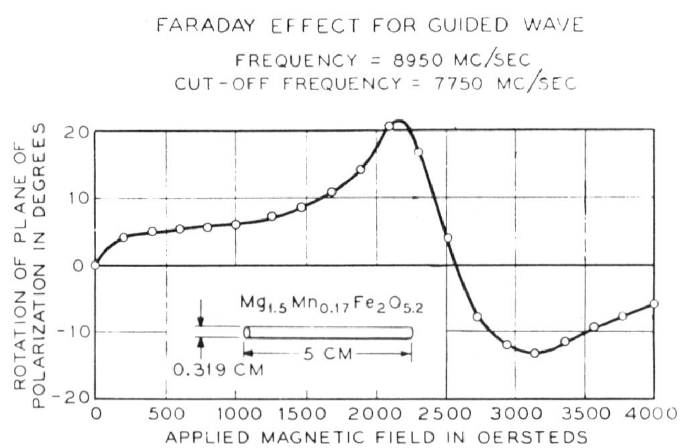

Fig. 15—Measured Faraday rotation of TE_{11} mode for a cylinder of ferrite in a circular x-band waveguide.

shows the rotation of the plane of polarization actually measured in a round waveguide carrying the dominant mode. It is seen that this is qualitatively identical to what one would predict from Fig. 9. In Fig. 16 is shown the measured attenuation of circularly polarized TE_{11} modes in a circular waveguide when they were propagated through a guide containing a small cylinder of ferrite, supported along the axis of the guide. It is seen that these data support the prediction of Fig. 10, that the medium should be practically lossless for both

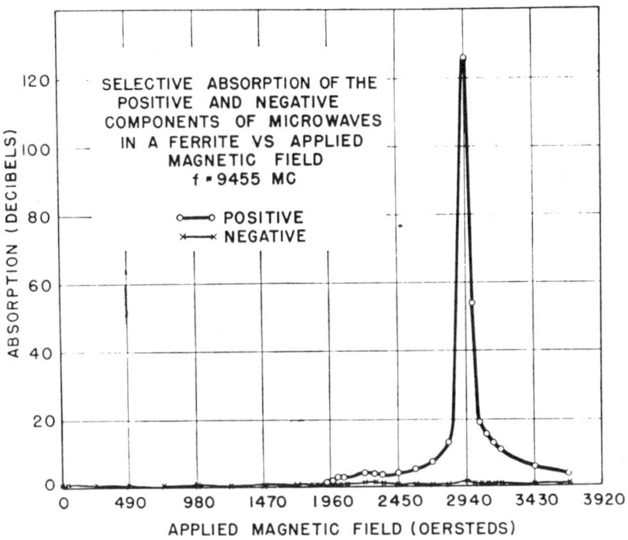

Fig. 16—Measured absorption of a positive and negative circularly polarized TE_{11} mode in a circular waveguide with a thin cylinder of ferrite located along the axis of the waveguide. The cylinder was 0.123 inch diameter and 0.815 inch long. The magnetic field was applied along the axis of the waveguide.

circular components, except in the region of the absorption line, where it should become very lossy for the positive circularly polarized component only.

Fig. 15 constitutes data taken on a magnesium manganese ferrite, and Fig. 16, a nickel zinc ferrite. Fig. 17 shows the experimental apparatus in which the rotation of the plane of polarization is measured.

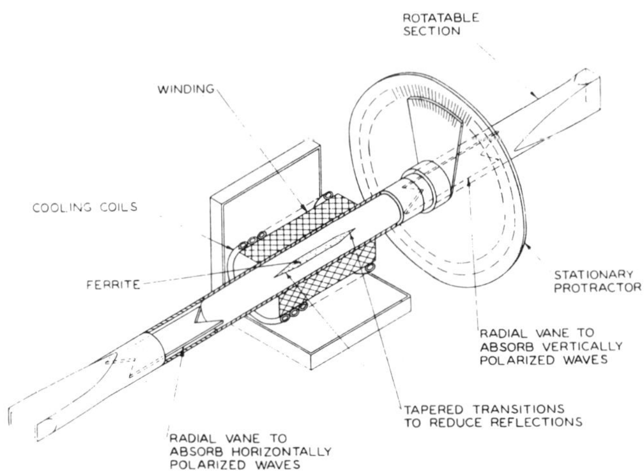

Fig. 17—Typical experimental set-up for measuring Faraday rotation.

For most practical applications the Faraday rotation is used in relatively weak applied magnetic fields. An inspection of Figs. 15 and 16 indicates why this is so. The rotation in weak fields is relatively large and the loss is relatively small. In this region, it is easy to derive a very fundamental equation for rotation per unit path length. Thus, the rotation per unit path length is obviously given by

$$\frac{\theta}{l} = \frac{\omega}{2}\left[\frac{1}{v_-} - \frac{1}{v_+}\right], \tag{64}$$

where

$v_\pm$ = velocity of positive and negative circularly polarized components.

Eq. (64) can be written in terms of the phase constant of the wave,

$$\frac{\theta}{l} = \frac{1}{2}\left[\beta_- - \beta_+\right]. \tag{65}$$

The phase constants can be evaluated from (51). If losses are neglected, (65) becomes

$$\frac{\theta}{l} = \frac{\omega}{2}\sqrt{\mu_0\epsilon}\left[\left(1 + \frac{4\pi M\gamma}{\gamma H_z + \omega}\right)^{1/2} - \left(1 + \frac{4\pi M\gamma}{\gamma H_z - \omega}\right)^{1/2}\right]. \tag{66}$$

In weak fields, $\gamma H_z \ll \omega$, and if the saturation moment is low enough or the frequency high enough so that

$$\frac{4\pi M\gamma}{\omega} \ll 1 \tag{67}$$

then (66) can be put into the extremely simple form

$$\frac{\theta}{l} = \frac{1}{2}\sqrt{\mu_0\epsilon}\left[4\pi M_s\gamma\right]\text{radians/meter.} \tag{68}$$

For assumed values

$$\epsilon = 10\epsilon_0, \quad 4\pi M_s = 3000 \text{ gauss},$$
$$\gamma = 17.6 \times 10^6 \text{ radians/sec/oersted},$$

(12) predicts rotations of 160 degrees/cm.

This rotation of the plane of polarization is known as the Faraday effect. It has long been known that the Faraday rotation of the plane of polarization in optics is antireciprocal. Its antireciprocal property distinguishes the Faraday effect from optical rotations caused by birefringent crystals, or by the Cotton-Mouton effect, which are reciprocal. That is, if a plane polarized light wave is incident upon a birefringent crystal in such a manner that the plane of polarization is rotated through an angle θ in passing through the crystal then this rotation will be cancelled if the wave is reflected back through the crystal to its source. In the Faraday rotation, however, the angle of rotation is doubled if the wave is reflected back along its path. Hence, if the length of path through the *active* material is adjusted so as to give a 90° original rotation, the beam on being reflected will have its plane of polarization rotated a total of 180° in passing in both directions through the material. This accounts for the one way transmission system which Lord Rayleigh discussed and which is shown in Fig. 3.

The antireciprocal property of the Faraday effect affords a means of realizing a microwave circuit element which is analogous to Tellegen's gyrator. Such a gyrator is illustrated in Fig. 18 along with diagrams which help explain its action. Beneath the gyrator are construction lines which indicate the plane of polarization of a wave as it travels through the gyrator in either direction. On each diagram is a dotted sine wave for reference only which indicates the constant plane of polarization of an unrotated wave. It is noticed that for propagation from left to right in Fig. 18, the screw rotation intro-

16

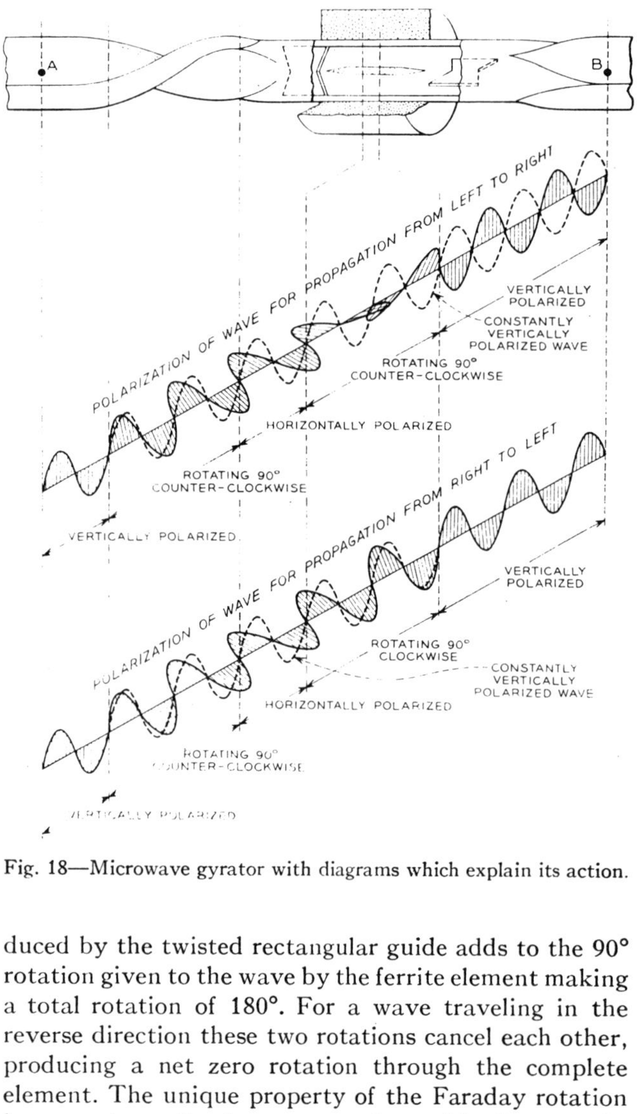

Fig. 18—Microwave gyrator with diagrams which explain its action.

duced by the twisted rectangular guide adds to the 90° rotation given to the wave by the ferrite element making a total rotation of 180°. For a wave traveling in the reverse direction these two rotations cancel each other, producing a net zero rotation through the complete element. The unique property of the Faraday rotation becomes immediately apparent from this diagram. In the case of the rotation induced by the twisted rectangular guide, the wave rotates in one direction in going from left to right through the twisted section, and rotates in the opposite direction when it traverses the section from right to left. For the case of the rotation induced by the ferrite element, the direction of rotation is indicated by the arrow in the upper figure for either direction of propagation. The important characteristic of the element is the time phase relation between two points such as A and B in the upper diagram. It is seen with the help of the diagrams illustrating the rotating waves that the field variations are in phase at points A and B for propagation from left to right and they are 180° out of phase for propagation from right to left. In other words the transmission line is an integral number of wavelengths long between A and B for propagation from left to right and is an odd integral number of half-wavelengths long for propagation from right to left.

If the rectangular wave guides on each side of the ferrite are rotated about their common axis so as to make an angle of 45° with each other, then a one-way transmission system can be created which is similar to Lord Rayleigh's one-way transmission system of optics. This one-way transmission system can be used, for example, to isolate the generator or detector from the wave guide in microwave systems. In this application it has the great advantage over the attenuators which are presently used for this purpose in that it can be made practically lossless for the direction of propagation which is desired but the reflected wave will be completely absorbed and hence more complete isolation can be effected.

This device which makes use of the 45° rotation has been called a circulator since it is essentially a more complex and more useful circuit element than its simple one-way transmission property would at first indicate. The word was coined by A. G. Fox. The circulator actually has four output branches corresponding to the two different polarizations at each end of the device. The polarizations of the four output branches are indicated schematically in Fig. 19(a). It is noticed that power sent into the polarization circulator with polarization a is turned into polarization b, also b is turned into c, c is turned into d, and d is turned into minus a. This property is indicated very clearly by the circuit symbol suggested in Fig. 19(a) the phase inversion between arms d and a being indicated by the minus sign between the d and a arms.

Another one-way transmission system can be created by combining the gyrator with two magic tees. This combination is indicated in Fig. 19(b). Since this device has all of the fundamental properties of the polarization circulator with the exception of the phase inversion between arms d and a it is suggested that it be called a circulator and the circuit symbol suggested which indicates its properties is also given in Fig. 19. The circulator can also be constructed by using other forms of the hybrid junction than the normal magic tees.

Kales[16] and his associates at the Naval Research Laboratories have shown that gyrators can be built in rectangular waveguides and by extending their reasoning to other guided wave systems, a very powerful approach is available for designing gyrators for many different types of guided wave systems. As an introduction to Kales' work on the partially loaded rectangu-

[16] M. L. Kales, H. N. Chait, and N. G. Sakiotis, "A non-reciprocal microwave component," *J. Appl. Phys.*, vol. 24, p. 816; 1953.

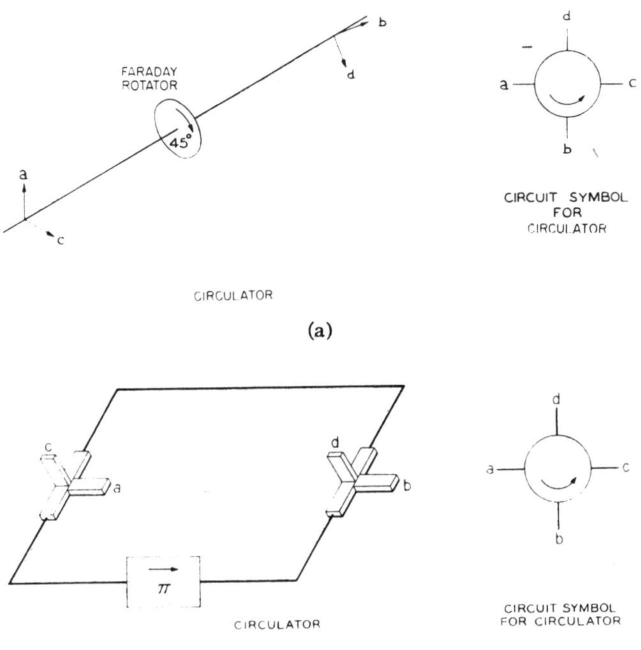

Fig. 19—(a) One type of Faraday rotation circulator, and (b) general circulator employing a gyrator and hybrid junctions.

lar waveguide, it is enlightening to examine the problem of the completely filled rectangular guide. This problem was first solved by Van Trier[17] by a rather general and very powerful mathematical method. For our purpose, we shall arrive at the solution by a less powerful but more direct route. In the first place we shall be interested in only the lowest or dominant mode that exists in a completely filled guide. The system of axes is illustrated in Fig. 20.

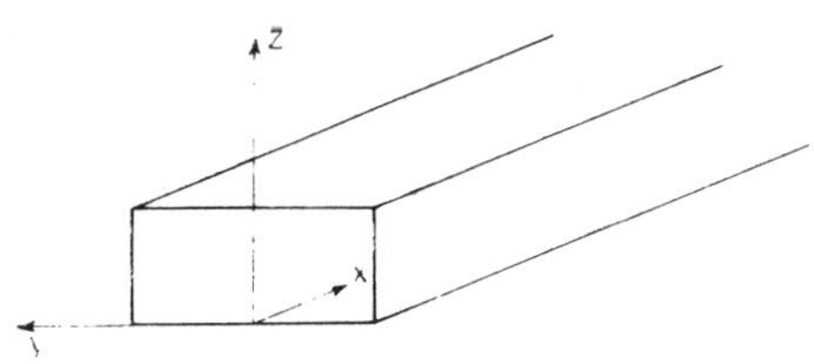

Fig. 20—System of axes for analysis of rectangular waveguide loaded with magnetized ferrite.

To find a solution to the problem of the completely filled guide, we begin with Maxwell's equations (mks).

$$\vec{\nabla} X \vec{E} = -\frac{\partial \vec{B}}{\partial t},$$
$$\vec{\nabla} X \vec{H} = \frac{\partial \vec{D}}{\partial t}. \tag{69}$$

[17] A. A. Th. M. Van Trier, "Guided electromagnetic waves in anisotropic media," *Appl. Sci. Research*, vol. B3, pp. 305–371; 1953.

To these we must add the relations

$$\vec{D} = \epsilon \vec{E} \tag{70}$$

and

$$\begin{aligned} b_x &= \mu_0(\mu h_x - j\kappa h_y) \\ b_y &= \mu_0(j\kappa h_x + \mu h_y) \\ b_z &= \mu_0 h_z. \end{aligned} \tag{71}$$

If now we assume that the fields in the guide are all proportional to exp $[-ay - bz - \Gamma x]$, we can derive from Maxwell's equations the explicit form of the wave equation which applies in this case. For a TE mode we require that $E_x = 0$. If this condition is placed on the set of equations, we are led to the result that $B = 0$. The resulting equation for the magnetic field vectors then becomes

$$\begin{aligned} &\left[\omega^2 \epsilon \mu_0 \mu + a^2 + \Gamma^2 + j\frac{\kappa}{\mu} a\Gamma\right] h_x \\ &\qquad + \left[-j\frac{\kappa}{\mu}(\omega^2 \epsilon \mu_0 \mu + \Gamma^2)\right] h_y = 0, \\ &\left[j\frac{\kappa}{\mu}(\omega^2 \epsilon \mu_0 \mu + a^2)\right] h_x \\ &\qquad + \left[\omega^2 \epsilon \mu_0 \mu + a^2 + \Gamma^2 - j\frac{\kappa}{\mu} a\Gamma\right] h_y = 0. \end{aligned} \tag{72}$$

In order for this set of equations to have a solution, the determinant must vanish. If the determinant is set to zero and then factored, two equations result, the solutions of which would allow wave propagation under the conditions which we have imposed. It can be shown that one equation leads to the trivial case that

$$\omega^2 \epsilon \mu_0 \mu = 0. \tag{73}$$

The other equation leads to the result

$$\omega^2 \epsilon \mu_0 \left(\frac{\mu^2 - \kappa^2}{\mu}\right) + a^2 + \Gamma^2 = 0 \tag{74}$$

which is the identical result for a dielectric loaded waveguide, except that the relative permeability of the medium is given by the expression in the parentheses above. In order to satisfy the boundary conditions on the edges of the guide (E_z must vanish), it is obvious that

$$a = j\frac{n\pi}{y_0}. \tag{75}$$

From here the solution of the problem is similar to that for guides loaded with isotropic material. The field components are

$$E_x = E_y = 0$$
$$E_z = E_0 \cos \frac{\pi y}{y_0} \exp [-j\beta x], \tag{76}$$

$$B_x = \frac{-j\pi}{\omega y_0} E_0 \sin \frac{\pi y}{y_0} \exp [-j\beta x]$$
$$B_y = \frac{-}{\omega} E_0 \cos \frac{\pi y}{y_0} \exp [-j\beta x] \tag{77}$$
$$B_z = 0$$

$$h_y = \frac{-E_0 \exp [-j\beta x]}{\omega\mu_0 \left(\frac{\mu^2 - \kappa^2}{\mu}\right)} \left[\beta \cos \frac{\pi y}{y_0} + \frac{\pi\kappa}{\mu y_0} \sin \frac{\pi y}{y_0}\right]$$
$$h_x = \frac{-jE_0 \exp [-j\beta x]}{\omega\mu_0 \frac{\mu^2 - \kappa^2}{\mu}} \left[\frac{\pi}{y_0} \sin \frac{\pi y}{y_0} + \frac{\beta\kappa}{\mu} \cos \frac{\pi y}{y_0}\right]. \tag{78}$$

Now the surface current density in the wall of the guide, necessary to terminate the tangential component of the magnetic field, is

$$\vec{K} = \vec{n} \times \vec{h} \tag{79}$$

where $\vec{n}$ is a unit vector normal to the surface.

For the top and bottom walls of the guide, it is possible to find an expression for the longitudinal (x component) and transverse (y component) current flowing in the guide walls from (78) and (79). From this one obtains

$$|K_y| \sim \frac{\pi}{y_0} \sin \frac{\pi y}{y_0} + \frac{\beta\kappa}{\mu} \cos \frac{\pi y}{y_0}, \tag{80}$$

provided the guide is above cutoff.

If a longitudinal slot is cut in the broad wall of the guide, it must be cut in a position where $|K_y|$ is zero in order to be a nonradiating slot. If we set (80) equal to zero and solve we find the value of y which gives the position of a nonradiating slot. This is

$$y = \frac{y_0}{\pi} \tan^{-1}\left(-\frac{\beta\kappa y_0}{\pi\mu}\right). \tag{81}$$

If the off-diagonal component of the tensor permeability κ is zero, the position of the nonradiating slot occurs at $y=0$ or the center line of the guide as usual. However, as the ratio of κ/μ increases, the position of the nonradiating slot moves over to one side of the guide. The direction which it moves depends upon the sign of β, or in other words, on the direction of propagation. Thus a slot can be cut in a ferrite filled guide that is radiating for one direction of propagation but not for the other.

The problem of the partially filled guide rectangular waveguide has been solved in part by many authors since Kales first showed that it could be made anti-reciprocal. The most complete solution is that due to B. Lax, K. J. Button, and L. M. Roth at Lincoln Laboratory. We shall not present their solution here but instead we shall present an intuitive explanation of the problem which can be extrapolated more rapidly to other configurations.

Fig. 21 shows a diagram of the field patterns associated with the dominant mode in rectangular guide.

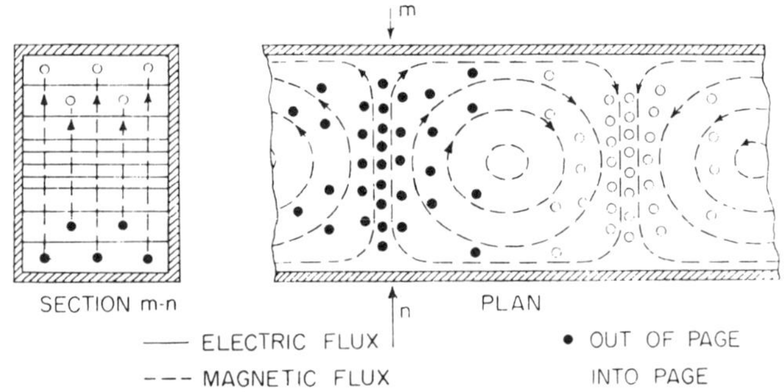

Fig. 21—Magnetic and electric field lines in a rectangular waveguide propagating the dominant mode.

The magnetic field associated with this pattern is usually decomposed into longitudinal and transverse fields, such that we can write

$$H_y = C\sqrt{\left(\frac{\omega}{\omega_c}\right)^2 - 1} \cos \frac{y\pi}{y_0} \sin (\beta x - \omega t),$$
$$H_x = C \sin \frac{\pi y}{y_0} \cos (\beta x - \omega t). \tag{82}$$

The field can be equally well decomposed into counter rotating circularly polarized vectors in the xy plane. Thus from (82) it is easy to show that the magnitudes of the circularly polarized fields that are equivalent to the fields given in (82) are

$$|H_{+cp}| = \frac{C}{2}\left|\left[\sqrt{\left(\frac{\omega}{\omega_c}\right)^2 - 1} \cos \frac{\pi y}{y_0} + \sin \frac{\pi y}{y_0}\right]\right|,$$
$$|H_{-cp}| = \frac{C}{2}\left|\left[\sqrt{\left(\frac{\omega}{\omega_c}\right)^2 - 1} \cos \frac{\pi y}{y_0} - \sin \frac{\pi y}{y_0}\right]\right|. \tag{83}$$

If these are plotted as a function of position in the guide, the results are as indicated in Fig. 22.

At points A and B in the rectangular guide, the magnetic field is purely circularly polarized but in opposite senses. If the direction of propagation is reversed, the pattern shown in Fig. 21 remains unchanged but the sense of rotation assigned to the two curves reverses. Thus an observer standing at point A would see a pure positive circularly polarized field for one direction of propagation and a pure negatively polarized field for the other direction of propagation. The position of point A or B can be easily calculated from (83), which yields the relation

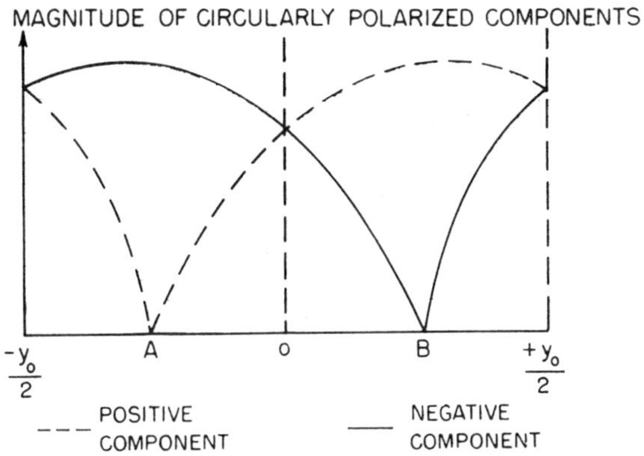

Fig. 22—Magnetic field components in a rectangular waveguide propagating the dominant mode.

$$\tan \frac{\pi y_A}{y_0} = \sqrt{\left(\frac{\omega}{\omega_c}\right)^2 - 1} \tag{84}$$

so that point A exists at the center of the guide when the frequency is just equal to the cutoff frequency and moves toward the side wall as the frequency is increased. As an example, a guide operating at 10,000 mc, whose cutoff frequency is 8000 mc, has the points of pure circular polarization at

$$y_A = 0.41\left(\frac{y_0}{2}\right). \tag{85}$$

If now a slab of ferrite were placed at point A and biased in the z direction, it would be expected that the slab would present the permeabilities given in Fig. 9, for the two different directions of propagation. Such a device is illustrated in Fig. 23.

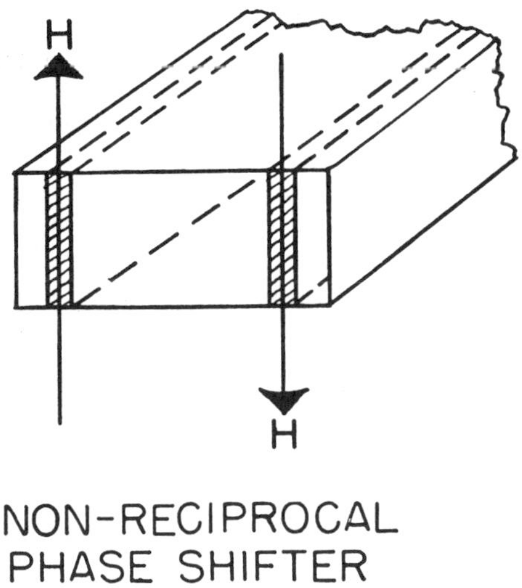

Fig. 23—Rectangular waveguide loaded with magnetized slabs of ferrite in a manner so as to make it nonreciprocal.

Actually it is not quite so simple, because the curves of Fig. 9 apply either for internal fields or for external fields if the ferrite has cylindrical symmetry about the z axis. Obviously, the slab given in Fig. 23 does not have cylindrical symmetry about the z axis. Hence, if it were placed in a region where the external fields were circularly polarized, the internal fields would be elliptically polarized and the sample would show resonance absorption for both directions of propagation. A closer approximation can be found by finding what conditions must be placed on the external fields so that the internal fields will be circularly polarized. This relation is very easily derived from (48). If the following requirement is placed on the internal fields (*i.e.*, they can be circularly polarized)

$$h_x = \pm j h_y \tag{86}$$

we are led to the following relation between the components of the external field

$$h_x^e = \pm j \frac{[1 + N_x X_{xx} \mp N_x X_{xy}]}{[1 + N_y X_{xx} \mp N_y X_{xy}]} h_y^e \tag{87}$$

19

where

$$X_{xx} = \frac{\mu - 1}{4\pi} = \frac{M\gamma\omega_0}{\omega_0^2 - \omega^2},$$

$$X_{xy} = \frac{K}{4\pi} = \frac{M\gamma\omega}{\omega_0^2 - \omega^2}.$$

If now the slabs shown in Fig. 22 are extremely thin, it was shown in the above discussion that the demagnetizing factors can be set approximately equal to

$$N_x = 0 \qquad N_y = 4\pi \qquad N_z = 0$$

regardless how long the slabs may be. (See Fig. 14.)

For this geometry, (87) reduces to:

$$h_x^e = \pm \frac{j h_y^e}{\mu \mp \kappa}. \tag{88}$$

Thus it is impossible to put the slab into the guide, in a manner such that it sees a pure positively rotating internal field for one direction of propagation and a negatively rotating internal field for the other direction unless the off diagonal component of the permeability tensor is zero. If the element is being used as a differential attenuator rather than a differential phase shifter, it is important that the ferrite see a pure negative circularly polarized wave for one direction of propagation so that there will be no absorption. For the other direction of propagation the internal field will be elliptically polarized, but that makes little difference. Thus, if the effective permeability for the negative circular component $(\mu - \kappa)$ is approximately two, then the relation between the external fields, so that the internal fields will be circularly polarized, is

$$h_x^e = \frac{j h_y^e}{2}, \tag{89}$$

which means the ferrite must be closer to the center than the point A in Fig. 22, if it is to have zero loss for the one direction of propagation. It is difficult, using this approach, to say anything about the point of maximum differential phase shift for a long thin slab. However, it is interesting to note that if in place of a long thin slab, extremely thin cylinders were placed in the guide with their axes parallel to the Z axis, then the position of maximum differential phase shift and maximum differential attenuation would occur when the cylinders were placed at point A or B in Fig. 22. The above analysis is valid only if the fields in the waveguide are undisturbed by the dielectric loading caused by inserting the ferrite. Experimental data indicates that if the slab goes all the way from top to bottom of the guide as indicated in Fig. 23 then the slabs must be thinner than 0.010 inch in order for this assumption to hold at X band. If, however, pieces of ferrite are placed against the upper and lower walls the pieces can be considerably thicker and the above analysis can be applied

Comparison of Various Microwave Devices

Differential Phase Shifter

The above theory makes it evident that several different types of nonreciprocal microwave devices can now be built in rectangular, round or square waveguide and it is often a perplexing problem for the component engineer to decide which particular geometry he should pick to realize a particular system function. Often the criterion which decides this for the engineer is a consideration which is not even included in the above theory, such as picking a particular geometry so that the ferrite can be kept in intimate contact with the waveguide walls, making it possible to dissipate the energy which it absorbs from the microwave fields rapidly enough so that it does not heat up above the Curie temperature. However, in spite of the complexity of the problem many general statements can be made about the various devices so that some guide can be given to the design engineer.

Thus a differential phase shifter operates best when the ferrite is biased with a weak magnetic field which is just sufficient to magnetize it. The strength of the field required is relatively independent of the frequency of operation of the device but since waveguide components get successively smaller as one goes to higher frequencies, the problem of providing the magnetic field for a differential phase shifter becomes increasingly easier as one operates at higher frequencies. For this reason alone, differential phase shifters look increasingly more attractive as one operates at frequencies above K band.

In addition, since (68) indicates that the differential phase shift (or Faraday rotation) per unit length is sensibly the same for two waveguides whose cross-sections contain the same amount of ferrite regardless of whether the guide is operating at 10,000 mc or 100,000 mc it turns out that the approximate length required to obtain a certain amount of differential phase shift is sensibly independent of the frequency of operation. The same equation indicates that a differential phase shifter can in principle be made as broad band as the waveguide itself.

As an example one can predict that if it is possible to build a 45° Faraday rotator which is 3 inches long with an applied field of 300 oersteds at a frequency of 24,000 mc then a 45° Faraday rotator to operate at 100,000 mc will be about 3 inches long and will require a magnetic field of approximately 300 oersteds.

The main advantages of a differential phase shifter can be summarized by stating that a very small biasing magnetic field relative to other microwave ferrite devices is required for its operation, and the ferrite need not necessarily absorb any microwave energy itself and hence the heating can be kept to a minimum. This turns out to be an important consideration for high average powers.

One disadvantage of the differential phase shifter is that it requires a great deal of other external circuitry (such as hybrid junctions) in addition to the phase shifter in order to build a simple isolator. (Of course if one is attempting to build a circulator then this extra circuitry is essential.) In addition as one goes to frequencies below 1000 mc the properties of a phase shifter deteriorate rapidly as shown in the paper by Lax.[18] Thus it is probably impossible to build a circulator to operate at 1000 mc with an insertion loss less than 1 db, and impossible to reduce the insertion loss below 2 db at 500 mc with any kind of ferromagnetic material which is known today regardless of whether it has ideal properties or not with respect to all its other characteristics.

This low frequency limitation, in itself, is really not too severe and if it were possible to build ferrites today whose other properties were ideal, differential phase shifters operating at 1000 mc and perhaps 500 mc would be available today. To realize this ultimate low frequency limit, we must develop materials which more nearly approach the ideal material assumed in the above discussion. The problems are well understood and research is now underway to develop materials which will make these devices operable. It is outside the realm of this paper to discuss these problems in detail. However it is felt that this paper would be incomplete without indicating the nature of the problem.

It has been shown by Polder and Smit[19] that demagnetized ferrites would be expected to have a large magnetic loss tangent between the frequencies of

$$\gamma H_a \leqq \omega \leqq \gamma(H_a + 4\pi M_s) \qquad (90)$$

where

$$H_a = \text{effective anistropy field} = \frac{2K_1}{M}.$$

[18] B. Lax, p. 1368, this issue.

[19] D. Polder, and J. Smit, "Resonance phenomena in ferrties," *Rev. Mod. Phys.*, vol. 25, p. 89; 1953.

(See Van Vleck.[20]) Thus it is to be expected that all demagnetized ferrites will have a large magnetic loss in this frequency region. However, when the ferrite is magnetized, this loss is not spread out over such a large band of frequencies but exists in a relatively narrow region (line width) around a frequency given by Kittel's equation.

If we assume a ferrite for which

$$\frac{2K_1}{M} = 400 \text{ oersteds} \tag{91a}$$

$$4\pi M_s = 2000 \text{ gauss} \tag{91b}$$

$$N_z = 0; \quad N_x = N_y = 2\pi \text{ (long cylinder)} \tag{91c}$$

and if we assume the wavelength of the radiation being considered is always much larger than the diameter of the ferrite (so that Kittel's equation can be applied), then one of the low frequency problems becomes immediately evident.

Eq. 90 indicates that this ferrite cylinder will be very lossy between 1120 mc and 6720 mc when it is demagnetized. When the ferrite is just magnetized (say in an applied field of 200 oersteds), the ferromagnetic resonance occurs at 3360 mc. If measurements are being taken at 4000 mc, a cylinder of the above ferrite will be at ferromagnetic resonance in an applied field of 430 oersteds. Thus, if the attenuation of a 4000 mc wave is measured as a function of the applied longitudinal field as the wave is propagated through a round waveguide containing the above cylinder two components are evident in the loss. First there is the initial loss which occurs in the demagnetized ferrite which approaches zero as the ferrite is magnetized, and then the ferromagnetic resonance of the magnetized cylinder which reaches a maximum in an applied field of 430 oersteds. This is indicated in Fig. 24.

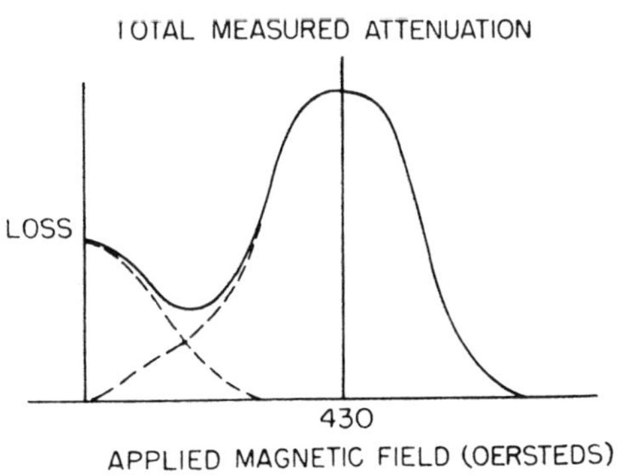

Fig. 24—Measured attenuation as a function of applied magnetic field for a cylinder of ferrite being used as a Faraday rotator at 4000 mc. (See text.)

Now if the saturation moment of the cylinder is reduced so that

$$(H_a + 4\pi M_s) < \omega, \tag{92}$$

then the initial loss in the demagnetized state will not occur for the frequency being used and the ferromagnetic resonance of the magnetized cylinder will not occur until higher magnetic fields are applied. Hence, the total loss in weak applied fields can be greatly reduced. Fig. 25 shows the measured loss and rotation of the plane of polarization for several samples of magnesium ferrite to which aluminum has been added to reduce the saturation moment of the material.[21] The data were

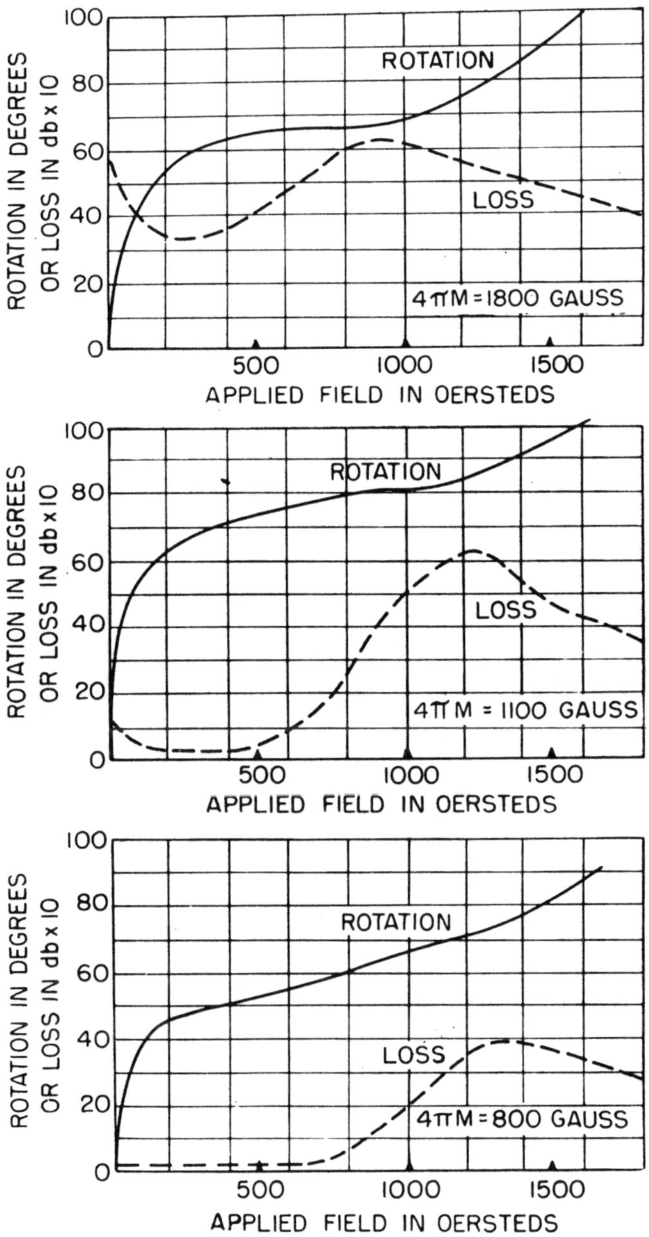

Fig. 25—Measured rotation and attenuation of cylinders of ferrite with various saturation moments when used as Faraday rotators at 4000 mc.

[20] J. H. Van Vleck, p. 1248, this issue.

[21] L. G. Van Uitert, J. P. Shafer, and C. L. Hogan, "Low loss ferrites for applications at 4000 mc/sec.," *J. Appl. Phys.*, vol. 25, p. 925; 1954.

taken at 4000 mc. It should be pointed out in this connection that the initial low-field loss does not completely disappear until the microscopic saturation moment of the individual ferrite grains has been reduced below that given in (92). Often the microscopic saturation moment is 10 per cent or 20 per cent larger than the macroscopic moment since the ferrites are usually not fired to maximum density. It can be easily shown, however, that the macroscopic moment is the quantity required in Kittel's equation, unless the ferrite is grossly nonhomogeneous.

Thus it is evident that if the major loss in a differential phase shifter is a magnetic loss it can be greatly reduced by reducing the saturation moment of the ferrite. (This is the major reason why some ferrites show an increase in the factor of merit when the temperature of the ferrite is increased.) However, as shown in (68), the rotation of the plane of polarization or the phase shift per unit length in a rectangular guide are proportional to the saturation moment of the ferrite, thus if the saturation moment is reduced too far, the physical size of the ferrite needed to obtain the desired result will become prohibitive. Thus a long cylinder of ferrite whose saturation moment is only 720 gauss will be at ferromagnetic resonance by the time it is magnetized, when it is subjected to a 1000 mc alternating field. Thus it appears that long cylinders are practical for a phase shifter only down to frequencies of the order of 1000 mc, unless one operates on the high field side of the resonance line. If on the other hand one uses a thin disc of ferrite rather than a cylinder, and magnetizes the disc perpendicular to the plane of the disc, then one can extend the low frequency limit of a particular ferrite. For instance, if the ferrite whose properties are described in (91a) and (91b) is cut into a thin disc instead of a cylinder, the demagnetizing factors for this geometry become

$$N_x = N_y = 0; \qquad N_z = 4\pi \tag{93}$$

and hence from Kittel's equation it is seen that the applied magnetic field necessary for resonance is

$$H_a = \frac{f(mc)}{2.8} + 4\pi M_s. \tag{94}$$

Thus if this ferrite is again used at 4000 mc ferromagnetic resonance will not occur in the disc until an applied field of

$$H_a = 1430 + 2000 = 3430 \text{ oersteds} \tag{95}$$

is reached.

On the other hand, the disc will almost completely saturate in an applied field of 2000 oersteds, and hence the low field loss will disappear long before the peak in the ferromagnetic resonance occurs. Thus, in place of Fig. 23, the loss vs applied field for the disc will be as indicated in Fig. 26.

It is seen by comparing Fig. 23 with Fig. 26 that the identical piece of ferrite might be usable at 4000 mc when cut into a thin disc but not when cut into a long cylinder, and hence part of the low frequency problem can be considered a geometry problem as well as a materials problem.

In addition, a polycrystalline ferrite has a ferromagnetic resonance absorption line width at least of the order of magnitude of the effective crystalline anisotropy field. Thus a polycrystalline nickel ferrite would have a resonance absorption line width of approximately 500 oersteds. Obviously, these wide lines hurt more and more as one attempts to operate at lower frequencies, and hence for extremely low microwave frequencies (below 1000 mc), it becomes increasingly important to obtain polycrystalline ferrites with low crystalline anisotropy, or in lieu of this, to obtain good single crystals.

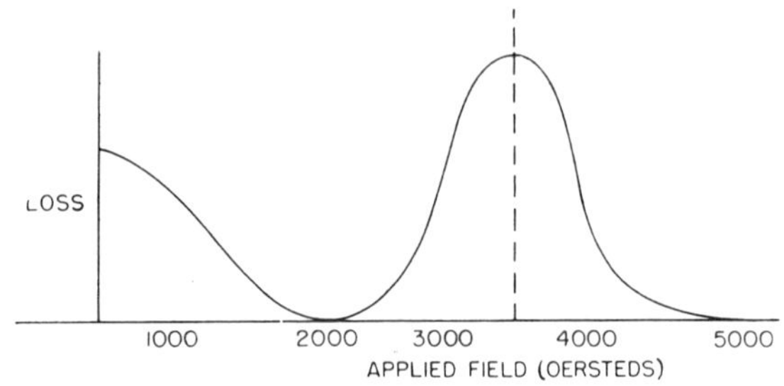

Fig. 26—Measured attenuation as a function of applied field of same ferrite as shown in Fig. 24. Here the ferrite is cut as a disc and located coaxially in a round waveguide or as thin slabs in a rectangular waveguide as shown in Fig. 27.

Resonance Absorption Devices

It is obvious that if a ferrite is placed in a waveguide and biased with a dc magnetic to its resonant frequency as given by Kittel's equation it will be extremely lossy provided the internal rf magnetic field vector in the ferrite has a positive circularly polarized component. It is on the basis of this phenomenon that microwave isolators can be built. Perhaps the simplest is that described above where a rectangular waveguide is loaded as shown in Fig. 23 and the magnetic field is adjusted to bring the ferrite to resonance. Usually in practice the slabs do not go all the way across the guide as shown in Fig. 23 but consist of small pieces placed in contact with the upper and lower walls of the guide.

This device has the advantages of constituting one of the simplest microwave ferrite devices from a circuit point of view and also having the best low frequency limit as shown by Lax.[22] As shown in that paper the back to front loss ratio in this device can be as high 10 db to 1 db for such a device operating at 200 mc. Certainly until ferrite materials are developed which overcome the problems discussed in the above section on differential phase shifters, resonance absorption devices are the only practical devices which can be built for operation at 1000 mc or lower today.

[22] Lax, *loc. cit.*

In reference to Figs. 24 and 26 it is evident that one should choose a geometry for the ferrite which maximizes the difference between the field required to magnetize and the field required for resonance as one goes to very low frequency operation (below 2000 mc). The simplest geometry which affords this result is with thin flat slabs in contact with the upper and lower walls of the guide as indicated in Fig. 27. This geometry also improves the thermal contact of the ferrite with the waveguide walls and thus allows the ferrite to dissipate the heat absorbed from the microwave energy more readily.

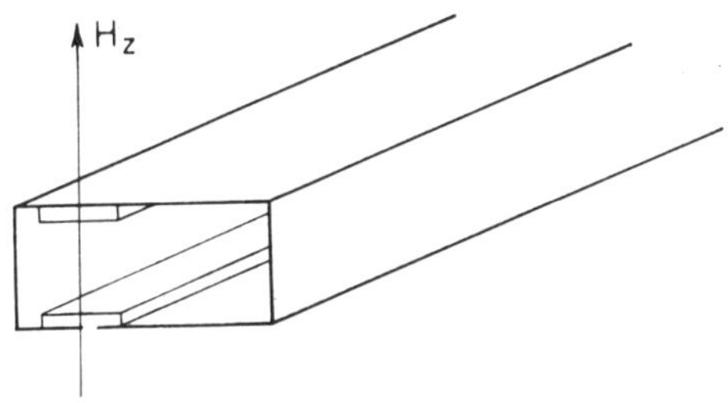

Fig. 27—Optimum geometry for loading a rectangular waveguide at very low microwave frequencies.

One of the main disadvantages of a resonance absorption isolator is that the ferrite, itself, must absorb all of the backward wave and in microwave systems which operate at large average power levels into loads with a high vswr this disadvantage can become disastrous if one does not resort to forced cooling. Another disadvantage of a resonance absorption isolator is that the magnetic field required for resonance becomes prohibitively high as one goes to very high frequency microwave systems even it one chooses optimum geometries. Thus if a resonance absorption isolator to operate at 56,000 mc is built with the geometry of Fig. 23 (optimum geometry for minimizing the dc magnetic field) an applied magnetic field of approximately 18,500 oersteds will be required if the ferrite has a saturation moment of 3000 gauss. It is doubtful if any permanent magnet today can produce this field over the volume required and at the same time have the high stability which is necessary.

Two considerations enter into determining the band width of a resonance absorption isolator. The first is related to the geometry of the guide itself. Thus to maximize the reverse loss and minimize the forward loss the ferrite should be placed in a position such that it sees a pure positive circular polarization for one direction of propagation and a negative circular polarization for the other direction of propagation. As we saw from (87) above this is in general impossible for a rectangular waveguide and even if one selects the optimum position, it is easily shown from (82) and (87) that this optimum position is in itself a function of frequency. Recently this problem has been partially overcome by dielectrically loading the waveguide. Even if the position of pure circular polarization does not change with frequency the resonance absorption isolator must have a bandwidth determined by the width of the resonance absorption line in Fig. 10 unless the ferrite is placed in a grossly inhomogeneous magnetic field.

By a combination of dielectric loading and inhomogeneous field, Melchor, Ayers, and Vartanian[23] at the Electronic Defense Laboratory have produced the broadest band resonance absorption isolator known to the author. The characteristics of this device are shown in Fig. 28. It seems apparent that slightly more development of this particular device would give front to back ratios in excess of 75 db to 1 db over a band from 8000 mc to 14,000 mc.

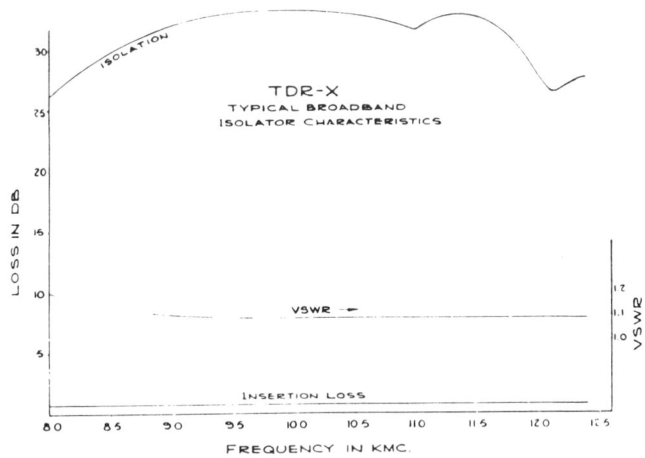

Fig. 28—Measured characteristics of dielectrically loaded ferrite isolator developed at the Electronic Defense Laboratory.

Field Displacement Devices

Field displacement devices are probably the most difficult to understand from an intuitive point of view since they cannot be completely understood either from an infinite plane wave analysis, or from perturbation theory of the loaded waveguide. However much can be said about the design of a field displacement isolator from the theory which has been developed above.

Thus if an infinite plane wave is incident upon a semi-infinite slab of ferrite magnetized in the direction of propagation of the wave as indicated in Fig. 29, some rather interesting aspects of the wave motion can be deduced from the above theory.

Thus as was shown in (51), the propagation constant for the two counter-rotating circularly polarized components of the wave can be written as

$$\Gamma_{\pm} = j\omega[\epsilon\mu_0\mu_{\pm}]^{1/2} \tag{96}$$

where

$$\mu_{\pm} = \mu \pm \kappa.$$

The real and imaginary parts of $\mu \pm \kappa$ are plotted in Figs. 9 and 10 for a particular set of values of the

[23] Private communication.

23

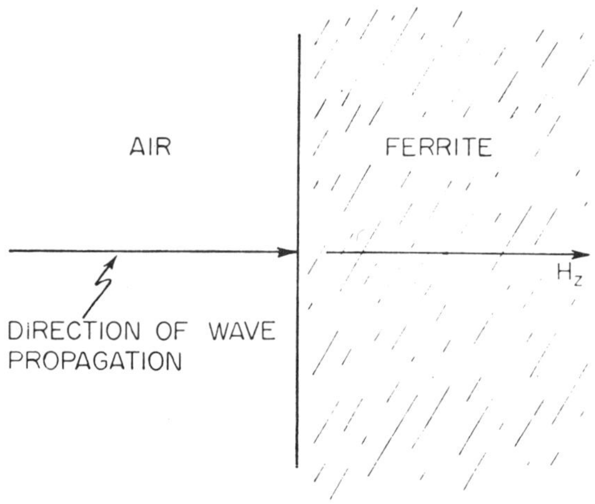

Fig. 29—Infinite plane wave incident upon a ferrite-air interface. Ferrite is magnetized in the direction of propagation of the wave.

parameters. Now if (96) is solved for the attenuation and phase constant under the assumption that the only loss is due to the imaginary part of the effective permeability shown in Fig. 10, one gets

$$\begin{aligned} \alpha_{\pm} &= \omega\sqrt{\epsilon\mu_0\mu_r{}^{\pm}} \\ \beta_{\pm} &= \omega\sqrt{\epsilon\mu_0\mu_L{}^{\pm}} \end{aligned} \tag{97}$$

where

$$\mu_r{}^{\pm} = \frac{[(\mu_{\pm}')^2 + (\mu_{\pm}'')^2]^{1/2} - \mu_{\pm}}{2}$$

$$\mu_L{}^{\pm} = \frac{[(\mu_{\pm}')^2 + (\mu_{\pm}'')^2]^{1/2} + \mu_{\pm}'}{2}.$$

If we refer to Fig. 9 we should expect some rather startling behavior for the positive circular component if the magnetic field is raised to the point where the effective permeability (*i.e.*, μ_+') is zero or negative. In fact if we neglect the losses associated with the *tail end* of the absorption line at first we see that the semi-infinite ferrite medium behaves like a waveguide beyond cutoff for the positive circular component. That is, neglecting all losses, when the magnetic field is adjusted to the zero permeability value for the positive component both α_+ and β_+ are zero. When the magnetic field is adjusted so that the effective permeability, μ_+', is negative we find that

$$\begin{aligned} \beta_+ &= 0 \\ \alpha_+ &= \omega\sqrt{\epsilon\mu_0 \mid \mu_+' \mid}. \end{aligned} \tag{98}$$

Thus we have an attenuation constant but no phase constant. This is obviously a reactive attenuation such as one encounters in a lossless waveguide beyond cutoff. In other words the wave is reflected from the air-ferrite interface. Thus we can define a skin depth which is the distance into the ferrite which the wave will penetrate before it is reduced to $1/e$ of its initial amplitude. On the basis of this

$$\delta_+ = \frac{1}{\omega\sqrt{\epsilon\mu_0 \mid \mu_+' \mid}}. \tag{99}$$

As the effective permeability for the positive circular polarization takes on larger and larger negative values (increasing the applied magnetic field) the depth of penetration becomes smaller and smaller. The effective depth of penetration assuming no loss can be easily computed from the above equation. Thus, for instance, if one assumes an operating frequency of 9000 mc, a relative dielectric constant of 10, and in addition assumes that the biasing magnetic field is adjusted so that the effective permeability of the medium for the positive circular component is minus 1, then (99) indicates that the effective depth of penetration of the wave into the medium is

$$\delta_+ = 1.7 \text{ mm.} = 0.067 \text{ inch.}$$

A similar effect occurs if the ferrite is placed in the waveguide at a position where it sees a positive circular polarization for one direction of propagation and a negative circular polarization for the other direction. For the direction of propagation in which the magnetic field vector is positively polarized the rf fields tend to be expelled from the ferrite loaded region and penetrate the ferrite to a depth indicated by (99). If the ferrite dimensions are large compared to the effective skin depth thus calculated, then appreciable field displacement effects should take place in the waveguide. More exact criteria can be established by the reader by the use of (97) and the theory developed above.

It will be recognized by the reader that the criteria given above are greatly oversimplified. Some field displacement takes place whenever β_+ is different from β_- in (97) and the effect is maximized by maximizing the difference. In practice, however, it is found the effect is not large unless the effective permeability for the positive circular polarization is zero or negative.

It is apparent that in order to build a good field displacement device it is necessary to bias the ferrite into the region where it has zero or negative permeability for the positive circular component but on the other hand the magnetic field must be low enough so that losses due to the resonance absorption line are negligible. These requirements of course put frequency limitations in which a practical field displacement device can be built. At too high a frequency (above K band) the magnetic field required to reach the zero permeability region is becoming prohibitively high. At too low a frequency (below L band) the ferrite will have considerable loss when the real part of the permeability is zero and like the cutoff in a very lossy guide, the field displacement effect deteriorates rapidly.

The behavior of a field displacement device is clearly illustrated in some calculations made by Lax and Button[24] and two of the modes which they calculated are shown in Fig. 30. The propagation constants for

[24] B. Lax and K. J. Button, "Theory of new ferrite modes in rectangular waveguide," *J. Appl. Phys.*, vol. 26, p. 1185; 1955.

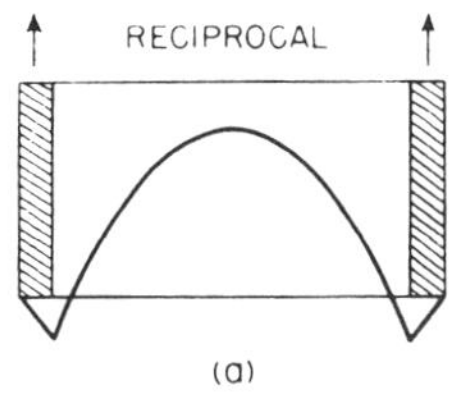

Fig. 30—Modes in a ferrite loaded rectangular waveguide. Magnetic field directions shown by arrows.

these modes are shown in Fig. 31. It is noticed that for very thin slabs of ferrite only mode A exists which is not radically different from the empty wave guide mode. However when the slabs are thick enough a strong field displacement toward one side of the waveguide takes place for one direction of propagation and toward the other side of the guide for the opposite direction of propagation. It is thus obvious that if a resistance card is placed on the surface of one of the slabs in Fig. 30(b) it will cause a very large attenuation for one direction of propagation and a very small attenuation for the other direction of propagation.

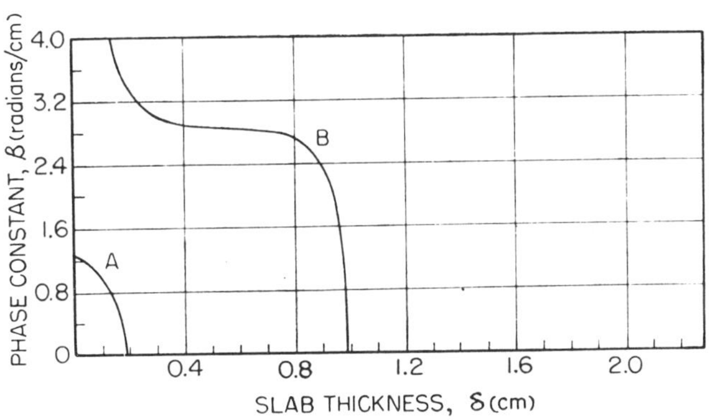

Fig. 31—Propagation constants for modes shown in Fig. 30 as a function of slab thickness. (In both cases the magnetic field is such that the ferrite has an intrinsic negative permeability for positive circular polarization.)

Commercial Ferrite Devices

In conclusion it is felt that this article would not be complete without indicating to the reader the types of commercial devices which have already come into wide use today. The list is not complete but is meant only to show the reader the extent of present day development.

Figs. 32 and 33 show an assembled and exploded view of a Faraday rotator isolator developed at Hughes Research and Development Laboratories. It has an isolation greater than 18 db and an insertion loss less than 0.5 db over a 10 per cent band centered around X band. It will handle a peak power of 300 kw and an average power of 300 watts with forced air cooling. The ferrite is magnetized by means of bar magnets on the outside wall of the waveguide. One of these magnets is shown in Fig. 33.

Fig. 34 shows an X-band circulator developed at Airtron, Inc. It is capable of handling a peak power of

Fig. 32—Assembled view of high power isolator developed at Hughes Research Laboratories.

Fig. 33—Exploded view of isolator shown in Fig. 32.

Fig. 34—Typical ferrite circulator using one folded hybrid and one Riblet hybrid.

25

500 kw and an average power of 500 watts without cooling. This circulator gives an insertion loss of 0.5 db and the cross coupling is down by 20 db.

Fig. 35 shows a typical rectangular waveguide resonance absorption isolator also developed at Airtron. This particular device gives an isolation greater than 10 db and an insertion loss less than 0.5 db while handling 2.5 megawatts of power at S band. It will work into a 2 to 1 load mismatch.

This list is not complete and there are available on the market today dozens of different types of microwave ferrite devices made by at least a dozen different concerns.

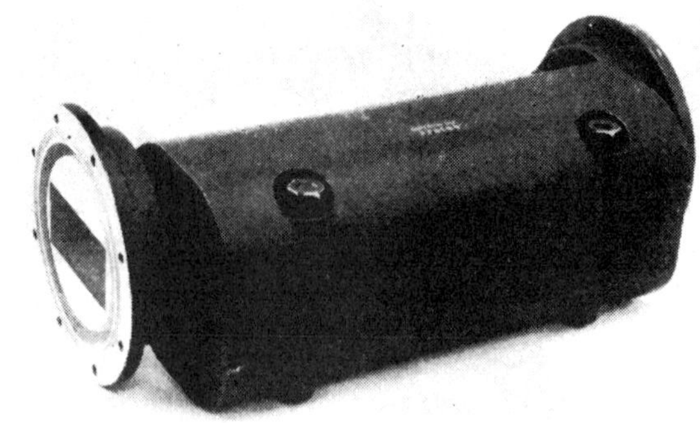

Fig. 35—High power S-band isolator developed by Airtron, Inc.

Application of Ferrites to the Microwave Antenna Problem*

N. G. Sakiotis, A. J. Simmons and H. N. Chait†

The design of rapid-scanning antennas is a problem of long standing in the field of microwave antennas. In many radar applications—such as target search and acquisition, gunfire and missile control, and homing—radiation beams of various widths and shapes must be moved rapidly through a given angle. When the antenna is small, as in airborne gun laying, the entire unit may be moved rapidly. In larger designs, however, this direct approach cannot be used, so that ingenious schemes have to be invented for shifting the beam position without moving the bulk of the antenna. Unfortunately, these schemes frequently result in mechanically complicated structures, which are not only hard to construct and to maintain mechanically but which also impose limitations on the electrical performance of the system.

A scanning antenna which requires no physical motion, either of the secondary focusing system (reflector or lens) or of the primary feeding source, has been a goal of the microwave antenna designer ever since the inception of the scanning problem. One attractive method of moving a beam without moving the antenna is to shift relative phase between the signals radiated by elementary antennas in an array. The array elements may be fed by tapping off a feed line at regular intervals, or by branching the feed line, or they may be field-fed from a parallel-plate "pillbox" source. In any case a phasing element may be inserted in the transmission line just before the radiating element. Rapid scanning by this means has been limited by the fact that the only available lossless phase shifters operate mechanically by means of a moving waveguide or dielectric section. Thus the disadvantages of mechanical scanning have not been entirely eliminated. An electronic means of shifting phase, if it could be made practical, would be highly desirable. It would possess the potentiality of producing a rapid-scanning antenna which is mechanically simple and quick-operating.

A possible way to shift phase electronically is by use of a material whose dielectric constant, ϵ, or whose permeability, μ, may be changed by means of external dc or low-frequency electric or magnetic fields. Such materials inserted in waveguide or coaxial line would, upon application of the external field, cause a change in electrical length of the transmission line, equivalent to a shift in phase. Materials of variable ϵ, such as the titanates, have proved too lossy for antenna applications, but materials of variable μ such as the ferrites—combinations of iron oxide (Fe_2O_3) with other bivalent metallic oxides—show some promise. These materials are used at ordinary radio frequencies for coil cores since they combine high permeability with low conductivity and thus have low losses. Because of the low conductivity of the ferrites, being composed of oxides rather than pure metals, skin-effect phenomena do not prevent the penetration of microwave fields into these materials. Thus they are still useful in these high-frequency ranges, even though the permeability values are much lower than those obtained at lower radio frequencies.

---

*The material contained in this article is Unclassified.

†Antenna Research Branch, Radio Division I

The characteristic of most interest at microwave frequencies is the variation of μ with an applied magnetic field. This variation takes place in an anisotropic fashion, as evidenced by the rotation of plane of polarization of a linearly polarized wave propagating in a direction parallel to the direction of applied magnetic field. This effect, the Faraday effect, may be explained by an anisotropy introduced between the μ for right-hand and that for left-hand circularly polarized waves.[1] The linearly polarized wave is thought of as a combination of right-hand and left-hand circularly polarized waves, and the ferrite anisotropy causes a relative phase shift between these two components, rotating the plane of polarization of the resultant linear wave through a prescribed angle when the dc field is applied. Because of this rotation of the plane of polarization, the ferrites can be utilized to advantage in a waveguide switch. With magnetizing current off, the output polarization is the same as input. With current on, the output polarization may be rotated 90° and passed through a different channel. Such a switch could have many applications such as scanning by
28 switching rapidly from one primary feed to another, or as a TR device separating

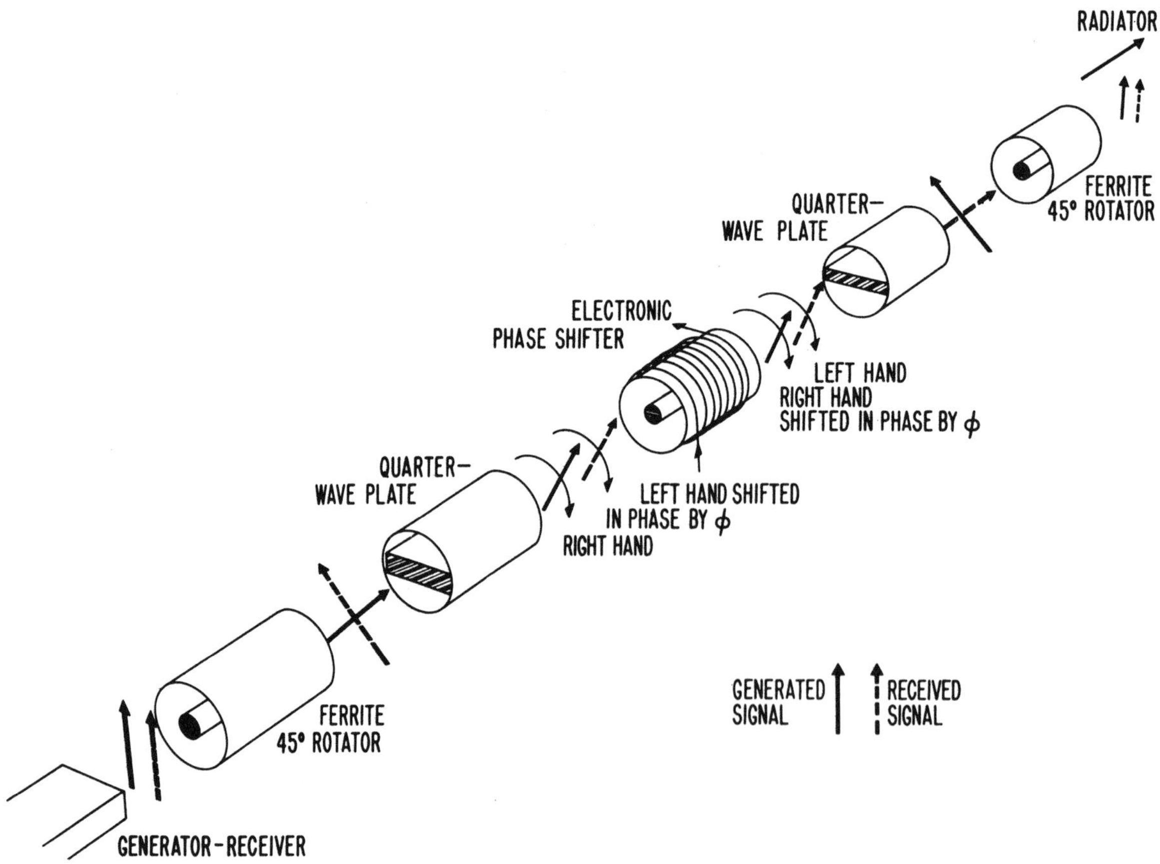

Figure 1 - Bilateral phase shifter

---

[1] D. Polder, "On the Theory of Ferromagnetic Resonance," Philosophical Magazine 40, 99-115 (1949)

receiver and transmitter functions in a radar set. For general scanning purposes, however, the phase-shifting properties due to the variation of μ are of major interest. For use as a phase shifter, a single circularly polarized wave may be passed through a section of hollow waveguide containing the ferrite; the phase of the output wave relative to the input is varied by application of the axial dc magnetic field.

An interesting property of these materials is that the reciprocity principle is not obeyed—a wave traveling in the direction of the external magnetic field is affected differently from one traveling in the opposite direction. For example, if the plane of polarization of a linearly polarized wave propagating in the same direction as the external magnetic field is rotated in a clockwise direction, a wave propagating in the opposite direction will be rotated counterclockwise. For scanning purposes this non-reciprocity is undesirable since the beam of the radar antenna on reception should be directed at the same point as the beam on transmission. This difficulty may be met in various ways, one of which is to combine transmission line sections containing ferrites in the manner shown in Figure 1. The solid vectors represent transmission, the dotted vectors, reception. The phase-shifting element will shift a right-hand wave traveling in one axial direction exactly as much as a left-hand wave traveling in the opposite direction.[2] By eliminating the 45^{0} rotator on the generator-receiver end, the received polarization is at 90^{0} to that transmitted, permitting the use of polarization switching as a TR device.

A research program is under way to investigate the properties of ferrites in waveguide at X-band, including phase shift, attenuation, and power-handling capacity as well as the variation of these properties with temperature and rf power level. Circular and rectangular waveguide are being investigated, each with circular and linear polarization and with longitudinal and transverse applied magnetic field. Some preliminary results on the phase shift and insertion loss of a circularly polarized wave passing through a cylindrical rod of Ferramic A[3] are shown in Figures 2 and 3. The results are plotted versus the magnetizing current in a 500-turn coil. The sample dimensions and the sample holder[4] are shown schematically in

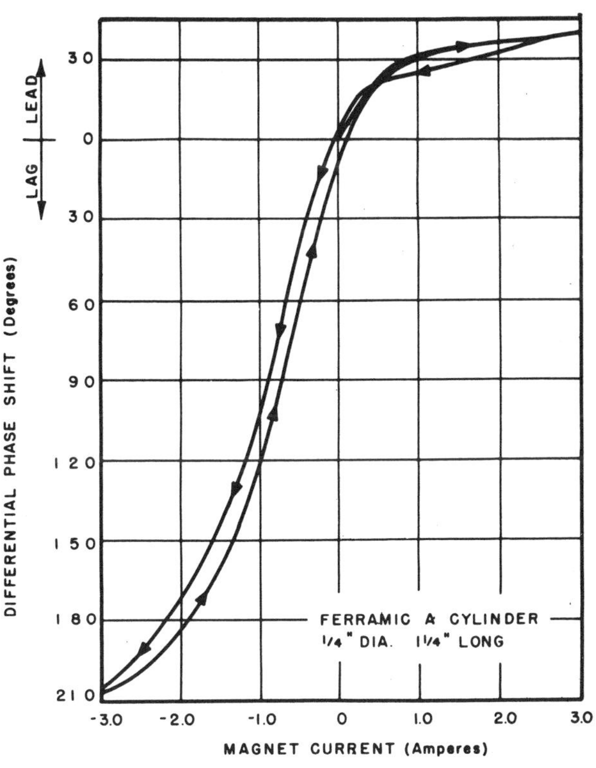

Figure 2 - Phase shift versus current

[2] The sense of rotation is defined according to IRE Standards as right hand = clockwise rotation looking in the direction of propagation; left hand, counterclockwise.

[3] Ferramic A is a trade name given to a sintered mixture of oxides (in ceramic form) manufactured by General Ceramic Company.

[4] This sample holder is modeled after one used by C. H. Luhrs (C. H. Luhrs and Co., Hackensack, N.J.) in a polarization switch.

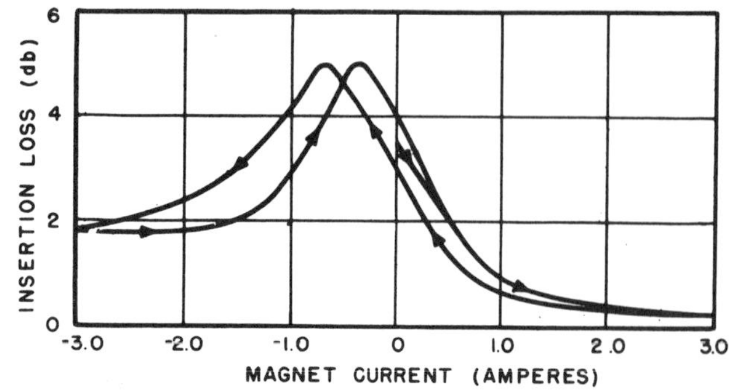

Figure 3 - Insertion loss versus current

Figure 4. The experimental setup is shown in Figure 5. This setup permits simultaneous measurement of phase shift, insertion loss, input VSWR and ellipticity as the dc magnetic field is applied. All elements are carefully matched to the line before the field is applied.

The effect of size and shape on the properties of the ferrites, as well as the theoretical propagation modes in ferrite-filled waveguide,

30

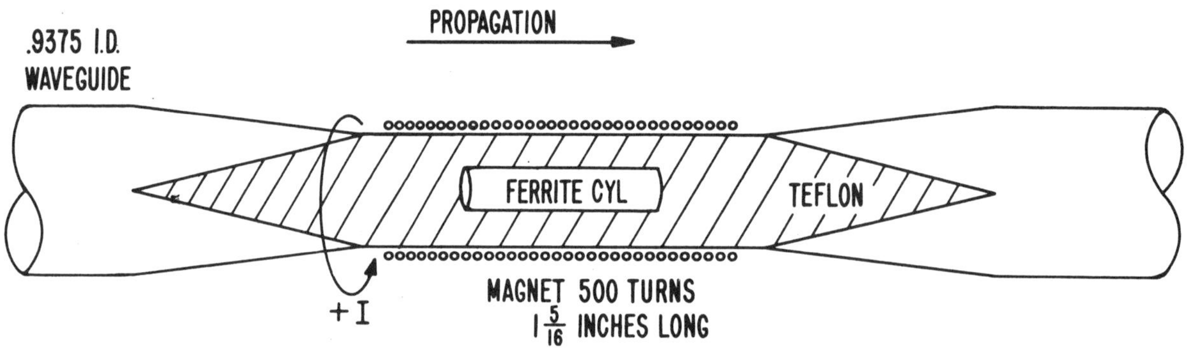

Figure 4 - Sample and holder (schematic)

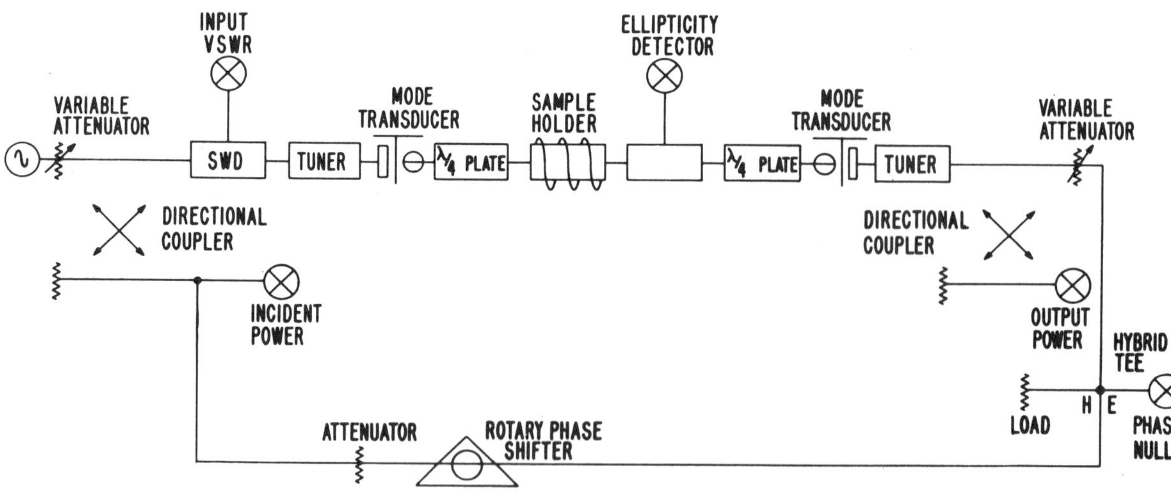

Figure 5 - Experimental setup (schematic)

are being studied. One result of tests so far shows that the phase shift per unit length for a given diameter rod is a function of length, increasing as length increases because of the change in demagnetization factor as the rods get longer. That is, for a given applied field the effective magnetizing field is of smaller magnitude for the shorter samples. A quantitative study of this and other effects is being undertaken preliminary to possible application to scanning antenna systems. It is expected that the information obtained from this study will be used in the design of many new-type microwave components such as electronic phase shifters, switches, duplexers, modulators, attenuators and polarization changers.

* * *

31

NRL Reprint No. 4-53

FERRITES AT MICROWAVES

H. N. Chait and N. G. Sakiotis

Reprinted from INSTITUTE OF RADIO ENGINEERS, January 1953

NAVAL RESEARCH LABORATORY

WASHINGTON, D.C.

Reprinted from the PROCEEDINGS OF THE I.R.E.
VOL. 41, NO. 1, JANUARY, 1953

PRINTED IN THE U.S.A.

Ferrites at Microwaves*

N. G. SAKIOTIS†, ASSOCIATE, IRE, AND H. N. CHAIT†

33

***Summary*—After a preliminary discussion of the propagation of waves in an unbounded ferromagnetic medium, and in waveguides containing ferrites, the results of an experimental study of the propagation characteristics of waveguides containing ferrites are presented for some samples of commercially available ferrites. Some of the applications of ferrites to microwave circuitry are briefly described.**

Introduction

Recently there have appeared on the microwave scene new materials under the generic name of "ferrites." Consisting of metallic oxides having a spinel structure and a general chemical formula XFe_2O_4, where X represents a bivalent metal such as magnesium, nickel, or cobalt, the ferrites furnish an interesting microwave propagation medium. The high resistivity of the oxide components together with the ferromagnetic properties of the iron component classify the ferrites as ferromagnetic dielectrics. Because of their high magnetic susceptance at low frequencies, the ferrites furnish a microwave medium whose properties may be varied over quite a large range by magnetizing the ferrite with a controlled static magnetic field.

Among their other ferromagnetic properties, the ferrites exhibit a strong Faraday Effect. This phenomenon was first observed by Faraday in 1848 when he passed a plane polarized light beam through a thin sheet of magnetic material magnetized in the direction of propagation. The following results were observed and have recently been obtained in the microwave region.[1]

1. The plane of polarization is rotated in passing through the medium, the angle of rotation being proportional to the thickness of material traversed.
2. The direction of rotation of plane of polarization depends only on the direction of the applied static magnetic field. Relative to an observer looking along the direction of the magnetic field, the sense of rotation of a linearly polarized wave will be the same whether the wave is traveling toward or away from him.
3. The angle of rotation is a function of the strength of the magnetic field.

The Infinite Ferromagnetic Medium

The observed phenomena may be explained[2] by considering the linearly polarized wave to be made up of two

* Decimal classification: R282.3. Original manuscript received by the Institute, September 29, 1952.
† Antenna Research Branch, Radio Division I, Naval Research Laboratory, Washington 25, D. C.

[1] F. F. Roberts, "A note on the ferromagnetic Faraday effect at centimetre wavelengths," *Le Jour. De Phys. et le Radium*, vol. 12, pp. 305–307; March, 1952.

[2] The following is based on the explanation given by M. Abraham and R. Becker, "Theorie der Elektrizitat," Taubner, Leipzig, vol. II, p. 27; 1933.

circular components, one right-handed and the other left. Upon application of the magnetic field the material becomes anisotropic in such a way as to have different indices of refraction for right- and left-hand circular polarization. Thus, in passing through the material, a relative phase shift is introduced between the two circularly polarized components which then recombine to give a rotated linear polarization at the output. The above explanation may be expanded in mathematical terms as follows:

Let a linearly polarized plane wave propagate through a lossless, semi-infinite medium in the direction of the applied field intensity $\bar{H}$. If we erect rectangular axes x, y, and z with the z-axis parallel to $\bar{H}$, the E-vector of the plane wave will lie in the x-y plane. We will specify the magnitude of $\dot{E}$ to vary as $\cos(\omega t-\beta_0 z)$:

$$|\bar{E}| = E_0 \cos(\omega t - \beta_0 z),$$

34 or if $\bar{i}_x$ and $\bar{i}_y$ are unit vectors along the x and y axes,

$$\bar{E} = [E_{0x}\bar{i}_x + E_{0y}\bar{i}_y]\cos(\omega t - \beta_0 z)$$
$$\bar{E} = |\bar{E}|\cos\theta\bar{i}_x + |\bar{E}|\sin\theta\bar{i}_y$$
$$\bar{E} = \mathrm{Re}\,(|\bar{E}|\,e^{j\theta})\bar{i}_x + \mathrm{Im}\,(|\bar{E}|\,e^{j\theta})\bar{i}_y.$$

We see from the last equation that we can derive the x- and y-components of $\bar{E}$ from the real and imaginary parts of the scalar $|\bar{E}|e^{j\theta}=E_0e^{j\theta}\cos(\omega t-\beta_0 z)$. Thus, $e^{j\theta}$ in this system of notation indicates spatial orientation rather than a time phase as is more usual. In this notation, then, a right-hand circularly polarized wave would be represented by $Ee^{j\omega t}$, a left-hand by $Ee^{-j\omega t}$, where E may be complex.

In accordance with the description of the Faraday effect given above we will let $\theta=az$, where θ is the angle rotation of the plane of polarization assuming the wave enters the medium aligned with the x-axis, at $z=0$. Then the expression for $|\bar{E}|e^{j\theta}$ is

$$E_0e^{j\theta}\cos(\omega t - \beta_0 z) = \frac{E_0}{2}e^{jaz}\left[e^{j(\omega t-\beta_0 z)} + e^{-j(\omega t-\beta_0 z)}\right]$$
$$= \frac{E_0}{2}\left[e^{j[\omega t-(\beta_0-a)z]} + e^{-j[\omega t-(\beta_0+a)z]}\right].$$

Let $(\beta_0-a)=\beta_+$ and $(\beta_0+a)=\beta_-$. Then,

$$|\bar{E}|\,e^{j\theta} = \frac{E_0}{2}e^{j(\omega t-\beta_+ z)} + \frac{E_0}{2}e^{-j(\omega t-\beta_- z)}$$

and

$$\bar{E} = \bar{i}_x\frac{E_0}{2}\cos(\omega t - \beta_+ z) + \bar{i}_y\frac{E_0}{2}\sin(\omega t - \beta_+ z)$$
$$+ \bar{i}_x\frac{E_0}{2}\cos(\omega t - \beta_- z) - \bar{i}_y\frac{E_0}{2}\sin(\omega t - \beta_- z).$$

It can be seen by inspection of the above expression that the rotating plane wave is equivalent to two circularly polarized plane waves traveling along the z-axis, with different phase velocities. One is circularly polarized in the same sense as the angle of rotation of the linearly polarized wave and has a phase constant $\beta_+=\beta_0-a$, while the other is circularly polarized in the opposite sense having $\beta_-=\beta_0+a$.

The rate of rotation a, and also the angle of rotation θ, can then be expressed as

$$a = \frac{\beta_- - \beta_+}{2},$$

$$\theta = \frac{\beta_- - \beta_+}{2}z. \tag{1}$$

It must be emphasized that the β_0 is the phase constant for a rotating linearly polarized wave in the medium and is not necessarily identical with the phase constant of an ordinary nonrotating, linearly polarized wave traveling through the medium. We may express β_0 in terms of β_+ and β_-,

$$\beta_0 = \frac{\beta_- + \beta_+}{2}.$$

Since θ is a function of the magnetic field, β_0, β_+, and β_- will then be functions of the magnetic field as well as of the material. These functions will have to be determined by an analysis of the interaction of the material and the applied fields.

The above model shows that in order that the Faraday effect take place the material must exhibit different phase velocities for the two senses of circularly polarized waves. Since the direction of rotation of the plane wave depends only on the direction of the applied static field, the phase velocity of each of the circularly polarized components will depend upon which sense of polarization the component has with respect to the applied field, not upon which sense it has with respect to the direction of propagation. Thus, in order to prevent confusion, we will have to abandon the usual nomenclature of right- and left-hand circular polarization since it depends on the direction of propagation. Instead, a wave circularly polarized in the sense that would advance a right-hand screw in the direction of the static field will by definition have a positive sense of polarization and will be characterized by a phase velocity β_+ irrespective of whether it propagates in a direction parallel or antiparallel to the static field. The opposite sense will be the negative sense with phase velocity β_-. Therefore a given circular polarization will change sense if the direction of the applied field is reversed with respect to the direction of propagation.

Polder[3] has shown that a ferromagnetic medium which is homogeneously magnetized to saturation is characterized by a tensor permeability. That is, RF magnetic field intensity $\bar{h}$ and flux density $\bar{b}$ are related by

$$\bar{b} = [T]\bar{h},$$

where $[T]$ is a tensor of the form

[3] D. Polder, "On the theory of ferromagnetic resonance," *Phil. Mag.*, vol. 40, pp. 99–115; January, 1949.

$$[T] = \begin{bmatrix} \mu & -j\alpha & 0 \\ j\alpha & \mu & 0 \\ 0 & 0 & \mu_0 \end{bmatrix},$$

where the tensor components, μ and α, are functions of the static field strength, flux density, and magnetization as well as the frequency of the microwave field.

Uniform plane-wave solutions of Maxwell's equations have been obtained[3,4] for such a saturated ferromagnetic medium magnetized in an arbitrary direction with respect to the direction of propagation. In general, two solutions are obtained, representing two elliptically polarized waves propagating in the same direction but with different velocities. In the case of the medium magnetized parallel to the direction of propagation, the two solutions represent contrarotating circularly polarized waves with phase velocities

$$\beta_- = \omega\sqrt{\epsilon(\mu + \alpha)}, \qquad \beta_+ = \omega\sqrt{\epsilon(\mu - \alpha)}.$$

In the case of the medium magnetized perpendicularly to the direction of propagation, the two solutions represent orthogonal linearly polarized waves. One wave has its E-vector polarized parallel to the direction of magnetization and travels with the phase velocity

$$\beta_{\parallel} = \omega\sqrt{\epsilon\left(\frac{\mu^2 - \alpha^2}{\mu}\right)},$$

while the other is polarized perpendicularly to the magnetization and travels with the phase velocity

$$\beta_{\perp} = \omega\sqrt{\epsilon\mu_0},$$

which is clearly not a function of the magnetization of the medium.

Ferrites in Waveguide

While the above discussion was based upon an infinite medium, the results have been found to be useful guides in qualitatively understanding the results obtained in the case of a circular waveguide with a small rod of ferrite placed on the axis of the waveguide. On the other hand, the infinite medium propagation constants have not been useful in understanding the data obtained in the case of a waveguide of rectangular cross section containing ferrite material.

The μ-tensor given above seems to be an inherent property of magnetized media and is therefore applicable to all analyses of ferrite propagation phenomena irrespective of the geometry of the medium. The phase velocity and the configuration of the fields, however, depend on the geometry involved. Thus, in order to analyze the propagation of energy through waveguides containing ferrite media, solutions must be found to Maxwell's equations with the ordinary scalar permeability replaced by the tensor permeability and subject to the waveguide boundary conditions.

[4] M. Born, "Optik," Springer, Berlin, pp. 363–365; 1933.

Formal solutions have been obtained by Kales[5] and by Suhl and Walker[6] for the modes in a completely filled circular waveguide magnetized in the axial direction. The rectangular waveguide case is also under investigation and some of the modes that may exist in the filled rectangular waveguide magnetized in the transverse plane have been obtained by Van Trier.[7] These theoretical investigations indicate that the modes which exist in a magnetized medium are quite different from the usual TE or TM waveguide modes. However, before the formal solutions can be used to obtain mode configurations and phase velocities, numerical solutions are necessary which require the knowledge of the μ-tensor components as functions of the applied field. This requires μ and α to be obtained experimentally for values of applied field at least below saturation. We are not at present aware of a satisfactory experimental method of measuring μ and α.

We have been engaged in an experimental study of 35
the propagation characteristics of both circular and rectangular waveguide containing pieces of standard commercial ferrites at 9,375 mc. Let us consider first the case of the circular waveguide. It can be seen from (1) that the angle of rotation of the plane of polarization may be obtained from measurements of the phase velocity of the circularly polarized components. The transmission loss, reflections and changes in ellipticity of a linearly polarized wave may also be deduced from measurements made upon circularly polarized waves. Because of the more fundamental nature of the information obtained by studying the propagation of circular polarization, most of the measurements are being made with circularly polarized waves using the apparatus shown in Fig. 1(a). A right-hand circularly

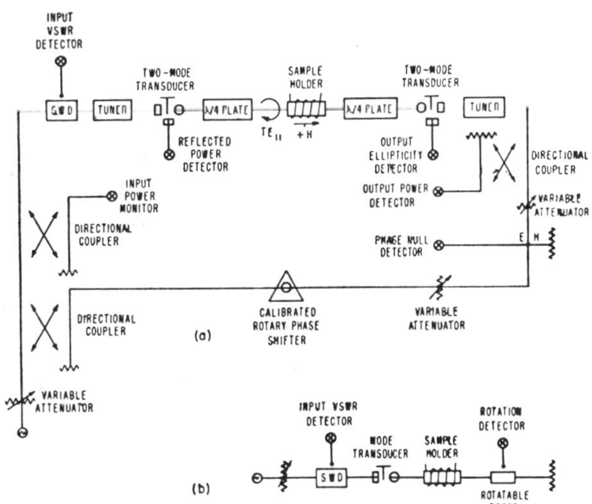

Fig. 1—Schematic diagram of the measuring apparatus.

[5] M. L. Kales, "Modes in Waveguides Containing Ferrites," NRL Report #4027, August 8, 1952.

[6] H. Suhl and L. R. Walker, "Faraday rotation of guided waves," *Phys. Rev.*, vol. 86, no. 1, pp. 122–123; April 1, 1952. (Letters to the Editor.)

[7] Van Trier, "Anomalous wave types in waveguides containing ferromagnetics," *Phys. Rev.*, vol. 87, no. 1, pp. 227–228; July 1, 1952. (Abstract of talk given at meeting of A.P.S. in Washington, D. C.)

polarized TE_{11} mode is incident upon a ferrite cylinder mounted coaxially in a length of 0.9375 inch diameter waveguide by means of a polyfoam spacer. The ferrite is magnetized in the axial direction by a solenoid wound around the waveguide. When the applied field is in the direction of propagation, the incident wave is by definition polarized in the positive sense. Upon reversal of the applied field, the incident field must be regarded as polarized in the negative sense.

36

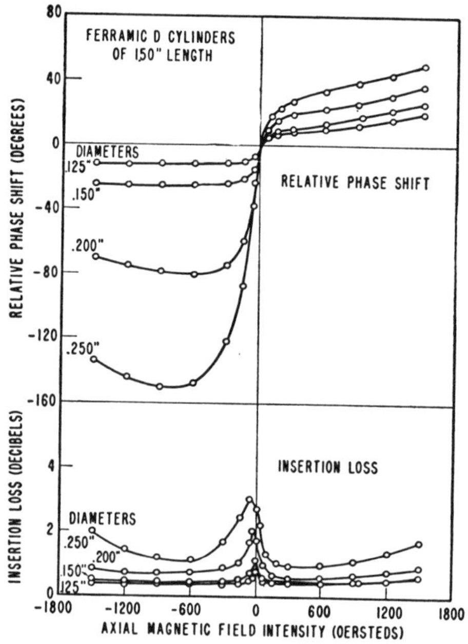

Fig. 2—Relative phase shift and insertion loss versus applied axial magnetic field intensity with the diameter of the sample as a parameter in the case of circularly polarized wave propagation.

As can be seen from the schematic, all reflections as well as changes in ellipticity are measured. The power absorbed by the ferrite can then be calculated from the measured insertion loss, the difference in readings of the input monitor, and the output power detector. It has been found, however, that for most samples the power absorbed is at least 90 per cent of the insertion loss.

The change in the phase constant $(\beta_{\pm}-\beta^0)$ due to the application of the magnetizing field was measured by observing the amount of phase shift that had to be introduced in the reference line to maintain a minimum reading at the phase null indicator. The phase shift was introduced by means of a calibrated rotary phase shifter capable of producing any change of phase within an accuracy of at least ± 2 degrees. In terms of ϕ, the difference in settings of the phase shifter for the magnetized and unmagnetized states, $\beta_{\pm}$, is given by

$$\beta_{\pm} l = \beta^0 l - \phi_{\pm},$$

where l is the length of the sample and the subscripts $+$, $-$ refer, as before, to the phase velocities for the positive and negative senses while the superscript, 0, refers to the phase velocity in the unmagnetized medium. The angle of rotation for a plane polarized wave can then be calculated from

$$\theta = \frac{\phi_{+} - \phi_{-}}{2} l.$$

The angle of rotation was measured directly in the apparatus shown schematically in Fig. 1(b) to an accuracy of at least ± 2 degrees and found to be in agreement with calculated values within experimental errors.

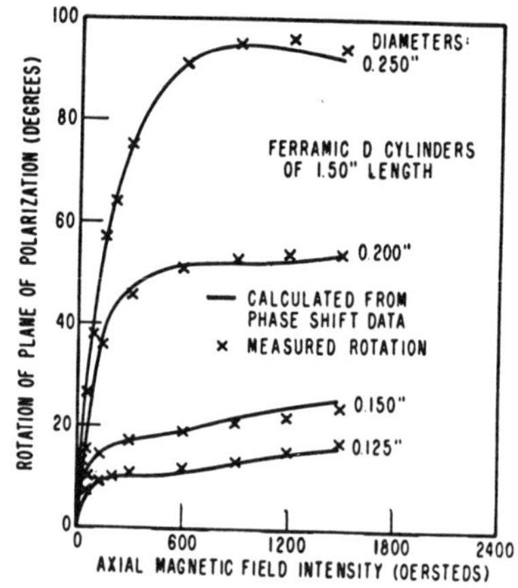

Fig. 3—Calculated and measured angle of rotation of the plane of polarization versus applied axial magnetic field intensity.

The curves of ϕ and insertion loss versus external applied field H are shown in Fig. 2 for samples of Feramic D of the same length but different diameters. The angle of rotation calculated from these curves as well as the measured rotation is shown in Fig. 3. The effect of changing only the length of the sample can be seen in Fig. 4. The corresponding curves of angle of

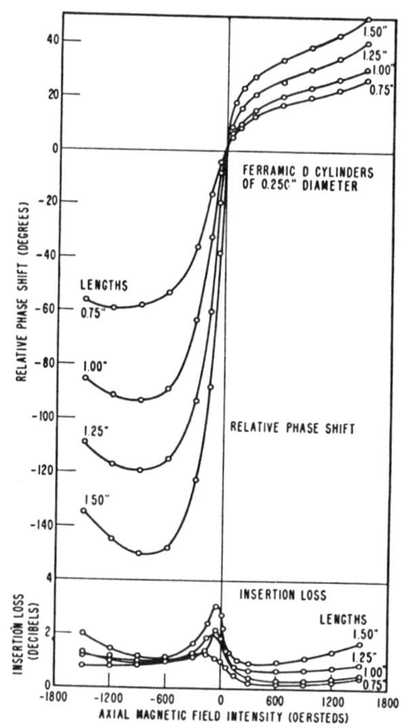

Fig. 4—Relative phase shift and insertion loss versus applied axial magnetic field intensity with the length of the sample as a parameter in the case of circularly polarized wave propagation.

rotation are presented in Fig. 5. The general shape of the curves is seen to be the same as the physical dimensions are changed, only the magnitude apparently being affected.

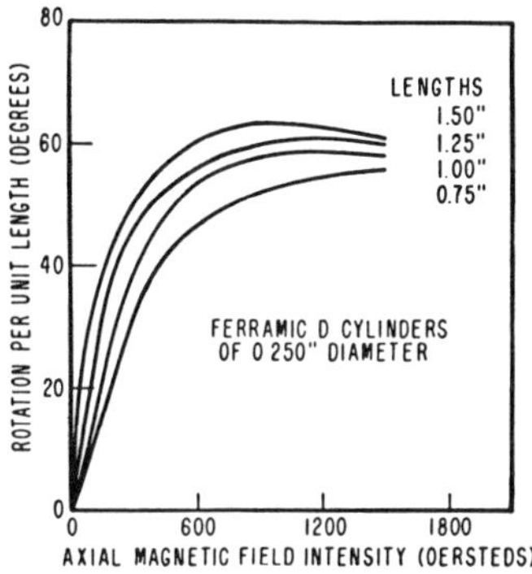

Fig. 5—Calculated angle of rotation of the plane of polarization versus applied axial magnetic field intensity.

The horizontal shift of the peaks of the phase and insertion loss curves as the length or diameter is changed may be a result of the change of the demagnetization factor as the dimensions are changed, resulting in different internal fields for the same applied field, as well as end effects producing lumped phase shifts at the ends of the cylinder. These factors may also explain why the rotation per unit length curves do not coalesce into a single curve as would be expected from (1).

The origin of the peak which is present in the insertion-loss curves is not yet fully understood. There are some indications that it may be a shape resonance since as one goes to much larger diameters in some ferrites more peaks appear. We have found, however, that if in addition to the axial field a magnetic field is applied transverse to the axis of the waveguide a substantial reduction can be made in the magnitude of the peak in many cases. This is shown in Fig. 6, where the parameter is the value of the transverse field, the dimensions of the sample being held constant.

Because of the limitations of space, only the curves for Ferramic D are shown. These were chosen since this particular ferrite exhibits the lowest absorption loss of the materials we have tested. The rest of the samples,

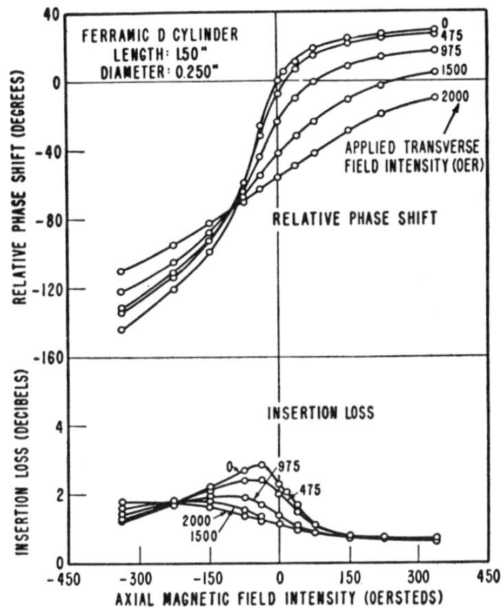

Fig. 6—Relative phase shift and insertion loss versus applied axial magnetic field intensity with the applied transverse magnetic field intensity as a parameter in the case of circularly polarized wave propagation.

with the exception of three, exhibit phase and insertion loss curves having the same general shape as those shown for Ferramic *D*, the difference between curves for individual materials of the same size being one of magnitude rather than of shape. A résumé of the values of the ϕ and insertion loss curves at the more interesting points is given in Table I for 0.250-inch$\times$1.50-inch cylinders of the materials tested. The three exceptions are Ferramic B, Ferramic H, and Ferroxcube 4B. These three materials give ϕ and insertion loss curves which seem to have the same general trend, but differ markedly in

TABLE I

Ferrite	Peak of ϕ_-	Magnitude of loss peak	Loss at $H=0$	ϕ_+ at $H=1000$ *oe*	Loss at $H=+1000$ *oe*	Loss at $H=-1000$ *oe*
Ferramic[8] A	−145°	3.4 db	2.6 db	43°	0.8 db	1.2 db
Ferramic A-34	−188°	2.1 db	0.6 db	95°	0.3 db	1.5 db
Ferramic C	−244°	11.5 db	6.0 db	35°	0.9 db	5.4 db
Ferramic D {D[1]	−150°	3.0 db	2.7 db	41°	1.2 db	1.2 db
Ferramic D {D[2]	−350°	4.4 db	3.2 db	35°	1.0 db	3.2 db
Ferramic E	−540°	12.5 db	3.0 db	35°	1.1 db	8.6 db
Ferramic G	−650°	26.0 db	5.5 db	30°	1.0 db	6.0 db
Ferramic H_1	−238°	26.0 db	3.5 db	98°	0.25 db	18.0 db
Ferramic I	−490°	20.0 db	5.5 db	123°	0.8 db	19.4 db
Ferramic J	−305°	11.5 db	6.0 db	40°	1.0 db	6.5 db
Ferroxcube[9] 4D	−312°	15.0 db	6.2 db	38°	1.2 db	9.7 db
Ferroxcube 4E	−170°	2.8 db	1.7 db	42°	0.5 db	0.8 db
Crolite[10]	−570°	21.0 db	5.5 db	35°	1.0 db	6.5 db

[8] General Ceramics and Steatite Corp., Keasbey, N. J.
[9] Ferroxcube Corporation of America, New York, N. Y.
[10] Henry L. Crowley and Co., West Orange, N. J.

37

form from the other materials. An unusual feature of these materials is that over the range of values measured ϕ_+ is less than ϕ_-; hence, negative rotation is obtained.

The data presented is intended to give a qualitative rather than a quantitative picture of what may be expected. In view of the nonuniformity of commercial ferrite materials with respect to their microwave properties, exactly reproducible results cannot in general be expected for materials having the same commercial designation. Thus, for example, we have been able to separate our Ferramic D samples into two groups from their phase- and insertion-loss characteristics which we designate as D^1 and D^2. The differences in the characteristics between these two forms of Ferramic D can be seen from Table I. The data presented applies to Ferramic D^1.

In the rectangular waveguide case, experiments
38 have been conducted with various physical dimensions and relative orientations of the waveguide and the ferrite. It is interesting to note that the transmission loss depends not only on the intrinsic properties of the ferrite material but also on the particular physical configuration used. For example, it has been found that when a small rectangular piece of ferrite is inserted in the center of a standard X-band waveguide the loss is much higher than if the same piece of ferrite were placed adjacent to the narrow wall of the waveguide.

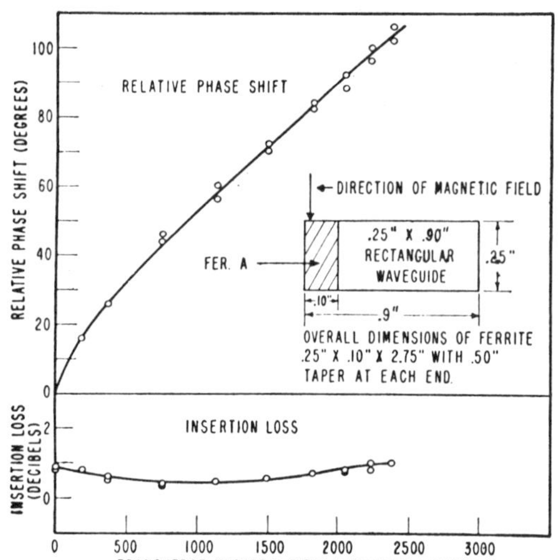

Fig. 7—Relative phase shift and insertion loss versus applied transverse magnetic field intensity for rectangular waveguide case.

On the other hand, the range of phase shift is not appreciably changed. A possible explanation is that the reduction in the loss is due to a reduction in the dielectric loss which results from placing the ferrite in a part of the waveguide where the electric field is small. Figs. 7 and 8 show the phase shift and loss as a function of transverse field obtained from two different physical arrangements.

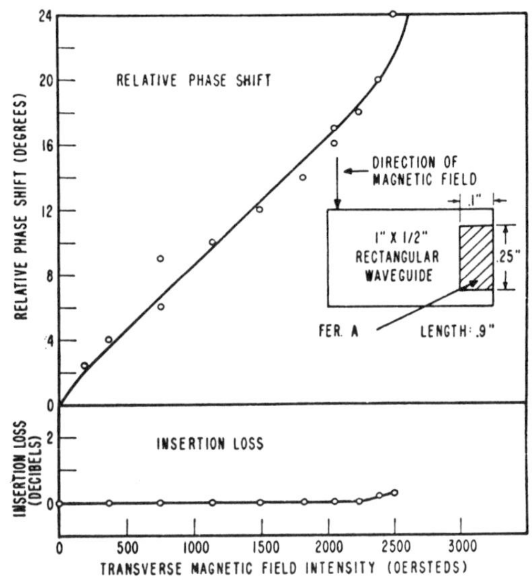

Fig. 8—Relative phase shift and insertion loss versus applied transverse magnetic field intensity for rectangular waveguide case.

Applications

Since it is possible, by means of ferrite devices, to vary electrically the phase or polarization of microwave fields, the ferrites should have many applications in microwave systems. Before such applications may be realized, however, the absorption losses of the ferrite materials must be minimized and further study must be made of their power-handling capabilities and the dependence upon frequency and temperature of their propagation characteristics. In view of the fact that commercially available ferrites were developed as relatively-low frequency transformer cores, it is not to be expected that they are optimum at microwave frequencies. Thus, it is quite possible that new ferrite materials can be developed with loss so low that they will be practical for many applications.

A direct application of the Faraday Rotation in ferrites has been made by Luhrs in his microwave switch.[11] Essentially, the Luhrs' switch consists of a section of dielectric-filled circular waveguide in which a piece of ferrite rod is axially located. A coil is wound on the circular waveguide to provide an axial magnetic field. The input to the ferrite section is a rectangular waveguide and the output another length of rectangular waveguide which is rotated 90° about its axis with respect to the input waveguide. Thus with the ferrite unmagnetized, there is about 50-db isolation between the crossed waveguides. When the proper current is supplied to the coil, the ferrite rotates the plane of polarization 90° and the input and output guides are coupled. The transmission loss in the "on" state is about 0.25 db. If a square waveguide having two outputs is used at the output, one could electrically switch between the two output channels. A number of two-channel switches,

[11] C. H. Luhrs and Co., Hackensack, N. J.

if cascaded, could form an organ-pipe scanner[12] with no moving parts. A rod of ferrite magnetized to produce 45° of rotation and located on the axis of a circular waveguide terminated at each end by a two-mode transducer produces a very interesting device called a "circulator," the properties of which are described by Hogan.[13] In the same paper, Hogan describes a one-way π phase shifter utilizing a 90° rotator. This device produces a phase shift of π for linearly polarized waves traveling in one direction. If the direction of propagation is reversed, no phase shift is introduced.

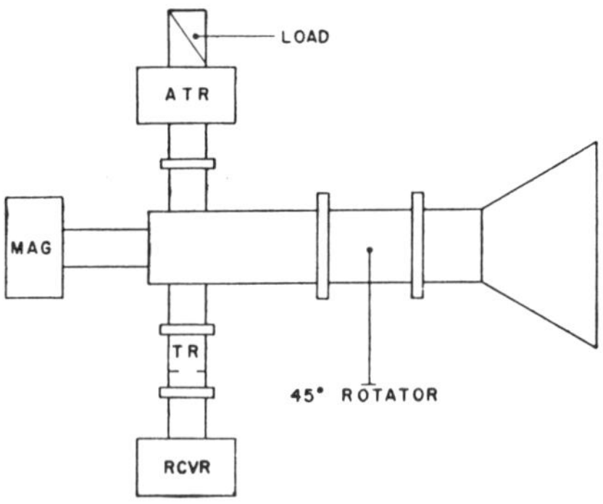

Fig. 9—Diagram of ferrite isolating device.

Another important possible application is the isolation of the generator from the load. If successful, it may make possible many antenna designs, particularly for scanning antennas, which in the past have had to be rejected because of the variable impedance presented to the generator. Fig. 9 shows a typical possible arrangement. When the magnetron fires, both the TR and ATR are closed. The output of the magnetron is rotated 45° in passing through the 45° rotator, and then proceeds to the antenna. If any power is reflected by the antenna, it comes back down the line, passes through the rotator, and is rotated another 45° in the same direction, making a total of 90°. The reflected wave therefore cannot enter the input and must proceed up one of the two side arms. Since the TR and ATR are still closed, the reflected wave proceeds to the upper arm and is absorbed in the load. By the time the target reflection reaches the antenna, both the TR and ATR are open and thus the target energy proceeds to the receiver instead of the load.

The electrical variation of phase made possible by the ferrites makes these materials especially useful in the design of scanning antennas, since one can scan a radar beam by controlling the phase distribution over the radiating aperture. We are at present developing an array whose beam can be electrically scanned. The array consists of shunt slots cut along the edge of a rectangular waveguide. Pieces of ferrite are located between the slots and energized by small electromagnets in such a way as to progressively change the phase at each slot, thus causing the radiated beam to scan.

A phase shifter utilizing the variation in phase velocity of circular polarization will be nonreciprocal since a wave traveling in one direction will suffer a phase shift which differs from that of the wave traveling in the reverse direction. However, such a device can be made reciprocal by the addition of a 45° ferrite rotator at each end.[14]

We have described only a few of the many possible applications of ferrites in present-day microwave systems. The field is relatively new and there is much more to be learned about ferrites at microwave frequencies.

Acknowledgments

The authors wish to thank Dr. M. L. Kales of the Naval Research Laboratory for his many helpful comments and suggestions and the many persons who ably assisted in taking the experimental data.

[12] K. S. Kelleher and H. H. Hibbs, "Organ pipe radar scanner," *Electronics*, vol. 25, pp. 126–127; May, 1952.

[13] C. L. Hogan, "The ferromagnetic Faraday effect at microwave frequencies and its applications—the microwave gyrator," *Bell Sys. Tech. Jour.*, vol. 31, pp. 22–26; January, 1952.

[14] N. G. Sakiotis, A. J. Simmons, and H. N. Chait, "Microwave-antenna ferrite applications," *Electronics*, vol. 25, pp. 156–166; June, 1952.

PROPERTIES OF FERRITES IN WAVEGUIDES

N. G. Sakiotis and H. N. Chait
Antenna Research Branch
Naval Research Laboratory
Washington 25, D. C.

40

*

REPRINTED FROM

TRANSACTIONS OF THE I.R.E.
Professional Group
on
Microwave Theory and Techniques

*

November, 1953

Volume MTT-1 Number 2

N. G. Sakiotis and H. N. Chait
Antenna Research Branch
Naval Research Laboratory
Washington 25, D. C.

In the past few years research with ferrites has shown that these materials possess unusual properties of considerable importance to microwave engineers. These properties result from the fact that at microwave frequencies and in the presence of a static magnetizing field, these non-conducting ferromagnetic media are characterized by an asymmetric tensor permeability whose components are functions of the static magnetizing field. Because of this one can, by means of a variable magneto-static field, vary the propagation characteristics, e. g., the phase, polarization, and transmission loss of a wave propagating in waveguide-containing ferrite materials. Since the static magnetizing field can be varied by purely electrical means, it is possible with the use of these materials to construct microwave components which will very rapidly vary either phase of polarization electrically. 41

At the Naval Research Laboratory, we have been engaged in a study of the propagation characteristics of the ferrites at X-Band and we would like to discuss a few of the results of that study.

Although it would be desirable to make measurements of the intrinsic properties of the ferrite materials with a view to ultimately being able to predict propagation characteristics in waveguides containing ferrites, such an approach entails many difficulties. Instead, we have chosen to measure the propagation characteristics directly, using ferrite-loaded waveguides. In this manner we measure the effects of the ferrite materials under conditions similar to those under which they will be used.

Clearly, the results of such measurements will depend not only on the intrinsic properties of the ferrites but also on the dimensions of the waveguide, the dielectric constant of the other media present in the waveguide, and the shape of the ferrite. Therefore, it should be kept in mind that the data which will be presented apply only to the geometry described, and may not, in general, be used to predict behaviour in other configurations.

In spite of these reservations, it is our hope that the results of our studies can be useful in understanding some of the intrinsic properties of the ferrite materials.

We would like to point out that the data to be presented were chosen as being typical of the behaviour of the ferrites we have examined and do not represent the best results which have been obtained as far as practical applications are concerned.

---

*Presented by N. G. Sakiotis at the URSI-IRE meeting at the National Bureau of Standards, April 27-30, 1953.

One of the properties with which we have been concerned is the Faraday effect in waveguide, that is, the rotation of the plane of polarization of a plane wave when it propagates through a section of circular waveguide containing a cylinder of ferrite magnetized along the direction of propagation. We are interested in obtaining a large amount of rotation per unit length of the ferrite, while at the same time maintaining the power absorption per unit length as small as possible.

Fig. 1 shows a curve of the Faraday rotation obtained with a cylindrical piece of Ferramic A-34, a magnesium-manganese ferrite, 1/4 inch in diameter and 1-1/2 inch in length, located axially in a circular waveguide and supported by means of a polyfoam cylinder. The operating frequency is 9375 mc. An axial magnetic field is applied by means of a solenoid wound about the waveguide.

This curve is typical of most ferrites we have examined, the angle of rotation of the plane of polarization increasing linearly and then leveling off to the saturation rotation when the material becomes saturated. As can be seen, the power absorbed by the ferrite is a function of the magnetizing field.

If one uses the ferrite to produce a fixed rotation, it is clear that it is very desirable to choose the dimension of the ferrite so that it produces the desired rotation when it is saturated, since in that condition the rotation is insensitive to variations in the magnetizing field. Accordingly, the data presented are restricted to ferrite polarization rotators operating in the saturation region. If the diameter of the ferrite cylinder is kept constant and the length increased, and if end effects could be neglected, it would be expected that the amount of rotation absorbed power would increase linearly with length. We have checked the dependence on length experimentally and, except for some end effects in the longer diameters, both the saturation loss in db and the saturation rotation vary directly as the length. Therefore the ratio of saturation loss to saturation rotation is independent of length.

Fig. 2 shows the saturation rotation per unit length as a function of sample diameter for samples of the same A-34 material. The rotation increases slowly with diameter until one reaches a diameter of about 1/8 inch and then increases more rapidly until a diameter of about 1/4 inch is reached. Beyond this, large fluctuations occur in the rotation and absorption curves. The diameter at which these fluctuations set in depends on the particular ferrite.

A significant factor in the design of a rotator is the absorption loss per degree of rotation. The dependence of this ratio upon diameter is also shown in Fig. 2. Note that the loss per degree of rotation decreases as the sample diameter is increased. Therefore, if one were to build, for example, a 45 degree rotator polarization rotator using a length of ferrite 1/8 inch in diameter, the absorption loss would be 1 db. If, however, the same 45 degree rotator utilized a cylinder 1/4 inch in diameter, the absorption loss would be reduced to 1/4 db.

The variation of the saturation rotation with frequency is shown in the next figure. In Fig. 3 can be seen the deviation of saturation rotation from the value it has at 9375 mc for three samples of different diameters and same length. The rotation increases with frequency for all three samples, the rate of increase being greater as the diameter increases. From these curves we see that for a fixed amount of rotation, the bandwidth of the rotator increases as the diameter of the sample is decreased. Therefore, as one increases the diameter of the sample to obtain less power absorption, he must pay for it by having a smaller bandwidth.

Another aspect of our work is concerned with a study of propagation in a ferrite-loaded rectangular waveguide. While for the rotation measurements we applied an axial magnetizing field, in this case we apply a magnetizing field transverse to the axis of the waveguide. The propagation characteristics, e. g., the phase velocity and the power absorption, depend of course upon the shape and placement of the sample in the waveguide.

The general geometry involved can be seen in Fig. 4. The magnetizing field is applied in a direction perpendicular to the broad face of the guide along the x-axis and the direction of propagation is along the y-axis.

One of the interesting features of this situation is that the ferrite and the magnetizing field can be so disposed as to obtain propagation characteristics which are either reciprocal or non-reciprocal. For example, it can be shown that for the arrangement shown in this figure, where the ferrite section is situated asymmetrically with respect to the central plane, the propagation characteristic will be non-reciprocal. On the other hand, if the ferrite is symmetrically placed, the propagation will be reciprocal.

Fig. 5 shows the absorption losses and phase shift for a symmetrical placement of the ferrite in the waveguide. As can be seen, the phase shift and loss is the same for the two directions of propagation. This arrangement is useful for a reciprocal variable phase-shifter. With this particular sample, a magnesium-manganese ferrite, we can obtain any phase shift between 0 and 80 degrees with an absorption loss of less than 1 db.

Fig. 6 shows the non-reciprocal behavior we get if we place the ferrite against the side wall. Here we have a phase velocity which is different for the two directions of propagation. For this case, it happens that the losses are about the same for the two directions of propagation for the values of the magnetizing field shown. This arrangement might be used for a non-reciprocal phase shifting device. With the sample shown one can obtain a 65 degree difference in phase shift between the two directions of propagation with an absorption loss of less than 1/2 db. By choosing the proper length, one can obtain any desired differential phase shift, 180 degrees, for example.

A particularly interesting application of the non-reciprocal characteristics of the asymmetrical configuration is the construction of a one-way transmission line which absorbs most of the power for one direction of propagation, but transmits most of the power in the other. An understanding of how this can be done can be obtained from the theory of gyromagnetic resonance. If the rf magnetic field is circularly polarized in a plane perpendicular to the magnetization of the ferrite, a large resonance absorption of power from the rf wave should occur at that value of the magnetizing field which brings the ferrite into gyromagnetic resonance, whenever the sense of the circular polarization is positive with respect to the magnetizing field. Such resonance absorption should not occur for the negative sense of circular polarization.

For a waveguide-propagating energy in the dominant TE_{10} mode, there are two parallel planes such that the H-vector is circularly polarized in one sense in one of these planes, and in the opposite sense in the other. This can easily be seen from Fig. 7. It will be recalled that the longitudinal and transverse components of H are in phase quadrature. The variation in amplitude across the guide of the two components is shown in the diagram. At two planes, equi-spaced about the center line, the amplitudes are equal, so that the total H-field is circularly polarized. The longitudinal component, however, has the reverse sign at one plane with respect to the other. Since the change in sign is equivalent to a 180 degree difference in phase of the longitudinal component at the two planes, the sense of circular polarization is positive at one plane and negative at the other.

If the direction of propagation is reversed, the sign of the longitudinal component changes at both planes, and the sense of polarization reverses at both planes. Therefore, one would expect that if a piece of ferrite were placed along the plane where the rf H-field is circularly polarized in the positive sense with respect to the applied magnetizing field, there would be a large resonance absorption of power at that value of magnetizing field which brings the ferrite into gyromagnetic resonance. If the magnetizing field is now kept at the value necessary for resonance and we reverse the direction of propagation, the rf magnetic field incident on the ferrite will be circularly polarized in the negative sense with respect to the magnetizing field, and therefore there will be no resonance absorption. Since the resonance process usually is accompanied by absorption losses of at least 20 db, we have here a one-way transmission device.

Fig. 8 shows some experimental verification. The curve represents absorption loss versus magnetizing field for each direction of propagation. For the wave propagating in the +y direction, the loss is very small until the magnetizing field approaches the resonance value. In the resonance region the absorption loss becomes very high, the transmitted power being down 26 db. The wave traveling in the -y direction suffers very small losses until it reaches the field required for resonance, where the absorption rises to 2 db. If we therefore maintain a constant current of 21 amperes through our magnet we have a one-way transmission line with a differential transmission of about 24 db.

Fig. 9 shows the variation of relative phase in the same arrangement. Again we can see the non-reciprocal nature of the configuration. The wave traveling in the +y direction suffers an increase in phase while the wave traveling in the opposite direction has its phase velocity decreased. It is interesting to note the typical change in sign of phase generally associated with resonance for the wave which undergoes the resonance absorption.

Acknowledgment

The authors wish to thank Dr. M. L. Kales of the Naval Research Laboratory for his many valuable contributions, as well as Messrs. T. Curry and C. Banks who assembled the data presented here.

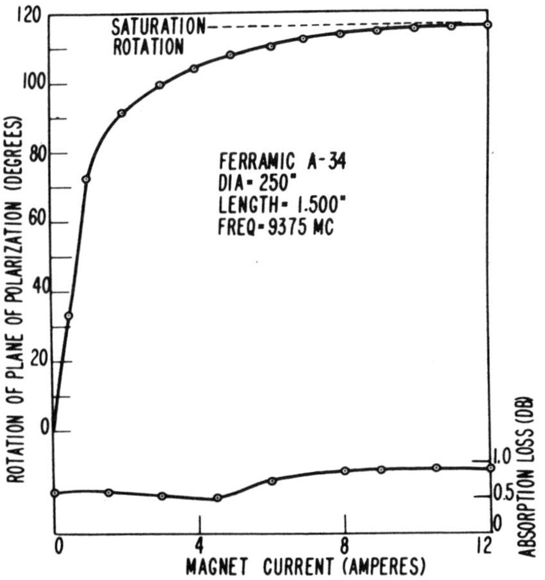

Fig. 1

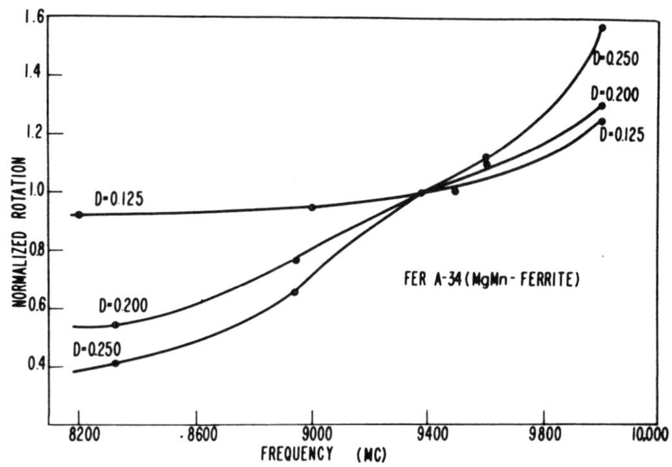

Fig. 3

45

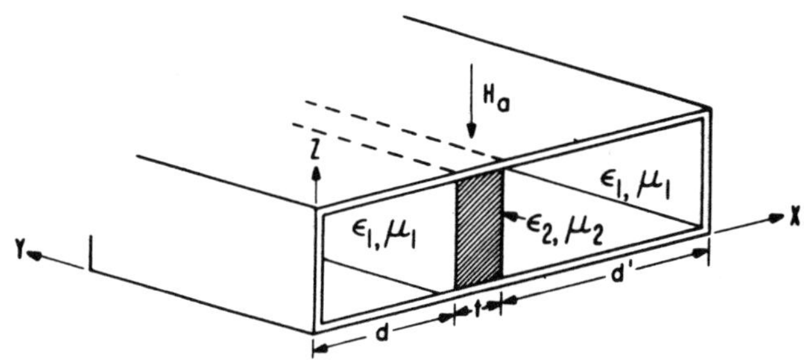

Fig. 4

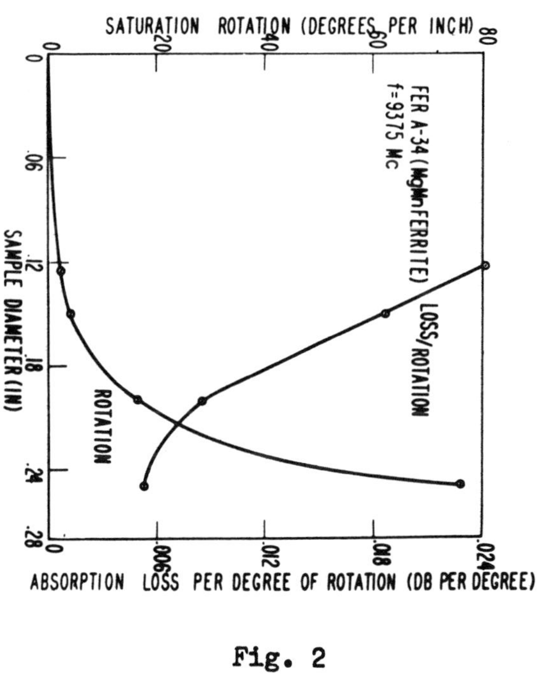

Fig. 2

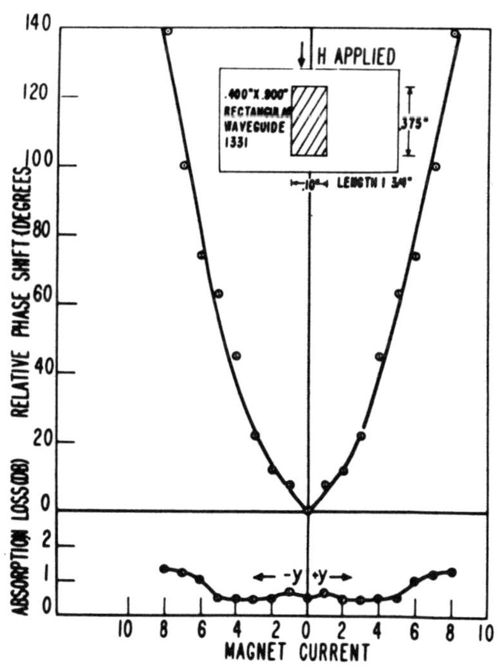

Fig. 5

46

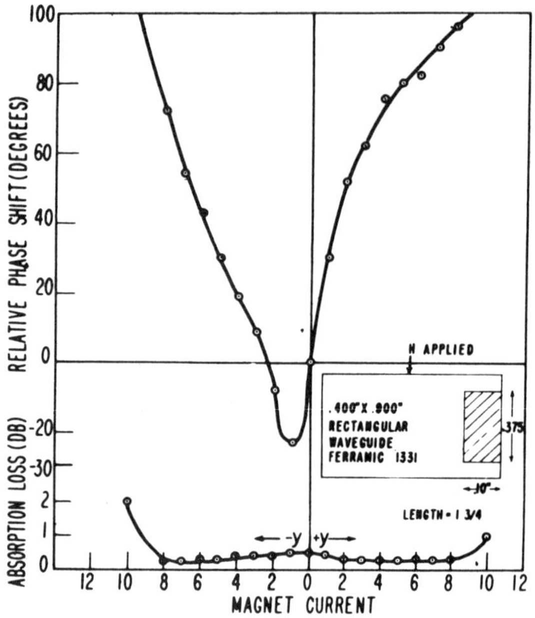

Fig. 6

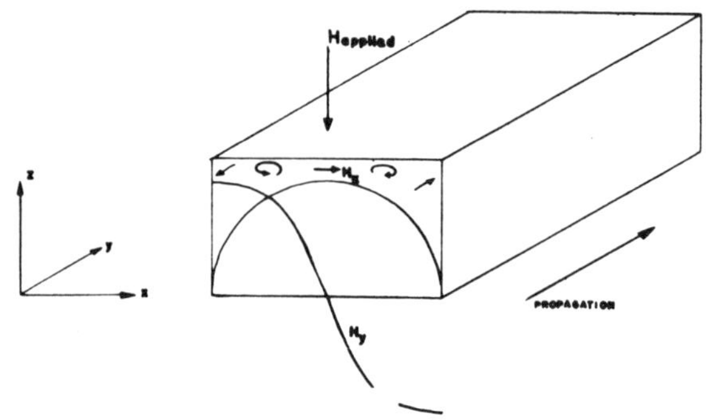

Fig. 7

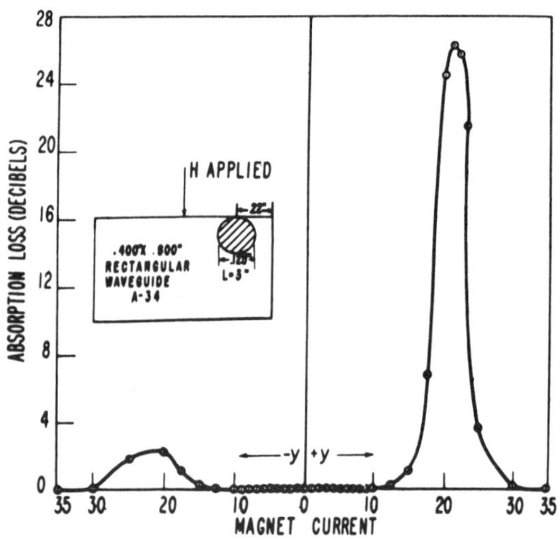

Fig. 8

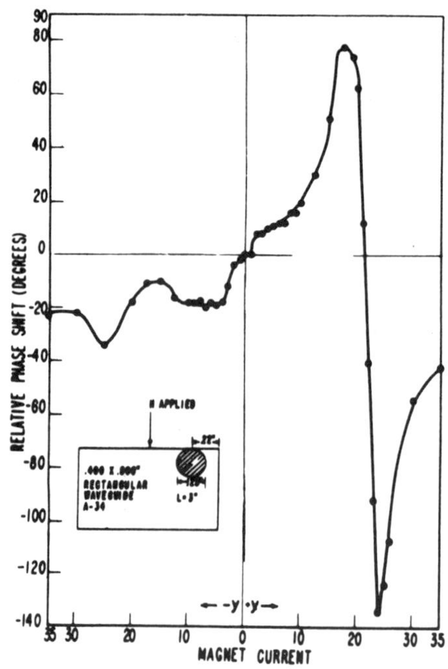

Fig. 9

Three New Ferrite Phase Shifters*

HOWARD SCHARFMAN†, ASSOCIATE MEMBER, IRE

Summary—**Three new devices for the electrical control of microwave phase shift are discussed. The first of these, the bucking rotator phase shifter, is a reciprocal unit. It employs two series ferrite rotators, the field coils of which are connected so as to produce opposing longitudinal magnetic fields. It is shown that the total rotation is zero independent of coil current, whereas the phase shift is a function of the current. The operation and limitations of the device are discussed. Curves of phase shift vs coil current for several configurations are presented.**

The other two devices are based on the use of a circular polarized phase shifter between quarter wave plates. In the new types, a reflection technique is described wherein only one quarter wave plate is required and the phase shift section is used twice. A nonreciprocal phase shifter and an electrically controllable short circuit based on this technique are described. The principles of operation are discussed, and the devices are compared with other types as to performance, size, and bandwidth.

Introduction

A NUMBER of techniques employing ferrites for obtaining essentially lossless microwave phase shift have recently been devised. Several varieties of phase shifters, based on the use of a transversely magnetized slab of ferrite located either in a region of local circular polarization or along the longitudinal axis in rectangular guide, have been successfully constructed.[1–4] A number of articles on the theory and application of such phase shifters may be found in the literature, and specific devices are available commercially. Another type of ferrite phase shifter which has received considerable attention uses a longitudinally magnetized ferrite rod in square or circular guide preceded and followed by circular polarizers. Detailed reports on this type may also be found in the literature.[4] Each type has advantages and disadvantages which suit it best for specific applications.

In this paper three new phase shift devices are presented. Their individual characteristics are described and compared with each other as well as with other devices previously investigated.

The Bucking-Rotator Phase Shifter

As has been shown by Sakiotis and Chait, both the rotation of the plane of polarization and the transmission phase shift through a ferrite rotator are functions of the left- and right-hand circular polarized phase constants. These phase constants are in turn functions of the applied longitudinal magnetic field. The rotation, θ, and phase shift, ϕ, can be expressed in terms of the phase constants by:

$$\theta = \frac{(\beta_- - \beta_+)L}{2} \tag{1}$$

$$\phi = \frac{(\beta_- + \beta_+)L}{2} \tag{2}$$

47

where

β_+ = phase constant of the circularly polarized wave rotating in the same sense as current causing the magnetic field.

β_- = phase constant of the circularly polarized wave rotating in the opposite sense of B_+.

L = length of ferrite rod.

The variation of the phase constants as functions of applied field for X-band rotators have been investigated previously. Experimental curves taken in our laboratories are shown in Fig. 1. Plotted in the same figure

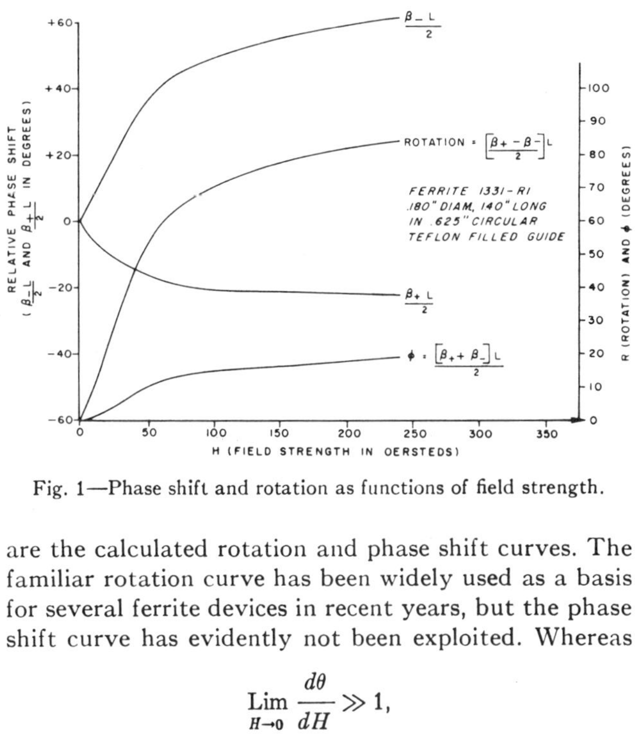

Fig. 1—Phase shift and rotation as functions of field strength.

are the calculated rotation and phase shift curves. The familiar rotation curve has been widely used as a basis for several ferrite devices in recent years, but the phase shift curve has evidently not been exploited. Whereas

$$\lim_{H\to 0} \frac{d\theta}{dH} \gg 1,$$

$$\lim_{H\to 0} \frac{d\phi}{dH} = 0.$$

As the field is increased, the phase shift increases slowly,

* Original manuscript received by the IRE, July 31, 1956.

† Raytheon Mfg. Co., Bedford, Mass.

[1] N. G. Sakiotis and H. N. Chait, "Ferrite at microwaves," Proc. IRE, vol. 41, pp. 87–93; January, 1953.

[2] H. N. Chait, "Nonreciprocal microwave components," 1934 IRE Convention Record, part 8, pp. 82–87.

[3] B. Lax, K. J. Button, and L. M. Roth, "Ferrite phase shifters in rectangular waveguide," *J. Appl. Phys.*, vol. 25, pp. 1413–1421; November, 1954.

[4] A. G. Fox, S. E. Miller, and M. T. Weiss, "Behavior and applications of ferrites in the microwave region," *Bell Syst. Tech. J.*, vol. 34, pp. 5–103; January, 1955.

then passes through an approximately linear region, and finally saturates. The linear region is of considerable interest, for it can be used to obtain linear, reciprocal phase shift proportional to the magnetic field. By using a dc biasing field sufficient to reach the center of the linear region, phase advance or delay may be achieved.

In order to use the phase shift phenomenon described above in a worth-while device, some arrangement must be made for the ouput waveguide to rotate (either mechanically or electrically) with the applied field. This can be done most easily by means of a second rotator, similar to the first but rotating the polarization in the opposite sense. If the coils of the two rotators are connected in series, the total rotation through the two rotators will be zero independent of current, but the phase shift will be twice that for each rotator.

48 This type of phase shifter is illustrated in Figs. 2, 3, and 4. As a two-port microwave device, it has reciprocal

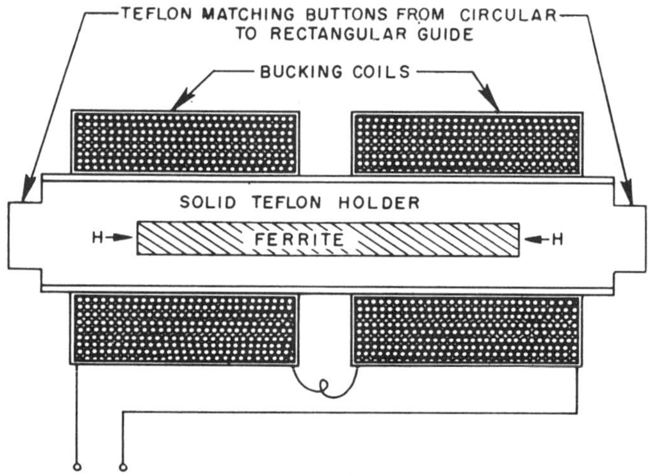

Fig. 2—Bucking rotator phase shifter.

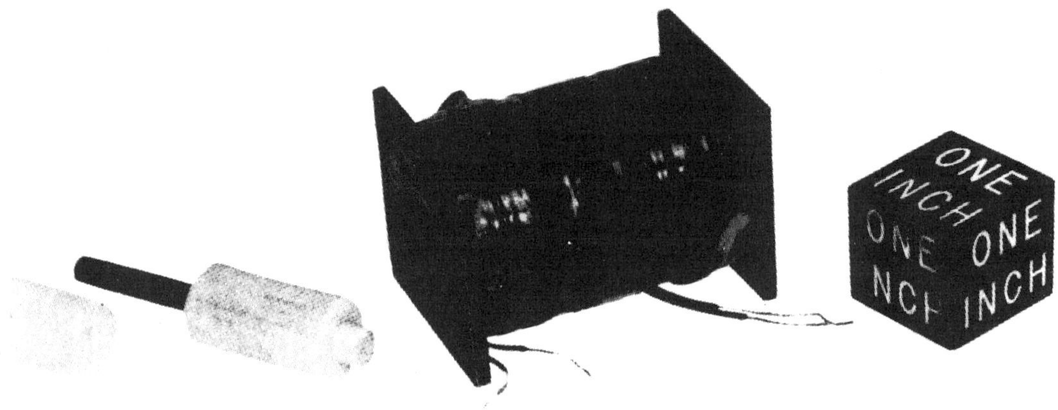

Fig. 3—Bucking rotator phase shifter, disassembled.

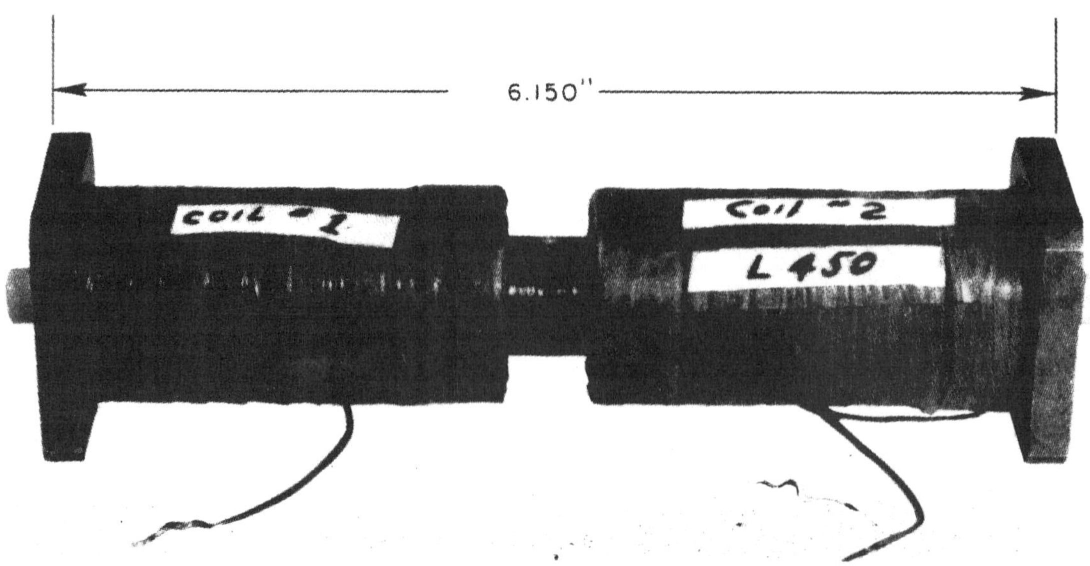

Fig. 4—Bucking rotator phase shifter.

properties for any value of current. With an appropriate dc biasing current applied, it can be operated to produce a microwave phase shift (either positive or negative) that is approximately linear with coil current. In addition, the insertion loss of the device is independent of applied coil current. Figs. 5, 6, and 7 show curves of phase shift vs applied field for three bucking-rotator phase shifters of different lengths and ferrite sizes. In each case the device operated over $2\frac{1}{2}$ per cent of the X band. A vswr of less than 1.1 was maintained when connected to 1 inch$\times\frac{1}{2}$ inch input and output waveguides.

As can be seen from the curves of Figs. 1, 5, 6, and 7 the maximum variation of phase shift of a bucking-rotator phase shifter is a function of ferrite length, diameter, and saturation magnetization. In principle, any arbitrarily large electrically variable phase shift could be obtained by increasing the length of the device. However, the practical difficulties inherent in maintaining precise symmetry in the two rotator sections become severe as the diameter and length of the ferrite rod is increased. In the latter case the rotations and rates of change of rotation of the plane of polarization in the individual rotator sections become quite large for small changes in applied coil current (*i.e.*, magnetic field). Consequently slight asymmetries may cause the rotations in each rotator section to be appreciably different with a resulting net rotation through the device. This results in the "capture" and eventual dissipation of some of the input energy with a corresponding spike of insertion loss. By the same token, any stray magnetic fields in the region of the device can cause asymmetries leading to similar effects. As a result, the device appears limited to phase variations up to ± 90 degrees unless special precautions are taken. The device is particularly well suited for applications requiring small linear phase variations with low constant insertion loss. It is simple in construction and small in size and weight. The longitudinal fields required can be applied from a low power driver over a wide frequency range provided proper precautions are taken to circumvent the shorted turn effect of the circular waveguide.

The Cross-Polarized Phase Shifter

As mentioned earlier in this paper, a longitudinally magnetized ferrite rod preceded and followed by circular polarizers has been used as a variable phase shifter.[4] such a device is illustrated schematically in Fig. 8 opposite. The energy is converted from linearly polarized to circular polarized in the first polarizer, is phase shifted in the ferrite section proportional to the applied longitudinal field, and converted back to linear polarization in the output polarizer. The resultant phase shift is proportional to the applied field (up to saturation of the ferrite).

Two improvements have been made on this device. Fig. 9 illustrates a method of doubling the phase shift for a given applied field (or doubling the maximum phase shift before saturation), Energy is fed through a two-mode transducer (thruplexer) in either round or square waveguide into a circular polarizer. The resultant circularly polarized field is phase shifted proportional to the applied longitudinal field in the ferrite section and hits the shorting plate. The reflected wave moving back through the ferrite experiences a second phase shift

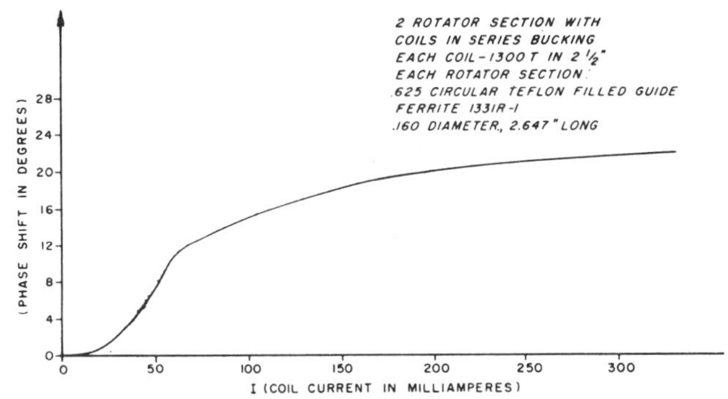

Fig. 5—Phase shift vs coil current.

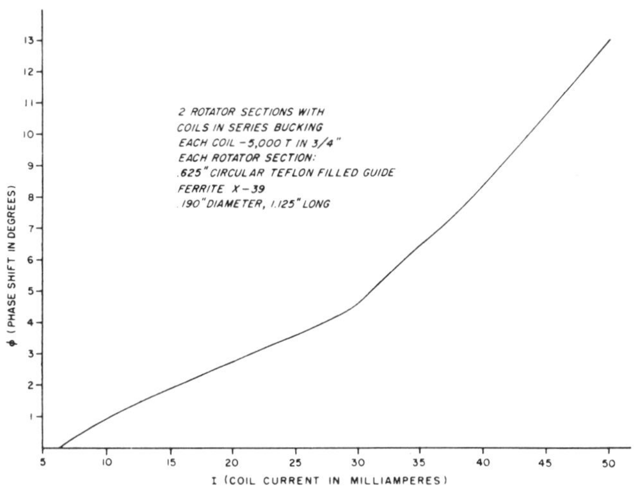

Fig. 6—Phase shift vs coil current.

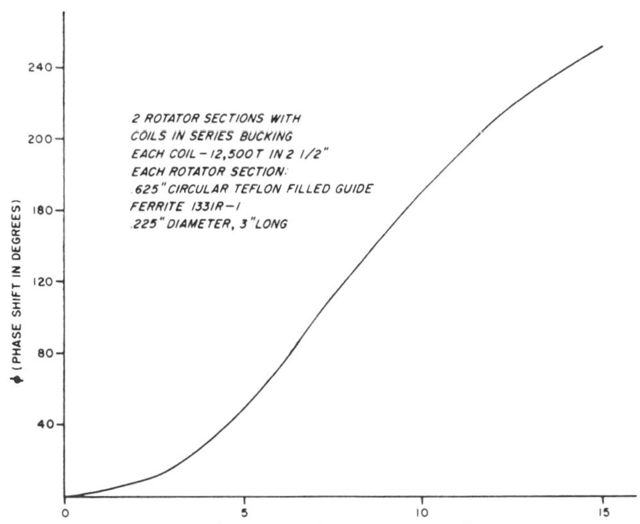

Fig. 7—Phase shift vs coil current.

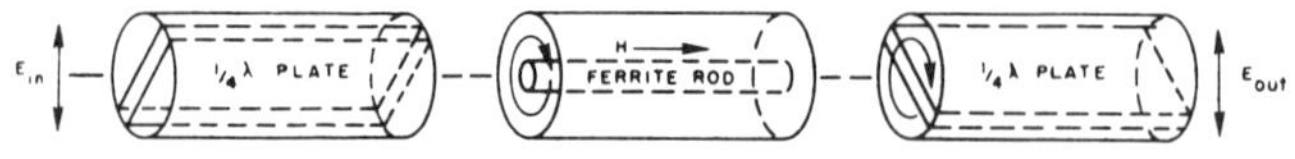

Fig. 8—Circular polarized phase shifter.

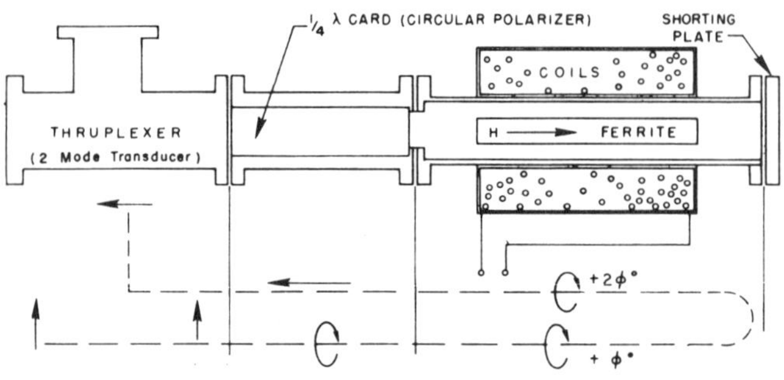

Fig. 9—Cross-polarized phase shifter.

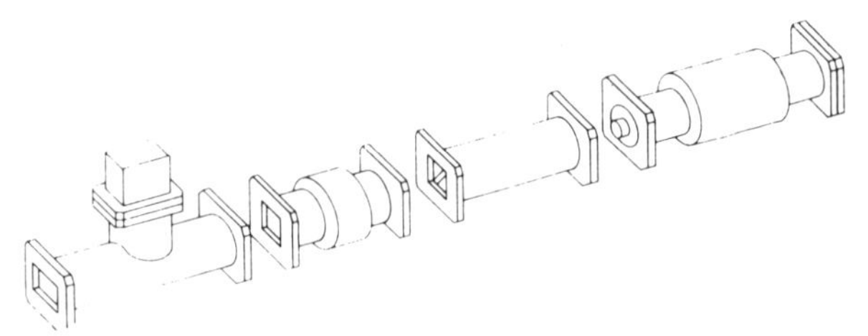

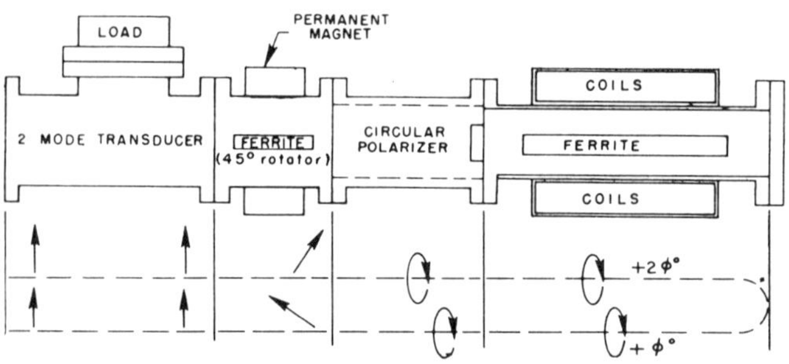

Fig. 10—Electrically controllable short circuit.

of the same sign as the first. Consequently the phase shift is doubled on the second trip through the ferrite. The action of the polarizer is to convert the circularly polarized reflected wave back to linear in a plane cross-polarized to the input field. The side arm of the thruplexer may therefore be used as the output of the device. It should be noted that this device, as the one from which it was derived, is also nonreciprocal in the sense that a phase advance is realized for one direction of propagation and a delay is obtained for the other direction. The cross-polarized phase shifter is capable of producing very large phase variations. It is considerably larger and more complex than the bucking-rotator type because of the required thruplexer and circular polarizer. These two types are supplementary rather than competitive. In contrast to the slab type and the original circularly-polarized type described above, the cross-polarized phase shifter produces greater phase variation per unit length. Another point to be considered, depending on system application and packaging, is the location of the output port close to the input port.

Electrically Controllable Short Circuit

As shown in Fig. 10 on the left, a modification of the cross-polarized phase shifter may be made for certain applications. In place of the thruplexer, a fixed 45 degree rotator (or isolator) is used preceding the polarizer. In this manner the input and output linearly polarized fields are in the same plane. The device is thus an electrically controllable short circuit.

Broad-Band Considerations

Although the three new devices introduced above were not specifically designed for broad-band operation, their limitations are readily apparent. The bucking-rotator phase shifter is completely symmetrical. Its satisfactory operation over any band is thus only a function of how well one can match the input and output guides into the ferrite rod section. The phase shift for a particular applied field will, of course, vary with frequency, but from a microwave viewpoint it appears that this device could be made very broad-band (*i.e.*, approximately 25 per cent bandwidth).

The broad-band limitations of the cross-polarized phase shifters and controllable short is a function of the broad-band properties of several components (polarizers, rotators, thruplexers, and 45 degree isolators). From this standpoint one would expect a gradual degradation in performance for bandwidths in excess of 10 per cent.

Conclusion

Three new types of electrically controllable ferrite phase shift devices have been presented. Their characteristics are supplementary to each other and in a general sense to other types which have been previously reported. The bucking-rotator type is particularly suitable for applications requiring small phase changes in a small space. The cross-polarized phase shifter and controllable short circuits have very high possible phase variation and in some applications can be more easily packaged in a system than other high phase variation types.

In order to obtain the necessary difference between β_+ and β_- at other than dc conditions, the longitudinal field devices require less driving power than the transverse field types. Therefore, all of the new types discussed herein employ a longitudinal magnetic field and have been successfully modulated using miniature tube drivers.

Acknowledgment

It is a pleasure to acknowledge the help of Dr. Ernest Wantuch and Dr. Peter Rizzi whose many suggestions and ideas contributed greatly to this work. Thanks are due also to C. G. Shaw for making many of the measurements.

Topics in Guided-Wave Propagation in Magnetized Ferrites*

MORRIS L. KALES†, MEMBER, IRE

Summary—**Contributions of various authors to the linear theory of propagation in magnetized ferrites are briefly reviewed. Modes of propagation are discussed for the cases where a static magnetic field is applied to a ferrite in directions parallel and perpendicular respectively to the direction of propagation of the microwave field. Various properties of propagation are described for waveguides of rectangular or circular cross section which are either partially or completely filled with ferrites.**

Emphasis is placed on those characteristics of propagation in ferrite media which are in contrast with those obtained in conventional media. Unusual types of mode "cutoffs" are described, as well as various nonreciprocal propagation properties resulting from the particular type of anisotropy which is present in a magnetized ferrite.

The review closes with a brief discussion of some results obtained by perturbation methods.

Introduction

IN ACCORDANCE with the linear theory of propagation, if a ferrite is magnetized in the direction of the z-axis by the application of a static magnetic field H_0, the relation between the rf components of $\vec{B}$ and $\vec{H}$ is given by the equations[1,2]

$$B_x = \mu H_x - j\kappa H_y$$
$$B_y = j\kappa H_x + \mu H_y$$
$$B_z = \mu_z H_z.$$

For a given unbounded ferrite medium, and at a given frequency, the components of the permeability tensor

$$\bar{\bar{\mu}} \equiv \begin{pmatrix} \mu & -j\kappa & 0 \\ j\kappa & \mu & 0 \\ 0 & 0 & \mu_z \end{pmatrix}$$

will depend on H_0.

Recent investigations have shown that propagation in a ferrite medium may be nonlinear at power levels normally encountered in practical applications.[3] The discussion that follows, however, is confined to the case where the propagation is linear.

Although the subject of this discussion is guided-wave propagation, it will be instructive to consider first the case of uniform plane waves in an unbounded ferrite medium. Polder has shown that for any direction of propagation there are two such waves.[1] In general these waves are elliptically polarized, and have unequal propagation constants. Two cases are of particular interest: they are the ones for which the directions of propagation are either parallel or perpendicular to the static magnetic field. In the first case, the waves are circularly polarized with opposite senses of rotation. Each of these waves is identical in structure with a circularly polarized plane wave in an isotropic, homogeneous medium having an appropriate permeability. For the wave rotating in the positive sense with respect to the magnetizing field the permeability is $\mu_+ = \mu - \kappa$. For the other wave the permeability is $\mu_- = \mu + \kappa$. These "effective" values of permeability depend only on the sense of rotation with respect to the magnetizing field and not on the sense of propagation. When two circularly polarizing plane waves of equal amplitudes and opposite senses of rotation propagate simultaneously in the direction of the magnetizing field, their sum is linearly polarized at every point in space, but the plane of polarization rotates as the wave advances. This is the well-known Faraday rotation. As a consequence of the fact that the propagation constants for each of the circularly polarized waves depend only on the sense of the magnetizing field, and not on the sense of the propagation direction, the rotation is nonreciprocal.

When the direction of propagation is at right angles to the magnetizing field, one of the two possible plane waves has a component of $\vec{H}$ in the direction of the z-axis only. For this wave only the μ_z component of the tensor permeability enters into consideration and, as a result, the wave is identical in form with a uniform plane wave in an isotropic, homogeneous medium of permeability μ_z. The second plane wave has no component of $\vec{H}$ in the direction of the z-axis. This wave has the $\vec{E}$-vector in the direction of the z-axis, and insofar as $\vec{E}$ and $\vec{B}$ are concerned, it is identical with a plane wave in an isotropic medium with permeability $\mu\{1-(\kappa/\mu)^2\}$. With respect to the vector $\vec{H}$, however, there is an important difference. In this case the $\vec{H}$-vector has a component in the direction of propagation and is elliptically polarized.

Ferrite-Loaded Waveguide Longitudinally Magnetized

We turn now to the consideration of propagation in a waveguide. Most of the studies of propagation in a ferrite-loaded waveguide have treated the case where the

* Original manuscript received by the IRE, May 21, 1956. Presented at Symposium on the Microwave Properties and Applications of Ferrites, Harvard Univ., Cambridge, Mass., April 2–4, 1956.

† Microwave Antennas and Components Branch, Electronics Div., Naval Res. Lab., Washington 25, D. C.

[1] D. Polder, "On the theory of ferromagnetic resonance," *Phil. Mag.*, vol. 40, pp. 99–115; 1949.

[2] C. L. Hogan, "The microwave gyrator," *Bell. Sys. Tech. J.*, vol. 31, pp. 1–31; January, 1952.

[3] N. G. Sakiotis, H. N. Chait, and M. L. Kales, "Nonlinearity of propagation in ferrite media," PROC. IRE, vol. 43, p. 1011; August, 1955.

51

magnetizing field is either parallel or transverse to the longitudinal axis of the waveguide.[4–9] We shall consider first the case where the magnetizing field is parallel to the longitudinal axis which is taken as the z-axis. It is assumed that the ferrite is uniformly magnetized, and that its cross section is uniform with respect to z. As in the case of an isotropic medium, all the field components of any mode can be expressed in terms of the longitudinal components E_z and H_z. These components satisfy a pair of equations of the form

$$\Delta^2 E_z + aE_z + b\,H_z = 0.$$

$$\nabla^2 H_z + cH_z + dE_z = 0.$$

The coefficients in these equations are functions of the propagation constant of the mode, and other physical constants of the medium. Formal solutions for the modes which depend on the solution of a transcendental
52 equation for the propagation constant can be obtained in the case of a round guide filled or partially filled with ferrite in an axially symmetric fashion. A complete study of the modes is very difficult, but certain general properties can be readily obtained. For example, it can easily be shown that no TE, TM, or TEM modes are possible. The modes fall into two groups; those having an $e^{in\theta}$ and those having an $e^{-in\theta}$ θ-dependence. The transverse field vectors of these modes are circularly polarized in opposite senses on the axis of the guide. (If $|n|>1$ this statement must be interpreted in a limiting sense, inasmuch as the transverse field components vanish on the axis in this case.) For a given direction of propagation, the propagation constant of a mode with an $e^{in\theta}$ dependence differs from one with an $e^{-in\theta}$ dependence. In other words, as in the case of the unbounded medium, modes which are circularly polarized in opposite senses have unequal propagation constants. This leads to a rotation of the plane of polarization when an incident linearly polarized mode passes through a ferrite waveguide section. This situation differs from the unbounded medium case, inasmuch as the counter-rotating circularly polarized modes not only have unequal propagation constants, but have different field configurations as well. It has been pointed out by some writers that two such modes do not combine to give a linearly polarized transverse field pattern which simply rotates as the wave progresses in the direction of propagation. While this is true within the ferrite medium itself, in the usual application one is not concerned with the field configuration within the ferrite medium, but rather with the relation between input and output waves on passing through a ferrite section of waveguide which joins two empty sections. From symmetry considerations, it is clear that an incident circularly polarized mode will emerge circularly polarized, in the same sense. However, the phase change on passing through the ferrite section will be different for two incident circularly polarized modes of opposite sense. Thus if a linearly polarized mode is incident on the ferrite section, the relative phase of its two circularly polarized components will be altered on passing through the section. If the insertion losses are equal for the two components they will recombine to form a linearly polarized mode, rotated with respect to the incident mode. If the amplitudes of the transmitted components are unequal, the resultant will be elliptically polarized, but the major axis of the ellipse will make the same angle with the direction of polarization of the incident wave as would be the case if the amplitudes were equal.

For the case of the completely filled circular waveguide, calculated values of the propagation constant for several special cases are given in a paper by Gamo.[5] Suhl and Walker[6] have made an extensive study of the behavior of the modes as a function of various physical constants of the medium. The results which they have obtained, as well as the analysis, are rather involved and we shall not attempt to summarize them here. Instead we shall single out for discussion just one of the interesting results which they present, which serves to illustrate some of the novel features of propagation in the ferrite medium. This has to do with the cutoffs of the modes. The analysis of Suhl and Walker shows that there are three different types of cutoffs. One of these types has been obtained by others, and is similar to the cutoff which occurs if the circular waveguide is filled with a homogeneous, isotropic medium. In such a waveguide, when the propagation constant β becomes zero, the field can be represented as a superposition of plane waves propagating at right angles to the longitudinal axis of the guide. It was pointed out earlier that in the unbounded ferrite medium two plane waves propagating at right angles to the direction of the magnetizing field were possible. With respect to the vectors $\vec{E}$ and $\vec{B}$ these waves are identical with plane waves in an isotropic medium. Since the boundary conditions in the filled guide involve only the tangential components of $\vec{E}$, it is clear that a cutoff condition can be obtained in the ferrite waveguide by a superposition of plane waves propagating at right angles to the axis. In calculating the cutoff one has only to use the effective value of μ, which is μ_z if the $\vec{H}$-vector of the plane waves is parallel to the axis, and $\mu(1-(\kappa/\mu)^2)$ if the $\vec{H}$-vector is perpendicular to the axis of the waveguide. Fig. 1 is a plot of the normalized propagation constant vs the normalized wave-

[4] M. L. Kales, "Modes in waveguides containing ferrites," *J. Appl. Phys.*, vol. 24, pp. 604–608; May, 1953.

[5] H. Gamo, "The Faraday rotation of waves in a circular waveguide," *J. Phys. Soc. Japan*, vol. 8, pp. 176–182; March–April, 1953.

[6] H. Suhl and L. R. Walker, "Topics in guided-wave propagation through gyromagnetic media, Part I—The completely filled cylindrical guide," *Bell Sys. Tech. J.*, vol. 33, pp. 579–659; May, 1954.

[7] H. Suhl and L. R. Walker, "Topics in guided-wave propagation through gyromagnetic media, Part II—Transverse magnetization and the non-reciprocal helix," *Bell. Sys Tech. J.*, vol. 33, pp. 939–986; July, 1954.

[8] H. Suhl and L. R. Walker, "Topics in guided-wave propagation through gyromagnetic media, Part III—Perturbation theory and miscellaneous results," *Bell Sys. Tech. J.*, vol. 33, pp. 1133–1194; September, 1954.

[9] A. A. Th. M. VanTrier, "Guided electromagnetic waves in anisotropic media," *Appl. Sci. Res.*, sec. B, vol. 3, pp. 305–371; 1953.

guide radius for a value of $\kappa/\mu=\frac{1}{2}$. (Modes are referred to as quasi-TE or quasi-TM if the limit of zero magnetizing field they reduce to TE or TM modes.) Note that the cutoff occurs at the same value of the radius for both the positive and negative circularly polarized modes. The dashed line is the asymptotic value of the propagation constant for the positive circularly polarized modes, as the waveguide radius becomes infinite. This value is that which would be obtained in the unbounded magnetized ferrite medium.

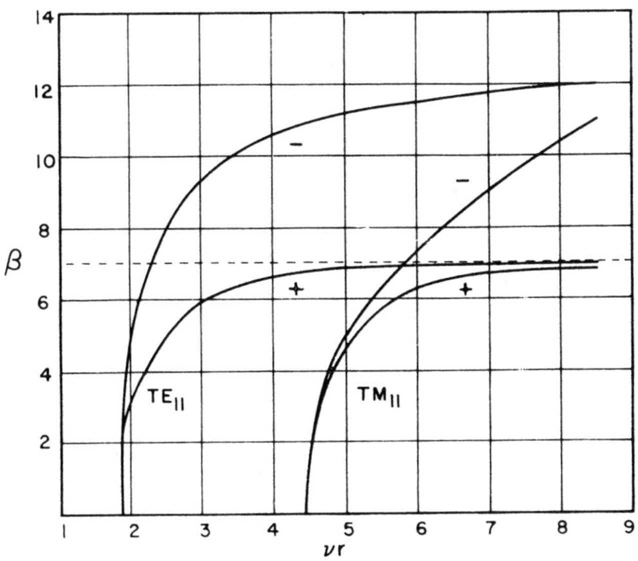

Fig. 1—Normalized propagation constant vs normalized radius. (Reproduced, with minor change in notation, from Gamo.[5])

The second type of cutoff, for which β is zero, is one for which the longitudinal components E_z and H_z both vanish. An examination of the equations which express the transverse components of the field in terms of the longitudinal components reveals that a nonzero solution is possible only if $\kappa/\mu=\pm 1$. It then turns out that $\vec{E}$ and $\vec{B}$ are both zero, and we are left only with a circularly polarized $\vec{H}$-vector rotating in planes perpendicular to the axis of the guide. For an $e^{in\theta}$ θ-dependence it can be shown that $\vec{H}=\nabla r^n e^{in\theta}$, where r and θ are the polar-coordinates of a point in the transverse plane. The striking feature of this solution is the existence of a time-varying $\vec{H}$-field, without an accompanying $\vec{E}$-field. Such a situation cannot exist in an isotropic medium, but is here possible because the appropriate effective μ, which is $\mu\pm\kappa$ when the $\vec{H}$-vector is circularly polarized in a transverse plane, vanishes. The occurrence of this type of cutoff can be seen in Fig. 2, where a normalized value of the propagation constant is plotted vs a parameter p for several values of a normalized radius. The parameter p is proportional to the magnetization of the ferrite. When $p=-1$, $\mu+\kappa=0$, which is a condition for cutoff. Note that the cutoff does not occur if the guide radius

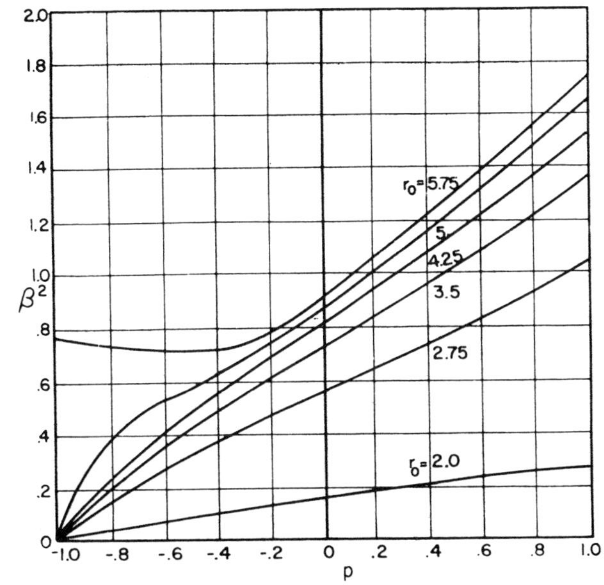

Fig. 2—Square of normalized propagation constant vs parameter p. (Reproduced from Suhl and Walker.[6])

is sufficiently large. The curves shown are for just one sense of polarization. Curves for the other sense may be obtained by reflection in the vertical axis.

The third type of cutoff is perhaps the most unusual in that, when it occurs, β is not zero. This cutoff occurs when the main diagonal component μ of the tensor permeability becomes zero. It is not clear from the analysis whether this mode still exists as a nonpropagating mode or simply ceases to exist when μ becomes negative. It appears to the writer that the method of analysis is such, that only those values of β^2 which are real can be obtained. The question then arises whether or not β can become complex, even though the medium is assumed to be lossless. (Continuity considerations would appear to rule out the possibility of an imaginary value of β.) Such a result would be in contrast with the homogeneous, isotropic case where for a lossless medium the propagation constant is always real or imaginary. (According to Adler this is true even for an inhomogeneous, isotropic medium.)[10]

We turn now to a recent publication by Chambers which deals with the case of a waveguide of *arbitrary* cross section completely filled with ferrite and magnetized in the longitudinal direction.[11] The author treats the case where the main diagonal components of the tensor μ and μ_z are equal. The longitudinal field components and the square of the propagation constant are represented as series in powers of κ/μ. When $\kappa/\mu=0$, these quantities reduce to the corresponding values appropriate to a particular mode which exists for the waveguide filled with the unmagnetized and therefore iso-

[10] R. B. Adler, "Properties of Guided Waves on Inhomogeneous Cylindrical Structures," Res. Lab. Elec. Tech. Rep. No. 102; 1949.
[11] L. L. G. Chambers, "Propagation in a ferrite-filled waveguide," *Quart. J. Mech. and Appl. Math.*, vol. VIII, Part 4; December, 1955.

tropic, ferrite medium. The determination of the coefficients requires the solution of an infinite set of pairs of wave equations. However, the solution of each pair depends only on the solutions of the preceding equations. Thus, in principle, it is possible to obtain any given number of coefficients. This writer is of the opinion that in its present form the development is not generally valid, although it may be possible to modify it so that it is. For example, the results which Chambers obtains indicate that when the ferrite is weakly magnetized the change in the propagation constant is of the second order in κ/μ, whereas it has been found by other methods to be of the first order in the case of the circularly polarized TE_{11} mode. Furthermore, it would appear from Chambers' results that a linearly polarized mode could
54 be obtained as a limiting form of a single mode in the weakly magnetized ferrite. On the other hand, it is known that if the unmagnetized ferrite medium supports a linearly polarized mode in a circular waveguide, as soon as a longitudinal magnetizing field is applied, the mode splits into *two* counter-rotating circularly polarized modes. These two modes have unequal propagation constants, and therefore cannot combine to form a single mode which would reduce to the linearly polarized mode in the limit of zero magnetizing field.

Ferrite-Loaded Waveguide Transversely Magnetized

The next problem which we shall consider is the one where the magnetizing field is transverse to the longitudinal axis of the guide. The problem which has received the most extensive theoretical treatment is the case of the rectangular waveguide with the magnetizing field perpendicular to the broad faces of the guide. If the z-axis is again taken in the direction of the magnetizing field it can be shown that modes which are independent of z can have only an $\vec{E}$ component in the z-direction, and that $\vec{H}$ is transverse to the z-direction. For the completely filled guide these modes can be found in the same way that they are frequently obtained in the isotropic case: that is, by a superposition of two uniform plane waves traveling in directions which are parallel to the broad faces and make a suitable angle $\pm\theta$ with the longitudinal axis. It will be recalled that, in the unbounded ferrite medium, uniform plane waves exist which propagate in any direction perpendicular to the magnetizing field, have their $\vec{E}$-vector in the direction of the magnetizing field, and insofar as $\vec{E}$ and $\vec{B}$ are concerned are identical in form with uniform plane waves in an isotropic, homogeneous medium of permeability $\mu\left[1-\left(\frac{K}{2}\right)^2\right]$. Since in the completely filled guide the boundary conditions involve only the tangential components of $\vec{E}$, it is obvious that for these z-independent modes the solutions are the same as in the empty waveguide insofar as $\vec{E}$ and $\vec{B}$ are concerned. It should be emphasized at this point that these remarks do not apply in the case where the ferrite only partially fills the cross section of the guide. In such a case the boundary conditions involve the tangential components of $\vec{H}$ at an interface between the ferrite and the air.

If we solve for the components of $\vec{H}$ in terms of those of $\vec{B}$ we find that

$$H_x = (\mu B_x + j\kappa B_y)(\mu^2 - \kappa^2)^{-1}$$

$$H_y = (-j\kappa B_x + \mu B_y)(\mu^2 - \kappa^2)^{-1}.$$

Thus we see that if the components of $\vec{B}$ are the same as they would be in an isotropic, homogeneous guide, the components of $\vec{H}$ must differ.

As a consequence of the distortion of the $\vec{H}$-field in the ferrite medium, if a normal waveguide mode is incident on a ferrite-filled section of waveguide an infinite number of modes must be excited in order to satisfy the boundary condition. This differs from the case of the isotropic medium where only modes of the same type are needed to satisfy the boundary conditions.

From the point of view of practical applications, the case where the ferrite only partially fills the cross section of the guide has received the greatest amount of study. The simplest case to analyze is that in which the ferrite consists of one or more slabs of rectangular cross section which fill the guide in height, and for which the modes are independent of z. In this case formal solutions which depend on the solution of a transcendental equation for the propagation constants can be obtained. An examination of the equation which determines the propagation constants reveals that when a single slab is asymmetrically located in the waveguide, the propagation constants of the modes depend upon the direction of propagation. The same thing is true for an asymmetrical distribution of several slabs. (Throughout this discussion a distribution will be considered asymmetrical even in the case where there is geometrical symmetry, if the geometrically similar portions are oppositely magnetized.) From the fact that the propagation constants are dependent on the direction of propagation, it may be concluded that the relation between the transmitted wave and the wave which is incident on such a section is dependent on the direction of propagation. In other words the ferrite section is "nonreciprocal." From a rigorous mathematical point of view this argument is not conclusive since, for either direction of propagation, an infinite number of modes are excited at the transitions between the empty and ferrite sections of guide. Intuitively, however, one might except nonreciprocal behavior, and this is borne out by experimental results.[12–14] The nonreciprocal character of the propaga-

[12] M. L. Kales, H. N. Chait, and N. G. Sakiotis, "A nonreciprocal microwave component," *J. Appl. Phys.*, vol. 24, pp. 816–817; June, 1953.

[13] N. G. Sakiotis and H. N. Chait, "Properties of ferrites in waveguides," Trans. IRE, vol. MTT-1, pp. 11–16; November, 1953.

[14] H. N. Chait, "Non-reciprocal microwave components," 1954 IRE Convention Record, Part 8, pp. 82–87.

tion can also be accounted for by a physical argument which is useful in analyzing other cases. When the ferrite is magnetized, the spin axes of the electrons in the ferrite have a natural sense and frequency of precession about the direction of magnetization. If the rf magnetic vector is circularly polarized, in a plane perpendicular to the magnetizing field, its interaction with the electron spins will depend on whether the sense of rotation is the same or opposite to the sense of precession of the spin axes. If the rf magnetic vector is elliptically polarized, it can be resolved into unequal circularly polarized components of opposite sense. The effect of the medium on propagation will depend on which sense of circular polarization is greater. Now let us consider the rf magnetic vector of the dominant mode in empty waveguide. This vector is elliptically polarized in planes parallel to the broad faces of the guide. In passing from one side of the guide to the other the sense of rotation of the vector reverses. If the direction of propagation is reversed, the senses of polarization on the two sides are interchanged. Thus, when the dominant mode is incident on the ferrite section for one direction of propagation one sense of circular polarization is greater, but when the direction of propagation is reversed the other sense of polarization is greater at the place where the ferrite is located. Here again the argument is not conclusive since it is the resultant field in the ferrite medium which is the determining factor. Again intuition suggests that if the polarization of the incident field is different for the two directions of propagation, the same will be true of the resultant field.

If the ferrite is symmetrically distributed, it follows from symmetry considerations that the relationship between incident and transmitted fields is independent of the direction of propagation. One of the interesting features of the ferrite medium is that the field distribution is asymmetrical even though the medium has physical symmetry. We have already seen that this is true of the $\vec{H}$-field in the case of the filled waveguide. When the waveguide is only partially filled, both the $\vec{E}$-field and $\vec{H}$-field may be asymmetrical. This fact has an important bearing on the design of ferrite devices. For example, if a scanning antenna is designed by cutting slots in the narrow wall of a ferrite-loaded guide, the field configuration at the slots depends on the direction of propagation. Thus the identity of the transmitting and receiving pattern that exists in the case of conventional antennas is destroyed. This violation of reciprocity is not surprising when it is considered that the slots on one side of the guide destroy the symmetry which otherwise leads to reciprocal propagation characteristics. Although the nonreciprocal field distribution is a disadvantage in the case of antennas, there are other applications which take advantage of this fact. One of these is the so-called field displacement isolator where the displacement of the field from one side of the guide to the other, when the propagation direction is reversed, is used to produce a much greater dissipation of energy for one direction of propagation than the other.[15]

Lax and his co-workers have made a study of the propagation constants and the mode configurations for a number of reciprocal and nonreciprocal cases.[16,17] We shall here review some of the results which appear to be of particular interest. The first deals with the case where slabs of equal thickness are placed adjacent to each of the narrow walls of the guide and magnetized in the same direction. The electric field distributions and a plot of the phase constant vs slab thickness for two propagating modes are shown in Figs. 3 and 4. The first

Fig. 3—Electric field patterns for ferrite modes in rectangular waveguide. (Reproduced from Lax and Button.[16,17])

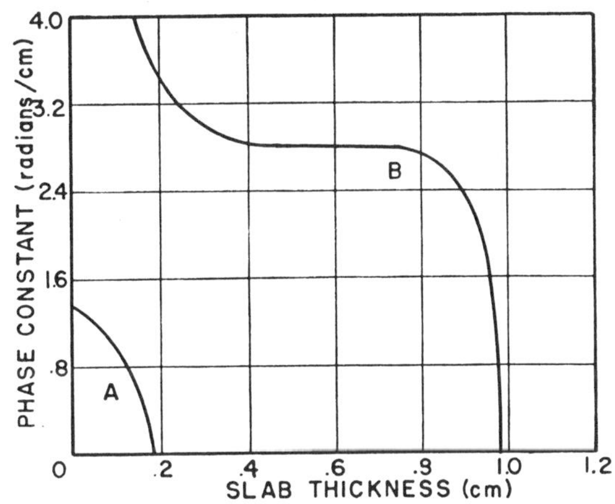

Fig. 4—Phase constant vs slab thickness. (Reproduced from Lax and Button.[16,17])

mode designated by "A" has an electric field distribution similar to that of the dominant mode of the empty guide, and in the limit of zero slab thickness this mode reduces to the dominant TE_{10} mode. On the other hand when the slab thickness exceeds a certain critical value this mode is cut off. At first glance, it seems rather surprising that the mode should be cut off when the slab thickness increases. One, perhaps, thinks of the case of pure dielectric slabs where, as the slab thickness increases, the electrical width of the waveguide increases

[15] A. G. Fox, S. E. Miller, and M. T. Weiss, "The behavior and applications of ferrites in the microwave region," *Bell Sys. Tech. J.*, vol. 34, pp. 6–103; January, 1955.

[16] B. Lax and K. J. Button, "Theory of new ferrite modes in rectangular wave guide," *J. Appl. Phys.*, vol. 26, pp. 1184–1185; September, 1955.

[17] B. Lax and K. J. Button, "New ferrite mode configurations and their applications," *J. Appl. Phys.*, vol. 26, pp. 1186–1187; September, 1955.

and therefore there is no tendency for a mode to be cut off. However, it will be recalled that for plane waves traveling at right angles to the direction of static magnetization, the effective permeability is given by $\mu_{eff}=\mu(1-(\kappa/\mu)^2)$. We see that μ_{eff} can be made arbitrarily small, and even negative, by making κ/μ close to unity. We can think of the electrical thickness of the slab as being given by $\omega\sqrt{\epsilon\mu_{eff}}\, t$, where t is the actual thickness. Thus, even though the dielectric constant of the ferrite is rather large, the electrical thickness of the slab can be made smaller than that of air, by making μ_{eff} small. If this is true, making the slabs thicker would narrow the electrical width of the guide, and tend to suppress the dominant mode.

The second mode which is designated by "B" does not make its appearance until the slab thickness reaches a certain minimum value, and it is cut off as soon as the thickness exceeds a critical value. This mode has been compared by Lax and Button with a so-called "ferrite dielectric" mode, shown in Fig. 5. The "ferrite dielectric"

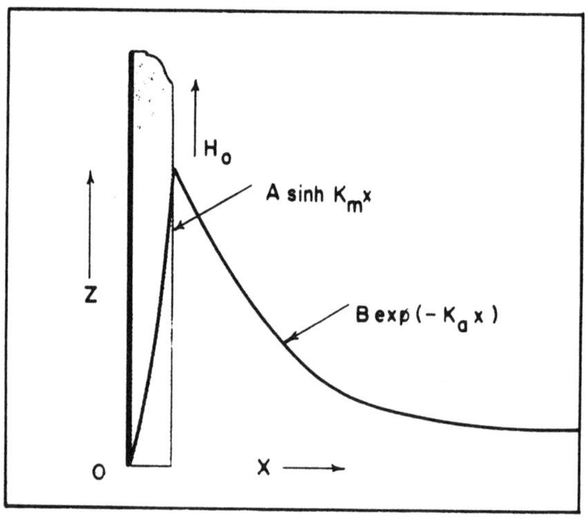

Fig. 5—Electric field distribution of the "ferrite dielectric" mode. (Reproduced from Lax and Button.[16,17])

mode is a bound wave which propagates in the semi-infinite space consisting of the space on one side of an infinite perfectly conducting plane, with an infinite magnetized slab adjacent to it. It is possible for a wave to be guided by this slab for one direction of propagation, but not the other. For this wave the electric field reaches a peak value at the interface between the ferrite and air, and is then attenuated exponentially with distance from the conducting plane. In the waveguide, the second slab and waveguide wall are in a region where the field of the mode guided by the first slab is weak. Presumably, therefore, the effect of the second wall and slab is to perturb the "ferrite dielectric" mode somewhat. If the direction of propagation is reversed, the second slab guides the mode, and thus the peak field intensity shifts from one side of the guide to the other.

If a single slab is placed in the waveguide, Lax and Button have shown that a situation may exist which has aroused wide interest. This is the fact that if the thickness and magnetizing field are properly chosen, a propagating mode will exist for one direction of propagation, but there will be no propagating modes for the other. From energy considerations it is known that if two ports of a junction are coupled by a lossless transmission path in such a way that energy can be transmitted from one to the other, then transmission in the reverse direction is possible. Ferrite circuits do exist which permit energy to be transferred from one port to a second, but not from the second to the first. In these cases, energy transmitted from the second port is transferred to a load, and the transmission path must be regarded as a lossy path for that direction of transmission. It might at first seem that the conditions described by Lax and Button, where a propagating mode is possible in one direction, but none is possible in the opposite, violate these energy principles. However, it should be kept in mind that in discussing a single mode, the waveguide must be considered to extend to infinity in both directions, and have no available ports. It is helpful, perhaps, to think of the waveguide as being in the form of a closed circle of infinite radius. It is then not difficult to think of energy circulating around this path in one sense, but not in the other, and no energy principles are violated. On the other hand if the ferrite section is finite, and leads at both ends into an empty section of waveguide, then an infinite number of modes must be excited at the transition planes. Now, even in the case of a conventional waveguide at a frequency for which all modes are cut off, it is possible to transmit energy through a finite length. This requires only that two properly phased modes of the same kind, and attenuated in reverse direction, be present. It therefore does not seem unreasonable to expect that propagation through the finite ferrite section is possible for either direction of propagation, when modes belonging to both directions are present simultaneously.

The modes which have been discussed thus far are those which have no z-dependence, and for which the magnetizing field is parallel to the z-axis. The more general problem of finding modes of the completely filled rectangular waveguide which are z-dependent has been solved by Mikaelyan.[18] The method used by Mikaelyan is a modified form of one which has been used in the case of conventional waveguides. A parallel plate ferrite medium is first considered with planes perpendicular to the z-axis. A pair of plane waves of the type discussed in the first section of this paper are combined to give a mode for which the components of $\vec{E}$, which are normal to the z-axis, have a sin $n\pi z/z_0$ dependence on z, where z_0 is the distance between the parallel plates, and n is an integer. It is found that for each integer n there are two such modes propagating in an arbitrary direction parallel to the plates. If a third perfectly conducting

[18] A. L. Mikaelyan, "Electromagnetic waves in a rectangular wave-guide filled with a magnetized ferrite," *Doklady, A. N., USSR*, vol. 98, pp. 941–944; 1954.

plane perpendicular to the first two is introduced, it is found that when one of these parallel plate modes is incident on the third plane, two are reflected. Therefore when the waveguide is completed by adding a fourth plane parallel to the third, two incident and two reflected waves are needed to satisfy the boundary conditions at both walls. These boundary conditions then lead to a transcendental equation for the determination of the propagation constants, and the modes are completely determined when the propagation constants are found.

Perturbation Theory

The final theoretical topic to be discussed is that of perturbation theory. A number of writers have made contributions in this area, but probably the most comprehensive treatment is that of Suhl and Walker.[8] Two types of problems have been treated. One of these is where the magnetization of the ferrite is of arbitrary magnitude, but because the appropriate dimensions of the ferrite are small the perturbation of the homogeneous, isotropic mode is small. The other is the case where the ferrite dimensions are unrestricted, but the magnetizing field is weak. This represents a perturbation of a mode corresponding to an isotropic, but not necessarily homogeneously filled waveguide. The perturbation theory has been applied principally to cavity techniques for determining the intrinsic physical constants of the medium. From the point of view of propagation studies, it is usually the case that when the perturbation is sufficiently small to permit accurate quantitative calculation, the desired effect is too small to be of practical value. Nevertheless, the perturbation theory is often useful in providing a qualitative guide to the behavior of various ferrite devices. Thus, for example, if the dominant linearly polarized mode in a circular waveguide is incident on a longitudinally magnetized ferrite section, the propagation constants of the positive and negative circularly polarized components in the ferrite section are given by formulas of the form

$$\beta_+ = \beta_0 + A\kappa + 0(\kappa^2)$$
$$\beta_- = \beta_0 - A\kappa + 0(\kappa^2).$$

Here β_0 is the propagation constant in the unmagnetized ferrite medium and A is a constant. If the ferrite is in the form of a cylinder which is long compared to its diameter, κ will vary linearly with the applied magnetizing field. If the effect of multiple reflections can be neglected, the angle of rotation of the transmitted wave is determined by $(\beta_+ - \beta_-)/2 = A\kappa + 0(\kappa^2)$ and the phase is determined by $(\beta_+ + \beta_-)/2 = \beta_0 + 0(\kappa^2)$. Thus we see that a ferrite section can be devised which, when second and higher order terms are neglected, produces a rotation of the incident wave which varies linearly with the magnetizing field, but for which the phase of the transmitted wave is independent of the magnetization. Such a device is useful when it is desired to produce an amplitude modulation without any accompanying phase modulation.

Conclusion

At the present time the exact theory of propagation in ferrite-loaded waveguides is inadequate and too involved for the treatment of practical problems. Nevertheless, much has been learned by using the theory as a qualitative guide in conjunction with experimental methods. We shall close with a discussion of a configuration which has been investigated experimentally, but for which a rigorous theory is not available. If, in the case of the transverse magnetizing field, the rectangular waveguide which can propagate only the dominant mode is replaced by a circular waveguide, an added degree of freedom is obtained. (Actually, for this added degree of freedom, it is only necessary that the waveguide be capable of supporting two orthogonally polarized modes.) We shall assume that the axis of the ferrite cylinder coincides with the axis of the waveguide. By analogy with the unbounded ferrite medium theory for the case of propagation perpendicular to the direction of magnetization, we find that what we can call an effective propagation constant depends on whether the $\vec{E}$-vector of an incident linearly polarized mode is parallel or perpendicular to the direction of the magnetizing field. Experimentally, it has been determined in these two cases, that after passing through the ferrite section, an incident linearly polarized mode emerges linearly polarized. If the $\vec{E}$-vector of the incident mode makes some other angle θ with the direction of the magnetizing field, the incident mode can be resolved into a parallel and perpendicular mode. After emerging from the ferrite section, the relative phase of these two component modes will depend on the length of the section, and the two modes can be recombined to form a single elliptically polarized mode. For proper lengths a half-wave or quarter-wave plate can be produced.

If the axis of the ferrite cylinder is displaced from the axis of the guide, it is possible to obtain either reciprocal or nonreciprocal propagation characteristics. In particular, as Turner has shown, a nonreciprocal phase shifter can be obtained.[19] Another possibility is a ferrite section which acts as a half-wave plate for one direction of propagation, and a quarter-wave plate for the other.[20]

The discussion in this paper has been limited principally to topics which have been treated theoretically. Although some of the experimental results have been touched upon, a great deal more has been presented in a number of published articles during the past few years. The knowledge gained from these theoretical and experimental studies of propagation in ferrite media has already led to the development of a number of important ferrite devices of both the reciprocal and nonreciprocal variety, and probably in time we shall see many more such developments.

[19] E. H. Turner, "A new nonreciprocal waveguide medium using ferrites," Proc. IRE, vol. 41, p. 937; July, 1953.
[20] N. G. Sakiotis, "Ferrite quarter-wave and half-wave plates at X-band," 1954 IRE Convention Record, Part 8, p. 88.

57

In this section, papers dealing with junction circulators are reprinted. The material has been broken into five areas. These include historical works, theory and practice for strip transmission line and waveguide junction circulators, switching circulators, microstrip circulators, and lumped constant circulators.

In the historic sub-section, the paper by Chait and Curry describes the first experimental results on circulators. Two important analytical papers are included, also. The first, by Auld, is concerned with the synthesis of symmetrical waveguide circulators. The second, by Bosma, refines analysis in one of his earlier papers (1) and provides a boundary value solution for the strip transmission line Y-circulator.

In the next sub-section, important papers dealing with theory and practice of strip transmission line and waveguide junction circulators are reprinted. The milestone paper by Fay and Comstock (2) which describes the operation of the junction circulator in terms of rotating modes is included here. This paper provides detail procedures for designing junction circulators and is widely used. Another important experimental paper by

JUNCTION CIRCULATORS

II

Simon (3) describes work performed at about the same time as the Fay and Comstock analysis and describes experimentally determined procedures for broadbanding junction circulators. A series of papers by Castillo and Davis are included. These papers are concerned with the computer-aided design of waveguide junction circulators.

The other sub-sections are concerned with physical realization of various configurations of junction circulators. These configurations are sub-divided as follows:

- Switching Circulators
- Microstrip Circulators
- Lumped Constant Circulators

At the end of this section, a paper by Kumar (4) is included which lists a rather extensive circulator bibliography covering the 1957-1969 time span.

References

1. H. Bosma, "On the Principle of Stripline Circulation," Proc. IEEE, 109, Suppl. 21, (1962) pp. 137-146.
2. C.E. Fay and R.L. Comstock, "Operation of the Ferrite Junction Circulator," MTT-13, Jan. 1965, pp. 15-27.
3. J.W. Simon, "Broadband Strip Transmission Line Y-Junction Circulators," MTT-13, May 1965, pp. 335-345.
4. R.C. Kumar, "Ferrite-Junction Circulator Bibliography," MTT-18, Sep. 1970, pp. 524-530.

ABSTRACT

It is well known that the field distribution in a rectangular waveguide containing a magnetized ferrite can be asymmetrical even though the physical configuration is symmetrical. By properly choosing the ferrite, its shape, and location, it is possible to concentrate the energy on one side of the waveguide. Reversing the field or changing the direction of propagation will cause the energy to concentrate on the other side of the waveguide. This phenomenon has been utilized to construct a new type of circulator. A ferrite post has been placed at the junction of three waveguides intersecting at 120 degrees. The dc field is applied transverse to the broad side of the waveguide. Circulator action can be obtained by using a ferrite rod of the correct size and applying a dc field of the proper magnitude. Similar results have been obtained by using single and double slab configurations as well as with wedges. A waveguide switch can be obtained by reversing the direction of the applied field. Results to date indicate that it is possible to build this Y-type circulator with a loss less than 0.5 db and isolations and reflections of less than 20 db over a 500-megacycle band at X-band. A four-port junction operating on this same principle has also been under investigation.

PROBLEM STATUS

This is an interim report; work continues on the problem.

AUTHORIZATION

NRL Problem R09-02
Project NE 120-000-27
BuShips No. S-1617

Manuscript submitted February 27, 1959

Reprinted by permission of the *Naval Research Laboratory Report,* No. 5297, May 1959, pp. 1-7

THE Y- OR JUNCTION-TYPE CIRCULATOR

INTRODUCTION

The Y-circulator consists of an H-plane junction of three identical waveguides joined to form a symmetrical Y-shaped figure. The region of the junction of the three waveguides contains a symmetrical distribution of ferrite which is transversely magnetized. The symmetry is threefold, designed to make the structure appear the same from any one of the three waveguides (Fig. 1). It is clear from the geometrical symmetry that if all the power into one of the guides can be made to come out of an adjacent guide, then this will be true of all three guides and the device will be a circulator.

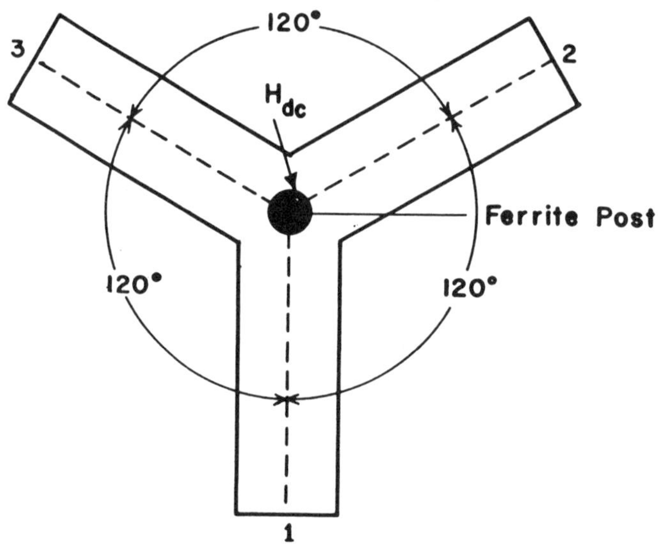

Fig. 1 - The Y-type circulator

Since the magnetized ferrite is an anisotropic medium, this configuration is not symmetrical from an electrical point of view even though it has geometrical symmetry. It is well known that when waves propagate through a section of ferrite-loaded rectangular waveguide which is transversely magnetized, the rf field distribution is asymmetrical even though the distribution of ferrite itself is symmetrical. This is shown in Fig. 2 for a rectangular waveguide filled with ferrite. The asymmetrical field distribution is a consequence of the fact that the rf magnetic field is elliptically polarized in planes parallel to the broad faces of the guide, and is of opposite sense on either side of the guide. Since the "effective" rf permeability of the magnetized ferrite depends on the sense of polarization, it is seen that the two sides of the ferrite-loaded guide are electrically dissimilar.

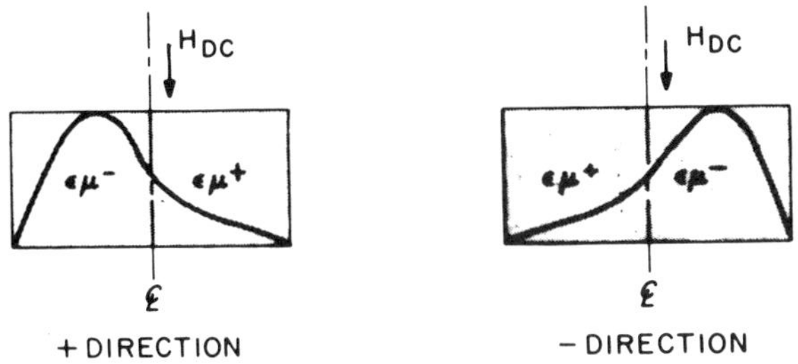

Fig. 2 - Field distribution in a rectangular waveguide filled with a magnetized ferrite

This fact suggests the possibility which has been verified experimentally of displacing the field to one side in such a way that all the power can be diverted into an adjacent waveguide in the junction region.

A number of different methods of guiding all the power from one waveguide to an adjacent one have been tried. One of these methods as shown in Fig. 1 makes use of a ferrite post in the junction region. Intuitively it seems unlikely that the ferrite post by itself would be able to deviate the energy entirely into an adjacent guide. One would expect some energy to be transmitted into the unwanted guide. Very likely, when the dimensions and field are properly chosen, reflections from the junction boundaries cancel the signal that might otherwise enter the third guide.

Another method uses slabs along each side of the three waveguides as shown in Fig. 3. 63
This can be done in such a way that almost all of the energy is trapped in one of the slabs in the so-called "ferrite dielectric mode" and is then guided continuously by the slab from one waveguide into an adjacent one. Lax and Button* have calculated the field distribution for symmetrical slabs of ferrite, transversely magnetized in the same direction. If the ferrite slabs are sufficiently thick a wave travelling in the forward direction will be guided by one slab whereas the wave travelling in the reverse direction will be guided by the other slab. Figure 4 shows the rf E- and H-fields under these conditions.

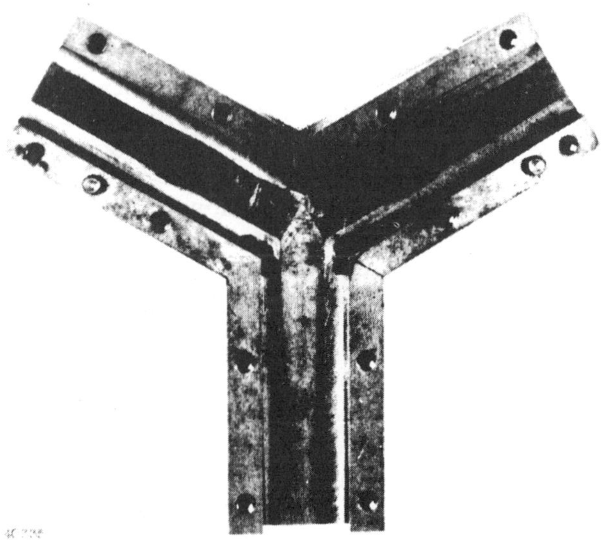

Fig. 3 - Ferrite slabs in the Y-circulator

*B. Lax and K. J. Button, J. Appl. Phys. 26:1184 (1955); 26:1186 (1955)

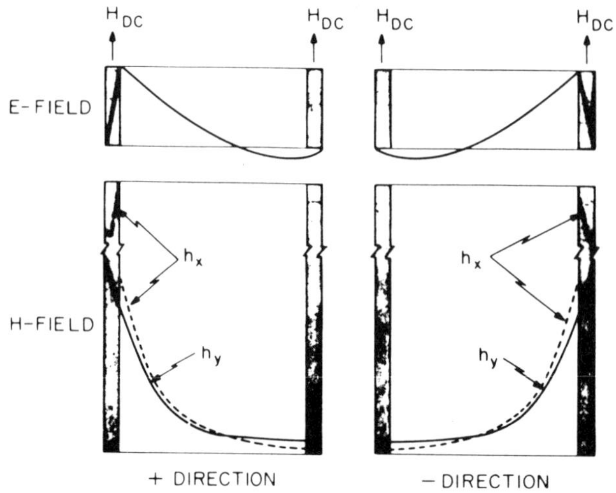

Fig. 4 - Field distributions of the "ferrite dielectric mode"

64

EXPERIMENTAL RESULTS

The first symmetrical configuration which was used experimentally consisted of a ferrite circular cylinder placed at the junction of three rectangular waveguides intersecting at 120 degrees. Circulator action was obtained by choosing the correct rod diameter and by applying a transverse field of proper magnitude. Rod diameters of from 0.125 inch up to 0.500 inch were tried and best results at 9375 Mc were obtained using a Ferramic R-1 rod 0.350 inch in diameter filling the 0.400 × 0.900 inch guide in height. The loss was less than 0.5 db and the isolation and reflection greater than 30 db over a narrow frequency band (about 50 Mc). The dc field was about 120 oersteds. By choosing a slightly different dc field and a different rod diameter, one could shift this band any place from 8.2 to 12.5 kMc.

Various other configurations having the threefold symmetry, which is required, were tried. Circulator action was also obtained using ferrite slabs and wedges. Figure 5 shows a few of the different ferrite configurations. In each case it is important that the ferrite dimensions be properly selected. The double-slab combination which is most amenable to theoretical evaluation was more lossy and had higher reflections than the post or wedge. This is probably a consequence of the relatively large volume of ferrite material which is required. At 9375 Mc, slabs 0.125 inch or more were necessary to get proper circulator performance. It should be possible to obtain better loss figures and more bandwidth from the slab configuration by the use of better ferrite material and special matching techniques. Work along these lines is in progress.

Figure 6 shows a typical set of data using a wedge of Ferramic R-1, 0.400 inch on a side filling the guide in height. The dc field was 189 oersteds. As shown on the curves the loss is less than 0.75 db, the reflections and isolation greater than 20 db from about 9200 to 9500 Mc. One can shift this frequency band by either selecting different ferrite dimensions or changing the dc field. In fact it has been found that a linear relationship exists between the length of the sides of the wedge and the operating frequency. Isolations and reflections of from 30 to 40 db are obtainable any place in the X-band. Limited high-power tests of this unit indicate that it will operate without breakdown at 50 kw peak. However, the power absorbed increased to 1.3 db because of the nonlinear characteristics of the ferrite material. Further tests using other ferrites will be carried out to reduce this loss.

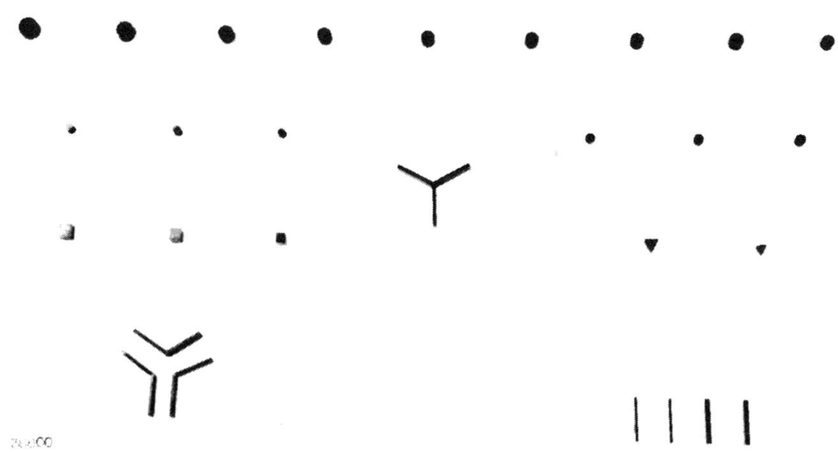

Fig. 5 - Ferrite configurations used in the Y- and junction-type circulators

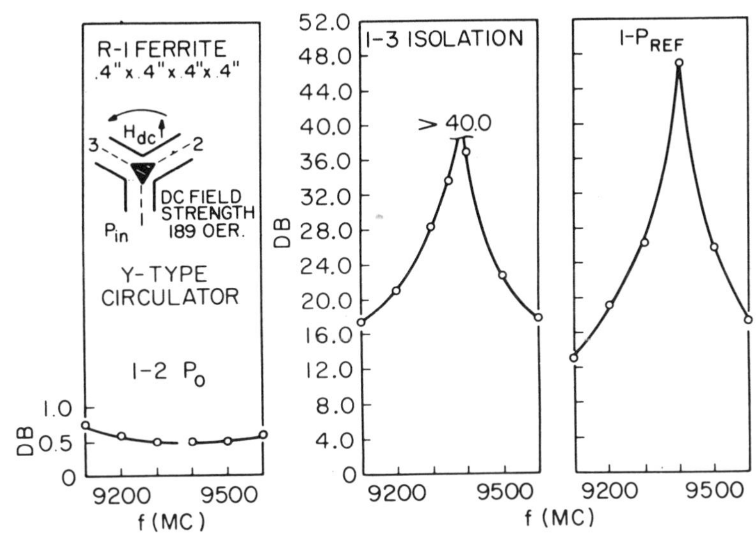

Fig. 6 - Data on ferrite wedge in Y-circulator (X-band)

Two of these three-port Y-circulators can be connected in series to produce a four-port circulator. Such a circulator was built where the Y-junctions were separated by about one-half guide wavelength. As shown in Fig. 7, the loss was less than 0.75 db and the reflection and isolations greater than 18 db from 9200 to 9400 Mc.

Another four-port circulator consisting of four waveguides intersecting at 90 degrees has been built and tested. Figure 8 shows some typical data on this type of circulator. N-port circulators of this type should also be feasible.

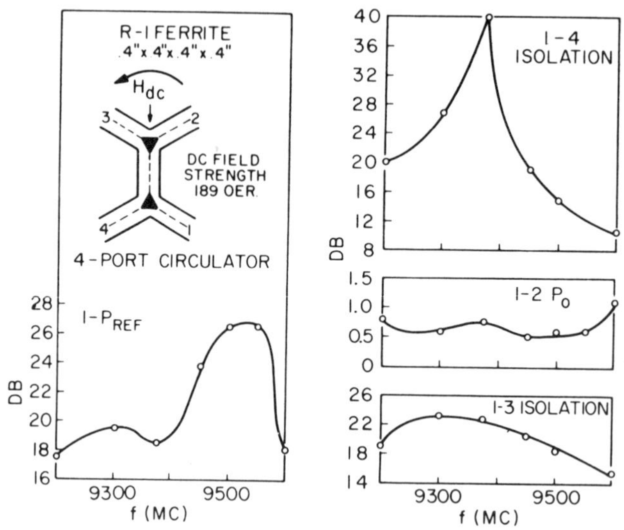

Fig. 7 - Data on ferrite wedge in two cascaded Y-circulators (X-band)

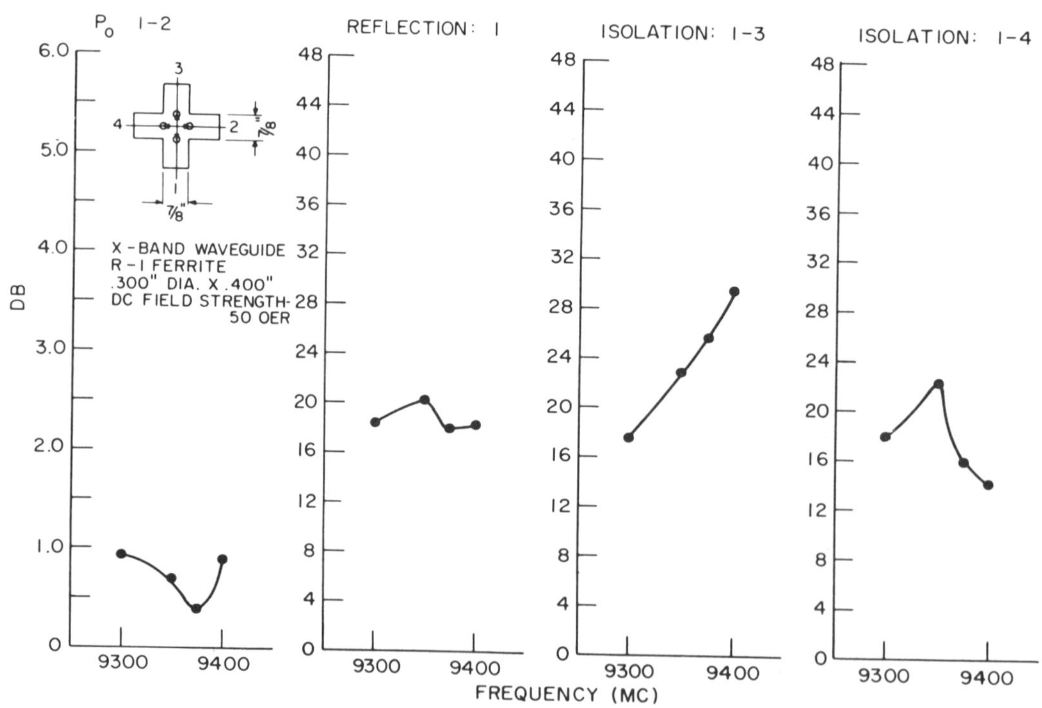

Fig. 8 - Data on cross circulator (X-band)

Although this technique had been applied and tested at X-band, it seemed that it should be possible to build circulators or switches of this type at higher and lower frequencies. Indeed, S-band models of the Y- and cross circulators have been built and tested. Figure 9 shows the results of some of these tests. Losses less than 0.25 db have been observed, making such a circulator ideally suited for various maser and parametric amplifier applications.

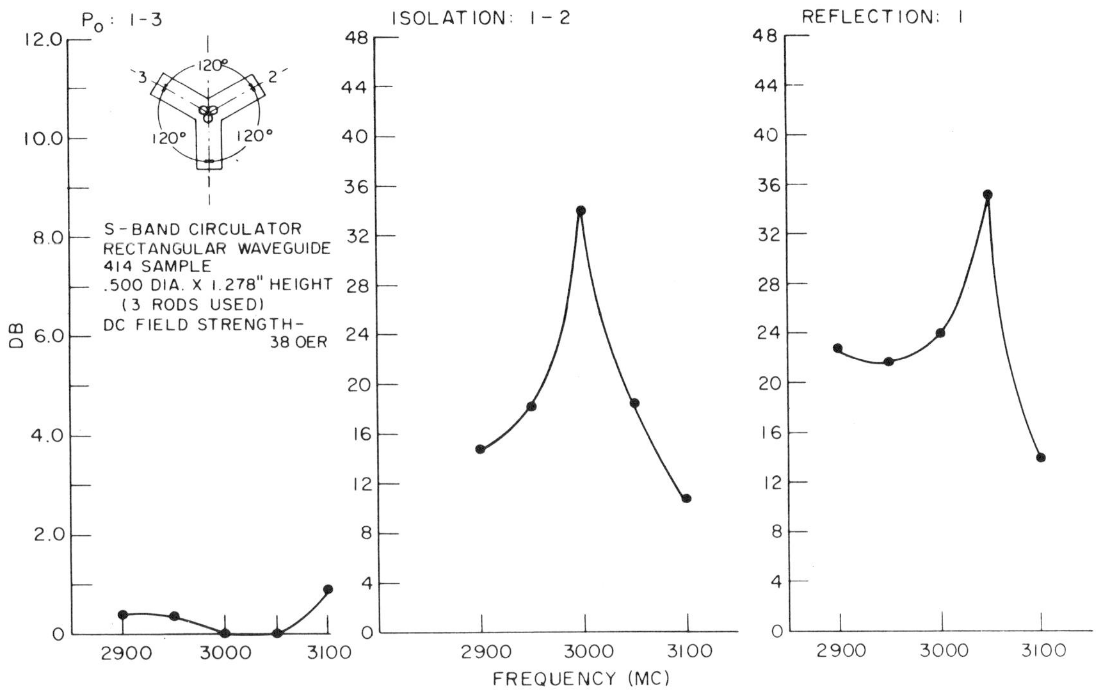

Fig. 9 - Data on Y-circulator (S-band)

CONCLUSIONS

The practical utility of the junction-type circulator has been demonstrated. It should prove very useful to the microwave engineer. It is smaller in size and lighter in weight than the Faraday rotation or rectangular waveguide differential phase type circulators. It requires no expensive hybrids thereby making it much cheaper to build than other types of circulators. At present the bandwidth is somewhat less than that of other circulators, however, it is adequate for most radar applications. With some further effort it should be possible to increase the bandwidth of this device.

Since this circulator is small and requires a low dc magnetic field it should be useful in many antenna and fast-switching applications.

A number of these circulators arranged in series, as shown in Fig. 10, can be used to produce a scanning array with separate control of the phase and amplitude of each radiator.

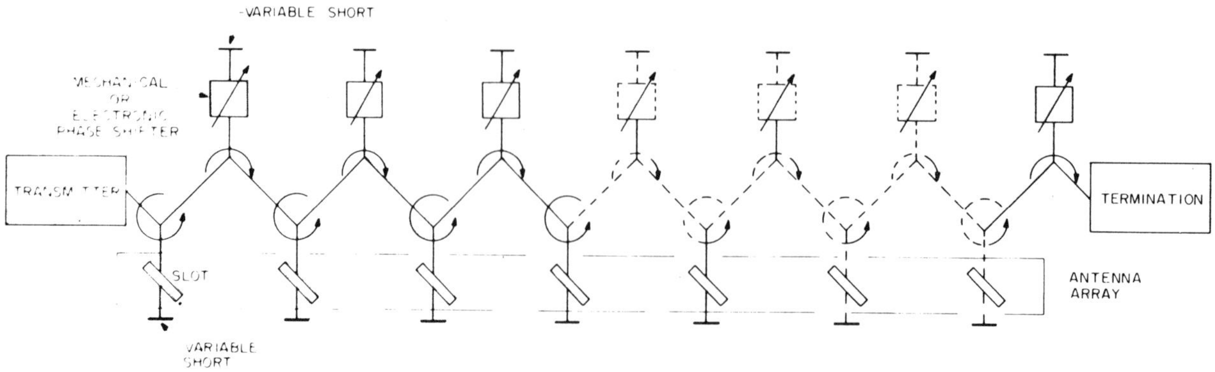

Fig. 10 - Series-fed array using Y-circulators

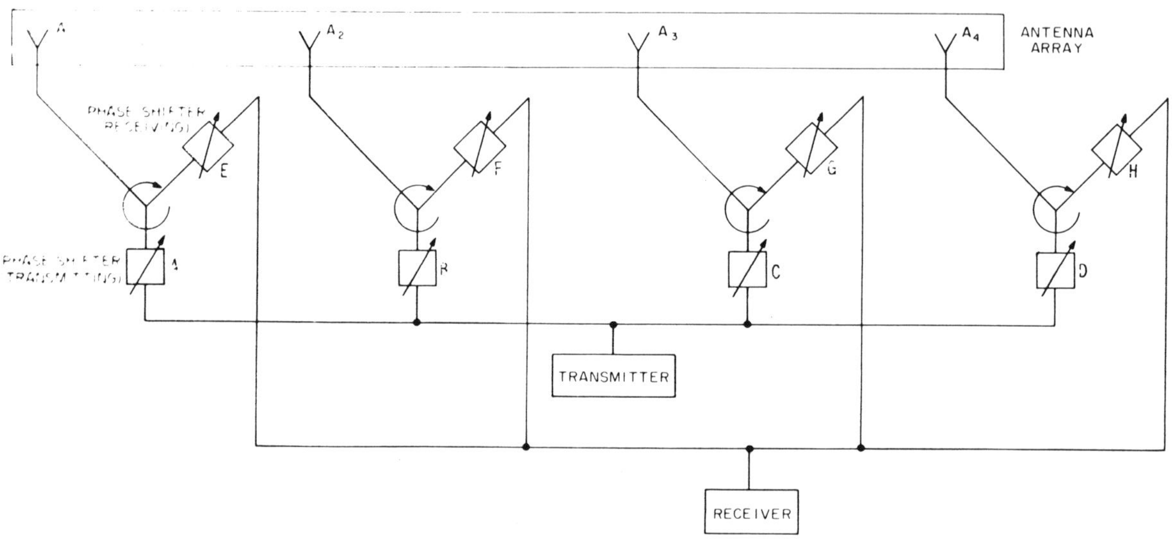

68

Fig. 11 - Parallel-fed array using Y-circulators

Shown in Fig. 11 is a scheme which can be used to control individually the positions of the transmitting and receiving beams of the antenna. Phase shifters A, B, C, and D control the position of the transmitting beam while E, F, G, and H will position the receiving beam.

These are just two of many possible applications for circulators in antenna systems. Because of its compactness and light weight, the junction-type circulator should make schemes such as these more practical and feasible.

* * *

The Synthesis of Symmetrical Waveguide Circulators*

B. A. AULD†

Summary—**A method for synthesizing symmetrical waveguide circulators by adjusting the eigenvalues of the scattering matrix is described. This procedure is particularly useful for the design of very compact circulators in the form of waveguide junctions containing ferrite obstacles. Permissible structural symmetries for a circulator are listed, and a standard form for the scattering matrix of a symmetrical circulator is defined. The synthesis procedure is then described in detail, stating the conditions to be imposed on the scattering matrix eigenvalues, and an expression is obtained for the changes in the eigenvalues due to the placing of anisotropic material within the junction.**

By applying the theory to Allen's 4-port turnstile circulator, it is shown that the use of a matched turnstile junction and a reflectionless Faraday rotator is not essential. The theory is also applied to the design of novel 3- and 4-port circulators, and two 6-port circulators, one of which may be used as a 5-position waveguide switch, are described. Some experimental results are presented for a compact 3-port circulator in the form of an *H*-plane *Y* junction, in 1 inch by ½ inch waveguide, containing a ferrite post obstacle. This circulator, which operates with a bias field of approximately 25 oersted, has a useful bandwidth of 3 per cent. Greater bandwidths would be expected in a Stripline or a fin-line version of this device.

INTRODUCTION

THE usual approach to the synthesis problem for microwave circuits is to use lumped-element network representations of the waveguide components. A lumped-element network having the desired characteristics is first obtained by conventional synthesis procedures, and a microwave structure exhibiting the same characteristics is then synthesized by interconnecting basic waveguide components, represented by lumped element networks and sections of transmission line. This method has been extended by Carlin[1] to the synthesis at a single frequency of non-reciprocal networks. At microwave frequencies the physical structure of the circuit is not defined by the network diagram as at lower frequencies, and for this reason there is often no particular advantage in working from an explicit network representation of the desired circuit.

A more direct approach is to consider the microwave circuit as a waveguide junction characterized by a scattering matrix S, without reference to a specific network representation. The scattering matrix coefficients are then fixed in accordance with the desired circuit characteristics by suitably adjusting the geometry of the boundary-value problem. It has been shown by Dicke[2] that this is a useful approach when the microwave circuit has structural symmetry, in which case it is convenient to regard the eigenvectors and eigenvalues of S, rather than the scattering coefficients themselves, as adjustable parameters of the circuit. Given a specific scattering matrix S_d to be realized by a symmetrical junction at a single frequency, the synthesis procedure is as follows:

a) Determine which structural symmetries are consistent with the form of the matrix S_d.

b) Find the eigenvectors and eigenvalues of S_d.

c) Select a junction having one of the symmetries found in a) and, by making symmetrical perturbations of the junction geometry, adjust the scattering matrix eigenvectors and eigenvalues so as to coincide with b).

69

As in the case of the equivalent network approach, the procedure outlined above does not lead to a specific microwave circuit. However, once one of the permissible symmetries in a) has been chosen, the *nature* of the adjustments required in c) is strongly suggested by symmetry. The *range* of adjustment required must be determined experimentally. This procedure is particularly useful in the design of very compact structures which perform rather complicated circuit functions, such as the magic-T hybrid and the waveguide Wheatstone bridge.[3] However, it is also applicable to the design of symmetrical circuits which comprise interconnections of distinct waveguide components. The procedure provides for the realization of the desired circuit characteristics at a single frequency only. Nevertheless, as is usual with microwave circuits designed at a single frequency, the circuit will function with somewhat degraded performance over a band of frequencies in the vicinity of the design frequency. The bandwidth depends on the minimum performance specifications and also on details of the circuit structure not under direct control of the synthesis procedure. For example, a Stripline or a fin-line structure would be expected to have greater bandwidth than a rectangular waveguide structure with the same symmetry. A disadvantage of the symmetry method of synthesis is that circuit losses are not easily taken into account. Consequently, it is mainly applicable to circuits in which the losses are small and incidental.

* Original manuscript received by the PGMTT, September 3, 1958. Revised manuscript received, December 23, 1958. The research reported in this paper was supported by the National Research Council of Canada.

† Dept. of Electrical Engineering, Univ. of British Columbia, Vancouver, B. C., Can.; now at the Microwave Laboratory, Stanford University, Stanford, Calif.

[1] H. J. Carlin, "Principles of Gyrator Networks," *Proc. Symp. on Modern Advances in Microwave Techniques*, Brooklyn Inst. Polytech., pp. 175–204; November, 1954.

[2] C. Montgomery, R. H. Dicke, and E. M. Purcell, "Principles of Microwave Circuits," McGraw-Hill Book Co., Inc., New York, N. Y., ch. 12; 1948.

[3] M. Chodorow, E. L. Ginzton, and J. F. Kane, "A microwave impedance bridge," Proc. IRE, vol. 37, pp. 634–639; June, 1949.

Although symmetrical circulators of several types have been described in the literature by Allen,[4] Fowler,[5] Vartanian,[6] and Kock,[7] symmetry methods have not been widely applied to non-reciprocal networks. However, Treuhaft[8] has described some of the general symmetry properties of circulators, and Fowler's circulator was designed by adjusting the eigenvalues of S, although full use was not made of the advantages of the method. The purpose of this paper is to apply the synthesis procedure outlined above to symmetrical waveguide circulators in general. In this connection the term "waveguide" is understood to include TEM structures such as coaxial line and Stripline.

Theory of Symmetrical Circulators

Permissible Symmetries

It has been shown by Dicke[2] that the scattering matrix S of any symmetrical junction must satisfy a set of commutation relations,

$$F_{g_i}S = SF_{g_i}, \tag{1}$$

70

which determine the restrictions imposed on S by the junction symmetry. In (1) the F_{g_i}'s are a set of independent generators for a group of "symmetry operators" F for the junction. The "symmetry operators" F are matrices which indicate how the terminal fields transform under the operations of the symmetry group of the junction, *i.e.*, rotations and reflections that carry the junction into itself. For a detailed discussion of the properties of the F-matrices the above-mentioned reference and others[9,10] may be consulted. In the present application it is sufficient to note that the group of F-matrices constitute a "representation" of the symmetry group of the junction. That is, if F_1 is the matrix corresponding to a symmetry operation T_1, and F_2 corresponds to T_2, then the matrix product F_1F_2 corresponds to the symmetry operation T_1T_2 (operation T_2 followed by operation T_1). For an n-port circulator, defined as a device whose scattering matrix S_{cn} performs a cyclic substitution on the n incident wave amplitudes, the group of F-matrices can include only 1) matrices contained in any one of the subgroups of the cyclic group generated by S_{cn}; and 2) the products of these matrices with $-I$.[8] Since a cyclic group possesses only cyclic subgroups,[8] the only possible F-groups for a circulator are those generated by the following sets of generators:

a) $F_1 = (S_{cn})^p$ (p a factor of n), and $F_2 = -I$

$$F_1^{n/p} = F_2^2 = I, \qquad F_1F_2 = F_2F_1,$$

b) $F_1 = (S_{cn})^p$ (p a factor of n)

$$F_1^{n/p} = I,$$

c) $F_2 = -I$

$$F_2^2 = I,$$

where case c) is trivial, since every $n \times n$ matrix commutes with $-I$. It is found by consulting tabulations of all the possible symmetry groups for a physical structure (3-dimensional point groups),[10,11] that only those symmetry groups listed in Table I, where the usual Schoenflies notation is used, can be represented by the groups of F-matrices a) and b). Accordingly, these are the only permissible non-trivial symmetries for an n-port circulator (Fig. 1). It should be noted that the boundaries of the junction may possess more than the permitted number of symmetries, these extra symmetries being removed by the anisotropic material placed within the junction in order to induce non-reciprocity. In Figs. 1(a), (b), (c), and (d) for example, the vertical planes of symmetry for the junction boundaries might be removed by a circular ferrite post placed along the axis of symmetry of the junction, with a bias field parallel to the symmetry axis. It is demonstrated in the theory of finite groups[12] that when a cyclic substitution on n symbols is raised to a power $p = n/q$ the result is a noncyclic substitution consisting of p cycles, each operating on a distinct set of q symbols. Consequently, when $p \neq 1$ the "symmetry operator" F_1 permutes the junction ports *within* sets of $q = n/p$ ports but does not interchange sets of ports, as in Figs. 1(d), (e), and (f).

TABLE I

Permissible Symmetries for an n-Port Circulator

	(p is a factor of n)
(a)	$C_{(n/p)h}$
(b)	$C_{n/p}$, $S_{n/p}$ (n/p even), $C_{(n/2p)h}$($n/2p$ odd), C_s ($p = n/2$)

Treuhaft's definition of a circulator does not specify uniquely the form of the scattering matrix, and for a given n-port circulator different systems of port numbering will lead to $(n-1)!$ distinct scattering matrices. In the case of a structurally symmetrical circulator the number of possibilities may be reduced, however, by specifying a standard numbering system. For an n-port

[4] P. J. Allen, "The turnstile circulator," IRE Trans. on Microwave Theory and Techniques, vol. MTT-4, pp. 223–228; October, 1956.

[5] H. Fowler, paper presented at Symposium on Microwave Properties and Applications of Ferrites, Harvard Univ., Cambridge, Mass.; April, 1956.

[6] P. H. Vartanian, "Theory and Applications of Ferrites at Microwave Frequencies," Sylvania Electronic Defense Lab. Rep. E 15, pp. 119–126; April, 1956.

[7] W. E. Kock, "Signal Routing Apparatus," U. S. Patent No. 2,794,172; May 28, 1957.

[8] M. A. Treuhaft, "Network properties of circulators based on the scattering concept," IRE Trans. on Circuit Theory, vol. CT-3, pp. 127–135; June, 1956. See also Proc. IRE, vol. 44, pp. 1394–1402; October, 1956.

[9] D. M. Kerns, "Analysis of symmetrical waveguide junctions," *J. of Research Nat. Bur. of Standards*, vol. 46, pp. 267–282; April, 1951.

[10] B. A. Auld, "Applications of Group Theory in the Study of Symmetrical Waveguide Junctions," MLR-157, Stanford University, Stanford, Calif.; March, 1952.

[11] F. D. Murnaghan, "The Theory of Group Representations," The Johns Hopkins Press, Baltimore, Md.; 1938.

[12] H. Hilton, "An Introduction to the Theory of Groups of Finite Order," Clarendon Press, Oxford, Eng., p. 8; 1908.

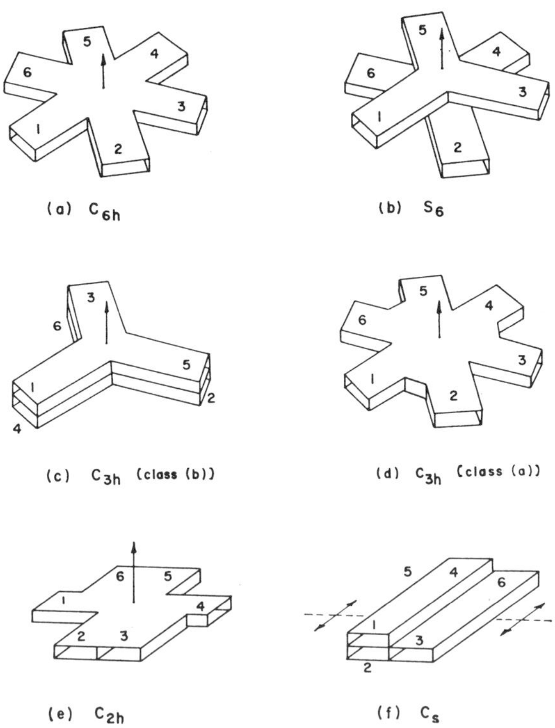

Fig. 1—Examples of some permissible symmetries for a 6-port circulator.

junction having any of the symmetries listed in Table I it is always possible to number the port, so that the generators of the F-group are $F_1 = K_n{}^p$ and $F_2 = -I$, where K_n is the matrix representing the cyclic substitution $(1, 2, 3, \cdots, n)$, and $K_n{}^p$ represents, therefore, the substitution[12]

$$(1, 1+p, 1+2p, \cdots, 1+[q-1]p) \cdots$$
$$\cdots (r, r+p, r+2p, \cdots, r+[q-1]p) \cdots$$
$$\cdots (p, 2p, 3p, \cdots, n).$$

This is, in fact, a natural system of port numbering for such junctions (Fig. 1). If the ports are numbered according to this system, the form of the scattering matrix S of the junction is determined by the commutation relation[13]

$$SK_n{}^p = K_n{}^p S, \tag{2}$$

and the scattering matrices S_{scn} of all symmetrical circulators are then restricted to those cyclic substitution matrices which satisfy (2). This may be taken as the standard form for the scattering matrix of a symmetrical circulator. When $p=1$, which will be shown to be the most useful case in practice, S_{scn} commutes with the *cyclic* substitution matrix K_n, and the standard form may be given explicitly. That is,

$$S_{scn} = K_n{}^f (f = 1, \text{ or is prime to } n), \tag{3}$$

since the only substitutions on n letters commutative with K_n are $K_n{}^f$, and $K_n{}^f$ is cyclic only if $f=1$ or is prime to n.[8,12]

The Synthesis Procedure

The procedure used for the synthesis of symmetrical circulators follows Dicke's work on symmetrical *reciprocal* junctions. In essence, the method requires finding the eigenvectors and eigenvalues of the desired scattering matrix, and then making symmetrical perturbations of the junction geometry until the eigenvectors and eigenvalues of the junction scattering matrix coincide with those of the desired scattering matrix. Each junction eigenvector v and its associated eigenvalue s specify a set of incident wave amplitudes v and reflected wave amplitudes sv at the terminal planes of the junction, and therefore describe uniquely a solution (eigensolution) to the junction boundary-value problem. For a lossless junction $|s| = 1$,[2] and each eigensolution is characterized by pure standing waves in the arms of the junction, the amplitudes of the fields in the arms being specified by the elements of the eigenvector, and the reflection coefficient at the terminal planes by the associated eigenvalue. In determining the *nature* of the perturbations required for adjusting the eigensolutions, it is necessary to make use of the symmetry properties of the eigensolutions. For this reason the synthesis procedure is of greatest practical use when the eigenvectors are completely determined by the symmetry of the junction, and the discussion will be limited to this case.

Let S be the scattering matrix of any lossless n-port junction having one of the permissible symmetries listed in Table I. If the ports are numbered appropriately, S satisfies the particular commutation relation (2), from which the symmetry restrictions on the eigenvectors and eigenvalues of S may be derived by means of the theory of group representations.[9,10] It is found that the eigenvectors are completely determined by symmetry only for symmetry groups C_{nh}, C_n, S_n (n even), or $C_{(n/2)h}$ ($n/2$ odd), that is, when $p=1$ in Table I. In this case, the general form of the scattering matrix eigenvectors is given by the set of column matrices u_a:

$$(u_a)_i = \frac{e^{-j2\pi a(i-1)/n}}{\sqrt{n}} \tag{4}$$

$$a = 0, \pm 1, \pm 2, \cdots, \left(\frac{n}{2}\right)$$

$$i = 1, 2, 3, \cdots, n,$$

and the associated eigenvalues s_a are, in general, nondegenerate. Since $p=1$, the standard form for the scattering matrix of a circulator with one of the above symmetries is $K_n{}^f$, and the first step in the synthesis procedure is to find the eigenvectors and eigenvalues of this matrix.

For the matrix K_n the characteristic equation is

$$|K_n - \lambda I| = |\lambda^n - 1| = 0$$

[13] The second commutation relation, $S(-I) = (-I)S$, is redundant, since it is satisfied by any matrix S.

from which the eigenvalues are

$$\lambda_a = e^{-j2\pi a/n}$$
$$a = 0, \pm 1, \pm 2, \cdots, \left(\frac{n}{2}\right). \tag{5}$$

The eigenvectors v_a are found by substituting (5) into

$$K_n v_a = \lambda_a v_a$$

from which it follows directly that

$$(v_a)_i = \epsilon^{-j2\pi a/n}(v_a)_{i-1}.$$

Taking $(v_a)_1 = 1/\sqrt{n}$, a set of normalized eigenvectors,

$$(v_a)_i = \frac{\epsilon^{-j2\pi a(i-1)/n}}{\sqrt{n}} \tag{6}$$

is obtained. These are also eigenvectors for the matrix $K_n{}^f$, with associated eigenvalues

$$\lambda_a{}^f = \epsilon^{-j2\pi af/n}. \tag{7}$$

Since the junction eigenvectors (4) already coincide with the required circulator eigenvectors (6), the final step in the synthesis procedure is to adjust the junction eigen-
72 values s_a in accordance with (7). It should be noted that the s_a's may all be rotated through an arbitrary angle θ by moving all the terminal reference planes through the electrical angle $\theta/2$. Consequently, only the relative phase angles of the eigenvalues are of importance, and the circulator conditions may be written

$$s_0 = \epsilon^{-j2\pi af/n} \qquad \left(a = \pm 1, \pm 2, \cdots, \left(\frac{n}{2}\right)\right), \tag{8}$$

where $f=1$ or is prime to n. When the eigenvalues have been adjusted so as to satisfy (8), the scattering matrix of the junction takes the desired form

$$S_{scn} = \epsilon^{j\theta} K_n{}^f$$

where the exponential factor allows for an arbitrary shift of the terminal reference planes.

Perturbation Theorem for the Scattering Matrix Eigenvalues

It has been shown that to realize a circulator in the form of a symmetrical junction it is necessary to perturb the junction geometry so as to fulfill condition (8). The perturbations take the form of deformations of the walls and the placing of isotropic and anistropic materials within the junction, due regard being paid to the preservation of symmetry in order that the eigenvectors (4) are not disturbed. Changes in the eigenvalues due to a particular perturbation are functions of both the nature of the perturbation and its position within the junction, and may be expressed as integrals of the eigensolution fields over the volume of the perturbation.

Assuming that the parameters within a volume V_1 inside the junction are changed from μ, ϵ (scalars) to $\overleftrightarrow{\mu}'$, $\overleftrightarrow{\epsilon}'$ (either scalars or tensors), let $\boldsymbol{E}^{(-a)}$, $\boldsymbol{H}^{(-a)}$ represent the electromagnetic field of the $-a$th eigensolution for the unperturbed junction, and let $\boldsymbol{E}^{(a)'}$ $\boldsymbol{H}^{(a)'}$ represent the electromagnetic field of the ath eigensolution for the perturbed junction. Then

$$\nabla \cdot (\boldsymbol{E}^{(-a)} \times \boldsymbol{H}^{(a)'} - \boldsymbol{E}^{(a)'} \times \boldsymbol{H}^{(-a)}) = j\omega(\boldsymbol{H}^{(-a)} \cdot \delta\overleftrightarrow{\mu}\boldsymbol{H}^{(a)'} - \boldsymbol{E}^{(-a)} \cdot \delta\overleftrightarrow{\epsilon}\boldsymbol{E}^{(a)'}) \tag{9}$$

where $\delta\overleftrightarrow{\mu} = \overleftrightarrow{\mu}' - \mu$ and $\delta\overleftrightarrow{\epsilon} = \overleftrightarrow{\epsilon}' - \epsilon$. Eq. (9) is integrated over the volume bounded by the junction and the left-hand side is converted in the usual manner into a sum of surface integrals over the terminal planes. Since the fields are eigensolutions, this sum may be expressed in matrix notation as

$$\sum_i \int_{S_i} (\boldsymbol{E}_i^{(-a)} \times \boldsymbol{H}_i^{(a)'} - \boldsymbol{E}_i^{(a)'} \times \boldsymbol{H}_i^{(-a)}) \cdot \boldsymbol{n}_i dS_i = 2(s_a' - s_{-a})\tilde{\boldsymbol{u}}_{-a}\boldsymbol{u}_a$$

where the symbol $\tilde{\boldsymbol{u}}_{-a}$ means the transpose of the column matrix $\boldsymbol{u}_{-a}$. From (4)

$$\tilde{\boldsymbol{u}}_{-a}\boldsymbol{u}_a = 1.$$

Hence,

$$s_a' - s_{-a} = j\frac{\omega}{2}\int_{V_1} (\boldsymbol{H}^{(-a)} \cdot \delta\overleftrightarrow{\mu}\boldsymbol{H}^{(a)'} - \boldsymbol{E}^{(-a)} \cdot \delta\overleftrightarrow{\epsilon}\boldsymbol{E}^{(a)'})dV. \tag{10}$$

For the ath and $-a$th eigensolutions of the unperturbed junction,

$$S\boldsymbol{u}_a = s_a\boldsymbol{u}_a$$
$$S\boldsymbol{u}_{-a} = s_{-a}\boldsymbol{u}_{-a}.$$

Multiplying the transpose of the second equation by $\boldsymbol{u}_a$ on the right and the first equation by $\tilde{\boldsymbol{u}}_{-a}$ on the left, and subtracting,

$$\tilde{\boldsymbol{u}}_{-a}(\tilde{S} - S)\boldsymbol{u}_a = (s_{-a} - s_a)$$

Since the unperturbed junction is assumed to contain only isotropic material, $\tilde{S} = S$ and it follows that $s_a = s_{-a}$ (reciprocity degeneracy).[9] Substituting this result into (10),

$$\delta s_a = s_a' - s_a = j\frac{\omega}{2}\int_{V_1} (\boldsymbol{H}^{(-a)} \cdot \delta\overleftrightarrow{\mu}\boldsymbol{H}^{(a)'} - \boldsymbol{E}^{(-a)} \cdot \delta\overleftrightarrow{\epsilon}\boldsymbol{E}^{(a)'})dV \tag{11}$$

which gives the change of the ath eigenvalue due to the perturbation.

Applications

Turnstile Circulators

A familiar example of a circulator with symmetry group C_4 is Allen's[4] 4-port circulator, based on the well-known 4-arm turnstile junction [Fig. 2(a)]. In this circulator a *matched* turnstile junction is used, and the circular arm is terminated by a Faraday rotator in front of an adjustable short-circuit plunger. Allen treated the device as an interconnection of two distinct

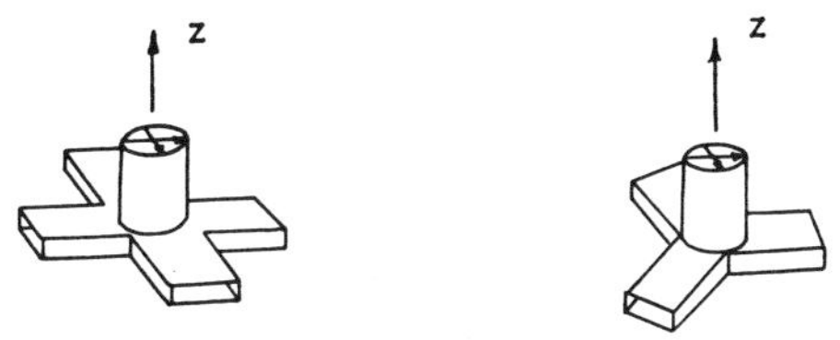

(a) (b)

Fig. 2—(a) Four-arm turnstile junction. (b) Three-arm turnstile junction.

components and, assuming that the Faraday rotator was reflectionless and did not extend into the fringing fields of the turnstile, demonstrated that circulator action can be achieved by suitably adjusting the plane of polarization and the phase of the reflected wave in the circular arm. An alternative approach would be to apply the symmetry theory outlined above, the scattering matrix eigenvalues being adjusted in accordance with (8) by means of the Faraday rotator and the sliding short-circuit. This approach, which might appear to be unnecessarily complicated, leads to several conclusions not readily apparent from Allen's analysis.

When applying symmetry theory to a composite junction, it is often convenient to express the composite eigensolutions as linear combinations of eigensolutions for the component parts of the junction, thereby arriving at expressions for the eigenvalues of the composite junction in terms of the eigenvalues of its component parts. In the case of the turnstile circulator, no initial assumption is made regarding the matching of the turnstile junction. The composite eigensolutions are constructed by taking linear combinations of the turnstile eigensolutions, subject to restrictions imposed by the load attached to the circular arm. Relations between the composite eigenvalues s_a and the turnstile eigenvalues s_{bt} are obtained, and certain circulator conditions on the s_{bt}'s are derived by applying the circulator conditions (8) to the s_a's.

Several general conclusions may be drawn from such an analysis.[14] In the first place, the ferrite element does not have to be matched. If the circular pipe propagates only the lowest mode and the fringing fields of the termination and the turnstile do not interact, the termination does not even have to be of the same symmetry as the turnstile. The ferrite element may, in fact, have any n-fold rotational symmetry ($n \neq 2$). Furthermore, matching of the turnstile is not essential. It is necessary only to insert a thin metal pin along the axis of symmetry, and this adjustment is not sufficient to match the turnstile.[15] By deliberately mismatching the turnstile, thereby converting the terminated circular pipe into a resonant cavity, a reduction in the volume of ferrite required may be obtained at the expense of bandwidth. Similar results are obtained in the case of Fowler's circulator, based on the 3-arm turnstile junction [Fig. 2(b)].

[14] B. A. Auld, "The Synthesis of Symmetrical Waveguide Circulators," Electrical Engineering Report, Univ. of British Columbia, Vancouver, Canada; Aug., 1958.
[15] Montgomery, Dicke, and Purcell, *op. cit.*, pp. 459–466.

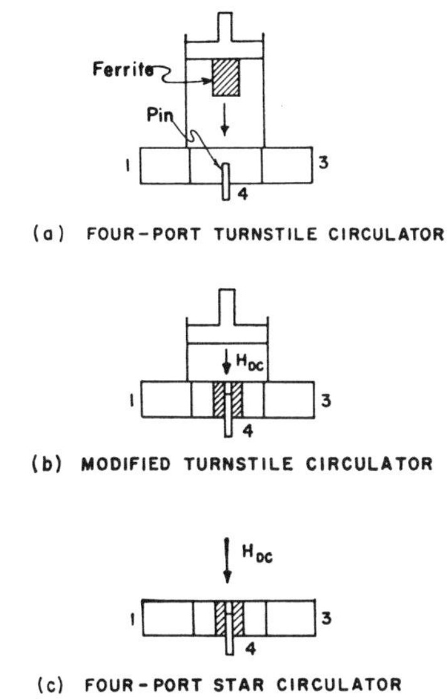

Fig. 3—Modifications of the 4-port turnstile junction.

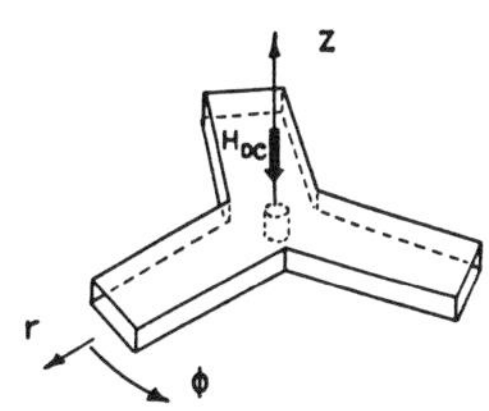

Fig. 4—Three-port star circulator.

The above discussion has been based on the explicit assumption that the fringing fields of the ferrite element and the turnstile do not interact. This restriction is not essential, and circulator action can still be realized even when the ferrite element is placed entirely within the junction region [Fig. 3(b)]. The advantage of this is that the short-circuit plunger can now be placed closer to the junction, shortening the electrical length of the side arm and reducing the frequency sensitivity of the circulator.[16] Since the short circuit plunger is placed outside the fringing fields, the composite junction theory may again be used. However, in this case the results are more complicated, owing to the nonreciprocity of the turnstile junction itself; and the adjustment procedure is consequently more difficult.

Star Junction Circulators

Further reduction of frequency sensitivity is achieved if the plunger and the ferrite element are both placed within the fringing fields, resulting in a star junction structure [Fig. 3(c)]. This structure cannot be treated as a

[16] The writer has recently learned that another method for increasing the bandwidth of a turnstile circulator has been used very successfully in the 3-port case by T. Schaug-Pettersen ("Novel Design of a 3-Port Circulator," Internal Rep., Norwegian Defence Res. Est., Bergen, Norway; January, 1958). In this method the phase-frequency response curves for all the scattering matrix eigenvalues are adjusted so as to approximately coincide over a band of frequencies, thus permitting the circulator conditions (8) to be approximately satisfied over this entire frequency band.

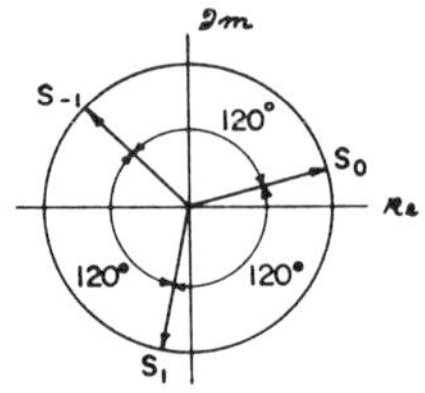

Fig. 5—Scattering matrix eigenvalues for a 3-port circulator.

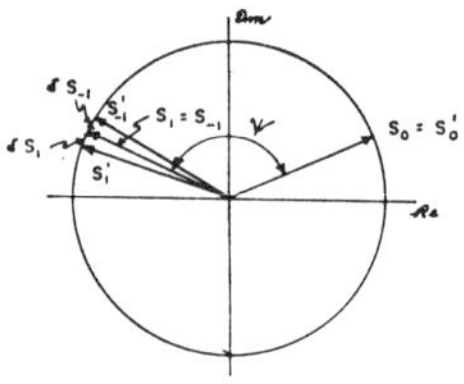

Fig. 6—Splitting of the reciprocity degeneracy due to the application of a small bias field.

composite junction, and the circulator conditions (8) are established by altering the geometry of the junction region. As an example of this type of circulator, a 3-port
74 circulator in the form of a symmetrical H-plane Y-junction with a ferrite post across the guide on the axis of symmetry (Fig. 4), will be considered.

The circulator conditions for a 3-port junction are, from (8),

$$s_1 = \epsilon^{\mp j2\pi/3} s_0$$
$$s_{-1} = \epsilon^{\pm j2\pi/3} s_0,$$

the upper signs corresponding to circulation in the direction 1→2→3 (Fig. 5), and the lower signs to circulation in the reverse direction. To satisfy these conditions, it is necessary to set only two of the junction parameters, the phases of s_1 and s_{-1} relative to s_0, and only two physical adjustments are required. These may be taken to be the diameter of the ferrite post and the strength of the bias field.

Let it be assumed that with no bias field applied, the eigenvalues are s_0 and $s_1 = s_{-1}$ (Fig. 6). Assuming no losses, these will lie on the unit circle in the complex plane with a relative angular spacing ψ, which is a function of the post diameter. The eigenvalues are, in effect, reflection coefficients and must have phase angles which increase with frequency. Consequently, s_0 and $s_1 = s_{-1}$ rotate around the unit circle in a counter-clockwise direction with increasing frequency, and, since the phase angles of s_0 and $s_1 = s_{-1}$ will not generally have the same frequency dependence, the angular spacing ψ will vary with frequency as well as with post diameter.

In general, the effect of applying bias field [$\overleftrightarrow{\delta\epsilon} = 0$ in (11)] is to alter the phases of all the eigenvalues. Because of this, the circulator conditions cannot be set up by a straightforward balancing procedure as in the case of the turnstile circulator. However, if only small perturbations are considered and second-order effects neglected, the results are sufficiently simple to serve as the basis for a fairly systematic adjustment procedure. When second-order terms are neglected, (11) becomes

$$\delta s_a = j \frac{\omega}{2} \int_{V_1} \boldsymbol{H}^{(-a)} \cdot \overleftrightarrow{\delta\mu} \boldsymbol{H}^{(a)} dV \qquad (a = 0, +1, -1). \tag{12}$$

If the internal bias field is assumed to be uniform, the biased ferrite medium is characterized by the Polder permeability tensor, and

$$\delta\mu = \begin{bmatrix} \mu - \mu_u & j\kappa & 0 \\ -j\kappa & \mu - \mu_u & 0 \\ 0 & 0 & \mu_z - \mu_u \end{bmatrix} \tag{13}$$

where μ_u is the scalar permeability of the unbiased ferrite. The terminal conditions for the eigensolution fields $\boldsymbol{E}^{(a)}$, $\boldsymbol{H}^{(a)}$ are defined by the scattering matrix eigenvectors, in this case

$$u_0 = \frac{1}{\sqrt{3}} \begin{bmatrix} 1 \\ 1 \\ 1 \end{bmatrix} \qquad u_1 = \frac{1}{\sqrt{3}} \begin{bmatrix} 1 \\ \epsilon^{-j2\pi/3} \\ \epsilon^{j2\pi/3} \end{bmatrix}$$

$$u_{-1} = \frac{1}{\sqrt{3}} \begin{bmatrix} 1 \\ \epsilon^{j2\pi/3} \\ \epsilon^{-j2\pi/3} \end{bmatrix}. \tag{14}$$

Accordingly, the electric fields of the eigensolutions must satisfy relations

$$E_z^{(a)}(r, \phi) = \epsilon^{j2\pi am/3} \quad E_z^{(a)}\left(r, \phi + \frac{2\pi m}{3}\right)$$
$$(a = 0, +1, -1), \qquad (m \text{ any integer})$$

where only the z component appears because the structure is uniform in the z direction. The eigensolution electric fields within the ferrite post must therefore take the form

$$E_z^{(a)} = \sum_{p=-\infty}^{\infty} A_p^{(a)} J_{|a+3p|}(kr) \epsilon^{-j(a+3p)\phi}$$

where $k = \omega\sqrt{\mu_u \epsilon}$ and the magnetic fields are of the form

$$\boldsymbol{H}^{(a)} = -\frac{\nabla \times k E_z^{(a)}}{j\omega\mu_u}$$
$$= \frac{-\epsilon^{-ja\phi}}{j\omega\mu_u} \sum_{p=-\infty}^{\infty} A_p^{(a)} \boldsymbol{f}_p^{(a)}(kr) \epsilon^{-j3p\phi}. \tag{15}$$

Substituting (13) and (15) into (12),

$$\delta s_a = -j \frac{b}{2\omega\mu_u^2} \sum_{p=-\infty}^{\infty} A_{-p}^{(-a)} A_p^{(a)}$$
$$\cdot \left\{ (\mu - \mu_u) \int_0^R \boldsymbol{f}_{-p}^{(-a)} \cdot \boldsymbol{f}_p^{(a)} r dr \right.$$
$$\left. + j\kappa \int_0^R \boldsymbol{k} \cdot \boldsymbol{f}_{-p}^{(-a)} \times \boldsymbol{f}_p^{(a)} r dr \right\} \qquad (a \neq 0)$$

$$\delta s_0 = -j \frac{b}{2\omega\mu_u^2} \sum_{p=-\infty}^{\infty} A_{-p}^{(0)} A_p^{(0)} (\mu - \mu_u)$$
$$\cdot \int_0^R \boldsymbol{f}_{-p}^{(0)} \cdot \boldsymbol{f}_p^{(0)} r dr \tag{16}$$

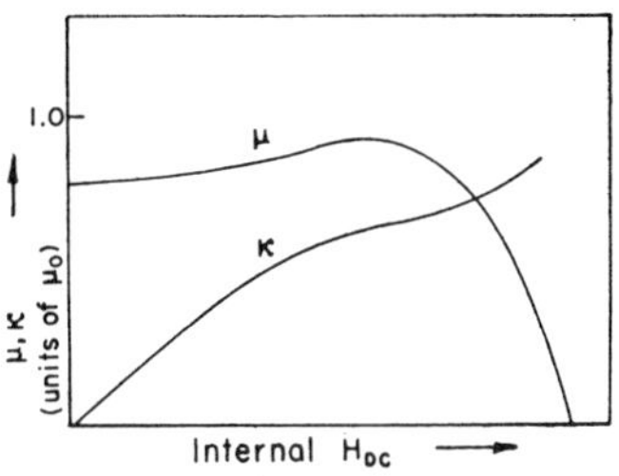

Fig. 7—Typical variation of ferrite permeability components as a function of internal bias field.

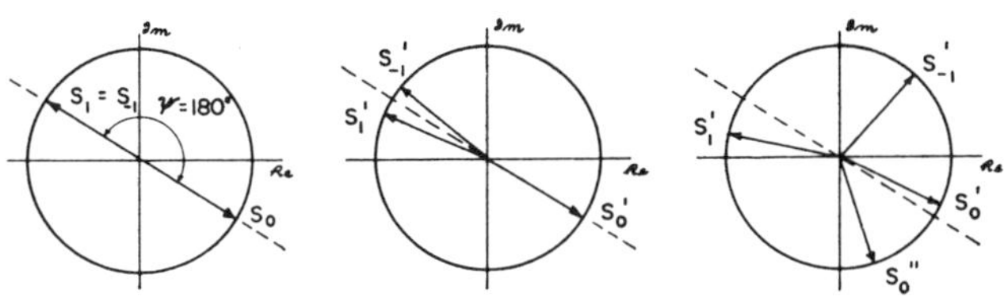

(a) (b) (c)

Fig. 8—Stages in the adjustment of a 3-port star circulator.

where R is the radius of the ferrite-post and b is the height of the waveguide.

For a typical ferrite, the components μ, κ of the Polder tensor vary with applied field as shown in Fig. 7. It is seen from this that for small bias fields $\mu - \mu_u = 0$, and the changes in the eigenvalues are $\delta s_0 = 0$ and $\delta s_1 = -\delta s_{-1}$ (Fig. 6). In attempting to establish the conditions for circulation, it is desirable, before applying the bias field, to adjust the post diameter so that $\psi = 180°$ [Fig. 8(a)]. (This condition is satisfied by determining experimentally the post diameter d_{opt} required for optimum matching of the unbiased junction; from the Appendix, $|S_{11}|_{opt} = \frac{1}{3}$.) Then, when a small bias field is applied, the reciprocity degeneracy is split symmetrically [Fig. 8(b)], the amount of splitting increasing with the strength of the bias field. However, as the separation of s_1' and s_{-1}' is increased by increasing the bias field, terms in $(\mu - \mu_u)$ can no longer be neglected in (16), and asymmetrical splitting, as well as a change in s_0, occurs. Higher order perturbation terms in (11) will contribute further to this effect. Consequently when the required 120° splitting of s_1', s_{-1}' has been reached, s_0' will not lie in the position required for circulation [Fig. 8(c)]. To obtain circulator action, the angle ψ must be adjusted, by varying either the frequency or the post diameter, so as to compensate for this effect. If a post diameter d_{opt} is used, circulation may be established by varying the frequency and the bias field until the junction is matched. Alternatively, the circulator may be adjusted at a fixed frequency by selecting various post diameters in the neighborhood of d_{opt} and varying the bias field until a combination is found for which the junction is matched. The adjustment procedure is simplified greatly if a third adjustment, a thin metal pin along the axis of symmetry, is added. Since the pin affects only the 0th eigensolution it

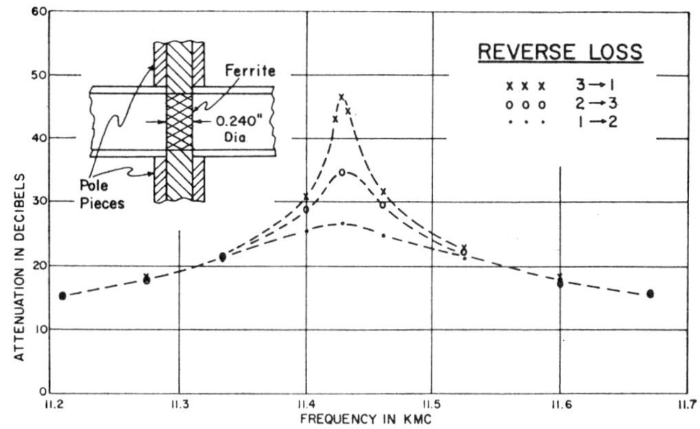

(a)

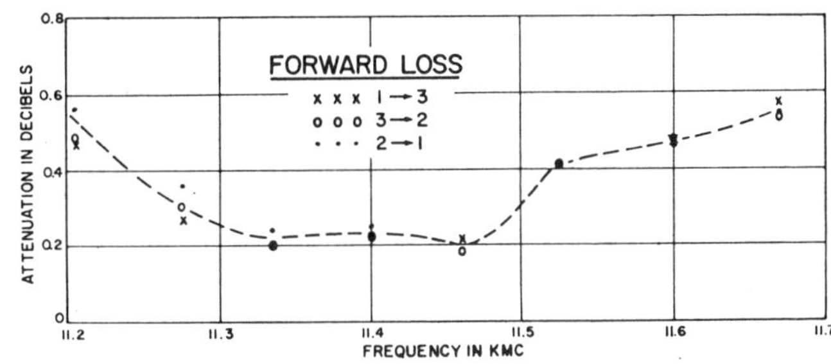

(b)

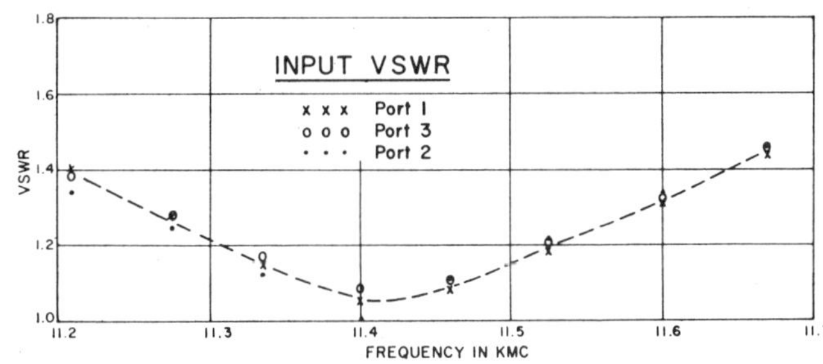

(c)

Fig. 9—Typical performance curves for a 3-port star circulator in 1 inch by $\frac{1}{2}$ inch waveguide. $H_{dc} \approx 25$ oersted in the empty gap.

may be used to shift s_0' in Fig. 8(c) to its correct position s_0'', without changing s_1' and s_{-1}'; and circulation is obtained by alternately adjusting the pin and the bias field for minimum reflection until a match is obtained.

In practice, ferrite circulator performance is degraded to some extent by incidental losses in the ferrite, which have not been taken into account in the above theory. Consequently, the synthesis procedure will be useful in practice only if the eigenvalues can be adjusted over a sufficiently large range without biasing the ferrite into the region of high losses. For the 8.2 to 12.4 kmc band the usefulness of the procedure has been demonstrated by the construction of 3-port star junction circulators in 1 by $\frac{1}{2}$ inch waveguide. In these circulators the ferrite element, a cylindrical post across the waveguide (Fig. 4), was placed directly between the magnet pole pieces

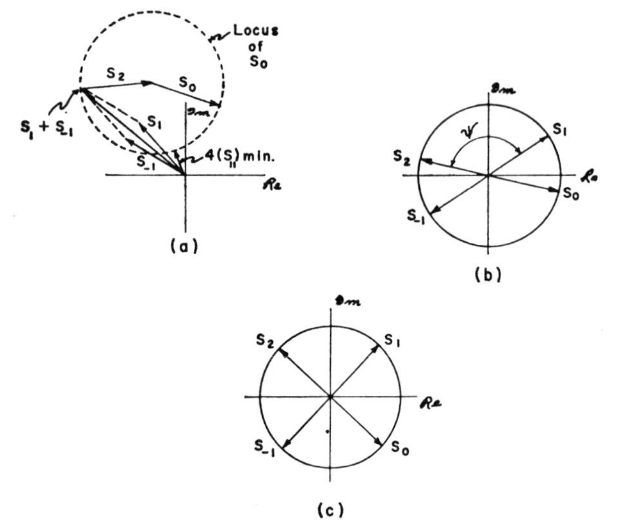

Fig. 10—Stages in the adjustment of a 4-port star circulator.

so as to form a closed magnetic circuit. With Ferramic
76 R-1 posts the bias field strengths required were found to be less than 100 oersted *in the empty gap*. Typical performance curves are presented in Fig. 9, showing the effect of small asymmetries in the structure. In the case of one of the circulators tested it was found that the effects of asymmetries could be partially cancelled by rotating the ferrite post to an optimum position. The peaks of the three reverse loss curves were then all greater than 40 db, with no deterioration of the forward loss or VSWR characteristics.[17]

The synthesis of a 4-port star circulator would proceed in similar fashion. For a 4-port junction the circulator conditions are given by (8) with $n=4$, Fig. 10(c). To meet these conditions, it is necessary to set three of the junction parameters, and three physical adjustments are required. As in the case of the 3-port, two of these adjustments may be the post diameter and the bias field strength. A convenient choice for the third adjustment is the length of a thin metal pin on the axis of symmetry. This perturbs only the 0th eigensolution and permits the setting of s_0 independently. Since the other two adjustments change the phase angles of all four eigenvalues, the adjustment procedure is necessarily somewhat involved. With a given post diameter, bias field is applied, splitting the degeneracy $s_1 = s_{-1}$; and $4|S_{11}| = |s_0+s_1+s_{-1}+s_2|$ (see Appendix) is minimized by adjusting the pin [Fig. 10a]. The splitting of s_1 and s_{-1} is increased in steps by increasing the bias field, the pin being adjusted for minimum $|S_{11}|$ at each step, until both $(S_{11})_{\min}$ and S_{13} vanish for the same pin setting. This can occur only when the splitting of s_1 and s_{-1} is 180°, so that $S_{11} = S_{13} = (s_0+s_2)/4$ (see Appendix), the null condition then being $s_2 = -s_0$ [Fig. 10(b)]. Unless the initial choice of post diameter has been very fortunate, the angle ψ will not be 90°, and the procedure must then be repeated for various post diameters until

$$S_{12} = \frac{s_0 + js_1}{2} = 0, \text{ or } S_{21} = \frac{s_0 - js_1}{2} = 0.$$

At this point, $s_1 = \pm js_0$ and circulation is established [Fig. 10(c)]. This result can be achieved only if the metal pin provides a sufficiently large range of adjustment of s_0. If this is not the case, other methods for adjusting s_0 must be considered.

The synthesis procedure may, in principle, be applied to circulators having any number of ports. In practice, however, the difficulties increase rapidly with the number of ports. Taking 5- and 6-port star circulators as examples, it is seen from the circulator conditions (8) that in both cases it is necessary to split two sets of reciprocity degenerate eigenvalues. This requires the use of two biased ferrite elements, and results in greatly increased complexity of the adjustment procedure. Fortunately, it may be shown that 6- and 8-port symmetrical circulators can easily be obtained without having recourse to the theory.

Symmetrical Ring Circulators

An alternative physical arrangement for an n-port circulator with symmetry group C_n is the symmetrical ring structure. Vartanian[6] has proposed a 3-port circulator of this type, using three gyrators, and a similar device is described by Kock.[7] Design formulas for the analogous 4-port symmetrical ring circulator, using four gyrators, are easily derived by regarding the circuit as a composite junction and applying symmetry theory, as in the case of the turnstile circulator. However, a more useful type of symmetrical ring circulator is obtained by simply interconnecting six 3-port star or turnstile circulators [Fig. 11(a)]. If the bias fields for the six circulators all have the same polarity, this structure functions as a 6-port circulator, with symmetry group C_6. A useful property of this circuit is that it degenerates into two disconnected circulators when symmetry is destroyed by reversing some of the bias fields. Accordingly, it may be used as a five position waveguide switch [Fig. 11(b)], the output being selected by applying an appropriate set of bias fields to the six circulators. Similarly, an 8-port symmetrical ring circulator *with symmetry group* C_4 is obtained by connecting four 4-port star or turnstile circulators in a square array.

Discussion

The synthesis procedure described in this paper is particularly useful for the design of *symmetrical* circulators in the form of waveguide junctions containing ferrite obstacles.[18] A typical example of a circulator of

[17] Experimental results for circulators of this type are also reported by H. N. Chait and T. R. Curry in a paper, "A new *Y*-type circulator," to be published in *J. Appl. Phys.*; March, 1959.

[18] An unsymmetrical circulator of this type, based on an *H*-plane Tee structure, has been described by W. E. Swanson and G. J. Wheeler in a paper, "Tee Circulator," 1958 IRE WESCON Convention Record, vol. 2, pt. 1, pp. 151–156. This device has no symmetries, except for the trivial reflection symmetry in the *H*-plane, and a matching device is required in the side arm of the tee.

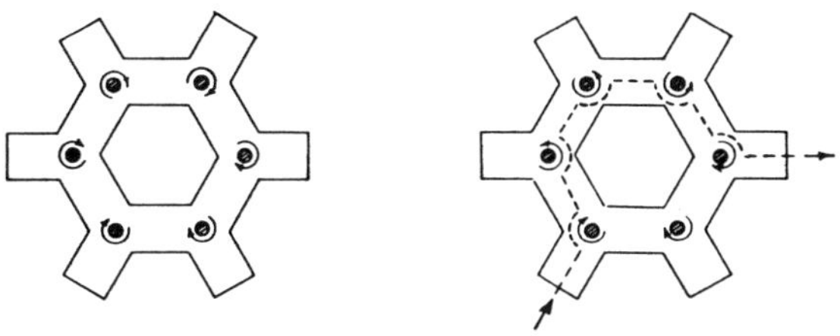

a b

Fig. 11—(a) Six-port ring circulator. (b) Five-position waveguide switch.

this type is the 3-port star circulator shown in Fig. 5. Such circulators are compact and structurally simple, and have the important practical advantage of requiring only a small volume of ferrite material.

The specific applications which have been discussed in this paper have all used C_{nh} (*class (a)*) or C_n symmetry. Other circulators of novel design may be devised by using some of the alternative symmetries listed on Table I. For example, a circulator having S_4 symmetry may in principle be realized by coupling two crossed guides with a ferrite-loaded aperture, the bias field being applied along the axis of symmetry (Fig. 12). The coupling arrangement might conveniently take the form of a ferrite post passing through the coupling hole and extending across the full height of both waveguides. Another example, with C_{3h} (*class (b)*) symmetry, is shown in Fig. 13. A wave with amplitude a incident at port 1 passes through the 3 db coupler and divides into incident waves with amplitudes $0.707a$ and $j.0.707a$ in arms 1 and 4, respectively. These waves pass through the star circulators into arms 5 and 2 and combine in the 3 db coupler to give a transmitted wave of amplitude a at port 2. Similarly, a wave incident at port 2 emerges at port 3, etc. The scattering matrix, therefore, performs a cyclic substitution on the incident wave amplitudes, and the junction functions as a circulator. This device is more compact than the 6-port ring circulator [Fig. 12(a)]. It also has the advantage of requiring only two ferrite posts, which may both be biased by the same magnet.

Applications of the synthesis procedure are not necessarily restricted to junctions of rectangular waveguide. For example, a 3-port star circulator might be realized in the form of an H-plane Y junction of fin-line waveguide. Such a structure would be expected to have a larger bandwidth than the corresponding rectangular waveguide structure. Furthermore, the volume of ferrite material required would be reduced, owing to the localization of the microwave fields. The adjustment procedure would be identical with that outlined above for the rectangular waveguide version. In similar fashion, 3-port circulators might be realized in the form of either Stripline or coaxial line Y junctions.

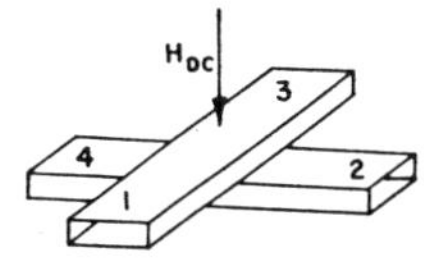

Fig. 12—Circulator with S_4 symmetry, using crossed guides coupled with a ferrite-loaded aperture.

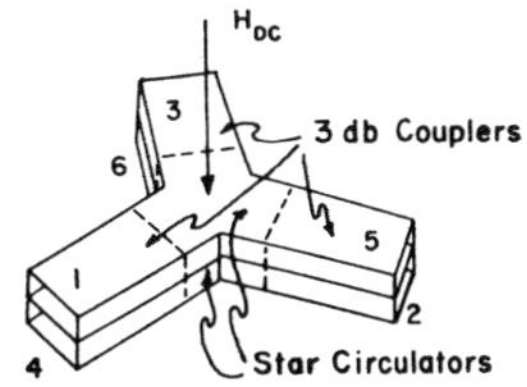

Fig. 13—Six-port circulator with C_{3h} (*class (b)*) symmetry, using two 3-port star circulators and three 3 db couplers.

Appendix

Let S be the scattering matrix of a lossless 3-port junction with C_3 or C_{3h} symmetry. S is diagonalized by a unitary matrix U having for its columns the normalized eigenvectors (14). Then,

77

$$S = U\begin{bmatrix} s_0 & 0 & 0 \\ 0 & s_1 & 0 \\ 0 & 0 & s_{-1} \end{bmatrix} U^{-1}$$

where s_0, s_1, s_{-1} are the eigenvalues of S. When the right-hand side of this equation is multiplied out, the scattering coefficients are obtained as functions of the eigenvalues.

$$S_{11} = S_{22} = S_{33} = \frac{s_0 + s_1 + s_{-1}}{3}$$

$$S_{12} = S_{23} = S_{31} = \frac{s_0 + \epsilon^{j2\pi/3}s_1 + \epsilon^{-j2\pi/3}s_{-1}}{3}$$

$$S_{13} = S_{32} = S_{21} = \frac{s_0 + \epsilon^{-j2\pi/3}s_1 + \epsilon^{j2\pi/3}s_{-1}}{3}.$$

Similarly, the scattering coefficients of a lossless 4-port junction with C_4, C_{4h}, or S_4 symmetry are given as functions of the eigenvalues s_0, s_1, s_{-1}, s_2 by the equations,

$$S_{11} = S_{22} = S_{33} = S_{44} = \frac{s_0 + s_1 + s_{-1} + s_2}{4}$$

$$S_{12} = S_{23} = S_{34} = S_{41} = \frac{s_0 + js_1 - js_{-1} - s_2}{4}$$

$$S_{13} = S_{31} = S_{24} = S_{42} = \frac{s_0 - s_1 - s_{-1} + s_2}{4}$$

$$S_{14} = S_{43} = S_{32} = S_{21} = \frac{s_0 - js_1 + js_{-1} - s_2}{4}.$$

On Stripline Y-Circulation at UHF*

H. BOSMA†

Summary—The simplified boundary-value problem of the circular strip-line Y-circulator is stated and, following experimental results, is solved approximately. The circulation parameters are calculated and discussed. The frequency characteristics are evaluated and a general method for broadbanding the device is presented. From the calculated and measured field distribution, an explanatory description of the circulation mechanism is given. Finally, other possible solutions and the features for use at UHF are discussed.

78

Introduction

THE STRIPLINE Y-circulator (Fig. 1) polarized above the resonance field has been found to have useful properties in the UHF region and, consequently, has received much attention.[1–5] In its basic form, it is a member of a larger family of nonreciprocal devices, the junction circulators. The general properties including the existence of and the tuning procedure for the symmetrical junction circulators have been studied by Auld[6] through the consideration of their scattering matrices and associated quantities. For the stripline Y-circulator in particular, the same has been done by Milano,[1] *et al.*

However, as far as the present author is aware, no consistent theory is yet available about the intrinsic circulator mechanism in terms of the electromagnetic field. The same can be said about the frequency characteristics of these types of circulators. It is true that several attempts have been made to explain the junction circulator in terms of Faraday rotation,[7,8] asymmetrical diffraction[9] or field displacement[5] but, although these efforts have confirmed, more or less, the intuitive thinking about the phenomenon, they have not led to a consistent and manageable theory. In a recent paper, Skomal[10] explained the junction circulator in terms of two contrarotating surface waves. He evaluated some of the design parameters. However, some of his results, *e.g.*, the dependence of applied field on magnetization, do not agree with the findings of the present paper. This may be due to the fact that some of the assumptions made by Skomal are not always tenable. In particular, the circumferential distance between adjacent ports is not equal to the effective wavelength in all cases[11] (see below).

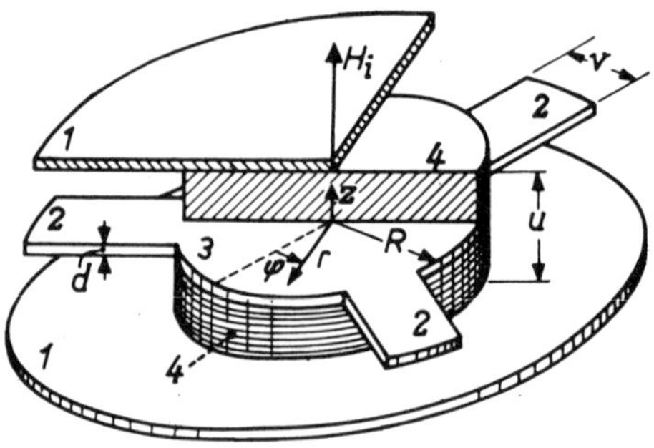

Fig. 1—The circulator configuration. (1) Outer conductors. (2) Inner conductors. (3) Center conductor. (4) Ferrite disks.

The investigation by Bosma,[5] which dealt with the 3-port stripline Y-circulator as this paper does, looked quite promising, but it has turned out that some of its results were not in agreement with experiment. Although the problem was stated correctly, the attempt, as is shown below, failed because a practically wrong (although mathematically nearly possible) solution was traced.

In this paper, the same problem is restated and, following the results of experiment,[12] an approximation of the problem is proposed. Then, again led by experimental results, a supposition for a practical solution is made. This solution is tested in the circulator equations and is shown to be consistent with other theoretical considerations. It also agrees well with all experimental results available.

From the solution, first order approximation, of the frequency dependence of the circulator characteristics are obtained. A general method of broadbanding the circulator is proposed. These results are also confirmed by experiment.

* Received April 5, 1963; revised manuscript received August 16, 1963.

† Philips Research Laboratories, N. V. Philips' Gloeilampenfabrieken, Eindhoven, The Netherlands.

[1] U. Milano, J. Saunders and L. Davis, "A Y-junction stripline circulator," IRE Trans. on Microwave Theory and Techniques, vol. 8, pp. 346–351; May, 1960.

[2] S. Yoshida, "J-Band stripline Y circulator," Proc. IRE, vol. 48, p. 1664; September, 1960.

[3] L. Freiberg, "Lightweight Y-junction stripline circulator," IRE Trans. on Microwave Theory and Techniques, vol. 8, p. 672; November, 1960.

[4] G. V. Buehler and A. F. Eikenberg, "Stripline Y-circulator for the 100 to 400 Mc region," Proc. IRE, vol. 49, p. 518; February, 1961.

[5] H. Bosma, "On the principle of stripline circulation," *Proc. IEE*, vol. 109, pt. B, suppl. no. 21, pp. 137–146; January, 1962.

[6] B. A. Auld, "The synthesis of symmetrical waveguide circulators," IRE Trans. on Microwave Theory and Techniques, vol. 7, pp. 238–247; April, 1959.

[7] P. J. Allen, "The turnstile circulator," IRE Trans. on Microwave Theory and Techniques, vol. 4, pp. 223–228; October, 1956.

[8] C. Bownes, "Discussion on microwave ferrites II," Proc. IEEE, vol. 109, pt. B, suppl. no. 21, p. 163; January, 1962.

[9] V. G. Feoktistov, "Diffraction model of Y-circulator," *Radiotekhnika i Elektronika*, vol. 7, pp. 1763–1768; October, 1962.

[10] E. N. Skomal, "Theory of operation of a 3-port Y-junction ferrite circulator," IEEE Trans. on Microwave Theory and Techniques, vol. 11, pp. 117–123; March, 1963.

[11] J. B. Davies and P. Cohen, "Theoretical Design of Symmetrical Junction Stripline Circulators," IEEE Trans. on Microwave Theory and Techniques, vol. MTT-11, pp. 506–512; November, 1963.

[12] Since this paper is dealing with the theory of the stripline circulator, the experiments and their results are not described in detail. They will probably be published elsewhere.

Furthermore, it will be found that circulator action is based on a slight distortion of the degeneracy of two first order resonances of the "isotropic" disk configuration caused by a small amount of gyrotropy of the disk material. Under the amount of gyrotropy, we understand the ratio κ/μ, where κ and μ are the elements of the Polder tensor [*cf.*, (12)]. The small amount of gyrotropy introduces a small asymmetric distortion of the isotropic field configuration. This asymmetry is necessary for a device to be nonreciprocal.

The smallness of the gyrotropy is an essential feature. As a consequence, the internal polarizing static magnetic field H_i is much larger than the internal resonance field[13] H_0. In most cases, the former is more than four times as large as the latter. In the opinion of the author, this is the main reason why the 3-port stripline Y-circulator is so effective in the UHF region.

Properties of the Scattering Matrix

Before the program outlined above is started, some remarks on the scattering matrix of a lossless, cyclic-symmetric 3-port may be made. The scattering matrix S may be written in the form

$$S = \begin{pmatrix} \alpha & \gamma & \beta \\ \beta & \alpha & \gamma \\ \gamma & \beta & \alpha \end{pmatrix}. \tag{1}$$

Carlin[14] has shown that a matched, lossless 3-port is a circulator and Thaxter and Heller[15] have pointed out that, if the reflection-coefficient α is small $|\alpha| \ll 1$, the equations

$$|\gamma| = |\alpha|, \qquad |\beta| = 1 - 2|\alpha|^2 \tag{2}$$

hold approximately. For a circulator which rotates in the opposite sense, γ is interchanged with β. An elegant description of these properties has been given by Butterweck.[16]

The elements of S satisfy the equations

$$|\alpha|^2 + |\beta|^2 + |\gamma|^2 = 1 \tag{3}$$

$$\alpha\beta^* + \beta\gamma^* + \gamma\alpha^* = 0. \tag{4}$$

By virtue of (3), $|\beta|$ can be expressed in terms of $|\alpha|$ and $|\gamma|$. The three terms of (4) can be interpreted as three vectors which span a triangle. Then, inequality relations like

$$|\beta|\,|\gamma| \leq |\alpha|(|\beta| + |\gamma|) \tag{5}$$

are valid. From these considerations Butterweck deduced that in the $|\alpha|$, $|\gamma|$ diagram the possible 3-ports are restricted to a region which is bounded by three ellipses. In Fig. 2 this region is indicated by the shaded area. The origin of the diagram $(|\alpha|, |\gamma|) = (0, 0)$ represents an ideal clockwise rotating circulator. Starting from this point, it is easy to show that, for $|\alpha| \ll 1$, $|\gamma|$ is given by the series

$$|\gamma| = |\alpha| + \theta|\alpha|^2 + \cdots, \quad -1 \leq \theta \leq 1. \tag{6}$$

Then, by virtue of (3), $|\beta|^2$ is given by the series

$$|\beta|^2 = 1 - 2|\alpha|^2 - 2\theta|\alpha|^3 \cdots. \tag{7}$$

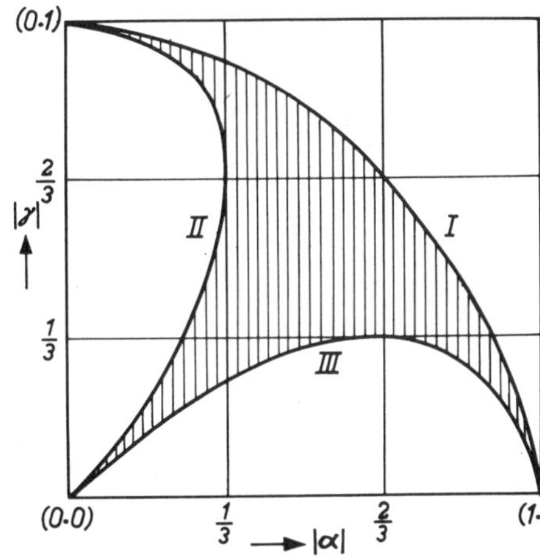

Fig. 2—The closed region of possible, lossless and cyclic-symmetric 3-ports, (Butterweck[16]).
(I) $|\alpha|^2 + |\alpha|\,|\gamma| + |\gamma|^2 - |\alpha| - |\gamma| = 0.$
(II) $|\alpha|^2 - |\alpha|\,|\gamma| + |\gamma|^2 + |\alpha| - |\gamma| = 0.$
(III) $|\alpha|^2 - |\alpha|\,|\gamma| + |\gamma|^2 - |\alpha| + |\gamma| = 0.$

The 3-Port Stripline Y-Circulator

In Fig. 1, a picture of the circulator is given. The cylindrical coordinate system, (r, ϕ, z), used further on, has been drawn into it and several of the important dimensions are presented.

The two ferrite disks are placed between the two earth plates (the outer conductors) and the center conductor. They are magnetized perpendicularly to the plane of the conductors by a static magnetic field. The value of the internal polarizing field (in the ferrite) is denoted by H_i.

In the elementary case, the ferrite disks and the center conductor both have circular form and all three have the same radius R. In practice this need not be so. The disks and the center conductor may have different sizes and forms. Other forms, such as triangles or clover[4] leaves, can be used if only a 3-fold rotation symmetry is present; this, in order to maintain the 3-fold cyclic symmetry of the circulator. Moreover, dielectric inserts may be present. For mathematical simplicity only the elementary configuration is investigated in this paper.

At equal distances around the edge of the center conductor, three inner conductors of stripline waveguides

[13] For reasons of simplicity, we presume that H_0 is given by $H_0 = \omega/\gamma$, where $\gamma = 2\pi 2.8$McOe, although in the configuration in question the ferromagnetic resonance occurs at an internal bias field which can be solved from $\omega = \gamma\sqrt{H(H+4\pi M)}$, where $4\pi M$ is the magnetization of the ferrite. H_0 is used as a reference quantity only.

[14] H. J. Carlin, "Principles of gryator networks," *Proc. Symp. on Modern Advances in Microwave Techniques*, Polytechnical Institute of Brooklyn, N. Y., p. 175–204; November, 1954.

[15] J. B. Thaxter and G. S. Heller, "Circulators at 70 and 140 kMc," Proc. IRE, vol. 48, pp. 110–111; January, 1960.

[16] H. J. Butterweck, "Der Y-zirkulator," *Archiv der Elektrischen Uebertrabung*, vol. 17, p. 163–176; April, 1963.

79

are connected. The centers of these connections are taken at ϕ-values $-\pi/3$, $\pi/3$ and π for the input, output and decoupled line, respectively. If v is the width of the stripline, a stripline width angle Ψ is defined by (see Fig. 3)

$$\Psi = \sin^{-1}(v/2R). \tag{8}$$

The thickness d of the inner and center conductors is assumed to be zero. The distance between the two outer conductors is u, twice the thickness of one ferrite disk.

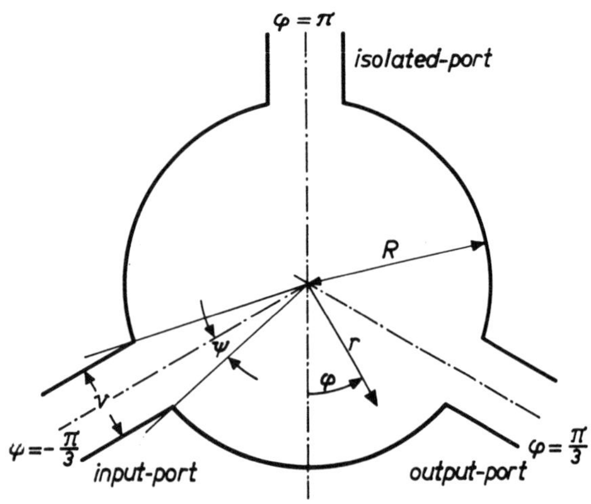

Fig. 3—The configuration of the center conductor.

Field Configuration in and Characteristic Impedance of the Striplines

In the cross section of the striplines, the field intensities are supposed not to vary over the width of the inner conductors and to be zero outside these conductors. The stray fields at the edges are neglected. However, especially in the neighborhood of the connections to the center conductor, rather large differences from the assumed ideal field configuration can be expected to exist. This supposition made about the fields in the stripline is clearly an approximation only.

The striplines carry TEM modes only. Hence, the electric field intensities are perpendicular to the conductors (in the z-direction) and the magnetic field intensities are parallel to them and perpendicular to the directions of propagation (in the ϕ-direction). Furthermore, the fields on the two sides of the inner conductors are 180° out of phase.

The characteristic impedance of the striplines is not important, because of the method of analysis presented. For convenience of matching to the other elements of the experimental setups, the characteristic impedances are made equal to 50 Ω in all practical cases. This is done by the proper choice[17] of u with respect to v (and d).

The ratio of the electric and magnetic field intensities of a traveling wave, the intrinsic wave impedance ζ, is of much more importance for the analysis. In the simple model of the circulator, no dielectric materials other than air are placed between the conductors of the striplines. Hence, ζ is given by

$$\zeta = \sqrt{\mu_0/\epsilon_0} = 120\pi\,\Omega. \tag{9}$$

Magnetic Field Intensity at the Edge of a Disk

As is the case with the fields in the striplines, the electric as well as the magnetic field intensities on either side of the center conductor are equal and oppositely directed at any instant. Hence, the field problem need be solved only for one disk in association with the fields on the appropriate sides of the inner conductors of the striplines.

Furthermore, it is supposed that the field in a disk does not depend on the z-coordinate, so that the problem can be reduced to two of three dimensions.

Except at the connections of the striplines, no radial current can flow from the edge of the center conductor. From boundary equations, it is known that, at the connections, the tangential component of the magnetic field intensity in the ferrite at its edge H_ϕ (R, ϕ) is equal to the magnetic field intensity in the corresponding stripline at the same place. Consequently, H_ϕ (R, ϕ) is constant over the stripline widths and it is zero elsewhere.[18] This can be expressed by

$$H_\phi(R, \phi) = \begin{cases} \text{a)} & -\pi/3 - \Psi < \phi < -\pi/3 + \Psi, \\ \text{b)} & \pi/3 - \Psi < \phi < \pi/3 + \Psi, \\ \text{c)} & \pi - \Psi < \phi < \pi + \Psi, \\ \text{d)} & \text{elsewhere.} \end{cases} \tag{10}$$

The fields at the different connections need not be in phase, so that in principle a, b and c are complex numbers.

The investigation in this paper will be restricted to values of v which are not too large in comparison with R; more precisely

$$v < R. \tag{11}$$

The smaller the ratio v/R is [*cf.*, (8)], the more accurate is the assumed distribution of $H_\phi(R, \phi)$, but the larger is the disturbing influence of the stray fields of the striplines. Hence, v must not be very small.

Electric Field Intensity in the Disk

The electric field intensity in the disk is assumed to have a z-component only. It is denoted $E_z(r, \phi)$. The specific permittivity of the ferrite is denoted by ϵ and the

[17] R. W. Peters *et al.*, "Handbook of Triplate Microwave Components," Sanders Associates, Inc., Nashua, N. H. 1956.

[18] Hereafter, it is shown that the electric field intensity at the edge of the disk is not equal to zero. Thus, an electric stray field which has a radial component is present. Consequently, the supposition $H_\phi(R, \phi) = 0$ outside the connections is not exact. However, from a theoretical estimation and also from experiments, it appeared that the resultant inaccuracy is not larger than those of the other approximate suppositions.

specific gyrotropic permeability by $\|\mu\|$. The latter has a tensor character and is given by

$$\| \mu \| = \begin{Bmatrix} \mu & -i\kappa & 0 \\ i\kappa & \mu & 0 \\ 0 & 0 & 1 \end{Bmatrix}. \tag{12}$$

An effective specific permeability μ_{eff} can be introduced by

$$\mu_{eff} = (\mu^2 - \kappa^2)/\mu \tag{13}$$

and an intrinsic wave number k by

$$k^2 = \omega^2 \mu_0 \epsilon_0 \mu_{eff} \epsilon. \tag{14}$$

With these definitions made, it is shown by Bosma[5] that $E_z(r, \phi)$ satisfies the homogeneous Helmholtz equation

$$\left[\frac{\partial^2}{\partial r^2} + \frac{1}{r}\frac{\partial}{\partial r} + \frac{1}{r^2}\frac{\partial^2}{\partial \phi^2} + k^2\right] E_r(r, \phi) = 0 \tag{15}$$

and that the tangential and radial components of the magnetic field intensity in the disk are related to $E_z(r, \phi)$ by, respectively,

$$H_\phi(r, \phi) = i\left[\frac{\partial E_r}{\partial r} + i\frac{\kappa}{\mu}\frac{1}{r}\frac{\partial E_r}{\partial \phi}\right] \Big/ \omega\mu_0\mu_{eff} \tag{16}$$

$$H_r(r, \phi) = -i\left[\frac{1}{r}\frac{\partial E_r}{\partial \phi} - i\frac{\kappa}{\mu}\frac{\partial E_r}{\partial r}\right] \Big/ \omega\mu_0\mu_{eff}. \tag{17}$$

If $H_\phi(r, \phi)$ is known at the edge [*cf.*, (10)], then, by virtue of (16), $E_z(r, \phi)$ is determined by oblique boundary conditions. Then, a Green's function $G(r, \phi; r', \phi')$ can be introduced[19] such that

$$E_z(r, \phi) = \int_{-\pi}^{\pi} G(r, \phi; R, \phi') H_\phi(R, \phi') d\phi'. \tag{18}$$

In most cases, one is interested in $E_z(r, \phi)$ at the edge only and, therefore, in order to simplify the equations, one may write

$$G(R, \phi; R, \phi') = G(\phi; \phi'). \tag{19}$$

From the distribution $H_\phi(R, \phi)$ given by (10), $E_z(R, \phi)$ can be evaluated at the centers of the connections.[20] For small values of Ψ, the following approximate equations result:

$$E_z(R, -\pi/3) = A = 2\Psi[G(-\pi/3; -\pi/3)a + G(-\pi/3; \pi/3)b + G(-\pi/3, \pi)c] \tag{20}$$

$$E_z(R, \pi/3) = B = 2\Psi[G(\pi/3; -\pi/3)a + G(\pi/3; \pi/3)b + G(\pi/3, \pi)c] \tag{21}$$

$$E_z(R, \pi) = C = 2\Psi[G(\pi; -\pi/3)a + G(\pi; \pi/3)b + G(\pi; \pi)c]. \tag{22}$$

Since E_z is parallel to the cylindrical air-ferrite interface, it is the same on both sides of the edge. Therefore, A, B and C are also the electrical field intensities in the corresponding striplines.

Eqs. (20)–(22) express the relations between the RF-em fields in the striplines (the incident and reflected waves) in terms of the properties of the field pattern in the disk. Hence, the scattering matrix elements of the 3-port circulator can be expressed in terms of $G(\phi; \phi')$.

Properties of $G(\phi; \phi')$

The Green's function can be interpreted as a transfer impedance function for the fields in the disk configuration. Before it is evaluated for the particular case of this paper, the conditions imposed upon it by general circulator characteristics are investigated.

A. Losslessness

If the field configuration at the boundary of the disk is known, by Poynting's theorem the average power which is radiated into or out of the disk can be calculated. For a lossless 3-port this must be zero. With the complex Poynting vector $\boldsymbol{P}$,

$$\boldsymbol{P} = \tfrac{1}{2}\boldsymbol{E} \times \boldsymbol{H}^*, \tag{23}$$

in which the asterisk denotes the complex conjugate, this is expressed by

$$\mathrm{Re}\left[\int_S \boldsymbol{P}\cdot\boldsymbol{n}dS\right] = 0, \tag{24}$$

where $\boldsymbol{n}$ is normal to the closed ferrite surface S. As the normal component of $\boldsymbol{P}$ is nonzero only on the cylindrical part of the ferrite surface, by virtue of (18), (19), (23) and (24) it is easy to show that losslessness of the disk implies that the equation

$$\int_{-\pi}^{\pi}\int_{-\pi}^{\pi} H_\phi^*(R, \phi)[G(\phi; \phi') + G^*(\phi'; \phi)] \cdot H_\phi(R, \phi') d\phi d\phi' = 0 \tag{25}$$

must be satisfied. Since this must be valid for all distributions $H_\phi(R, \phi)$, the condition for losslessness is

$$G(\phi'; \phi) = -G^*(\phi; \phi') \tag{26}$$

which is analogous to the relation $Z_{ji} = -Z_{ij}^*$ which is valid for lossless $2n$-ports in the impedance matrix theory.

B. Cyclic Symmetry

The cyclic symmetry of a 3-port requires the condition

$$G(\phi + 2\pi/3; \phi' + 2\pi/3) = G(\phi; \phi') \tag{27}$$

to be fulfilled.

[19] D. van Dantzig and H. A. Lauwerier, "The North Sea problem," *Proc. Koninklijke Nederlandse Akademie van Wetenschappen*, series A (mathematical sciences), vol. LXIII, p. 170–180; 1960.

[20] Instead of evaluating $E_z(R, \phi)$ at the centers of the striplines, the average values of $E_z(R, \phi)$ over the stripline widths could have been determined. This more complicated method has been carried out by Bosma,[5] but its results are not much more accurate.

C. Resonances of the Disk

The disk configuration with disconnected striplines has two types of resonances.

1) *Electric wall:* If the edge of the center conductor is short-circuited to the outer conductors, the resonance condition is that $E_z(R, \phi)$ is identically zero. Hence, the resonance condition is

$$G(\phi; \phi') = 0. \tag{28}$$

For the subject matter of this paper, this type of resonances is of no importance.

2) *Magnetic wall:* If the edge of the disk constitutes a magnetic wall, the resonance condition is

$$G(\phi; \phi') = \text{infinity}. \tag{29}$$

Simplified Boundary Equations

At this stage, it is advantageous to introduce two quantities ξ and Φ defined by

$$i\xi = 2\Psi G(-\pi/3; -\pi/3) \tag{30}$$

$$\Phi = 2\Psi G(-\pi/3; \pi/3). \tag{31}$$

The condition for losslessness (26) implies that ξ is real. In general, ϕ is complex. The application of the cyclic symmetry condition (27) to the coefficients of the right hand members of (20)–(22) and the substitution of (30) and (31) into these equations result in

$$A = i\xi a + \Phi b - \Phi^* c \tag{32}$$

$$B = -\Phi^* a + i\xi b + \Phi c \tag{33}$$

$$C = \Phi a - \Phi^* b + i\xi c. \tag{34}$$

Matching Condition

A good method of investigating the more special properties of the circulator is to terminate the output and the isolated port with reflectionless loads. Then, the parts of a wave incident upon the input port, which are dissipated in these terminations, are measures of the transmission coefficients. The reflected part of that wave is related to the reflection coefficient. In an earlier paper,[5] this has been called the matching condition.

In this situation, only outward traveling waves are present in the output and isolated striplines so that, by virtue of (9), the relations

$$B/b = C/c = -\zeta \tag{35}$$

hold. Substitution into (32)–(34) and introduction of the complex quantity Θ given by

$$\Theta = \zeta + i\xi \tag{36}$$

result in the relations

$$A = \left(i\xi + \frac{\Phi^3 - \Phi^{*3} + 2\Phi\Phi^*\Theta}{\Phi\Phi^* + \Theta^2}\right) a \tag{37}$$

$$b = \frac{\Phi^2 + \Phi^*\Theta}{\Phi\Phi^* + \Theta^2}\, a \tag{38}$$

$$c = \frac{\Phi^{*2} - \Phi\Theta}{\Phi\Phi^* + \Theta^2}\, a. \tag{39}$$

Scattering Matrix

The scattering matrix S of a cyclic symmetric 3-port is given by (1). If the electric and magnetic field intensities of the wave incident upon the input port are A_i and a_i, respectively, the following relations are valid:

$$A_i = \zeta a_i \tag{40}$$

$$A = (1 + \alpha) A_i \tag{41}$$

$$a = (1 - \alpha) a_i \tag{42}$$

$$\beta = B/A_i \tag{43}$$

$$\gamma = C/A_i. \tag{44}$$

After some algebraic manipulation, the following expressions are found for α, β and γ:

$$\alpha = \frac{\Phi^3 - \Phi^{*3} + \Phi\Phi^*(2\Theta - \Theta^*) - \Theta^2\Theta^*}{\Phi^3 - \Phi^{*3} + 3\Phi\Phi^*\Theta + \Theta^3} \tag{45}$$

$$\beta = -\frac{(\Phi^2 + \Phi^*\Theta)(\Theta + \Theta^*)}{\Phi^3 - \Phi^{*3} + 3\Phi\Phi^*\Theta + \Theta^3} \tag{46}$$

$$\gamma = -\frac{(\Phi^{*2} - \Phi\Theta)(\Theta + \Theta^*)}{\Phi^3 - \Phi^{*3} + 3\Phi\Phi^*\Theta + \Theta^3}. \tag{47}$$

All the elements of the scattering matrix are now expressed in terms of the Green's function $G(\phi; \phi')$ and the properties of the medium[21] of the striplines ζ.

Circulation Condition

From the considerations made above about cyclic symmetric 3-ports, such a 3-port is an ideal clockwise rotating circulator, if the conditions

$$\alpha = \gamma = 0 \tag{48}$$

$$|\beta| = 1 \tag{49}$$

are fulfilled. It will be clear that this only can be so if all media are lossless and all conductors are perfect. These conditions can be satisfied for given values of R and $4\pi M$ only with certain values of ω and H_i. Such a combination of the parameters R, ω, H_i and $4\pi M$ could be called a circulation adjustment.

The condition $\gamma = 0$ is conceptually the same as the requirement $c = 0$ for all a [*cf.*, (39)]. In the latter form, it was earlier called[5] the circulation condition. Since

$$\Theta + \Theta^* = 2\zeta \neq 0. \tag{50}$$

[21] The formulas given apply only if the stripline medium is lossless.

application of $\gamma=0$ to (47) yields the circulation equation

$$\Theta = \Phi^{*2}/\Phi. \tag{51}$$

Substitution into (45) and (46) shows that $\alpha=0$ and

$$\beta = -\Phi/\Phi^{*} \tag{52}$$

which satisfy the other circulation requirements posed by (48) and (49). At first sight, one might wonder why $\gamma=0$ is a sufficient condition for circulator action, but it has been shown above that this is implied by losslessness and cyclic symmetry.

Rewriting the circulation equation (51) in its real and imaginary parts yields the two circulation equations (54) and (55) derived by Bosma.[5] Therefore, the elements of the latter can be explained now in terms of the real and imaginary parts of the transfer wave impedances of the disk configuration and of the intrinsic wave impedance of the stripline medium.

Green's Function

To solve the circulation equation in a particular case, Φ and Θ must be expressed in the characterizing quantities R, Ψ, ω, H_i and $4\pi M$. To that end $G(\phi;\phi')$ will now be deduced.

In the present case, the general solution of the Helmholtz equation (15) is a series in which the general term is given by

$$E_{z,n}(r,\phi) = a_n J_n(kr)e^{in\phi}, \tag{53}$$

where J_n is the nth order Bessel function. The corresponding term of the series for $H_\phi(r,\phi)$ is, by virtue of (16),

$$H_{\phi,n}(r,\phi) = i\frac{a_n}{\zeta_{\text{eff}}}\left[J_n'(kr) - \frac{\kappa}{\mu}\,\frac{nJ_n(kr)}{kr}\right]e^{in\phi}, \tag{54}$$

where ζ_{eff} is the effective intrinsic wave impedance of the ferrite as given by

$$\zeta_{\text{eff}} = \sqrt{\mu_0\mu_{\text{eff}}/\epsilon_0\epsilon}. \tag{55}$$

Inversely, if

$$H_{\phi,n}(R,\phi) = A_n e^{in\phi} \tag{56}$$

is the nth Fourier component of $H_\phi(R,\phi)$ at the edge, $r=R$ the corresponding term of the Bessel-Fourier series of $E_z(r,\phi)$ is given by

$$E_{z,n}(r,\phi) = iA_n\frac{\zeta_{\text{eff}}J_n(kr)e^{in\phi}}{\dfrac{\kappa}{\mu}\,\dfrac{nJ_n(x)}{x} - J_n'(x)}, \tag{57}$$

where

$$x = kR = \omega\sqrt{\mu_0\epsilon_0\mu_{\text{eff}}\epsilon}\,R. \tag{58}$$

Let $H_\phi(R,\phi)$ be everywhere zero at the edge except at an azimuth ϕ' over a small angle $\Delta\phi'$, where it is $H_\phi(R,\phi')$. Then,

$$A_n = (1/2\pi)H_\phi(R,\phi')e^{-in\phi'}\Delta\phi' \tag{59}$$

and the corresponding contribution to $E_z(r,\phi)$ is

$$\Delta E_z(r,\phi) = \frac{i\zeta_{\text{eff}}H_\phi(R,\phi')\Delta\phi'}{2\pi}\sum_{n=-\infty}^{\infty}\frac{J_n(kr)e^{in(\phi-\phi')}}{\dfrac{\kappa}{\mu}\,\dfrac{nJ_n(x)}{x} - J_n'(x)}. \tag{60}$$

Comparing this with the defining relation (18) after some algebra, one obtains

$$G(r,\phi;R,\phi') = -\frac{i\zeta_{\text{eff}}J_0(kr)}{2\pi J_0'(x)} + \frac{\zeta_{\text{eff}}}{\pi}\sum_{n=1}^{\infty}\frac{\dfrac{\kappa}{\mu}\,\dfrac{nJ_n(x)}{x}\sin n(\phi-\phi') - iJ_n'(x)\cos n(\phi-\phi')}{\{J_n'(x)\}^2 - \left\{\dfrac{\kappa}{\mu}\,\dfrac{nJ_n(x)}{x}\right\}^2}J_n(kr). \tag{61}$$

Resonances

Only the resonances of the disk configuration with magnetically short-circuited edge have importance for the present investigation. From (29), it is seen that such a resonance occurs if the denominator of one of the terms of the series (61) is zero,

$$J_n'(x) - \frac{\kappa}{\mu}\,\frac{nJ_n(x)}{x} = 0 \tag{62}$$

where n may be any positive or negative integer. From (54), it is easy to show that for these resonances $H_{\phi,n}(R,\phi)$ is identically zero. Moreover, for $n>0$ the field pattern is rotating to the right and for $n<0$ to the left. If the disk medium is isotropic, $\kappa/\mu=0$, the resonance conditions of the clockwise and anticlockwise rotating modes are identical

$$J_n'(x) = 0 \tag{63}$$

and, hence, these resonances are degenerate. When the gyrotropy increases, $\kappa/\mu\neq 0$, such a degenerate pair splits giving different x-values. For a disk R at a particular magnetic adjustment κ/μ and μ_{eff}, these two resonances have different frequencies which can be calculated from (58). With (62) and the recurrence relations of Bessel functions, it is easy to deduce that for a small amount of gyrotropy $\kappa/\mu \ll 1$ the difference of x-values of a pair of resonances is determined by

$$(\Delta x)_{n,j} = 2 \frac{n x_{n,j}}{x_{n,j}^2 - n^2} (\kappa/\mu), \tag{64}$$

where $x_{n,j}$ is the jth solution of the nth order equation (63). n denotes the order of the resonant modes and j the number of the pair. In Fig. 4, the κ/μ-dependence has been sketched for several of these modes.

84

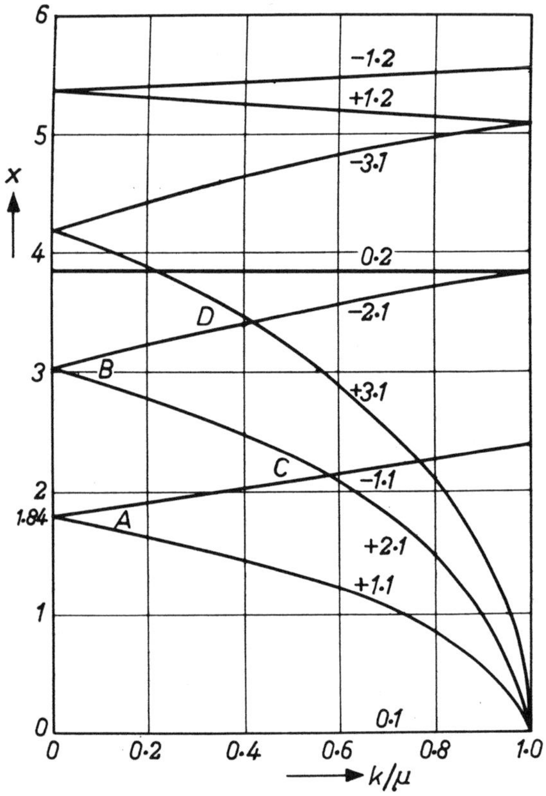

Fig. 4—The dependence on κ/μ of the x-values of several resonant disk modes. + sign denotes a clockwise rotating mode. − sign denotes an anticlockwise rotating mode. The first index refers to the order and the second one to the pair-number of a mode.

From experiments, it appeared that the practical circulation adjustment is situated between the first pair 1st-order resonances. In Fig. 4, this is in the neighborhood of A. Hence, the value of x is about equal to $x_{1,1}$. Therefore, it is supposed that the relation

$$J_1'(x_{1,1}) = 0, \qquad x_{1,1} = 1.84 \tag{65}$$

is exactly valid for the circulator.

From the measured frequency difference between the first pair 1st-order resonances at the (magnetic) circulation adjustment and the value of $x_{1,1}$ calculated from it with (58), it appeared, by virtue of (64), that $\kappa/\mu = 0.125$. The other parameters, determined experimentally, of this circulation adjustment are $f = 450$ MHz, $R = 3.14$ cm, $H_i \approx 850$ Oe, $4\pi M = 1750$ Oe and $\Delta H \approx 150$ Oe.

The experimental value of κ/μ is very small, much smaller than would have been presumed intuitively. It is the opinion of the author that this is the main reason why an explanation of the junction circulation mechanism has been awaited for so long.

Approximate Circulation Adjustment

Remembering the experimental data above, it is easy to see that the term for $n = 1$ of the Green's function (61) is large compared with the other terms. Therefore, $G(r, \phi; R, \phi')$ will be approximated by retaining the term for $n = 1$ only. A numerical estimation made afterwards showed that the error introduced in this way amounts to a few per cent only.

Substituting (65) into (61), by virtue of (19), for $G(\phi; \phi')$, the formula

$$G(\phi; \phi') = -\frac{x_{1,1}\zeta_{eff}}{\pi(\kappa/\mu)} \sin(\phi - \phi') \tag{66}$$

is found. Then, by virtue of (30), (31) and (36), Θ and Φ are found to be

$$\Theta = \zeta \tag{67}$$

$$\Phi = \frac{\sqrt{3}x_{1,1}\zeta_{eff}}{\pi(\kappa/\mu)} \Psi. \tag{68}$$

In this particular approximated case, both these quantities are real. Due to (9) and (55), substitution into the circulation equation (51) yields

$$\kappa/\mu = \frac{\sqrt{3}x_{1,1}}{\pi} \Psi \sqrt{\mu_{eff}/\epsilon}. \tag{69}$$

As $\omega\sqrt{\mu_0\epsilon_0} = 2\pi/\lambda$, λ being the free space wavelength, (58) yields

$$R/\lambda = \frac{x_{1,1}}{2\pi\sqrt{\mu_{eff}\epsilon}} \cdot \tag{70}$$

For small values of Ψ, this factor[22] in the right hand member of (69) can be replaced by $(v/2R)$. Then, elimination of R from (69) and (70) yields the relation

$$\frac{\kappa/\mu}{\mu_{eff}} = \sqrt{3}\frac{v}{\lambda} \cdot \tag{71}$$

Ferrite Properties

In order that the parameters R and H_i can be calculated at a given frequency, the electromagnetic properties of the ferrite, ϵ, κ, μ and μ_{eff} have to be known. Since the ferrite is biased far above the magnetic resonance

[22] If, for the electrical field intensities at the ports, the averaged values and not the values of $E_z(R, \phi)$ at the centers of the striplines had been taken, the factor in Ψ in (69) would have been[5] sin (Ψ). Then, the replacement by $(v/2R)$ is exact for all values of Ψ.

field $\kappa/\mu \ll 1$, the real parts of κ and μ need be considered only. These are given by

$$\mu = 1 + \frac{hm(h^2 - 1)}{(h^2 - 1)^2 - s^2} \tag{72}$$

$$\kappa = \frac{m(h^2 - 1)}{(h^2 - 1)^2 + s^2}, \tag{73}$$

where

$$m = 4\pi M/H_0 \tag{74}$$

$$h = H_i/H_0 \tag{75}$$

$$s = \Delta H/H_0. \tag{76}$$

For the circulators considered, the approximations $h^2 \gg 1$ and $h^2 \gg s$ can be made, so that (73) and (74) become

$$\mu = \mu_{eff} = (h + m)/h \tag{77}$$

$$\kappa = m/h^2. \tag{78}$$

Circulation Parameters

Insertion of (77) and (78) into (70) and (71) yields

$$R/\lambda = \frac{x_{1,1}}{2\pi\sqrt{\epsilon}}\sqrt{\frac{h}{h + m}}, \tag{79}$$

$$h = \sqrt{\frac{\lambda m}{\sqrt{3}v}} - m, \tag{80}$$

while elimination of h from these equations results in

$$R/\lambda = \frac{x_{1,1}}{2\pi\sqrt{\epsilon}}\sqrt{1 - \sqrt{\sqrt{3}\,\frac{mv}{\lambda}}}. \tag{81}$$

From the last relation, one might conclude that small circulators can be obtained with large magnetizations and stripline widths. However, calculation of h shows that this quantity then becomes small (or even negative), which from considerations of resonance losses and magnetic saturation is not practical.

In general, it may be stated that in the UHF region h must be as large as possible. From (81), it is easily seen that h attains its maximum value for

$$m_{h_{max}} = h_{max} = (1/4\sqrt{3})\,\frac{\lambda}{v}. \tag{82}$$

By virtue of (81), the corresponding relative disk radius is given by

$$(R/\lambda)_{h_{max}} = \frac{x_{1,1}}{2\pi\sqrt{2\epsilon}} = 0.207e^{-1/2}. \tag{83}$$

The inverse proportionality with $\sqrt{\epsilon}$ is self-evident. Increase of h_{max} by diminution of v (and corresponding choice of m) is not unrestricted, as the resulting stray fields become undesirably large.

Eqs. (80) and (81) can be rewritten in the form

$$H_i = \sqrt{\frac{\lambda}{\sqrt{3}v}\,H_0 \cdot 4\pi M} - 4\pi M \tag{84}$$

$$R/\lambda = \frac{x_{1,1}}{2\pi\sqrt{\epsilon}}\sqrt{\frac{H_i}{H_i + 4\pi M}}. \tag{85}$$

As H_0 is inversely proportional with λ, for the same stripline width and magnetization (v and $4\pi M$ constant), H_i and R/λ do not depend on frequency. Hence, with the latter increasing, H_i and R become too small with respect to H_0 and v, respectively. Consequently, v must be chosen smaller for higher frequencies. A good estimation is obtained by the assumption that H_i must be four times as large as H_0 at least, *i.e.*, $h > 4$. This number is rather arbitrary and it may be chosen smaller for the higher frequencies and small line widths. Due to (83), v is restricted in accordance with

$$v < \lambda/30, \tag{86}$$

so that for $\epsilon \approx 16$ a restriction of the kind

$$v/R < 0.75 \tag{87}$$

is valid, which is in agreement with (11).

Most of the experiments are performed with a manganese ferrite aluminate, the significant properties of which are $4\pi M = 1750$ Oe, $\Delta H = 150$ Oe and $\epsilon = 14.2$. The geometrical properties of the circulator were $v = 15$ mm and $u = 11$ mm. With these parameters, it can be calculated from (84) and (85) that $H_i = 935$ Oe and $R/\lambda = 0.046$. At a frequency of 450 MHz, one then finds theoretically $h = 5.82$, $\kappa/\mu = 0.112$, $R = 3.07$ cm and $v/R = 0.48$.

All approximations made above about h, κ/μ and v/R are justified. The agreement of these theoretical results with the experimental values quoted above is good. This good correspondence between theory and experiment is also seen in Fig. 5, where experimental and

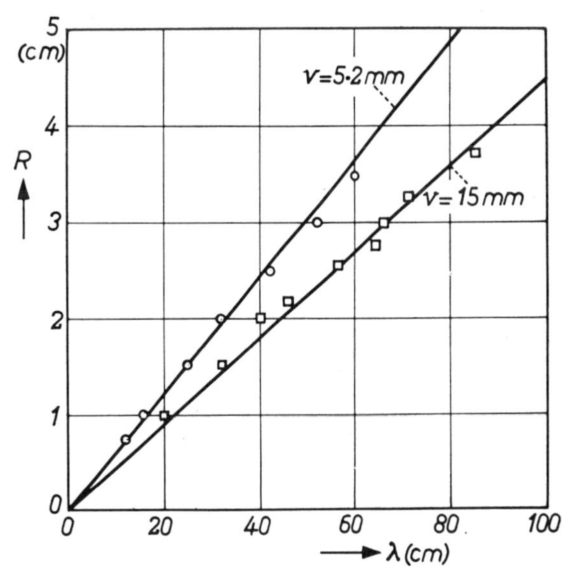

Fig. 5—Theoretical and experimental disk radii against free space wavelength with the stripline width as parameter. The drawn lines denote the theoretical results, (○) measured radii for $v = 5.2$ mm and (□) measured radii for $v = 15.0$ mm.

theoretical values of R are given as a function of λ, with values of v 15 mm and 5.2 mm, respectively. Values of H_i are not inserted into Fig. 5 as they were not measured accurately because of concentration of the static magnetic flux in the disks in most cases.

Frequency Dependence

For any ferrite disk R, there is a circulation adjustment for one frequency only. This frequency will be called the circulation frequency ω_0. If all other parameters are kept constant, the circulator characteristics are functions of frequency. Analogously, H_i can be varied in order to have a tunable circulator. The latter case will not be investigated here.

If $\omega \neq \omega_0$, $G(\phi; \phi')$ is a function of ω. Neglecting all terms for $n \neq 1$ again, from (61) it is

$$G(\phi; \phi') = \frac{Z_{\text{eff}}\left[\dfrac{\kappa}{\mu}\dfrac{J_1(x)}{x}\sin(\phi-\phi') - iJ_1'(x)\cos(\phi-\phi')\right]J_1(x)}{\pi\left[\{J_1'(x)\}^2 - \left\{\dfrac{\kappa}{\mu}\dfrac{J_1(x)}{x}\right\}^2\right]}. \tag{88}$$

For $x = x_{1,1}$ and $\omega = \omega_0$ it is easy to deduce that μ does not depend on ω and that the relations

$$\frac{d(\kappa/\mu)}{d\omega} = \frac{\kappa/\mu}{\omega_0}, \qquad \frac{dx}{d\omega} = \frac{x_{1,1}}{\omega_0} \tag{89}$$

hold. Using them together with recurrence relations of Bessel functions and (65) and (66), $G(\phi; \phi')$ can be developed into a Taylor-series. Neglecting 2nd and higher order terms of $\delta\omega$, where

$$\delta\omega = \omega - \omega_0, \tag{90}$$

this series is given by

$$G(\phi; \phi') = -\frac{\zeta_{\text{eff}}}{\pi}\left[\frac{x_{1,1}}{\kappa/\mu}\sin(\phi-\phi') + i\frac{x_{1,1}(x_{1,1}^2-1)}{(\kappa/\mu)^2\omega_0}\cos(\phi-\phi')\delta\omega\right]. \tag{91}$$

Then, by virtue of (30), (31) and (36), the quantities Θ and Φ are found to be

$$\Theta = \zeta(1 - i2\eta) \tag{92}$$

$$\Phi = \zeta(1 + i\eta), \tag{93}$$

where

$$\eta = \frac{x_{1,1}^2 - 1}{\sqrt{3}(\kappa/\mu)} \cdot \frac{\delta\omega}{\omega_0}. \tag{94}$$

Substitution into (45)–(47) yields, for the elements of the scattering matrix,

$$\alpha = \frac{-3\eta^2 + i2\eta}{(4 - 9\eta^2) - i6\eta} \tag{95}$$

$$\beta = \frac{(-4 + 6\eta^2) + i2\eta}{(4 - 9\eta^2) - i6\eta} \tag{96}$$

$$\gamma = \frac{b\eta^2 + i2\eta}{(4 - 9\eta^2) - i6\eta}. \tag{97}$$

Eqs. (3) and (4) are easily verified. For small η or, what is the same, for small frequency shifts, $\delta\omega/\omega_0 \ll 1$, α, β and γ can be approximated by

$$\alpha = \gamma = i\eta/2 = i\frac{x_{1,1}^2 - 1}{2\sqrt{3}(\kappa/\mu)} \cdot \frac{\delta\omega}{\omega_0} \tag{98}$$

$$1 - |\beta|^2 = \frac{1}{2}\eta^2 = \frac{1}{2}\left\{\frac{x_{1,1}^2 - 1}{\sqrt{3}(\kappa/\mu)}\right\}^2\left(\frac{\delta\omega}{\omega_0}\right)^2. \tag{99}$$

α and γ measure the return losses and the isolation, while $(1 - |\beta|^2)$ is an expression for the forward attenuation. Due to dissipative losses, the latter is much larger in practice. The relation $\alpha = \gamma$ is easy to verify experimentally.

Bandwidth

The bandwidth of a circulator is not an unambiguous quantity. It can be defined in several ways. A practically useful method is the specification of a maximal value for the modulus of the reflection coefficient. Thus,

$$|\alpha| \leq \rho_{\max}. \tag{100}$$

The isolation is limited at the same time. Substitution of (98) yields, for the relative bandwidth,

$$\frac{\Delta\omega}{\omega_0} = \frac{4\sqrt{3}}{x_{1,1}^2 - 1}(\kappa/\mu)\rho_{\max}. \tag{101}$$

For $\rho_{\max} = 0.1$ (which is the same as a VSWR of 1.22, an isolation of 20 db and a minimal insertion loss of 0.1 db) at 450 MHz, a bandwidth of 3.2 per cent is found. Experimentally, it appeared to be 4.1 per cent. Why the experimental bandwidth is larger than the theoretical one can be understood from the fact that the dissipative losses, which tend to increase the bandwidth, are not considered in the theory.

From (101), it can be seen that the bandwidth is proportional to the amount of gyrotropy κ/μ. This can be understood from an inspection of (62) or (64). The disturbance of the original isotropic resonance is proportional to κ/μ. The larger κ/μ is, the more distant are the resonant clock- and anticlock-wise rotating modes. Hence, in the frequency region between the latter, the dependence of several quantities on frequency will be smaller, resulting in a larger bandwidth. Although, in most cases, a large bandwidth is desirable, κ/μ cannot be chosen large, as a small H_i would be the result which is not permitted because of the losses due to it. However, it is shown below that another straightforward method exists to obtain large bandwidths.

Input Impedance

Let the characteristic impedance of the striplines be Z and the output and isolated ports be reflectionlessly terminated. Then, upon neglecting higher order terms of $\delta\omega/\omega_x$ again, the input impedance Z_i is easily derived. The result is

$$Z_i = R_i + iX_i \tag{102}$$

with

$$R_i = Z \tag{103}$$

$$X_i = Z\frac{x_{1,1}^2 - 1}{\sqrt{3}(\kappa/\mu)}\cdot\frac{\delta\omega}{\omega_0}\cdot \tag{104}$$

The real part of Z_i, R_i does not depend on frequency, but the imaginary part X_i does. Remembering that the time dependent factor of the field is $\exp(-i\omega t)$, one can see that, for $\omega<\omega_0$, X_i is inductive and, for $\omega>\omega_0$, capacitive. This parallel resonant circuit behavior is confirmed by experiment.

Bandwidth Enlargement

From the dependence of the input impedance on frequency, a method of enlarging the bandwidth enormously can be proposed. For a lossless, cyclic-symmetric 3-port circulator, the isolation is related directly to the reflection coefficient [*cf.*, (7)]. When the latter is zero, the former is infinite. Now, α can be made zero over a rather large bandwidth and, hence, so can γ. This can be accomplished with a lossless series circuit of an inductance L and a capacitance C. The series impedance of such a circuit is

$$Z_{\text{series}} = i\left(\frac{1}{\omega C} - \omega L\right). \tag{105}$$

Choosing L and C so that the relations

$$\omega_0 = 1/\sqrt{LC}, \tag{106}$$

$$2\omega L_0 = Z\frac{x_{1,1}^2 - 1}{\sqrt{3}(\kappa/\mu)} \tag{107}$$

hold and connecting such resonant LC-circuits in series with the three ports, the reflections are eliminated over a rather large bandwidth centered at the circulation frequency.

In this way, with small coils and capacitors, it was found possible to make a circulator having a bandwidth of about 30 per cent. From 310 MHz up till 420 MHz, it had a maximal VSWR of 1.2, a minimal isolation of 19.5 db and a maximal insertion loss of 1.1 db. This last figure is a little high, but it is believed that it can be improved by using a better ferrite.

Field Distribution

It is easy to show that at the circulation adjustment the relations

$$b = a, \qquad c = 0 \tag{108}$$

hold [*cf.*, (10)] and that, upon substitution into (18), they yield, for the electric field intensity in the disk, the 1st order approximation

$$E_z(r,\phi) = \frac{2x_{1,1}Z_{\text{eff}}\sin\Psi J_1(kr)\sin\phi}{\pi(\kappa/\mu)J_1(x_{1,1})}a. \tag{109}$$

Insertion of this result into (16) and (17) gives for the magnetic field intensity in the disk the approximate expressions

$$H_\phi(r,\phi) = -\frac{2x_{1,1}\sin\Psi}{\pi(\kappa/\mu)J_1(x_{1,1})}\cdot\left[\frac{\kappa}{\mu}\frac{J_1(kr)}{kr}\cos\phi - iJ_1'(kr)\sin\phi\right]a \tag{110}$$

$$H_r(r,\phi) = -\frac{2x_{1,1}\sin\Psi}{\pi(\kappa/\mu)J_1(x_{1,1})}\cdot\left[\frac{\kappa}{\mu}J_1'(kr)\sin\phi + i\frac{J_1(kr)}{kr}\cos\phi\right]a. \tag{111}$$

Inspection of these formulas shows that the magnetic field intensity is anticlockwise rotating and elliptically polarized. The axis ratio is κ/μ. Calculation of the high frequency part of the magnetization shows that it is clockwise rotating and also elliptically polarized with an axis ratio $\kappa/\mu/(\mu_{\text{eff}}-1)$.

The electric field has been measured, too. The results of this experiment are sketched in Fig. 6, in which the lines of equal amplitude and the constant phase lines have been drawn. Comparison of the experimental results with the expression given above shows that the agreement is good, except at those places where the latter is small; that is, along the diameter $\phi=0, \pi$, which is symmetrically situated between the input and the output port. This is quite understandable as in those places the neglected terms for $n\neq 1$ give a non-negligible contribution to the total electric field intensity and spread the phase lines.

Retaining 18 terms of the series involved, a solution of the same problem, as it is stated by Bosma,[5] has been determined with an electronic computer. Indeed, the field distribution so obtained agrees much better with the experimental result of Fig. 6. The circulation parameters of this "accurate" solution are $x=1.852$, $h=6.034$, $H_i=970$ Oe, $4\pi M=1750$ Oe, $\kappa/\mu=0.109$, $f=450$ MHz and $R=2.90$ cm. These data, too, correspond well to those quoted above.

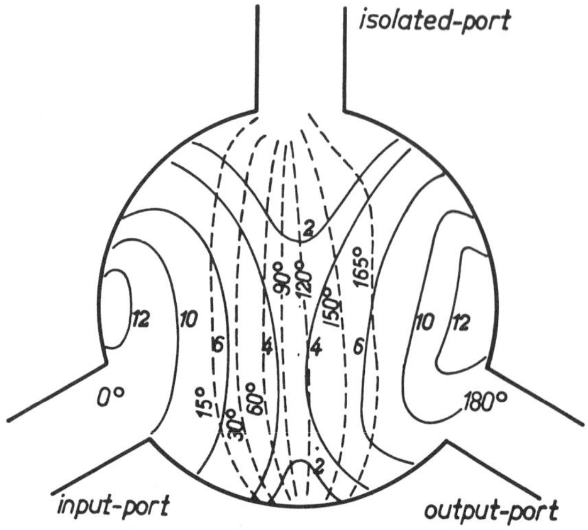

Fig. 6—Measured amplitude and phase of the electric field in the circulator disks. Drawn lines indicate the lines of equal amplitude. Broken lines are equal phase-lines. (The amplitude is given in arbitrary units.)

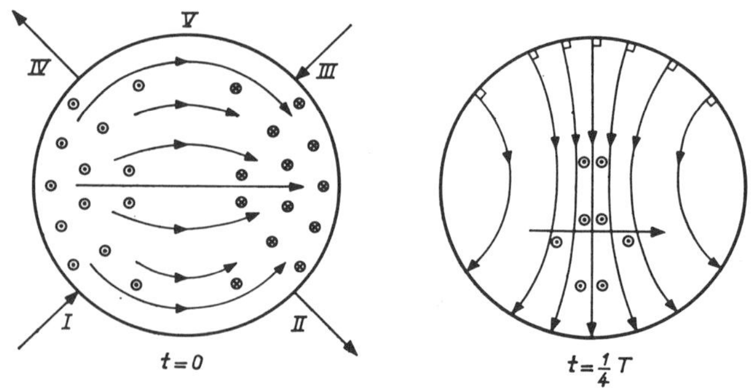

(a) (b)

Fig. 7—The electromagnetic field in the circulator disks at times (a) $t=0$ and (b) $t=T/4$, respectively. The circles denote the electric field intensity, the drawn lines the magnetic field intensity and the straight arrows the power flow.

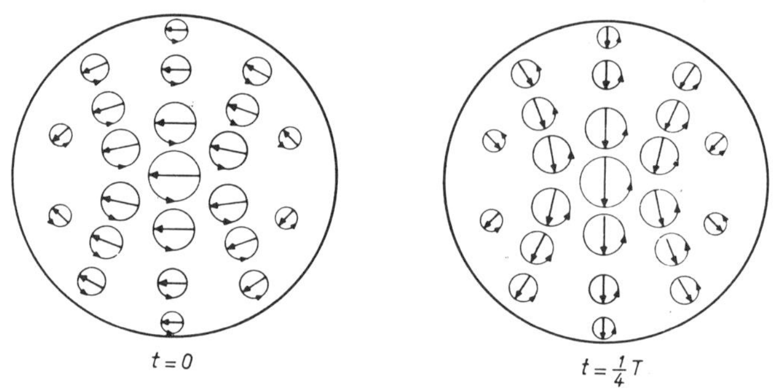

(a) (b)

Fig. 8—The magnetic dipole distribution in the circulator disks at times (a) $t=0$ and (b) $t=T/4$, respectively.

The Circulation Mechanism

The primary condition for circulation seems to be that the adjustment is in the neighborhood of the degeneracy of two resonances of the disk configuration (with disconnected striplines). If these resonances form a pair of the same order, as is the case with the solution found here, the implication of (64) is that κ/μ is small. This is also confirmed by Butterweck[16] who studied a model of the 3-port waveguide junction circulator, consisting of a circular-cylindrical cavity with an axially biased, concentric, thin and full height post of ferrite and three rectangular waveguides coupled weakly to the cavity at equal distances around the cylindrical wall. Butterweck also found that circulator action was possible near the degeneracy of the two first-order oppositely rotating modes. Although, in his case, κ/μ in the ferrite is not small, the effective gyrotropy, as averaged over the whole cavity volume, is.

The first order resonance of the isotropic disks has a distribution of the electric field as sketched in Fig. 7(a) at $t=0$. It is antisymmetric with respect to the symmetry diameter. A quarter of a period later $t=t/4$, the electric field is identically zero but, then, the magnetic field attains its maximum. The lines of the latter have been drawn in Fig. 7(b). If, instead of being isotropic, the disks are weakly gyrotropic, the magnetic field present at $t=T/4$ will induce a magnetic dipole distribution. This has been sketched in Fig. 8(b). As the disks are polarized perpendicularly, these dipoles rotate to the right. A quarter of a period earlier $t=0$, they were directed as it has been indicated in Fig. 8(a). The magnetic induction then being zero, they cause a demagnetizing field which, too, has been sketched in Fig. 7(a). The latter is in phase (or counterphase) with the electric field and it is symmetric with respect to the symmetry diameter. Hence, at the edge $r=R$, a real, radial power flow exists. In four regions, this power flow is alternately inwardly and outwardly directed. In Fig. 7(a), this has been indicated with four radial arrows. Because of the resonant character of the isotropic (symmetric) part of the magnetic field, the reactive power flow at the edge is zero.

If, in two successive regions (I and II), waveguides (striplines) are connected, the power radiated inwardly and outwardly, respectively, can be transported via these guides. If, in the other two regions (III and IV), no guides are present, no radial current can flow from the center conductor, no power can be radiated and, consequently, the azimuthal component of the magnetic field must there be zero. This can be so if only other modes are excited. As these modes are not resonant, they do not contribute substantially to the electric field at the edge, but they do near the symmetry diameter. The (small) electric field excited in this way will be in phase with the magnetic field there [Fig. 7(b)], so that across the symmetry diameter a power flow occurs which transports the energy from region I towards region II.

Due to the excitation of the other modes, $n\neq 1$ everywhere at the edge except at the regions I and II, the azimuthal magnetic field is zero. Because near V [Fig. 7(a)] the electric field is also zero, connection of a waveguide at that place does not disturb the fields and, hence, no power exchange occurs there. This guide is isolated.

Finally, it will be clear that the wave impedances of the guides at I and II have to be matched to the ratios of the electric and magnetic field intensities in the ferrite at those places. If R is chosen such that the isotropic disks are resonant, κ/μ and μ_{eff} have to be adjusted by means of H_i so that the guides are matched to the disks. Mathematically, (79) is the expression of the former condition and (80) that of the latter.

It will be clear that the effective electrical length of the circulator is equal to half a wavelength.

Application in the UHF Region

Three reasons exist why the 3-port stripline Y-circulator polarized above resonance is advantageously used at frequencies below 1000 Mc/sec.

First, it is known from resonance isolators at the lower kMc-frequencies that thin, flat forms of ferrite magnetized perpendicularly are favorable with respect to saturation and, hence, for small low-field losses. The same is valid for the circulator in question.

Second, nearly all known nonreciprocal devices are of the same order of largeness as the free space wavelength, at least in one dimension. This signifies that in the UHF region the construction would become big and voluminous. However, the diameter of the stripline Y-circulator is determined mainly by the wavelength in the ferrite. By virtue of (84) $\epsilon = 14$–16, the latter is an order of magnitude smaller than the free space wavelength as is, therefore, the circulator as well.

The last and most important reason is the fact that the circulator in question requires essentially a very high bias field $h > 4$. It is much higher than it is in the case of any one of the well-known principles of nonreciprocal action. And, as has been argued above, this high bias field is very favorable for small insertion losses, which are usually difficult to obtain in the UHF region.

Other Solutions of the Problem

Any solution of the circulation equation (51) may result in conditions for circulator adjustment. Such solutions can be found in the neighborhood of degeneracies of disk resonances. This has been confirmed by Cohen and Davies[11] who computed electronically, from (54) of Bosma,[5] curves in the κ/μ, x-plane along which circulator action is possible. The straight line $x = 1.84$ for $\kappa/\mu < 0.25$ coincides nearly with one of these curves. Other curves join at the degeneracies B, C and D (Fig. 4) and are between the mode pairs $(-2,1—2,1)$, $(-1,1—2,1)$ and $(-2,1—3,1)$, respectively. To give a complete picture, for small values of κ/μ, they have also found a curve within the region ABC, which seems not to be related to a degeneracy. Moreover, the separate curves are linked crossing over a resonant mode curve somewhere between two degeneracies, *e.g.*, over the curve $-1,1$ between A and C. However, it may appear that these solutions do not apply to practical circulators and, in many cases, they do not.

In fact, circulator action has been detected experimentally somewhere near B (Fig. 4). However, using the second order resonance is not a practical proposition, since due to the large x-value the radius is large. Furthermore, relative to the transmitted power, much of its field energy is stored in the disk, so that the insertion loss is relatively large. It is the opinion of the author that many rectangular waveguide junction circulators can be improved because, in many of them, higher order resonances are employed.

The solution evaluated by Bosma[5] is near the degeneracy of the first anticlockwise rotating 1st order mode and the first clockwise rotating 2nd order mode which is in the neighborhood of C in Fig. 4. The mathematical difficulties of the evaluation of this solution are much greater than of the solution presented in this paper, because the gyrotropy is not small ($\kappa/\mu \approx 0.5$) and it is not permitted to neglect all terms of the series involved except the two resonant terms. Besides, it has turned out that the approximation $J_0(x) = 0$ made by Bosma[5] is not quite justified. But, although the solution near C (Fig. 4) is mathematically nearly possible and would require a small disk radius, it is not a practical proposition since κ/μ is not very small; hence, as h is not large, $h \approx 1.5$–2.0 in the UHF region, resonance (and perhaps low field) losses are not small. In fact, the resonances of the disk configuration could not be detected even at values of κ/μ larger than about 0.3, because they were masked by losses. The agreement between the theoretical and the experimental radii as a function of frequency found by Bosma[5] must be looked upon as merely an awkward coincidence. If at the time the field distribution in the disk had been measured, the practically false character of the solution would have been observed. Now, the conclusions drawn from it may be considered obsolete.

The theory developed in this paper applies also to circulators which are biased below resonance, if the section about the magnetic properties of the ferrite and its consequences are restated. Moreover, an analogous analysis can probably be carried out for the rectangular waveguide junction circulator.

Acknowledgment

The author wishes to express his gratitude to G. de Vries for his lasting and stimulating interest, for the fruitful discussions held with him; K. Kegel for the many accurately performed experiments; and H. J. C. A. Nunnink and J. Vlietstra for the trouble taken by them to find, despite the many singularities of the problem, a practical solution with an electronic computer. This solution has contributed in great measure to the attainment of the simplified analysis presented in this paper.

Four-Port Crossed-Waveguide Junction Circulators*

L. E. DAVIS, M. D. COLEMAN† AND J. J. COTTER‡

Summary—**A detailed account is given of the experimental investigation which has led to the design of 4-port circulators at frequencies from 2.5 to 25 Gc. The complex behaviour of these devices is clarified by establishing the modes of circulation in any particular waveguide passband. The modes are defined in terms of the static magnetic field (*H*) required, the microwave frequency (*f*) and the direction of circulation. The shape of the mode characteristic depends upon the ferrite/brass configuration in the center of the junction and the mode charts are given for two possible ferrite arrangements. The fixed-field bandwidths of the two configurations are, respectively, 2–4 per cent and 4–8 per cent. The performances are: isolation (ports 1–4)$>$20 db, cross-coupling (ports 1–3)$>$20 db, VSWR$<$1.2 and insertion loss (ports 1–2), respectively, $<$0.5 db and $<$1.0 db for the two configurations. The tunable bandwidth depends upon the mode of circulation and varied from 5 to 15 per cent. The positions of the modes in the waveguide pass-band can be adjusted by changing the diameter of the center ferrite/brass post. This shifting of the center frequency is described in detail for both ferrite arrangements and, for particular modes of circulation, figures are given of the frequency shift per 0.001″ change in ferrite and brass diameter.**

I. Introduction

IN 1957, ALLIN AND SMITH[1] proposed that circulator action could be obtained with a symmetrical 4-port *H*-plane waveguide junction containing a central composite post of polarized ferrite and a conducting material, *e.g.*, brass. In 1958, Chait and Curry[2] demonstrated that circulator action was possible in a symmetrical ferrite-loaded 3-port *H*-plane junction. Since these proposals, much work has been carried out on 3-port circulators but comparatively little on the 4-port device. If their bandwidth can be improved, these latter devices will have the advantage of greater compactness in applications where at present two 3-port devices are required. However, a symmetrical 3-port circulator requires the correct adjustment of only two independent parameters for circulation at one particular frequency, whereas a 4-port component requires three.[3] This additional parameter increases the problems of broadbanding and considerably complicates the behavior of the device.

If a symmetrical 4-port circulator has the ports numbered cyclicly 1-2-3-4 in the direction of power flow, the following terms are used to express its performance:

isolation (db)	ports 1-4
cross-coupling (db)	ports 1-3
insertion loss (db)	ports 1-2
VSWR	ports 1-1.

It is clear that the insertion loss consists not only of dissipation within the ferrite material but also of power scattered to ports 3, 4 and back into 1. A symmetrical, loss-less, nonreciporcal 3-port junction can be made a circulator by matching the three ports and, therefore, the bandwidth is limited only by the bandwidth of the reflection coefficient. But, in a near-perfect 4-port device, the isolation is (to a 1st order) independent of the reflection coefficient. The signals scattered to port 1 and port 3 are equal and interdependent. By reducing the reflection coefficient to zero the cross-coupling is also reduced to zero and this power instead of being lost is circulated to port 2. By matching in this way, the insertion loss can be reduced by twice the reflective loss of the unmatched junction. However, the isolation remains unaffected and is dependent only on the properties of the central ferrite/brass geometry. Therefore, a broadband 4-port junction circulator is possible only if a central configuration can be found which gives a broadband isolation characteristic and which can then be matched. This point is effectively demonstrated in the performance curves of a 3 Gc circulator shown in Fig. 1. The insertion loss, VSWR and cross-coupling curves are interdependent and broadband but the useful bandwidth of the device is limited by the narrow bandwidth of the isolation curve.

Two arrangements of ferrite and brass have been investigated and these are shown in Fig. 2. When a static magnetic field is applied perpendicular to the broad waveguide wall, the junction becomes asymmetrical and circulator action occurs. Although no simple physical explanation of this behavior exists, it may be regarded as asymmetrical scattering from an anisotropic dielectric rod. It cannot be regarded as a problem of nonreciprocal phase-shift giving signal cancellation at the unwanted ports. This approach can be used to describe the action of the 3-port circulator but it cannot be extended to the 4-port device; yet, the 4-port and 3-port circulators are manifestations of the same principle.

The behavior of 4-port junction circulators is complex but it can be clarified by establishing the various modes of circulation. Each mode is defined in terms of the diameters of the ferrite (d_f) and brass (d_b), the

* Received May 20, 1963; revised manuscript received September 30, 1963.

† Mullard Research Laboratories, Redhill, Surrey, England.
‡ Mullard Equipment Ltd., Crawley, Sussex, England.

[1] P. E. V. Allin and F. W. Smith, "Waveguide Circulators," British patent specification No. 852, 751; application date December 2, 1957.

[2] H. N. Chait and T. R. Curry, "Y circulator," *J. Appl. Phys.*, supplement to vol. 30, pp. 1525–1535; April 1959.

[3] B. A. Auld, "The synthesis of symmetrical waveguide circulators," IRE Trans. on Microwave Theory and Techniques, vol. 7, pp. 238–246; April, 1959. See p. 245.

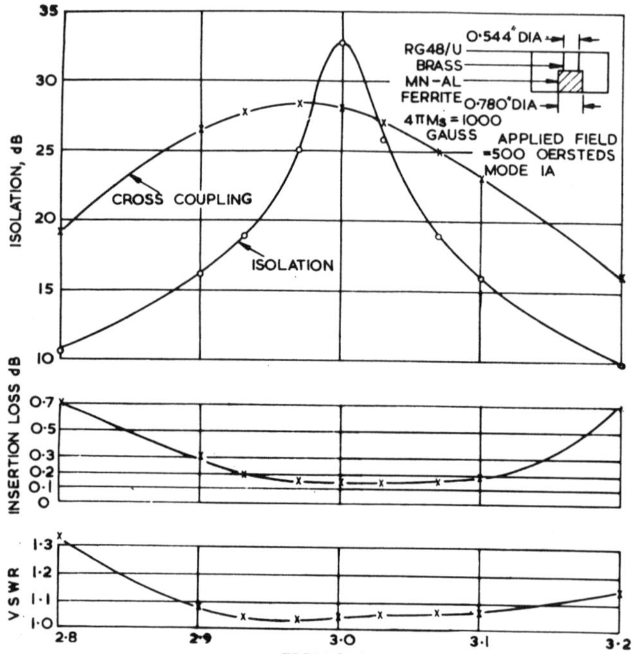

Fig. 1—The performance of a 4-port circulator for operation at 3.0 Gc.

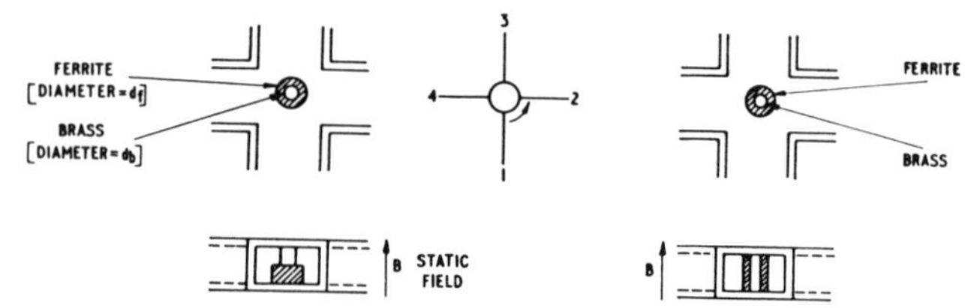

(a) (b)

Fig. 2—4-port H-plane waveguide junction circulator. (a) Half-height configuration. (b) Full-height configuration.

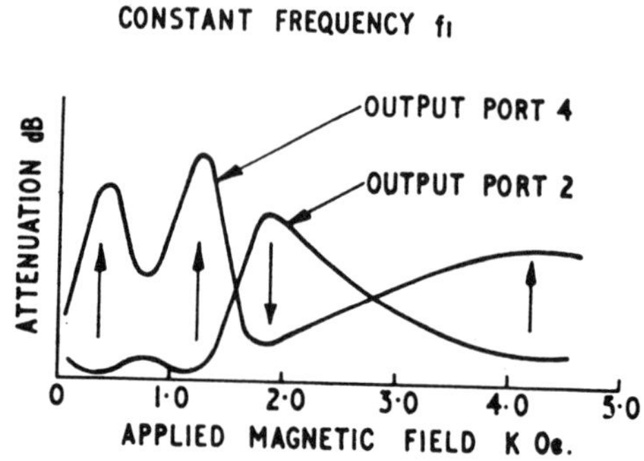

(a)

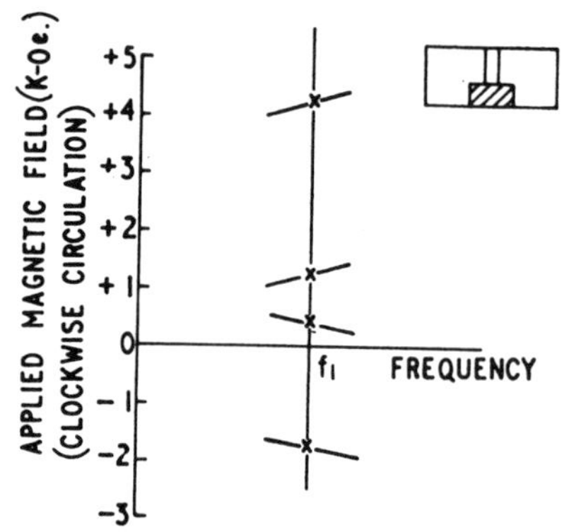

(b)

Fig. 3—(a) The output from ports 2 and 4 as the magnetic field is varied. (b) Modes of circulation. The negative and positive magnetic field values, required for circulation in a constant direction, plotted against frequency.

applied static magnetic field (H), the microwave frequency (f) and the direction of circulation.

II. The Half-Height Configuration

A. The Modes of Circulation

This configuration is shown in Fig. 2(a) and consists of a central polarized ferrite post, which is half the waveguide height, on top of which is a cylindrical brass post. These cylinders make firm flat contact with each other and with the waveguide walls. To obtain circulation at a specified frequency, the three independent variables chosen were the ferrite diameter, brass diameter and the magnetic field. To distinguish the modes of circulation the microwave frequency (f) was increased in discrete steps and at each frequency the outputs from ports 2 and 4 were monitored as the applied magnetic field (H) was swept from zero to approximately 3000 oersteds. A typical sketch of the output at f_1 from ports 2 and 4 is shown in Fig. 3(a) in which two features are apparent: that not all modes are equally useful and that at a single frequency circulation may be obtained in either direction by adjusting the value of the applied field. This information may be presented in another way by plotting H against f, with the circulation in a reverse sense represented by a negative applied field. In this way, the indicated circulation is always in the same sense and the graph is self-consistent [Fig. 3(b)]. By repeating this plot each time as the frequency is increased in discrete steps, a mode chart (H vs f) may be built up as shown in Fig. 4. Thus, a picture is obtained of the regions of circulation which are possible in any waveguide pass-band, but without any indication of performance. Although the modes have been arbitrarily numbered, the distinction in direction has been made by giving all modes in one sense odd numbers and in the other sense even numbers. The best performance is usually obtained on the sloping parts of the mode 1 and 1A characteristics and the fixed field bandwidth is about 3 per cent. Only poor performance is obtained over the flat regions of the mode 1 characteristic where broadband operation might be expected. At a constant value of applied field ($\pm H$), circulation at two very different frequencies ($f_1 f_2$) can be obtained but in opposite directions. This should not be confused with the diplexing action reported by Brown and Clarke.[4]

[4] J. Brown and J. Clark, "A unique solid state diplexer." IRE Trans. on Microwave Theory and Techniques, (*Correspondence*), vol. 10, p. 298; July, 1962.

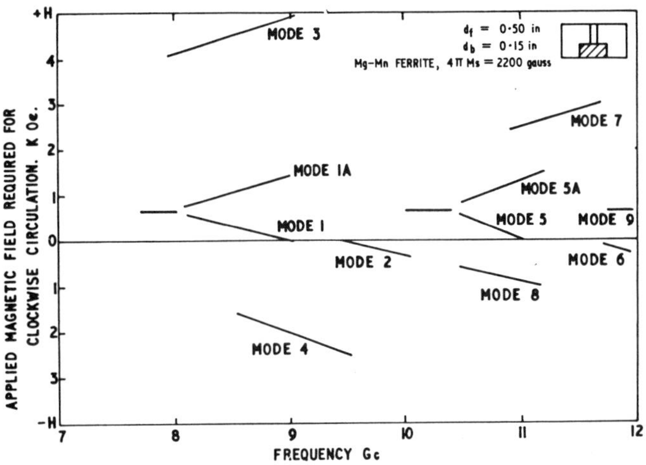

Fig. 4—The complete mode chart, 8–12 Gc and 0–4 kOe, for the "half-height" configuration.

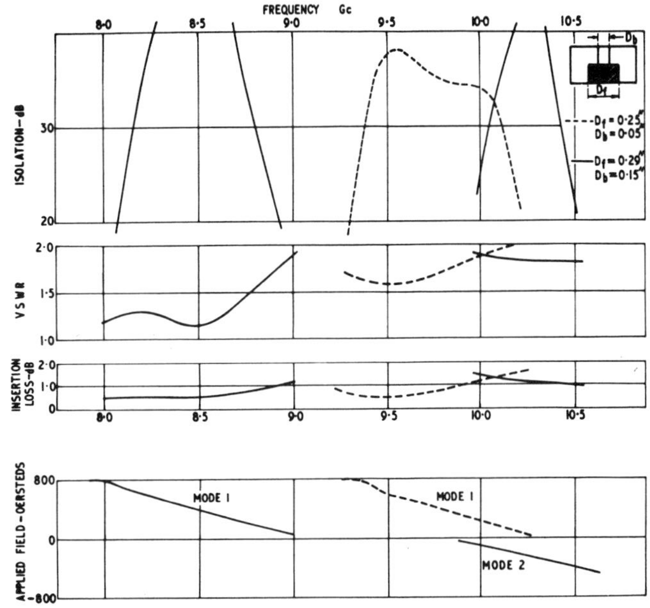

Fig. 5—A comparison of performance of two similar unmatched configurations. The need to distinguish the mode and direction of circulation is clear. Mn-Mg ferrite, $4\pi M_s = 2200$ gauss. To avoid confusion the performance of mode 1A is not shown.

They obtained opposite directions of circulation in a stripline circulator at two frequencies, because the circulator was below resonance at one frequency and above resonance at the other. However, from Fig. 4, it can be seen that with the "half-height" 4-port circulator opposite directions of circulation at two frequencies can occur with the magnetic field not only below resonance but also below saturation. Some circulation has been measured with only the remanent flux in a 0.25″ diameter post of manganese-magnesium ferrite with a 0.20″ diameter post of brass. The isolation was 12 db, insertion loss 4 db, cross-coupling 12 db, VSWR 2.5. Another interesting feature of the mode chart is the demonstration of higher-order modes of circulation, as shown by the repetition of modes 1–4 at the higher frequencies. The performance in these modes has not yet been investigated in detail.

TABLE I

FREQUENCY SHIFT PER 0.001″ CHANGE ON DIAMETER FOR HALF-HEIGHT CONFIGURATION

Frequency	W/g Size	Ferrite	Ferrite Δf	Brass Δf
(*Gc*)		4πMs(Gauss)	(Mc)	(Mc)
3.0	RG48/U	Mn−Al 1000	−2	−0.5
9.0	RG52/U	Mn−Mg 2200	−20	−5
22.0	RG53/U	Ni−Zn 5000	−160	−30

93

B. Circulator Design

The mode chart shown in Fig. 4 shows a large number of modes because the diameter of the ferrite post was large, 0.5″. In practice, many of these modes are eliminated and the behavior is simplified by using a smaller ferrite diameter. If the diameter of the ferrite or conducting post is changed, the operating frequency of the circulator is altered. Decreasing the diameter, increases the frequency; increasing the diameter, decreases the frequency. Thus, the mode chart may be shifted along the frequency axis. By reducing the diameter of the ferrite post in waveguide RG-52/U from 0.5″ to 0.30–0.25″, modes 1–4 are shifted up to the center of the waveguide pass-band and modes 5–8 are eliminated. If now the performances of ferrites of 0.25″ diameter and 0.29″ diameter are compared in Fig. 5, we see not only a simplification in the mode pattern (see Fig. 4) but also a possible source of confusion. The 0.29″ ferrite operates in mode 2 over a similar frequency band to the 0.25″ ferrite operating in modes 1 and 1A. That is to say, these two similar configurations have similar performances over a common frequency band but in opposite directions. Therefore, the two performances can only be compared properly when accompanied by a knowledge of the mode structure. The approximate figures of frequency shift per 0.001″ change in ferrite and brass diameter when operating in mode 1 are given in Table I. They were obtained by plotting the mode 1 characteristic for various ferrite diameters with a constant brass diameter and by plotting the characteristics for various brass diameters with a constant ferrite diameter. Then, at a level of constant applied field where the circulation was optimum, a line was drawn across each set of characteristics intersecting each mode at a different frequency. The graph of these frequencies vs their associated ferrite or brass diameters was in general a straight line, the slope of which gave the frequency shift per 0.001″ change in diameter.

By adjusting the composite post to obtain the desired center frequency in mode 1 or 1A, 4-port circulators with the half-height configuration have been completed in the frequency ranges of around 3, 9 and 22 Gc. The performance at 3 Gc is shown in Fig. 1. In realizing a practical device, it may be possible to shift the desired mode to the required frequency by adjusting only the ferrite or the brass diameter, but in general it is necessary to adjust both in order to optimize both the isolation and the VSWR. Also, it may not be sufficient to consider only the shift of the modes along the frequency axis. It is important to consider the position of the optimum operating frequency within each mode. The position of the best performance on the characteristics may change as the modes are shifted along the frequency axis. Although a characteristic may be shifted up in frequency by decreasing the diameter of the composite post, the optimum position on the characteristic may be moved down by a larger amount. Thus, decreasing the diameter may lower the optimum frequency instead of increasing it.

A circulator using polycrystalline yttrium iron garnet has been operated successfully in liquid nitrogen. The insertion loss at 77°K was not significantly different to that at 293°K, provided that the applied magnetic field was readjusted. The magnitude of the insertion loss depended upon the mode of operation. Modes 1 and 2 appeared to be promising with losses of less than 0.40 db, *i.e.*, a noise contribution of less than 10°K at 77°K over tunable bandwidths of 10 per cent.

III. The Full-Height Configuration

With this configuration, shown in Fig. 2(b), the brass post is the full height of the waveguide and is surrounded by a closely fitting ferrite tube of the same height. The static magnetic field is again applied perpendicular to the broad walls. Since it can now be assumed that the microwave fields do not vary in this direction, this type of 4-port circulator is more amenable to theoretical solution. The structure has other advantages over the half-height configuration: the brass may be used for pinning the ferrite and it may be hollow to pass a coolant. It would also be less prone to high peak-power breakdown, particularly if the top and bottom and inner surfaces of the ferrite tube are plated integrally with the waveguide junction.

To obtain circulation at any specified frequency, the three variables were again the ferrite, brass diameter and the applied magnetic field. Using the technique described earlier for the half-height configuration, the mode chart shown in Fig. 6(a) was established for a circulation in RG-52/U *X*-Band. As before, all modes in one sense have even numbers and in the other sense, odd numbers. Keeping the brass diameter constant, 0.10″, and increasing the ferrite diameter from 0.3″ to 0.40″, modes 5 and 3 are shifted to lower frequencies until they are eliminated as shown in Fig. 6(b) and (c). With a ferrite diameter of 0.45″, mode 4 is introduced at the higher frequencies, as shown in Fig. 6(d). Good

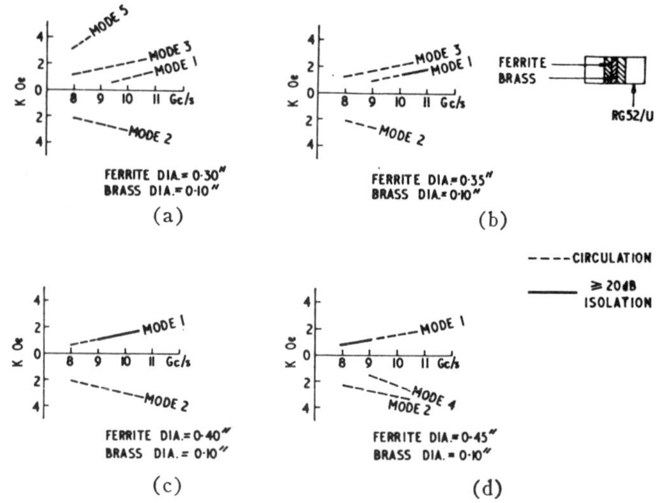

Fig. 6—The mode charts for "full-height" configurations. Mn-Mg ferrite, $4\pi M_s = 2200$ gauss. Heavy lines indicate isolation >20 db. Brass diameter, $d_b = 0.10$." Ferrite diameter (d_f) in (a) 0.30″, (b) 0.35″, (c) 0.40″ (d) 0.45″.

circulation can be obtained in mode 1 (VSWR∼1.7) with ferrite diameters 0.3–0.4″ or in mode 3 (VSWR ∼1.9) with ferrite diameters near 0.2″. Mode 1 has the advantage that smaller applied field values are required. The regions of circulation in each mode move from higher to lower frequencies as the ferrite diameter is increased. If the brass diameter is increased with the ferrite diameter held constant, the effect is equal and opposite. The operating frequency increases. That is to say, it is the thickness not the diameter of ferrite which governs the circulating frequency. With manganese-magnesium ferrite, a 0.32″ diameter ferrite with 0.050″ diameter brass post has a performance similar to a 0.38″ diameter ferrite with a 0.100″ brass post. The operating frequency shifts −20 Mc per 0.001″ change in diameter. It is surprising that a specific increase (or decrease) in the ferrite thickness produces approximately the same frequency shift whether it occurs on the inside or outside radius. With ferrite posts larger than about 0.50″ outer diameter, the mode chart becomes complex; up to eight modes have been identified but circulation in mode 1 has remained appreciably better than in any other mode. In general, the larger the brass diameter, the larger the reflection coefficient. The usefulness of decreasing the brass diameter to lower the optimum frequency appears to reach a limit with a diameter of approximately 0.030″. Reducing the brass diameter further does very little to shift the center frequency, but if it is removed completely the performance deteriorates abruptly.

The performance of an unmatched "full-height" 4-port circulator is shown in Fig. 7. The broader band isolation provides isolation > 20 db over the frequency range 9.3–10.0 Gc. The ferrite material is nickel-manganese ferrite, $4\pi M_s = 2800$ gauss. If the VSWR is re-

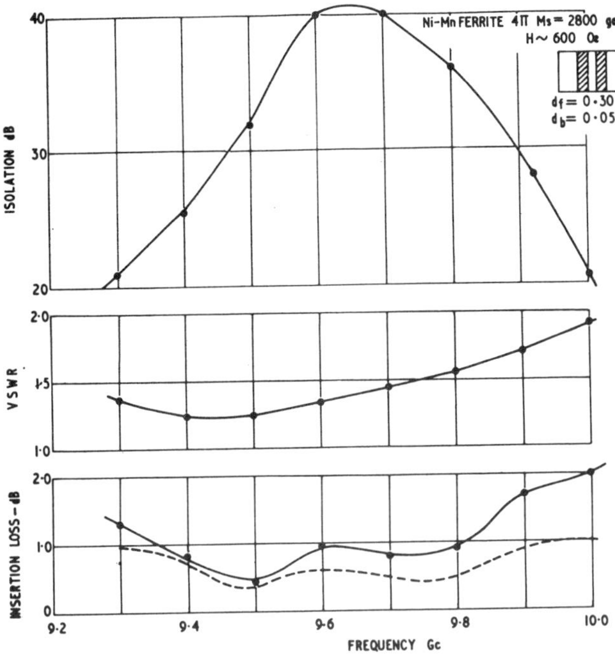

Fig. 7—The performance of an unmatched 4-port circulator. The isolation curve shows an 8 per cent bandwidth. The realizable insertion loss after matching is shown by the broken line.

duced by external matching elements until it makes a negligible contribution to the insertion loss as described in the introduction, the realizable insertion loss with this material will be approximately as shown by the dotted curve. It should be emphasized that this theory applies only to a near-perfect circulator. It has been found in practice that, when a 4-port circulator with a VSWR ≥ 1.5 is matched, the improvement in insertion loss is rather better than that predicted by this theory.

Since the correct adjustment of three variables is required in a symmetrical 4-port junction for circulation at a particular frequency, at least one more variable must be introduced for operation over a broader band. The parameters which have been introduced in turn are the height of the ferrite and an additional coaxial tube of dielectric around the ferrite. Reducing the height of the ferrite shifted the mode curves to higher levels of magnetic field, due to the increased demagnetizing factor. The addition of dielectric (up to $K=4$) around the ferrite lowers the frequency of circulation as shown in Fig. 8, but the effect upon the bandwidth is not yet clear, since the brass, ferrite and dielectric diameters should be optimized with each configuration.

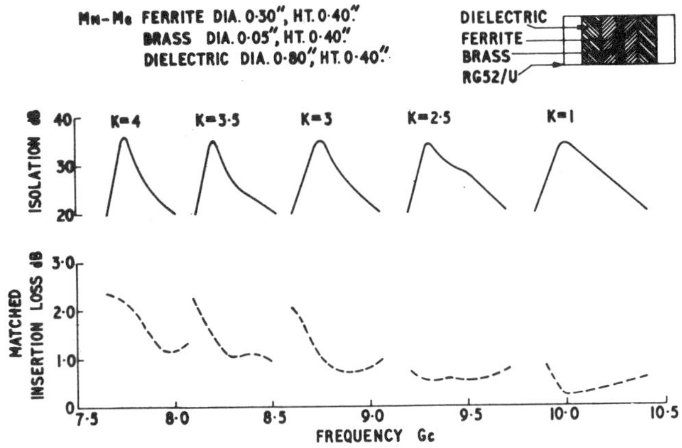

Fig. 8—The effect of loading the circulator with coaxial dielectric cylinders.

IV. Conclusions

The circulator action of a symmetrical ferrite-loaded 4-port H-plane waveguide junction has been demonstrated. The complex behavior of these devices has been clarified by establishing the modes of circulation within a waveguide pass-band. These modes depend entirely on the particular configuration of ferrite and conducting material at the center of the junction and a technique by which these modes may be distinguished has been described in detail. Two composite ferrite-brass posts have been investigated and their mode charts have been established. Design data have been compiled which enable these circulators to be operated in their optimum modes at frequencies about 9 Gc with the "full-height" configuration and at frequencies about 3, 9 and 22 Gc with the "half-height" configuration.

It has been established that the bandwidth of a 4-port junction circulator depends upon the bandwidth of the isolation curve. The isolation is substantially independent of the reflection coefficient; thus the bandwidth is entirely governed by the properties of the central composite ferrite post. The broadest bandwidth obtained with the two 4-port configurations described above is 8 per cent. If this can be improved, these devices offer the advantage of compactness where otherwise two 3-port circulators would be required.

V. Acknowledgment

The authors would like to thank J. Sapp who made many of the waveguide junctions and C. V. George who ground the ferrite material.

Operation of the Ferrite Junction Circulator

C. E. FAY FELLOW, IEEE, AND R. L. COMSTOCK MEMBER, IEEE

96

Abstract—The operation of symmetrical circulators is described in terms of the counter-rotating normal modes (fields varying as exp $n\phi$) of the ferrite-loaded circuits. The rotating modes, which are split by the applied magnetic field, form a stationary pattern which can be rotated in space to isolate one of the ports of the circulator. A detailed field theory of the strip-line Y-junction circulator operating with $n=1$ is presented. Experiments designed to confirm the validity of the rotating normal mode description of circulator action in the Y-junction circulator also are presented; these include measurements of mode frequencies and electric field patterns. The results of the field theory are used in a design procedure for quarter-wave coupled strip-line circulators. The results of the design procedure are shown to compare adequately with experimental circulators. Higher mode operation of strip-line circulators is described. The operation of waveguide cavity circulators is shown to depend on the rotating ferrite-loaded cavity modes.

Introduction

A VERSATILE MICROWAVE DEVICE which is becoming one of the most widely used is the ferrite junction circulator. Its versatility is indicated by the fact that in addition to its use as a circulator, it also can be used as an isolator or as a switch. The three-port version of the ferrite junction circulator, usually called the Y-junction circulator, is most commonly used. It can be constructed in either rectangular waveguide or strip line. The waveguide version is usually an H-plane junction, although E-plane junction circulators also can be made. The strip-line ferrite junction circulator is usually made with coaxial connectors and is principally applicable to the UHF and low-microwave frequencies.

Manuscript received July 13, 1964; revised September 8, 1964.
C. E. Fay is with Bell Telephone Labs., Inc., Murray Hill, N. J.
R. L. Comstock is with Lockheed Missiles and Space Co., Research Labs., Palo Alto, Calif., He was formerly with Bell Telephone Labs., Inc., Murray Hill, N. J.

The ferrite junction circulator was in use for a number of years before its theory of operation received much attention in the literature. Early experimenters found that waveguide T junctions having a transversely magnetized ferrite slab suitably placed in the junction could, with proper matching and adjustment of the magnetic field, be changed into circulators. The bandwidth of such devices was very narrow. Refinements producing greater symmetry were found to increase bandwidth so that useful devices were obtained [1]–[4].

More recently a number of papers have appeared in the literature which bear on the theory of operation of the ferrite junction circulator. Auld [5] has considered the theory of symmetrical junction circulators in terms of the scattering matrix of the device. He has shown the necessary relations of the eigenvectors of the matrix, and has indicated how these relations may be obtained. Milano, Saunders, and Davis [6] have applied these concepts in the design of a Y-junction strip-line circulator. Bosma [7], [8], has made an analysis of the strip-line Y-junction circulator in terms of the normal modes of the center disk structure. In his second paper, he shows that the circulation condition is near a degeneracy of a pair of resonances of the disk structure. Butterweck [9] has considered the case of the waveguide junction circulator and has given the waveguide equivalent of Bosma's explanation. Others have attempted explanations based on field displacement, scattering from a post, or surface waves. All of these, if used with proper boundary conditions, conceivably could lead to the same conclusions.

This treatment will consist of an extension of Bosma's approach to the problem. First, we shall present a phenomenological description of the operation of the strip-line Y-junction circulator followed by a development of the field theory applicable to this form of circulator, and then some experimental evidence of the validity of the approach will be submitted. Then, the theory is applied to provide a detailed design procedure for quarter-wave coupled strip-line circulators. Finally, we shall discuss operation in higher modes, operation of a four-port junction, and operation in rectangular waveguide.

I. Phenomenological Description

The strip-line Y-junction ferrite circulator consisting of two ferrite cylinders filling the space between a metallic conducting center disk and two conducting ground plates presents the simplest geometrical arrangement and, therefore, is the easiest to treat analytically. We shall present a development of the strip-line Y-circulator starting from more familiar concepts. This circulator in its basic form is illustrated in Fig. 1. The connections to the center disk are in the form of three strip-line center conductors attached to the disk at points 120° apart around its circumference. A magnetizing field is applied parallel to the axis of the ferrite cylinders.

It was found, experimentally, that the Y-circulator had some but not all of the properties of a low-loss transmission cavity. At its resonant frequency, it was well matched but slightly undercoupled, and a standing wave existed in the structure. The maximum isolation occurred almost at the frequency at which the insertion loss was minimum. The isolation at the third port and the return loss from the input port correspond quite well as the frequency is changed. The above experimental evidence suggests a resonance of the center disk structure as being an essential feature of the operation of the circulator. The lowest frequency resonance of the circular disk structure of Fig. 1 is the dipolar mode in which the electric field vectors are perpendicular to the plane of the disk and the RF magnetic field vectors lie parallel to the plane of the disk. This mode, as excited at port 1, is illustrated in the unmagnetized case by the standing-wave pattern of Fig. 2(a). In the case of an isolated disk, the RF H lines would curve over the edge of the disk and continue back on the underside. With ferrite cylinders on each side of the disk, the bottom cylinder behaves as a mirror image of the top one, and the analysis of the device need only be concerned with one cylinder. Ports 2 and 3, if open-circuited, will see voltages which are 180° out of phase with the input voltage, and about half of the value of the input voltage. If the standing-wave pattern is rotated, as in Fig. 2(b), then port 3 is situated at the voltage null of the disk and the voltages at ports 1 and 2 are equal. The device is equivalent to a transmission cavity between ports 1 and 2, and port 3 is isolated. The standing-wave pattern of Fig. 2(b) in which all fields vary as $e^{j\omega t}$ can be generated by two counter-rotating field patterns of the same configuration. Each of these patterns would involve an

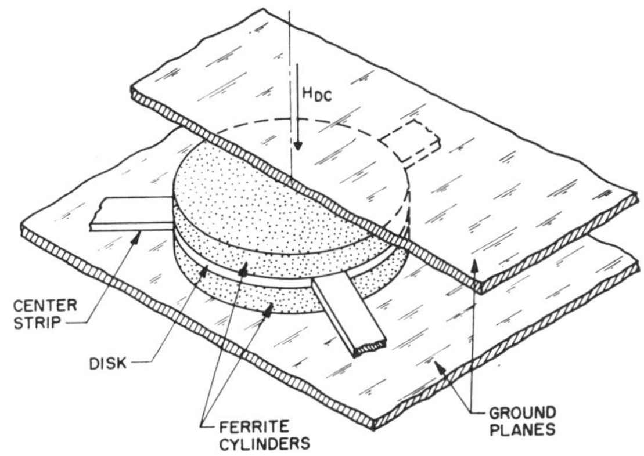

Fig. 1. Strip-line Y-junction circulator.

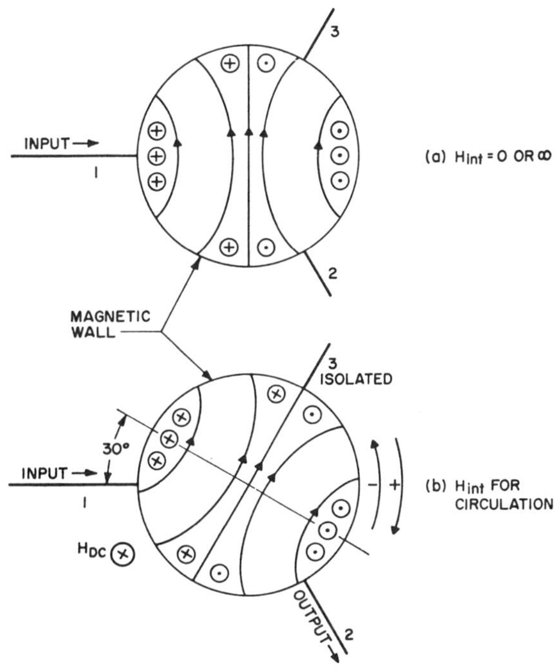

Fig. 2. (a) Dipolar mode of a dielectric disk. $H_{int}=0$ or ∞. (b) Analogous pattern of a magnetized disk. H_{int} for circulation. The pattern for the magnetized disk has been rotated to isolate port 3. The magnetic fields of higher modes than the dipolar mode are needed to achieve the pattern shown in (b).

RF magnetic field pattern which is circularly polarized at the center of the disk, becomes more elliptical as the radius increases, and is linearly polarized at the edge of the disk. If a magnetic biasing field (H_{dc}) is applied in the direction of the axis of the disk, the two counter-rotating patterns are no longer resonant at the same frequency. It is necessary when considering circulator action in the magnetized disk to include all of the normal modes, of which the dipolar field is but one, especially the nonrotating ϕ-independent mode ($n=0$ in the notation of Section II). This is the case since, as shown by Auld [5], circulator action must involve at least three normal modes. For the magnetized disk, the magnetic field pattern is a net pattern composed of many modes (the mode components of this pattern

97

will be derived under specialized assumptions in Section II). The pattern rotating, in the same sense as the electron spins of the ferrite cylinder tend to precess under the influence of the biasing field, will have an effective scalar permeability which we shall call μ^+ at the center of the cylinder, and an effective permeability μ_e for linear polarization at the edge of the disk. The pattern rotating in the opposite sense from the electron spin precession will have an effective scalar permeability μ^- at the center of the cylinder, and an effective permeability μ_e at the edge of the cylinder. Accordingly, we shall designate these split modes as "+" (plus) or "−" (minus) depending on their sense of rotation compared to that of the electron spin precession. If we excite the system of Fig. 2(b) at a frequency intermediate between the resonant frequencies of the split modes, the impedance of the + mode, which has the higher resonant frequency, will have an inductive reactance component, and that of the − mode, which has the lower resonant frequency, will have a capacitive reactance component. If the frequency chosen is such that the capacitive reactance component of one equals the inductive reactance component of the other, the total impedance will be real at the operating frequency f_0, Fig. 3. If the degree of splitting is adjusted such that the phase angles of the impedances of the two modes are each 30° ($|X/R| = \tan 30°$) at the operating frequency, then the standing-wave pattern will be rotated 30° from that which obtains with no splitting of the modes. This is illustrated in Fig. 2(b). The + mode, which has the inductive reactance component, will have its voltage maximum leading the current (or H-field) maximum at the input port by 30° in time phase, which is also 30° in space phase since for this mode there is one revolution per cycle. The − mode will have its voltage maximum lagging the current maximum at the input port by 30°. Therefore, the two voltage maxima will coincide at a point 30° away from the input port as shown in Fig. 2(b). The rotation of the *pattern* is in the direction of rotation of the + mode. This 30° rotation brings the E-field null of the pattern to port 3 so that no voltage exists at this port. However, for the power entering at port 1, the device acts as a transmission cavity with power leaving at port 2. Thus, the direction of circulation is the direction of rotation of the − mode for $H_{dc} < H_{res}$ (ferromagnetic resonance field); for $H_{dc} > H_{res}$, the circulation is in the opposite direction. Since the device has complete symmetry, any power entering at port 2 will set up a new set of similar modes which will result in this power leaving at port 3, and port 1 will be isolated. Similarly, power entering at port 3 is transmitted to port 1, and port 2 is isolated.

The sum of the impedances of the two counter-rotating modes gives the impedance of the stationary mode. This stationary resonance will have the same Q as each of its components, and will be the one to be considered when the loaded Q and the matching of the circulator are discussed. It is shown in the next section that the approximate equivalent circuit of the resulting stationary mode is a shunt resonator as indicated in Fig. 3.

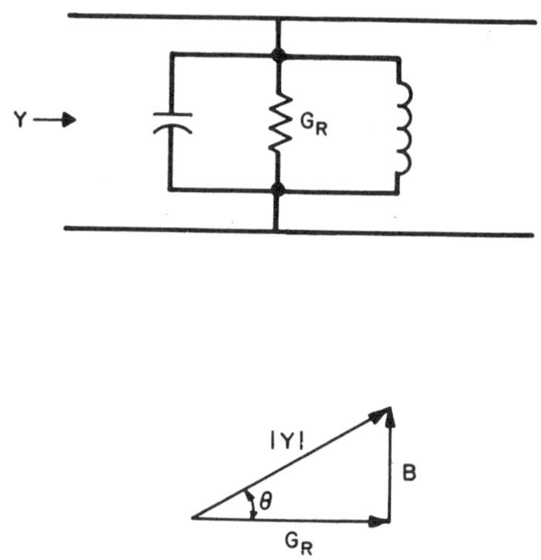

Fig. 3. Lumped element equivalent circuit of the stationary mode resonator.

We assume the resonator is lossless in this analysis so that the loaded Q, Q_L is described completely by the stored energy in the resonator and the power radiated into the connecting striplines. The field patterns of Fig. 2 are only those of the standing wave. The power transfer must be through a traveling wave whose fields are such as to result in an outgoing Poynting flux at port 2. In transmission cavities of large Q, the fields of the standing wave predominate. However, as Q_L becomes small this picture becomes inaccurate but still useful as will be shown by experimental results.

II. Field Theory of the Ferrite Junction Circulator

In the preceding section, a phenomenological description of circulator operation was presented. Here we will use the Maxwell equations, and the equations of motion of the magnetization to show the validity of the phenomenological picture and to arrive at some useful results relative to the design of this class of circulators. The boundary value problem which we shall discuss is the same as that considered by Bosma [7]. The circulator geometry consists of two ferrite cylinders separated by a disk center conductor fed by symmetrical transmission lines (see Fig. 1). The analysis is expected to predict the behavior of many similar circulators, e.g., the waveguide junction with a ferrite post.

Transverse electric (TE) waves with no variation along the dc bias field are considered. For these waves the electric field in the ferrite ($\boldsymbol{E} = \boldsymbol{a}_z E_z$) satisfies the homogeneous Helmholtz equation in cylindrical coordinates

$$\left[\frac{\partial^2}{\partial r^2} + \frac{1}{r}\frac{\partial}{\partial r} + \frac{1}{r^2}\frac{\partial^2}{\partial \phi^2} + k^2\right] E_z = 0 \qquad (1)$$

with time dependence $e^{j\omega t}$ and

$$k^2 = \omega^2 \epsilon\epsilon_0 \mu_{\text{eff}} \mu_0 = \omega^2 \epsilon_0 \epsilon \mu_0 (\mu^2 - \kappa^2)/\mu \quad (2)$$

where μ and κ are the Polder tensor components appropriate to a normally magnetized disk [10]. Equation (1) has solutions, depending on the value of n, given by

$$E_{zn} = J_n(x)(a_{+n}e^{jn\phi} + a_{-n}e^{-jn\phi}) \quad (3)$$

where

$$x = kr$$

Maxwell's equations then give for the ϕ-component of $\vec{H}_n$

$$H_{\phi n} = jY_{\text{eff}} \left\{ a_{+n}e^{jn\phi} \left[J_{n-1}(x) - \frac{nJ_n(x)}{x}\left(1 + \frac{\kappa}{\mu}\right) \right] + a_{-n}e^{-jn\phi} \left[J_{n-1}(x) - \frac{nJ_n(x)}{x}\left(1 - \frac{\kappa}{\mu}\right) \right] \right\} \quad (4)$$

with

$$Y_{\text{eff}} = \sqrt{\frac{\epsilon\epsilon_0}{\mu_0 \mu_{\text{eff}}}}$$

As stated in the preceding section, it has been found useful to treat the circulator as a tightly coupled resonator. With this in mind, it is then of interest to find the character of the normal modes in the uncoupled case ($\psi = 0°$, where 2ψ is the angle subtended by the strip lines, Fig. 4). Even in the uncoupled case, it is difficult to record the exact boundary conditions. Bosma [8] has argued that the normal modes resulting from the boundary condition given by

$$H_\phi(r = R) = 0 \quad (5)$$

are sufficient for describing the field in the disk in the uncoupled case (Fig. 4 for $\psi = 0°$). Justification for this boundary condition is the exponential falloff of fields outside the disk boundary. The normal modes, which result when the boundary condition given by (5) is used with (4), have resonant frequencies given by the roots of

$$+ \text{ mode}, \quad J_{n-1}(kR) - \frac{nJ_n(kR)}{kR}\left(1 + \frac{\kappa}{\mu}\right) = 0$$

$$- \text{ mode}, \quad J_{n-1}(kR) - \frac{nJ_n(kR)}{kR}\left(1 - \frac{\kappa}{\mu}\right) = 0 \quad (6)$$

Several of the roots of these equations have been given by Bosma [8] in the form $(kR)_{m,n}^{\pm}$ vs. κ/μ, where m is the order of the root for a given value of n. For fixed value of m and n, there are two modes whose frequency separation depends on κ/μ and for small κ/μ is proportional to κ/μ. Fig. 5 shows a curve of κ/μ vs. σ where $\sigma = |\gamma| H_{\text{int}}/\omega$ with $p = |\gamma| 4\omega M_s/\omega$ as a parameter (H_{int} is the internal field and $4\pi M_s$ is the saturation induction). From an inspection of these curves, it is seen that $|\kappa/\mu|$ and hence the splitting is largest for a fixed value of p when $\sigma < 1$. This is shown, experimentally, to be the case in Section III. Also, it is evident that the splitting will vary rapidly with p (and hence $4\pi M_s$) in this region. The lowest root of (6) will be arrived at graphically after equations are derived relating to the coupled disk.

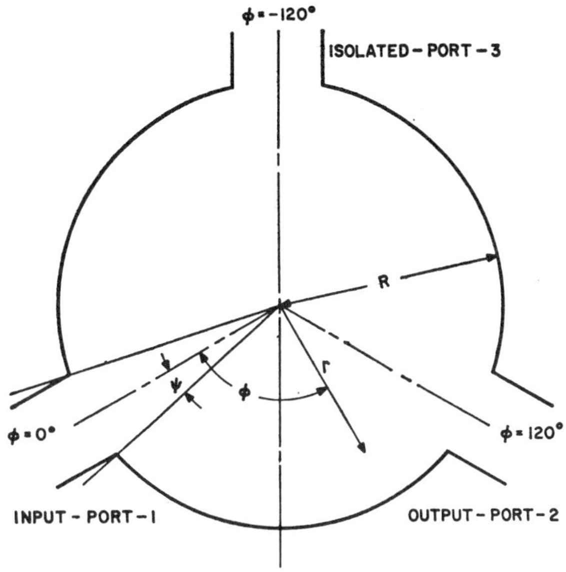

Fig. 4. Schematic diagram of strip-line circulator.

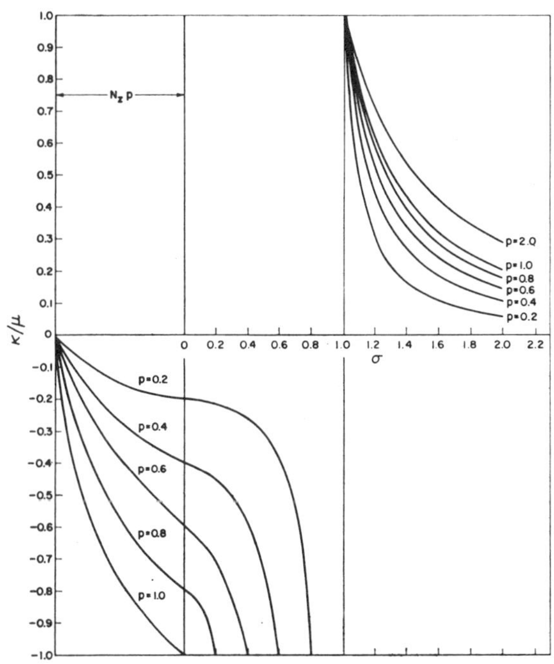

Fig. 5. κ/μ vs. σ, where $\sigma = |\gamma| H_{\text{int}}/\omega$.

The procedure we will use for treating the coupled case ($\psi \neq 0$), where the normal modes are driven by the TEM waves on the strip lines, is to assume initially that the system is functioning as a circulator and then to find the conditions on the rotating normal modes to satisfy the assumption. The conditions for circulator action are perhaps most easily described in terms of the scattering matrix [5] but, in the field theory discussed here, it is necessary to state the conditions in terms of the electric and magnetic fields. We follow Bosma [7] in assuming H_ϕ to be constant over the width of the strip lines, and take for the boundary conditions at $r=R$

$$\left.\begin{aligned} -\psi<\phi<\psi, &\qquad H_\phi = H_1\\ 120^\circ-\psi<\phi<\psi+120^\circ, &\qquad H_\phi = H_1\\ -120^\circ-\psi<\phi<\psi-120^\circ, &\qquad H_\phi = 0\\ \text{elsewhere}, &\qquad H_\phi = 0\end{aligned}\right\} \tag{7}$$

and

$$\left.\begin{aligned} \phi = 0, &\qquad E_z = E_1\\ \phi = 120^\circ, &\qquad E_z = -E_1\\ \phi = -120^\circ, &\qquad E_z = 0\end{aligned}\right\} \tag{8}$$

It is necessary for both E_z and H_ϕ to be zero at the isolated port ($\phi=-120^\circ$) since the boundary condition there is an admittance boundary condition, i.e., the ratio of H_ϕ to E_z at this port must be the wave admittance.

To compute the complete function $E_z(\phi, R)$, it is assumed as an approximation that E_z has a sinusoidal
100 distribution with one period around the periphery of the disk, i.e., only the $n=1$ mode is included for the electric field. The justification for this approximation is twofold: First, the amplitudes of the electric fields of the higher modes will be small since only the $n=1$ mode is near resonance; and second, measurements of the electric field described in Section III conform closely to such a sinusoidal distribution, at least for narrow-band circulators. Equation (3) with the boundary conditions (8) then gives for the mode amplitudes

$$a_+ = \frac{E_1}{2J_1(kR)}\left(1+\frac{j}{\sqrt{3}}\right)$$

and

$$a_- = \frac{E_1}{2J_1(kR)}\left(1-\frac{j}{\sqrt{3}}\right) \tag{9}$$

so that the total electric field is in the form of a sinusoidal standing wave

$$E_z = E_1\frac{J_1(kr)}{J_1(kR)}\left(\cos\phi - \frac{\sin\phi}{\sqrt{3}}\right) \tag{10}$$

For the magnetic field, we expand the function specified in (7) into a Fourier series, as was done previously by Bosma [7], so that

$$H = H_1\left[\frac{2\psi}{\pi} + \sum_{n=1}^{\infty}\frac{\sin n\psi}{n\pi}\cos n\phi + \sqrt{3}\frac{\sin n\psi}{n\pi}\sin n\phi\right] \tag{11}$$

which for the $n=1$ mode can be written

$$H_{\phi 1} = H_1\frac{\sin\psi}{2\pi}\left[(1-j\sqrt{3})e^{j\phi} + (1+j\sqrt{3})e^{-j\phi}\right] \tag{12}$$

However, the $n=1$ component of H does not satisfy the circulation conditions. These conditions are satisfied only by the entire set of modes with the amplitudes given above.

Another solution for $H_{\phi 1}$ can be obtained from (4) by setting $n=1$ and using the values of the coefficients a_+ and a_- given in (9). The result

$$\begin{aligned} H_{\phi 1} = jY_{\text{eff}}\frac{E_1}{2J_1(kR)} &\cdot\left\{\left(1+\frac{j}{\sqrt{3}}\right)\left[J_0(kR) - \frac{J_1(kR)}{kR}\left(1+\frac{\kappa}{\mu}\right)\right]e^{j\phi}\right.\\ &\left.+\left(1-\frac{j}{\sqrt{3}}\right)\left[J_0(kR) - \frac{J_1(kR)}{kR}\left(1-\frac{\kappa}{\mu}\right)\right]e^{-j\phi}\right\}\end{aligned} \tag{13}$$

is only compatible with (12) if

$$J_0(kR) - \frac{J_1(kR)}{kR}\left(1-\frac{\kappa}{\mu}\right) = -\left[J_0(kR) - \frac{J_1(kR)}{kR}\left(1+\frac{\kappa}{\mu}\right)\right]$$

or

$$J_0(kR) = \frac{J_1(kR)}{kR} = 0 \tag{14}$$

of which the first root is

$$(kR)_{1,1} = 1.84 \tag{15}$$

This result is exact under the present boundary conditions and not approximate as assumed by Bosma [8]. By comparing (15) with the normal mode resonances in the uncoupled case (6), it is seen that $(kR)_{1,1}$ lies somewhere between $(kR)_{1,1}{}^+$ and $(kR)_{1,1}{}^-$. The exact value of operating frequency must be determined from a solution of (15) with k given by (2). In Fig. 6 is shown a graph of μ_{eff} vs. σ with p as a parameter for a thin disk. This curve is useful in conjunction with (15) in evaluating the disk radius R, given M_s, H_0, and ω. To illustrate the relationship of the operating frequency to the normal mode resonant frequencies, we show in Fig. 7 a plot of (6) ($n=1$) and (14) vs. kR.

One condition for circulation is that the operating frequency lie between the resonant frequencies for the $n=\pm 1$ modes. This is not the only condition since it is also required that the input admittance at port 1 and the output admittance at port 2 be adjusted to match the admittance at the terminals. Equating the values of $H_{\phi 1}$ given by (12) and (13), and taking account of the condition given by (14), leads to an expression for the input wave admittance of the circulator at resonance

$$Y_w = G_R = \frac{H_1}{E_1} \approx \frac{Y_{\text{eff}}\,|\kappa/\mu|}{\sin\psi} \tag{16}$$

which can be reduced to (69) of Bosma [7] for small ψ. The loaded Q of the circulator can be derived from the

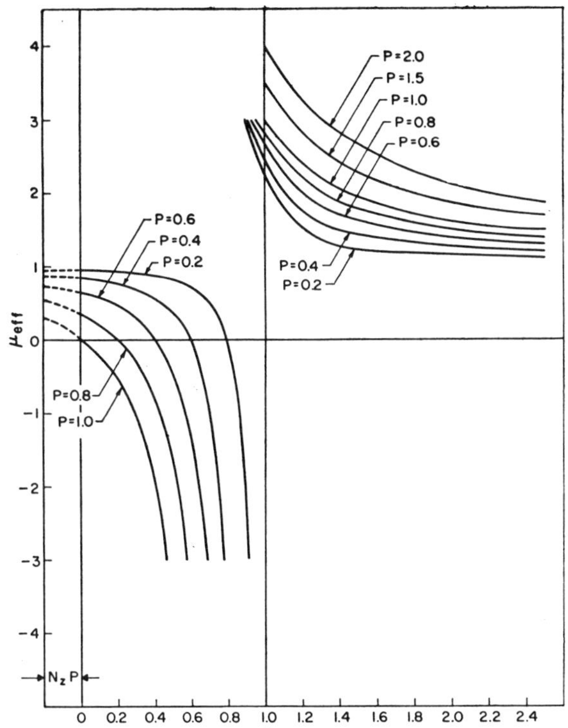

Fig. 6. μ_{eff} vs. σ

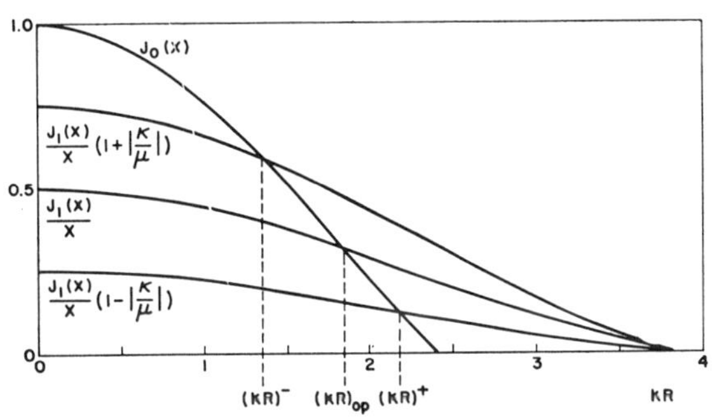

Fig. 7. Graphical solution of (9) and (17) for $n=1$. The operating frequency ($kR=1.84$) lies between the two normal mode resonances $(kR)^{\pm}$.

fields given in this section by starting from the definition

$$Q_L = \frac{\omega U}{P_{out}} \tag{17}$$

where U is the stored energy in the two cylinders and P_{out} is the power radiated out the strip lines. For U, we have

$$\begin{aligned} U &= \frac{\epsilon\epsilon_0 d}{J_1^2(kR)} \int_0^{2\pi} \int_0^R E_m^2 J_1^2(kr) \cos^2 \phi' r dr d\phi' \\ &= 1.11\, E_m R^2 d\epsilon\epsilon_0 \end{aligned} \tag{18}$$

where E_m is the maximum electric field at the periphery of the disk, d is the cylinder height, and ϕ' has its zero reference at $\phi = -30°$. The total radiated output power is given by

$$P_{out} = d^2 E_m^2 G_R \cos^2 30° \tag{19}$$

where G_R is the conductance at the output strip lines and also the conductance looking into the resonator disk. Combining (17), (18), and (19), and evaluating the Bessel functions results in

$$Q_L = 1.48 \frac{\omega R^2 \epsilon\epsilon_0}{G_R d} \tag{20}$$

This result is physically satisfying because a large G_R is realized by making ψ large, which increases the coupling to the resonator.

It is possible to calculate the insertion loss of the ferrite cylinder and metallic disk circulator by calculating the unloaded Q, Q_0 of the structure as the result of including magnetic loss and dielectric loss. Since the expression for the resonant frequency and standing-wave pattern at the operating point involved only μ_{eff}, [(14) and (2)], if we assign a magnetic loss component to μ_{eff} so that

$$\mu_{eff} = \mu'_{eff} - j\mu''_{eff}$$

then the magnetic Q, Q_μ is just

$$\frac{\mu'_{eff}}{\mu''_{eff}}$$

From [11], we find

$$\mu_{eff} = \frac{1-(p+\sigma+j\alpha)^2}{1-(\sigma+j\alpha)(p+\sigma+j\alpha)} \tag{21}$$

where the loss has been introduced by substituting for σ, $\sigma+j\alpha$, where $\alpha=\gamma\Delta H/2\omega$ is the resonance linewidth in the usual reduced units. Separating the real and imaginary parts of (21) we find, taking their ratio and neglecting terms in α^4, that

$$\begin{aligned} Q_\mu &= \frac{\mu'_{eff}}{\mu''_{eff}} \\ &= \frac{(p\sigma+\sigma^2-1)[(p+\sigma)^2-1]+\alpha^2(2\sigma^2+3\sigma p+p^2+2)}{\alpha p[(p+\sigma)^2+1]} \end{aligned} \tag{22}$$

In the case of $\sigma=0$, this expression reduces to

$$Q_\mu = \frac{1-(\kappa/\mu)^2}{[1+(\kappa/\mu)^2]\alpha\kappa/\mu} \tag{23}$$

For most circulators operated below resonance, p and κ/μ are $\ll 1$ and the unloaded Q can be approximated by

$$Q_\mu = \frac{2\omega^2}{\gamma^2 4\pi M_s \Delta H} \tag{24}$$

Likewise, the dielectric loss can be taken into account if we know the dielectric loss tangent of the ferrite. For this we have $Q_\epsilon = \epsilon'/\epsilon'' = 1/\tan\,\delta$. Then the total unloaded Q, Q_0 neglecting conductor losses is given by

$$\frac{1}{Q_0} = \frac{1}{Q_\mu} + \frac{1}{Q_\epsilon} \tag{25}$$

The insertion loss in dB is then, at band-center

$$\text{IL (dB)} = 10 \log_{10}\left(1 - \frac{Q_L}{Q_0}\right) \tag{26}$$

This represents a lower limit since the loss is assumed to be due to the fields of the standing wave only, and the conductor losses are not included.

The analysis up to this point has resulted in equations for the loaded Q of a single resonator. To relate the properties of this resonator to those of the two rotating normal modes, we express the electric field from (3) in terms of the normal mode fields, at $r=R$.

$$E_z = E^+ e^{-j\phi} + E^- e^{j\phi} \tag{27}$$

where, from (9)

$$E^\pm = \frac{E_1}{2}\left(1 \pm \frac{j}{\sqrt{3}}\right)$$

When both sides of (27) are divided by H_1, the driving magnetic field, and evaluated at $\phi=0$, we find expressions for the impedances for the two normal modes at port 1 at the operating frequency ω_{op}

$$Z^\pm = \frac{1}{2G_R}\left(1 \pm \frac{j}{\sqrt{3}}\right) \tag{28}$$

The input impedance is given by

$$Z = Z^+ + Z^- = \frac{1}{G_R} \tag{29}$$

These impedances can be represented by the lumped constant resonators shown in Fig. 8(a). The resonant frequencies of the resonators ($\omega^\pm$) are given by the roots of (6). The impedances of the + and − resonators at ω_{op} have phase angles of $\pm 30°$ as described in Section I. The operating frequency is approximately midway between the + and − mode resonant frequencies, i.e.,

$$\omega_{op} \approx \frac{\omega^+ + \omega^-}{2} \tag{30}$$

A detailed analysis of this equivalent circuit shows that in the vicinity of band-center, the two resonant circuits can be replaced by the single shunt resonator shown in Fig. 3 with the conductance on resonance and Q determined by (16) and (20).

Since the individual rotating mode resonators have 30° phase angles at band-center, a definite relation exists between the splitting $2\delta' = (\omega^+ - \omega^-)/\omega$, and the loaded Q, i.e.,

$$2\delta' = \frac{\tan 30°}{Q_L} = \frac{1}{\sqrt{3}Q_L} \tag{31}$$

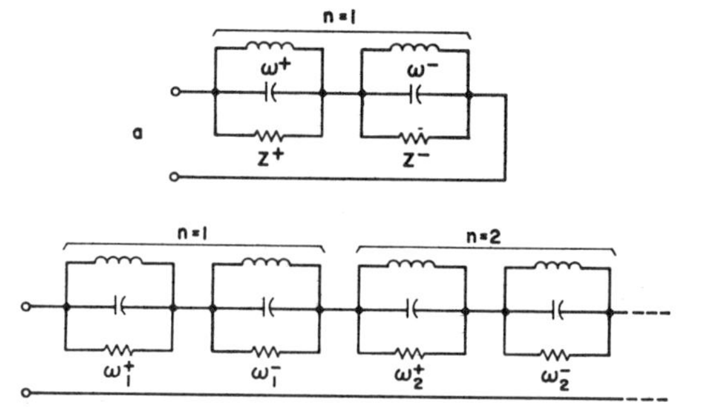

Fig. 8. Equivalent circuit of rotating normal modes ($\omega^\pm$). The $n=1$ mode resonances are shown in (a), and the more general circuit is shown in (b).

Equation (31) has been derived for the waveguide circulator by Butterweck [9] and is expected to be a general result for all three-port circulators which operate on the principle of the splitting of normal modes. A generalization of the above arguments for the higher order modes of the disk leads to the equivalent circuit shown in Fig. 8(b). In Sections III and IV it is shown that circulator action is possible between the resonances of the higher modes as represented in the equivalent circuit by the additional resonators. For example, if $\omega_{op} \approx (\omega_2^+ + \omega_2^-)/2$, the $n=1$ and all other modes are effectively short circuits, and we have the equivalent circuit similar to that shown in Fig. 8(a).

In Section IV, these results will be applied to the design of a stripline circulator. In the next section, some pertinent experimental results will be quoted.

III. Experimental

In this section, experimental evidence in support of the concepts developed in Sections I and II is presented. The experimental data have been taken on an L-band strip-line circulator which has dimensions large enough to allow probing of the electric fields at the edge of the center disk with sufficient mechanical accuracy so that relatively smooth plots are obtained.

A mode plot of the circulator is given in Fig. 9, which relates the mode resonances to the external field, H_{dc}. For the purposes of this plot, the center disk structure is lightly coupled to the 50-ohm strip lines by means of quarter-wave sections of high impedance line. This is done so that the various modes of the disk structure will be lightly loaded, and the resonances sharp and easily distinguished. The plot of resonant frequency as a function of applied magnetic field in Fig. 9 shows the $n=\pm 1$ modes, the $n=\pm 2$ modes, and parts of the $n=0$ and $n=+3$ modes. For circulator purposes, we are interested principally in the $n=1$ modes since these involve the smallest size of the disk structure for a given frequency and, therefore, presumably the least loss in the structure. The points of circulation are shown on the plot. These occur near the degeneracies of modes of like n. The sense of circulation, also indicated, is for the H_{dc} vector pointing toward the viewer. It will be noted that the sense of circulation for a given set of modes

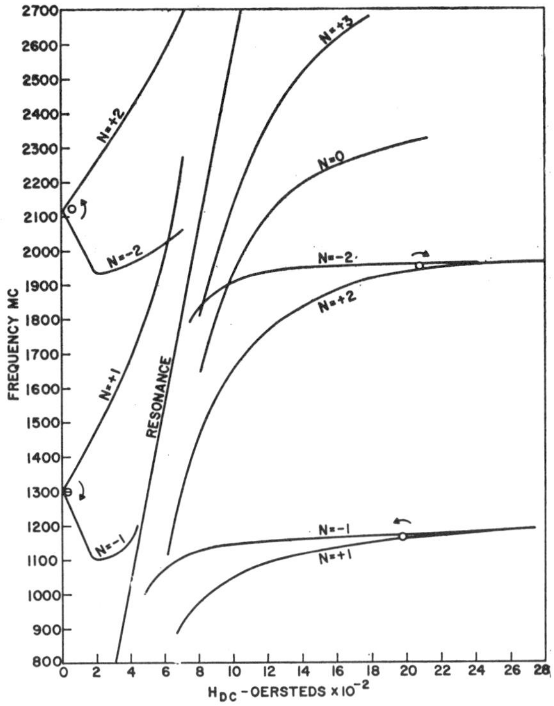

Fig. 9. Mode chart of experimental L-band circulator. Circulation is observed at the positions of the circles in the directions shown.

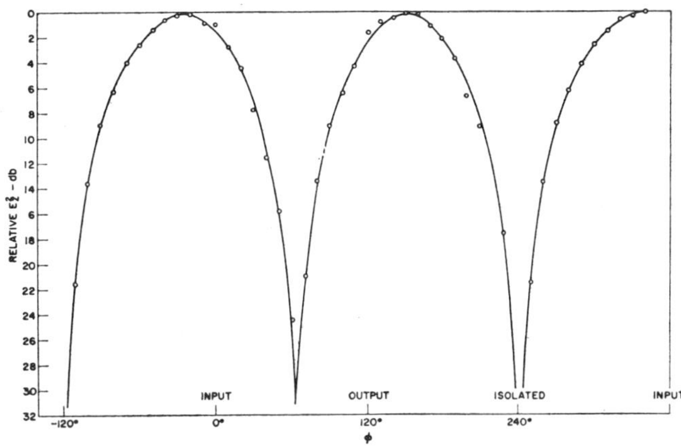

Fig. 10. Measured electric field at the periphery of a loosely coupled circulator. The measured curve is predicted accurately by the function 20 log (cos ϕ+sin ϕ/tan 120°).

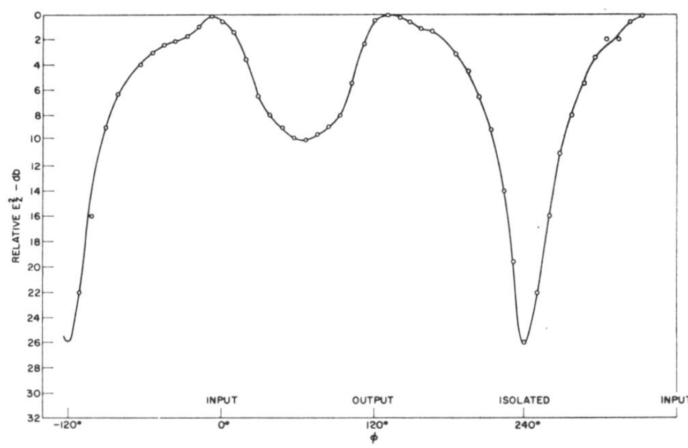

Fig. 11. Measured electric field at the periphery of a tightly coupled circulator.

above ferrimagnetic resonance is opposite to that for the same set of modes below ferrimagnetic resonance. The reason for this is that the − mode is the lower frequency one below resonance, and the higher frequency one above resonance. Therefore, the sign of the reactance of this mode at the operating frequency changes from the below resonance to the above resonance case, and hence, the sense of circulation will change if the reasoning given in Section I is followed. For the $n=2$ mode, the sense of circulation in each case is opposite to that of the corresponding $n=1$ mode case. The reason for this will be considered in the discussion of higher mode circulators.

A plot of the relative electric field, E_z, at the edge of the disk for the lightly coupled circulator is shown in Fig. 10. The circulator was adjusted for the below resonance point of circulation near the convergence of the $n=1$ modes, and the electric field probed at points around the circumference of the disk. The distribution is very nearly sinusoidal as predicted, with a slight distortion at the points where the high impedance feed lines are connected. The field maxima are seen to be approximately 30° away from the input and output ports. The circulator is very narrow-band in this case, and the required biasing field quite small, ~20 oersteds.

A more heavily coupled case is shown in Fig. 11. Here the same center structure is coupled to the 50 ohm strip lines by means of low impedance quarter-wave sections. This provides a more representative circulator which instead of being extremely narrow band, as in the previous case, has a>25 dB isolation bandwidth of approximately 15 per cent. The frequency of operation is nearly the same as before, but the required biasing field has increased to about 200 oersteds. This implies a greater separation or splitting of the modes as indicated in Fig. 9. The field plot of Fig. 11 shows a greater distortion in the vicinity of the connections to the center disk, as might be expected, since these connections are now considerably wider than in the previous case ($\psi=22°$). The other result to be noted is that the minimum in the electric field between the input and output ports is much shallower than previously. Aside from the distortions discussed above, it is obvious that the standing-wave pattern in this case is maintaining the 30° rotation predicted but is requiring a greater mode separation to do so. This greater mode separation is required in order to maintain the 30° admittance phase angles with the lower loaded Q of the modes which results from the heavier coupling to the strip lines.

IV. Y-Circulator Design Considerations

The design of a strip-line circulator employing the circular disk and ferrite cylinder form can be carried out using the principles discussed in Sections I and II.

The result will yield a reasonable approximation to the desired operating characteristics.

The use of the circulation condition between the resonances of the $n=1$ modes seems most desirable since this results in a minimum diameter of the metallic disk and ferrite cylinders. There seems to be no advantage in using the higher modes for a simple Y-circulator. The design will be based on the use of a quarter-wave transformer for coupling the circulator to the strip lines.

We must first choose the isolation bandwidth of the circulator. This is usually one of the things specified for the particular application of the circulator. Since the isolation of a Y-circulator usually corresponds quite closely to the return loss at the input port, and the return loss at the input port can be converted to VSWR (voltage standing-wave ratio), we define the following:

δ = fractional frequency deviation from f_0 (band-center) $= \frac{1}{2}$ bandwidth

VSWR = voltage standing-wave ratio at input port at frequencies $f=f_0(1\pm\delta)$

Y_R = admittance of disk resonator at $f=f_0(1\pm\delta)$

G_R = conductance of disk resonator

θ = phase angle of Y_R

Q_L = loaded Q of the disk resonator

δ' = fractional frequency splitting required for an admittance phase angle of 30° at the resonator

The resonator is assumed to have the equivalent circuit of Fig. 3. If a quarter-wave transformer or equivalent means is used for matching the resonator to the connecting strip lines, the susceptance of the transformer at $f=f_0(1\pm\delta)$ can be used to cancel that of the resonator at these frequencies. For these frequencies, it can then be shown that for small δ

$$\text{VSWR} \approx \frac{|Y_R|^2}{G_R^2} = \sec^2\theta \tag{32}$$

which allows us to determine θ. Then

$$Q_L = \frac{\tan\theta}{2\delta} \tag{33}$$

and we can find δ' from (31) using Q_L found previously. Since δ' relates to the amount of "splitting" required of the two $n=1$ modes, using (6), the value of κ/μ can be determined. Since the relation of δ' to κ/μ is approximately linear for small splitting, we find an approximate relation

$$\frac{\kappa}{\mu} = 2.46\delta' = \frac{0.71}{Q_L} \tag{34}$$

that is useful for design purposes. The value of the constant in the above equation is appropriate for values of κ/μ in the range 0.25 to 0.5 which should cover the most practical cases. Having found the value of κ/μ for the desired bandwidth, we find from Fig. 5 that we can obtain it with the biasing field in the ferrite either above or below that required for resonance, i.e., $\sigma>1$ for above resonance, or $\sigma<1$ for below resonance, where $\sigma=\gamma H_{\text{int}}/\omega$. Above resonance operation requires a somewhat smaller disk diameter, a higher saturation, p, where $p=\gamma 4\pi M_s/\omega$; and a higher biasing field, hence, a larger magnet. Below resonance operation requires a minimum field and a relatively low-saturation material, ($p<1$). Most Y-circulators in use at the present time employ the below resonance type of operation. For minimum field requirements it is convenient to operate at $\sigma=0$ which implies that the ferrite is just saturated or $H_{\text{dc}}\approx N_z 4\pi M_s$. From Fig. 5 and 6, it can be seen that at $\sigma=0$, $p=\kappa/\mu$ and $\mu_{\text{eff}}=1-(\kappa/\mu)^2$. It is now possible to calculate R, the disk radius, from (2) and (15) assuming the value of ϵ is known.

It is of course acceptable to use a lower value of p than that determined previously at the expense of increased biasing field. On the other hand, if too low a value of p is chosen, operation too close to $\sigma=1$ may be required and resonance losses will be appreciable. If higher values of p are used, the ferrite is operated below saturation, and hence, low-field loss may be experienced.

The remaining dimension to be found is the height of the ferrite cylinder, d. Equation (20) may be rewritten as

$$G_R d = \frac{1.48\omega R^2\epsilon\epsilon_0}{Q_L} \tag{35}$$

where everything on the right-hand side is already known. Thus, if we choose a value of G_R, the height d is determined. However, G_R cannot be specified until the matching is provided.

So far it has been assumed that a quarter-wave transformer will be used for matching and that its susceptance at the band edges will cancel the resonator susceptance at these points. In the admittance Smith chart of Fig. 12, there are shown two radii which represent the loci of admittances at $f_0(1+\delta)$ and $f_0(1-\delta)$ looking into a terminated transformer which is a quarter wavelength long at f_0, the band-center. Also shown are dotted curves representing loci of admittances having phase angles θ from (32). The example of Fig. 12 is taken for a bandwidth of 15 per cent ($\delta=0.075$) with 30 dB isolation (VSWR = 1.065, $\theta=14.3°$). The intersections of the radii and dotted loci define a resonator having the desired Q_L. The value of G_R and the characteristic admittance of the transformer, Y_T, now are determined if we specify the characteristic admittance, Y_0, of the strip line. Assuming for the example that $Y_0=0.02$ mho (50 ohm), then $Y_T=0.048$ mho and $G_R=0.115$ mho. The resonator admittance characteristic is seen to be wrapped into a tight loop at the end of the transformer and then transformed to a match at f_0, and to approximately the desired VSWR at the crossover point, $f_0(1\pm\delta)$.

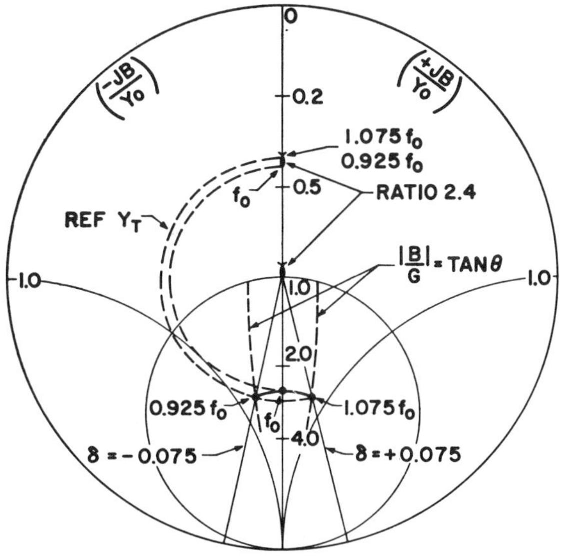

Fig. 12. Smith chart showing quarter-wavelength transformation to achieve a wide-band match. In this case the circulator is matched at midband with a 30 dB isolation bandwidth of 15 per cent.

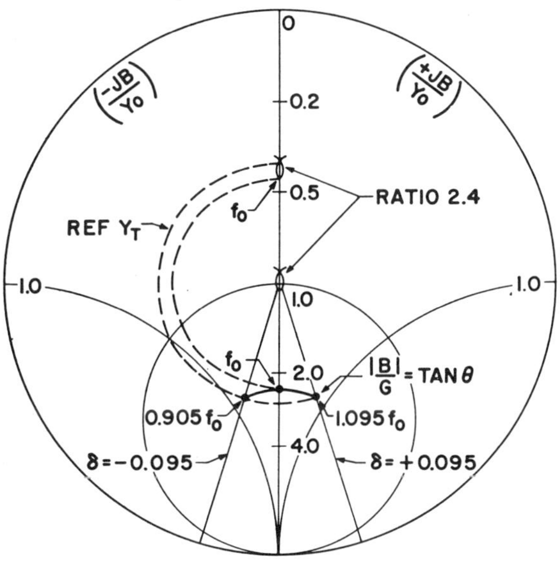

Fig. 13. Quarter-wavelength transformation for wide-band match with ripple in the isolation response. The 30 dB isolation bandwidth is 19 per cent in this case.

On inspection of Fig. 12, it will seem obvious that if either the transformer ratio or G_R of the resonator were changed slightly, the loop could be moved so that the origin was at its center, and the VSWR would be more uniform over the band. On the other hand, a larger loop could be used, still keeping all points within the specified VSWR. In Fig. 13, we show the latter case with the loop laid out so the midband point and the band edges all lie on a 1.065 VSWR circle. Using the same transformer ratio and working backward as compared to the previous procedure, it is found that G_R is now 0.108 mho and $\delta \approx 0.095$ with $\tan\theta = 0.310$, $Q_L = 1.63$. The value of G_R found in the foregoing can be used in (35) to find the thickness of the ferrite cylinder, d.

TABLE I

Comparison of Design and Actual Circulators

Design quantity	From fig. or eq.	*L*-Band Circulator		*C*-Band Circulator	
		19% B-W design	actual	19% B-W design	actual
Q_L	(33)	1.63	1.72	1.63	1.2
κ/μ	(34)	0.435	~0.5	0.435	~0.59
p	Fig. 5	0.435	0.52	0.435	0.28
μ_{eff}	Fig. 6	0.81	*	0.81	~0.6
ϵ		14	*	14	14
R cm	(15)	2.01	2.10*	0.655	0.76
$4\pi M_s$ gauss	$\frac{p\omega}{\gamma}$	200	240	625	400
G_R mho	Fig. 13	0.108	0.112	0.108	0.082
Y_T mho	Fig. 13	0.048	0.049	0.048	0.041
d cm	(35)	0.342	0.310	0.112	0.133
H_{dc} oers	$N_z 4\pi M_s$	145	~200	450	~870
Insertion loss dB	(26)	0.15	~0.1*	0.15	~0.17
30 dB Bandwidth	Fig. 13	0.19	0.185	0.19	0.12

* This structure has a 1.35-inch diam. ferrite cylinder inside a 1.75-inch-diam. alumina ring, hence μ_{eff} and ϵ will not be uniform and there is less material having magnetic loss than assumed in the design.

The result of the foregoing 19 per cent bandwidth design for 30 dB isolation compared to actual data on two circulators is shown in Table I. 105

Both circulators listed in the table were designed before the procedure outlined here was developed. The *L*-band circulator is observed to approximate fairly closely the design outlined.

V. Higher Mode Circulators and Other Forms

In this section we briefly sketch the operation of other strip-line circulators and also some waveguide circulators.

Y Circulator Using the $n=2$ Modes

As indicated previously, a *Y* circulator can be made using the $n=2$ modes (also $n=4$ and higher, excluding those having threefold symmetry). The resonator pattern rotation required for the $n=2$ mode is shown in Fig. 14. The required rotation in this case is 15° geometrically which corresponds to 30° electrically since this mode has two cycles around the circumference of the disk. The direction of rotation of the pattern is the same as the sense of circulation for this circulator.

The Four-Port Single Junction Circulator in Strip Line

The four-port single junction or *X*-junction circulator [12] seems attractive from the standpoint of compactness. An additional parameter is required [5] for this circulator as compared to the three-port junction. In this case, $n=1$ modes are split as in the three-port circulator and, in addition, the $n=0$ mode, which cannot split, is tuned to the operating frequency of the split $n=1$ modes. The resonator standing-wave patterns are illustrated in Fig. 15. The $n=1$ mode standing-wave pattern is rotated 45° toward the output port. With the

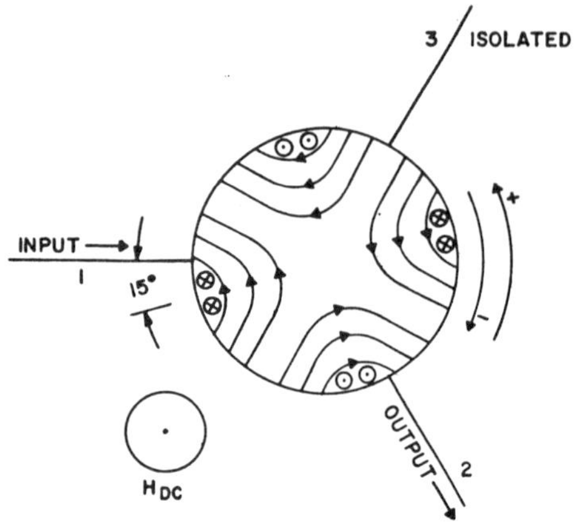

Fig. 14. Mode pattern for $n=2$ mode operation of disk circulator.

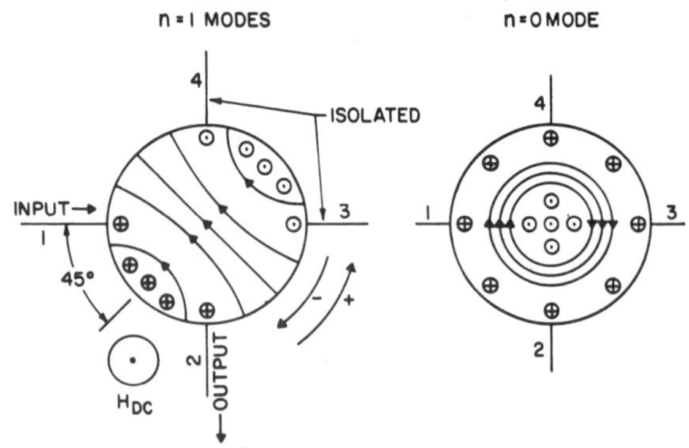

Fig. 15. Four-port circulator synthesized from $n=1$ and $n=0$ modes.

$n=0$ mode excited by the input port and with equal E-fields for the two patterns at the input port, the fields are also equal and in phase at the output, port 2. However, at ports 3 and 4, the fields are in opposite phase and will cancel. Thus we have transmission from port 1 to port 2, and ports 3 and 4 are isolated. The $n=0$ mode, which is normally about twice the frequency of the $n=1$ modes can be tuned lower in frequency by means of a capacitive post extending from the ground plane through a hole in the center of the ferrite cylinder to close proximity to the center of the metallic center disk. This post also raises the frequency of the $n=1$ modes somewhat so that the two resonant frequencies converge more rapidly. An $n=0$ mode of the cavity enclosing the ferrite strip-line assembly can be used in place of the $n=0$ disk mode. Therefore, this mode may be found to be more easily brought to the desired operating frequency.

The Y-Junction Circulator in Rectangular Waveguide—The H-Plane Junction

The three-port waveguide circulator usually consists of a 120° H-plane junction having a post of ferrite at the center of the junction as shown in Fig. 16. The biasing magnetic field is applied in the direction of the

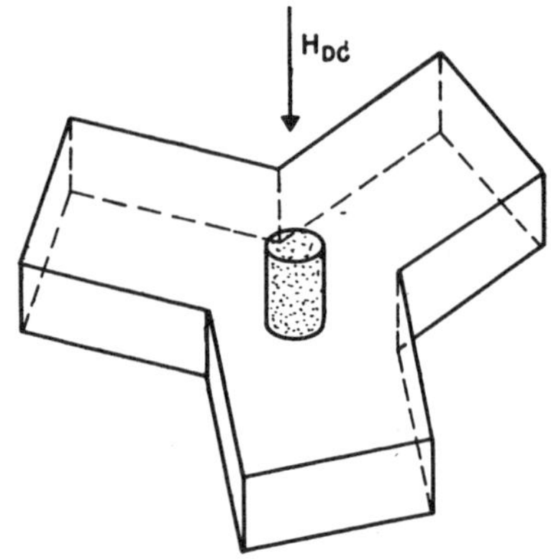

Fig. 16. Three-port H-plane waveguide circulator.

cylindrical axis of the post. With rectangular waveguide operating in the dominant TE_{10} mode, and with the ferrite post being considerably smaller in diameter than the width of the waveguide, the ferrite is excited mainly by the transverse RF H-field of the wave.

The operation of the waveguide circulator can be developed in a manner similar to that of the strip-line circulator [9]. First let us consider a lightly coupled case in which the ferrite post is contained in a cylindrical cavity which is coupled to the branching waveguides by small irises as shown in Fig. 17. If we assume this cavity is resonant at the operating frequency in the TM_{110} mode, the field pattern will be similar to that shown in the figure. Such a standing-wave pattern can be generated by two similar counter rotating patterns. The RF magnetic field will be almost circularly polarized in the ferrite for each sense of rotation. As soon as magnetic biasing field is applied to the ferrite, the resonance splits in a manner similar to that of the strip-line circulator. If the bias is adjusted so that each of the counter-rotating modes has a 30° phase angle, the resultant standing-wave pattern will have the orientation shown in Fig. 17. Here it is obvious that two ports are coupled to the standing-wave pattern and the third port lies at a null. This is equivalent to the $n=1$ mode circulation of the strip-line junction. Calculation of the frequency of resonance of this cavity is possible if small coupling is assumed. Writing down expressions for the fields in the ferrite and air portions, and equating E_z, B_r, and H_θ at the ferrite air boundary will yield an equation which could be solved for the resonant frequency. However, in the case of the actual circulator, the coupling irises are enlarged to the full size of the waveguide so that the cavity boundaries are no longer well defined. This is the equivalent of the broadbanding of the stripline circulator by tighter coupling to the feed lines. The calculation of the operating frequency becomes difficult in this case. Additional broadbanding can be obtained by impedance transforming means. One often used method is to make the circulator in reduced height, hence low-impedance waveguide, and to match to normal waveguide by stepped transformer

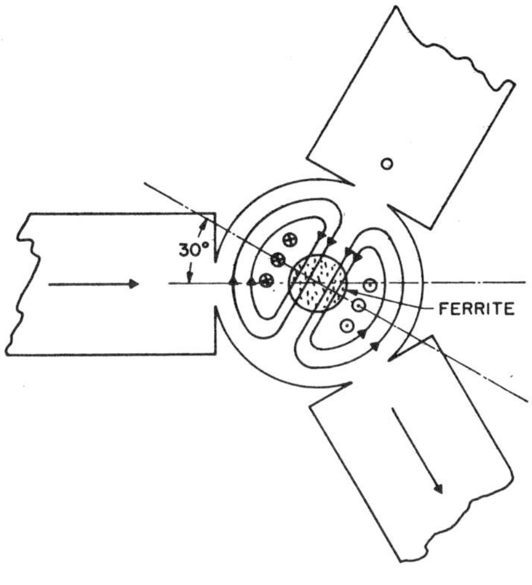

Fig. 17. Mode pattern for H-plane waveguide circulator using TM_{110} mode.

sections. Another method is to use triangular shaped ferrites whose points are in the centers of the waveguides. The points of the triangles are probably in regions of nearly linear polarization so that they are acting more like dielectric tapers than ferrimagnetic material. A partial height ferrite post compensated for by increased diameter also may be used, as well as a reduced diameter post with dielectric sleeve to maintain the proper resonant frequency.

The waveguide circulator can operate in the $n=2$ mode also in a manner similar to the strip-line case. However, the larger ferrite required, with its probable higher loss, makes this mode unattractive.

The E-Plane Y-Junction Waveguide Circulator

An E-plane waveguide Y-junction circulator can also be made [13]. The ferrite post in this case is mounted perpendicular to the narrow wall junction and is excited by the longitudinal RF H-fields. The cavity mode involved in this case is a TE_{111} mode as illustrated in Fig. 18. It is advantageous here to remove the center portion of the ferrite post leaving only two short cylinders at the ends of the cavity, since the center of the cavity is not active magnetically and ferrite here is of little value as well as a possible source of some loss. There is no strip line equivalent to this device since there are no longitudinal H-fields in TEM mode line.

The E-plane junction circulator has not been used to any extent. It might be advantageous in high power applications since the ferrite can be confined to regions of low E-field. It has some disadvantage in that a large air gap in the magnetic biasing circuit is inherent.

The Four-Port X-Junction Circulator in Waveguide

A few four-port waveguide junction circulators or X-junction circulators, have been described [14], [15]. These operate in a manner similar to the strip-line four-port junction. The TM_{110} cavity mode is operated with

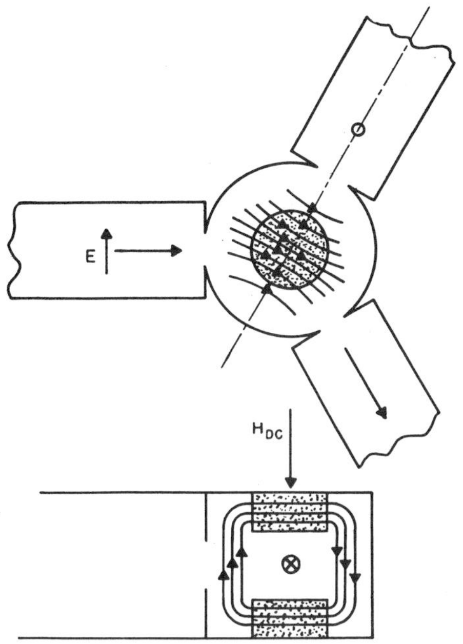

Fig. 18. E-plane waveguide circulator using TE_{111} mode.

107

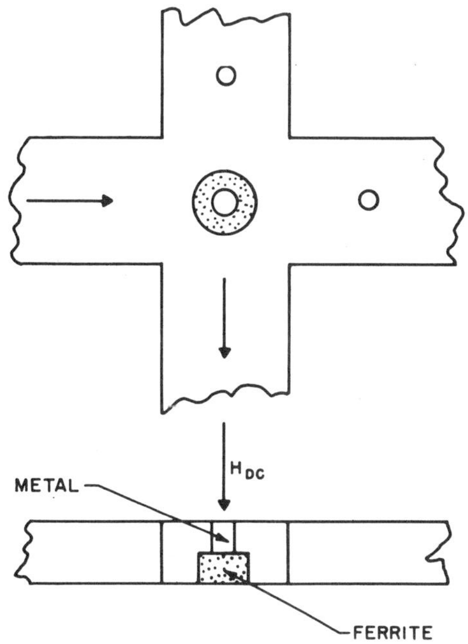

Fig. 19. Waveguide four-port circulator.

a 45° rotation of the standing-wave pattern and the TM_{010} cavity mode tuned to resonate at the same frequency by means of a central metallic post as illustrated in Fig. 19. The ferrite post and the metallic post are usually each partial height.

Conclusion

The operation of the ferrite junction circulator has been explained on the basis of a split resonance of the junction in which the standing-wave pattern of the resonator is rotated to produce a null or no coupling at the isolated port. The derivation of the field relations necessary to provide the above condition for the strip-line case has been carried out and experimental verification of the normal modes of the resonator and the postulated field distribution for circulation have been shown. A design procedure using the quarter-wave transformer for matching a strip-line circulator has been outlined.

While the explanation and analysis of the strip-line circulator given here has been based on a circular disk and cylindrical ferrite structure for simplicity, it seems reasonable that the same basic principles should apply to other structures having a threefold symmetry. These principles are:

1) The symmetrical microwave circuit must support two counter-rotating modes which in the absence of a magnetic biasing field are degenerate.
2) The degeneracy must be removed by a magnetic biasing field directed along the axis of symmetry.
3) Circulation then will occur at a frequency between the resonant frequencies of the nondegenerate modes.
4) The splitting of the nondegenerate modes must be adjusted to approximate the desired bandwidth, which is determined by the degree of external coupling to the microwave circuit, (31).

The basic resonator can, in principle, extend well beyond the ferrite loaded portion, since the ferrite is useful only where the RF H fields approach circular polarization. Other symmetrical junctions which have been used are simple 120° strip-line junctions without a circular disk, and triangular junctions of both ferrite and center conductor. Application of these principles to waveguide circulators has been outlined.

Acknowledgment

The authors are indebted to L. K. Anderson and B. A. Auld for many constructive suggestions and criticisms.

References

[1] Fowler, H., presented at Symposium on Mircowave Properties and Applications of Ferrites, Harvard University, Cambridge, Mass., Apr, 1956.
[2] Schauge-Pettersen, T., Novel design of a 3-port circulator, Norwegian Defense Establishment Rept. 1958.
[3] Swanson, W. E., and G. J. Wheeler, Tee circulator, *IRE WESCON Conv. Rec.*, pt. 1, 1958, pp 151–156.
[4] Chait, H. N., and T. R. Curry, A new type *Y*-circulator, *J. Appl. Phys.*, suppl. vol 30, Apr 1959, pp 152S–153S.
[5] Auld, B. A., The synthesis of symmetrical waveguide circulators, *IRE Trans. on Microwave Theory and Techniques*, vol MTT-7, Apr 1959, pp 238–246.
[6] Milano, U., J. Saunders, and L. Davis, A *Y*-junction stripline circulator, *IRE Trans. on Microwave Theory and Techniques*, vol MTT-8, May 1960, pp 346–351.
[7] Bosma, H. On the principle of strip line circulation, *Proc. IEE*, vol 109, pt B, suppl 21, Jan 1962, pp 137–146.
[8] Bosma, H., On strip line *Y*-circulation at UHF, *IEEE Trans. on Microwave Theory and Technique*, vol MTT-12, Jan 1964, pp 61–72.
[9] Butterweck, H. J., Der *Y* Zirkulator, *AEU*, vol 17, Apr 1963, pp 163–176.
[10] Lax, B., and K. J. Button, *Microwave Ferrites and Ferrimagnetics*, New York: McGraw-Hill, 1962.
[11] Suhl, H., and L. R. Walker, Topics in guided wave propagation through gyromagnetic media, *Bell Sys. Tech. J.* vol 33, May 1954, pp 579–659; Jul 1954, pp 939–986; Sep 1964, pp 1133–1194.
[12] Editorial, *Microwaves*, vol 2, p 92, Aug 1963.
[13] Yoshida, S., *E*-Type T circulator, *Proc. IRE* (*Correspondence*), vol 47, Nov 1959, p. 2018.
[14] Yoshida, S., *X* circulator, *Proc. IRE* (*Correspondence*), vol 47, Jun 1959, p. 1150.
[15] Davis, L. E., M. D. Coleman, and J. J. Cotter, Four-port crossed waveguide junction circulators, *IEEE Trans. on Microwave Theory and Techniques*, vol MTT-12, Jan 1964, pp 43–47.

Reprinted from JOURNAL OF APPLIED PHYSICS, Vol. 36, No. 3 (Part 2), 1253–1258, March 1965

Performance and Ferrimagnetic Material Considerations in Cryogenic Microwave Devices

R. L. COMSTOCK*
Lockheed Missiles and Space Company, Research Laboratories, Palo Alto, California
AND
C. E. FAY
Bell Telephone Laboratories, Incorporated, Murray Hill, New Jersey

Circulators which can be operated at cryogenic temperatures are discussed. Both symmetrical Y-junction circulators using polycrystalline ferrimagnetic garnets and hybrid-gyrator circulators using single crystals are evaluated. The temperature dependence of the ferromagnetic resonance losses of polycrystalline aluminum-doped and undoped yttrium iron garnet is reported. Device performance is evaluated in terms of the measured material properties.

1. INTRODUCTION

IN this paper we will discuss the application of ferrimagnetic insulators to low-temperature microwave transmission devices. Low-temperature microwave magnetic devices have become important in connection with low-noise microwave preamplifiers, e.g., resonance isolators in traveling-wave masers, and circulators in recent experiments with low-temperature parametric amplifiers. The problems associated with the operation of microwave ferrimagnetic devices at cryogenic temperatures are critically related to the properties, e.g., ferromagnetic resonance losses, of the ferrimagnetic garnets at low temperatures (Sec. 2). The low-temperature properties of microwave circulators[1] are emphasized in this paper but many of the conclusions concerning material properties will be generally applicable to the maser isolator and other low-temperature devices. Two broad classes of solutions to the low-temperature circulator problem have evolved: symmetrical Y-junction circulators as previously developed for room-temperature operation with polycrystalline ferrimagnetic materials, and gyrator-hybrid circulators using single-crystal ferrimagnets. The properties of Y-junction circulators at low temperatures will be discussed in Sec. 3, single-crystal hybrid-gyrator circulators in Sec. 4, and the conclusions summarized in Sec. 5.

Cryogenic Circulator—Parametric Amplifiers

The need for cryogenic temperature circulators to be used in conjunction with cryogenic temperature parametric amplifiers can be established by considering the noise temperatures of the two microwave receivers outlined in block diagram form in Fig. 1. Each receiver consists of two circulator-coupled reflection-type parametric amplifiers operated at 4.2° and 77°K, respectively, and a room-temperature mixer with an effective input noise temperature of 2000°K. In the first receiver the insertion loss of the circulators, assumed to be 0.4 dB per pass, takes place at the same temperatures as the respective parametric amplifiers, while in the second receiver, the circulators are operated at room temperature (290°K). The circulators are assumed to have infinite isolation. If thermal noise is the dominant noise contribution, calculations indicate that for gallium arsenide diodes the noise temperatures of the two parametric amplifiers are 1° and 20°K, respectively.[1a] The effective input noise temperatures of the receivers are calculated to be 3°K with cooled circulators and 30°K with room-temperature circulators. If the room-temperature circulators have 0.2-dB insertion loss, the effective input noise temperature is still 15°K. Uenohara and Josenhans have constructed a two-stage parametric amplifier using a liquid-helium temperature circulator (to be described in Sec. 3) in the first stage and a nitrogen temperature circulator in the second stage.[1a] The receiver was similar to that shown in the block diagram in Fig. 1. The average effective input noise temperature measured on the complete receiver was 6°±3°K. The receiver bandwidth was approximately 3% at 4 Gc/sec. The above noise temperatures are competitive with those observed in microwave masers.

The low-noise temperatures available with cryogenic circulator-coupled parametric amplifiers justifies the development of suitable low-temperature circulators. The most critical part of the circulator at low temperatures is the ferrimagnetic material. The relevant properties of available materials are discussed in the next section.

2. FERRIMAGNETIC MATERIAL CONSIDERATIONS

The most critical property of ferrimagnetic insulators for use at low temperatures is the ferromagnetic resonance loss. For this reason only the ferrimagnetic garnets are considered since they have been observed to have lower linewidths than spinels or other ferrimagnets.

* Formerly with Bell Telephone Laboratories, Murray Hill, New Jersey.

[1] An n-port circulator transmits power incident on port k to port $(k+1)$. An isolator can be made from an $n=3$ port circulator by terminating one port.

[1a] M. Uenohara and J. Josenhans, Summaries of papers of International Conference on Microwaves, Circuit Theory and Information Theory, p. 133, Inst. Elec. Comm. Engrs., Tokyo, Japan (1964).

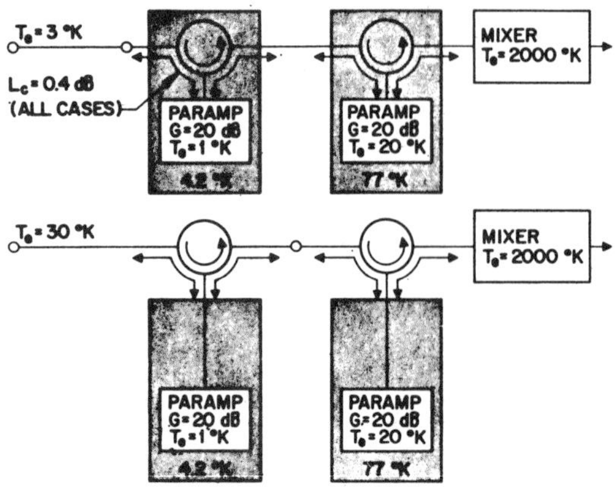

FIG. 1. Microwave parametric amplifiers with cooled and uncooled circulators. The theoretical thermal input noise temperatures of the two receivers are 3° and 30°K, respectively.

Yttrium Iron Garnet

The narrow ferromagnetic resonance linewidth (ΔH) of yttrium iron garnet (YIG) has made it an important material for room-temperature devices.[2] The same is true for low-temperature devices. The resonance losses of moderate purity single crystals of YIG are dominated at low temperatures by impurities, the most important of which are silicon and the rare earths. Divalent iron resulting from charge compensation with tetravalent silicon results in a peak in the single-crystal linewidth at ~50°K (~6 Oe).[3,4] For polycrystals the linewidths have contributions from impurities, crystalline anisotropy, and porosity and are, in general, at least an order of magnitude larger than for the corresponding single crystal. In certain cases, the measured ferromagnetic resonance line in polycrystals is asymmetrical. In all the measurements reported here the asymmetry is negligible and the characterization of the losses in terms of a single parameter (the linewidth at the half absorption points) is justified. The results of a measurement of the temperature dependence of ΔH for a dense polycrystalline sample of YIG are shown in Fig. 2 (points). All linewidth measurements on the polycrystalline garnets were made at 5.5 Gc/sec. The YIG material was supplied by Trans-Tech, Inc. (G–113). The samples were unpolished since the surface linewidth is much smaller than the volume porosity linewidth.[5] Previous measurements of ΔH for $77° < T < 300°$K on a less-dense sample of polycrystalline YIG showed larger linewidths at all temperatures, but the same temperature dependence as our sample was observed in the above range.[6] The peak in the linewidth at ~20°K has not been previously reported for polycrystalline YIG and occurs near the temperature at which the linewidth in single-crystal YIG has a similar peak.[3] In order to interpret the observed linewidth, the two-magnon scattering theory developed by Geschwind and Clogston[7] (referred to as GC) will be used. The GC theory predicts an inhomogeneous ferromagnetic resonance linewidth narrowed by dipolar forces having anisotropy,[8] and porosity contributions[9]

$$\Delta H = \Delta H_{\text{ani}} + \Delta H_{\text{pores}}$$
$$= 1.08 J (H_A^2/4\pi M) + 0.5 J 4\pi M p. \quad (1)$$

The shape factor J is given in Ref. 7, the anisotropy field H_A is given by $2|K_1|/M$, and p is the porosity. The porosity of the sample used in obtaining the data for Fig. 2 was determined from the observed and theoretical densities (ρ_{ob} and ρ_{th}) as $p = 1 - (\rho_{ob}/\rho_{th}) = 0.6\%$. The resonance linewidth from these contributions increases with decreasing temperature since the factor J increases with increasing magnetization, especially at low saturating fields, and the anisotropy field increases much faster than the magnetization.[10] The observed linewidth also may have homogeneous contributions which can be measured on single crystals of the same material; this linewidth contribution will be referred to here as $\Delta H_{s.c.}$. Previous measurements of K_1 (Refs. 11, 12) and M (Ref. 10) for single-crystal YIG lead to predictions of linewidths larger than observed; e.g., the anisotropy field measured by Dillon at 1.5°K ($2|K_1|/M = 250$ Oe) by itself predicts a linewidth at this temperature of 130 Oe. Part of this dis-

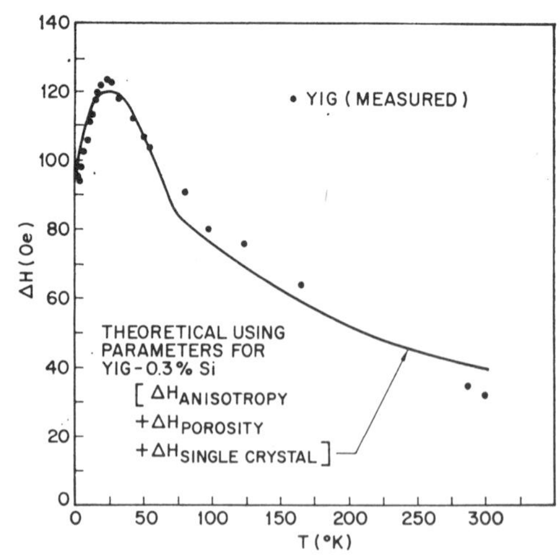

FIG. 2. Measured (points) and theoretical linewidth of polycrystalline YIG at 5.5 Gc/sec. The theoretical curve is based on Dillon's measurements of YIG–0.3% Si.

[2] E. G. Spencer and R. C. LeCraw, Proc. I.E.E. **109**, Part B Suppl. 21 (1962).

[3] E. G. Spencer and J. P. Remeika, International Conference on Nonlinear Magnetics 12-1-2, Washington, D. C., April 1964.

[4] E. G. Spencer, J. P. Remeika, and P. V. Lenzo, Appl. Phys. Letters **4**, 171 (1964).

[5] M. Sparks, *Ferromagnetic-Relaxation Theory* (McGraw-Hill Book Company, Inc., New York, 1964).

[6] P. E. Seiden and J. G. Grunberg, J. Appl. Phys. **34**, 1696 (1963).

[7] S. Geschwind and A. M. Clogston, Phys. Rev. **108**, 49 (1957).

[8] E. Schlömann, J. Phys. Chem. Solids **6**, 242 (1958).

[9] Calculation by E. Schlömann described in Ref. 6.

[10] M. A. Gilleo and S. Geller, Phys. Rev. **110**, 73 (1954).

[11] J. F. Dillon, Jr., Phys. Rev. **105**, 759 (1957).

[12] G. P. Rodrigue, H. Meyer, and R. V. Jones, J. Appl. Phys. **31**, 375S (1960).

crepancy may be attributed to the (supposedly) different amounts of rare-earth impurities in the yttrium oxide from which the two garnets were made. Recently, it has been suggested[13] that the reduction of anisotropy observed in silicon-doped single-crystal YIG by Dillon[14] may explain the reduction of anisotropy observed in polycrystalline YIG. The addition of silicon to the garnet lattice results in ferrous iron by valence compensation, while ferrous iron is present in polycrystalline YIG because of oxygen deficiency resulting from sintering. The anisotropy contribution to the linewidth of polycrystalline YIG using the first term in Eq. (1) with Dillon's measurements[13] of K_1 in YIG doped with 0.3% silicon is shown in Fig. 3(A). The anisotropy contribution is shown added to the porosity contribution in Fig. 3(B). The single-crystal linewidth measured on the same sample (averaged over the principal axes) is shown in Fig. 3(C). The sum of all three contributions to the linewidth is shown as the curve in Fig. 2. It is not known if the amount of ferrous iron is the same in the polycrystal and single crystal; however, the agreement between the measured and theoretical linewidths suggests that ferrous ions may reduce polycrystalline linewidths as long as the reduction in anisotropy is larger than the increase in homogeneous relaxation.

Low Magnetization Ferrimagnets

In certain applications, to be described in Sec. 3, ferrimagnetic materials are needed which have smaller values of M than YIG. One successful approach to the problem of obtaining low-magnetization materials has been to replace ferric ions in YIG with nonmagnetic ions. Gilleo and Geller[10] have shown that nonmagnetic Ga^{+++} and Al^{+++} can replace Fe^{+++} in YIG in sufficient amounts to reduce the 1.5°K value of $4\pi M$ to less than one third the value for YIG. Spencer and LeCraw[15] have reported room-temperature linewidths

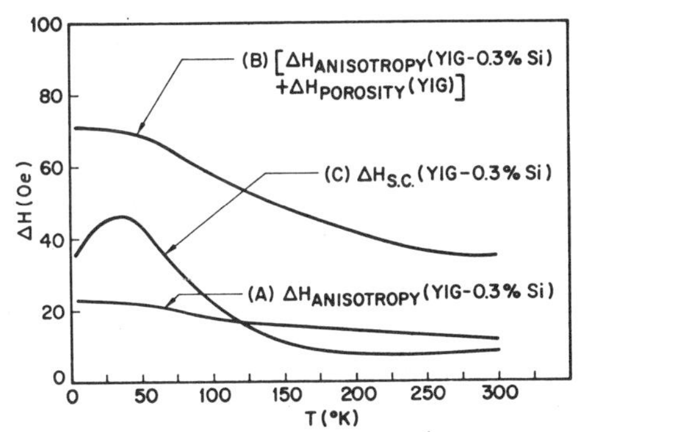

FIG. 3. Anisotropy linewidth (A), anisotropy linewidth+porosity linewidth (B), and single-crystal linewidth (C) of silicon-doped YIG based on the dipole-narrowed linewidth theory.

[13] J. F. Dillon and R. L. Comstock (unpublished).
[14] J. F. Dillon, Bull. Am. Phys. Soc. **6**, 160 (1961).
[15] E. G. Spencer and R. C. LeCraw, Bull. Am. Phys. Soc. **5**, 58 (1960).

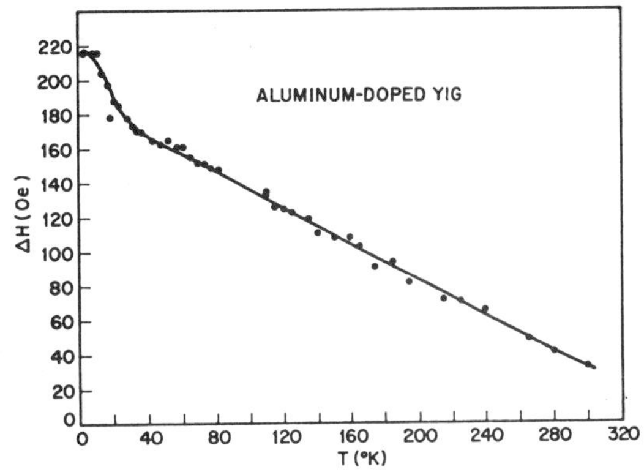

FIG. 4. Measured linewidth of aluminum-doped YIG [$4\pi M$ (300°K) = 500 G] at 5.5 Gc/sec.

of 1.2 Oe for single-crystal YIG–Ga ($4\pi M = 400$ G) as compared to 0.375 Oe for YIG made from starting materials of the same purity. Linewidths of polycrystalline YIG–Ga at room temperature have been reported by Harrison and Hodges,[16] Saunders and Green,[17] and Anderson, Cunningham, McDuffie, and Stauder.[18] It is found that ΔH for YIG–Ga increases with gallium substitution [25 Oe for YIG to 90 Oe for 20% substitution of Ga^{+++} for Fe^{+++}].[18] The temperature dependence of ΔH in single crystals of YIG–Ga or YIG–Al are both similar to the undoped YIG.[15] The temperature dependence of ΔH in a dense polycrystalline sample of YIG–Al is shown in Fig. 4. The value of $4\pi M$ is 495 G at 300°K. The linewidths at all temperatures are larger than for YIG because of the smaller dipolar narrowing and the larger anisotropy. In the YIG–Al, the homogeneous relaxation from the ferrous iron, which undoubtedly is present, apparently is swamped out by the large anisotropy contribution to the linewidth (the porosity was small).

The magnetization of YIG also can be reduced by substituting magnetic ions on the yttrium site, e.g., relatively dilute concentrations (<35%) of Gd. Magnetization measurements of

$$3[(1-x)Y_2O_3 \cdot xGd_2O_3]5Fe_2O_3$$

from 300° to 4.2°K have been made by Morris and Miller[19] for $x = 0.05$, 0.10, 0.15. These measurements indicate a substantial reduction in magnetization from YIG and for $x = 0.10$ a possible improvement in linewidth at 4.2°K over the undoped YIG as evidenced by an improvement in the operation of a maser isolator at 2 Gc/sec.

Other compounds which have lower magnetization than YIG are the recently discovered class of garnets

[16] G. R. Harrison and L. R. Hodges, Jr., J. Am. Ceram. Soc. **44**, 214 (1961).
[17] J. H. Saunders and J. J. Green, J. Appl. Phys. **32**, 161S (1961).
[18] E. E. Anderson, J. R. Cunningham, Jr., G. E. McDuffie, Jr., and F. R. Stauder, J. Phys. Soc. Japan **17**, Suppl. B-1, 369 (1962).
[19] L. C. Morris and D. J. Miller, I.E.E.E. Trans. on Microwave Theory and Techniques **12**, 421 (1964).

111

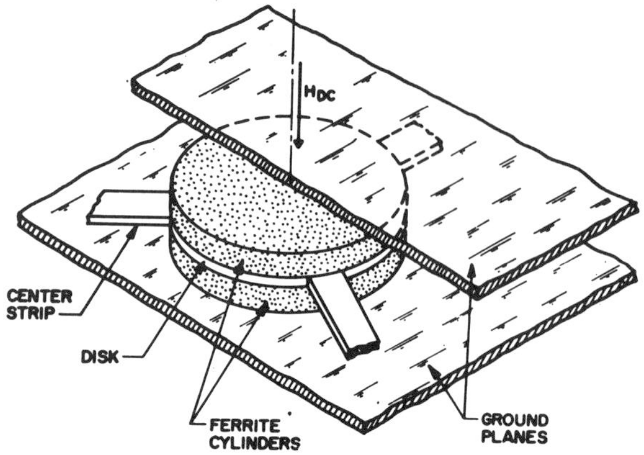

FIG. 5. Schematic diagram of 3-port Y-junction circulator.

containing no yttrium or rare earths.[20] As an example, the calcium–vanadium–bismuth garnet

$$\{Ca_{2.5}Bi_{0.5}\}[Fe_2](Fe_{1.75}V_{1.25})O_{12}$$

has $4\pi M$ equal to 600 G at 4.2°K. Recent measurements by Nilsen and Spencer show narrow linewidths and low anisotropy at low temperatures for a calcium–vanadium–bismuth garnet.[21]

3. SYMMETRICAL Y-JUNCTION CRYOGENIC CIRCULATORS

Design Considerations

It is the purpose of this section to relate the measurements of material parameters discussed in the last section to the theoretical operation of circulators which use polycrystalline ferrimagnets. The most commonly used circulator for room-temperature operation is the ferrimagnetically loaded symmetrical junction. The most common junctions are strip lines, which are used for frequencies from uhf to low X band (8 Gc/sec), and rectangular waveguides which are used at X band and above. Only the strip-line junction circulator has been evaluated at liquid-helium temperature and the detailed discussion will be limited to this case. Ferrimagnetic single crystals can be used in symmetrical junction circulators if they are oriented so as not to destroy the junction symmetry. The results of measurements on junction circulators indicate that polycrystals should be adequate for most applications.

A typical 3-port strip-line circulator is shown schematically in Fig. 5. Ferrimagnetic cylinders are placed between the circular center conductor and ground plane of the strip-line junction. The operation of the "Y-junction" circulator can be explained by reference to Fig. 6, which shows the electric and magnetic fields of the dipolar mode which is excited at port #1, with and without an applied dc magnetic field adjusted for circulation.[22] The *stationary* pattern (fields varying as $\cos\phi$, where ϕ is the azimuthal coordinate) shown in Fig. 6(A) results from the combination of the two rotating modes (fields varying as $e^{\pm j\phi}$), whose resonant frequencies are "split" by the internal magnetic field (H_{int}). The phase angles of the impedances of the two rotating modes are ±30°, resulting in a spatial rotation of the dipolar mode of 30°, which isolates port #3. Two cases of circulator operation can be distinguished: "below resonance" ($H_{int}<\omega/\gamma$) in which the pattern is oriented as shown in Fig. 6 and "above resonance" ($H_{int}>\omega/\gamma$) which results when 30° rotation in the opposite sense obtains and port #2 is isolated.

In many respects the junction circulator is similar to a symmetrical transmission cavity, e.g., if lossless, it is matched, and the fields are represented by a stationary mode. The loaded Q of the circulator mode is given by

$$Q_L = 1.5(\omega R^2\epsilon/G_R d), \tag{2}$$

where d is the ferrite disk thickness, R the disk radius, and G_R is the load conductance referred to the edge of the disk. For circulation it is also required that

$$Q_L = 1/(2\sqrt{3}\delta), \tag{3}$$

where δ is the fractional splitting of the rotating modes for 30° phase shift. The loaded Q, and hence the splitting δ, is not expected to vary appreciably with temperature [Eq. (2)]. The splitting is given in terms of the Polder tensor components (κ, μ) by $\kappa/\mu = 2.46\delta$. In order to keep δ constant with increasing M, which results when the temperature is lowered, it is necessary to decrease the internal field for $H_{int}<\omega/\gamma$ and increase it for $H_{int}>\omega/\gamma$. The center frequency of the circulator (ω_r) is given by

$$\omega_r = 1.84/R(\epsilon\mu_{eff})^{\frac{1}{2}}, \tag{4}$$

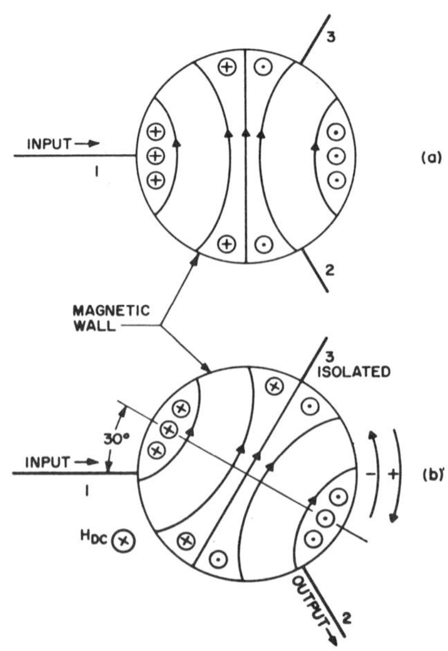

FIG. 6. Dipolar mode used in Y-junction circulator (a) $H_{int}=0$ or ∞, (b) H_{int} for below resonance circulation. The circles indicate the electric fields.

[20] S. Geller, G. P. Espinosa, H. J. Williams, R. C. Sherwood, and E. A. Nesbitt, Appl. Phys. Letters **3**, 60 (1963).

[21] W. G. Nilsen and E. G. Spencer (to be published).

[22] All of the theoretical discussion in this section is derived from: C. E. Fay and R. L. Comstock, Trans. I.E.E.E. Trans. Microwave Theory Tech. (to be published).

where μ_{eff} is the scalar microwave permeability. The change in dc field required to maintain a constant value of δ also changes μ_{eff} and hence the required operating frequency. The resulting operating field and frequency can be found graphically if the change in magnetization is known, by using curves of κ/μ and μ_{eff} vs H_{int} and M. In general, for $H_{int}<\omega/\gamma$ the required field decreases and the operating frequency increases with increasing M, and for $H_{int}>\omega/\gamma$ the opposite is true. For below resonance circulators operated at magnetic saturation the required magnetization can be found from the value of κ/μ with $H_{int}=0$,

$$4\pi M=(0.71/Q_L)(\omega/\gamma) \tag{5}$$

at 4 Gc/sec with $Q_L=2$ the required value of $4\pi M$ is 500 G. With larger values of $4\pi M$ low field losses may become excessive and with smaller values ferromagnetic resonance losses may become excessive. For above resonance circulators the required $4\pi M$ is larger and the cylinder diameter is reduced.

The magnetic part of the unloaded Q of the circulator mode for below resonance operation is approximately given by

$$Q_0=2\omega^2/\gamma^2 4\pi M\Delta H. \tag{6}$$

In most cases the magnetic losses dominate the losses of the circulator, in which case the insertion loss (in dB) is given by $20\log(1-Q_L/Q_0)$. The loaded resonator Q is specified by the reverse isolation (VSWR) bandwidth and is typically in the range $1<Q_L<4$. In order to keep the insertion loss less than 0.5 dB it is necessary for $Q_0>20Q_L$, resulting in a minimum value for Q_0 of 20. From Eq. (6) with $Q_0=20$, $4\pi M=500$ G, at 4 Gc/sec, the linewidth is restricted to $\Delta H<400$ Oe. The polycrystalline materials described in Sec. 1 (YIG–Al) all have linewidths at 4.2°K smaller than 200 Oe. Similar considerations for the above resonance mode lead to similar restrictions on ferromagnetic resonance linewidth.

Experimental Results

An above-resonance strip-line circulator was designed using the principles outlined in the previous section and a photograph of the assembled circulator is shown

FIG. 7. Photograph of cryogenic Y-junction circulator with superconducting magnet.

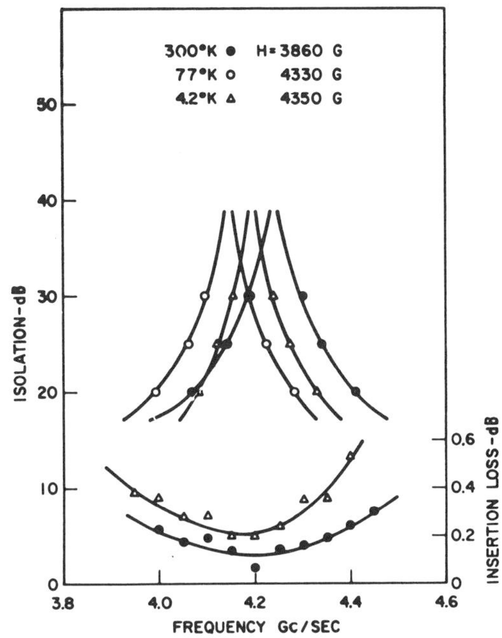

FIG. 8. Experimental results for above resonance circulator at 300°, 77°, and 4.2°K. The insertion loss at 77°K was almost identical to that at 4.2°K.

as Fig. 7. The dc magnetic field was supplied by a superconducting magnet (niobium–zirconium alloy wire) mounted on the circulator body. The ferrimagnetic material was polycrystalline yttrium iron garnet (see Fig. 2) in the form of disks ($2R=0.350$ in., $d=0.052$ in). The circulator was mounted on coaxial stainless steel transmission lines and the temperature runs were made with a stainless steel double Dewar with the experimental results shown in Fig. 8. The temperature dependence of the operating field and frequency are approximately predicted by the variation in μ_{eff} with κ/μ constant. The bandwidth of the circulator isolation could be improved with the addition of more sophisticated impedance-matching circuits [only a single quarter-wavelength line was used to match the low impedance (~20 Ω) of the circulator mode to the 50-Ω input lines]. The measured insertion loss was compatible with the linewidth measured on spheres of the same material (Fig. 2). The circulator was combined with a parametric amplifier and was operated at 4.2°K. The experimental results were described in the introduction.

Below resonance circulators also have been made with results similar to those obtained with the above resonance mode.[23] The below resonance mode of circulator operation is preferred to the above resonance mode because of the lower values of magnetic field. However, for Y-junction circulators requiring the lowest possible values of insertion loss it may be preferable to choose the above resonance mode, which allows the use of YIG for frequencies in the vicinity of 4 Gc/sec.

[23] J. A. deGruyl, W. E. Heinz, and S. Okwit, Proc. I.E.E.E. (Correspondence) **51**, 947 (1963).

FIG. 9. Single-crystal YIG gyrator. The 0.072-in. YIG sphere is located between the crossed semiloops.

4. SINGLE-CRYSTAL HYBRID-GYRATOR CIRCULATOR

Multicomponent circulators can be made by combining two hybrids and a gyrator in a circuit first described by Fox.[24] Recently, the gyrator element in a multicomponent circulator was made using a small single crystal of YIG.[25] Single-crystal YIG of adequate purity is ideally suited to low-temperature applications
114 since its linewidth is lower at 4.2°K than at 300°K.[4] The single-crystal gyrator was made using the ferrimagnetically coupled strip-line circuit first described by DeGrasse[26] and analyzed in detail by Comstock.[27] The strip-line gyrator is shown in Fig. 9. Two half-wave strip-line resonators with centrally located semiloops are crossed-over and magnetically coupled by the dipolar fields of the ferrimagnetic sphere (d= 0.072 in.) excited in the uniform precession resonance. The resonators are coupled to the external terminals by high impedance quarter-wave lines. The nonreciprocal phase shift of the gyrator was measured as $\pm(90°\pm1.5°)$ over a 200-Mc/sec bandwidth centered at 4 Gc/sec. A photograph of the four-port circulator is shown in Fig. 10. The performance of the circulator at 4.2°K is shown in Fig. 11. The reverse isolation from ports 2 to 1 had a bandwidth at the 30-dB isolation points of approximately 50 Mc/sec, while the forward insertion loss was less than 1 dB over the same range. With adjustment of impedances near the end of the strip-line resonators all of the four ports operated at the same center frequency. The bandwidth of the circulator is believed to be limited by spurious reflections within the circulator.

FIG. 10. Hybrid-gyrator circulator.

5. CONCLUSIONS

The low-temperature properties of ferrimagnets are of increasing technical importance in microwave transmission circuits. We have discussed in this paper the cryogenic microwave circulator since for this device detailed comparisons are available relating material properties to device performance Studies have shown that presently available polycrystalline ferrimagnetic materials, principally the magnetic garnets, are adequate for all but the most severe Y-junction circulator applications. For extremely low loss using the below resonance mode of circulator operation it may be necessary to develop new low loss materials. These materials will have the following characteristics: (i) magnetization at 4.2°K less than 500 G, (ii) anisotropy fields ($2|K_1|/M$) less than 100 Oe, and (iii) single-crystal linewidths comparable to YIG (<5 Oe).

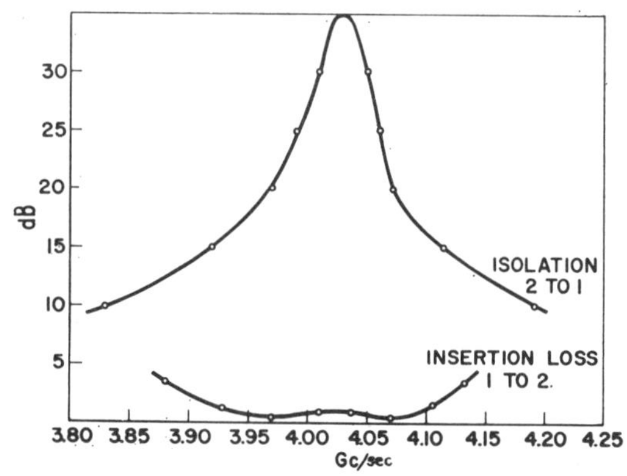

FIG. 11. Experimental results for single-crystal hybrid-gyrator circulator at 4.2°K.

An alternative solution to the cryogenic circulator problem consisting of a hybrid-gyrator circulator with a single-crystal gyrator was discussed. This four-port circulator has similar performance to the Y-junction circulator but is not believed to possess the potential for wideband operation.

ACKNOWLEDGMENT

The authors wish to acknowledge the assistance of M. Loewy in the construction of the devices and in taking much of the data reported here.

[24] R. H. Fox, Lincoln Lab. Tech. Rept. No. 68, (1954).
[25] R. L. Comstock, Proc. I.E.E.E. (Correspondence) **51**, 1768 (1963).
[26] R. W. DeGrasse, J. Appl. Phys. **30**, 155S (1959).
[27] R. L. Comstock, in Trans. I.E.E.E. Trans. on Microwave Theory Tech. MTT-12, 599 (1964).

Circulator Synthesis

JERALD A. WEISS, SENIOR MEMBER, IEEE

***Abstract*—A symmetrical three-port ring network composed of reciprocal *T* junctions and nonreciprocal phase shifters is analyzed theoretically to determine conditions under which it exhibits perfect circulation. All physically realizable *T* junctions are considered. It is found that many such junctions, combined with appropriate phase shifters specified by the theory, form perfect circulators. Among these are many cases for which the internal wave amplitudes are small and which require only very small amounts of nonreciprocal phase shift. Circulators designed in accordance with this model may offer appreciable advantages in insertion loss and bandwidth, as well as in mechanical characteristics such as size and weight, and in the possibility of adapting the design for special applications such as high-power capability, high-speed switching, etc. The nature of the model and the method of calculation are summarized.**

Manuscript received August 20, 1964; revised October 12, 1964. This work was performed under the sponsorship of the Array Radars Group of Lincoln Laboratory.

The author is with the Lincoln Laboratory, Massachusetts Institute of Technology, Lexington, Mass. (operated with support from the United States Air Force), and the Worcester Polytechnic Institute, Worcester, Mass.

THEORETICAL CONSIDERATION of the junction circulator has taken the form of group-theoretical treatment of the characteristic modes of symmetrical junctions [1] and analysis of the spatial configurations of the modes [2] under certain simplifying assumptions regarding the structure of the junction. Although a considerable advance in the quality of *Y*-junction circulators has taken place during the same period, it has not been possible to apply the results of the theory to the design effort. The mechanical improvements, which include the shaping of the ferrite, use of composite dielectric-ferrite elements, addition of tuning elements at the ports, and, in strip-line versions, shaping of the center conductor have increased the complexity of the device to the point where simplified theoretical models can serve, at most, only as a qualitative guide.

So long as circulator design remains essentially empirical in nature, there exists the possibility that alternative styles of design offering substantial advantages in one or more specifications may be overlooked. As a familiar example, we may consider the question: how "efficiently" is the volume of ferrite used in present devices? It is now well known that dielectric loading of the junction by the ferrite plays an important part in its performance; designers have found it advantageous, in fact, to replace a considerable fraction of the ferrite by simpler dielectric materials. This naturally leads to speculation as to whether still further improvements might be made if the ferrite and dielectric elements were shaped and distributed in other ways.

As the term synthesis suggests, we attempt to decompose the circulator action into its elementary parts and to specify a combination of these elements which, in a sense, is optimum. If we choose as our optimum, for example, a structure which requires a minimum volume of ferrite, then the object is to characterize the nonreciprocal action separately from the overall scattering by the junction, and to synthesize the structure in such a way that this action is minimized. There is not an
116 obvious reason to suppose in advance that this procedure will lead to results which are of practical significance. We shall show, however, that there do indeed exist circulators which demand only extremely small nonreciprocal effects. Our estimate of losses indicates that there is no penalty in loss in these designs. Estimates of bandwidth are difficult to make without introducing special assumptions regarding the dispersive properties of the components, but there is no indication in the model to be considered that a bandwidth penalty is associated with these designs.

This paper considers a network model in which two aspects of circulator performance are singled out as fundamental: nonreciprocal differential phase shift, regarded as taking place in a distributed manner in the region between the ports; and scattering at the ports, regarded as localized and reciprocal. The relevance of this idealized model to existing circulator designs is based on the observations, 1) that broadband circulation seems to occur only when the electrical distance between ports is at least an appreciable fraction of a wavelength; 2) that coupling to the ports, visualized as localized scattering, must involve internal reflections as a fundamental part of circulator action. No attempt is made here, however, to establish a detailed connection between the network model and a rigorous field theory of junction circulators. Reduction of the complex junction circulator structure to a few disjoint components must, of course, be regarded as a purely conceptual simplification. On the other hand, the model may be regarded as a basis for the synthesis of novel types of circulators. The results of the calculation indicate that some important advantages in performance may be inherent in the ring network to be described.

The ring-network representation of the three-port Y circulator to be considered in this paper is illustrated in Fig. 1, in which the elements L are nonreciprocal phase shifters, and the elements T are symmetrical, reciprocal T junctions. The idea of synthesizing a circulator by joining three ferrite phase shifters has been advanced by others [3], [4], but has not received much attention because it has not appeared to offer any practical advantage over existing junction designs. In the present paper, the network is examined more rigorously. Specifically, the role of reflections at the T junctions in determining the characteristics of the circulator is incorporated in a systematic way.

The most significant result of the theory is the discovery that the ring network may exhibit perfect circulation even when the amount of phase differential for propagation around the ring in the two-clock senses is very small. This result by itself is not unexpected: a familiar circulator theorem asserts that *any* nonreciprocal three-port junction can be made to circulate by the addition of obstacles at the ports. It is found, however, that the amplitudes of the standing waves excited within the ring are surprisingly small in many cases. Some examples are presented in Tables I and II. As will be explained in the subsequent discussion and appendices, the example of Table I requires nonreciprocal phase shifters having a differential phase of only about 8° per sector; yet, the amplitude of the standing waves in the sectors adjoining the isolated port is only a little over twice the amplitude of the incident signal. A measure of merit of such a circulator may be formulated in terms of the amount of nonreciprocal phase and the amplitudes of the internal standing waves. This quantity turns out to be quite insensitive to the amount of nonreciprocal phase and, in fact, slightly lower (corresponding to lower insertion loss) in many of the low-phase cases than in those of higher phase. This means that in the designs prescribed by the theory for which the required differential phase is small, the saving in insertion loss due to the requirement of only a small volume of nonreciprocal medium (ferrite), more than compensates for the penalty in insertion loss resulting from the relatively large amplitudes of the interval waves.

The theory also yields a prescription for the frequency dependence of the scattering at the T's, such that circulation may be made to persist over a band of any width (with, of course, increasing complexity of structure as the bandwidth specification is increased). Thus, if the frequency dependence of the intrinsic reciprocal and nonreciprocal phases of the ferrite-loaded junction is known, the characteristics of the corresponding T junctions required for circulation over the entire band are completely determined. For the types of differential phase shifter characteristics commonly encountered in waveguide and coaxial device practice, it is reasonable to expect that the required T's can be synthesized by the application of known filter principles, resulting in a rigorously flat broadband Y-junction circulator, which exhibits only incidental loss and which may be built with extremely efficient use of ferrite. Depending on the

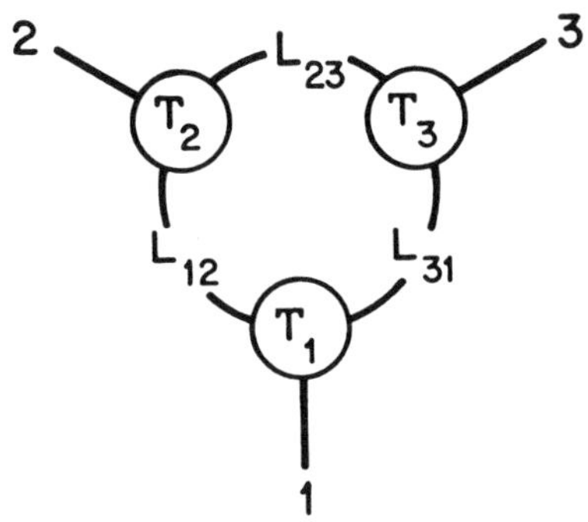

Fig. 1. The ring network. T and L denote symmetrical T junctions and nonreciprocal phase shifters, respectively.

TABLE I

THE RING-NETWORK CIRCULATOR, CASE (27°; 631B)

	Magnitude	Phase, degrees
r	0.254	170.17
s	0.878	58.68
r_d	0.820	14.55
s_d	0.405	135.00
ϵ	1.00	58.12
δ	1.00	3.97
ϕ_+	—	62.09
ϕ_-	—	54.15
$\Delta\phi$	—	7.94
E_1	0	—
E_2	1.00	9.69
E_3	0	—
C_{12}	1.56	153.82
C_{23}	1.12	−79.71
C_{31}	1.12	27.38
D_{21}	1.20	22.25
D_{32}	1.12	−152.69
D_{13}	1.12	108.16
M	4.94	—

TABLE II

THE RING-NETWORK CIRCULATOR

Summary of ten cases, identified as (γ; 631b) with $\gamma=4.50°, \cdots, 45.00°$. Listed with γ are the nonreciprocal phase parameter arg δ, the amplitude of the internal wave C_{31}, and the parameter of merit $M=|C_{31}|^2$ arg δ. The magnitudes of the scattering coefficients of the T junctions are also listed. Additional details of the case $\gamma=27°$ are given in Table I.

γ	4.50	9.00	13.50	18.00	22.50	27.00	31.50	36.00	40.50	45.00
arg δ	0.137	0.539	1.18	2.02	2.98	3.97	4.88	5.56	5.91	5.85
$\lvert C_{31}\rvert$	5.92	2.98	2.02	1.55	1.28	1.12	1.02	0.962	0.948	0.971
M	4.79	4.80	4.81	4.84	4.88	4.94	5.03	5.15	5.31	5.52
$\lvert r\rvert$	0.379	0.367	0.347	0.321	0.290	0.254	0.217	0.183	0.156	0.146
$\lvert s\rvert$	0.922	0.917	0.910	0.900	0.889	0.878	0.869	0.861	0.855	0.854
$\lvert r_d\rvert$	0.994	0.976	0.947	0.910	0.866	0.820	0.777	0.740	0.716	0.707
$\lvert s_d\rvert$	0.078	0.155	0.227	0.294	0.354	0.405	0.446	0.476	0.494	0.500

application contemplated, this efficiency may be manifested in some combination of compact size, low-loss, high-peak and average-power capacity, high-speed reversal of the direction of circulation, etc.

The method of carrying out the network analysis is as follows. The scattering coefficients of a symmetrical T junction are expressed in terms of the characteristic modes of the junction. The derivation follows the standard method [5] of the theory of group representations, but the result is obtained with full generality not found in standard treatments of the subject so as to represent all physically realizable T junctions satisfying reciprocity, energy conservation, and T symmetry. The scattering of a matched, lossless, nonreciprocal phase shifter (denoted by L in Fig. 1) is represented by a reciprocal, or average, phase factor $\epsilon=\exp i(\phi_+ + \phi_-)/2$, and a nonreciprocal, or differential phase factor $\delta=\exp i(\phi_+ - \phi_-)/2$, where ϕ_+ and ϕ_- refer to clockwise and counterclockwise propagation, respectively. The overall scattering by the network is then calculated in terms of the parameters of the T's and L's. We obtain the scattering coefficients E_1, E_2, and E_3 denoting, respectively, reflection, transmission, and leakage in response to a unit signal incident on port 1 of the ring.

The conditions for perfect circulation are now imposed on the overall scattering coefficients E_1, E_2, E_3. They are: input match, $E_1=0$; perfect isolation, $E_3=0$; lossless transmission, $|E_2|=1$. These conditions are, of course, not mutually independent; on the contrary, the last condition implies the other two. It turns out to be convenient, however, to use the isolation condition $E_3=0$; solving this relation to find the reciprocal phase factor ϵ and the nonreciprocal phase factor δ, we obtain a biquartic, algebraic equation for ϵ and an expression for δ in terms of ϵ, both relations involving the scattering coefficients of the T's as parameters. The problem now becomes one of computation.

A program was written for the MIT Lincoln Laboratory IBM 7094-C computer, in which values were assigned to the scattering coefficients of the T's in sets (cases) covering the range of physically realizable T's in a convenient number of steps. For each case, the problem was solved to find (a) the phase factors ϵ and δ if they exist; (b) the amplitudes and phases of the internal waves excited by a unit wave incident on port 1 of the ring; (c) the phase of the transmitted wave E_2; (d) the

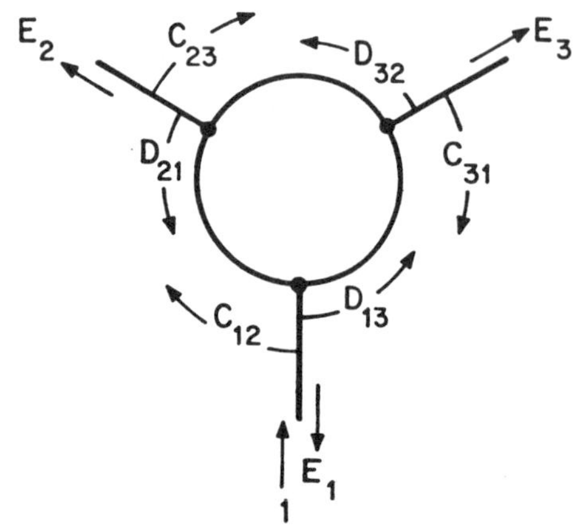

Fig. 2. Internal and scattered waves in the ring network.

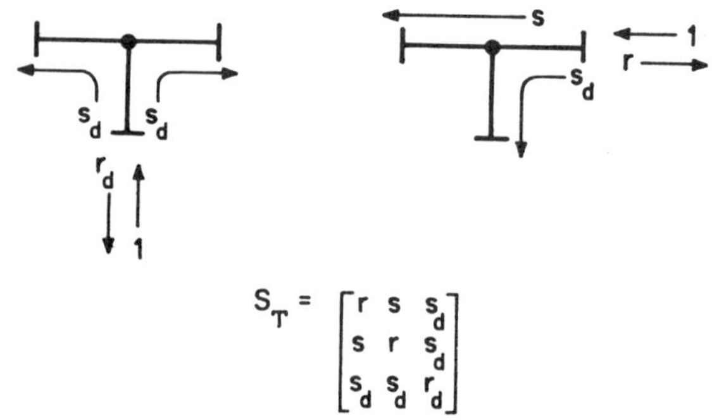

Fig. 3. Definition of the scattering matrix of a symmetrical T junction.

quantity M, defined subsequently, which provides an estimate of the relative insertion loss. In all, about 700 cases were considered. Of these, well over half gave solutions satisfying all requirements. As an example, the case identified as (27°; 631b) is summarized in Table I.
118 The scattering coefficients r, s, r_d, and s_d of the T's are defined in Fig. 3, and the internal partial waves C and D are defined in Fig. 2.

The table first lists the four scattering coefficients of the T junctions. They specify what would ordinarily be considered "poorly" matched junctions; for example, the coefficient r_d specifying reflection at the accessible port d (Fig. 3) has the magnitude 0.820 corresponding to a VSWR (voltage standing-wave ratio) of 10/1 for the T junction if its other ports were terminated in matched loads. When combined to form the ring, these T's, together with the corresponding L's yield, of course, a perfectly matched circulator.

The next series of entries specifies the characteristics of the phase shifters. The most significant number is the parameter δ which defines the differential phase $\Delta\phi$ through the relation $\Delta\phi = 2 \arg \delta$. In this example, $\Delta\phi$ is only 7.94° (per sector). The parameter ϵ specifying the average phase per sector is of less significance, since it can be "traded" with the phases of the scattering coefficients of the T's without affecting δ or the amplitudes of the internal waves.

The overall scattering coefficients E_1, E_2, and E_3 are listed next. They verify that perfect circulation indeed does occur and specify, through the phase of E_2, the overall phase of transmission through the ring.

The amplitudes of the C's and D's, the six internal wave amplitudes (Fig. 2) are listed. The total standing-wave amplitudes in the sectors 2–3 and 3–1 are the sums $C_{23}+D_{32}$ and $C_{31}+D_{13}$, respectively. In this example, they are equal to 2.24 times the amplitude of the signal incident on port 1.

A measure of the merit of the network with respect to its dissipative losses may be obtained from these data, notwithstanding the fact that dissipative effects have been neglected in the model. We make the plausible assumption that the dominating contribution to loss is magnetic in origin, and take the loss, as characterized by the Poynting theorem, to be proportional to the product of $\arg \delta$ (which measures the required length of the differential phase shifters) times $|C_{31}|^2$ (which measures the power level of the internal standing-wave configuration). The value of this product M for the present example is $M = 4.94$. The significance of this value of M may be seen when the case presented in Table I is compared with others in a series, as shown in Table II.

Table II lists a series of ten cases, showing values of $\arg \delta$ ranging from less than one degree to about 6°, with a corresponding decrease in $|C_{31}|$ from about six to about one. A number of details of the solutions have been omitted in order to emphasize the most significant features. The competition between $\arg \delta$ and $|C_{31}|$ results in a value of the parameter M which remains quite stable but shows a slight increase as $\arg \delta$ increases. This suggests that the very small-phase ring circulator designs are actually of higher merit (lower M) as far as magnetic losses are concerned, although other sources of loss might overbalance this tendency when the wave amplitudes become large.

In Appendices I and II, some further details are presented relating to the evaluation of the scattering matrix of the T junction, and the solution of the ring network problem.

Appendix I

The Scattering Matrix of a Symmetrical T Junction

The logic of the ring-circulator theory, whereby we begin with an assumed scattering matrix S_T for the three T junctions and calculate the corresponding requirements on the nonreciprocal phase shifters, demands that we have a completely general formulation of S_T, such that all physically realizable T junctions are represented. Since this is more general than that found in the standard treatments of the problem, we summarize here the derivation of S_T.

The T junctions are characterized by T symmetry, reciprocity, and losslessness. Reciprocity and energy conservation are represented, respectively, by symmetry and unitarity of the scattering matrix

$$\tilde{S}_T = S_T$$
$$S_T^\dagger S_T = I$$

where $\sim$ denotes the transpose, $\dagger$ the Hermitean conjugate, and I the unit matrix. T symmetry is represented by invariance of S_T with respect to a reflection which interchanges the two symmetrical ports of the junction. Let P denote the reflection operator represented by the matrix

$$P = \begin{bmatrix} 0 & 1 & 0 \\ 1 & 0 & 0 \\ 0 & 0 & 1 \end{bmatrix}$$

Then

$$P^{-1} S_T P = S_T$$

or

$$S_T P - P S_T = 0$$

that is, S_T commutes with P. Since the two matrices commute, there exists a unitary transformation matrix M which simultaneously diagonalizes S_T and P. We find M by carrying out the diagonalization of the reflection matrix P. The result is

$$M^\dagger P M = \begin{bmatrix} -1 & 0 & 0 \\ 0 & +1 & 0 \\ 0 & 0 & +1 \end{bmatrix}$$

where

$$M = \frac{1}{\sqrt{2}} \begin{bmatrix} 1 & e^{i\phi_b} \cos\gamma & e^{i\phi_c} \sin\gamma \\ -1 & e^{i\phi_b} \cos\gamma & e^{i\phi_c} \sin\gamma \\ 0 & \sqrt{2}\, e^{i\phi_b} \sin\gamma & -\sqrt{2}\, e^{i\phi_c} \cos\gamma \end{bmatrix}$$

in which ϕ_b, ϕ_c, and γ are arbitrary real angles. The dependence of M on γ is a consequence of the degeneracy in the eigenvalues of P. We may assume, for the diagonal representation of S_T, the form

$$M^\dagger S_T M = \begin{bmatrix} s_a & 0 & 0 \\ 0 & s_b & 0 \\ 0 & 0 & s_c \end{bmatrix}$$

where s_a, s_b, and s_c are complex numbers of unit magnitude but otherwise arbitrary. Reversing this transformation, we obtain

$$S_T = \begin{bmatrix} r & s & s_d \\ s & r & s_d \\ s_d & s_d & r_d \end{bmatrix}$$

where

$$r = \tfrac{1}{2}(s_a + s_b \cos^2\gamma + s_c \sin^2\gamma)$$
$$s = \tfrac{1}{2}(-s_a + s_b \cos^2\gamma + s_c \sin^2\gamma)$$
$$r_d = s_b \sin^2\gamma + s_c \cos^2\gamma$$
$$s_d = \frac{1}{\sqrt{2}}(s_b - s_c)\cos\gamma \sin\gamma$$

All T junctions satisfying T symmetry, reciprocity, and energy conservation are characterized by this set of scattering coefficients. For computing purposes, we express s_a as exp $(i\sigma_a)$, and similarly for s_b and s_c. We assign values of σ_a and σ_b in the range $0 \leq \sigma_a, \sigma_b < 2\pi$; σ_c can be held fixed at zero without loss of generality. We assign values of γ in the range $0 < \gamma \leq \pi/4$.

Appendix II

The Ring Circulator Problem

With the scattering coefficients of the T junctions (Fig. 3) characterized as discussed in Appendix I, we may analyze the scattering within the ring network (Fig. 2) as follows: the scattered wave C_{12}, for example, is composed of contributions due to transmission of the unit incident signal into the sector 1–2, the transmission of the wave C_{31} from sector 3–1 into sector 1–2, and the reflection of D_{21} at the T junction and we get

$$C_{12} = s_d + s C_{31} e^{-i\phi_+} + r D_{21} e^{-i\phi_-} \tag{1}$$

Let

$$e^{-\frac{1}{2}i(\phi_+ + \phi_-)} = \epsilon \tag{2}$$
$$e^{-\frac{1}{2}i(\phi_+ - \phi_-)} = \delta \tag{3}$$
$$r\epsilon = R \tag{4}$$
$$s\epsilon = S \tag{5}$$

Using (2), $\cdots$, (5) in (1), we have

$$C_{12} = s_d + \delta S C_{31} + \delta^* R D_{21} \tag{6}$$

Figure 4 illustrates schematically the composition of (6). Similarly, we may construct six relations connecting the C's and D's. In matrix form they are

$$\begin{bmatrix} \delta^* R & -1 & & & & \delta S \\ -1 & \delta R & \delta^* S & & & \\ & \delta S & \delta^* R & -1 & & \\ & & -1 & \delta R & \delta^* S & \\ & & & \delta S & \delta^* R & -1 \\ \delta^* S & & & & -1 & \delta R \end{bmatrix} \begin{bmatrix} D_{21} \\ C_{12} \\ D_{32} \\ C_{23} \\ D_{13} \\ C_{31} \end{bmatrix} = -s_d \begin{bmatrix} 1 \\ \\ \\ \\ \\ 1 \end{bmatrix} \tag{7}$$

In (7), null elements have been omitted for clarity. Expressions for the waves E_1, E_2, and E_3 scattered out of the ring at the three ports may be formed in the same way. We obtain

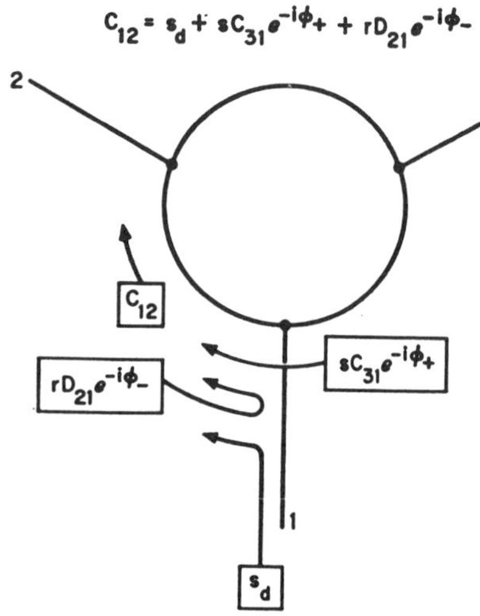

Fig. 4. Illustrating the composition of the scattering equations for the ring network.

$$E_1 = r_d + \epsilon s_d(\delta C_{31} + \delta^* D_{21}) \tag{8}$$

120

$$E_2 = \epsilon s_d(\delta C_{12} + \delta^* D_{32}) \tag{9}$$

$$E_3 = \epsilon s_d(\delta C_{23} + \delta^* D_{13}) \tag{10}$$

Solving the system (7), we obtain for the C's,

$$C_{12} = -\frac{s_d}{\Delta}\{[R(R-S)(R^2-S^2-1)+(1-R^2)] + \delta^{*3}S^2(R-S)\} \tag{11}$$

$$C_{23} = -\frac{s_d}{\Delta}\{\delta^{*2}S[R-(R-S)^2(R+S)] - \delta S(R^2-RS-1)\} \tag{12}$$

$$C_{31} = -\frac{s_d}{\Delta}\{\delta^*[(R-S)^3(R+S)^2 - R(2R^2-RS-S^2-1)] + \delta^2 S^2\} \tag{13}$$

where

$$\Delta = (R^2-S^2)^3 - 3R^2(R^2-S^2-1) + (\delta^3+\delta^{*3})S^3 - 1 \tag{14}$$

The D's may be found similarly by direct calculation, but we note that they are related to the C's in a simple way as a consequence of the symmetry of the network, as manifested in the form of the system (7). Under interchange of δ and δ^*, that is, $\delta \leftrightarrow \delta^*$, the C's and D's are interchanged according to

$$C_{12} \leftrightarrow D_{13}$$
$$C_{23} \leftrightarrow D_{32}$$
$$C_{31} \leftrightarrow D_{21}$$

Now, using the C's and D's in (8), (9), and (10), we obtain the desired expressions for the scattered waves

$$E_1 = r_d - \epsilon\frac{s_d^2}{\Delta}\{2(R-S)[(R^2-S^2)^2 - R(2R+S)] + (\delta^3+\delta^{*3})S^2 + 2R\} \tag{15}$$

$$E_2 = -\delta\epsilon\frac{s_d^2}{\Delta}\{(R-S)^3(R+S) - 2R(R-S) + \delta^{*3}S[1-(R-S)^2] + 1\} \tag{16}$$

$$E_3 \leftrightarrow E_2 \quad \text{under} \quad \delta \leftrightarrow \delta^* \tag{17}$$

We now impose a requirement for perfect circulation,

$$E_3 = 0 \tag{18}$$

describing perfect isolation of port 3. For a lossless three-port junction, this condition implies either perfect circulation, $|E_2| = 1$, or complete reflection at the input, $|E_1| = 1$. We discard the latter solutions. With E_3 given by (17), (18) yields

$$0 = (R-S)^3(R+S) - 2R(R-S) + \delta^3 S[1-(R-S)^2] + 1 \tag{19}$$

Using (4) and (5), we obtain from (19)

$$(a_4\epsilon^4 + a_2\epsilon^2 + a_0) + (a_3\epsilon^3 + a_1\epsilon)\delta^3 = 0 \tag{20}$$

where

$$\left.\begin{aligned} a_4 &= (r-s)^3(r+s) \\ a_3 &= -s(r-s)^2 \\ a_2 &= -2r(r-s) \\ a_1 &= s \\ a_0 &= 1 \end{aligned}\right\} \tag{21}$$

Solving (20) for δ in terms of ϵ, we have

$$\delta^3 = -\frac{a_4\epsilon^4 + a_2\epsilon^2 + a_0}{a_3\epsilon^3 + a_1\epsilon} \tag{22}$$

With the requirements from (2) and (3) that

$$|\delta| = |\epsilon| = 1 \tag{23}$$

we obtain from (22) the following biquartic equation in ϵ:

$$A_8\epsilon^8 + A_6\epsilon^6 + A_4\epsilon^4 + A_2\epsilon^2 + A_0 = 0 \tag{24}$$

where

$$\left.\begin{aligned} A_8 &= A_0^* = a_4a_0^* \\ A_6 &= A_2^* = a_4a_2^* + a_2a_0^* - a_3a_1^* \\ A_4 &= |a_4|^2 + |a_2|^2 + |a_0|^2 - |a_3|^2 - |a_1|^2 \end{aligned}\right\} \tag{25}$$

Equation (24) is the basis for the numerical solution of the problem. To solve a single case we select a set of values for the angles σ_a, σ_b, σ_c, and γ characterizing the scattering matrix S_T as discussed in Appendix I. Using these, we compute and record the four scattering coefficients r, s, r_d, and s_d of the T junctions. In terms of the coefficients r and s, we successively evaluate the coefficients $a_0, \cdots, a_4$, (21), and the coefficients $A_0, \cdots, A_8$, (25). We then solve (24) for ϵ and use (22) to evaluate δ. We test to ascertain that the magnitudes of ϵ and δ fulfill the requirement (23); if they do, the case yields a physically realizable circulator. We proceed to evaluate the C's and D's specified by (11), (12), and (13), and the

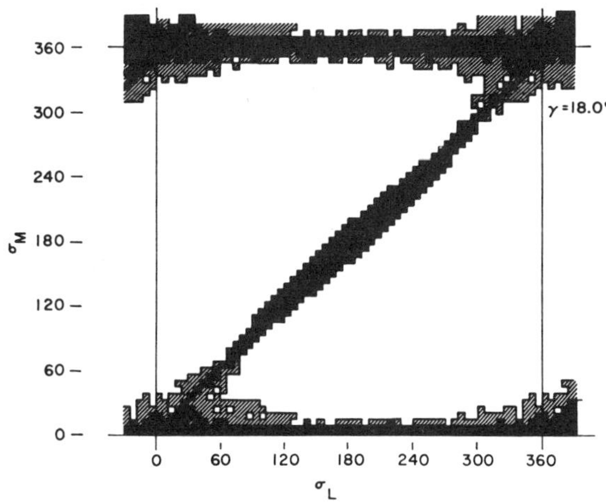

Fig. 5. A map of the σ_L, σ_M plane for $\gamma = 18.0°$.
White: physically realizable ring circulators;
Black: no ring circulator exists;
Shaded: borderline cases, uncertain due to rounding-off errors.

overall transmission coefficient E_2, (9). Finally, the parameter of merit M is evaluated.

A map illustrating the range of soluble cases is shown in Fig. 5. The particular range of cases shown there is for the value $\gamma = 18.0°$ and $\sigma_c = 0$, with σ_a and σ_b ranging from 0° to 360° (σ_L and σ_M are the same as σ_a and σ_b, respectively). The white areas represent physically realizable cases; the black, inadmissible cases. The diagonally shaded regions are borderline cases which are ambiguous because of rounding-off errors in the computation. A series of eleven such maps, for $\gamma = 0°$, 4.5°, · · · , 45.0°, show that well over half of the "parameter space," whose coordinates are σ_a, σ_b, σ_c, γ, is occupied by physically realizable cases.

Acknowledgment

The support and encouragement of D. H. Temme and other members of the group is hereby acknowledged. I also wish to acknowledge the invaluable assistance of J. Tabasky and C. Work of the Computer Facility of the Lincoln Laboratory.

References

[1] Auld, B. A., The synthesis of symmetrical waveguide circulators, *IRE Trans. on Microwave Theory and Techniques*, vol MTT-7, Apr 1959, pp 238–246.
[2] Bosma, H., On the principle of stripline circulation, *Proc. IEE* (London), vol 109B, suppl 21, 1962, p 137.
——, On stripline *Y*-circulation at UHF, *IEEE Trans. on Microwave Theory and Techniques*, vol MTT-12, Jan 1964, pp 61–72.
[3] Grace, M., and F. R. Arams, Three-port ring circulators, *Proc. IRE (Correspondence)*, vol 48, Aug 1960, p 1497.
[4] Vartanian, P. N., The theory and applications of ferrites at microwave frequencies, Sylvania EDL Rept. DDC-101, 888, Sylvania Defense Lab., Mountain View, Calif., Apr 1956.
[5] Montgomery, C. G., R. H. Dicke, and E. M. Purcell, *Principles of Microwave Circuits*, M.I.T. Rad. Lab. Ser., vol 8, ch. 12, New York: McGraw-Hill, 1948.

121

Broadband Strip-Transmission Line *Y*-Junction Circulators

J. W. SIMON, STUDENT MEMBER, IEEE

***Abstract*—Broadband circulators have been developed which operate below resonance and cover octave bandwidths in the frequency regions 0.6–8.0 Gc. The interrelationship of voltage standing-wave ratio (VSWR) and isolation between various ports is established without imposing symmetry. The fact that building a three-port circulator is essentially a matching process is used as a basis for a successful experimental development procedure. Empirically determined criteria for broadbanding are reported with particular emphasis placed on the discussion of optimizing saturation magnetization, linewidth, strip-transmission line parameters, and magnetic biasing field. The importance of having a ferrite with low dielectric loss tangent is also discussed. The empirical data are considered in light of theoretical work done in this laboratory and in light of the papers by Bosma.**

122

Introduction

A NUMBER of papers [1], [2], [3] have been given which present in a general nature the methods utilized in making narrow-band circulators. Papers also have been presented which contributed to the fundamental understanding of junction circulators [4], [5], [6] but which do not deal specifically with the problem of broadbanding these devices. It is the aim of this paper to present useful information from these papers along with a large amount of empirical data which have been derived in the course of developing a number of broadband circulators.

These procedures have been applied successfully to the development of broadband strip-transmission-line *Y*-junction circulators for use over octave bandwidths in the frequency ranges of 0.6–1.2, 1.0–2.0, 1.5–3.0, 2.0–4.0, and 4.0–8.0 Gc.

The basis for this successful development procedure was suggested by the Humphrey and Davies paper [7], in which it was shown that the bandwidth could be extended by matching the ferrite element symmetrically using external reciprocal reactive networks. Using this approach, it was discovered that certain parameters (of the microwave structure and of the ferrite itself) could be varied to permit matching to be carried out with greater ease. Of greatest importance, an empirical evaluation of the impedance vs. frequency characteristics of the ferrite loaded junction revealed that adjustment of the saturation magnetization of the material could indeed be useful in creating the type of impedance response which realistically could be matched using microwave reactive structures. This evaluation was carried out in each band and resulted in the selection of a series of ferrite materials producing impedance characteristics suitable for broadband matching.

Thus, the operating bandwidth has been extended by utilizing external matching networks applied to ferrite loaded junctions that have been adjusted and optimized to reduce the difficulty of matching.

Empirical data are presented which indicate the relation between the saturation magnetization of the ferrite and the impedance vs. frequency characteristics in the bands previously mentioned. Also discussed in this regard are the effects of linewidth, width of the strip-transmission line, and the magnetic biasing field upon the impedance characteristics. Useful curves are presented which relate optimum ranges of $4\pi M_s$ and ferrite disk diameter to frequency. Methods of matching into the ferrite loaded junction are discussed, and the results are indicated for each band.

Technical Considerations

Broadband Approach

An important factor in the broadbanding effort was the recognition of the fact that the isolation of any port is dependent on the impedance match at the preceding port. It was verified experimentally, for a circulator operating in the low-loss mode port 1→port 2→port 3 →port 1 (1→2→3→1), that, when the power entered port 1, the isolation between port 1 and port 3 could be expressed by the following expression:

$$\text{Isolation } (1 \rightarrow 3) = 10 \log \frac{1}{\Gamma_2^2}$$

(provided that the normal situation exists in which power is not radiated), where Γ_2 = reflection coefficient of port 2.

This relation was found to hold accurately for VSWR values at least as high as 1.50 and does not require symmetrical matching.[1]

From this, it is clear that the basic problem of producing high isolation and low loss reduces to that of finding methods for matching into the ferrite loaded junction over a broad frequency spectrum.

Manuscript received June 5, 1964; revised January 26, 1965. This work was performed by Sperry Microwave Electronics Co. under Contract DA-36-039-SC-89214 sponsored by the U. S. Army Electronics Research and Development Lab., Fort Monmouth, N. J.

The author is with Scientific Atlanta, Inc., Atlanta, Ga. He was formerly with Sperry Microwave Electronics Co., Clearwater, Fla.

[1] A formal proof of this experimental result was derived using scattering matrix formulation and is provided in Appendix I.

This problem was approached in two ways: First, the impedance characteristics of the ferrite loaded junction were studied, while various parameters of the ferrite and microwave structure were varied. The result desired from this study was information leading to the development of techniques for simplifying the matching problem. Secondly, matching structures were developed to be used in conjunction with the forementioned characteristics.

Materials Selection

Since it was desired to discover what effect, if any, the saturation magnetization, $4\pi M_s$, of the ferrite material had upon the circulator impedance vs. frequency characteristics, it was necessary to choose a class of materials in which the dielectric constant, g factor, and linewidth ΔH could be held essentially constant while large changes in $4\pi M_s$ could be effected.

The basic material chosen for this evaluation was an aluminum substituted yttrium iron garnet. The properties of this garnet are well suited to this kind of test because of the following three factors:

1) The $4\pi M_s$ values can be varied from 1780 to 133 gauss for aluminum substitutions ranging from 0 to 27 per cent.
2) The material has a narrow linewidth that is approximately constant for aluminum substitutions from 0 to 27 per cent.
3) The value of the dielectric constant ϵ and the g factor change less than 10 per cent for aluminum substitutions from 0 to 25 per cent.

The particular $4\pi M_s$ values that were used for this test were 133, 200, 400, 500, 670, and 1005 gauss, respectively, which are listed in Table I along with properties and compositions.

TABLE I
ALUMINUM SUBSTITUTED YTTRIUM IRON GARNETS OF THE FORMULA $3Y_2O_3 \cdot 5[X\%Al_2O_3 \cdot (1-X)\%Fe_2O_3]$

Material No.	X	ΔH*	ϵ	$4\pi M_s$	°C Curie temp.	g factor	tan δ (less than)
1	27	40	14.25	133	136°	2.10	0.0002
2	25	40	14.38	200	140°	2.03	0.0002
3	20	40	16.0	500	150°	2.024	0.002
4	20	52	15.1	400	150°	2.03	0.0001
5	15	39	14.20	670	170°	2.02	0.0001
6	10	44	15.5	1005	200°	2.02	0.0001

* ΔH measured at 10 Gc.

PARAMETERS INVESTIGATION

Effects of $4\pi M_s$ in L and S Bands

Testing of ferrimagnetic materials with various values of $4\pi M_s$ was accomplished in the frequency bands 0.6–1.2, 1.0–2.0, 1.5–3.0, 2.0–4.0, and 4.0–8.0 Gc. For purposes of brevity, only the tests that were made in the 1.0–2.0 and 2.0–4.0 Gc bands will be described in detail here. Since the procedure is typical of all bands, only the final results will be presented in the 0.6–1.2, 1.5–3.0, and 4.0–8.0 Gc.

Material Evaluation, L-Band Region: The yttrium iron garnets listed in Table I as 1, 2, 4, and 5, respectively, were tested in this frequency band using a housing as shown in Fig. 1. These materials were tested at several magnetic fields, and the impedance characteristics were plotted on Smith charts.

The results of these tests are shown in Figs. 2 through 5. An interesting observation, which can be made from these Smith chart plots, is that, as the saturation magnetization of the garnet is increased, the impedance plot begins to coalesce. This suggests that the use of materials possessing high $4\pi M_s$ values would be advantageous. Because of low field loss considerations, however, only a limited increase in the saturation magnetization can be tolerated. From the measurements taken on these materials in L band, low field losses were apparent only for material no. 5, which exhibited a forward loss of 4.5 dB at 1.1 Gc. 123

Thus, in this band, material no. 4, whose operational characteristics produced a well-grouped impedance plot while operating in the low loss region, was considered the most promising material for the application of broadband matching techniques.

Material Evaluation, S-Band Region: The work done in the S-band region consisted of testing ferrimagnetic materials with various values of $4\pi M_s$ in a manner similar to the tests accomplished in the L-band region.

The $4\pi M_s$ values that were used for these tests were 200, 400, 670, and 1005 gauss which correspond to materials 2, 4, 5, and 6 listed in Table I.

All tests were conducted using a housing as shown in Fig. 1. Smith chart plots of the circulator using these materials were taken at various magnetic fields and are presented in Figs. 6 through 9.

These data agree with the results obtained in L band in that, as the $4\pi M_s$ is increased, the impedance plots tend to coalesce. Measurements of the loss characteristics indicated that the material with the highest $4\pi M_s$ which could be used, and yet avoid low field losses, was number 6. In both bands, then, a high saturation magnetization was required for producing the type of tight impedance plot grouping which effectively permits broadband matching.

A similar materials evaluation was made covering the remaining portion of the 0.6–8.0 Gc band. The results obtained were similar to those presented in the L- and S-band tests, in that, the higher the $4\pi M_s$, the more tightly grouped were the impedance characteristics for a particular frequency range. Also, as the frequency increased, the required $4\pi M_s$ for tight grouping became higher. The results of this investigation are presented in Fig. 10, which shows the optimum values of $4\pi M_s$ and ferrite disk diameters for each frequency.

DIMENSION	L-BAND	S-BAND
A	0.550	0.300
B	0.500	0.265
C	0.063	0.025

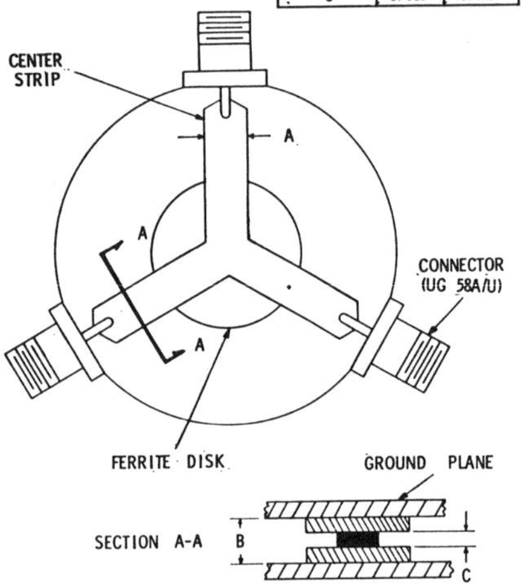

124

Fig. 1. Strip-transmission-line housing and ferr te configuration for the L- and S-band regions.

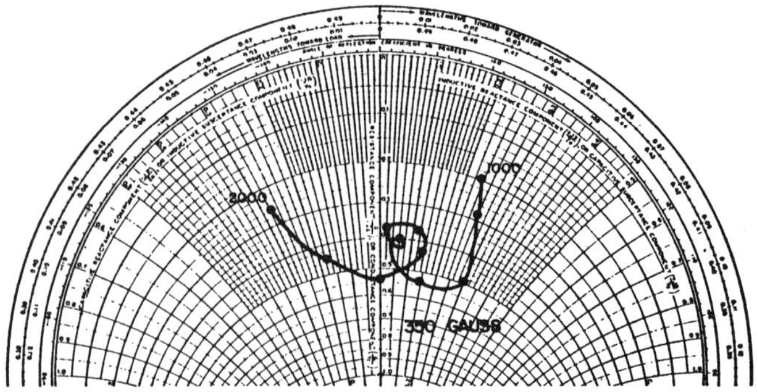

Fig. 4. L-band circulator impedance plot using material no. 4 ($4\pi M_s=400$ gauss).

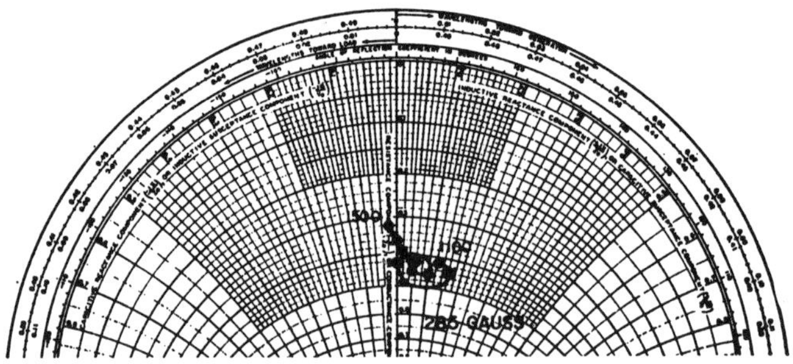

Fig. 5. L-band circulator impedance plot using material no. 5 ($4\pi M_s=670$ gauss).

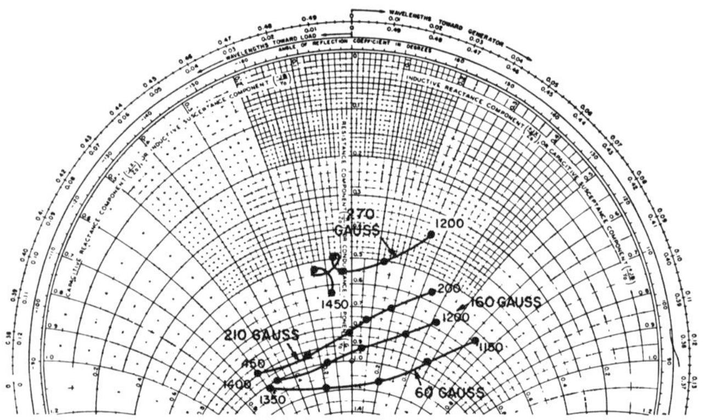

Fig. 2. L-band circulator impedance plot using material no. 1 ($4\pi M_s=133$ gauss).

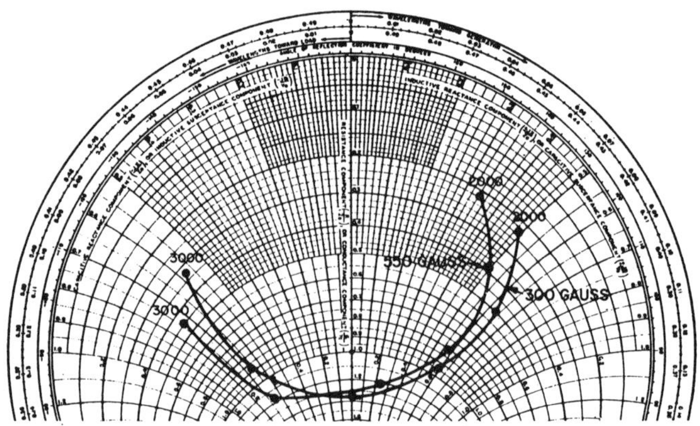

Fig. 6. S-band circulator impedance plot using material no. 2 ($4\pi M_s=200$ gauss).

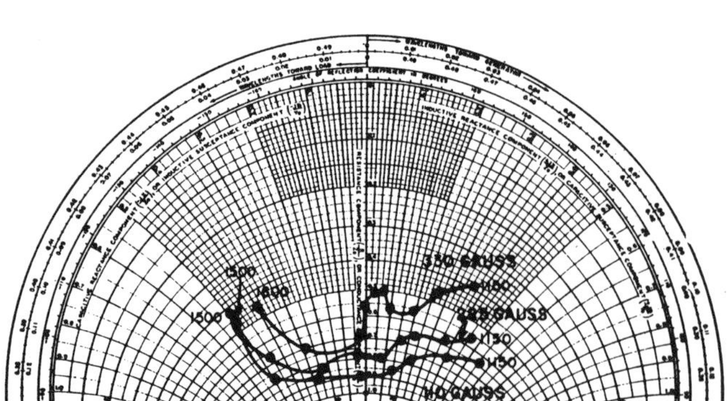

Fig. 3. L-band circulator impedance plot using material no. 2 ($4\pi M_s=200$ gauss).

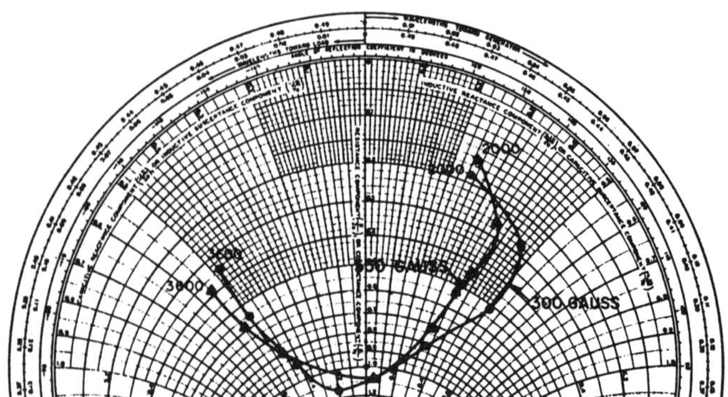

Fig. 7. S-band circulator impedance plot using material no. 4 ($4\pi M_s=400$ gauss).

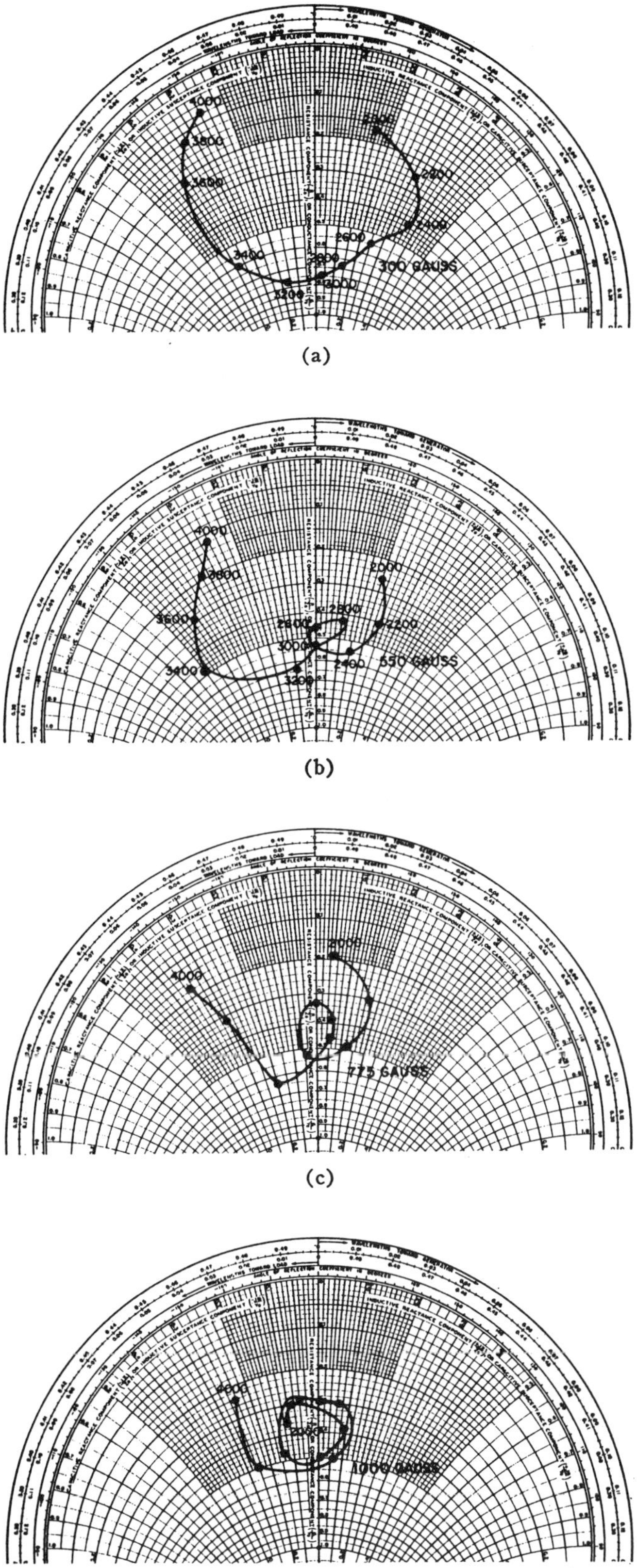

Fig. 8. *S*-band circulator impedance plot vs. magnetic field using material no. 5 ($4\pi M_s = 670$ gauss).

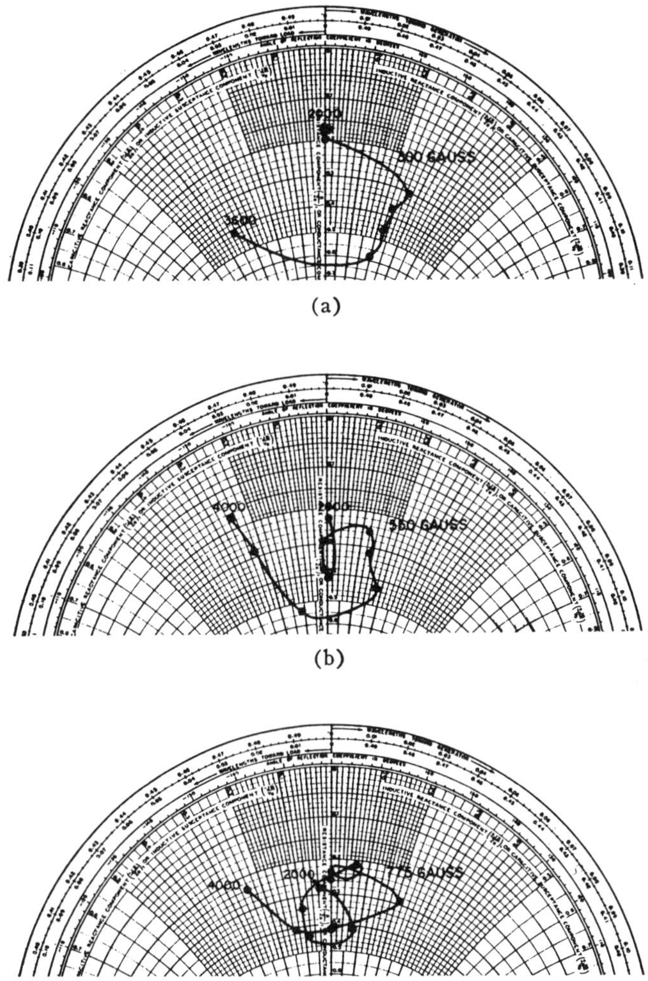

Fig. 9. *S*-band circulator impedance plot using material no. 6 ($4\pi M_s = 1005$ gauss).

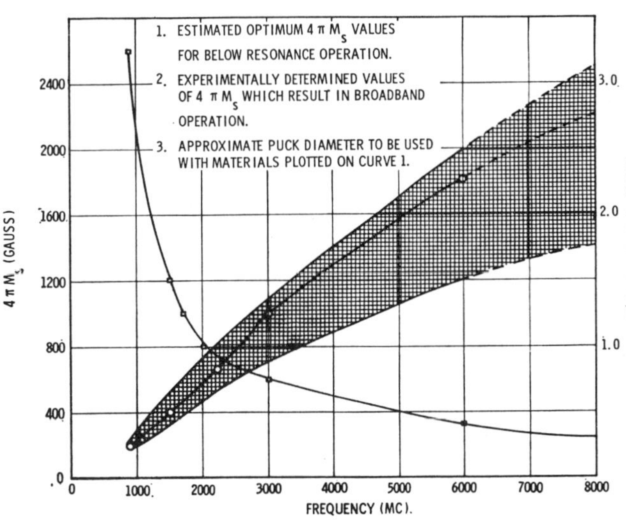

Fig. 10. Envelope of practical $4\pi M_s$ values and puck diameters for below resonance broadband operation.

Effects of Magnetic Field

The effects of varying the magnetic field are clearly illustrated in the impedance plots previously presented and are typical of all bands. It can be seen, for example, in Fig. 8(a) and 8(b) that the impedance becomes more tightly grouped as the applied field is increased; and at even higher fields, as shown in Fig. 8(c) and 8(d), the impedance plot continues to coalesce. However, as the field is increased the resonant frequency also increases, and a point is reached where high losses become evident on the lower end of the frequency band. In any given case, the resonance loss limits the extent to which the impedance plot can be coalesced by raising the applied field. In each band the optimum field was found to be approximately equal to the $4\pi M_s$ of the ferrite material.

Effects of Ferrite Geometry

Two examples are cited in Figs. 3 and 11 in which the impedance characteristics of the triangular- and disk-shaped garnet materials can be compared. In both cases, the similarity of the plots is surprisingly close.
126 In the tests to date, no significant difference in the impedance characteristics for the disk or triangle has been found for any of the frequency bands or materials.

The disk-shaped ferrite, being easier to fabricate and smaller in size, was used in the development of these devices.

Effects of Dielectric Loss

The importance of using ferrimagnetic materials having low dielectric loss tangents, tan δ, can be simply illustrated through the use of experimental data taken recently in the L-band region. The materials used were numbers 3 and 4, respectively, whose only significant differences are in their dielectric loss tangents. The circulators containing the two materials were tuned and tested under similar conditions, and the results are plotted in Fig. 12(a) and 12(b). While the isolation and VSWR are quite similar, a noticeable drop of about 0.5 to 0.6 dB in the forward loss is evident for the material with the lower loss tangents in Fig. 12(b).

Thus, for circulators operating below resonance utilizing low saturation magnetization materials, and hence large disk diameters, dielectric loss is an important loss mechanism.

Effects of Linewidth

In choosing materials for possible below resonance operation, careful consideration must be given to the linewidth and frequency of operation. At the lower frequencies, where the field required for resonance becomes smaller and smaller and the tail of the resonance loss curve approaches the low field loss curve, the possible range of magnetic field values for low-loss operation becomes diminishingly small. This also becomes true at the higher frequencies where it is desirable to use the highest possible $4\pi M_s$ and biasing field. In either case, a narrow linewidth material has an appreciably larger operating region between the low field loss and the resonant loss region. From this standpoint, materials having the narrowest possible linewidths are desirable.

Materials have been tested which have all properties similar except linewidth. The results of these tests indicate that the linewidth has no measurable effect on the impedance characteristics of a circulator while operating in the low-loss region.

On the basis of these results, the linewidth of the material should be chosen to be as narrow as possible to permit the use of the highest possible saturation magnetization.

Effects of Strip-Transmission-Line Parameters

Only a limited amount of experimental data has been taken to study the specific function that stripwidth, strip thickness, characteristic impedance, etc., play in the bandwidth properties of circulators.

The strip-line configurations, which were used in the development of the circulators in the 1.0–2.0, 1.5–3.0, 2.0–4.0, and 4.0–8.0 Gc bands, are basically similar to Fig. 1. The ground plane spacings, stripwidth, and strip thickness were chosen to produce 50-ohm lines and also to suppress the propagation of higher order modes. This symmetrical Y-center conductor was used throughout the development described in the preceding paragraphs.

In the final stages of development of the 1.5–3.0 and 2.0–4.0 Gc bands, difficulty was encountered in obtaining a sufficiently good impedance grouping when using the material selected from the foregoing evaluation in a standard junction. Attempts were made to obtain an improved impedance plot by altering the center conductor geometry in the region of the junction. It became readily apparent from the experimental trials made that any *large* deviation from the straight Y configuration was detrimental to the impedance grouping. After a number of trials, some improvement was obtained by adding a slight taper to each leg entering the junction. Through this method, the necessary bandwidth was acquired for both bands.

In the 600–1200 Mc unit, a center conductor with a shielding disk [5] was found to be very useful for grouping the impedance characteristics. In this band, using material no. 2, at a constant magnetic biasing field of 190 gauss, a series of Smith chart plots was made for the following shield diameters: Straight Y (no shield), 0.800, 1.750, 2.000, 2.700, and 3.220 inches. The effects of the shield diameters are illustrated in Fig. 13. The two-inch-diameter shield was chosen to be used as the final design and later resulted in an octave bandwidth circulator.

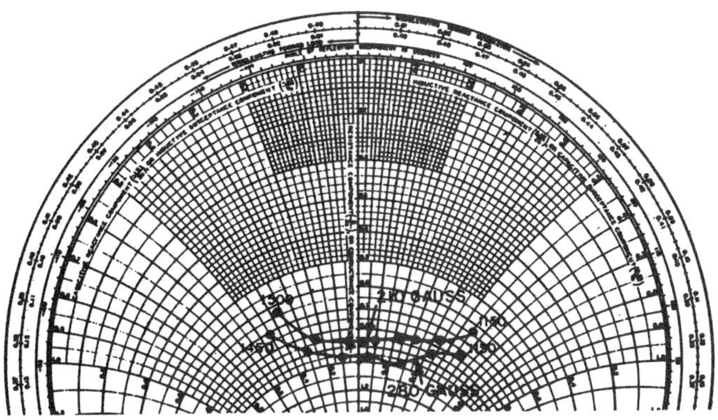

Fig. 11. *L*-band circulator impedance plot using material no. 2 (triangular-shaped ferrite element).

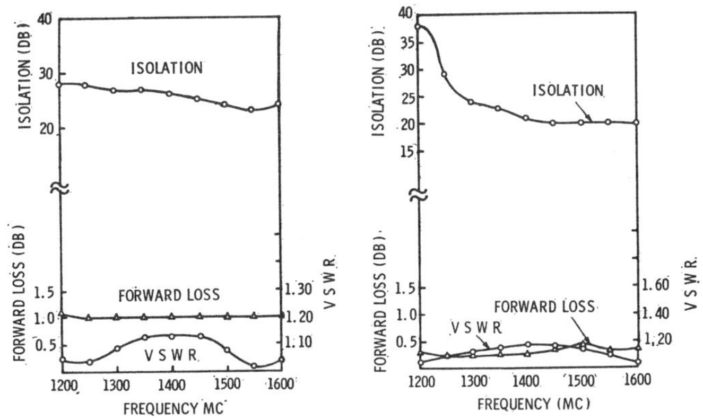

(a) (b)

Fig. 12 (a) Loss characteristics of an *L*-band circulator using material no. 3 (tan $\delta = 0.002$). (b) Loss characteristics of an *L*-band circulator using material no. 4 (tan $\delta = 0.0001$).

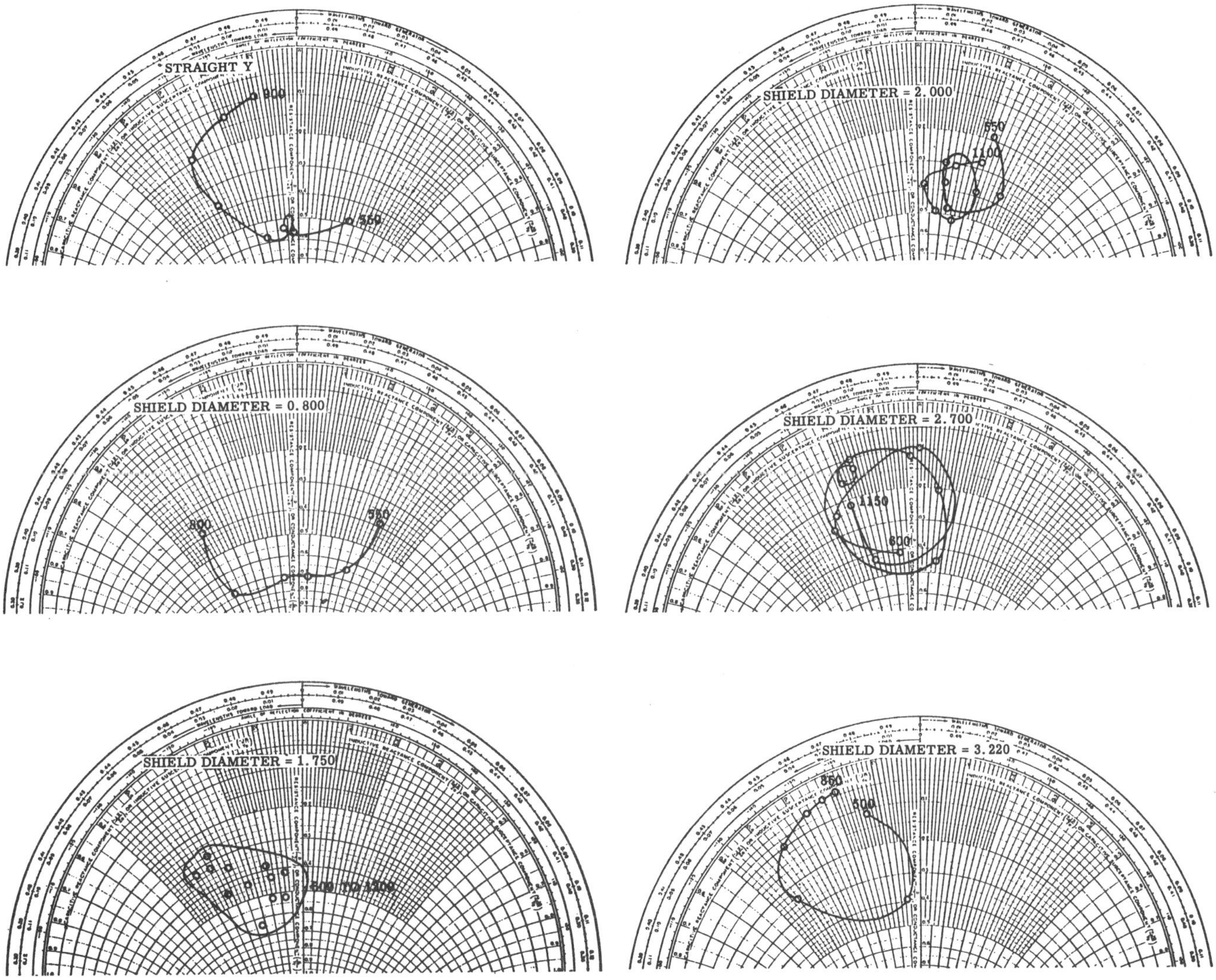

Fig. 13. UHF circulator impedance plots using a straight *Y*, 0.800, 1.750, 2.000, 2.700, and 3.220-inch shield diameters (material no. 2, biasing field = 190 gauss).

Octave Bandwidth Circulator Designs

Design Generation

From the results of the study just outlined, sufficient information is available for designing a circulator with impedance characteristics most favorably grouped for broadband matching. The steps to be followed in choosing optimum parameters are outlined below:

1) The choice of $4\pi M_s$ and ferrite disk diameter may be taken directly from Fig. 10.
2) The magnetic biasing field should be approximately equal to the $4\pi M_s$ of the material previously chosen, but can be adjusted in test for producing the best performance.
3) A ferrite material having a very low dielectric loss tangent is needed if low insertion loss is to be maintained.
4) The linewidth of the material must be narrow (approximately 50 to 100 Oe) if resonance losses are to be avoided.
5) In the strip-transmission-line structure, many combinations of stripwidth, strip thickness, and ground plane spacings may result in a broadband device. It is felt that the main consideration should be maintaining a 50-ohm line and preventing the propagation of higher order modes. Some guide in choosing the stripwidth, with relation to the ferrite disk diameter, can be obtained from Fig. 1.

After the optimization of the foregoing parameters, it becomes necessary to match these characteristics over octave bandwidths. The techniques that have been developed for all bands utilize common matching circuits. A description of some of these circuits and the results obtained are given later.

Specific Matching Techniques

In Fig. 4 an impedance plot of a 1.0–2.0 Gc circulator in the unmatched condition is shown. The grouping of this plot, as it stands, does not appear particularly good for broadband matching. However, if the output ports are matched to prevent re-reflections within the circulator, a surprisingly well-grouped plot results.

With the use of a double stub tuner inserted in the output arm (port 2), as shown in Fig. 14, the re-reflections between ports can essentially be eliminated by adjusting for maximum isolation between ports 1 and 3.

Using this technique for each of the frequency points, results were obtained as illustrated in Fig. 15. This plot now represents more accurately the impedance characteristic for which to design the matching element, since in the matched condition these internal reflections are very nearly canceled. The impedance response now looks very nearly like a low impedance resistive load.

The type of matching structure which has been used with greatest success, thus far, has been a single-element quarter-wavelength transformer located symmetrically in each arm. The function of this transformer is to match the 50-ohm line to the low impedance ferrite junction. With the use of this dielectric transformer, it was possible to obtain a good impedance match for 1.5–3.0, 2.0–4.0, and the 4.0–8.0 Gc units over an octave bandwidth.

In the development of the two lowest frequency units, difficulty was encountered in obtaining the full octave bandwidth using the quarter-wavelength matching transformers. The results of both units, when matched with the single transformer, are typically illustrated by the impedance plot shown in Fig. 16. This impedance plot is one which can be further improved by the addition of a resonant element having an impedance with an equal magnitude and an opposite phase slope. This type of circuit can be attained by the use of an open-circuited, quarter-wavelength stub in series with the strip-line center conductor. A stub of this type is equivalent to a series resonant circuit whose resonant frequency is controlled by the length of the stub. This circuit was built and tested in a separate housing, the results of which are shown in Fig. 17. With the addition of the resonant stub to the circulator, considerable improvement was obtained as shown in Fig. 18. A sketch of this resonant stub is shown in Fig. 19, and a photograph of the circuit built into the circulator is shown in Fig. 20.

Similar techniques were employed in improving the matched characteristics of the 1.0–2.0 Gc unit. The only parameter that required adjustment in using this stub in the 1.0–2.0 Gc band was the length of the wire stub.

Thus, in general, for achieving a sufficiently good match in any of the frequency bands, from 0.6–8.0 Gc, several techniques have been developed and demonstrated to be useful in matching the type of response which is characteristic of the below resonance circulator. With the use of these matching techniques and the information previously listed, octave bandwidth designs may be derived.

Electrical Performance

Typical data for the 0.6–1.2, 1.0–2.0, and 2.0–4.0 Gc units are shown in Figs. 21, 22, and 23. The performance of the 1.5–3.0 and 4.0–8.0 Gc units is quite similar to the 2.0–4.0 Gc circulator and will not be reproduced here. In some cases the relations between VSWR and isolation derived earlier do not hold in these illustrations. This variance is due to the ports being slightly unsymmetrical.

It should be noted that the high loss characteristics at the lower frequencies in Fig. 21 are probably a result of low field losses. With available ferrimagnetic materials, this type of loss mechanism is extremely difficult to avoid at frequencies as low as 0.600 Gc.

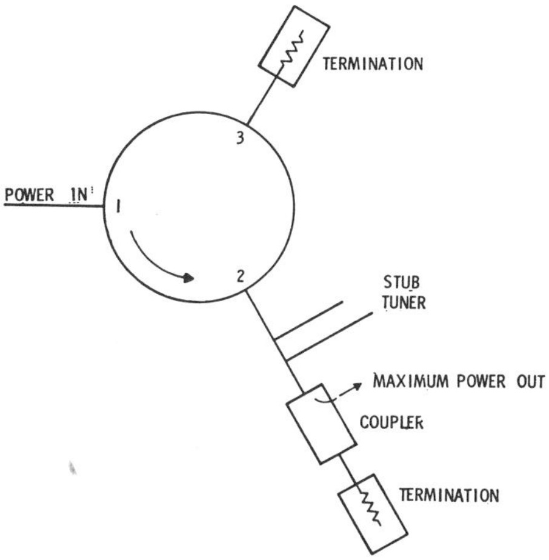

Fig. 14. The use of the double stub tuner for preventing re-reflections in circulator measurements.

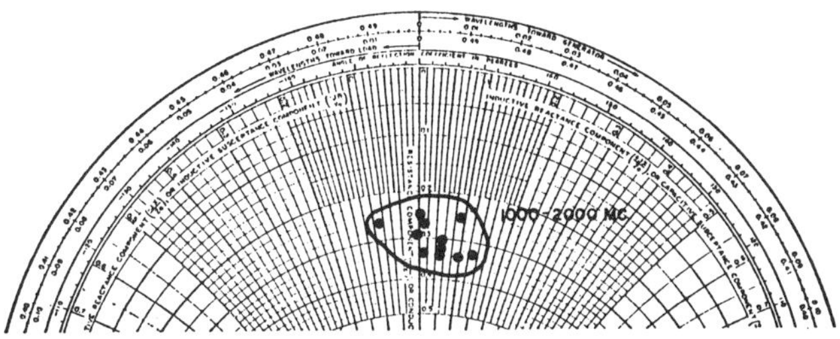

Fig. 15. *L*-band circulator impedance plot using material no. 4 with tuner in output arm.

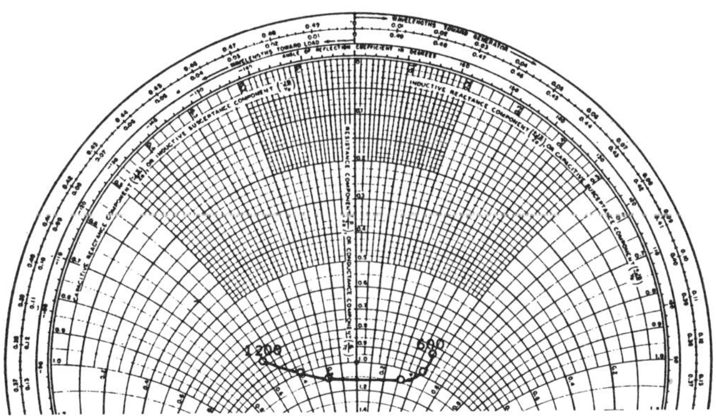

Fig. 16. Typical impedance characteristics of a 0.6–1.2 Gc circulator when matched with single-element dielectric transformers.

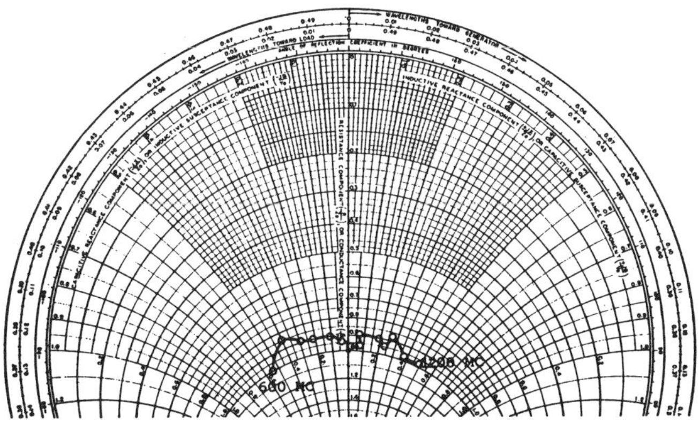

Fig. 17. Impedance characteristic of an open-circuited quarter-wavelength stub when tested in a 50-ohm line.

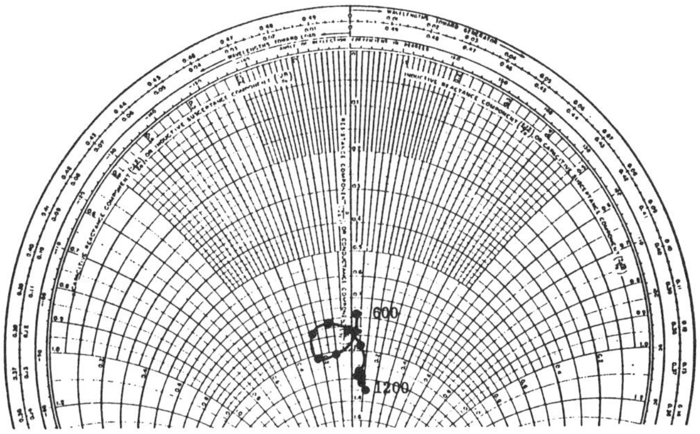

Fig. 18. Impedance characteristic of a 0.6–1.2 Gc circulator when matched with a resonant stub.

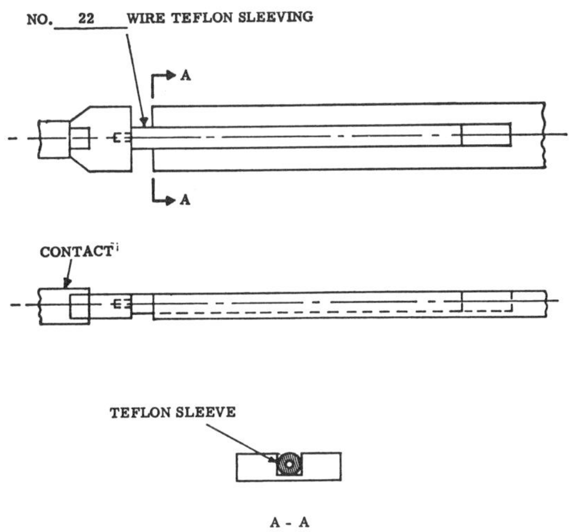

Fig. 19. Sketch of the open-circuited stub mounted in series with the strip-line center conductor.

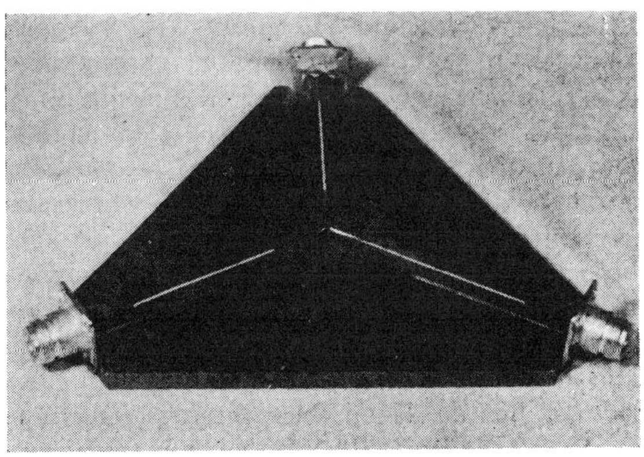

Fig. 20. Photograph illustrating the location of the resonant circuit.

129

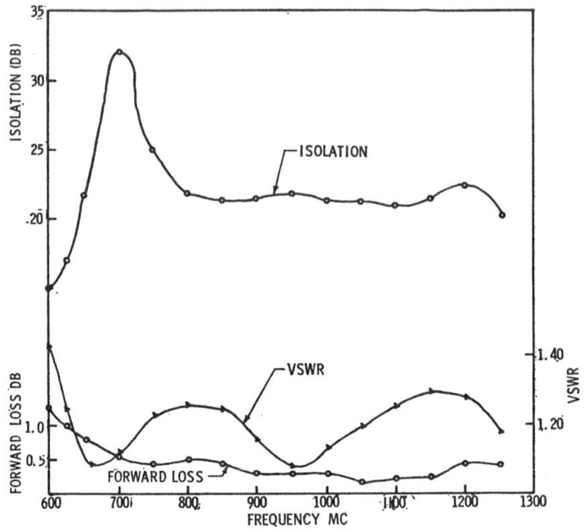

Fig. 21. UHF circulator characteristics.

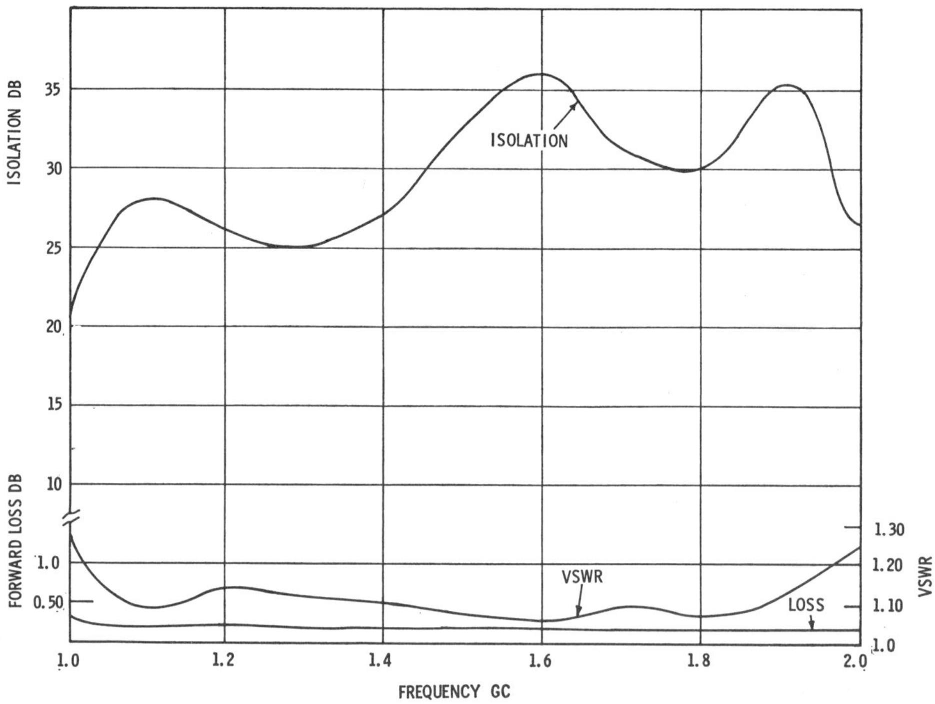

Fig. 22. Octoave bandwidth operating characteristics of the 1.0–2.0 Gc circulator.

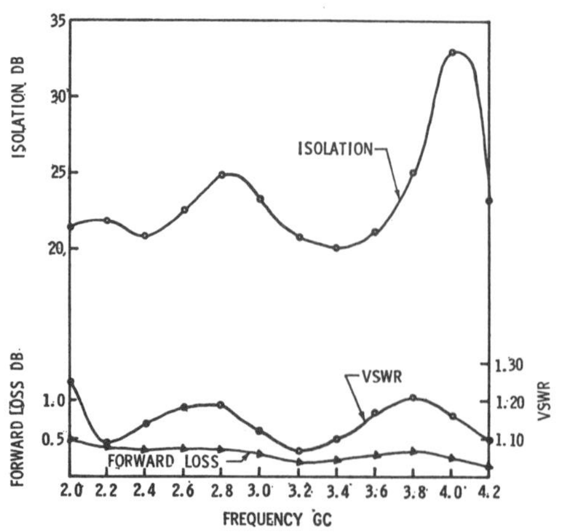

Fig. 23. Characteristics of a typical *S*-band circulator.

SECTION II

Conclusions

The most important single factor in contributing to the development of the broadband circulators was the discovery of the strong dependence of the impedance characteristics upon the saturation magnetization. Once the optimum $4\pi M_s$ is established, the parameters of secondary importance such as magnetic field, strip thickness, stripwidth, ground plane spacing, matching techniques, etc., can be carefully adjusted for attaining maximum bandwidth.

Although a great deal has been accomplished in broadbanding circulators by empirically optimizing the saturation magnetization and developing the appropriate matching techniques, it is presently felt that greater bandwidths are attainable. A likely means of accomplishing this end is through the use of multiple-element matching transformers and through improved center conductor geometries that can be developed experimentally.

Appendix I

An interesting theoretical result that confirms what has been observed experimentally for some time with regard to the relation between isolation and VSWR in a circulator is derived in this section.

It is well known [9] that a lossless *reciprocal* three-port junction cannot be matched out. This is easily shown as follows: If the junction is matched and reciprocal, the scattering matrix must be of the form:

$$\mathcal{S} = \begin{pmatrix} 0 & S_{21} & S_{31} \\ S_{21} & 0 & S_{32} \\ S_{31} & S_{32} & 0 \end{pmatrix}. \qquad (1)$$

If, in addition, it is lossless, the following equations must hold:

$$S_{21}S_{32*} = S_{21}S_{31*} = S_{31}S_{32*} = 0. \qquad (2)$$

$$|S_{21}|^2 + |S_{31}|^2 = |S_{21}|^2 + |S_{32}|^2 = 1. \qquad (3)$$

There does not exist a set of values satisfying (2) and (3) since this would require one of the elements to be zero and, at the same time, have an absolute value of unity, which obviously cannot be. Hence, a reciprocal lossless three-port junction cannot be matched.

One can arrive at essentially the same result from a different point of view. If the three-port junction is required to be symmetric, then the scattering matrix can be of the form:

$$\mathcal{S} = \begin{pmatrix} S_{11} & S_{31} & S_{21} \\ S_{21} & S_{11} & S_{31} \\ S_{31} & S_{21} & S_{11} \end{pmatrix}. \qquad (4)$$

If the junction is also to be matched, then the matrix must be of the form:

$$\mathcal{S} = \begin{pmatrix} 0 & S_{31} & S_{21} \\ S_{21} & 0 & S_{31} \\ S_{31} & S_{21} & 0 \end{pmatrix} \qquad (5)$$

and if the junction is lossless, then it is seen that

$$|S_{21}| = 1 \quad \text{and} \quad S_{31} = 0 \qquad (6)$$

or

$$S_{21} = 0 \quad \text{and} \quad |S_{31}| = 1. \qquad (7)$$

Condition (6) gives a junction that circulates in the mode, port 1→port 2→port 3→port 1 (1→2→3→1), and condition (7) gives one that circulates in the mode (1→3→2→1). Hence, a matched symmetric, lossless three-port junction is necessarily a circulator. Also, this shows that the procedure of matching out ports in the laboratory is precisely equivalent to building a circulator; a result that was also obtained from the solution of the boundary-value problem using Bosma's boundary conditions [5].

It has been observed for some time that, if one determines the VSWR for each of the ports of a Y junction and if the VSWR's are relatively small (e.g., less than 1.5), then the isolation of port 3, when power is put in port 1, will be equal to the VSWR of port 2 when converted to decibels. This is true provided the junction is tending to circulate in the 1→2→3→1 mode. If circulation is in the mode 1→3→2→1, obvious changes in the foregoing statement are required. This result can be obtained theoretically as follows. Consider the scattering matrix for a general three-port junction:

$$\mathcal{S} = \begin{pmatrix} S_{11} & S_{12} & S_{13} \\ S_{21} & S_{22} & S_{23} \\ S_{31} & S_{32} & S_{33} \end{pmatrix}. \qquad (8)$$

If the VSWR's of all ports are small, this requires

$$|S_{ii}| \ll 1 \qquad \text{for } i = 1, 2, 3. \qquad (9)$$

If the Y junction is tending to circulate in the 1→2→3→1 mode, then

$$|S_{31}|, \quad |S_{12}|, \quad |S_{23}| \ll 1 \qquad (10)$$

and

$$|S_{21}| \doteq |S_{32}| \doteq |S_{13}| \doteq 1. \qquad (11)$$

It is necessary to assume the junction has low loss in order to obtain (11). If it has low loss, then

$$S_{11}S_{12*} + S_{21}S_{22*} + S_{31}S_{32*} \doteq 0. \qquad (12)$$

Since $S_{11}S_{12*}$ is small to the second order and $|S_{21}| \doteq |S_{32}| \doteq 1$ to first order, it follows that to first order we have

$$|S_{22}| = |S_{31}|. \qquad (13)$$

According to (13), if power $|A|^2$ enters port 1, then power $A^2|S_{31}|^2$ will come out port 3, and if power A^2 is put in port 2, power $|A|^2|S_{22}|^2=|A|^2|S_{31}|^2$ will be reflected. This is precisely the result sought.

A special case of this result, obtained by Aitken and McLean [8], states that, if the junction is symmetric, the power reflected at port 1 and the power out port 3 will be equal. This follows since, if the junction is symmetric, $S_{11}=S_{22}$ and $|A|^2|S_{31}|^2=|A|^2|S_{11}|^2$.

The fact that the derivation of (13) did not require symmetry makes the result particularly useful to the experimentalist since he will, in general, be dealing with junctions that are not truly symmetric.

Acknowledgment

The author wishes to acknowledge the assistance of W. Passaro and D. Landry in the development of these devices and Dr. G. Burdick, who contributed greatly in providing the theoretical background and guidance throughout the program.

References

[1] Milano, U., J. H. Saunders, and L. Davis, Jr., A Y-junction strip-line circulator, *IRE Trans. on Microwave Theory and Techniques*, vol MTT-8, May 1960, pp 346–350.
[2] Clark, J., Perturbation techniques for miniaturized coaxial Y-junction circulators, *J. Appl. Phys.*, suppl. to vol 32, Mar 1961, p 323s–324s.
[3] Clark, J., and J. Brown, Miniaturized, temperature stable coaxial Y-junction circulators, *IRE Trans. on Microwave Theory and Techniques*, vol MTT-9, May 1961, pp 267–269.
[4] Auld, B. A., The synthesis of symmetrical waveguide circulators, *IRE Trans. on Microwave Theory and Techniques*, vol MTT-7, Apr 1959, pp 238–246.
[5] Bosma, H., On the principle of stripline circulation, *Proc. IEE (London)*, vol 109, pt B, suppl. no 21, Jan 1963, p 137–146.
[6] ——, On stripline Y-circulation at UHF, *IEEE Trans. on Microwave Theory and Techniques*, vol MTT-12, Jan 1964, pp 61–72.
[7] Humphreys, B. L., and J. B. Davies, The synthesis of N-port circulators, *IRE Trans. on Microwave Theory and Techniques*, vol MTT-10, Nov 1962, pp 551–554.
[8] Aitken, F. M., and R. McLean, Some properties of the waveguide Y-circulator, *Proc. IEE (London)*, Feb 1963, pp 256–260.
[9] Montgomery, C., R. H. Dicke, and E. M. Purcell, *Principles of Microwave Circuits.* New York: McGraw-Hill, 1948, ch 12.

Reprinted from IEEE TRANSACTIONS
ON MICROWAVE THEORY AND TECHNIQUES
Volume MTT-13, Number 3, May, 1965
Pp. 335-345

A High-Power UHF Circulator

YOSHIHIRO KONISHI, SENIOR MEMBER, IEEE

Abstract—**The insertion loss, the bandwidth ratio, and the nonlinearity of a high-power UHF circulator are discussed generally with regard to the characteristics, volume, and filling factor of the ferrite. Theory and experiment are made on the high-power circulator with ferrite, where either surface of the ferrite comes into contact with air. A wideband technique in improving the narrowband that is essentially the result of the filling factor of ferrite is also described.**

To avoid the center conductor heating effect, a circulator without a center conductor is described.

Experiments have proven that, for ferrite nonlinearity, the threshold power by spinwave occurs in a polycrystal for CW power even above resonance and is changed by a external field strength, whereas the nonlinearity is not observed in a single crystal.

I. Introduction

THE ORDINARY Y-stripline circulator has been developed and studied by many authors.[1]–[5] However, in the case of a high-power CW circulator important factors for practical use such as temperature rise, ferrite nonlinearity, bandwidth, and insertion loss have not been adequately explained. As an example, nonlinearity occurs in a polycrystal at a comparatively low-power CW even under an above resonance operation, although such a phenomenon is not observed in a single crystal.

First, the general relation between several electrical characteristics of a circulator is studied in Section II. As a result, it becomes clear that the ferrite of the volume necessary for linearity should come into contact with as wide a surface as possible for good heat conductivity. It is also clarified that insertion loss depends only on ηQ values of the ferrite material reaching a minimum value at the optimum external field, and further, that the bandwidth ratio is related to the filling factor of the ferrite.

Second, to satisfy the general requirements already mentioned, analysis is made for the circulator in which either surface of a ferrite plate comes into contact with a small dielectric constant and permeability, such as air, to create a wide surface and volume of ferrite. Wideband techniques to improve the narrowband, essentially caused by the filling factor of the ferrite, are described. For practical use, construction that offsets rising temperature as a result of the center conductor is also considered.

Finally, several experiments confirming both theory and practical use are described.

Manuscript received May 17, 1967; revised August 7, 1967.

The author is with the Technical Research Laboratories, Nippon Hoso Kyokai, Tokyo, Japan.

II. General Considerations

The approximate values of the characteristics of a circulator such as insertion loss L (decibels), bandwidth ratio w, and available maximum power P_{crit}, are obtained by relating the volume τ and the filling factor k_f (= the time average magnetic energy $\tilde{W}_m$/time average total magnetic energy $\tilde{W}_{mt}$) of the ferrite contained in a symmetrical 3 port under the following assumptions:

First, it is assumed that there exists no magnetic energy inside the ferrite under the same-phase excitation. This assumption is satisfied for the lumped element Y circulator[7] and is nearly satisfied for a stripline circulator as described in Section III-B.

Second, it is assumed that tensor permeability takes the positive and negative polarized permeability under positive and negative rotational excitation, respectively.

Let ω_3 and ω_2 be the angular eigenfrequencies at which each of the terminals are at open impedance under positive and negative excitations, and let $\tilde{\omega}$ be the eigenfrequency corresponding to rotational excitation when the ferrite is assumed to be an isotropic medium with a permeability of $(\mu_+ + \mu_-)/2 = \tilde{\mu}$.

Since ω_3 and ω_2 are considered to be the perturbed values of $\tilde{\omega}$ caused by perturbation of permeability $\tilde{\mu}$ to $\tilde{\mu}_+$ and $\tilde{\mu}_-$, we get

$$\frac{\omega_2 - \omega_3}{\tilde{\omega}} = \frac{(\mu_+ - \mu_-)\iiint_\tau |H|^2 dv}{2\tilde{\mu}\iiint_V |H|^2 dv} = \frac{\mu_+ - \mu_-}{2\tilde{\mu}} k_f = \eta k_f. \tag{1}$$

The eigenvalues of admittance corresponding to a negative and positive rotational excitation y_2 and y_3 should take $-j(1/\sqrt{3}R)$ and $+j(1/\sqrt{3}R)$ when the reference plane is defined so that the eigenvalue y_1 corresponding to a same-phase excitation is infinite, where R is the characteristic impedance of the junctions. Since y_2 and y_3 are the frequency variations from the center frequency $\tilde{\omega}$ to ω_2 and ω_3, they are connected to the total reactive energy $\tilde{W}_t$ at $\tilde{\omega}$ in the following:[7]

$$\delta y = y_3 = -y_2 = j\frac{2}{3}\delta\omega \cdot \tilde{W}_t = j\frac{\omega_2 - \omega_3}{3}\tilde{W}_t = j\frac{1}{\sqrt{3}R} \qquad \delta\omega = \omega_2 - \tilde{\omega} = \tilde{\omega} - \omega_3 \tag{2}$$

thus

$$(\omega_2 - \omega_3)\tilde{W}_t = \frac{\sqrt{3}}{R}. \tag{3}$$

As the external Q, Q_e takes the value of

$$Q_e = \frac{\tilde{\omega}\tilde{W}_t R}{3}. \tag{4}$$

From (1), (3), and (4)

$$k_f = \frac{2}{\sqrt{3}Q_e}\frac{\tilde{\mu}}{\mu_+ - \mu_-} = \frac{1}{\sqrt{3}Q_e\eta}. \tag{5}$$

Since the voltage at each terminal for the positive rotational excitation with a power of P_{crit} is $1/3\sqrt{RP_{crit}}$, the intensity of an RF magnetic field at the center of the ferrite h_{crit} takes the value of

$$h_{crit} = \tfrac{1}{3}\sqrt{RP_{crit}}h_{max}$$

thus

$$P_{crit} = \frac{9h^2_{crit}}{Rh^2_{max}} \tag{6}$$

where h_{max} is the RF magnetic field intensity for a unit voltage excitation. Considering the relation

$$\frac{1}{2}\tilde{W}_t k_f = \frac{\tilde{\mu}}{2}\iiint_\tau H^* H dv = K\frac{\tilde{\mu}h^2_{max}}{2}\tau \tag{7}$$

where $K=1$ in the case of a uniform field in ferrite, and from (4) through (7), we get

$$P_{crit} = \frac{9h^2_{crit}K\tilde{\omega}(\mu_+ - \mu_-)\tau}{2\sqrt{3}} = kh^2_{crit}(\mu_+ - \mu_-)\tau\omega. \tag{8}$$

Next, the bandwidth ratio w is obtained. The deviation of y_2 and y_3 in amounts of δy_2 and δy_3 corresponding to the deviation of frequency $\delta\omega$ causes the leakage voltage $|S''|$ at the isolated terminal, and is expressed by

$$|S''| = \frac{R}{2}\sqrt{|\delta y_2|^2 + |\delta y_3|^2 - |\delta y_2|\cdot|\delta y_3|}. \tag{9}$$

Substituting into (9) the relation obtained from (2) and (4) and approximating $\delta y_2 \doteqdot \delta y^3$ under $c \ll 1$, we get

$$w = 2\sqrt{3}|S''|k_f\eta. \tag{10}$$

Last, the insertion loss L (decibels) is expressed by the magnetic energy for rotational excitations and the corresponding quality factors of $\mu_\pm$, $Q_\pm$, and a quality factor of the dielectric constant of ferrite Q_c as in (11). (See Appendix I.)

$$L = 2.5\frac{1}{\eta}\left(\frac{1}{Q_+} + \frac{1}{Q_-} + \frac{2}{Q_c}\right) \doteqdot \frac{5}{\eta}\left(\frac{1}{Q_{eff}} + \frac{1}{Q_c}\right). \tag{11}$$

Equations (8), (10), and (11) show the relation between P_{crit}, w, L and τ, η, Q. Equation (8) shows that a high P_{crit} requires a large volume. It is also understood from (10) that the bandwidth ratio w, which becomes narrower for a smaller filling factor k_f required for a high-power circulator mentioned later and for a smaller η under a stronger dc external magnetic field, is not concerned with the volume τ. From (11) it is also understood that insertion loss L depends only on the $Q\eta$ product of material and not on τ and k_f. Further, an optimum dc magnetic field exists which satisfies the minimum insertion loss under the maximum $Q\eta$ value.

III. Partially Filled Coaxial Y Circulator

Increasing volume τ and surface of the ferrite required for a high-power circulator is possible by making either surface of the ferrite come into contact with an air medium. In this Section the approximate values of eigenvalues are theoretically obtained as a function of the ratio of the width of ferrite to that of an air medium, and to the diameter of the ferrite. Based upon these results, the design considerations for this type of circulator are presented.

A. Analysis

The TM mode exists in a direction perpendicular to the surface of the ferrite where the tangential component of electrical field changes in sinusoidal and exponential variation inside the ferrite and air region, respectively, that is, if the ferrite is assumed to be an isotropic medium.

Using the potential function of the TM mode for a trial function, the eigenfrequencies corresponding to eigenvectors are calculated by the variational technique, and the eigenvalues are obtained by the calculated values of the energies together with corresponding eigenfrequencies.

However, there are variations of (12) corresponding to a trial electric field E, magnetic field H, electric current J, and magnetic current M inside a cavity resonator.

$$\langle EJ^* \rangle_{V+S} + \langle HM^* \rangle_{V+S} = 0 \tag{12}$$

where $\langle \ \rangle$ denotes the sum of the volume integral of scalar product of the continuous region V and the surface integral on discontinuous surface S. The value of ω obtained by solving (12) takes approximate values of the second-order error.[6],[8] Assuming that the ferrite is an isotropic medium with a proper isotropic permeability, we can also obtain the angular frequency ω' by (12) using the same trial function.

However, it has been verified that ω can be obtained by ω' and the perturbational term (13). (See Appendix II.)

$$\omega = \omega' \left\{ 1 - \frac{\iiint_\tau H \cdot (\boldsymbol{\mu}^* - \mu) H^* dv}{2 \iiint_V H \cdot \boldsymbol{\mu}^* H^* dv} \right\}. \tag{13}$$

1) Same-Phase Excitation: The eigenfrequency satisfying the zero impedance at each terminal under a same-phase excitation with unit current is obtained. The trial potential function assumed in (14) corresponds to the first three terms of the infinite series solution and is for air and ferrite, respectively,

$$\begin{aligned}\varphi_1 &= C_1 \cos k_{Z1}(h - Z)\{J_0(k_\rho \rho) + C_3 \cos 3\phi \cdot J_3(k_\rho \rho) \\ &\qquad + C_6 \cos 6\phi \cdot J_6(k_\rho \rho)\}, \\ \varphi_2 &= C_2 \cos k_{Z2} Z \{J_0(k_\rho \rho) + C_3 \cos 3\phi \cdot J_3(k_\rho \rho) \\ &\qquad + C_6 \cos 6\phi \cdot J_6(k_\rho \rho)\}\end{aligned} \tag{14}$$

where

$$C_2 = C_1 \frac{\cos k_{Z1}(h - t)}{\cos k_{Z2} t}$$

from the continuity on the ferrite surface.

Since we have the boundary condition on the periphery of the center conductor

$$\begin{aligned} &E_Z = 0, \qquad \phi = \frac{2\pi}{3} m \qquad m = 0, 1, 2, \\ &H_\phi = 0, \qquad \frac{\theta}{2} + \frac{2\pi}{3} m < \phi < -\frac{\theta}{2} + \frac{2\pi}{3}(m+1) \\ &\qquad\qquad\qquad\qquad\qquad\qquad m = 0, 1, 2 \end{aligned} \tag{15}$$

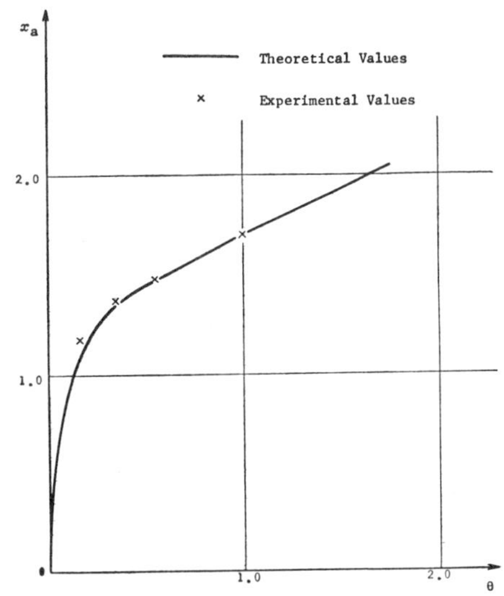

Fig. 1. Values of x_a.

the values of C_3 and C_6 may be selected to satisfy

$$E_Z = 0 \text{ at } \phi = \frac{2\pi}{3} m,$$

and

$$H_\phi = 0 \text{ at } \phi = \frac{2\pi}{6} + \frac{2\pi}{3} m.$$

However, since (14) satisfied Maxwell's equation inside the ferrite and the boundary condition on the broad surface of the ferrite, trial electric currents J_s and the magnetic current M_s may be considered only on the surface $\rho = a$. In this case, J_s flows toward the Z direction to satisfy (15) and M_s flows toward the ϕ direction to connect the mode of (14) with the dominant stripline mode of the feeders. Therefore, substituting the calculated values of J_s and M_s into (12), we get

$$J_0 J_0' + \frac{C_3^2}{2} J_3 J_3' + \frac{C_6^2}{2} J_6 J_6' - 2J_0' \left\{ C_3 \left[\frac{\sin \frac{3}{2}\theta}{\frac{3}{2}\theta} - 1 \right] J_3 + C_6 \left(\frac{\sin 3\theta}{3\theta} - 1 \right) J_6 \right\} = 0 \tag{16}$$

where

$$J_n = J_n(x_a), \qquad J_n' = J_n'(x_a).$$

The results of calculation of (16) are illustrated in Fig. 1. In the case where $\theta \ll 1$ and where ϕ contains the higher-order terms, x_a may instead be obtained from the infinite series [see (14)] by an approximate integral method which results in

$$J_0(x_a) + 2J_1(x_a) \cdot x_a \cdot C_i\left(\frac{3\theta}{2}\right) = 0. \tag{17}$$

Equation (17) shows that x_a approaches zero as θ approaches zero. However, the values obtained from (16) are almost measurable because θ is larger than 0.5 radian in practical cases.

Next, from the continuity of the surface impedance the following is obtained:

$$\frac{k_{Z1}}{\epsilon_1}\tan k_{Z1}(h-t)+\frac{k_{Z2}}{\epsilon_2}\tan k_{Z2}t=0 \tag{18}$$

and where one can also obtain the relation between wave-numbers

$$\begin{aligned} k\rho^2 + k_{Z1}^2 &= k_1^2,\\ k\rho^2 + k_{Z2}^2 &= k_2^2 \end{aligned} \tag{19}$$

and

$$k_{Z1}^2 - k_{Z2}^2 = k_1^2 - k_2^2 \qquad k_1^2=\omega^2\mu_1\epsilon_1,\ k_2^2=\omega^2\mu_2\epsilon_2. \tag{20}$$

After obtaining k_{Z1} and k_{Z2} from (18) and (20) and substituting them into (19) using x_a indicated in Fig. 1 we can get ω'. The calculation is further simplified by taking the first two terms of Fourier expansion of the tangent which is sufficient for good approximation in a practical case because

$$\begin{aligned} &k_{Z1}(h-t),\\ &k_{Z2}t<1. \end{aligned} \tag{21}$$

The results are

$$\omega'=\frac{x_a}{a\sqrt{\epsilon_2\mu_{\rm eff}}}\sqrt{\frac{2\left\{t'+\frac{\epsilon_{2s}}{\epsilon_{1s}}(1-t')\right\}}{\left\{1+\sqrt{1+F^2}\right\}\left\{t'+\frac{\mu_{1s}}{\mu_{{\rm eff},s}}(1-t')\right\}}},$$

$$F=\frac{2\sqrt{2}}{\sqrt{3}}\,\frac{t'(1-t')\left\{t'+\frac{\epsilon_{2s}}{\epsilon_{1s}}(1-t')\right\}^{1/2}}{\left\{t'+\frac{\mu_{1s}}{\mu_{{\rm eff},s}}(1-t')\right\}\left\{\frac{\epsilon_{1s}}{\epsilon_{2s}}t'+(1-t')\right\}^{1/2}}\,\frac{hx_a}{a}$$

$$\epsilon_{1s}=\mu_{1s}=1 \qquad \text{(for air medium)}$$

$$\epsilon_i=\epsilon_0\epsilon_{is},\ \mu_{\rm eff}=\mu_0\mu_{{\rm eff},s}. \tag{22}$$

To obtain the compensating term of (13), we define the vector

$$a^+ = i_\rho - ji_\phi, \qquad a^- = i_\rho + ji_\phi \tag{23}$$

and express $H_t = i_\rho H_\rho + i_\phi H_\phi)$ by

$$H_t = a^+a^+ + a^-a^- \tag{24}$$

where

$$a^+=\frac{H_\rho+jH_\phi}{2}, \qquad a^-=\frac{H_\rho-jH_\phi}{2}$$

$$H_\rho=\frac{1}{\rho}\frac{\partial\varphi}{\partial\phi}, \qquad H_\phi=-\frac{\partial\varphi}{\partial\rho}.$$

where

$$\xi=\frac{2\mu_0}{\mu_+ + \mu_-}\,\frac{\cos^2 k_{Z2}t\int_t^h\cos^2 k_{Z1}(h-Z)dZ}{\cos^2 k_{Z1}(h-t)\int_0^t\cos^2 k_{Z2}Z\,dZ}$$

$$\fallingdotseq \frac{2(1-t')}{t'}\,\frac{1}{\tilde{\mu}_s} \qquad \tilde{\mu}_s=\frac{\mu_+ + \mu_-}{2\mu_0}=\frac{\mu_{+s}+\mu_{-s}}{2}.$$

If $\mu_{\rm eff}$ is used instead of μ in (13) and (24) is substituted into the second term of (13) we get (25) which is a good approximation in the case where $\eta^2 \ll 1$.

$$\frac{\delta\omega}{\omega'}=\frac{\eta\iiint_T\{\mu_+|a^+|^2-\mu_-|a^-|^2\}dv}{2\iiint_V\{\mu_+|a^+|^2+\mu_-|a^-|^2\}dv}=\frac{\eta^2}{2}\,\frac{1}{1+\xi} \tag{25}$$

Therefore, (13) can be shown as (26) by using (22) and (25).

$$\omega=\frac{x_a(\theta)}{a\sqrt{\epsilon_2\mu_{\rm eff}}}\sqrt{\frac{2\{t'+\epsilon_{2s}(1-t')\}}{\left\{t'+\frac{1}{\mu_{{\rm eff},s}}(1-t')\right\}\{1+\sqrt{1+F^2}\}}}\cdot\left\{1-\frac{\eta^2}{2(1+\xi)}\right\}. \tag{26}$$

In an ordinary circulator (26) becomes (27) by substituting $t'=1$

$$\omega=\frac{x_a(\theta)}{a\sqrt{\epsilon_2\mu_{\rm eff}}}(1-\eta^2). \tag{27}$$

The approximate values of a total reactive energy for the constant unit current same-phase excitation $\tilde{W}_t$ is obtained from a trial function

$$\tilde{W}_t = \frac{6.75}{\omega}\frac{h\lambda}{a^2}\left(\frac{t'}{\epsilon_{2s}} + 1 - t'\right)\frac{I}{J_1^2(xa)} \tag{28}$$

where

$$I = 2\pi\int_0^{x_a} J_0^2(x)xdx + \pi C_3^2\int_0^{x_a} J_3^2(x)\cdot xdx + \pi C_6^2\int_0^{x_a} J_6^2(x)xdx.$$

The energy $\tilde{W}_t'$, as in the case of Fig. 6(b) described later, takes on the values of

$$\tilde{W}_t' = 4\tilde{W}_t. \tag{29}$$

2) *Negative and Positive Rotational Excitation:* Assuming isotropic permeabilities μ_- and μ_+ by tensor quantity, the potential function adopted for air and ferrite, respectively, is

$$\varphi_1 = C_1' \cos k_{Z1}(h - Z)\cdot e^{\pm j\phi}J_1(k_\rho\rho),$$
$$\varphi_2 = C_2' \cos k_{Z2}Z\cdot e^{\pm j\phi}J_1(k_\rho\rho). \tag{30}$$

From the boundary condition $H_\phi = 0$ at $\rho = a$

$$J_1'(x_a') = 0$$

thus

$$x_a' = k_\rho a = 1.841. \tag{31}$$

The second term of (13) expressed by $a^\pm$ substituting $\mu_\mp$ instead of μ, results in

$$\frac{\delta\omega}{\omega'} = \frac{\pm(\mu_+ - \mu_-)\iiint_\tau |a^\pm|^2 dv}{2\iiint_V \{\mu_+|a^+|^2 + \mu_-|a^-|^2\}dv} \tag{32}$$

where $a^\pm$ can be calculated by (24) and (30). Therefore, the approximate values of $\omega_\mp$ obtained from the above results are

$$\omega_\mp = \frac{1.841}{a\sqrt{\epsilon_2\mu_\mp}} \cdot \sqrt{\frac{2\left\{t' + \frac{\epsilon_{2s}}{\epsilon_{1s}}(1-t')\right\}}{\left\{t' + \frac{\mu_{1s}}{\mu_{\mp s}}(1-t')\right\}\{1+\sqrt{1+F^2}\}}} \cdot \left\{1 \mp \frac{0.063\eta}{0.787\left(1+\frac{1}{\tilde{\mu}_s}\right) \mp 0.66\eta\frac{1-t'}{t'}}\right\}. \tag{33}$$

The total reactive energy $\tilde{W}_{t\mp}'$ is also obtained as

$$\tilde{W}_{t\mp}' \doteqdot 0.0725\frac{a^2}{\omega h\lambda\left(1 - t' + \frac{t'}{\epsilon_{2s}}\right)}. \tag{34}$$

The eigenvalues of admittance for negative and positive phase excitations are obtained from (2) as

$$y_\mp = j(\omega - \omega_\mp)\tfrac{2}{3}W_{t\mp}'$$
$$= j0.0484\frac{a^2}{h\lambda\left(1 - t' + \frac{t'}{\epsilon_{2s}}\right)}\frac{\omega - \omega_\mp}{\omega} \tag{35}$$

where $\omega_\mp$ takes the values of (33).

B. Design Consideration for the Circulator

1) *Determination of a Transformer Ratio:* A circulator realized by inserting a proper parallel reactance and ideal transformer at each terminal is considered. As the difference of admittance eigenvalues $y_\mp$ should be $y_+ - y = j(2/\sqrt{3}R)$ when the eigenadmittance for a same-phase excitation is infinite, we get from (35) 137

$$0.0484\frac{a^2}{h\lambda\left(1 - t' + \frac{t'}{\epsilon_{2s}}\right)}\frac{\omega_- - \omega_+}{\omega}n^2 = \frac{2}{\sqrt{3}R}. \tag{36}$$

Substituting (33) into $\omega_\mp$ and considering the relation $\mu_\mp \doteqdot \tilde{\mu}(1\mp\eta)$ and the condition $\eta \ll 1$, one obtains

$$n^2 = 1.63\frac{h}{a}\frac{(1 - t' + \tilde{\mu}_s t')^{3/2}\left(1 - t' + \frac{t'}{\epsilon_{2s}}\right)^{1/2}}{t'\eta\tilde{\mu}_s} \tag{37}$$

which is a required transformer ratio.

2) *Reactance To Be Inserted at Each Terminal:* As discussed in Section III-A, x_a takes different values depending on θ whereas x_a' is independent. Therefore, a capacitive reactance is required under the condition $x_a < x_a'$, which corresponds to $\theta < 1.3$ radians, although no reactance is needed in the case of $x_a = x_a'$.

3) *Values of P_{crit}, w, and L:* Denote the RF magnetic field at center by h_{crit} for operation with a power P_{crit}

$$h_{crit} = h_\rho = \frac{1}{\rho}\frac{\partial\varphi}{\partial\phi_{\rho=0}} = C_2'\frac{x_a'}{2a}.$$

However, C_2' is obtained from the terminal voltage determined by P_{crit} as

$$C_2' = \frac{n\sqrt{R P_{crit}}}{3\omega\mu_+ h\left\{t' + \frac{1}{\mu_{+s}}(1-t')\right\}J_1(x_a')}.$$

Substituting (36) and (26) into n and ω of the preceding equations yields

$$P_{\rm crit} = 0.716 \left\{\frac{1 + (\mu_{+s} - 1)t'}{1 + (\tilde{\mu}_s - 1)t'}\right\}^2 h_{\rm crit}{}^2 \tau\omega(\mu_+ - \mu_-). \quad (38)$$

4) Bandwidth Ratio w: Equation (9) can also be expressed by the frequency variation of eigenvalues of reflection coefficient as

$$|S''| = \frac{1}{3}\sqrt{\sum_{i=1}^{3} |\delta S_i|^2 - \sum_{i\neq j} |\delta S_i| \cdot |\delta S_j|}. \quad (39)$$

Since we have the relation of $|\delta S_1| = (4/3R)\tilde{W}_{t1}|\delta\omega|$ and $|\delta S_{\frac{2}{3}}| = (|\delta\omega|/Y)\tilde{W}_{t\mp}'$, substituting (28) into $\tilde{W}_{t1}$ and (34) into $\tilde{W}_{t\mp}$, we get

$$\left|\frac{\delta S_1}{\delta S_i}\right|_{i\neq 1} = 0.087\left(\frac{h\lambda}{a^2}\right)^2 \frac{\left(1 - t' + \dfrac{t'}{\epsilon_{2s}}\right)^2 I}{J_1{}^2(x_a)} \ll 1. \quad (40)$$

It is understood that the eigenvalue for the same-phase excitation is very insensitive for a frequency variation other than that of rotational excitation. So, putting $\delta S_1 = 0$ we get

$$w = \frac{2|\delta\omega|}{\omega} = \frac{6|S''|}{n^2\omega\tilde{W}_t R}.$$

If the relation $n^2\omega\tilde{W}_t = \{\omega/(\omega_- = \omega_+)\}(\sqrt{3}/R)$ which is obtained from (34) and (36), is substituted into above equation, we get

$$w = \frac{6|S''|\eta t'\tilde{\mu}_s}{\sqrt{3}\{1 + (\tilde{\mu}_s - 1)t'\}}. \quad (41)$$

On the other hand, the filling factor k_f is calculated from a trial function as

$$k_f = \frac{\tilde{W}_m}{\tilde{W}_{tm}} = \frac{\tilde{\mu}_s t'}{\{1 + (\tilde{\mu}_s - 1)t'\}}. \quad (42)$$

Therefore,

$$w = 2\sqrt{3}|S''|\eta k_f$$

which is the same relation as (10) obtained by general consideration.

5) Insertion Loss: Under the consideration $\tilde{W}_{t1} \ll \tilde{W}_{t\mp}$, the same results can be obtained as in (11).

6) Volume: From the relation $\tau = \pi h t' a^2$ and (26), one obtains

$$\tau \doteqdot \frac{x_a{}^2}{2\pi} h\lambda^2 \frac{\left\{1 - \left(1 - \dfrac{1}{\epsilon_{2s}}\right)t'\right\} t' \left\{1 - \dfrac{\eta^2}{2(1+\xi)}\right\}}{\{1 + (\mu_{{\rm eff},s} - 1)t'\}\{1 + \sqrt{1 + F^2}\}}. \quad (43)$$

7) Rising Temperature: The heat generated per unit volume is inversely proportional to τ, and the rising temperature T is highest on the boundary surface between air and ferrite and is proportional to the square of the ferrite width if heat escapes only through the surface contacting the metal plate. Therefore we get the relation

$$T \sim \frac{t^2}{\tau} = \frac{t}{a^2} = \frac{ht'}{a^2} \quad (44)$$

which shows T is proportional to a width and inversely proportional to a surface.

8) One Design Procedure: Considering the several conditions previously mentioned one example of design procedure can be shown as follows:

a) $\mu_{\rm eff}$, μ_+, and μ_- are determined to minimize the insertion loss by (11);

b) τ, h, t', and a are determined from $P_{\rm crit}$ of (38) and T of (44) and (26);

c) w is obtained from (41).

C. Consideration of the Wideband

Since the $t' < 1$ required for high power makes the bandwidth narrower from (41), it is preferable to use the wideband technique. However, the frequency variation for same-phase excitation is more insensitive than that of the rotational excitation as shown in (40) and results in the narrow bandwidth. On the other hand, the frequency variations of rotational excitation $\delta S_{2,3}$ are essentially connected to k_f by the following equation derived from (39), (4), and (6):

$$|\delta S_i|_{i\neq 1} = \frac{\delta\omega}{Y}\tilde{W}_{ti} = \frac{\delta\omega}{Y}\frac{3Q_e}{\tilde{\omega}R} = \frac{\delta\omega}{\tilde{\omega}}\frac{\sqrt{3}}{k_f\eta}.$$

Therefore by increasing reactive energy only for the same phase of excitation it is most effective to make a wideband until satisfying

$$|\delta S_i|_{i\neq 1} = |\delta S_1| = \frac{4}{3R}\tilde{W}_{t1}|\delta\omega| = \frac{|\delta\omega|}{Y}\tilde{W}_{ti(i\neq 1)}$$

thus

$$\tilde{W}_{t1} = \frac{3R^2}{4}\tilde{W}_{ti}. \quad (45)$$

One example of the techniques of coupling only to a same-phase excitation is illustrated in Fig. 2(a). The equivalent networks of each excitation are shown in Fig. 2(b) where $L_c'C_c'$ and $L_c''C_c''$ can be neglected in a practical case because their resonant frequencies differ greatly from ω.

Next, the most important factor is the rising temperature on the center conductor; this heats the ferrite surface. This matter could be improved by using the reentrant cavity

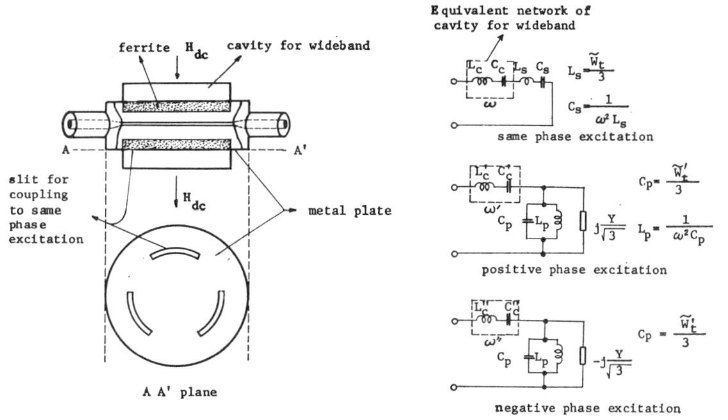

(a) (b)

Fig. 2. Wideband technique and the equivalent network. (a) An example of coupling cavity for the wideband to the same-phase excitation (b) Equivalent network of each eigenvalue. ω: center angular frequency of circulator; $\omega' \neq \omega$; $\omega'' \neq \omega$.

coupled with the coaxial junction as shown in Fig. 6(b) where ferrite is inserted at the narrow space. Since, for the rotational excitation, the TE_{11} mode in the coaxial part inside cavity couples to the quasi TM_{11} mode in ferrite, the length of coaxial part is chosen at a quarter of the TE_{11} mode wavelength. This contributes to an open boundary condition at the periphery of the ferrite disk which is same as an ordinary circulator. For a same-phase excitation, the cavities are nonresonant, and this contributes to maintaining its eigenvalue at a zero impedance which is the same as that of the aforementioned stripline circulator.

IV. Experiments

A. Confirmation of the Theoretical Eigenvalues

The experiment was made with a ferrite disk of 150 mm in diameter, 7 mm in width, where $\epsilon_{2s} = 14$ and under the state of $t' = 0.425$ and $\theta = 0.6$. The measured value of $\mu_{eff,s}$ is 3.2; μ_+ and μ_- are calculated from $\mu_{eff,s}$.

The theoretical values of y_1, y_2, and y_3 calculated from the above data are plotted by a dotted line on Smith chart of Fig. 3; the experimental values are also indicated by lines, which show good coincidence with theoretical values.

B. Nonlinearity of Ferrite

The experiment was made with a lumped element circulator operated at above resonance to use small samples of polycrystals and single crystals of YIG. As indicated in Fig. 4(a), the threshold power P_{crit} occurs at comparably small CW power for polycrystal and it becomes higher for a stronger dc magnetic field. This threshold power was not observed in the single crystal and nonlinearity takes place immediately after power is applied, as the temperature of the ferrite cannot be raised. For pulsed power, the nonlinearity did not appear in the experiment below 2 kW for both polycrystal and single crystal as shown in Fig. 4(b).

This phenomenon is supposed to be caused by local heating owing to spinwave generation around pore regions of polycrystals where the internal dc field is below resonance because of the demagnetizing and anisotropic field in their vicinities; and local heating increases in CW power. The increase of P_{crit} for a stronger dc field should be caused by the statistical diminish of the amount of such a grain. From the above reasons, it is preferable to use materials with small values of $\omega_M/\Delta H$, although an excessive value of ΔH of polycrystal should be avoided to keep the insertion loss small.

C. Performance of the Circulator

1) Wideband Characteristics: The characteristics of a trial circulator at 670 MHz with $t' = 0.4$ is 25 MHz in bandwidth and 0.2 dB in insertion loss. This becomes wideband, as shown in Fig. 5 by the construction of Fig. 2(a).

2) High-Power Operation: The circulator [Fig. 6(b)] with ferrite of 150 mm in diameter and 5 mm in width was tested by 15 kW CW operation at 670 MHz. The rising temperature of the ferrite surface was 10 degrees higher than the cavity which was cooled by water. The nonlinearity did not occur.

Appendix I

Showing the absolute values of eigenvalues of reflection coefficients S_1, S_2, and S_3 by $e^{-\alpha_1}$, $e^{-\alpha_2}$, and $e^{-\alpha_3}$, and assuming no energy in the same-phase excitation, the insertion loss L (decibels) is

$$L = 8.67 \ln \left(1 - \frac{\alpha_2 + \alpha_3}{3}\right) = \frac{8.67}{3} \sum_{i=2}^{3} \alpha_i. \quad (46)$$

However,

$$|S_i| = \left|\frac{Y - y_i}{Y + y_i}\right| = 1 - \frac{2Yy_i''}{Y^2 + y_i'^2} \qquad y_i = jy_i' + y_i''.$$

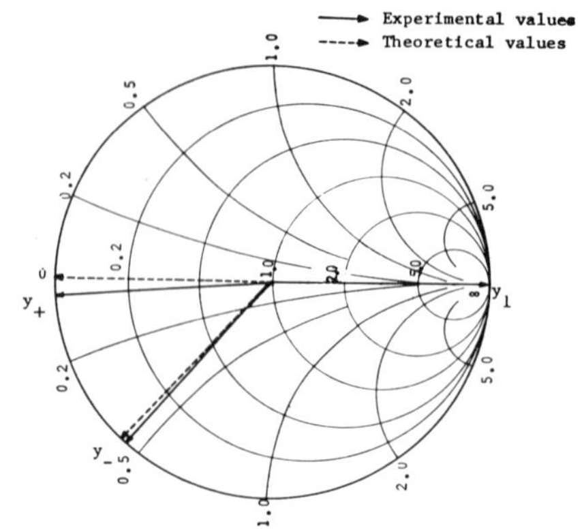

Fig. 3. Comparison of the theoretical values of the eigenvalues with experimental values on the normalized admittance Smith chart.

140

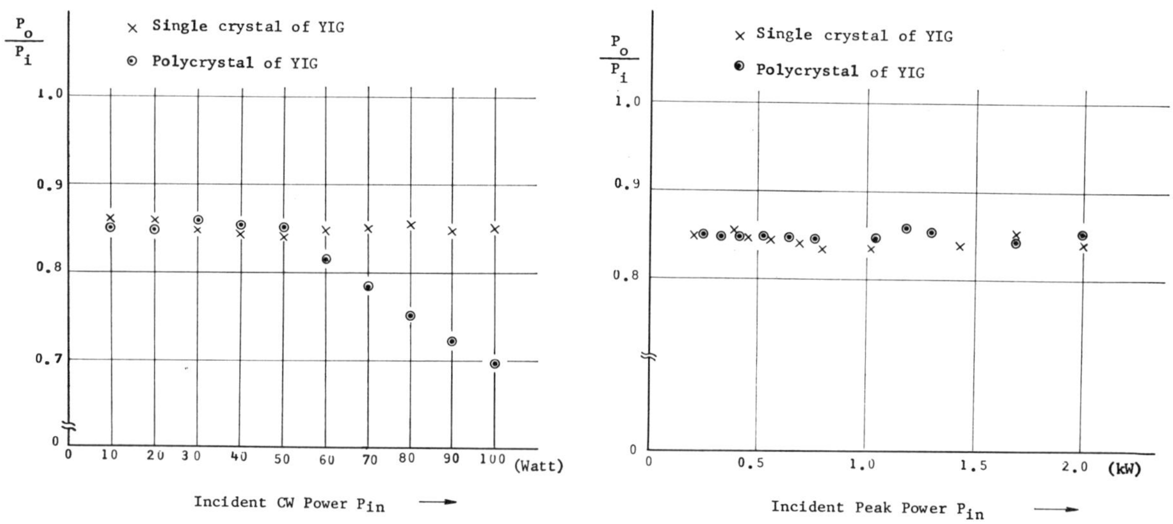

(a) (b)

Fig. 4. Measured values of linearity of circulator with single crystal and polycrystal. P_0 is the output power from the circulator. (a) Linearity for CW power. (b) Linearity for pulsed power. Pulsewidth: 3 μs; repetition frequency: 300 Hz.

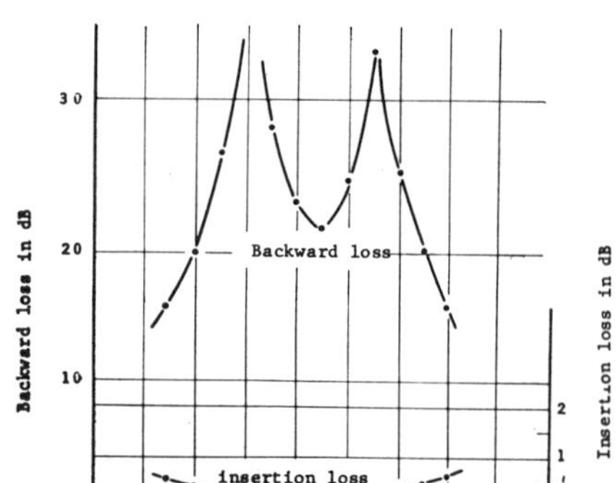

Fig. 5. Wideband characteristics of high-power UHF circulator.

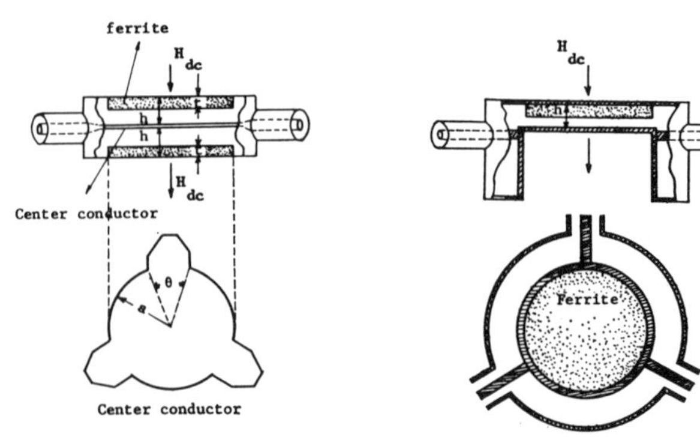

(a) (b)

Fig. 6. Example of high-power circulator.

Substitute

$$y_2' = -\frac{Y}{\sqrt{3}}, \qquad y_3' = \frac{Y}{\sqrt{3}},$$

$$\alpha_i = 1 - |S_i| = \frac{3}{2}\frac{y_i''}{Y} \tag{47}$$

since

$$y_i = j\frac{2}{3}\omega(\tilde{W}_{e_i} - \tilde{W}_{m_i})$$

$$y_i'' = \frac{2}{3}\left(\frac{\omega W_{m_i}}{Q_i} + \frac{\omega \tilde{W}_{e_i}}{Q_c}\right) \fallingdotseq \frac{2}{3}\omega\tilde{W}_{m_i}\left(\frac{1}{Q_i} + \frac{2}{Q_c}\right). \tag{48}$$

Substituting (47) and (48) into (46)

$$L = 2.9\omega\tilde{W}_t k_f R\left(\frac{1}{Q_+} + \frac{1}{Q_-} + \frac{1}{Q_c}\right),$$

$$Q_2 = Q_-, \qquad Q_3 = Q_+. \tag{49}$$

Since the relation

$$\tilde{W}_t k_f = \frac{\sqrt{3}}{\tilde{\omega}R}\frac{\tilde{\mu}}{\mu_+ - \mu_-} = \frac{\sqrt{3}}{2\tilde{\omega}R}\frac{1}{\eta}$$

is derived from (4) and (6), substituting this relation into (49) under the approximation $\omega \fallingdotseq \tilde{\omega}$, we get

$$L = 2.51\frac{1}{\eta}\left(\frac{1}{Q_+} + \frac{1}{Q_-} + \frac{2}{Q_c}\right). \tag{50}$$

Appendix II

We have the field relation

$$\langle E_c J^*\rangle_{V+S} + \langle H_c M^*\rangle_{V+S} = 0 \tag{51}$$

where E_c and H_c are correct fields.[6] Maxwell's equations

$$\nabla \times H - j\omega\epsilon E = J, \qquad \nabla \times E + j\omega\boldsymbol{\mu}H = -M \tag{52}$$

and boundary conditions

$$n \times [H_2 - H_1] = J, \qquad [E_2 - E_1] \times n = M \tag{53}$$

should be satisfied between trial field and current inside the medium with the tensor permeability $\boldsymbol{\mu}$. If the same trial fields are used inside the medium with a scalar permeability μ, Maxwell's equations should be changed to

$$\nabla \times H - j\omega'\epsilon E = J',$$
$$\nabla \times E + j\omega'\mu H = -M' \qquad \omega' = \omega + \delta\omega \tag{54}$$

whereas boundary conditions are the same on discontinuous surfaces because of the same fields. Substituting the trial current obtained from (52) and (54) into (51),

$$\langle E_c \cdot \nabla \times H^*\rangle_V - \langle H_c \cdot \nabla \times E^*\rangle_V + j\omega\epsilon\langle E_c \cdot E^*\rangle_V + j\omega\langle H_c \cdot \boldsymbol{\mu}^* H^*\rangle_V + \langle E_c J_S^*\rangle_S + \langle H_c M_S^*\rangle_S = 0 \tag{55}$$

$$\langle E_c \cdot \nabla \times H^*\rangle_V - \langle H_c \cdot \nabla \times E^*\rangle_V + j\omega'\epsilon\langle E_c E^*\rangle_V + j\omega'\langle H_c \cdot \mu^* H^*\rangle_V + \langle E_c J_S'^*\rangle_S + \langle H_c M_S'^*\rangle_S = 0. \tag{56}$$

Subtracting (55) from (56) and taking account of $J_S = J_S'$ and $M_S = M_S'$ because of the same boundary condition

$$\frac{\delta\omega}{\omega'} = \frac{-\langle H_c \cdot \mu^* H^*\rangle_V + \langle H_c \cdot \boldsymbol{\mu}^* H^*\rangle_V}{\langle E_c \cdot \epsilon E^*\rangle_V + \langle H_c \cdot \boldsymbol{\mu}^* H^*\rangle_V}$$

so that

$$\omega = \omega'\left\{1 - \frac{\iiint_T H \cdot (\boldsymbol{\mu}^* - \mu)H^* dv}{2\iiint_V H \cdot \boldsymbol{\mu}^* H^* dv}\right\}. \tag{57}$$

Acknowledgment

The author wishes to express his gratitude and appreciation to the Director of the Laboratories, Dr. M. Komai, for his encouragement and discussion in connection with this study. Thanks are also due to Mr. Hoshino for recording the experimental data.

141

References

[1] U. Milano, J. H. Saunders, and L. Davis, Jr., "A Y-junction strip-line circulator," *IRE Trans. Microwave Theory and Techniques*, vol. MTT-8, pp. 346–351, May 1960.

[2] H. Bosma, "On stripline Y-circulation at UHF," *IEEE Trans. Microwave Theory and Techniques*, vol. MTT-12, pp. 61–72, January 1964.

[3] C. V. Beuhler and A. F. Eikenberg, "A VHF high-power Y-circulator," *IRE Trans. Microwave Theory and Techniques* (*Correspondence*), vol. MTT-9, pp. 569–570, November 1961.

[4] B. A. Auld, "The synthesis of symmetrical waveguide circulators," *IRE Trans. Microwave Theory and Techniques*, vol. MTT-7, pp. 238–246, April 1959.

[5] J. B. Davies and P. Cohen, "Theoretical design of symmetrical junction stripline circulators," *IEEE Trans. Microwave Theory and Techniques*, vol. MTT-11, pp. 506–512, November 1963.

[6] Y. Konishi, "VHF-UHF Y circulators," *Tech. J. Japan Broadcasting Corp.*, vol. 17, pp. 87–121, 1965. Also Nippon Hoso Kyokai, Tokyo, Japan, Tech. Monograph 6, May 1965.

[7] ——, "Lumped element Y circulator," *IEEE Trans. Microwave Theory and Techniques*, vol. MTT-13, pp. 852–864, November 1965.

[8] ——, "Characteristics of ferrite components and methods of approximate calculation for design," *Tech. J. Japan Broadcasting Corp.*, vol. 19, pp. 296–312, 1967.

Computer-Aided Design of Three-Port Waveguide Junction Circulators

JOSEPH B. CASTILLO, JR., MEMBER, IEEE, AND LIONEL E. DAVIS, SENIOR MEMBER, IEEE

Abstract—**The complete performance of a lossless three-port H-plane waveguide junction loaded coaxially with various inhomogeneous ferrite cylinders has been evaluated over the waveguide bandwidth and compared with experiment. Qualitative agreement between the predicted and measured performance was generally good using only the first three modes, $n=0\pm1$. It has been shown theoretically and verified experimentally that if the $4\pi M_s$ of a homogeneous rod or the internal field is increased, the circulation frequency f_0 increases; conversely, if the permittivity is increased, f_0 decreases. These conflicting effects are modified when the magnetization $4\pi M_s$ and permittivity ε_r are inhomogeneous. For example, if the $4\pi M_s(r)$ is small at the outer surface of the rod (with permittivity held constant), the effect on f_0 is very small; but if $4\pi M_s(r)$ approaches zero for r small, then f_0 may decrease significantly. On the other hand, if $\varepsilon_r(r)$ approaches unity near the outer surface of the rod, f_0 may increase significantly; but if $\varepsilon_r(r)$ approaches unity near the center of the rod, f_0 is affected relatively little. The inhomogeneous structure has also shown that decreasing the ferrite volume may improve the performance, and high-power applications are suggested. With a conducting pin down the center of the ferrite, relative bandwidths of 40–50 percent are predicted.**

I. Introduction

COMPARATIVELY few guidelines exist for the design of waveguide circulators, and this paper will describe some initial work to determine the usefulness of numerical analysis techniques. Present design techniques permit some prior selection of "best" ferrite properties for a particular application, but detailed prediction of performance, as is possible in filter design for example, has not so far been attempted. This is due to the complexity of the problem and to the fact that almost all of the empirically determined successful designs are not amenable to theoretical analysis. Our attention has been concentrated on those structures that can in principle be analyzed rigorously but which do not necessarily have good performance, rather than those which have excellent performance and can only be modeled very approximately. Therefore, this work was primarily to check the feasibility of this approach and to clarify the effects of some important parameters, in the hope that this will develop better insight into the behavior of circulators, and lead eventually to superior components. For example, it is not clear to what extent the waveguide itself limits the bandwidth of the waveguide circulator. Assuming it is the primary limitation, what is the maximum possible bandwidth in a ridged-waveguide, stripline, microstrip or lumped-constant circulator?

Another objective is to try to replace the expensive and time-consuming cut-and-try development in the laboratory with analysis in the computer, thereby permitting more efficient use of the engineer's time. An experienced technician would then check experimentally, finally adjust, and compare a few "optimized" designs suggested by the computer. He would then select a design according to other criteria such as size, cost, simplicity, ruggedness, etc. Ultimately, the goal is circulator synthesis rather than analysis. For this, we must have a successful theoretical model and so this work is seen as only a first step in that direction. However, much

Manuscript received March 13, 1969; revised July 14, 1969. This work was supported principally by the U. S. Army Electronics Command, Fort Monmouth, N. J., Contract DAAB-07-67-C-0491; and also by NASA Grants NsG-6-59 and NGL-44-006-033).

The authors are with the Department of Electrical Engineering, Rice University, Houston, Tex. 77001.

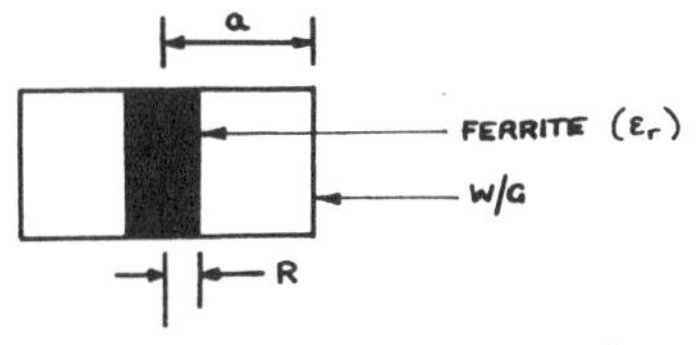

Fig. 1. Ferrite post structure.

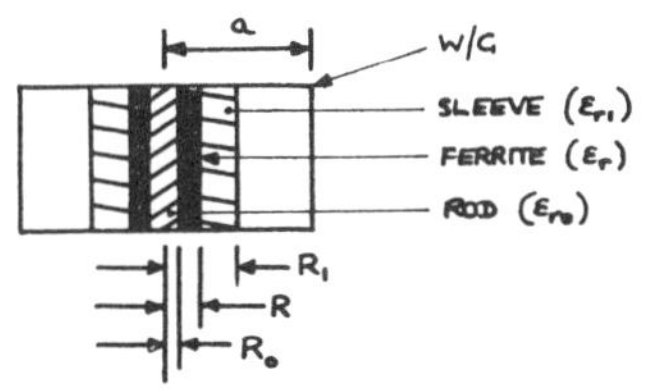

Fig. 2. Ferrite tube, dielectric rod, dielectric sleeve structure.

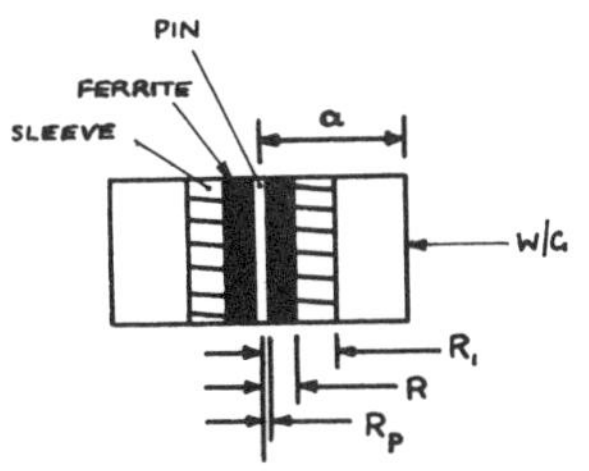

Fig. 3. Ferrite post, metal pin, dielectric sleeve structure.

research and development remains before we even begin to approach the degree of success achieved in filter synthesis.

In 1959 a paper by Auld [1] considered the external properties of waveguide junctions in terms of their degrees of symmetry, but although this paper showed what properties theoretically were possible it could not describe how these properties could be realized. In 1962 a paper by Davies [2] showed how the junction symmetry could be used in the boundary value problem to determine the scattering matrix of a simple ferrite rod circulator. This paper describes the application of Davies' method to various inhomogeneous cylindrical ferrite structures which extend the full height of the waveguide, as shown in Figs. 1–3, and compares theory with experiment.

The structures investigated were

1) the solid ferrite rod [2],
2) the ferrite tube, with an outer dielectric sleeve and a solid dielectric center,
3) the ferrite rod, with a thin conducting central pin and an outer dielectric sleeve [1], [3]–[8].

The simple solid rod is regarded as the basic design against which the more complicated structures may be compared. It is well known to have narrow bandwidth, but it has the advantage of simplicity. The tube was chosen for its versatility, and the "thin-pin" structure for its potential bandwidth and high-power capability.

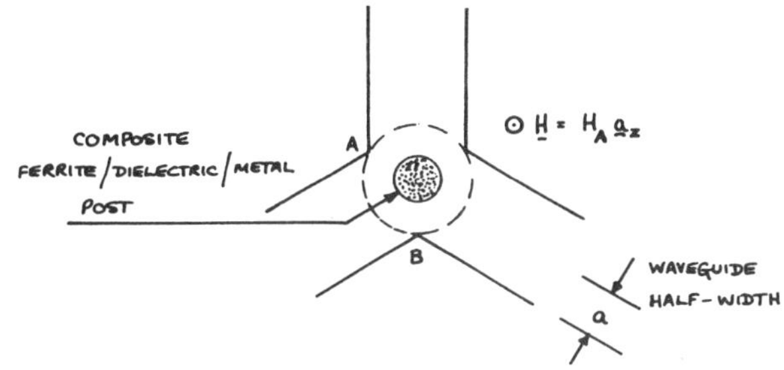

Fig. 4. Three-port waveguide junction (broken circle defines circulator "cavity" region).

II. Analysis

The details of the analysis will be kept to a minimum, but sufficient information will be given to enable computer programs to be written for any of the structures. The theory will be developed from the internal to the external point of view, i.e., from the fields within the junction to the scattering matrix which describes the external properties.

A. Field Theory

The waveguide junction is shown in Fig. 4, and all fields within the junction, and the ferrite magnetization are assumed independent of z. The permeability tensor for the lossless ferrite is given by [9]

$$[\mu] = \begin{bmatrix} \mu & -j\kappa & 0 \\ j\kappa & \mu & 0 \\ 0 & 0 & \mu_0 \end{bmatrix} \tag{1}$$

where

$$\begin{aligned} \frac{\mu}{\mu_0} - 1 &= f_m f_0 [f_0^2 - f^2(1-\alpha^2)] / \{[f_0^2 - f^2(1+\alpha^2)]^2 + 4f_0^2 f^2 \alpha^2\} \\ \frac{\kappa}{\mu_0} &= f_m f [f_0^2 - f^2(1+\alpha^2)] / \{[f_0^2 - f^2(1+\alpha^2)]^2 + 4f_0^2 f^2 \alpha^2\} \end{aligned} \tag{2}$$

and f = frequency (MHz), $f_0 = \gamma H_0$, $|\gamma| = 2.8$ MHz/Oe if the spectroscopic splitting factor $g = 2.0$, H_0 is the internal magnetic field (Oe), $f_m = \gamma 4\pi M_s$, $4\pi M_s$ is the saturation magnetization (gauss), $\alpha = \gamma \Delta H / 2f$, and ΔH is the 3-dB ferrimagnetic resonance linewidth.

All the inhomogeneous structures are combinations of the following elements: a dielectric tube or rod, a ferrite (or garnet) tube or rod, and a perfectly conducting pin. Expressions for the field components in each region, E_z, H_ϕ, H_r, may be derived, but we shall only need E_z and H_ϕ to match the fields at the interfaces. These may be written using nth-order cylindrical Bessel functions of the first and second kinds (J_n, Y_n) as follows.

Dielectric Tube:

$$E_z = \sum_{n=-\infty}^{\infty} \{A_n J_n(\gamma_1 r) + B_n Y_n(\gamma_1 r)\} \exp(-jn\phi)$$

$$H\phi = -j\frac{\sqrt{\epsilon_{r1}}}{\eta_0}\sum_{n=-\infty}^{\infty}\{A_nJ_n'(\gamma_1 r) + B_nY_n'(\gamma_1 r)\} \cdot \exp(-jn\phi) \tag{3}$$

where intrinsic impedance of free space $\eta_0 = (\mu_0/\epsilon_0)^{\frac{1}{2}}$, ϵ_{r1} is the relative permittivity of the tube, and $\gamma_1^2 = \omega^2\mu_0\epsilon_0\epsilon_{r1}$.

Dielectric Rod: The field equations are written as in (3) but with $B_n = 0$.

Ferrite Tube [10], [11]:

$$E_z = \sum_{n=-\infty}^{\infty}\{L_nJ_n(\gamma r) + M_nY_n(\gamma r)\}\exp(-jn\phi)$$

$$H_\phi = \frac{-j\mu\gamma}{\omega(\mu^2-\kappa^2)}\sum_{n=-\infty}^{\infty}\left\{[L_nJ_n'(\gamma r) + M_nY_n'(\gamma r)] + \frac{\kappa n}{\mu\gamma r}[L_nJ_n(\gamma r) + M_nY_n(\gamma r)]\right\}\exp(-jn\phi) \tag{4}$$

Ferrite Rod (Fig. 1):

$$\rho_n = -\{D_n[J_n(\gamma_0 R)/J_n(\gamma R)] - J_n'(\gamma_0 R)\} / \{D_n[Y_n(\gamma_0 R)/J_n(\gamma R) - Y_n'(\gamma_0 R)\} \tag{5}$$

where

$$D_n = \left[J_n'(\gamma R) + \frac{\kappa n}{\mu\gamma R}J_n(\gamma R)\right]\gamma_0\epsilon_r/\gamma,$$

$\gamma_0^2 = \omega^2\mu_0\epsilon_0$, and the other parameters are defined in previous equations and Fig. 1. This equation is the same as (14) derived by Davies [2] with $\rho_n = C_n/B_n$.

Ferrite Tube, Dielectric Sleeve (Fig. 2):

$$\rho_n = -\frac{[U_nJ_n'(\gamma_0 R_1) - V_nJ_n(\gamma_0 R_1)]}{[U_nY_n'(\gamma_0 R_1) - V_nY_n(\gamma_0 R_1)]} \tag{6}$$

where

$$\left.\begin{aligned} U_n &= J_n(\gamma_1 R_1) + \frac{D_n}{C_n}Y_n(\gamma_1 R_1) \\ V_n &= \sqrt{\epsilon_{r1}}\left\{J_n'(\gamma_1 R_1) + \frac{D_n}{C_n}Y_n'(\gamma_1 R_1)\right\} \end{aligned}\right\} \tag{7}$$

$$\frac{D_n}{C_n} = -\frac{[S_nJ_n'(\gamma_1 R) - T_nJ_n(\gamma_1 R)]}{[S_nY_n'(\gamma_1 R) - T_nY_n(\gamma_1 R)]} \tag{8}$$

$$\left.\begin{aligned} S_n &= J_n(\gamma R) + \frac{B_n}{A_n}Y_n(\gamma R) \\ T_n &= \frac{\gamma_1\epsilon_r}{\gamma\epsilon_{r1}}\left\{\left[J_n'(\gamma R) + \frac{B_n}{A_n}Y_n'(\gamma R)\right] + \frac{\kappa n}{\mu\gamma R}\left[J_n(\gamma R) + \frac{B_n}{A_n}Y_n(\gamma R)\right]\right\} \end{aligned}\right\} \tag{9}$$

$$\frac{B_n}{A_n} = -\frac{\left\{\frac{J_n(\gamma R_0)}{J_n(\gamma_{r0}R_0)}J_n'(\gamma_{r0}R_0) - \frac{\gamma_{r0}\epsilon_r}{\gamma\epsilon_{r0}}\left[J_n'(\gamma R_0) + \frac{\kappa n}{\mu\gamma R_0}J_n(\gamma R_0)\right]\right\}}{\left\{\frac{Y_n(\gamma R_0)}{J_n(\gamma_{r0}R_0)}J_n'(\gamma_{r0}R_0) - \frac{\gamma_{r0}\epsilon_r}{\gamma\epsilon_{r0}}\left[Y_n'(\gamma R_0) + \frac{\kappa n}{\mu\gamma R_0}Y_n(\gamma R_0)\right]\right\}}. \tag{10}$$

where $\gamma^2 = \omega^2\epsilon_0\epsilon_r(\mu^2-\kappa^2)/\mu$.

Ferrite Rod: The field equations are written as in (4), but with $M_n = 0$. Note that if $\kappa = 0$ and $\mu = \mu_0$, the ferrite equations (4) reduce to the dielectric equations (3). Hence, for small values of the ratio κ/μ the magnetized ferrite represents a small perturbation of the dielectric situation.

Any combination of these cylindrical regions can now be considered, and by matching E_z, H_ϕ at the interfaces all but two of the coefficients can be eliminated. By working outwards from the center of the junction, one is left with the ratio ρ_n of the coefficients in the outermost dielectric sleeve (usually air) at the limits of the junction "cavity" region (Fig. 4). Davies [2] has shown how these fields are related to the scattering matrix, and this is summarized in a later section.

The results of this elimination for the three structures shown in Figs. 1–3 may be summarized as follows.

The parameters are defined in previous equations and Fig. 2.

Ferrite Post, Metal Pin, Dielectric Sleeve (Fig. 3): ρ_n is given by (6)–(8), and in (9) we substitute

$$\frac{B_n}{A_n} = -\frac{J_n(\gamma R_p)}{Y_n(\gamma R_p)}. \tag{11}$$

At this point it is relatively easy to include the effect of mechanical tolerances on the pin and the hole. An annular dielectric space can be introduced between it and the ferrite. In this case ρ_n is given by (6)–(8), and in (9) we substitute

$$\frac{B_n}{A_n} = \frac{\left\{M_nJ_n(\gamma R_0) - L_n\left[J_n'(\gamma R_0) + \frac{\kappa n}{\mu\gamma R_0}J_n(\gamma R_0)\right]\right\}}{\left\{M_nY_n(\gamma R_0) - L_n\left[Y_n'(\gamma R_0) + \frac{\kappa n}{\mu\gamma R_0}Y_n(\gamma R_0)\right]\right\}} \tag{12}$$

where

$$\left.\begin{aligned} L_n &= J_n(\gamma_{r0}R_0) + \frac{\alpha_n''}{\alpha_n'} Y_n(\gamma_{r0}R_0) \\ M_n &= \frac{\gamma\epsilon_{r0}}{\gamma_{r0}\epsilon_r}\left\{J_n'(\gamma_{r0}R_0) + \frac{\alpha_n''}{\alpha_n'} Y_n'(\gamma_{r0}R_0)\right\} \end{aligned}\right\} \quad (13)$$

$$\frac{\alpha_n''}{\alpha_n'} = -\frac{J_n(\gamma_{r0}R_p)}{Y_n(\gamma_{r0}R_p)}. \quad (14)$$

In these equations $\gamma_{r0}^2 = \omega^2\mu_0\epsilon_0\epsilon_{r0}$, and the other parameters are defined in previous equations and Fig. 3.

Having determined the internal fields, these may now be used to estimate the scattering matrix.

B. The Scattering Matrix

The "external" performance of a three-port device can be described by a scattering matrix $[S]$. The input and output column vectors, $\{a\}$ and $\{b\}$, respectively, are related by the equation $\{b\} = [S]\{a\}$, where, if the junction has physical symmetry about its axis, we may write

$$[S] = \begin{bmatrix} \alpha & \beta & \gamma \\ \gamma & \alpha & \beta \\ \beta & \gamma & \alpha \end{bmatrix}$$

where

$$\begin{aligned} \alpha &= A \exp(j\phi_A) = s_{11} \\ \beta &= B \exp(j\phi_B) = s_{12} \\ \gamma &= G \exp(j\phi_G) = s_{13}. \end{aligned} \quad (15)$$

Throughout the analysis the circulator is assumed free of any dissipation, and therefore all calculated "losses" are due to scattering to "unwanted" ports. Also, since the circulator is lossless, the scattering matrix is unitary. Clearly A, B, G are the reflection coefficient, isolation, and insertion loss, respectively, and in a perfect circulator B (or G) would be unity, and A together with G (or B) would be zero.

The eigenvalue equation for $[S]$ can be written in the form

$$[S]v = sv \quad (16)$$

where v is an eigenvector of $[S]$, and s is an eigenvalue. The eigenvalues will be complex in general and will have unit magnitude if the junction is lossless. There are three eigenvalues for a three-port circulator, and when it is symmetrical they may be related to the elements of the scattering matrix as follows [1]:

$$\begin{aligned} \alpha &= \tfrac{1}{3}(s_0 + s_1 + s_{-1}) \\ \beta &= \tfrac{1}{3}[s_0 + s_1 \exp(j2\pi/3) + s_{-1}\exp(-j2\pi/3)] \\ \gamma &= \tfrac{1}{3}[s_0 + s_1 \exp(-j2\pi/3) + s_{-1}\exp(j2\pi/3)]. \end{aligned} \quad (17)$$

Therefore, if the eigenvalues can be calculated or measured, the amplitude and phase of the scattering elements α, β, γ, can be determined.

The link between the eigenvalues of the scattering matrix and electromagnetic fields within the m-port junction has been provided by Davies [2]. He has shown that, neglecting evanescent modes, the eigenvalues $s_i = \exp(j\theta_i)$ of a lossless symmetrical m-port junction shown in Fig. 4 can be evaluated approximately using

$$\tan \tfrac{1}{2}(\theta_i + \pi) = \frac{D_n^m + \rho_n E_n^m}{F_n^m + \rho_n G_n^m}. \quad (18)$$

In (18), ρ_n is the ratio of coefficients obtained in (5) and (6), and expressions for the coefficients D, E, F, G may be obtained from Davies' paper. They are not repeated here to conserve space. To summarize the analysis, ρ_n may now be computed for any one of the structures (with $m=3$, $n=0, \pm 1$ for any set of structural parameters, then substituted in (18) to obtain θ_i to use in (17) to obtain $[S]$.

III. Results and Discussions

All measurements were taken with conventional swept oscillator equipment incorporating broad-band directional couplers, a ratio meter, and an X-Y recorder driven from the oscillator. The junction was milled from brass and a threaded plug in the center of the lid facilitated the interchange of ferrite posts. The magnetic field was obtained with an adjustable-shunt permanent magnet; the pole gap was 2 cm, the diameter of the pole pieces was 5 cm, and the magnetic field was calibrated in the center of the junction without the ferrite post, using a Hall probe. The external field H_A required to produce the correct internal field H_0 depends upon the saturation magnetization $4\pi M_s$ and the demagnetizing factor N_z:

$$H_0 = H_A - 4\pi M_s \cdot N_z \begin{cases} H_A, H_0 & \text{in oersteds} \\ 4\pi M_s & \text{in gauss} \\ N_z & \text{dimensionless.} \end{cases} \quad (19)$$

The demagnetizing factor N_z for a ferrite rod magnetized parallel to the long axis depends upon the ratio of length to diameter, and values for the circulator rods were obtained from Bozorth [12].

A. The Ferrite Post

Four different rod structures with promising predicted performance were selected from several sets of computed data. Experimental tests gave results generally in good agreement with the predictions, and analysis and measurement are compared for one of the posts in Fig. 5. The material properties were supplied by the manufacturers, and the applied field, which was not necessarily a best fit, was set at the value calculated from the above demagnetization equation to give an internal field of 200 Oe. The isolation occurred at a frequency 2 percent below the predicted value. The measured reflection coefficient characteristic exhibits a slight "double-hump" at and below the circulation frequency which is a departure from prediction. This feature was apparent with all the solid rod structures to a greater or lesser extent and it is attributed to excitation of modes that do not contribute to the circulation. This is being investigated further. The small spikes seen in the experimental curves are spurious responses attributed to imperfect contact

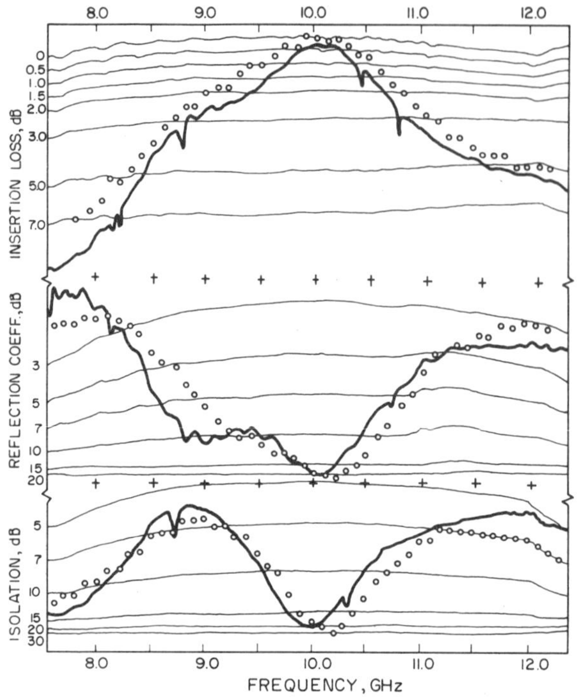

Fig. 5. Performance of G1002 rod, $R=0.300$ cm. Solid lines: measured values. Circles: computed values.

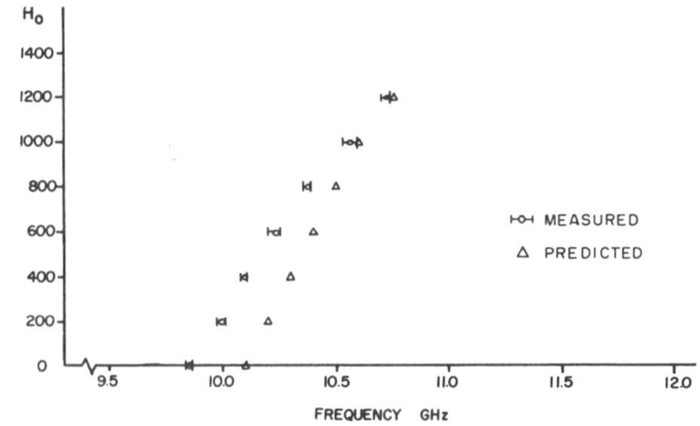

Fig. 6. Center frequency versus internal field. G 1002 rod, $R=0.300$ cm.

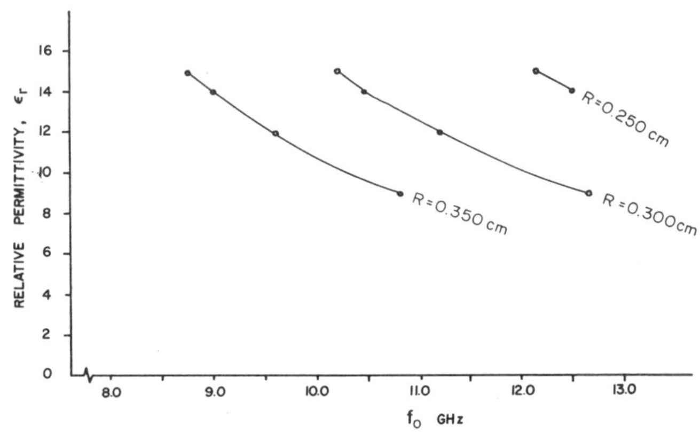

Fig. 7. Predicted center frequency versus relative permittivity.

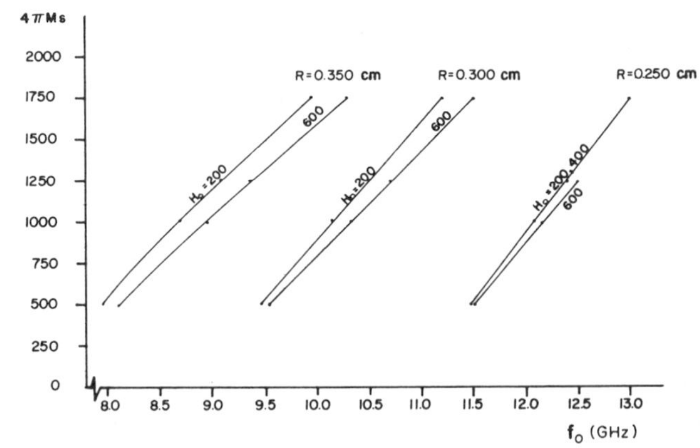

Fig. 8. Center frequency versus saturation magnetization.

between the ferrite post and the walls of the junction. It was also worthwhile to investigate whether the variable H_0 used in the computations bore any resemblance to the values of internal field calculated using the demagnetizing factor and used experimentally to produce the measured performance. The results of this comparison for the circulator described by Fig. 5 are shown in Fig. 6 by plotting center frequency of circulation f_0 versus the calculated internal field required. We see that the discrepancy decreased as the field required increased. Of the other three posts, one (TT1-109) exhibited slightly less discrepancy, and two (TT2-130 and G-1002) rather more. In particular, the TT2-130, a nickel ferrite, showed a marked deviation that is unexplained at present: the circulation frequency f_0 was 5–10 percent above that predicted, and the unpredicted "double-hump" reflection coefficient characteristic was most pronounced. The fact that this ferrite has a somewhat high Landé g factor of approximately 2.5 was taken into account in the computations.

The general behavior of this simple ferrite rod configuration is straightforward and is summarized in Figs. 7 and 8. Fig. 7 shows that if the ferrite relative permittivity ϵ_r is increased, the circulation frequency f_0 decreases. Fig. 8 shows that if the saturation magnetization $4\pi M_s$ or the internal bias field H_0 is increased, f_0 increases. Both figures show that as the ferrite rod radius R is increased, f_0 decreases. These trends may all be explained by the "equivalent resonator" point of view. If the central junction region is regarded as a low-Q cavity, with constant dimensions, an increase in ϵ_r will cause the "resonant" frequency f_0 to decrease. If the $4\pi M_s$ or H_0 is increased, the effective relative permeability is decreased and the "resonant" frequency f_0 increases (assuming the internal field is below that required for ferrimagnetic resonance). Similarly, if R is increased, and the material properties and internal field are unchanged, the permittivity and permeability of the whole junction region are increased and hence f_0 is decreased.

B. The Composite Ferrite–Dielectric Rod

This structure facilitates comparison of the theoretical and experimental effects of replacing parts of the ferrite rod with nonmagnetic dielectric. In this way, "dielectric effects" and "magnetic effects" can be distinguished and discussed. A convenient ferrite and rod diameter was selected from a range of computed results; then the effect of replacing the outer region and/or the inner region (core) of the ferrite with dielectric was examined, keeping the overall diameter of the composite rod constant. The ferrite tube with a dielectric core and an outer dielectric sleeve might offer advantages in high-power structures by surrounding the ferrite or garnet with dielectrics of high thermal conductivity. Putting it another way, it might enable the least amount of ferrite to be placed where it will do the most good.

The specific composite posts which were considered are shown in Fig. 9. The overall dimensions were kept constant, but the following combinations of materials were investigated:

(a) solid ferrite post, $R=0.350$ cm;
(b) solid ferrite post, $R=0.300$ cm, with a dielectric sleeve, $R_1=0.350$ cm and $\epsilon_{r1}=13.0$ or 1.0;
(c) thick ferrite tube, $R=0.350$ cm, with a central dielectric rod $R_0=0.150$ cm, and $\epsilon_{r0}=13.0$ or 1.0;
(d) thin ferrite tube, $R=0.300$ cm, $R_0=0.150$ cm, with a dielectric sleeve, $R_1=0.350$ cm, $\epsilon_{r1}=13.0$, and dielectric rod, $R_0=0.150$ cm, $\epsilon_{r0}=13.0$ or 1.0.

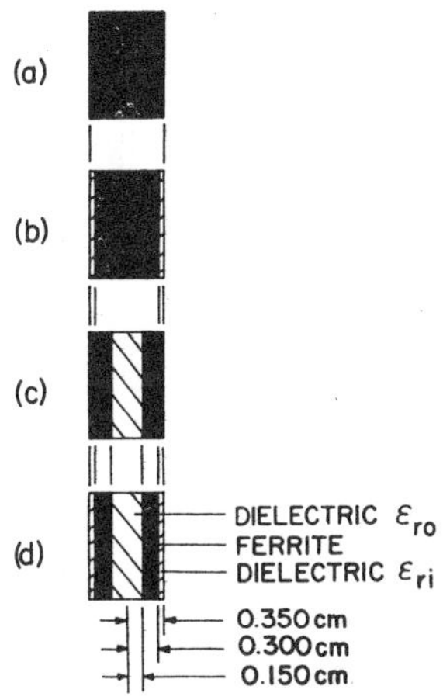

MATERIAL PARAMETERS

	$4\pi M_s$	ΔH	ε_r
TTI-390	2212	540	12.7
TTI-109	1317	135	11.7
DIELECTRIC SLEEVE	—	—	$\varepsilon_{ri}=12.9$ or 1.0
DIELECTRIC SLEEVE	—	—	$\varepsilon_{ro}=12.9$ or 1.0

Fig. 9. Composite ferrite–dielectric posts used to distinguish magnetic and dielectric effects.

We shall be concerned principally with the effects produced by various changes rather than with comparison of theory and experiment. For the present it is sufficient to note that the agreement between prediction and experiment was good enough for this discussion to be carried out in terms of either. Where there are significant discrepancies these will be mentioned. Two ferrite materials were used, TT 1-390 and TT 1-109, which have similar values of permittivity, but different values of saturation magnetization. The main features of the behavior with the two materials were similar, so only the performance with the TT 1-109 will be discussed in detail and some differences with the TT 1-390 will be noted parenthetically.

Fig. 10 shows the performance of a solid TT 1-109 rod, $R=0.350$ cm with a calculated internal field $H_0=200$ Oe. The general shapes of the theoretical and experimental characteristics agree satisfactorily. The frequency of best isolation ("circulation frequency") f_0 is 10 GHz, and a slight double hump again appears in the measured reflection coefficient, ρ, just below f_0. Also, the insertion loss increases where the reflection loss increases. The TT 1-390 rod, with a higher value of $4\pi M_s$, circulated at 11.0 GHz and did not exhibit such a large double hump in the "ρ-curve." Fig. 11 shows the performance of the solid rod of TT 1-109 with the outer region replaced with dielectric, $\epsilon_{r1}=12.9$, but with $H_0=200$ Oe and the overall diameter maintained. This diminution of the magnetic region by 26 percent (by volume), while retaining the dielectric effect, produced a negligible frequency shift in the insertion loss and reflection characteristics and produced a 3–4 percent shift in the isolation curve. The theory also produced virtually no change. However, we note that the discrepancy in the "ρ curve" is larger. An identical replacement on the TT 1-390 rod also produced virtually no change. Then, in the center of the original TT 1-109 rod R=0.350 cm, we replaced 18 percent (by volume) of the ferrite material by dielectric of similar permittivity, $\epsilon_{r0}=12.9$, $R_0=0.15$ cm, and the performance shown in Fig. 12 was obtained. In the absence of data on demagnetizing factors for tubes, the applied field was adjusted to 360 Oe to give a best fit to the theoretical curves computed on the basis of $H_0=200$ Oe. We see that the effect of substituting a nonmagnetic core is to lower the frequency f_0 from 10 GHz to 9.1 GHz. The level of isolation and reflection loss is lower than predicted, and an unpredicted peak now occurs in the isolation instead of the reflection loss characteristic. With TT 1-390 ($4\pi M_s$ $=2212$ gauss), the downward shift of f_0 was larger, approximately 1500 MHz from 11.0 GHz. To summarize: the effect of reducing the saturation magnetization of the rod depends strongly on *where* the reduction occurs. Reducing the magnetic volume by 26 percent at the periphery had negligible effect independent of $4\pi M_s$; reduction by 18 percent at the center lowered the center frequency f_0 by 9 percent and 15 percent depending on the $4\pi M_s$ of the material (1237 and 2212 gauss, respectively).

Then we investigated the effect of reducing the permittivity simultaneously with the magnetization. Reducing the relative permittivity ϵ_{r1} of the sleeve from 12.9 to 1.0 simply reduced the diameter of the solid ferrite rod from $R=0.350$ cm to $R=0.300$ cm. With $H_0=200$ Oe, this reduction shifted the circulation frequency up from 10 GHz to 11.6 GHz and deteriorated the performance. This was not checked experimentally since our earlier work with solid posts had shown that the effects of such simple structural changes are fairly accurately predicted [13].

The same reduction using TT 1-390 shifted the frequency up from 11 GHz to approximately 12.6 GHz. Thus we see that removing 26 percent (by volume) of the dielectric from the periphery increases the circulation frequency f_0 by approximately the same amount with both materials since their relative permittivities are similar in spite of the fact that their values of $4\pi M_s$ are markedly different. The next step was to reduce the relative permittivity of the central region from 12.9 to unity, i.e., to remove the central dielectric rod. The effect of this is shown in Fig. 13. The measured f_0 increases by approximately 200 MHz to 9.3 GHz and the predicted f_0 increases by 300 MHz to 9.5 GHz. So we see good correlation between theory and experiment in this respect. But the small shift in f_0 was an unexpected result. It is perhaps rather

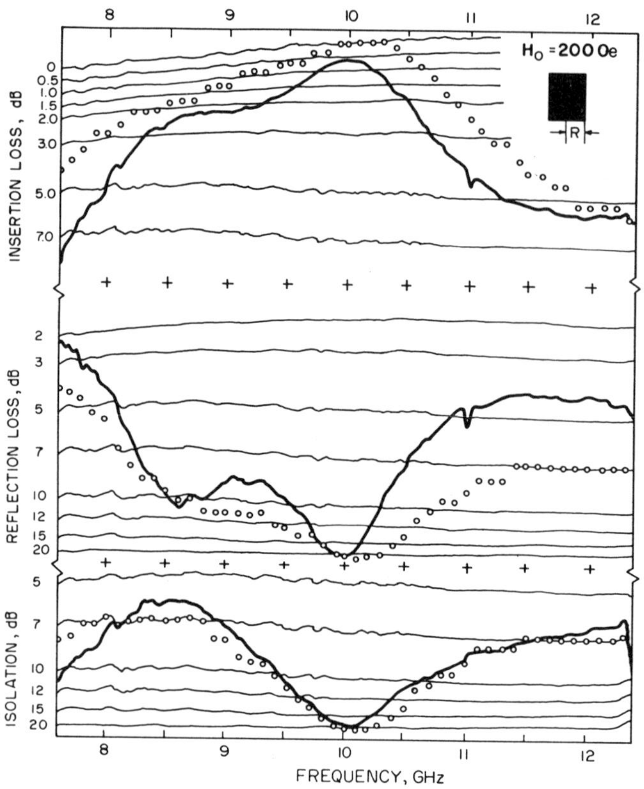

Fig. 10. Performance of a TT 1-109 ferrite rod. $R=0.350$ cm. Solid line: measured values. Circles: computed values.

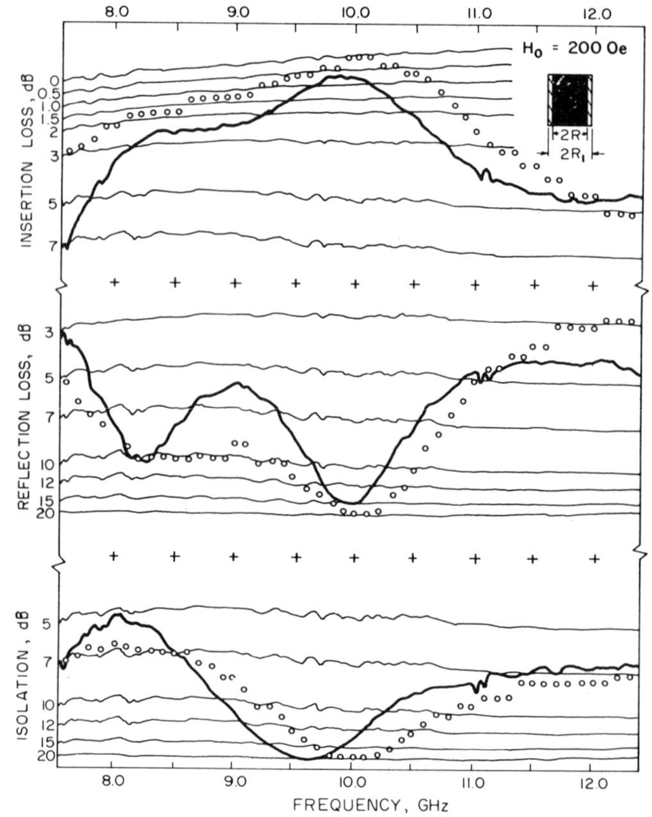

Fig. 11. Performance of a TT 1-109 ferrite rod with dielectric sleeve. $R=0.300$ cm, $R_1=0.350$ cm, $\epsilon_1=12.9$. Solid line: measured values. Circles: computed values.

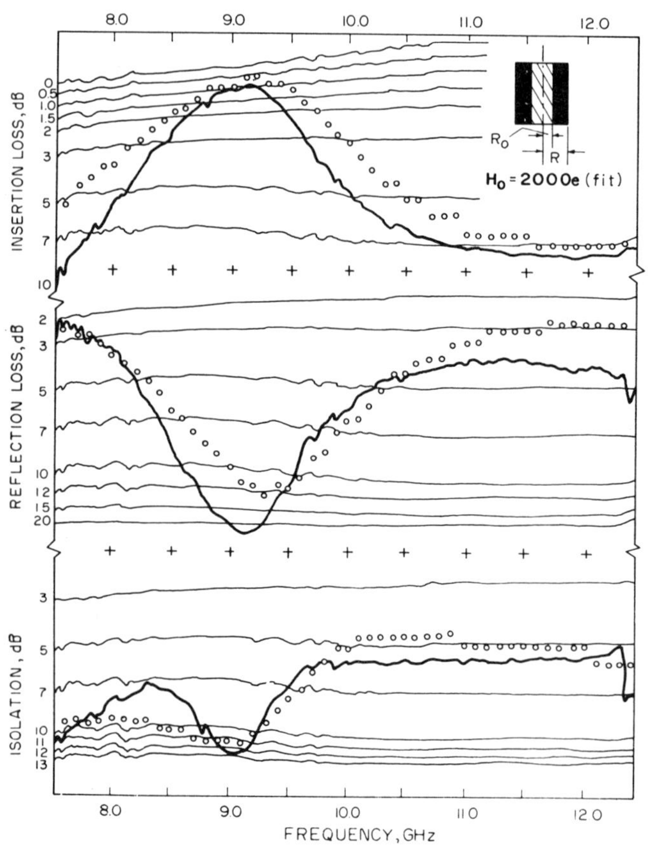

Fig. 12. Performance of a TT 1-109 ferrite tube with a dielectric rod. $R=0.350$ cm, $R_0=0.150$ cm, $\epsilon_{r0}=12.9$. Solid line: measured values. Circles: computed values.

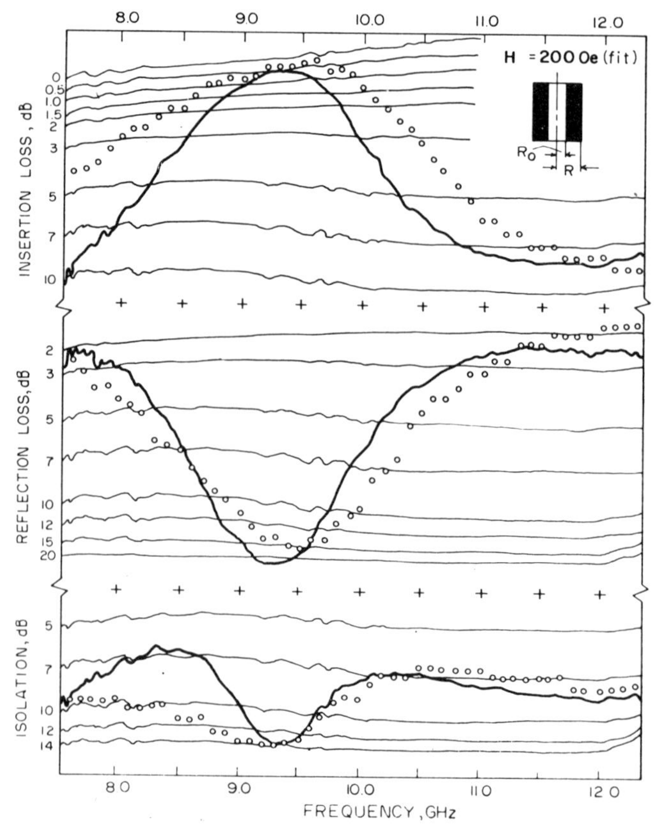

Fig. 13. Performance of a TT 1-109 hollow ferrite tube. $R=0.350$ cm, $R_0=0.150$, $\epsilon_{r0}=1.0$. Solid lines: measured values. Circles: computed values.

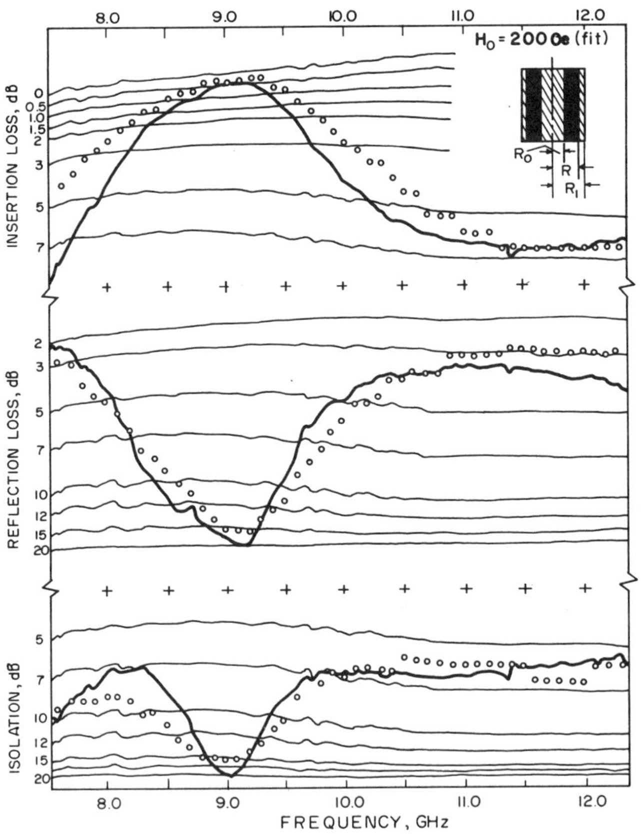

Fig. 14. Performance of a TT 1-109 ferrite tube, a central dielectric rod, and a dielectric sleeve. $R_1=0.350$ cm, $R=0.300$ cm, $R_0=0.150$ cm, $\epsilon_{r1}=\epsilon_{r0}=12.9$. Solid lines: measured values. Circles: computed values.

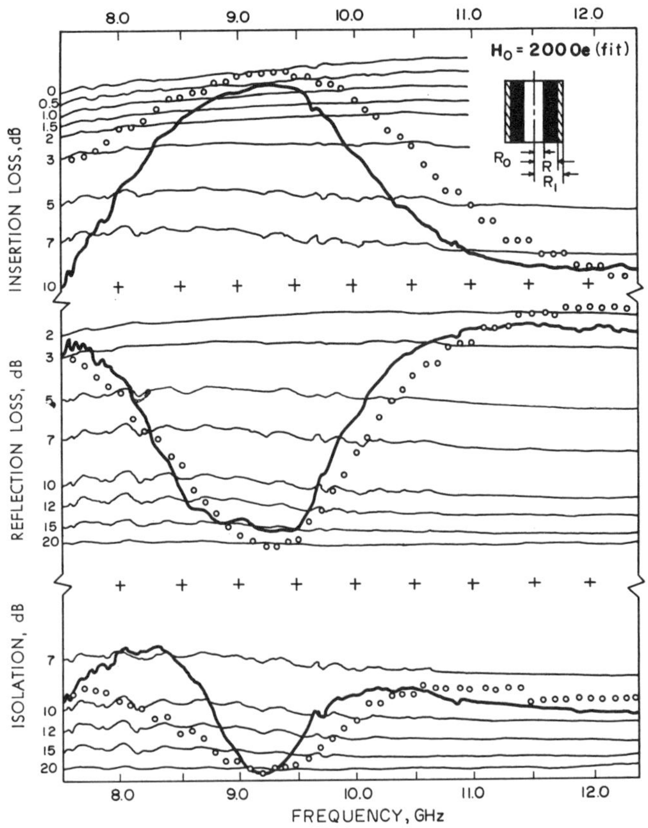

Fig. 15. Performance of a TT 1-109 hollow ferrite tube with a dielectric sleeve. $R_1=0.350$ cm, $R=0.300$ cm, $R_0=0.150$ cm, $\epsilon_{r1}=12.9$, $\epsilon_{r0}=1.0$. Solid lines: measured values. Circles: computed values.

surprising that f_0 hardly changes whether or not the central dielectric rod (diameter$=0.120$ inch, $\epsilon_{r0}\approx 13.0$) is present within the 0.280-inch-diameter ferrite rod. With TT 1-390 the effect of removing an identical central dielectric rod was to increase the measured and predicted f_0 by 500 MHz to 9.9 GHz. Thus we see that the effect of reducing the permittivity depends on *where* the reduction occurs. Reducing the dielectric volume by 26 percent at the periphery increased the circulation frequency f_0 by approximately 15 percent, but reduction of the dielectric volume by 18 percent at the center produced only 2–5 percent increase in f_0. This relative insensitivity to permittivity in the center may be a useful feature. It was seen first in the computed results and was one of the factors that prompted the investigation of composite posts.

These results emphasize the fact that to compare directly the performance of a solid ferrite rod with that of a hollow ferrite tube would be to overlook two interesting and opposing effects: 1) to decrease the magnetization decreases the frequency of circulation; 2) to decrease the permittivity increases the frequency of circulation. In the former effect, the frequency shift is largest if the magnetization is changed in the center of the rod. In the latter effect, the shift is largest if the permittivity is changed at the periphery of the rod.

Since introducing a dielectric sleeve on the solid ferrite rod did not produce a large frequency shift, it was used again with the ferrite tubes to see if the ferrite volume could be further reduced.

Fig. 14 shows the effects of including both a dielectric sleeve and a dielectric rod in the ferrite rod. The predicted results are for $H_0=200$ Oe, and the "best fit" measurements were taken with an applied field of 600 Oe. Comparing Fig. 14 with Fig. 12 we see that adding the sleeve, i.e., reducing the ferrite volume, does not shift the frequency, but it *improves* the performance. It is interesting to see that an improvement was predicted as well as measured, but in both Figs. 12 and 14 the measured isolation and reflection loss are better than the predicted. Figs. 13 and 15 permit the same type of comparison to be made for the hollow tube. With less ferrite (Fig. 15) a better reflection coefficient and isolation are predicted, and the measurements of isolation confirm this trend. But the measured reflection coefficient deteriorated. The ferrite volume in these structures is 56 percent that of the solid rod, $R=0.350$ cm.

Finally, we note that with both the hollow tube and the dielectric-filled tube, the effect of replacing some of the peripheral ferrite with the dielectric is to lower the circulation frequency slightly, since this increases the permittivity of the composite rod. The relative permittivity of TT 1-109 is 11.7, whereas that of the sleeve is 12.9. Also, we have seen that again the largest discrepancy between theory and experiment is the unpredicted peak just below the circulation

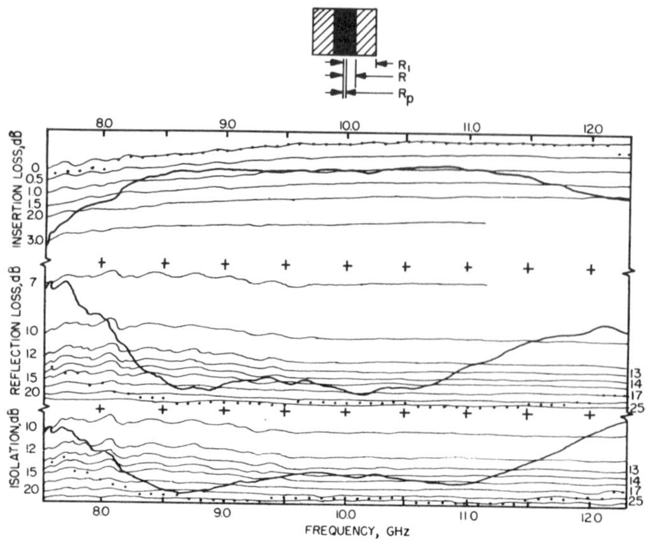

Fig. 16. Performance of a G-113 YIG rod with a metal pin and a dielectric sleeve. $R_1=0.700$ cm, $R=0.260$ cm, $R_p=0.025$ cm, $\epsilon_{r1}=2.08$. Solid lines: measured values. Dotted lines: predicted values.

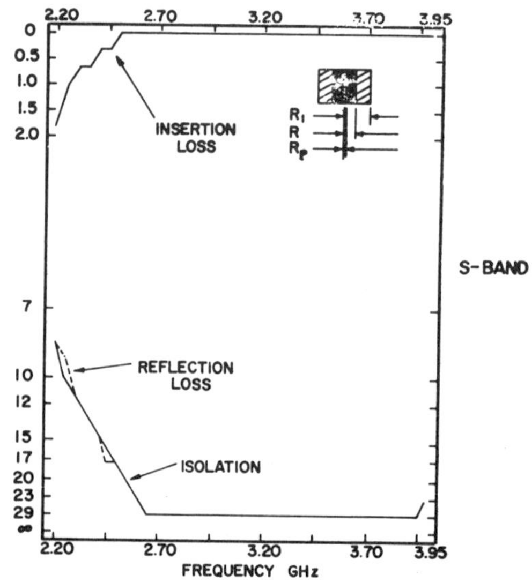

Fig. 17. Computed performance of a G610 garnet rod with a metal pin and a dielectric sleeve. $H_0=0$, $4\pi M_s=680$ gauss, $\Delta H=40$ Oe, $\epsilon_r=14.50$, $R_1=2.120$ cm, $R=0.790$ cm, $R_p=0.047$ cm, $\epsilon_{r1}=2.30$, WR-284.

frequency; however, with this material the measured peak is largest in the reflection loss characteristic with the rod structures and in the isolation characteristic with the tubes.

C. Ferrite Post, Metal Pin, and Dielectric Sleeve

This type of configuration has yielded predicted isolation of greater than 20 dB over 3–5 GHz in X-band waveguide, i.e., almost the waveguide bandwidth. Design guidelines have been discussed elsewhere [7], and experimental modifications of these predicted designs have yielded bandwidths of approximately 3 GHz in X-band waveguide. These broadband structures require a 0.012-inch-diameter metal pin down the center of the ferrite post, which poses some problems, but the important point is that this design probably would not even have been available for consideration but for the computer-aided search.

We have carried out preliminary experimental checks with three thin-pin structures. They all demonstrated greater bandwidth than designs without a central pin, but in each case it was inferior to the computed ideal performance. For example, the results for a YIG rod are shown in Fig. 16. The garnet post was contained in a close-fitting Teflon sleeve, and a copper wire was forced into the hole which was nominally 0.025 cm in radius. The measured bandwidth is 8.5–11.0 GHz for an insertion loss of the order of 1 dB and an isolation of 12–14 dB without external matching. It has been demonstrated elsewhere that with experimental modification this performance can be improved significantly. Following some suggestions made by one of us (L.E.D.) in May 1966, an independent experimental investigation of these structures carried out by Simon *et al.* [14] has resulted in excellent static and latched circulators with relative bandwidths approaching 30 percent. This figure compares favorably with predicted values of 40–50 percent. Experimental results based on computations have also been reported by Parsonson *et al.* [6], and Temme [15] has also tested a thin-pin C-band circulator from 50 to 300 kW peak power without electrical breakdown or increase in insertion loss in this power range.

In an X-band circulator the maximum predicted bandwidth is achieved with a pin diameter of approximately 0.012 inch. With smaller (larger) pin diameters the higher (lower) frequency performance deteriorates. If the pin diameter is made larger than about 0.020–0.030 inch, the predicted 20-dB isolation relative bandwidth decreases to about 30 percent, and simpler structures can already provide this performance. Therefore, the most promising applications of these "thin-pin" circulators would appear to be in C band and S band where scaled dimensions would result in thicker pins, and in high-power applications where the full-height geometry is very favorable and a smaller bandwidth requirement would permit thicker-than-optimum pins. An example of a promising theoretical design for S band is given in Fig. 17.

IV. Conclusions

Measurements on several different structures and comparison with theory have shown that the main features in the behavior of these circulators may be predicted fairly accurately over the entire waveguide bandwidth using only three modes, $n=0, \pm 1$, in the waveguide junction region. It has been shown theoretically and verified experimentally that if the saturation magnetization $4\pi M_s$ of a homogeneous rod is decreased, the circulation frequency f_0 decreases; conversely, if the permittivity ϵ_r is increased, f_0 decreases.

These effects also occur when the $4\pi M_s$ and ϵ_r are radially inhomogeneous. The way in which the magnitudes of the effects vary with the position has been brought out quantitatively, and the theory and experiments showed good agreement. If the $4\pi M_s(r)$ decreases as r increases (with permittivity held constant), the effect on f_0 is very small. But if $4\pi M_s(r)$ decreases as r decreases, then f_0 may decrease significantly. On the other hand, if $\epsilon_r(r)$ approaches unity near the outer surface of the rod, f_0 may increase significantly; but if $\epsilon_r(r)$ approaches unity near the center of the rod, f_0 is affected relatively little. In fact, with one tube (TT 1-109, outer radius $R=0.350$ cm, inner radius $R_0=0.150$ cm) circulation occurred at $f_0 \approx 9.3$ GHz whether or not the central region was filled with dielectric ($\epsilon_{r0}=12.9$ or 1.0). It has also been shown that replacing some of the ferrite at the center and the circumference simultaneously may improve the performance level. It is suggested that high-power circulators may be designed with relatively small volumes of ferrite surrounded by thermally conducting dielectric. This structure has not been studied for optimum bandwidth.

Another promising structure is the ferrite rod with a thin central metal pin and a dielectric sleeve. Relative bandwidths of 40–50 percent have been predicted, and values of approximately 30 percent have been obtained experimentally by other workers. The problem with these structures is principally mechanical, in that it is difficult to incorporate a very thin pin 0.010 inch in diameter down the center of a ferrite rod 0.400 inch long. Therefore, the most promising applications appear to be in high-power C-band and S-band devices where thicker pins would be required. A typical theoretical design for S band has been given.

A particular type of discrepancy between prediction and experiment frequently occurred in the structures without a central metal pin in the form of a peak in the reflection loss or isolation characteristic at a frequency just below the circulation frequency. The discrepancy is large when the outer radius of the solid (or composite) post is 0.350 cm. With smaller radii the effect is much smaller or negligible. This point may be resolved with further analysis.

From this preliminary investigation [16] numerical analysis appears to be very promising. Future work should include attempts to model some of the currently promising experimental designs in an attempt to obtain a better understanding of their limitations. Four-port and five-port single junction circulators should also be investigated since even moderate bandwidths in these devices would find immediate applications.

Acknowledgment

We would like to thank J. R. Bacon, a Rice University undergraduate, who wrote all of the computer programs used in this work.

References

[1] B. A. Auld, "The synthesis of symmetrical waveguide circulators," *IRE Trans. Microwave Theory and Techniques*, vol. MTT-7, pp. 238–246, April 1959.
[2] J. B. Davies, "An analysis of the *m*-port symmetrical *H*-plane waveguide junction with central ferrite post," *IRE Trans. Microwave Theory and Techniques*, vol. MTT-10, pp. 596–604, November 1962.
[3] L. E. Davis, M. D. Coleman, and J. J. Cotter. "Four-port crossed-waveguide junction circulators," *IEEE Trans. Microwave Theory and Techniques*, vol. MTT-1, pp. 43–47, January 1964.
[4] J. B. Davies, "Theoretical design of wideband waveguide circulators," *Electronics Letters*, vol. 1, pp. 60–61, May 1965.
[5] L. E. Davis, "Theoretical design of static and latching ferrite 3-port and 4-port symmetrical waveguide circulators," *Digest of 1966 Internatl. Microwave Symp.*, pp. 281–285.
[6] C. G. Parsonson, S. R. Longley, and J. B. Davies, "The theoretical design of broadband 3-port waveguide circulators," *IEEE Trans. Microwave Theory and Techniques* (Correspondence), vol. MTT-16, pp. 256–258, April 1968.
[7] L. E. Davis, "Central-pin tolerances in broadband 3-port waveguide circulators," *Electronics Letters*, vol. 4, No. 15, July 26, 1968.
[8] J. E. Pippin, "Ferrites—1968," *Microwave J.*, April 1968. See In particular Figs. 5(a)–(c).
[9] B. Lax and K. J. Button, *Microwave Ferrites and Ferrimagnetics.* New York: McGraw-Hill, 1962.
[10] P. S. Epstein, "Theory of wave propagation in a gyromagnetic medium," *Rev. Mod. Phys.*, vol. 28, pp. 3–17, January 1956.
[11] H. E. Bussey, and L. A. Steinert, "An exact solution for a cylindrical cavity containing a gyromagnetic material," *Proc. IRE* (Correspondence), vol. 45, pp. 693–694, May 1957.
[12] R. Bozorth, *Ferromagnetism.* Princeton, N. J.: Van Nostrand, 1951.
[13] J. B. Castillo, "The *H*-plane 3-port waveguide circulator with central ferrite post," M.S. thesis, Dept. Elec. Engrg., Rice University, Houston, Tex., 1968.
[14] J. W. Simon, W. K. Alverson, and J. E. Pippin, "Broadband latching switches and circulators" USAECOM Contract DA 28-043-AMC-02445(E), Repts. 1–4, March 1967–March 1968.
[15] D. H. Temme, M.I.T. Lincoln Laboratory, Lexington, Mass., private communication.
[16] L. E. Davis and J. B. Castillo, "Computer-aided design of 3-port waveguide circulators," Final Rept. USAECOM Contract DAAB-07-67-C-0491, October 1968, Fort Monmouth, N. J., contains further details, including computer programs, automatic plotting of computed impedance, a brief discussion of optimization, more results and analytical detail.

Identification of Spurious Modes in Circulators

Abstract—**The spurious, and sometimes troublesome spikes in circulator characteristics are shown by analysis and experiment to be due to hybrid resonant modes of the ferrite post open-resonator structure. They can be excited by axial nonuniformities such as imperfect contact between the ferrite and the conducting walls, but they cannot exist in axially uniform structures with axially uniform excitation.**

The unwanted spikes in circulator characteristics are well known to experimental designers, and examples of some selected transmission characteristics of narrow-band (<5 percent) waveguide devices are shown in Fig. 1. In a potentially broad-band (>25 percent) waveguide or stripline device, their occurrence often limits the useful bandwidth. They can be excited by very small air gaps between the ferrite and the adjacent conducting surfaces, and they may be suppressed using soft metal shims to eliminate the gaps.

These spurious modes have been referred to as "higher order mode resonances" [1], but they have not been identified. It is the purpose of this correspondence to show that they are indeed distinct resonant modes of the ferrite post open resonator (FPOR) formed by the post (or disk) between two conducting planes. The structure is shown in Fig. 2 (and, depending upon the ratio of diameter to height, it can be interpreted as the center of a waveguide or stripline structure).

The FPOR was first investigated by Steier and Coleman [2], [3], but their analysis was restricted to axially symmetric modes. In 1967 Godtmann and Haas published [4] tuning curves for the asymmetric as well as symmetric modes of the FPOR for values of internal magnetic field above and below resonance, but their curves in the below-resonance region for very low internal fields did not permit sufficient resolution for this investigation. More recently, Courtney [5] has used a dielectric post open resonator to measure the relative permeability and permittivity of demagnetized garnets with considerable accuracy.

A Burroughs B-5500 computer program for the FPOR modes was written based on the theory of Kales [6] with the usual ferrite parameters, the internal magnetic field (to compute the permeability tensor), and the ferrite dimensions as input data. The analysis did not take account of the gap needed to excite the modes; the gap was considered to be only a small perturbation required for the experiment. This is similar to calculating the spectrum of a resonant cavity and choosing to neglect the effect of the coupling hole required to excite the cavity. Instead of exciting an axially nonuniform structure with an axially uniform field, a more rigorous experiment would have been to excite a uniform structure with a nonuniform field. The computed results were checked by reproducing the results of Godtmann and Haas. It should be noted that the material is assumed to be *saturated* and to have a small *uniform* internal field.

MgMnAl and MgMn ferrites were used in the experiments and the material specifications were as follows: TT1-109, saturation magnetization given by $SM_{sat}=1.03\times10^5$ A/m (i.e., 1300 Oe),[1] $\epsilon_r=11.8$; and TT1-390, $SM_{sat}=1.71\times10^5$ A/m (i.e., 2150 Oe),[1] $\epsilon_r=12.7$. The mode charts for the range of frequencies and radii used in Fig. 1 and an internal field, $H_0=2\times10^5$ A/m, are plotted in Figs. 3 and 4, showing only the first few modes. They are identified by their indices (n, m, l) where n is the order of the Bessel function, and hence indicates the azimuthal variation ($\cos n\Phi$); m indicates the mth root of the Bessel function and hence the radial dependence; and l is the number of half wavelengths in the z direction. We see from [4] that these are HE hybrid modes; H_0 is well below the value required for resonance.

Manuscript received June 22, 1970; revised September 3, 1970. This work was supported in part by NASA Grants NsG-6-59 and NGL-44-006-033.

[1] Although the IEEE requires MKS units, the units of saturation magnetization are controversial at present. A detailed background to the whole question of units may be found in *Systems of Units in Electricity and Magnetism* by L. Young (London: Oliver and Boyd, 1969). The specific problem of $4\pi M_s$ units is discussed by the same author in "SI units and magnetisation—A proposal," to appear in IEEE SPECTRUM. Young's notation is used above to quantify saturation magnetization. The problem arises from using $B=\mu_0[H+M]$, in which M and H have the same dimensions and units, and then trying to reconcile this with the MKS and the 3- and 4-dimensional cgs systems.

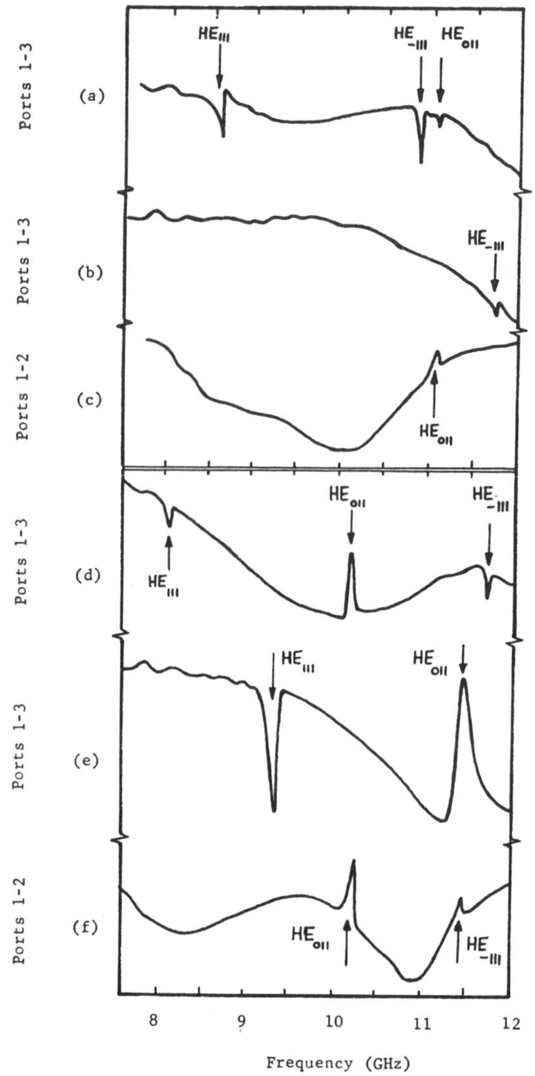

(a) 4-port, TT1-109, $R=0.350$ cm.
(b) 4-port, TT1-109, $R=0.300$ cm.
(c) 3-port, TT1-109, $R=0.350$ cm.
(d) 4-port, TT1-390, $R=0.350$ cm.
(e) 4-port, TT1-390, $R=0.300$ cm.
(f) 3-port, TT1-390, $R=0.350$ cm.

Fig. 1. Transmission characteristics of some 3- and 4-port circulators, selected to show unwanted "spikes." Vertical axis is attenuation in arbitrary units; internal bias field $H_0=200$ Oe.

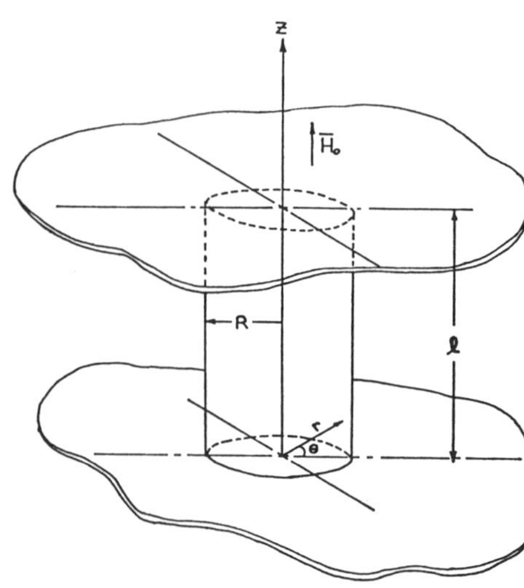

Fig. 2. Ferrite post open resonator (FPOR).

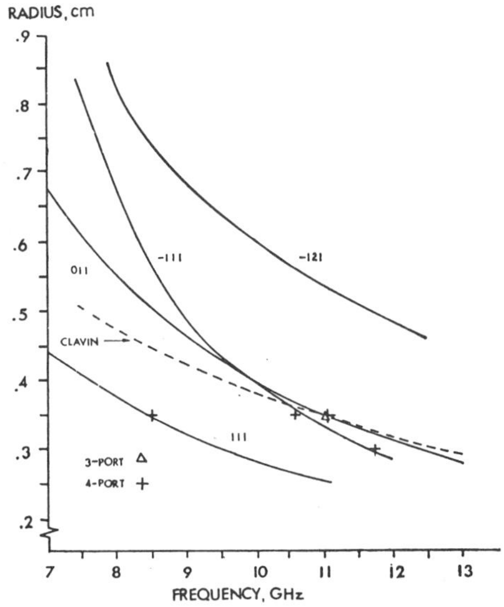

Fig. 3. Resonance frequency versus ferrite radius, for TT1-109 FPOR hybrid modes. $H_0 = 2\times10^5$ A/m; $L = 1.016$ cm.

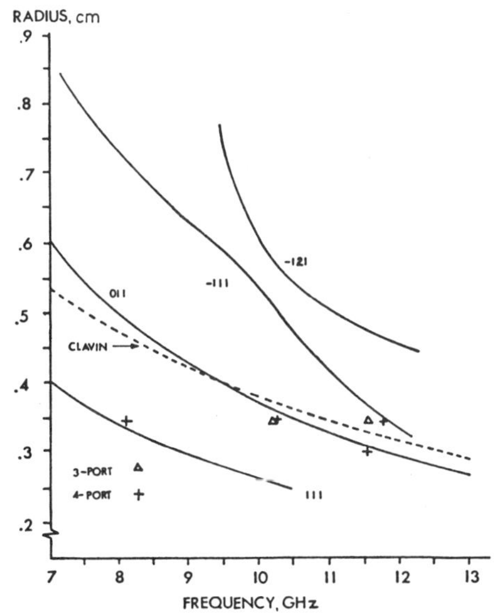

Fig. 4. Resonance frequency versus ferrite radius for TT1-390 FPOR hybrid modes. $H_0 = 2\times10^5$ A/m; $L = 1.016$ cm.

An important point is that if the relative permeability and permittivity are greater than zero and finite, resonances do not exist for $l=0$. This is to say, if 1) there is no longitudinal variation of structure between the conducting plates (no air gaps), 2) the sample is uniformly magnetized, and 3) the excitation is z independent, then open resonator modes cannot occur. For this reason the term "resonance" in connection with waveguide and stripline circulators should be used with caution, if at all. Of course, a field distribution exists with $l=0$, and it is this which causes circulator action, but the field is not resonant. The (0, m, 0), (1, m, 0) and (−1, m, 0) cylindrical modes frequently used in circulator theory are each the lowest order mode of an infinite set, where each set corresponds to one of the three eigensolutions of a 3-port circulator. Being the dominant modes, they provide a reasonable approximation to the circulator action. Also, it should be noted that the fields inside the ferrite may vary radially according to Bessel functions of the first kind (volume modes), or modified Bessel functions of the first kind (surface modes) [7]. The latter solutions are undesirable in waveguide circulators, since the fields tend to be excluded from the ferrite.

The spikes in Fig. 1 have been plotted on Figs. 3 and 4, and they lie sufficiently close to the mode characteristics to permit positive identification. The small discrepancies may be due to 1) the proximity of the corners of waveguide junction to the ferrite surface, within $0.3\lambda_0$–$0.4\lambda_0$; 2) the small gaps needed to excite the modes; 3) the slight axial nonuniformity of the ferrite magnetization.

Finally, it is interesting to compare these results with some experimental data on "higher-order mode resonances" in stripline circulators, reported by Clavin in 1961 [1]. The bias field was just sufficient to saturate the material; five different commercial materials were used; and the frequency range was 1.0–7.0 GHz. Plotting the frequency f(GHz) at which the resonance occurred against the diameter d(in), he obtained a good fit of the experimental points to the curve of the equation, $fd=3$. This equation is plotted as a broken line in Figs. 3 and 4 and it very strongly suggests that Clavin's unidentified spurious mode was the HE_{011}, the lowest order symmetrical mode of the ferrite post open resonator.

A straightforward extension of this analysis would yield the open resonator modes of more complex radially inhomogeneous structures such as the thin-pin circulator.

153

J. B. Castillo
L. E. Davis[2]
Dep. Elec. Eng.
Rice University
Houston, Tex. 77001

References

[1] A. Clavin, "Higher-order mode resonances in strip-line Y-junction circulators," *IEEE Trans. Microwave Theory Tech.* (Corresp.), vol. MTT-9, Nov. 1961, p. 575.
[2] W. H. Steier and P. D. Coleman, "Dipolar magnetodynamic ferrite modes," *J. Appl. Phys.*, vol. 30, 1959, pp. 1454–1455.
[3] ——, "Theory and application of dipolar ferrite modes," *J. Appl. Phys.*, vol. 31, Suppl. to no. 5, 1960, pp. 995–1005.
[4] H.-D. Godtmann and W. Haas, "Magnetodynamic modes in axially magnetized ferrite rods between two parallel conducting sheets," *IEEE Trans. Microwave Theory Tech.* (Corresp.), vol. MTT-15, Aug. 1967, pp. 478–480.
[5] W. E. Courtney, "Analysis and evaluation of a method of measuring the complex permittivity and permeability of microwave insulators," *IEEE Trans. Microwave Theory Tech.*, vol. MTT-18, Aug. 1970, pp. 476–485.
[6] M. L. Kales, "Modes in waveguides containing ferrites," *J. Appl. Phys.*, vol 24, 1953, pp. 604–608.
[7] F. W. Schott and T. F. Tao, "On the classification of electromagnetic waves in ferrite rods," *IEEE Trans. Microwave Theory Tech.* (Corresp.), vol. MTT-16, Nov. 1968, pp. 959–961.

[2] Now on leave to the University College London, London, England.

Reprinted from IEEE TRANSACTIONS ON *MICROWAVE THEORY AND TECHNIQUES*
Volume MTT-19, Number 1, January, 1971
pp. 112-113

Short Papers

A Higher Order Approximation for Waveguide Circulators

JOSEPH B. CASTILLO AND LIONEL E. DAVIS

***Abstract*—Earlier work on the analysis and experimental performance of 3-port and 4-port waveguide circulators over the waveguide bandwidth has shown some discrepancies. The computed performance tended to be too optimistic. It is shown that an extension of Davies' analysis to include higher order cylindrical modes and evanescent modes in the waveguide arms yields excellent correlation between theory and practice.**

I. Introduction

Extending the work of Auld [1] and Davies [2], the behavior of radially inhomogeneous waveguide circulators has recently been reported [3]. The ferrite/dielectric/metal structures uniformly extend the full height of the waveguide (Fig. 1) and the rectangular waveguide excitation is assumed to be the TE_{0p}, i.e., no field variation in the direction of the static magnetic field in the ferrite. This assumption permits an analysis in terms of cylindrical modes in the ferrite rod that are TM_z. The results of this investigation were very en-
154 couraging using azimuthal variations of order $n=0$, ± 1, but there were two noticeable types of discrepancy. The first was the occasional appearance of unpredicted spikes in the measured performance; the second was an unpredicted hump in the curve of measured reflection coefficient or isolation, with an associated increase in insertion loss. The spikes, which are well known to experimental designers, have since been identified as resonances of the ferrite post open resonator (FPOR) hybrid modes [4]. They are trapped modes and cannot participate in circulator action. They may be excited by an axial nonuniformity, and perhaps the most frequent cause is imperfect contact between the ferrite rod and the adjacent conducting surfaces. The "hump" discrepancy could not be explained using the FPOR modes, and it will be shown below that by including higher order circulating modes, i.e., cylindrical modes with $|n|>1$ and zero z variation, and the first two evanescent waveguide modes (TE_{0p}), the overall agreement is improved significantly.

II. Analysis

The general technique consists of solving the boundary value problem inside the geometrically symmetrical junction, which then enables the junction eigenvalues to be calculated [2]. From the eigenvalues, the elements of the scattering matrix of the m-port junction can be determined using the results in the appendix of Auld's paper [1].

The fields within the junction region, i.e., within the central dotted region of Fig. 1, are described by an infinite set of cylindrical modes. The fields in each waveguide near the junction are also described by an infinite set. If the m-port junction is excited with a particular eigenvector of the scattering matrix, the electric field of the eigensolution differs only by a phase factor for each port. Because of this fact, and using superposition of the eigensolutions, it is sufficient to discuss the fields in one waveguide together with those in the cylindrical region between it and the junction axis. Consequently, the problem may be solved by matching each cylindrical mode exactly to the complete set of rectangular waveguide modes. An infinity of equations results (one for each cylindrical mode) each involving an infinity of unknowns (the amplitudes of the waveguide modes):

Manuscript received December 29, 1971.
J. B. Castillo is at 4047 Majestic Lane, Fairfax, Va. 22030.
L. E. Davis is with the Electrical Engineering Department, Rice University, Houston, Tex. 77001.

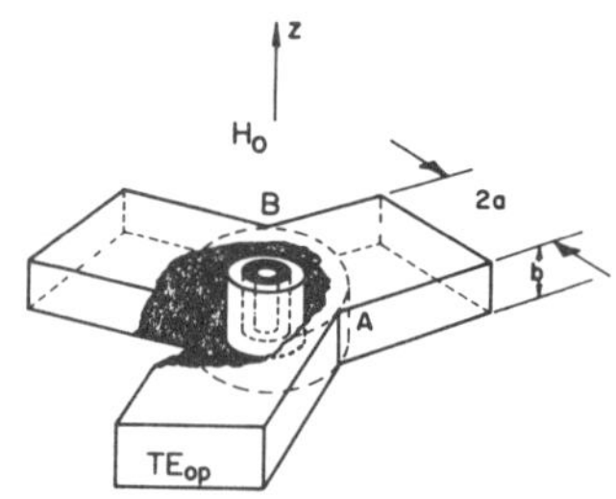

Fig. 1. Symmetrical waveguide junction with radially inhomogeneous composite ferrite post. The central rod may be a metal pin or a dielectric rod. The shaded inner annulus is a ferrite tube which is surrounded by a dielectric sleeve.

$$\sum_{p=-1}^{\infty} A_p(q)G_p = 0, \qquad q = 0, \pm 1, \pm 2 \cdots. \tag{1}$$

A_p and G_p are given in [2]. The calculation of the eigenvalues requires the ratio of the reflected and incident dominant waveguide modes, $p=-1$ and $p=1$. Using Cramer's Rule, the $(r+1)$th iterate for the successive approximate solution is given by Davies' equation (29), and Davies' first approximation took the form $\lambda_j = -A_1(q)/A_1{}^*(q)$ where q was not fixed, but chosen to minimize the expression $n=|mq+j|$. That is, he considered only the lowest possible order of cylindrical modes. In an attempt to improve the agreement between predictions and experiment, especially as related to the "hump" discrepancy, the next higher approximation was taken.

Successive approximate solutions to the infinite system can be derived from one of the linear equations, from three equations, and so on, the number of equations required always being odd. Thus the second approximate solution takes the form

$$\frac{G_{-1}}{G} = -\frac{\begin{vmatrix} A_1(1) & A_2(1) & A_3(1) \\ A_1(0) & A_2(0) & A_3(0) \\ A_1(-1) & A_2(-1) & A_3(-1) \end{vmatrix}}{\begin{vmatrix} A_{-1}(1) & A_2(1) & A_3(1) \\ A_{-1}(0) & A_2(0) & A_3(0) \\ A_{-1}(-1) & A_2(-1) & A_3(-1) \end{vmatrix}}$$

where q must have the values 0, ± 1 in order to obtain three equations. The inclusion of the terms $A_2(q)$ and $A_3(q)$ means physically that the fields of the first two evanescent waveguide modes ($p=2$, 3) are now included in the boundary value problem.

This extension differs slightly from Davies' approach in that he intuitively let q vary to obtain the lowest order mode associated with each eigensolution. We incremented q through the appropriate range of values without minimizing $|mq+j|$. Thus a comparison of the first and second approximations in terms of the number of waveguide and cylindrical modes included in the calculation yields: for the first approximation (FA), the $n=0$, ± 1 for a 3-port and $n=0$, ± 1, ± 2, for a 4-port; for the second approximation (SA), the incident and reflected dominant waveguide modes, the first two evanescent waveguide modes, $n=0$, ± 1, ± 2, ± 3, 4, 5 for a 3-port and $n=0$, ± 1, ± 2, ± 3, ± 4, 5, 6, 7 for a 4-port.

III. Experimental Results

The higher approximation was used to compute the performance of some 3-port devices for which the experimental behavior and the FA predictions were already available [3]. There was an overall improvement in the prediction, and the correlation with experiment was generally excellent. TT1-390 ferrite was used in the experiments and the nominal parameters of this MgMn material are: $M_s = 1.71$

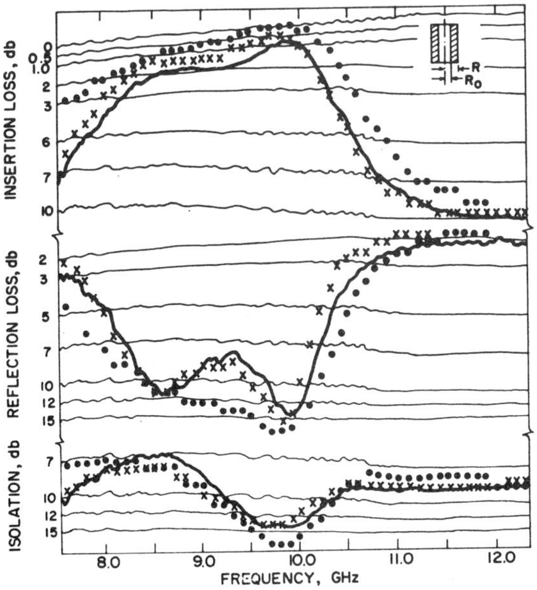

Fig. 2. 3-port performance of a hollow TT1-390 ferrite tube. $R=0.350$ cm, $R_0=0.150$ cm, $\epsilon_{r0}=1.0$, $H_0=2\times10^5$ A/m. (———, measured; · · ·, first approximation; ××× second approximation.)

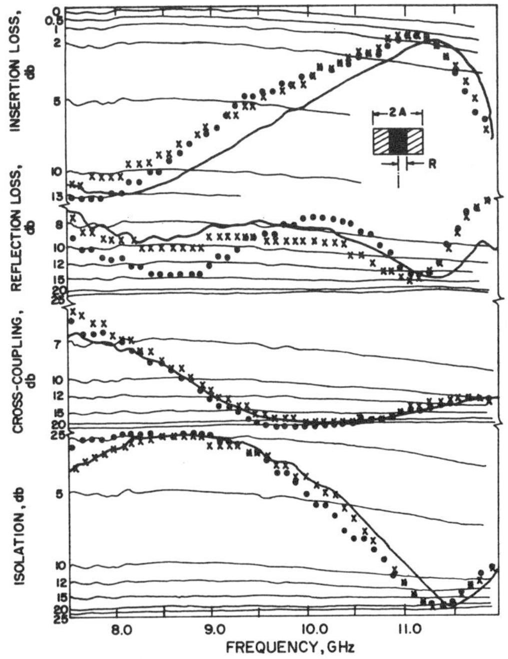

Fig. 4. 4-port performance of a TT1-390 ferrite rod, $R=0.350$ cm. (———, measured; · · ·, first approximation; ×××, second approximation.)

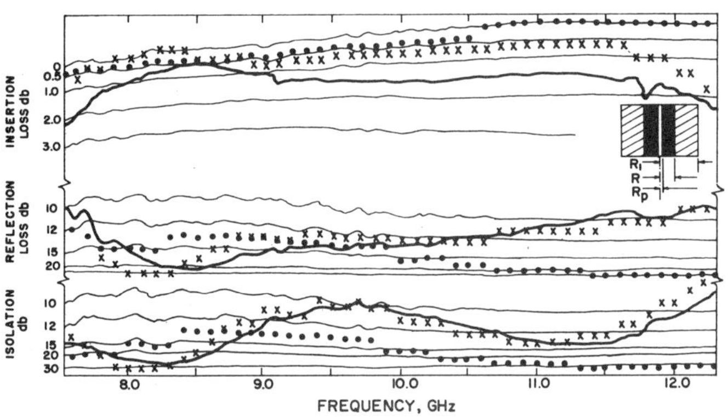

Fig. 3. 3-port performance of TT1-390 ferrite rod, a metal pin, and dielectric sleeve. $R_1=0.775$ cm, $R=0.275$ cm, $R_p=0.025$ cm, $\epsilon_{r1}=2.08$. (———, measured; · · ·, first approximation; ×××, second approximation.)

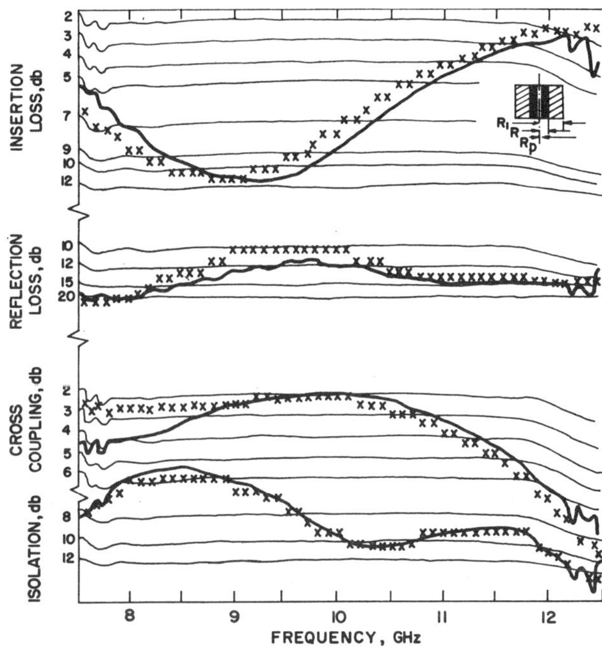

Fig. 5. 4-port performance of a TT1-390 ferrite rod, metal pin, and dielectric sleeve. $R_1=0.775$ cm, $R=0.275$, $\epsilon_{r1}=2.08$. (———, measured; ×××, second approximation.)

155

$\times10^5$ A/m, $H=5.4\times10^5$ A/m, $\epsilon_r=12.7$. However, upon request, the manufacturer supplied the following more accurate value of saturation magnetization. $M_s=1.76\times10^5$ A/m. In the figures, the closed circles mark the FA, the crosses mark the SA, and the continuous lines show the measured performance.

The main feature of the improvement in Fig. 2 for the hollow TT1-390 ferrite tube is that the "hump" discrepancy in the reflection loss curve is resolved. There is also much better agreement at both ends of the frequency range. The thin-pin structure of the type shown in Fig. 3 has yielded theoretical relative bandwidths of 40–50 percent using the FA [3], and experimental bandwidths of nearly 30 percent in X-band waveguide have been measured [5]. However, Fig. 3 shows that the FA can indicate a performance that is too optimistic. The SA shows a performance that is somewhat inferior to the FA, but which agrees more closely with the measured characteristics. It is particularly satisfying that reflection loss agreed well with the measurements, because 3-port circulator design is frequently based upon inspection of the reflection coefficient (or input impedance) characteristic as a function of frequency.

In order to test the 4-port theory, but not necessarily to make a circulator, two of the 3-port composite posts were tested in a 4-port waveguide junction. In Fig. 4 the FA and SA are shown with the measured results. There is relatively little difference between the two sets of computed curves, except for the reflection loss, and the agreement between the second computations and experiment is good. It should be noted that the insertion loss is minimum, while reflection loss and isolation (ports 1–4) are maximum at 11.5 GHz. But the cross-coupling (ports 1–3) curve reaches a minimum in the range 9.5–11.0 GHz, and has risen considerably at 11.5 GHz. This is the type of behavior which restricts the bandwidth of these devices. As another example, the thin-pin configuration provides good agreement with the SA theory in Fig. 5, and it appears that circulator action might be occurring about 12.5 or 13 GHz, but there is no indication of other than narrow-band performance.

Finally, two unsuccessful results should be mentioned briefly. Some results of empirical 4-port circulator design using full-height structures are available in the literature [6], [7], but attempts to predict these results were unsuccessful. This is attributed to a lack of reliable ferrite data, and the sensitivity of 4-port performance to small changes in material properties and dimensions. Also, an unsuccessful attempt was made to obtain a good (predicted) 4-port circulator using the design procedure suggested by Helszajn and Buffler [8]. Further investigations are under way.

IV. Discussion and Conclusions

Davies' theory of the symmetrical m-port nonreciprocal waveguide junction has been successfully extended to a higher approximation for $m=3$ and 4. The SA agrees very well and resolves some of the previously unexplained discrepancies displayed by the FA. The SA provides some of the "fine structure" and provides a more realistic estimate of bandwidth and loss between ports. This is particularly useful for the potentially broad-band thin-pin 3-port structures.

The penalty we have paid for improved calculations is an increase in computer time necessary to make them. The increase in the number of terms in n requires many more calculations and, especially for the more complex structures, the increase in computation time is marked. As examples let us consider a simple junction, the 3-port with a ferrite post, and then the more complicated 4-port with a ferrite post, metal pin, and dielectric sleeve. With the 3-port, the FA and SA take approximately 0.55 and 1.65 min, respectively, for a set of computations over the waveguide bandwidth. With the 4-port, the times are approximately 1.63 and 4.22 min, respectively. These times include approximately 0.180 min for the graphical printout in the form of a simulated swept-frequency plot. The programs might be made more efficient, and the machine was a Burroughs B-5500.

It should be emphasized that we now have a proven analysis technique for a particular class of waveguide structures; we do not have a circulator design program. Any successful design that is developed using this program is still likely to be the result of experience just as it would be in the laboratory, but the difference is that it would have been found in less time or it would be the best of a larger number of attempts. However, one would still not have any information as to whether this was the "best" design in any sense of the word.

Two avenues may now be followed. A performance figure of merit could be defined and optimization routines developed that would automatically maximize the figure of merit. This is probably the most straightforward extension, and one that will fill a real need. It will, however, be complicated by the many parameters present in all potentially attractive structures, the complexity of the behavior with parameter changes (particularly in the 4-port), and the fact that a global maximum is required.

The authors feel that while analysis and optimization are the only good design procedures within reach, the real requirement is for a successful circulator synthesis procedure.

References

[1] B. A. Auld, "The synthesis of symmetrical waveguide circulators," *IEEE Trans. Microwave Theory Tech.*, vol. MTT-7, pp. 238–246, Apr. 1959.

[2] J. B. Davies, "An analysis of the m-port symmetrical H-plane waveguide junction with central ferrite post," *IEEE Trans. Microwave Theory Tech.*, vol. MTT-10, pp. 596–604, Nov. 1962.

[3] J. B. Castillo and L. E. Davis, "Computer-aided design of three-port waveguide junction circulators," *IEEE Trans. Microwave Theory Tech.*, vol. MTT-18, pp. 25–34, Jan. 1970.

[4] ——, "Identification of spurious modes in circulators," *IEEE Trans. Microwave Theory Tech.* (Corresp.), vol. MTT-19, pp. 112–113, Jan. 1971.

[5] J. E. Pippin, "Ferrites—1968," *Microwave J.*, Apr. 1968; see fig. 5(a)–(c).

[6] L. E. Davis, M. D. Coleman, and J. J. Cotter, "Four-port-crossed-waveguide junction circulators," *IEEE Trans. Microwave Theory Tech.*, vol. MTT-12, pp. 43–47, Jan. 1964.

[7] G. S. Sidhu and O. P. Gandhi, "A four-port waveguide junction circulator and effects of dielectric loading on its performance," *IEEE Trans. Microwave Theory Tech.* (Corresp.), vol. MTT-13, pp. 388–389, May 1965.

[8] J. H. Helszajn and C. R. Buffler, "Adjustments of the 4-port single junction circulator," *Radio Electron. Eng.*, vol. 35, no. 6, 1968.

The Adjustment of the m-Port Single-Junction Circulator

JOSEPH HELSZAJN, MEMBER, IEEE

Abstract—The purpose of this paper is to present a systematic procedure for the adjustment of the m-port single-junction circulator in terms of the scattering-matrix eigenvalues of the junction. The theory used starts by defining the initial location of the scattering-matrix eigenvalues before any adjustment to the junction is made and also by defining the eigenvalue arrangement of an ideal circulator. The initial set of eigenvalues is then perturbed in a systematic way by adjusting the phases of $(m-1)$ of the eigenvalues one at a time until the final set of eigenvalues is obtained. This procedure leads to $(m-1)$ distinct scattering matrices which can be obtained with $(m-1)$ symmetric perturbations of the junction. This approach also leads to $(m-1)$ distinct boundary-value problems each involving one independent variable instead of the more usual single-boundary problem involving $(m-1)$ independent variables.

Manuscript received October 27, 1969; revised June 6, 1970.
The author is a consultant, 6 Kirkwood Crescent, Leeds 16, England.

I. Introduction

One method of synthesizing symmetrical waveguide junctions consists of adjusting the eigenvalues of the scattering matrix by symmetric perturbations of the junction until the required junction scattering coefficients are satisfied [1]. In the case of the m-port single-junction circulator the scattering matrix eigenvalues must be adjusted until they coincide with those of an ideal circulator [2]. For an ideal m-port circulator these eigenvalues lie equally spaced on a unit circle in the complex plane. This method has been applied in the construction of the three-port circulator [3]. To obtain the eigenvalue arrangement of an ideal circulator, it is necessary to adjust the phases of $(m-1)$ of the eigenvalues with respect

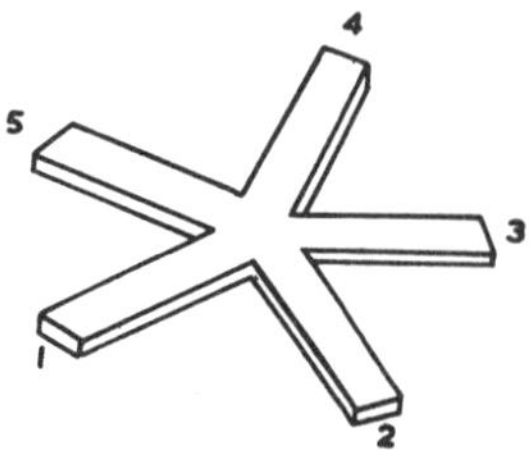

Fig. 1. Five-port single-junction circulator.

to one of them, and therefore $(m-1)$ physical perturbations of the junction are necessary. This leads to $(m-1)$ distinct scattering matrices which can be calculated once the eigenvectors of the junction are known. Since the eigenvectors of the junction are determined from symmetry conditions only, they remain unchanged under a symmetric perturbation of the junction. Hence the junction eigenvectors already coincide with those of an ideal circulator. Fig. 1 shows a waveguide junction with the symmetry considered in this paper for the case of $m=5$. The synthesis procedure then consists of adjusting the phases of $(m-1)$ of the eigenvalues one at a time by symmetric perturbation of the junction until the coefficients of the scattering matrix appropriate to each required eigenvalue arrangement are obtained. The purpose of this paper is to give a systematic way in which this can be done. In this approach it is necessary to know the initial and final locations of the eigenvalues in the complex plane. The final location of the eigenvalues are found from the scattering matrix of an ideal circulator. These lie equally spaced on a unit circle. The initial location of the eigenvalues can be obtained by forming the scattering matrix of the junction, assuming that all the junction modes are not resonant. The matrix formed in this way has all of its transmission coefficients zero and all of its reflection coefficients unity. This matrix is simply a unit matrix for which all the eigenvalues are equal. This can be seen immediately by setting the spur of the unit matrix, defined as a sum of the diagonal elements of the matrix, equal to the sum of the eigenvalues. Once these two diagrams are known, a systematic adjustment of the eigenvalues which will perturb the initial set of eigenvalues to the final set is possible. The order in which the eigenvalues are perturbed is determined by the choice of the physical perturbation of the junction. From the initial arrangement in which the eigenvalues have m-fold degeneracy one obtains the important result that rotating an eigenvalue through 180° in the complex plane minimizes the reflection coefficient. This statement provides one definition of an electromagnetic resonance in a cavity resonator for what follows.

The approach used also leads to $(m-1)$ distinct boundary-value problems. The electric-field amplitude at the terminals of the junction is also given after each perturbation of the junction. The approach described in this paper has been successfully used to construct a four-port circulator [5].

II. The Scattering-Matrix Eigenvalues and Eigenvectors of an Ideal m-Port Circulator

The scattering matrix of an m-port junction is defined by

$$\bar{b} = \bar{S}\bar{a} \tag{1}$$

where $\bar{S}$ relates the complex amplitude and phase of the transverse electric field a_i of the incident wave at the ith port to the emergent complex wave b_i at the same port. It is assumed that a_i and b_i are normalized in such a way that $\frac{1}{2}a^*a$ is the average input power and $\frac{1}{2}b^*b$ is the average emergent power. The principle of the conservation of energy requires that the scattering matrix be unitary. The matrix $\bar{S}$ has the form

$$\bar{S} = \begin{bmatrix} S_{11} & S_{12} & \cdots & S_{1m} \\ S_{21} & S_{22} & \cdots & S_{2m} \\ S_{m1} & S_{m2} & \cdots & S_{mm} \end{bmatrix}. \tag{2}$$

The eigenvalue equation of the square matrix $\bar{S}$ is

$$s_n \bar{U}_n = \bar{S}\bar{U}_n \tag{3}$$

where $\bar{U}_n$ is an eigenvector and s_n is an eigenvalue. By comparison with (1) it can be seen that $\bar{U}_n$ represents a possible field excitation in the junction with the fields at the terminal planes proportional to the elements of the eigenvector, and s_n represents a reflection coefficient measured at each terminal plane. Equation (3) has a nonvanishing value for $\bar{U}_n$ provided that

$$\det \left| \bar{S} - s_n \bar{I} \right| = 0. \tag{4}$$

The solution of the above characteristic equation determines the eigenvalues of the matrix $\bar{S}$. In the complex plane the eigenvalues lie on a unit circle because $\bar{S}$ is unitary. For an ideal m-port circulator at a suitable reference plane

$$s_n = e^{-j2\pi n/m}, \qquad n = 0, \pm 1, \pm 2, \cdots, m/2. \tag{5}$$

For m odd there is one real eigenvalue and $(m-1)$ complex eigenvalues. For m even there are two real eigenvalues and $(m-2)$ complex eigenvalues. Hence for an ideal circulator the eigenvalues lie equally spaced on a unit circle. In terms of the eigenvalue arrangement given by (5), the synthesis procedure consists of adjusting the phases of $(m-1)$ of the eigenvalues on the unit circle and hence requires $(m-1)$ independent physical variables. This is a well-known result.

The eigenvectors of the junction are determined by the symmetry properties of the junction only. In the case of the star-junction circulator shown in Fig. 1 a set of eigenvectors is given by the set of column matrices $\overline{U}_n$ [2]

$$(\overline{U}_n)_i = \frac{e^{-j2\pi n(i-1)/m}}{\sqrt{m}}, \quad n = 0, \pm 1, \pm 2, \cdots, m/2$$
$$i = 1, 2, 3, \cdots, m. \quad (6)$$

That the above eigenvectors satisfy the eigenvalue equation can be seen by direct substitution into (3). Each eigenvalue and eigenvector represents a unique eigensolution to the junction boundary-value problem. Since the eigenvectors already coincide with the required circulator eigenvectors, the adjustment procedure continues by making symmetrical perturbations of the junction geometry until the coefficients of the scattering matrix appropriate to each eigenvalue adjustment is obtained. The relation between the scattering-matrix eigenvalues and the coefficients of the scattering matrix is obtained by diagonalizing $\overline{S}$.

III. The Diagonalization Procedure

In order to determine independently the $(m-1)$ scattering matrices for obtaining the eigenvalue arrangement of an ideal circulator, it is necessary to diagonalize the matrix $\overline{S}$. This can be done by a matrix $\overline{U}$ having for its columns the eigenvectors of $\overline{S}$.

$$\overline{S} = \overline{U}\bar{\lambda}\overline{U}^{-1} \quad (7)$$

where $\bar{\lambda}$ is a diagonal matrix with the eigenvalues of $\overline{S}$ along its main diagonal, and $\overline{U}^{-1}$ is the inverse of $\overline{U}$. If the eigenvectors of $\overline{S}$ are obtained from the eigenvalue equation of an ideal circulator one has [6]

$$\overline{U}^{-1} = (\overline{U}^*)^T \quad (8)$$

where $(\overline{U}^*)^T$ is the transpose of the complex conjugate of $\overline{U}$. The relation between the eigenvalues and the coefficients of the scattering matrix is obtained by multiplying out (7).

The diagonalization procedure will now be developed for a five-port circulator. This gives the relation between the eigenvalues and the scattering coefficients of the scattering matrix. Using (6) the eigenvectors of the junction are

$$\overline{U}_0 = \frac{1}{\sqrt{5}}\begin{bmatrix}1\\1\\1\\1\\1\end{bmatrix} \quad (9)$$

$$\overline{U}_1 = \frac{1}{\sqrt{5}}\begin{bmatrix}1\\e^{-j2\pi/5}\\e^{-j4\pi/5}\\e^{-j6\pi/5}\\e^{-j8\pi/5}\end{bmatrix} \quad (10)$$

$$\overline{U}_{-1} = \frac{1}{\sqrt{5}}\begin{bmatrix}1\\e^{+j2\pi/5}\\e^{+j4\pi/5}\\e^{+j6\pi/5}\\e^{+j8\pi/5}\end{bmatrix} \quad (11)$$

$$\overline{U}_2 = \frac{1}{\sqrt{5}}\begin{bmatrix}1\\e^{-j4\pi/5}\\e^{-j8\pi/5}\\e^{-j12\pi/5}\\e^{-j16\pi/5}\end{bmatrix} \quad (12)$$

$$\overline{U}_{-2} = \frac{1}{\sqrt{5}}\begin{bmatrix}1\\e^{+j4\pi/5}\\e^{+j8\pi/5}\\e^{+j12\pi/5}\\e^{+j16\pi/5}\end{bmatrix}. \quad (13)$$

The matrix $\overline{U}$ having the eigenvectors of $\overline{S}$ as columns is

$$\frac{1}{\sqrt{5}}\begin{bmatrix}1 & 1 & 1 & 1 & 1\\ 1 & e^{-j2\pi/5} & e^{+j2\pi/5} & e^{-j4\pi/5} & e^{+j4\pi/5}\\ 1 & e^{-j4\pi/5} & e^{+j4\pi/5} & e^{-j8\pi/5} & e^{+j8\pi/5}\\ 1 & e^{-j6\pi/5} & e^{+j6\pi/5} & e^{-j12\pi/5} & e^{+j12\pi/5}\\ 1 & e^{-j8\pi/5} & e^{+j8\pi/5} & e^{-j16\pi/5} & e^{+j16\pi/5}\end{bmatrix}. \quad (14)$$

In the case of the five-port circulator there are two degenerate pairs of eigenvalues and one nondegenerate pair. The diagonal matrix $\bar{\lambda}$ with the eigenvalues of $\overline{S}$ is therefore

$$\bar{\lambda} = \begin{bmatrix}s_0 & 0 & 0 & 0 & 0\\ 0 & s_1 & 0 & 0 & 0\\ 0 & 0 & s_{-1} & 0 & 0\\ 0 & 0 & 0 & s_2 & 0\\ 0 & 0 & 0 & 0 & s_{-2}\end{bmatrix}. \quad (15)$$

Substituting the above quantities into (7) and making use of (8) give the following relation between the scattering coefficients and the eigenvalues.

$$5S_{11} = s_0 + s_1 + s_{-1} + s_2 + s_{-2} \quad (16)$$

$$5S_{12} = s_0 + s_1 e^{+j2\pi/5} + s_{-1}e^{-j2\pi/5} + s_2 e^{+j4\pi/5} + s_{-2}e^{-j4\pi/5} \quad (17)$$

$$5S_{13} = s_0 + s_1 e^{+j4\pi/5} + s_{-1}e^{-j4\pi/5} + s_2 e^{+j8\pi/5} + s_{-2}e^{-j8\pi/5} \quad (18)$$

$$5S_{14} = s_0 + s_1 e^{+j6\pi/5} + s_{-1}e^{-j6\pi/5} + s_2 e^{+j12\pi/5} + s_{-2}e^{-j12\pi/5} \quad (19)$$

$$5S_{15} = s_0 + s_1 e^{+j8\pi/5} + s_{-1}e^{-j8\pi/5} + s_2 e^{+j16\pi/5} + s_{-2}e^{-j16\pi/5}. \quad (20)$$

If the adjustment procedure is to be made in terms of the reflection coefficient S_{11} only, it is not necessary to diagonalize the matrix $\overline{S}$. In this instance S_{11} can be obtained quite simply by equating the spur of the scattering matrix to the sum of the eigenvalues.

IV. Electric Field Amplitudes

If the eigenvalues of (16) have five-fold degeneracy, one obtains the important result that rotating an eigenvalue through 180° in the complex plane minimizes the reflection coefficient. This can be taken as one definition of an electromagnetic-field resonance in a cavity resonator. If the electric-field pattern of the electromagnetic resonance is known, the nature of the physical adjustment is strongly suggested. The electric-field amplitudes established in this way at the terminals of the junction are assumed to be proportional to the corresponding eigenvectors.

To compute the total electric-field pattern at the edge of the ferrite disk, it is assumed as an approximation that $E_z(\phi)$ is proportional to a linear combination of the eigenvectors of the scattering matrix.

160

$$E_z(\phi) = a_0 + a_1^+ e^{j\phi} + a_1^- e^{-j\phi} + a_2^+ e^{j2\phi} + a_2^- e^{-j2\phi}. \quad (21)$$

Since the eigenvectors given by (9)–(13) are not unique, one assumes the following general form for the coefficients:

$$a_1^\pm = a_1' \pm jb_1'$$
$$a_2^\pm = a_2' \pm jb_2'.$$

This gives

$$E_z(\phi) = a_0 + a_1 \cos(\tau_1 + \phi) + a_2 \cos(\tau_2 + 2\phi) \quad (22)$$

where

$$\tau_{1,2} = \tan^{-1}\left(\frac{b_{1,2}'}{a_{1,2}'}\right)$$
$$a_{1,2} = \sqrt{(a_{1,2}')^2 + (b_{1,2}')^2}.$$

The electric-field amplitudes at the terminals are therefore

$$E_z(\phi = 0) = a_0 + a_1 \cos\tau_1 + a_2 \cos\tau_2 \quad (23)$$

$$E_z\left(\phi = \frac{2\pi}{5}\right) = a_0 + a_1 \cos\left(\tau_1 + \frac{2\pi}{5}\right) + a_2 \cos\left(\tau_2 + \frac{4\pi}{5}\right) \quad (24)$$

$$E_z\left(\phi = \frac{4\pi}{5}\right) = a_0 + a_1 \cos\left(\tau_1 + \frac{4\pi}{5}\right) + a_2 \cos\left(\tau_2 + \frac{8\pi}{5}\right) \quad (25)$$

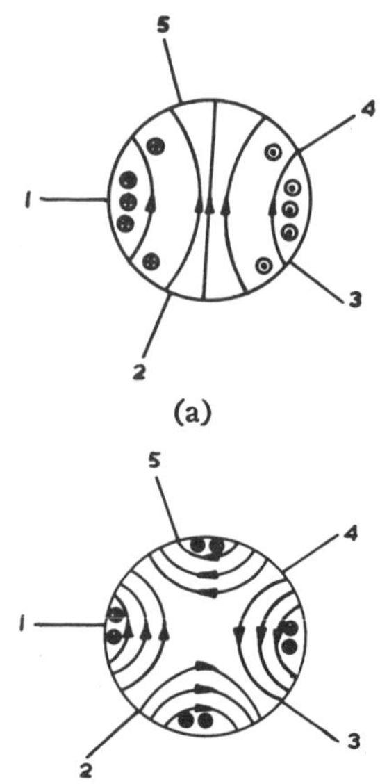

Fig. 2. (a) Field patterns for $n = \pm 1$ modes in ferrite disk. (b) Field patterns for $n = \pm 2$ modes in ferrite disk.

$$E_z\left(\phi = \frac{6\pi}{5}\right) = a_0 + a_1 \cos\left(\tau_1 + \frac{6\pi}{5}\right) + a_2 \cos\left(\tau_2 + \frac{12\pi}{5}\right) \quad (26)$$

$$E_z\left(\phi = \frac{8\pi}{5}\right) = a_0 + a_1 \cos\left(\tau_1 + \frac{8\pi}{5}\right) + a_2 \cos\left(\tau_2 + \frac{16\pi}{5}\right). \quad (27)$$

The electric-field pattern given by (22) with $\tau_1 = \tau_2 = 0$ consists of a linear combination of electromagnetic-field patterns whose electric fields have cosinusoidal variations around the periphery of the ferrite disk. Such field patterns have been described in the literature [7]–[9]. Fig. 2(a) and (b) shows the $n = \pm 1$ and $n = \pm 2$ field patterns of a ferrite disk. A suitable nonresonant field pattern (not shown) is obtained with $n = 0$. A five-port circulator can now be synthesized from these field patterns by rotating the two field patterns by the angles τ_1 and τ_2. This is done by splitting the degeneracy of each of the field patterns by magnetizing the junction.

The necessary rotation of the field patterns is obtained by applying the boundary conditions for a perfect circulator at the five ports in (23)–(27). The results are $\tau_1 = 54°$, $\tau_2 = 18°$, $a_0 = 0$, $a_1 = 0.4707$, and $a_2 = 0.7607$. It is observed that the angles τ_1 and τ_2 are those of the corresponding admittance eigenvalues.

Fig. 3(a) and (b) shows the field patterns shown in Fig. 2(a) and (b) rotated through 54° and 18°, respectively. If the magnitude of the E_z fields for the two patterns are equal at the five ports, then circulation occurs between ports 1 and 2, and all the other ports are isolated.

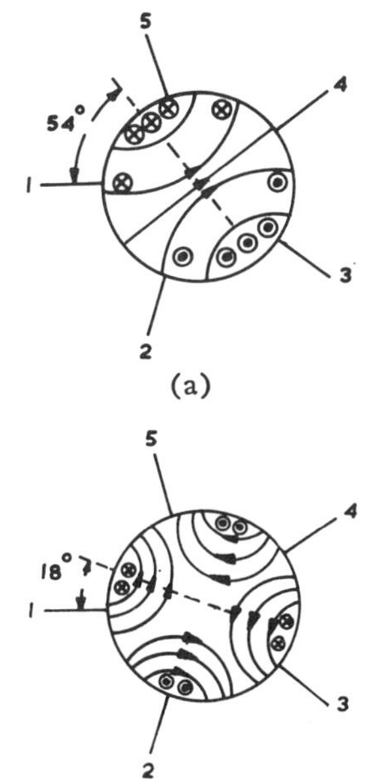

Fig. 3. (a) Field patterns for rotated $n=\pm 1$ modes in ferrite disk. (b) Field patterns for rotated $n=\pm 2$ modes in ferrite disk.

It is the purpose of this paper to relate the above four physical perturbations of the junction to measured quantities at the external terminals of the junction.

V. Circulation Adjustment for the Five-Port Circulator

In this section the synthesis procedure will be used to describe the adjustment of the five-port circulator. The physical perturbation described in this section applies to the stripline junction only. This will be done by rotating the eigenvalues in the complex plane one at a time until all the eigenvalues coincide with those of the scattering matrix of an ideal circulator. If all the electromagnetic fields are nonresonant, the scattering coefficients are

$$S_{11} = -1$$

$$S_{12} = S_{13} = S_{14} = S_{15} = 0$$

which correspond with the reference plane for which $S_{12}=-1$ for an ideal circulator. The eigenvalues corresponding to these scattering coefficients are

$$s_0 = s_1 = s_{-1} = s_2 = s_{-2} = -1.$$

Hence all the eigenvalues lie together on the unit circle. This is shown in Fig. 4.

The first perturbation of the junction is now obtained by rotating either the s_1 and s_{-1} or s_2 and s_{-2} degenerate pairs of eigenvalues through 180° on the unit circle. This decision is not unique but is determined by the choice of the ensuing physical adjustment. In what follows the s_2 and s_{-2} eigenvalues are first rotated through 180° on the unit circle by adjusting the ferrite disk ap-

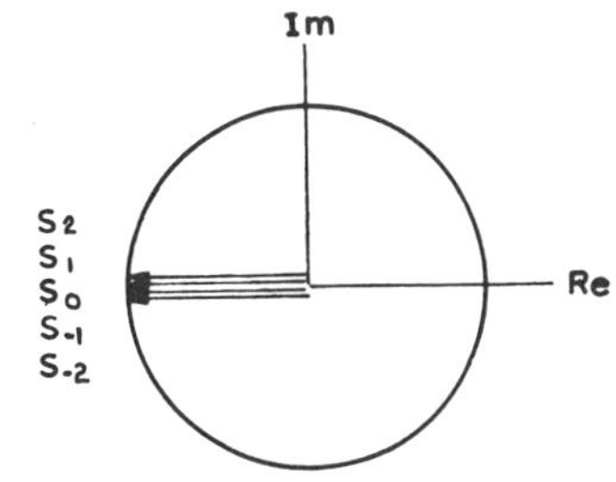

Fig. 4. Scattering-matrix eigenvalues for five-port junction having all its transmission coefficients zero and all its reflection coefficients unity.

propriate to the $n=\pm 2$ field patterns. The second physical perturbation of the junction then consists of also rotating the s_1 and s_{-1} eigenvalues through 180° without disturbing the s_2 and s_{-2} eigenvalues. One possible physical perturbation here is an inductive post at the center of the ferrite disk which establishes the $n=\pm 1$ field patterns without detuning the $n=\pm 2$ patterns. The adjustment procedure then continues by removing the degeneracy between the two pairs of eigenvalues as will be described shortly. An alternative approach, which leads to a different set of scattering coefficients, consists of first rotating the s_1 and s_{-1} eigenvalues through 180° on the unit circle. This first adjustment now leads to a different ferrite geometry appropriate to the $n=\pm 1$ field patterns. Then the s_2 and s_{-2} eigenvalues are also rotated by 180° on the unit circle. One physical variable here is a slotted-plate center conductor (in the case of stripline) which has been shown to tune the $n=\pm 2$ field patterns to the same frequency as the $n=\pm 1$ without detuning the latter [10]. The adjustment procedure again continues by splitting the degeneracy between the degenerate pairs of eigenvalues.

The first circulation adjustment is now obtained by rotating the degenerate pair of eigenvalues s_2 and s_{-2} through 180° on the unit circle. This leads to the arrangement between the eigenvalues shown in Fig. 5:

$$s_0 = s_1 = s_{-1}$$

$$s_0 = -s_2 = -s_{-2}.$$

Substituting the above eigenvalues into (16)–(20) gives the following scattering coefficients.

$$S_{11} = \frac{-1}{5}$$

$$S_{12} = \frac{-3.2360}{5}$$

$$S_{13} = \frac{+1.2360}{5}$$

$$S_{14} = \frac{+1.2360}{5}$$

$$S_{15} = \frac{-3.2360}{5}.$$

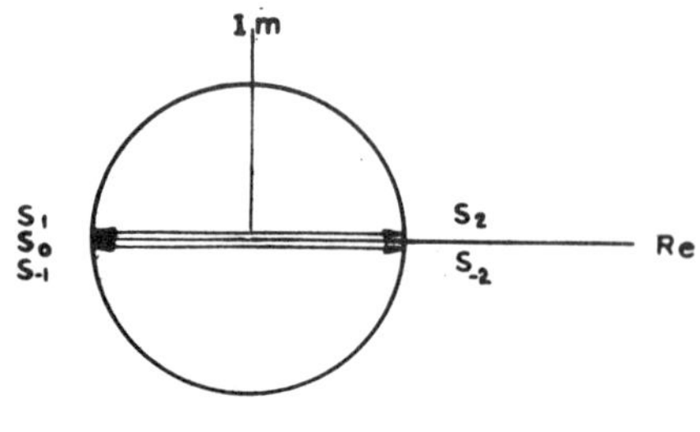

Fig. 5. Scattering-matrix eigenvalue after first circulation adjustment.

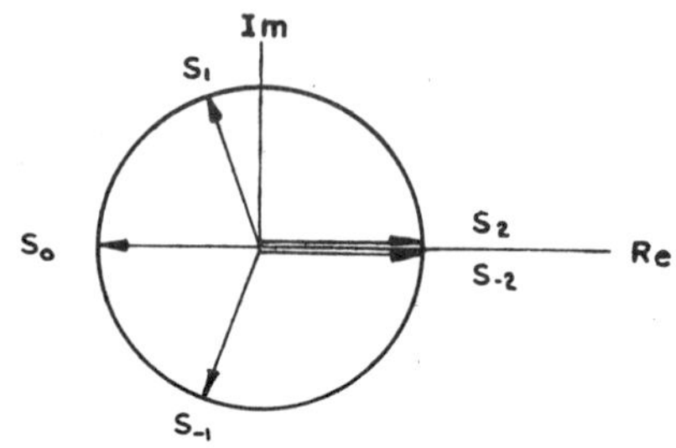

Fig. 7. Scattering-matrix eigenvalue after third circulation adjustment.

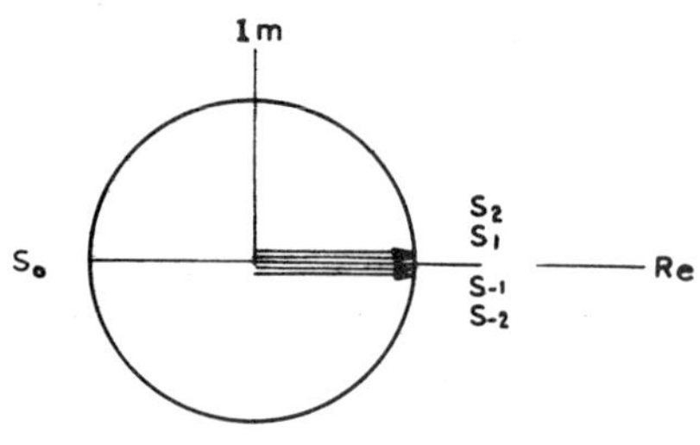

Fig. 6. Scattering-matrix eigenvalue after second circulation adjustment.

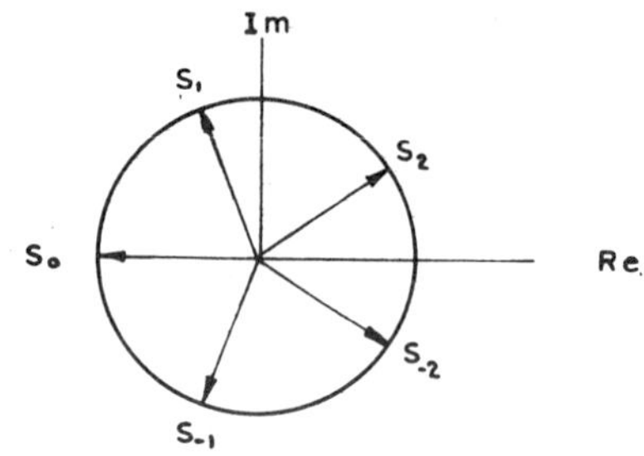

Fig. 8. Scattering-matrix eigenvalue after fourth circulation adjustment (ideal circulator).

162 Substituting the above boundary conditions into (23)–(27) gives for the electric field

$$E_z(\phi) = a_2 \cos 2\phi$$

where $a_2 = 0.80$.

The physical arrangement is the diameter of the ferrite disk appropriate to the $n = \pm 2$ field patterns. This field pattern is shown in Fig. 2(b).

The second circulation adjustment is now obtained by rotating the degenerate pair of eigenvalues s_1 and s_{-1} through 180° on the unit circle. This gives the eigenvalue diagram shown in Fig. 6:

$$s_0 = -s_2 = -s_{-2} = -s_1 = -s_{-1}.$$

This eigenvalue arrangement is obtained when the scattering coefficients take on the following values:

$$S_{11} = \tfrac{3}{5}$$
$$S_{12} = -\tfrac{2}{5}$$
$$S_{13} = -\tfrac{2}{5}$$
$$S_{14} = -\tfrac{2}{5}$$
$$S_{15} = -\tfrac{2}{5}.$$

Substituting the above boundary conditions into (23)–(27) gives for the electric field

$$E_z(\phi) = a_1 \cos \phi + a_2 \cos 2\phi$$

where $a_1 = 0.80$ and $a_2 = 0.80$.

The second perturbation consists of tuning the $n = \pm 1$ field patterns to the same frequency as the $n = \pm 2$ patterns. One possible physical perturbation is an inductive post at the center of the ferrite disk. The physical variable is obtained by noting that the $n = \pm 1$ field configuration has most of its magnetic energy at the center of the disk, whereas the $n = \pm 2$ field pattern has no energy there. In order not to tune the $n = 0$ mode, which has a maximum electric field at the center of the junction with this adjustment, it is necessary to short-circuit this mode with a thin metal post.

The third circulation adjustment is now obtained by splitting the degeneracy between the s_1 and s_{-1} eigenvalues without affecting the degeneracy between s_2 and s_{-2}:

$$s_2 = s_{-2} = -s_0$$
$$s_1 = s_0 e^{-j2\pi/5}$$
$$s_{-1} = s_0 e^{+j2\pi/5}.$$

This eigenvalue arrangement is shown in Fig. 7. It is obtained when

$$S_{11} = \frac{0.3820}{5}$$
$$S_{12} = \frac{-4.6180}{5}$$
$$S_{13} = \frac{-1}{5}$$
$$S_{14} = \frac{+1.2360}{5}$$
$$S_{15} = \frac{-1}{5}.$$

Substituting the above boundary conditions into (23)–(27) gives for the electric-field distribution

$$E_z(\phi) = a_1 \cos(\phi + \tau_1) + a_2 \cos 2\phi$$

where $a_1=0.4707$, $a_2=0.80$, and $\tau_1=54°$. The circulation adjustment rotates the $n=\pm 1$ field patterns by 54° as shown in Fig. 3(a).

The third physical perturbation is obtained by splitting the degeneracy between the $n=\pm 1$ field patterns. Here, a magnetic field at the center of the disk should split the degeneracy of the $n=\pm 1$ field patterns without affecting the degeneracy between the $n=\pm 2$ field patterns, which have little magnetic energy there.

The final eigenvalue adjustment, shown in Fig. 8 is now obtained by removing the degeneracy between the eigenvalues s_2 and s_{-2}:

$$s_1 = s_0 e^{-j2\pi/5}$$
$$s_{-1} = s_0 e^{+j2\pi/5}$$
$$s_2 = s_0 e^{-j4\pi/5}$$
$$s_{-2} = s_0 e^{+j4\pi/5}.$$

This eigenvalue arrangement is obtained when

$$S_{11} = S_{13} = S_{14} = S_{15} = 0$$
$$S_{12} = -1$$

which corresponds to an ideal circulator.

As already shown in Section IV, these boundary conditions give for the electric field the following:

$$E_z(\phi) = a_1 \cos(\phi + \tau_1) + a_2 \cos(2\phi + \tau_2)$$

where $a_1=0.4707$, $a_2=0.7607$, $\tau_1=54°$, and $\tau_2=18°$. This adjustment rotates the field patterns of the $n=\pm 2$ modes by 18° as shown in Fig. 3(b).

The fourth and last physical perturbation is a ring magnet at the edge of the ferrite disk where most of the magnetic energy of the $n=\pm 2$ field patterns resides.

VI. Conclusions

This paper has described a synthesis procedure for the construction of the *m*-port circulator which leads to $(m-1)$ distinct scattering matrices at the terminals of the junction. The approach developed here allows the $(m-1)$ independent physical perturbations of the junction to be carried out one at a time. The physical perturbations can be obtained by solving the boundary problems associated with each eigensolution. This approach therefore also leads to $(m-1)$ distinct electromagnetic problems, each involving only one independent variable instead of a single problem involving $(m-1)$ independent variables.

References

[1] C. Montgomery, R. H. Dicke, and E. M. Purcell, *Principles of Microwave Circuits*. New York: McGraw-Hill, 1948.

[2] B. A. Auld, "The synthesis of symmetrical waveguide circulators," *IRE Trans. Microwave Theory Tech.*, vol. MTT-7, pp. 238–246, April 1959.

[3] U. Milano, J. H. Saunders, and L. Davis, Jr., "A *Y*-junction strip-line circulator," *IRE Trans. Microwave Theory Tech.*, vol. MTT-8, pp. 346–351, May 1960.

[4] J. B. Davis, "An analysis of the *m*-port symmetrical *H* plane waveguide junction with central ferrite post," *IRE Trans. Microwave Theory Tech.*, vol. MTT-10, pp. 596–604, November 1962.

[5] J. Helszajn and C. R. Buffler, "Adjustment of the 4-port single junction circulator," *Radio Electron Eng.*, vol. 35, pp. 357–360, June 1968.

[6] H. J. Carlin and A. B. Giordano, *Network Theory, an Introduction to Reciprocal and Nonreciprocal Networks*. Englewood Cliffs, N. J.: Prentice-Hall, 1964.

[7] H. Bosma, "On stripline *Y*-circulation at UHF," *IEEE Trans. Microwave Theory Tech.*, vol. MTT-12, pp. 61–72, January 1964.

[8] C. E. Fay and R. L. Comstock, "Operation of the ferrite junction circulator," *IEEE Trans. Microwave Theory Tech.*, vol. MTT-13, pp. 15–17, January 1965.

[9] J. Watkins, "Circular resonant structures in microstrip," *Electron. Lett.*, vol. 5, pp. 524–525, October 1969.

[10] G. E. Fay and W. A. Dean, *1966 IEEE G–MTT Intern. Symp. Dig.*, p. 286.

Section II. C
Switching Circulators

Reciprocal and Nonreciprocal Switches Utilizing Ferrite Junction Circulators*

The symmetrical ferrite junction circulator has assumed a position of prominence in recent years due to its small physical size and excellent electrical characteristics. The device was described theoretically by Auld,[1] and other investigators have contributed to the design of improved electrical characteristics.[2–4] Due to their low loss, these circulators have found wide use with parametric and tunnel diode amplifiers. Other applications include duplexers, multiplexers and load isolators.

As with all types of circulators, if the sense of the magnetic field is reversed, the direction of circulation reverses, making the design of modulators and switches possible. It is the purpose of this communication to emphasize these properties of junction circulators and discuss possible circuits for both reciprocal and nonreciprocal switching. Some of the experimental characteristics of these devices will also be presented.

The junction circulator is represented schematically as indicated in Fig. 1(a) where the arrow indicates the direction of circulation. The implication is that power entering at Port A leaves at Port B, or power entering at Port B leaves at Port C, etc. A switchable circulator is depicted in Fig. 1(b). The schematic has both a solid and dashed arrow representing the direction of circulation in the two states of the applied magnetic field. In one position, the direction of the circulation is ABC and in the switched position, represented by the dashed arrow, the direction of circulation is ACB. It is obvious that the circuit of Fig. 1(b) represents a single pole-double throw switch; that is, an input at Terminal A can be

(a) (b)

Fig. 1—(a) Circulator. (b) Switchable circulator.

Fig. 2—Switchable circulator for the 2.0-kMc band. The unit performs the function of a nonreciprocal single pole-double throw switch.

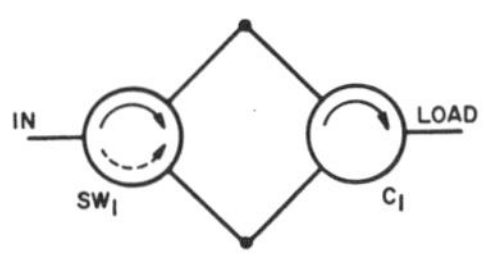

Fig. 3—Reactive, single pole-single throw reciprocal switch.

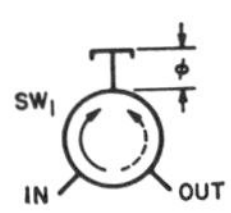

Fig. 4—Reactive balanced modulator or switch.

switched to Terminal B or Terminal C by alternating the direction of applied magnetic field. A photograph of such a switch designed for the S-band region is shown in Fig. 2.

There are several properties of this type of junction switch that should be pointed out. First, the device is nonreciprocal; second the impedance is not constant during switching; and third, fast switching times are difficult to achieve.

The first property is apparent from a study of Fig. 1(b). When the switch is connected between A and B, B is connected to C instead of A and duplexing of the switched signal cannot be accomplished. Furthermore, the isolation between the two loads at B and C is determined, not only by the circulator, but by how well the loads are matched to the transmission line.

The second item can be understood by considering the nature of the circulator at zero field. In this condition the device is a *reciprocal* 3-port (circulation occurs in both directions with equal amplitude) and cannot be matched bilaterally from port to port. The VSWR can be as high as 2:1 during switching. If this condition is intolerable to the system, it can be overcome by the addition of an isolator at the input of the switch.

The difficulty in achieving fast switching times in these devices is caused by demagnetizing fields and eddy currents. The strong demagnetizing fields in the circulator are due to the disk-shaped ferrite geometries used. These fields oppose the applied field and more switching power is required. The demagnetizing field can be reduced by making the ferrite longer with respect to its diameter.

Eddy currents are present in the coil forms as well as in the ground planes and the center strip of the circulator transmission lines. These currents are difficult to reduce in practice. "Thin wall" techniques have been tried with some success; however, it is felt that 100 μsec seems a practical limit for these devices using reasonable switching power. One μsec switching does not appear practical at this stage of development.

In the discussion to follow, a number of different switching circuits using 3-port junction circulators will be presented. These circuits should cover most switching requirements, both reciprocal and nonreciprocal.

Single Pole-Single Throw (Reciprocal)

A single pole-single throw reciprocal switch can be obtained by combining the switchable circulator with a nonswitchable 3-port circulator. The arrangement utilized for this function is shown in Fig. 3. When the switchable circulator is in the state 165
shown by the solid arrow, the input is connected to the load, and the load is similarly connected to the input. In the position indicated by the dashed arrow, power flows from the input around the loop containing C_1 and back out the input without being connected to the load. Similar power flow occurs looking into the load terminal. This circuit, therefore, is reactive in nature and the energy is either transmitted or reflected from the input in the two switched positions. This circuit is the basic building block of some of the switches that follow.

Balanced Switch (Reciprocal)

A balanced switch having single pole-single throw characteristics is shown in Fig. 4. In this device, one of the three ports is terminated in an adjustable length short circuit. In one direction of circulation, indicated by the solid arrow, the power at the input travels to the short circuit, is reflected and emerges from the output. In the other direction of circulation, indicated by the dashed arrow, the power at the input feeds directly to the output. Therefore, in *both* switched positions, the input is connected to the output. At zero field the direction of circulations are opposite and equal in magnitude. It is then possible to adjust the phase length (ϕ) of the short circuit so that at the output the two paths will be 180° out of phase and the energy will be reflected back to the input. When this balance is realized, the switch has total reflections at zero field and is transmitting in either direction of the applied field. This device, besides acting as a switch, will provide the function of balanced modulation if its solenoid is driven by a symmetrical periodic waveform. That is, the output will contain an upper and lower sideband spectrum with a suppressed carrier. For broad-band performance, ϕ should have the smallest possible value.

* Received February 19, 1963.

[1] B. A. Auld, "The synthesis of symmetrical circulators," IRE Trans. on Microwave Theory and Techniques, vol. MTT-7, pp. 238–246; April, 1959.

[2] L. Davis, U. Milano, and J. Saunders, "A strip line L-band compact circulator," Proc. IRE (*Correspondence*), vol. 48, pp. 115–116; January, 1960.

[3] J. B. Thaxter and G. S. Heller, "Circulators at 70- and 140-kMc," Proc. IRE (*Correspondence*), vol. 48, pp. 110–111; January, 1960.

[4] J. Clark, "Miniaturized, temperature stable, coaxial Y-junction circulators," IRE Trans. on Microwave Theory and Techniques (*Correspondence*), vol. MTT-9, pp. 267–269; May, 1961.

Single Pole-Double Throw (Reciprocal)

If it is desired to produce reciprocal single pole-double throw action, this can be done efficiently by using one nonreciprocal switchable circulator and two fixed circulators as shown in Fig. 5.

The action of the switch can be seen by following the path of the power flow. Power enters SW_1 and proceeds in the sense of the solid arrow through C_1 to Load 1. Power returned from Load 1 proceeds through C_2 and SW_1 back to the input. When SW_1 is switched so that the dashed arrow applies, power entering at the input proceeds through C_2 to Load 2. Power returned from Load 2 travels to C_1 and then to SW_1 and back to the input. This switch then produces reciprocal power flow. The forward- and reverse-phase in this switch may be different although it could be adjusted by choosing proper line lengths. The insertion loss in the two directions are not equal since, in the forward direction one suffers the loss of two circulators, while in the reverse direction the loss is due to passing through three circulators.

A reciprocal single-pole N-throw switch can be obtained by combining single pole-double throw switches, either in a parallel or series arrangement. A series arrangement
166 is shown in Fig 6.

Double Pole-Double Throw (Nonreciprocal)

A nonreciprocal double pole-double throw switch can be designed using two switchable junction circulators. This circuit is shown in Fig. 7. When the two switches are electrically ganged together, the switching action can be seen as follows. First, when the solid arrow applies, In_1 is connected to Out_1 and In_2 is connected to Out_2 by the energy passing directly through SW_1 and SW_2, respectively. When the circulators are switched and the dashed arrows apply, In_1 is connected to Out_2 by energy passing through both SW_1 and SW_2. Similarly. In_2 is connected to Out_1. The switch is nonreciprocal since the following sequences apply.

Solid Arrow	Dashed Arrow
$In_1 \rightarrow Out_1$	$In_1 \rightarrow Out_2$
$In_2 \rightarrow Out_2$	$In_2 \rightarrow Out_1$
$Out_1 \rightarrow In_2$	$Out_1 \rightarrow In_1$
$Out_2 \rightarrow In_1$	$Out_2 \rightarrow In_2$

Double Pole-Double Throw (Reciprocal)

The switch of Fig. 8 can be made reciprocal by the addition of two circulators as shown schematically in Fig. 8.

The reader can verify that the following switching sequences take place.

Solid Arrow	Dashed Arrow
$In_1 \rightarrow Out_1$	$In_1 \rightarrow Out_2$
$In_2 \rightarrow Out_2$	$In_2 \rightarrow Out_1$
$Out_1 \rightarrow In_1$	$Out_1 \rightarrow In_2$
$Out_2 \rightarrow In_2$	$Out_2 \rightarrow In_1$

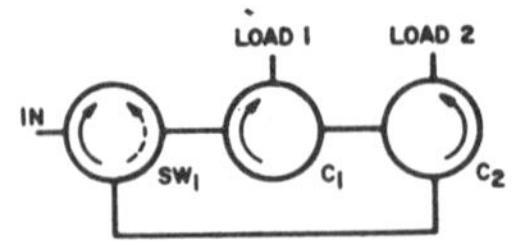

Fig. 5—Reciprocal single pole-double throw switch.

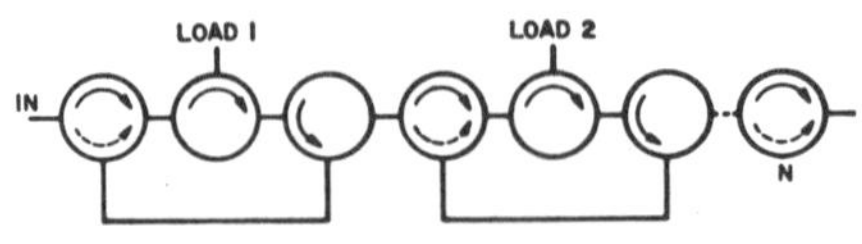

Fig. 6—Series arrangement of single pole-double throw reciprocal switches to make single pole-N throw switch.

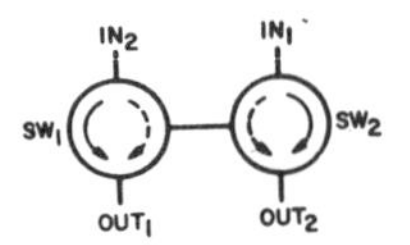

Fig. 7—Double pole-double throw nonreciprocal switch.

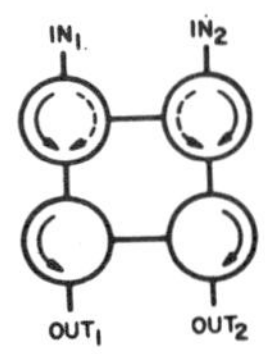

Fig. 8—Double pole-double throw reciprocal switch.

Experimental Characteristics

The three-port circulator shown in Fig. 2, having broad-band tuning, was investigated experimentally to observe its switching characteristics. The instantaneous bandwidth with a fixed field is shown in Fig. 9. The insertion loss vs coil current for several frequencies is shown in Fig. 10. An important characteristic should be noted from Fig. 10. It is difficult to overdrive this type of switch. The isolation and insertion loss show little change when one applies additional current to the unit. This is obviously not true in Faraday rotation switches, or switches utilizing phase shift principles. The fact that the switches are not current sensitive is an advantage to the systems engineer who need not worry about the design of a precision driver. Hysteresis measurements have shown this switch to have a maximum separation of 2 db between increasing and decreasing currents after driving the switch into saturation. Some of the hysteresis is due to the iron coil-forms used.

The coil current required is somewhat arbitrary and can be controlled by the number of turns and wire size. For fast switching requirements, inductance must also be controlled and some freedom of choice is lost. The holding power can be made extremely small by careful design. For example, a switch designed for space applications drew a current of 10 ma at 1 3/4 volts for proper operations.

Several arrangements for combining nonreciprocal, three-port switches with three-port circulators to produce reciprocal switches have been described. Single pole-single throw, single pole-double throw, double pole-double throw as well as

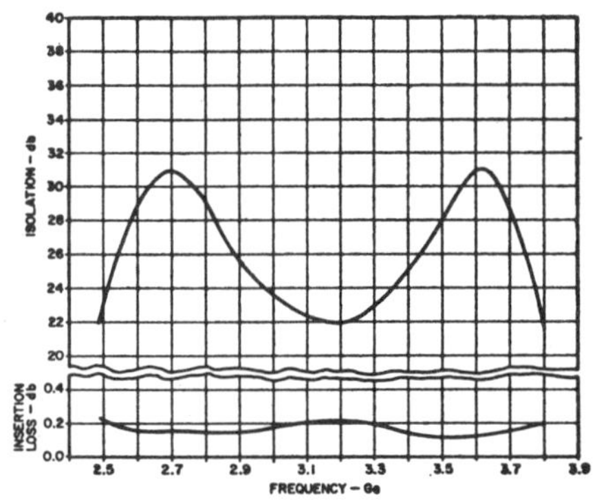

Fig. 9—Instantaneous isolation and insertion loss bandwidth of 3-port switchable junction circulator with fixed field.

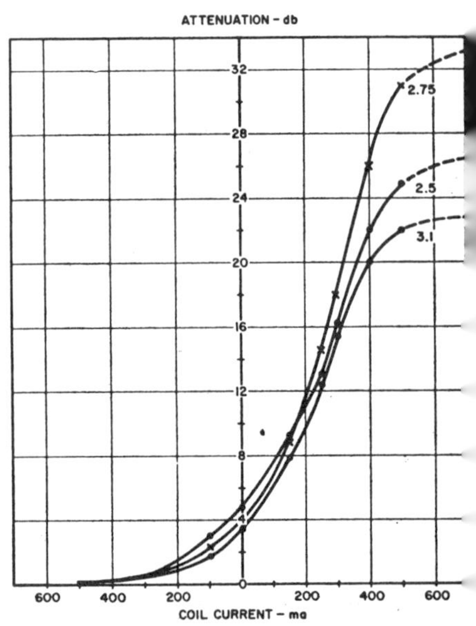

Fig. 10—Attenuation vs coil current for various fre quencies for a switchable junction circulato Only one-half of the frequency band is shown.

balanced switches can be designed. Thes switches do not require exact values c switching current to obtain their best per formance, and should find system use due t their small physical size. Some problem may be encountered with these devices du to their nonconstant input impedance dur ing switching, as well as the difficulty i obtaining very rapid switching times.

Alvin Clavi
RANTEC Corp
Calabasas, Cali

Reprinted from IEEE TRANSACTIONS
ON *MICROWAVE THEORY AND TECHNIQUES*
Volume MTT-14, Number 12, December, 1966
pp. 665-669

A Switching Circulator: S-Band; Stripline; Remanent; 15 kilowatts; 10 microseconds; Temperature-Stable

F. BETTS, MEMBER, IEEE, D. H. TEMME, MEMBER, IEEE, AND J. A. WEISS, SENIOR MEMBER, IEEE

***Abstract*—A stripline, three-port remanence circulator switch has been designed for high-speed switching of time delay in a phased array radar at S-band (2.9 GHz). Special attention was devoted to minimizing switching time and energy through design of the magnetic circuit and suppression of eddy currents. Temperature stabilization of insertion phase was accomplished by means of a flux regulating magnetic circuit. Switching performance: time: less than 10 microseconds; energy: 450 microjoules. Circulator performance: bandwidth for >26 dB isolation, 8.9 percent; insertion loss, 0.35 dB. Temperature stability of insertion phase: one electrical degree per 10°C. Peak RF power: 15 kW. The discussion includes details of the junction design and performance, techniques of eddy current suppression, temperature stabilization, and the method of switching energy measurements.**

Manuscript received July 1966.
F. Betts and D. H. Temme are with Lincoln Laboratory, Massachusetts Institute of Technology, Lexington, Mass. (Operated with support from the U. S. Advanced Research Projects Agency.)
J. A. Weiss is a Consultant at Worcester Polytechnic Institute, Worcester, Mass.

INTRODUCTION

THIS PAPER reports the development of a ferrite switching circulator suitable for time delay switching applications in phased array radars. The re-

Fig. 1. Interior of switching circulator.

TABLE I

Frequency	2.90 GHz
Input impedance level	50 ohms
Stripline ground plane spacing	0.250 inch
Ferrite: saturation magnetization	400 gauss
remanent magnetization	240 gauss
dielectric constant	14
ΔH (X-band)	60 oersteds
Ferrite disk diameter	0.705 inch
Matching transformer impedance	33 ohms

168

quirements are high speed and minimum switching energy, combined with good isolation and match, and low insertion loss over a moderate band. Stability of transmission phase is also an important requirement. Performance must be maintained in the presence of changes in operating conditions such as ambient temperature, RF power, and switching rate. Data are presented here on design considerations, performance, and methods of measurement.

To the best of the authors' knowledge the first latching circulator in TEM transmission line to achieve good microwave performance was built by Western Microwave Laboratories, Inc. [1]. This device operated the ferrite at a high dc magnetic bias with consequent high switching energy. In the present device the ferrite is operated at essentially zero internal field with consequent reduction of switching energy. Good temperature stability is maintained with a flux regulating magnetic circuit.

I. Circulator Design

The circulator design is based on the work of Bosma [2] as further developed by Fay and Comstock [3] with the modification that the ferrite is operated in the remanent state. The circulator with the top ground plane removed is shown in Fig. 1 to illustrate the microwave construction. The design is conventional with a circular center conductor, ferrite disks between it, and the ground plane and dielectric rings which form the quarter-wave matching transformers. A summary of the design is given in Table I.

TABLE II

Frequency	2.90 GHz
Maximum isolation	>40 dB
Bandwidth at 26-dB isolation	240 MHz (8.9 percent)
VSWR at 26-dB isolation	1.11
Minimum insertion loss	0.35 dB
Switching time	<10 μs (see text)
Switching energy delivered to the circulator	450 μJ
Temperature variation of transmission phase	1° phase per 10°C (to 60°C)
Average power capability (Dielectric melted at 400 watts)	300 watts (water cooling)
Peak RF power capability (RF arcing at 40 kW)	15 kW (0.1-dB rise in insertion loss

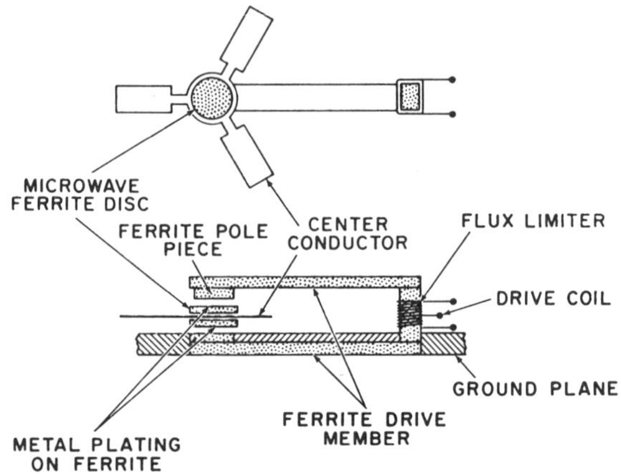

Fig. 2. Details of magnetic circuit.

A preliminary test model with copper center conductor was built to verify the design. It was operated wit an electromagnet, giving an insertion loss of 0.15 d and a bandwidth of 200 MHz, at a flux density of 21 gauss. In the final design, incorporating an integral magnetic yoke and operating at remanence, the higher resistivity conductors used to achieve good switching performance raised the insertion loss to 0.35 dB and the bandwidth to 240 MHz. The microwave and switching performance is summarized in Table II.

In defining switching time, it is important to consider the special requirements of the phased array application. Since insertion phase is the most crucial parameter, the appropriate definition is the time required for the phase of the transmitted signal to reach within a specified tolerance of its steady-state value. For an arbitraril chosen $\pm 2°$ tolerance, the switching time is found to be less than 10 microseconds. The device can be switche somewhat faster but with considerable increase in switching energy.

II. Magnetic Circuit

Details of the magnetic circuit are shown in Fig. 2. The magnetic circuit is brought through the ground planes by cylindrical pole pieces and the magnetic circuit completed outside the ground planes. The center conductor is photoetched out of 0.5 mil stainless ste with those portions outside the magnetic circuit copper plated to reduce microwave loss. Surfaces of the ferrite disks near the ground planes are plated with 0.2 mils of tin to maintain RF continuity across the ferrite pole pieces. Thickness (somewhat more than a skin depth) and resistivity of the platings are such that eddy currents are kept small with only moderate microwave loss. The ground plane is slit from the pole piece to the flux limiter so that it will not act as a shorted turn linking the switched flux. The ferrite platings are insulated from the ground planes with 0.5 mil FEP teflon film to preserve the open circuit for eddy currents created by the slot.

III. Temperature Stabilization

Temperature stability of the circulator is obtained through the flux regulating properties of the magnetic circuit. The principles and design of such circuits have been described by Stern and Ince [4] and will only be reviewed briefly. Referring to Fig. 2, the magnetomotive force necessary to maintain flux in the microwave ferrite is supplied by the drive members. In operation of the circuit these members are driven on a minor hysteresis loop. Ferrite material chosen for these pieces need not have good microwave properties but should have square hysteresis loop and coercive force that varies little over the operating temperature range. The drive coil is wound on the flux limiter which is of such cross-sectional area that it is driven on a major loop and sets the peak flux level in the circuit. Therefore the material chosen for the limiter should have a saturation magnetization that varies little over the operating temperature range. With the majority of the MMF supplied by a temperature-stable square loop material and the peak flux set by the temperature-stable limiter, dependence of the operating flux on properties of the microwave material is greatly reduced.

The degree of temperature stabilization afforded by the flux regulating magnetic circuit is indicated in Figs. 3 and 4. Figure 3 shows remanence flux, normalized to the room temperature value, both for a toroid of the microwave ferrite and for the flux regulating circuit. As the remanence magnetization in the microwave material tends to drop sharply, the driver ferrite supplies additional MMF to maintain the net flux at a high level. At 100°C the flux has fallen to 85 percent of the room temperature value, a variation which is essentially due to the variation of $4\pi M_s$ of the limiter ferrite. Near the Curie temperature of the microwave material, the driver loses control and the flux drops rapidly.

The extent to which the microwave performance has been stabilized is shown in Fig. 4. Bandwidth for 26 dB isolation is plotted vs. temperature, showing essentially constant bandwidth up to around 55°C. At 65°C the bandwidth has fallen to 80 percent of the room temperature value and falls off very rapidly above about 70°C.

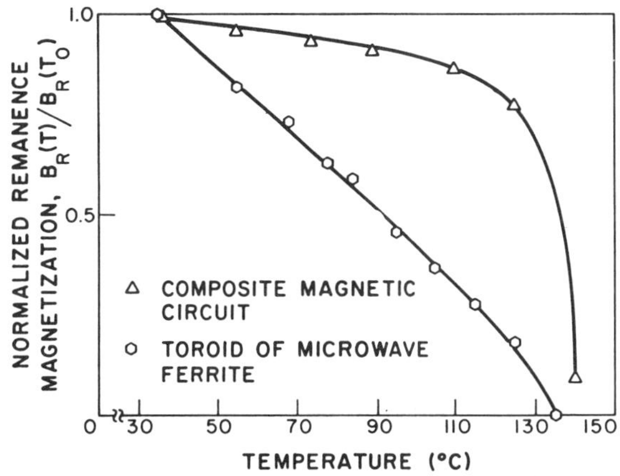

Fig. 3. Remanence flux vs. temperature.

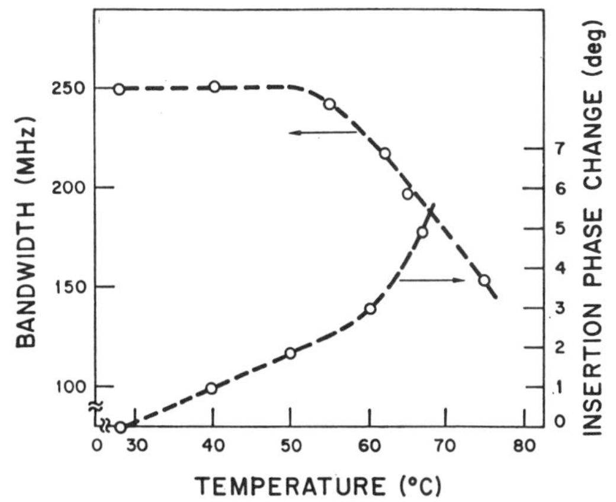

Fig. 4. Microwave performance vs. temperature.

The variation of insertion phase is also shown with changes of about 1° phase per 10°C temperature change up to 60°C where it begins to vary much more rapidly.

IV. Eddy Current Evaluation

The magnetic circuit used to evaluate eddy current effects is shown in Fig. 5. There are two pick-up coils on the pole pieces, one at the base of the bottom pole piece, which is linked by flux that closes around the whole circuit, and a second at the face of the upper pole piece near the metal surface being tested. A large drive current is applied to the coil on the left member, switching the circuit rapidly. Since the left member is virtually saturated by the drive, the flux due to eddy currents closes chiefly near the metal. Since this flux cannot be controlled by the drive, circulator performance will not stabilize until eddy currents have died out. The voltages from the two coils are subtracted and integrated to give a voltage proportional to eddy current flux. The eddy current time constant for the tin platings and for the stainless steel center conductor was found with this device to be of the order of 3 microseconds or less.

169

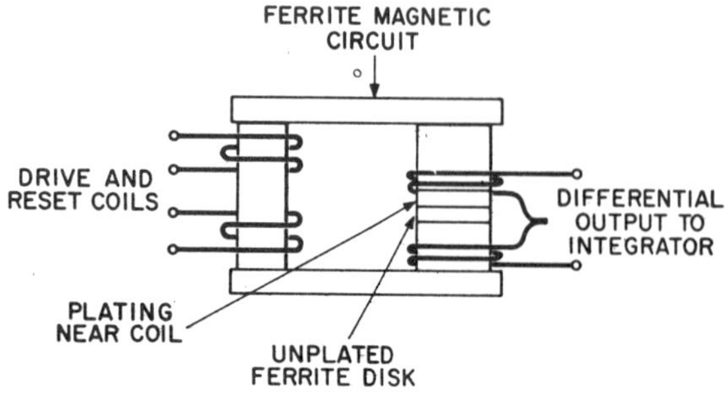

Fig. 5. Eddy current test circuit.

V. Switching Energy Measurement

A method of measuring switching energy was used, which gives accurate results and is faster and more convenient than the usual method; viz., integrating the product of the time-dependent voltage and current.

The switching energy measurements are made from flux vs. MMF loops of the magnetic circuit as follows. Referring to the equivalent circuit of Fig. 6, switching energy is defined as

$$W_s = \int_0^{T_s} eidt = \int_0^{T_s}\left(N\frac{d\phi}{dt} + L\frac{di}{dt} + Ri\right)idt.$$

170 T_s is switching time, L is leakage inductance, and R is the resistance of the drive circuit. Since the initial and final values of current are zero, and R is generally negligible, there remains only the term

$$W_s = \int_0^{T_s} N\left(\frac{d\phi}{dt}\right)idt = \int_{\phi_1}^{\phi_2} Nid\phi.$$

Referring to Fig. 6 and thinking of an integral as the area under a curve, it can be seen that the value of the integral is equal to the area bounded by the ϕ vs. Ni curve and the ϕ axis. That is, W_s equals half the area of the flux-MMF loop.

The loss contribution of magnetic hysteresis can be distinguished from that due to eddy currents by comparing ϕ-MMF loops obtained under pulsed switching with those observed at low frequencies. Several pulsed loops are illustrated in Fig. 7; the three frames show flux vs. applied MMF when eddy currents are absent, moderate, and severe, respectively. Switching time is approximately 10 microseconds. For the present circulator, the switching energy of 430 microjoules is composed of a magnetic hysteresis loss of 130 microjoules and an eddy current loss of 300 microjoules.

The equipment used to generate pulsed hysteresis loops is shown in Fig. 8. The pulse current is sensed with a Tektronix current probe and the flux signal is generated by means of a Tektronix-type "O" operational amplifier used as an integrator. The loop display is photographed and the area measured with a planimeter.

VI. Peak Power Effects

At sufficiently high RF peak power, the insertion loss of the circulator increases above the low power value due to nonlinear effects. Although in general the loss begins to rise at a more or less well defined power level, with certain ferrite materials the device can be operated considerably above this level with only modest increase in loss. Insertion loss vs. peak power data for the circulator with several different materials is shown in Fig. 9. The performance specified in Table II was obtained with a commercially available material having $4\pi M_s = 400$ and $\Delta H = 60$ Oe. It gives an increase in loss of 0.1 dB at 15 kW, and at 24 kW the insertion loss rise is about 0.2 dB. An experimental material with $4\pi M_s = 425$ and $\Delta H = 104$ gave a smaller rise in insertion loss at high power, but the loss at low power was higher by about 0.05 dB.

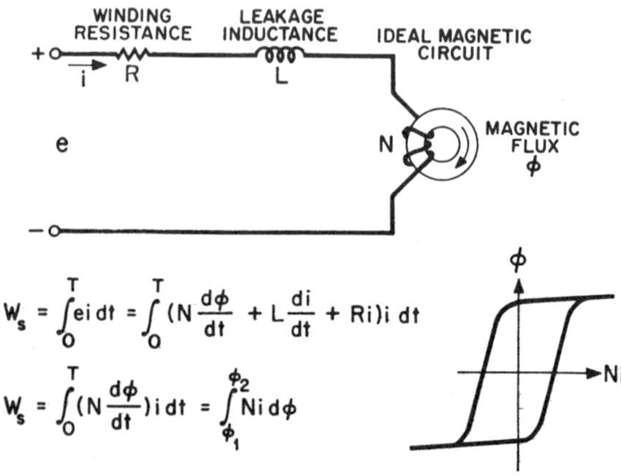

Fig. 6. Equivalent circuit, switching energy measurement.

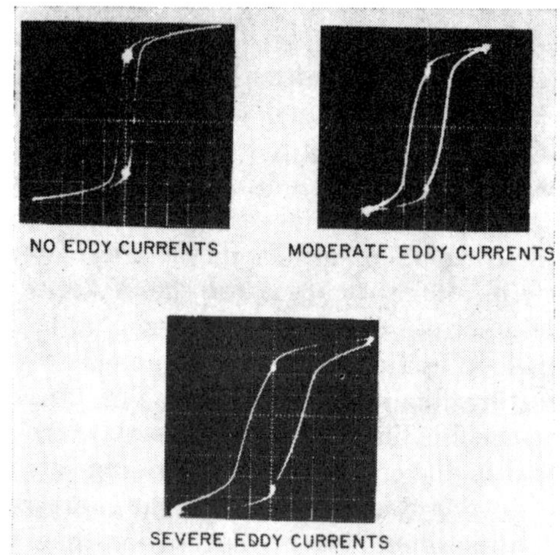

Fig. 7. Pulsed hysteresis loops showing eddy current effects.

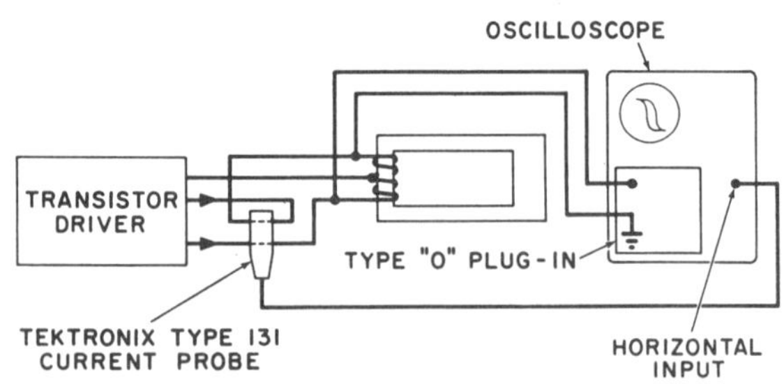

Fig. 8. Equipment to generate pulsed hysteresis loops.

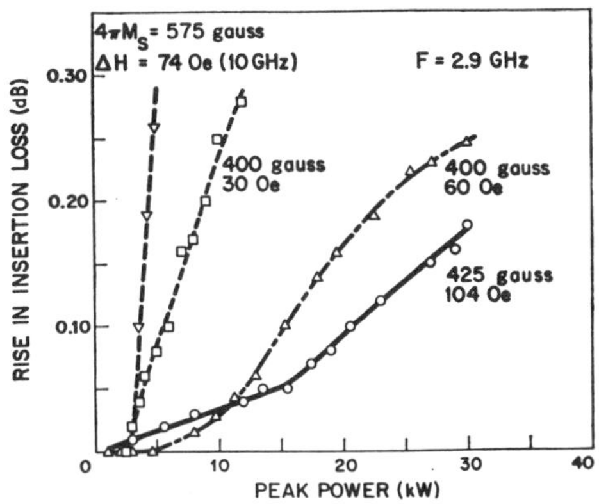

Fig. 9. Rise in insertion loss vs. peak power.

Acknowledgment

The interest of and helpful discussions with E. Stern are gratefully acknowledged.

References

[1] "Development of a fast switching, high power digital phase-shifter at X-band," Air Research and Development Command, Rome Air Development Center, Griffiss AFB, N.Y., Final Rept. on Contract AF 30(602)-3496, June 1965, by Western Microwave Labs., Santa Clara, Calif.

[2] H. Bosma, "On stripline Y-circulation at UHF," *IEEE Trans. on Microwave Theory and Techniques*, vol. MTT-12, pp. 61–72, January 1965.

[3] C. E. Fay and R. L. Comstock, "Operation of the ferrite junction circulator," *IEEE Trans. on Microwave Theory and Techniques*, vol. MTT-13, pp. 15–27, January 1965.

[4] E. Stern and W. J. Ince, "Temperature stabilization of unsaturated microwave ferrite devices," presented at the 1965 11th Conf. on Magnetism and Magnetic Materials, San Francisco, Calif.

171

A 35-GHz Latching Switch

W. C. PASSARO, MEMBER, IEEE, AND J. W. McMANUS, MEMBER, IEEE

172

Abstract—**The development of a fast switching, small, lightweight latching three-port ferrite circulator is outlined. Geometrical configurations, as relating to the toroidal ferrite element and their apparent effects on operating characteristics, are presented.**

Considerations relating to the proper selection of ferromagnetic materials, compatible with latching applications, are discussed. The finalized device operates at 35 GHz with an instantaneous bandwidth of 5 percent. Total weight is less than 0.6 oz, while total volume is less than 0.750 cubic inch.

Performance characteristics are presented, showing a maximum insertion loss of 0.50 dB and a minimum isolation of 15.0 dB, while switching times of less than 0.3 μs have been achieved (under dynamic operating conditions). The unit exhibits highly stable characteristics over the temperature range of −60°C to +100°C.

Introduction

THIS PAPER describes the development of a lightweight, fast switching, 35-GHz latching ferrite circulator, with an instantaneous bandwidth of 5 percent and switching time of less than 0.30 μs. The device is designed for use over the temperature range of −60°C to +100°C under dynamic operating conditions.

This latching circulator is particularly well suited for use as a switching element in applications where high reliability, small size and weight, and fast switching time are required. Total weight of the device is less than 0.60 ounce, and total volume is less than 0.75 cubic inch (see Fig. 1).

Manuscript received May 31, 1966; revised September 6, 1966.
The authors are with the Sperry Microwave Electronics Company, a division of the Sperry Rand Corporation, Clearwater, Fla.

Fig. 1. 35-GHz latching switch.

Design Considerations

Three-port junction circulators using externally applied magnetic biasing fields have been reported by a number of investigators [1]–[5]. Application of these devices as switching elements requires the use of cumbersome electromagnets, application of continuous holding power, and closely regulated power sources. In addition, these devices have been too slow for many proposed RF switching applications.

The use of latching ferrite elements in the construction of an S-band differential phase-shift circulator has been described by Levy and Silber [6], while more recently, the novel application of latching ferrites to Y-junction circulators has been reported by Goodman [7].

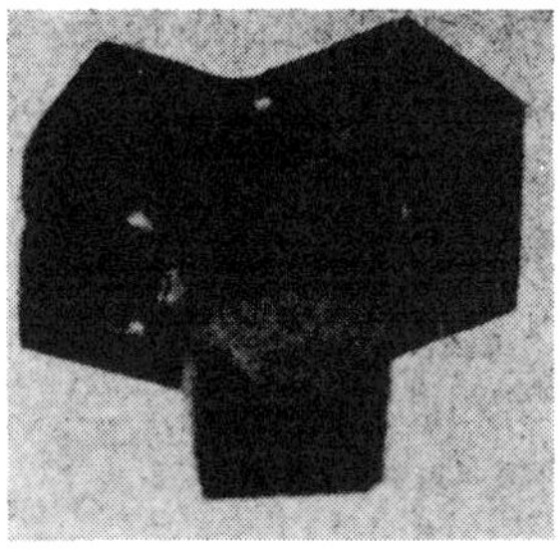

Fig. 2. Three-port latching element (similar to the design used by Goodman).

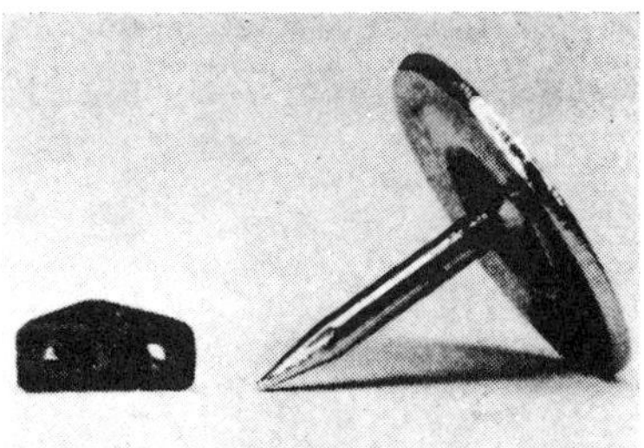

Fig. 3. Triangular ferrite latching element used in 35-GHz switch (size can be compared to normal size thumb tack).

The waveguide junction circulator, which Goodman has described, uses a ferrite toroidal structure similar to the one shown in Fig. 2, which is the microwave RF ferrite element and complete magnetic circuit combined. This general type of construction overcomes the disadvantages inherent in devices using externally applied biasing fields.

The latching V-band circulator described in this paper also utilizes an internal ferrite toroidal structure, but of a different design than that of Goodman's. The ferrite is constructed basically in the shape of a triangular prism as shown in Fig. 3.

The main advantage in using the triangular shaped latching element is that it appears to exhibit better broadband characteristics. The element is positioned so that its pointed edges are in the center of the waveguide and, as suggested by Fay and Comstock [5], are regions of nearly linear polarization and as such act more like dielectric tapers rather than ferrimagnetic material.

Circulator Design

The basic design considerations of the junction circulator is based on the work of Bosma [4], as further extended by Fay and Comstock [5]. Minor modifications in the design procedure are incorporated to allow for operation in the remanent state, since the remanent magnetization of the material appears to provide the necessary magnetic biasing of the ferrite.

The selection of materials was initially guided by conventional design criteria, as relating to three-port junction circulators [9]. Though these materials exhibited satisfactory results when using external magnetic biasing field, they behaved rather poorly when used as latching elements. Therefore, it was necessary to improve the hysteresis properties of the material. This was accomplished by improved processing techniques, during the preparation of the material. The final material chosen for this device was a nickel-zinc-manganese-ferrite with a remanence ratio (R_r) of 0.52 and a coercive force (H_c) of 0.90 oersted.

Fig. 4. Bottom half of waveguide junction showing ferrite element with charging wire.

The latching ferrite element, which occupies a volume of 0.024 cm^3, is shown in Fig. 4, positioned in place by rigid Styrofoam triangular spacers, directly above and below it. The physical shape and size of these spacers, as well as the RF characteristics of the material used, were arrived at empirically. In addition to acting as positioning devices, these pieces also perturb the junction characteristics so as to aid in the impedance matching of the device.

The element shown in Fig. 4 is designed to provide a closed magnetic path entirely within the material. The charging wire, which carries the current pulse required for switching, can be seen as it enters the waveguide junction, encircles the center leg of the latching element and emerges at the same point from which it entered. Extreme care was necessary in maintaining the wire in a plane parallel to the RF H-plane so as to minimize the possibility of RF magnetic coupling.

To initially activate the switch, it is necessary to apply a specified current pulse at the ends of the charging wire. Depending upon the polarity of the pulse, the switch is placed in a "latched" condition and the incoming RF energy is directed to either of the two output ports. Reversal of the sense of magnetization by changing the polarity of the switching pulse, results in reversing the direction of circulation. Since the magnetization in toroidal configurations can be reversed rapidly, the resultant device lends itself very well to fast-switching applications.

The fabrication of the ferrite latching element was performed using conventional grinding and ultrasonic cutting techniques. However, the rather strict tolerances inherently characteristic of millimeter wave devices necessitated extreme care. It was observed that a 0.005-inch variation in any direction (i.e., length, width, or thickness) of the ferrite element resulted in shifting the optimum operating frequency approximately 1 GHz.

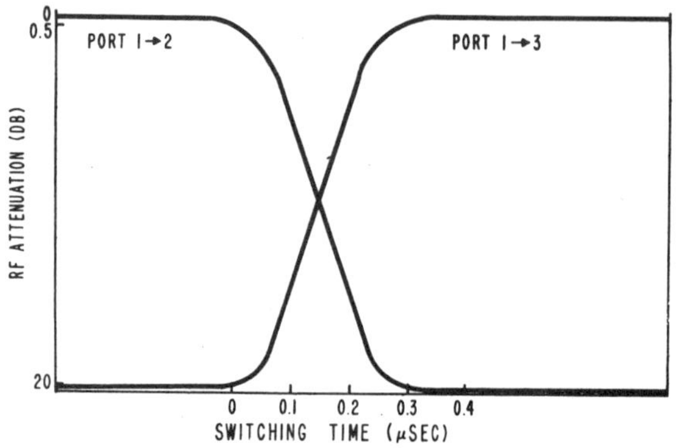

Fig. 5. Switching time.

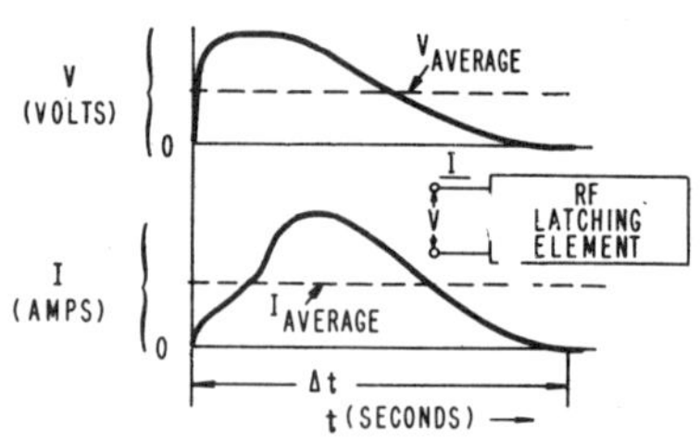

Fig. 6. Typical voltage and current waveforms observed during fast switching of the ferrite element.

Operating Characteristics

Operating characteristics of the device at room temperature are shown in Fig. 7, where it can be seen that the insertion loss has been kept to 0.50 dB maximum, while isolation in excess of 15.0 dB has been maintained. Fig. 8 displays the performance of the device at −60°C, while Fig. 9 shows the operating characteristics at +100°C.

To assure detection of any "spikes" which may be present, all microwave performance data were recorded on swept frequency test systems using permanent record recorders.

Techniques similar to those described in this paper have been used to develop a single-junction four-port latching circulator operating in X-band and three-port devices in both waveguide and stripline at S-, C-, and X-bands.

Final matching of the device was accomplished through the use of external reactive elements, combined with the junction perturbation provided by the Styrofoam spacers, as mentioned earlier.[1]

Switching Characteristics

Switching data are presented in Fig. 5, where it can be seen that switching times of less than 0.30 μs have been achieved, at a specified switching rate of 1 KHz. The switching time is defined as the time required for the isolation of the off port to reach a value of 15.0 dB.

[1] The reader is referred to the following paper which may be used as an excellent guide on matching Y-junction circualtors. J. W. Simon [10].

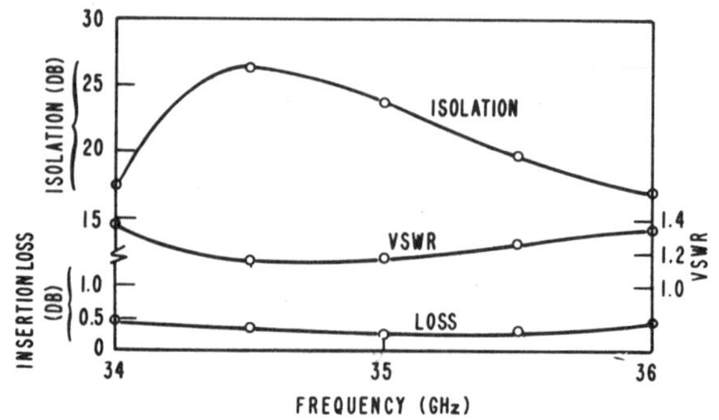

Fig. 7. Operating characteristics at room temperature (26°C).

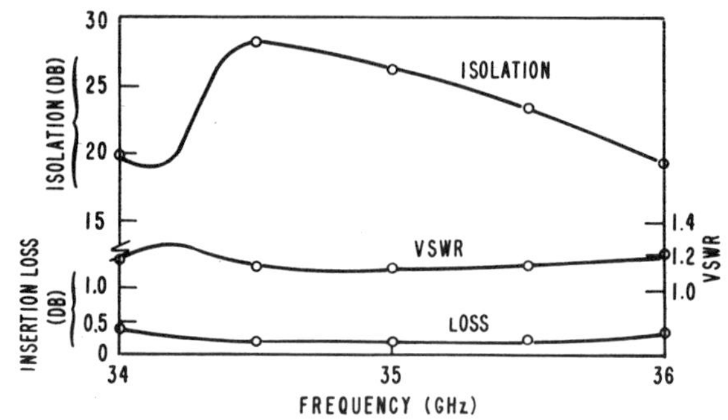

Fig. 8. Operating characteristics at −60°C.

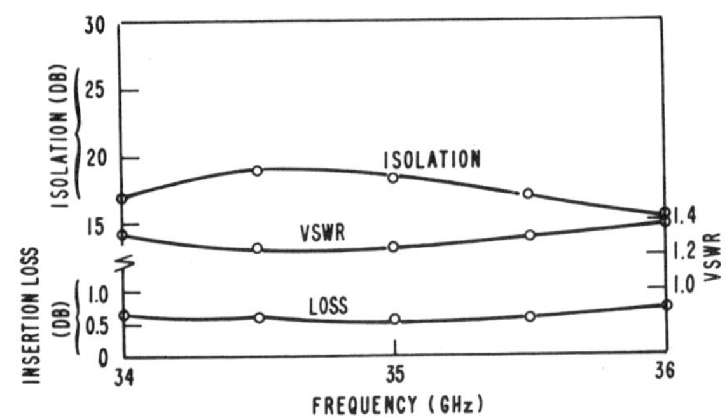

Fig. 9. Operating characteristics at +100°C.

Switching Energy

For this device, the switching energy (u_s) is defined as the amount of energy required to switch the unit from one state of operation to the other. The method used to determine the switching energy is outlined as follows.

The solid traces shown in Fig. 6 depict the voltage pulse across and the current pulse through the charging wire of the RF latching element. These traces are typical of oscillograms obtained using fast-responding voltage and current probes. The switching energy then may be calculated from the following equation [8].

$$\mu_s = V_{av.} I_{av.} (\Delta t) \text{(joules)}$$

where

$V_{av.}$ = average voltage (volts)
$I_{av.}$ = average current (amps)
(Δt) = pulse duration (seconds).

For this device the following data were recorded:

$$V_{av.} = 4.2 \text{ volts}$$
$$I_{av.} = 2.1 \text{ amps}$$
$$\Delta t = 2.0 \ \mu s.$$

Therefore,

$$\mu_s = 17.6 \ \mu J.$$

In accordance with the above criteria, the switching energy (u_s) associated with this device appears to be in the vicinity of 17.6 μJ.

References

[1] U. Milano, J. Saunders, and L. Davis, "A Y-junction strip-line circulator," *IRE Trans. on Microwave Theory and Techniques*, vol. MTT-8, pp. 346–351, May 1960.

[2] H. N. Chait and T. R. Curry, "Y-circulator," *J. Appl. Phys.*, vol. 30, p. 152 S, 1959.

[3] B. A. Auld, "The synthesis of symmetrical waveguide circulators," *IRE Trans. on Microwave Theory and Techniques*, vol. MTT-7, pp. 238–246, April 1959.

[4] H. Bosma, "On stripline Y-circulation at UHF," *IEEE Trans. on Microwave Theory and Techniques*, vol. MTT-12, pp. 61–72, January 1964.

[5] C. E. Fay and R. L. Comstock, "Operation of the ferrite junction circulator," *IEEE Trans. on Microwave Theory and Techniques*, vol. MTT-13, pp. 15–27, January 1965.

[6] L. Levy and M. Silber, "A fast switching S-band circulator utilizing ferrite toroids," *1960 IRE WESCON Conv. Rec.*, pt. 1, pp. 11–20.

[7] P. C. Goodman, "A latching ferrite junction circulator for phased array switching applications," presented at the 1965 IEEE G-MTT Symp., Clearwater, Fla.

[8] "Advanced ferrimagnetic materials applied to digital phase shifters," Air Research and Development Command, Rome Air Development Center, Griffiss AFB, N.Y., Contract AF 30(602)-3490.

[9] "Broadband junction circulators," U. S. Army Electronic Conv., prepared by Sperry Microwave Electronics Company, Clearwater, Fla, Final Rept. DA-36-039-SC-89214, May 1965.

[10] J. W. Simon, "Broadband strip-transmission line Y-junction circulators," *IEEE Trans. on Microwave Theory and Techniques*, vol. MTT-13, pp. 335–345, May 1965.

A New Type of Latching Switchable Ferrite Junction Circulator

W. W. SIEKANOWICZ MEMBER, IEEE, AND WILLIAM A. SCHILLING, MEMBER, IEEE

Abstract—**This paper presents an approximate theory and initial performance data for a new type of latching, switchable, ferrite junction circulator that is well suited for applications requiring fast (fractions of or several microseconds), low-energy (tens of microjoules in the microwave region) switching. The novelty of the device is the use of oppositely magnetized ferrite cylinder-and-ring assemblies within the region of circulation. The approximate theory yields the radius for circulation, bandwidth, and input impedance as functions of material properties and frequency. Operation of latched circulators has been demonstrated on assemblies operating in the region of 7.3 and 5.4 GHz. Insertion losses from 0.25 to 0.4 dB, 20-dB isolation bandwidths from 2.4 to 8.3 percent, and switching energies from 15 to 30 μJ have been obtained.**

Introduction

UNTIL RECENTLY, ferrite junction circulators[1],[2] required an externally applied dc magnetic field provided by permanent magnets or electromagnets. Designs employing electromagnets could also be used as switches, but switching required relatively large amounts of power and could be performed only at slow rates[3] (every 10 to 50 μs, approximately).

Recently, significant progress has been made in the development of switchable, latching, single-junction ferrite circulators in waveguide[3]–[5] and coaxial circuits.[6] In the waveguide configuration,[3],[4] latched operation has been obtained by use of a closed dc magnetic path within a triangular ferrite element; as a result, dc magnetic bias is provided by the remanent magnetization. In this configuration, circulation occurs primarily within a unidirectionally magnetized inner portion of the ferrite element, and closure of the dc magnetic circuit is completed through three arms projecting from the central region of that element. The direction of circulation can be switched by reversing the polarity of the current pulse through a wire loop that surrounds the inner portion of the ferrite element and passes through the openings in the arms of the ferrite member. In the coaxial configuration,[6] operation at remanence is achieved by closure of the dc magnetic circuit through a ferrite yoke outside the region of circulation.

This paper describes the theory and performance of a new type of three-port, stripline, latching, ferrite junction circulator, illustrated in Fig. 1. This circulator employs two latched ferrite elements placed between a central conductor and two outer conductors. Each element consists of an inner ferrite cylinder surrounded by a concentric ferrite ring.[5] The cylinder and the ring are separated by a small nonmagnetic gap, in which a circular wire loop is supported by two ceramic rings having a dielectric constant approximately equal to that of the ferrite. The upper and lower portions of the cylinder–ring assembly are in contact with ferrite or high-remanence metal disks. The circulator is put into operation by passing a pulse of direct current through the wire loop and thereby producing a magnetizing field, indicated approximately by the arrows in Fig. 1. When the current is turned off, the ferrite operates at remanence $4\pi M_r$. Operation at remanence is achieved because the magnetic path is closed entirely within the magnetic material and, as a result of this closure, the demagnetizing effects are absent. Switching of the latched circulator is accomplished by reversing the polarity of the current pulse.

The material requirements for the latched circulator (low coercive force, high remanence ratio, low electric and magnetic losses, etc.) are similar to those necessary for other types of latched microwave devices, such as phase shifters.[7]–[10]

The advantages of the design shown in Fig. 1 over state-of-the-art circulators are elimination of permanent magnets and provision of switching. The all-ferrite assembly is suitable for applications requiring fast (fractions of or several microseconds) and low-energy (tens of microjoules) switching. Reversal of the dc magnetic flux in the design employing high-remanence metal discs for flux transport can be accomplished only at the expense of losses arising from eddy currents in the metal disks; therefore, this design is best suited for applications requiring only occasional switching (tens of microseconds or more).

The significant differences between the latched circulator shown in Fig. 1 and previous configurations[1]–[4],[6] are the placement of the entire dc magnetic circuit within the region of circulation, and achievement of circulation by use of cylinder–ring assemblies having opposite directions of dc magnetization. Theoretical analysis and experiments show that the external diameters of such assemblies are comparable to those of the unidirectionally magnetized ferrite disks employed in state-of-the-art circulators. Preservation of small diameter and location of the entire magnetic circuit within the region of circulation result in a very compact design for this new type of circulator. Experiments also show that the RF performance (i.e., isolation and RF losses) of such a circulator is as good as that of state-of-the-art circulators employing externally produced magnetic fields, except for maximum attainable bandwidth.

Manuscript received May 17, 1967; revised August 30, 1967. This paper was presented at the 1967 G-MTT Symposium, Boston, Mass., May 8–11, 1967.

W. W. Siekanowicz is with Electronic Components and Devices, Radio Corporation of America, Princeton, N. J.

W. A. Schilling was with Electronic Components and Devices, Radio Corporation of America. He is now with RCA Frequency Engineering Laboratories, Inc., Farmingdale, N. J.

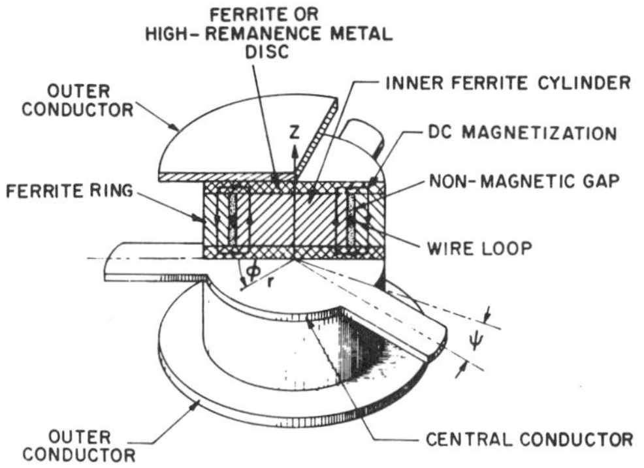

Fig. 1. Schematic representation of the latching circulator.

This paper presents the results of an approximate analysis of the cylinder–ring configuration having opposite directions of magnetization, describes the experimental results and comparison with the theory, and gives the conclusions. Details of the theoretical analysis are presented in the Appendix.

Theoretical Operation and Design

Existing field theories[1],[2],[11] for ferrite circulators apply only if the direction and magnitude of the dc flux density throughout the cross section of the ferrite element are constant. In the configuration described in this paper, the directions of the dc magnetic flux densities in the cylinder-and-ring regions are opposite. Three simplifying assumptions are made for analysis of this configuration: 1) the dc magnetic flux densities are directed either in the $+z$ or $-z$ directions, and their magnitudes are equal and constant throughout both regions; 2) the width of the nonmagnetic gap is negligible; and 3) the effects of dc magnetization in the disks is neglected. The theoretical analysis is described in some detail in the Appendix; the following paragraphs present the results.

The method of analysis of the latched circulator is similar to that described by Bosma[1] and Fay and Comstock.[2] As in the case of the unidirectionally magnetized disk circulator, operation of the latched cylinder–ring configuration is based on the splitting[12] of the resonances of the normal modes that exist in the absence of external lines ($\psi=0$, Fig. 1). The modes considered are the transverse electric (TE) waves having no variation in the z direction of the dc magnetic flux. The analysis consists of three steps: demonstration of the resonance split, derivation of conditions for circulation, and determination of the input admittance of the circulator. Knowledge of these parameters permits theoretical design of the circulator.

It is shown in the Appendix that the resonant frequencies of the normal modes for the uncoupled case can be obtained from the following expression:

$$\left[J_n'(x_2) \pm \frac{Kn}{\mu x_2} J_n(x_2)\right]\left[1 \pm \frac{\pi K n}{\mu} J_n(x_1) Y_n(x_1)\right] \mp \frac{\pi K n}{\mu} J_n^2(x_1)\left[Y_n'(x_2) \pm \frac{Kn}{\mu x_2} Y_n(x_2)\right] = 0. \quad (1)$$

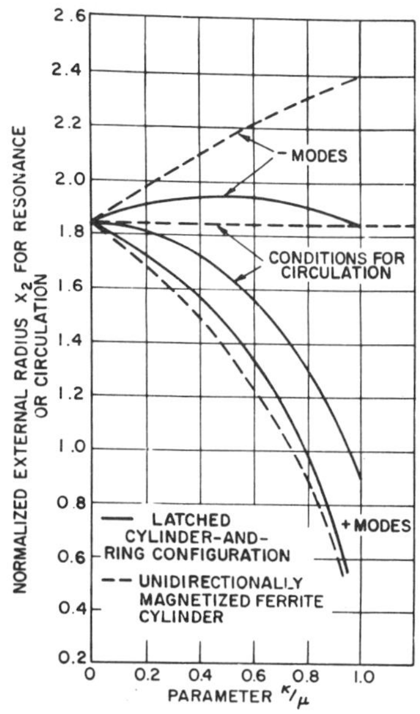

Fig. 2. Normalized outside radii for resonance and circulation of oppositely and unidirectionally magnetized ferrite elements for the $n=1$ mode as functions of parameter K/μ.

In this expression, J_n and Y_n denote Bessel functions of order n of the first and second kinds, respectively. The upper signs are used for the "+ mode" ($e^{+jn\phi}$) and the lower signs for the "− mode" ($e^{-jn\phi}$); n is a positive integer, μ and K denote the components of the Polder permeability tensor,[13] x_1 is the normalized[12] radius of the inner ferrite cylinder, and x_2 is the normalized radius of the ferrite ring. $x=kr$ and $k=\omega\sqrt{\epsilon\epsilon_0\mu_{eff}\mu_0}$, where $\omega=2\pi$ times frequency, ϵ_0 and μ_0 are, respectively, the dielectric constant (8.854×10^{-12} F/m) and permeability (1.257×10^{-6} H/m) of free space, and μ_{eff} is the effective permeability and ϵ the relative dielectric constant of the ferrite, respectively.

The upper and lower solid curves in Fig. 2 show the roots (x_2) of (1) for the $n=1$ mode as a function of K/μ. These roots were calculated for the following condition:

$$x_2 = \sqrt{2}x_1. \quad (2)$$

This condition is derived from the equality of the dc magnetic flux and flux densities in the cylinder-and-ring regions. The existence and inequality of the real roots of (1) provide theoretical proof for the splitting of the resonant frequency of the normal modes in the latched configuration and, as a result, provide the possibility of achieving circulation. For comparison, the upper and lower dashed curves in Fig. 2 show the roots of (1) for the unidirectionally magnetized ferrite disk employed in state-of-the-art circulators. These curves were obtained by making x_1 equal to x_2 in (1). The equation, in combination with simplifying relations for Bessel functions,[14] is then reduced to the forms given by Fay and Comstock.[12] Inspection of Fig. 2 shows that, for a given value of K/μ, the splitting (i.e., the separation) between the resonant modes of the latched configuration is smaller than that of the unidirectionally magnetized ferrite disk. This reduction of mode splitting is a consequence of

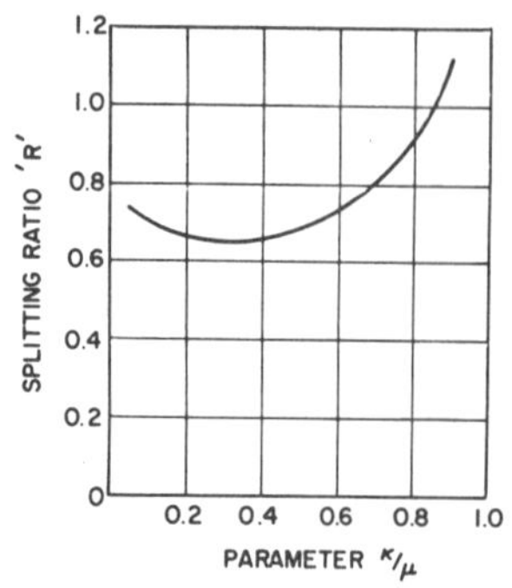

Fig. 3. Ratio of mode splitting in the oppositely and unidirectionally magnetized ferrite circulators as a function of parameter K/μ.

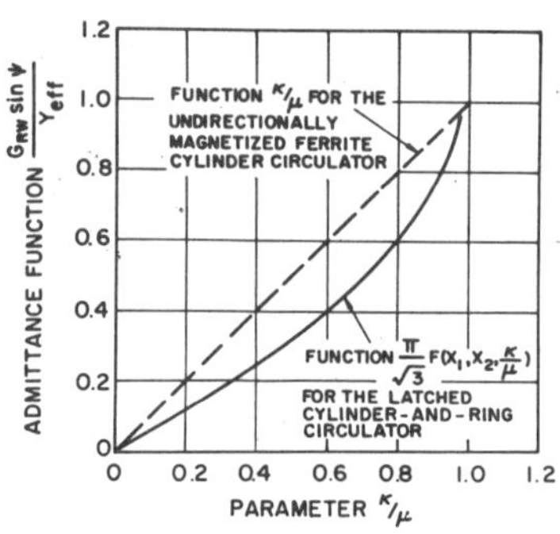

Fig. 4. Dependence of admittance functions for oppositely and unidirectionally magnetized ferrite elements on parameter K/μ.

the opposite directions of dc magnetizations in the cylinder-and-ring regions, and, as will be shown later, results in reduction of bandwidth in comparison with the unidirectionally magnetized ferrite disk.

For the latched configuration, the condition for circulation employing the $n=1$ modes is obtained from the following characteristic equation:

$$J_1'(x_2)J_1(x_2) + \frac{\pi^2K^2}{\mu^2} J_1^2(x_1)[Y_1(x_1)J_1'(x_2) - Y_1'(x_2)J_1(x_1)] \cdot [J_1(x_1)Y_1(x_2) - J_1(x_2)Y_1(x_1)] = 0. \quad (3)$$

The solid curve in the center of Fig. 2 shows the normalized outside radius x_2 as a function of K/μ, obtained by combination of (2) and (3). For comparison, the dashed curve in the center of Fig. 2 shows the circulation condition for the unidirectionally magnetized ferrite disk. This curve was obtained by making x_1 equal to x_2 in (3), which is then reduced to the form given by Fay and Comstock.[15] For the latched circulator, the theoretical normalized outside radius x_2 decreases with increasing K/μ, while the corresponding radius for the unidirectionally magnetized disk is equal to 1.84, independently of the value of K/μ. The solid curves in Fig. 2 show that the theoretical frequency for the circulation of the latched configuration lies between the resonant frequencies of the + and − modes, as in the case of the unidirectionally magnetized disc circulator.[16]

The bandwidth of a circulator is proportional to the splitting $2\delta'$ defined by the following expression:

$$2\delta' = \frac{\omega^- - \omega^+}{\omega_{circ}} \quad (4)$$

where ω^- and ω^+ are the resonant frequencies (proportional to x_2^- and x_2^+, shown, respectively, by the upper and lower curves in Fig. 2), and ω_{circ} is the frequency for circulation (proportional to x_{2circ}, shown by the curves in the center of Fig. 2). Fig. 3 shows the ratio R of the splitting in the latched circulator to that in the unidirectionally magnetized disk circulator as a function of K/μ. In the typical range of design ($0.2<K/\mu<0.5$, approximately), the theoretical bandwidth of the latched design is approximately 30 percent smaller than that of state-of-the-art circulators. This reduction of bandwidth is a consequence of the magnetic flux reversal in the region of circulation.

The input wave admittance of the latched circulator at resonance is given by the following expression:

$$G_{RW} = \frac{\pi}{\sqrt{3}\sin\psi} Y_{eff} F\left(x_1, x_2, \frac{K}{\mu}\right) \quad (5)$$

where Y_{eff} has been given by Fay and Comstock[17] and F is expressed as follows:

$$F\left(x_1, x_2, \frac{K}{\mu}\right) = \frac{\left[J_1'(x_2) - \frac{K}{\mu x_2}J_1(x_2)\right]\left[1 - \frac{\pi K}{\mu}J_1(x_1)Y_1(x_1)\right] + \frac{\pi K}{\mu}J_1^2(x_1)\left[Y_1'(x_2) - \frac{K}{\mu x_2}Y_1(x_2)\right]}{J_1(x_2) - \frac{\pi K}{\mu}J_1(x_1)[J_1(x_2)Y_1(x_1) - Y_1(x_2)J_1(x_1)]}. \quad (6)$$

The solid curve in Fig. 4 shows the normalized wave conductance $G_{RW}\sin\psi/Y_{eff}$ as a function of the parameter K/μ for the conditions given by (2) and the solid curve in the center of Fig. 2. For comparison, the dashed curve shows the corresponding conductance for the unidirectionally magnetized ferrite disk. For such a disk, this function[18] is very nearly equal to K/μ, and can be obtained from (5) and (6) by making x_1 equal to x_2 and employing simplifying relations for Bessel functions.[14]

For the coaxial configuration shown in Fig. 1 and small values of angle ψ, the input conductance G_R at the theoretical frequency of circulation is given by the following expression:

$$G_R \simeq \frac{4R_2\psi}{h} G_{RW} \quad (7)$$

where h is the height of a single ferrite assembly, and R_2 the external radius of the ferrite ring.

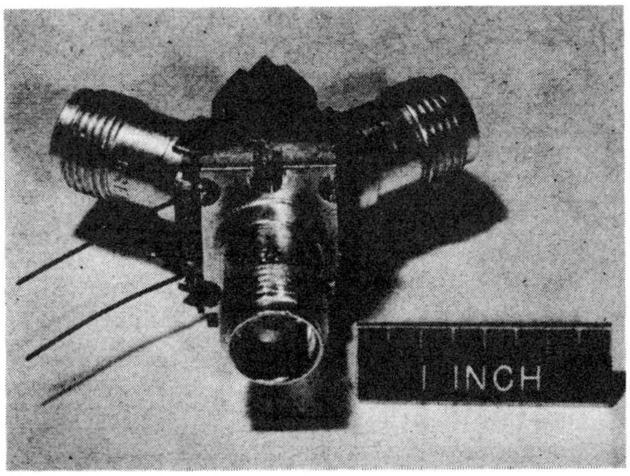

Fig. 5. Latching switchable circulator operating in the vicinity of 7.3 GHz.

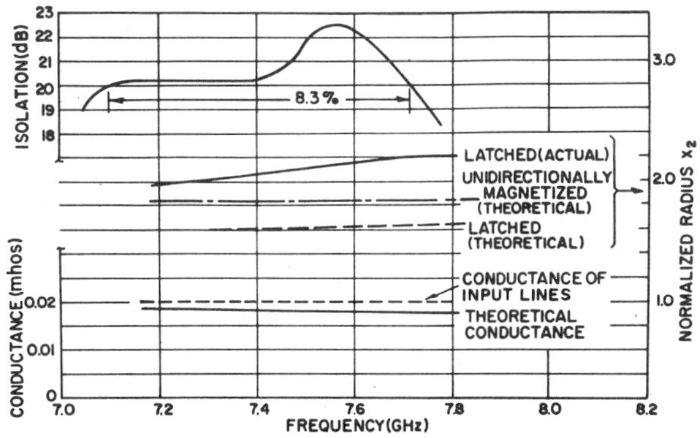

Fig. 6. Isolation, normalized radii, and input conductance as functions of frequency for a latching circulator operating in the vicinity of 7.3 GHz and employing high-remanence metal disks and fixed stub tuners.

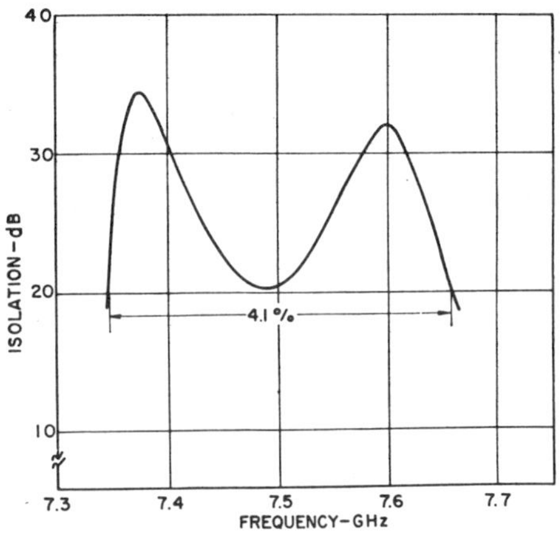

Fig. 7. Isolation as a function of frequency for a latching circulator employing ferrite disks for flux transport and fixed stub tuners for RF-impedance matches.

Experimental Results

Fig. 5 is a photograph of a latched switchable circulator operating in the vicinity of 7.3 GHz.

The purpose of the first experiment was the demonstration of circulation in a latched cylinder–ring assembly having opposite directions of dc magnetization. The cross-sectional areas of the ferrite ring and cylinder were made equal. In this experiment, high-remenence metal disks were employed for flux transport, and the ferrites were fed directly from 50-ohm coaxial lines. The 20-dB isolation bandwidth for this assembly was approximately 4 percent, the center frequency was 7.325 GHz, and the minimum insertion loss was 0.25 dB. This bandwidth was increased to 8.3 percent by use of fixed stub tuners for RF impedance matches, as shown in the upper portion of Fig. 6. The solid and dashed curves in the center of Fig. 6 show, respectively, the actual and theoretical external normalized radii as functions of frequency. The theoretical normalized radius for circulation was obtained from Fig. 2. The actual normalized radius was obtained by multiplying the physical radius by the parameter k defined by Fay and Comstock.[19]

The curves in the lower portion of Fig. 6 show the comparison between the theoretical conductance and the input conductance of 0.02 mho. The theoretical conductance was calculated by use of (5) and (7) and Fig. 4.

It was assumed in the above calculations that in the latched circulator, the internal dc magnetic field H is zero. This assumption leads to the following simplifying relations: $\mu=1$; $K=2.8(4\pi M_r)/f$ where $4\pi M_r$ is the remanence in gauss; and the frequency f is in megahertz.

Fig. 7 shows the isolation as a function of frequency for an assembly employing ferrite disks for flux transport and fixed stub tuners for RF impedance matches. The 20-dB isolation bandwidth for this condition was 4.1 percent. The switching time and energy for this circulator were 0.2 μs and 10 μJ, respectively.

Fig. 8 shows the isolation, actual and theoretical normalized radii, and theoretical and actual input conductances as functions of frequency for a C-band latched circulator employing high-remanence metal discs for flux transport and direct feed from 50-ohm coaxial lines. The linear dimensions of this assembly were approximately 25 percent greater than those of the circulator shown in Fig. 5. The 20-dB isolation bandwidth for the C-band assembly was 4.85 percent, and the insertion loss 0.4 dB. The theoretical value for the bandwidth was 24 percent. At the center frequency, the actual external normalized radius was 2.1. This value is 30 percent greater than the theoretical value, and the conductance of the feeding lines is 18 percent smaller than the theoretical values obtained from Fig. 4 and (5) and (7). The figure of 30 percent includes a finite thickness of the nonmagnetic gap (Fig. 1), which was not considered in the theory. At the optimum circulation frequency, the actual normalized radius of the ferrite cylinder was 1.36. This value is approximately 19 percent greater than the theoretical value. The latter figure is more significant in comparing experiments with theory.

Fig. 9 shows the isolation as a function of frequency for the C-band latched circulator when the high-remanence metal disks were replaced by ferrite disks and quarter-wave coaxial transformers were used for RF impedance matches. For this design, the 20-dB isolation occurred at 5.330 and 5.446 GHz (approximately 2.4 percent bandwidth), and the

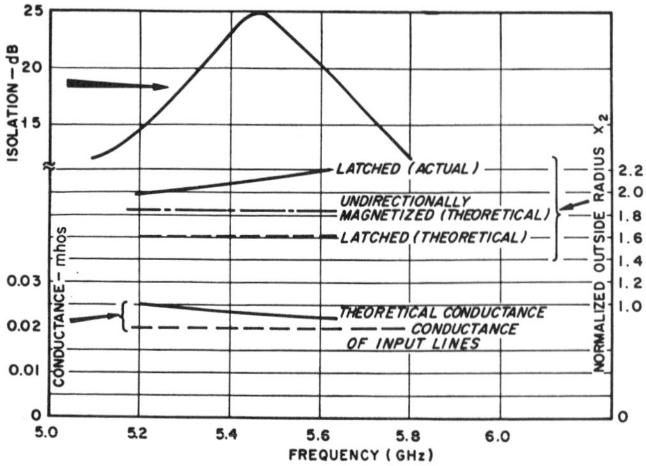

Fig. 8. Isolation, normalized radii, and input conductance as functions of frequency for a C-band latching circulator employing high-remanence metal disks for flux transport.

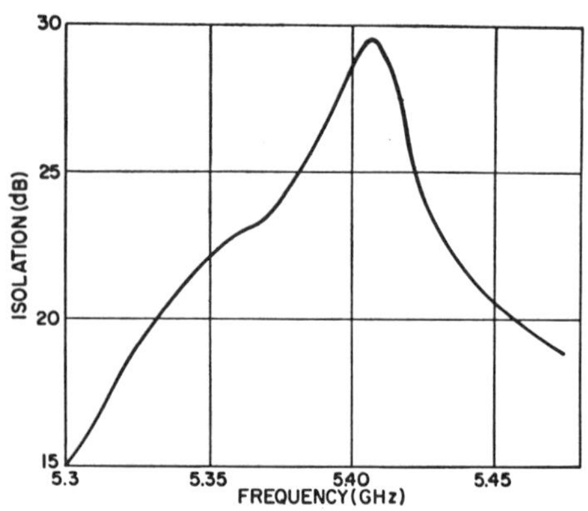

Fig. 9. Isolation as a function of frequency for a C-band latching circulator employing all-ferrite elements and quarter-wavelength transformers for RF-impedance matches.

insertion loss was 0.4 dB. The theoretical value for the bandwidth was 21 percent. This further reduction of bandwidth is attributed to increased distortion of the z component of dc magnetic induction, caused by replacement of high-remanence metal disks by ferrite disks. However, impedance measurements indicate that the bandwidth of the all-ferrite assembly can be extended to 10 percent by addition of reactive matching elements.

The switching time and energy for the all-ferrite assembly was 0.2 μs and 15 μJ, respectively. This switching energy was lower than that for the previous circulator because the coercive force of the ferrite and, therefore, the area of the B-H loop was significantly smaller.

Conclusions

This paper has presented the theory and operation of a new type of compact, latching, switchable ferrite junction circulator that is well suited for applications requiring fast (fractions of microsecond), low-energy (tens of microjoules in the microwave region) switching. The novelty of the device is the placement of a magnetically closed ferrite assembly within the RF interaction region, and utilization of the major portion of this assembly for circulation. Such an assembly consists of a ferrite cylinder surrounded by an oppositely magnetized ferrite ring, both elements being in contact with ferrite or high-remanence metal discs. The results of an approximate analysis for such an element were determinations of radius for circulation, bandwidth, and input impedance as functions of material properties and frequency.

Operation of latched circulators was demonstrated on assemblies operating in the vicinity of 7.3 and 5.4 GHz. When high-remanence metal disks were employed for flux transport and the ferrites were fed directly from 50-ohm coaxial lines, the 20-dB isolation bandwidth was approximately 4 percent and the insertion loss ranged from 0.25 to 0.4 dB. This bandwidth was increased to 8.3 percent by use of fixed stub tuners.

Comparison of experiments with theory showed that the actual radius of the ferrite cylinder was approximately 19 percent greater than the theoretical radius, and the input conductance ranged from 10 percent above to 18 percent below the calculated values.

When the metal disks were replaced by ferrite disks and quarter-wavelength transformers were employed for RF-impedance matching, the 20-dB isolation bandwidth was 2.4 percent. This bandwidth was increased to 4.1 percent by use of fixed stub tuners.

The switching time and energy for the C-band assembly were 0.2 μs and 15 μJ, respectively.

Appendix

IELD THEORY OF THE LATCHING FERRITE CIRCULATOR

The method of analysis of the latched circulator is similar to that described by Bosma[1] and Fay and Comstock.[2] The analysis of the device is confined to only one cylinder-and-ring assembly because the other assembly is a mirror image of its neighbor.

In the inner ferrite cylinder (Fig. 1) the electric field has a solution of the following form:[20]

$$E_{zn} = J_n(x)\,(a_{+n}e^{jn\phi} + a_{-n}e^{-jn\phi}) \tag{8}$$

where a_{+n} and a_{-n} are constants and x is the normalized radial coordinate.[17]

When the dc magnetization is in the same direction as the positive z axis, the ϕ component H_ϕ of the RF magnetic field is given by

$$H_\phi = \frac{j}{\omega\mu_0\mu_{\text{eff}}}\left(\frac{\partial E_z}{\partial r} + j\frac{K}{\mu r}\frac{\partial E_z}{\partial \phi}\right). \tag{9}$$

Substitution of (8) in (9) provides the following expression for the RF magnetic field in the inner ferrite cylinder:

$$H_{\phi n} = jY_{\text{eff}}\left\{\left[J_n'(x) - \frac{Kn}{\mu x}J_n(x)\right]a_{+n}e^{jn\phi} + \left[J_n'(x) + \frac{Kn}{\mu x}J_n(x)\right]a_{-n}e^{-jn\phi}. \right. \tag{10}$$

$$a_{\pm 1} = \frac{E_1\left(1 \pm \frac{j}{\sqrt{3}}\right)}{2\left\{J_1(x_2) \pm \frac{\pi K}{\mu} J_1(x_1)\left[J_1(x_2)Y_1(x_1) - Y_1(x_2)J_1(x_1)\right]\right\}}. \tag{15}$$

The solution for the electric field E_{zn} in the ferrite ring (Fig. 1) is given by the following expression:

$$E_{zn} = J_n(x)(b_{+n}e^{jn\phi} + b_{-n}e^{-jn\phi}) + Y_n(x)(c_{+n}e^{jn\phi} + c_{-n}e^{-jn\phi}) \tag{11}$$

where b_{+n}, b_{-n}, c_{+n}, and c_{-n} are constants. The ϕ component H_ϕ of the RF magnetic field in the ferrite ring can be obtained by changing the sign of K in (9), as required by reversal of dc magnetization. Combination of (11) with the modified (9) results in the following expression for the RF magnetic field in the ferrite ring:

$$\begin{aligned} H_{\phi n} = jY_{\text{eff}}\Bigg\{&\left[\left(J_n'(x) + \frac{Kn}{\mu x}J_n(x)\right)b_{+n} + \left(Y_n'(x) + \frac{Kn}{\mu x}Y_n(x)\right)c_{+n}\right]e^{jn\phi} \\ &+ \left[\left(J_n'(x) - \frac{Kn}{\mu x}J_n(x)\right)b_{-n} + \left(Y_n'(x) - \frac{Kn}{\mu x}Y_n(x)\right)c_{-n}\right]e^{-jn\phi}. \end{aligned} \tag{12}$$

The constants b_{+n}, b_{-n}, c_{+n}, and c_{-n} can be expressed in terms of a_{+n} and a_{-n} by imposing the condition that the electric field E_z and the RF magnetic field H_ϕ be continuous at $x=x_1$. This condition, together with simplifying expressions for Bessel functions,[14] yields the following relationships:

$$\begin{aligned} b_{\pm n} &= \left[1 \pm \frac{\pi Kn}{\mu} J_n(x_1)Y_n(x_1)\right]a_{\pm n} \\ c_{\pm n} &= \mp \frac{\pi Kn}{\mu} J_n^2(x_1)a_{\pm n}. \end{aligned} \tag{13}$$

As stated by previous authors,[17] it is useful to determine the normal modes for the uncoupled case ($\psi=0°$, where 2ψ is the angle subtended by the striplines, as shown in Fig. 1). These modes are obtained by use of the following condition:

$$H_\phi(x = x_2) = 0. \tag{14}$$

Combination of (10) and (12) through (14) shows that the resonant frequencies of these modes are given by the roots of (1).

Following Fay and Comstock's procedure,[17] the conditions for circulation are derived by assuming that the device shown in Fig. 1 functions as a circulator. These conditions, given by equations (7) and (8) in Fay and Comstock,[2] apply at normalized radius x_2 for the present device. For the latching circulator, the constants $a_{\pm 1}$ are given by the following expression:

When $x_2=x_1$ (i.e., for the case of the unidirectionally magnetized disk), (15) reduces to Fay and Comstock's equation (9).[2]

Combination of (12), (13), and (15) provides the following expression for $H_{\phi 1}$ at $x=x_2$:

$$H_{\phi 1} = jY_{\text{eff}}\frac{E_1}{2}\left[\left(1+\frac{j}{\sqrt{3}}\right)F\left(x_1, x_2, -\frac{K}{\mu}\right)e^{j\phi} + \left(1-\frac{j}{\sqrt{3}}\right)F\left(x_1, x_2, \frac{K}{\mu}\right)e^{-j\phi}\right]. \tag{16}$$

Following Fay and Comstock's procedure,[15] the RF magnetic field satisfies the condition for circulation when the following condition is satisfied:

$$F\left(x_1, x_2, \frac{K}{\mu}\right) = -F\left(x_1, x_2, -\frac{K}{\mu}\right). \tag{17}$$

Combination of (6) and (17) results in the characteristic equation (3) for circulation.

Equation (5) for wave conductance G_{RW} at the circulation frequency is obtained by combination of (16) and (17) and Fay and Comstock's equation (12).[2]

Acknowledgment

The authors thank H. K. Jenny and Dr. F. Sterzer for guidance in the course of the project, and I. Bardash (now with Sedco Systems, Inc., Farmingdale, N. Y.) for cooperation during initial stages of the project.

References

[1] H. Bosma, "On stripline Y-circulation at UHF," *IEEE Trans. Microwave Theory and Techniques*, vol. MTT-12, pp. 61–72, January 1964.

[2] C. E. Fay and R. L. Comstock, "Operation of the ferrite junction circulator," *IEEE Trans. Microwave Theory and Techniques*, vol. MTT-13, pp. 15–27, January 1965.

[3] P. C. Goodman, "A latching ferrite junction circulator for phased array switching applications," presented at the Symp. on Microwave Theory and Techniques, Clearwater, Fla., May 6, 1965.

[4] W. C. Passaro and J. W. McManus, "A 35 GHz latching switch," presented at the Internat'l Microwave Symp., Palo Alto, Calif., May 19, 1966.

[5] L. E. Davis, "Computed phase-shift and performance of a latching 3-port waveguide circulator," *NEREM Rec.*, p. 96, 1966.

[6] F. Betts, D. H. Temme, and J. A. Weiss, "A switching circulator: *S*-band; stripline; remanent; 10 kW; 10 microsecond; temperature-stable," presented at the Internat'l Microwave Symp., Palo Alto, Calif., May 19, 1966.

[7] W. J. Ince and E. Stern, "Computer analysis of ferrite digital phase shifters," *IEEE Internat'l Conv. Rec.*, pt. 5, pp. 32–38, March 21–25, 1966.

[8] E. Schlömann, "Theoretical analysis of twin-slab phase shifters in rectangular waveguide," *IEEE Trans. Microwave Theory and Techniques*, vol. MTT-14, pp. 15–23, January 1966.

[9] W. W. Siekanowicz, W. A. Schilling, T. E. Walsh, I. Bardash, and I. Gordon, "Design and performance of a 20-kilowatt latching nonreciprocal X-band ferrite phase shifter," *RCA Rev.*, vol. 26, pp. 574–589, December 1965.

[10] L. R. Whicker and R. R. Jones, "Design guide to latching phase shifters," *Microwaves*, pp. 43–47, December 1966.

[11] J. B. Davies, "An analysis of the m-port symmetrical H-plane waveguide junction with central ferrite post," *IRE Trans. Microwave Theory and Techniques*, vol. MTT-10, pp. 596–604, November 1962.

[12] Fay and Comstock, [2] p. 18.

[13] B. Lax and K. J. Button, *Microwave Ferrites and Ferrimagnetics.* New York: McGraw-Hill, 1962, pp. 150–151.

[14] N. W. McLachlan, *Bessel Functions for Engineers.* London: Oxford University Press, 1934 (example 64, p. 156, in combination with eq. 74, p. 162).

[15] Fay and Comstock, [2] p. 19, eq. (14).

[16] *Ibid.*, p. 19.

[17] *Ibid.*, p. 18.

[18] *Ibid.*, p. 19, eq. (16).

[19] *Ibid.*, p. 18, eq. (2).

[20] *Ibid.*, p. 18, eq. (3).

A Latching Ring-and-Post Ferrite Waveguide Circulator

WIESLAW W. SIEKANOWICZ, MEMBER, IEEE, ROBERT W. PAGLIONE, AND THOMAS E. WALSH

Abstract—**This paper presents the performance and normalized design parameters for a latching ring-and-post ferrite circulator in waveguide. A *C*-band circulator has provided an insertion loss of 0.35 dB and a 20-dB isolation bandwidth of 17 percent. When the circulator was matched for higher maximum isolation (50 dB) but narrower bandwidth (10 percent) at room temperature, the 20-dB isolation bandwidth was 7.8 percent across the −40° to +75°C temperature range.**

Low-loss operation was obtained at pulsed powers up to 7.5 kilowatts, and at least 20 dB of isolation was maintained up to 100 kilowatts. This performance, in conjunction with a switching speed of a fraction of a microsecond, permits the use of these circulators for transmitting–receiving functions in high-reliability radars.

Introduction

SEVERAL ferrite configurations can be used for latching, single-junction circulators employing internal return paths. These configurations include *Y*-shaped and triangular elements [1], [2], and a ring-and- post assembly [3]–[5]. A common feature of internal-return-path latching circulators is the use of a microwave ferrite element that has a magnetic path closed entirely within the region of microwave interaction. Such circulators can be switched in a fraction of a microsecond; the switching energy is in the range of tens of microjoules.

The preceding configurations are different from external-return-path designs in which part of the dc magnetic circuit is outside the microwave region [6], [7]. These latter designs provide wider bandwidths, but require longer switching times (typically several microseconds) and higher switching energies (typically hundreds of microjoules).

The performance of *Y*-shaped and triangular ferrite elements in waveguide circuits has been reported previously [1], [2], [8]. The feasibility and initial performance for ring-and-post elements were obtained in coaxial circuits [4], [5]. This paper presents results for the ring-and-post configuration in waveguide circuits.

Description of the Ring-and-Post Latching Circulator

A diagram of the latching ring-and-post circulator in waveguide is shown in Fig. 1. The assembly consists of an inner ferrite post concentric with two thin ceramic rings and an external ferrite ring. The upper and lower surfaces of the ring-and-post assembly are in contact with ferrite discs. A

Manuscript received July 15, 1969; revised October 27, 1969.
The authors are with the David Sarnoff Research Center, RCA Laboratories, Princeton, N. J.

wire loop is formed around the ferrite post and supported by the ceramic rings; its ends are brought out to the waveguide walls through two small holes in the ferrite ring. At the waveguide walls the ends of the latching wire are connected to capacitive feedthroughs which provide dc insulation between the wires and the waveguide.

The purpose of the ceramic rings, which have a dielectric constant approximately equal to that of the ferrite, is to provide a dc nonmagnetic gap, supply dielectric continuity at microwave frequencies between the ferrite post and ring, and furnish mechanical support for the loop.

For operation of the circulator, a pulse of direct current is passed through the wire loop to produce a magnetizing field indicated approximately by the dashed line in Fig. 1. When the current is turned off, the ferrite operates at remanence $4\pi Mr$. Operation at remanence is achieved because the magnetic path is closed entirely within the ferrite, as indicated by the dashed line in Fig. 1. The direction of circulation is determined by the direction of the magnetizing field in the ferrite post [9]; reversing the polarity of the dc pulse reverses the direction of circulation.

For continuity of dc magnetic flux, the cross-sectional areas of the ferrite ring and post are equal, and the thickness

of each ferrite disc is equal to one half the radius of the post.

Design Considerations

An approximate theory for the ring-and-post circulator was presented previously [4], [5]. One of the important results of this theory was the derivation of the normalized external radius of the ferrite ring for negligible thickness of ceramic rings as a function of the ratio κ/μ, where κ and μ are the components of the Polder permeability tensor [10]. The normalized radius [11] is given by kr, where r is the actual radius, and the parameter k is calculated from

$$k = \frac{2\pi}{\lambda_0}\sqrt{\epsilon\mu_{\text{eff}}}. \tag{1}$$

In this expression, λ_0 is the free-space wavelength, ϵ is the relative dielectric constant of the ferrite, and μ_{eff} is the effective permeability of the ferrite. For a latched ferrite device the internal magnetizing field H_o is zero. The general expression [12] for μ_{eff} then reduces to the simple form

$$\mu_{\text{eff}} = 1 - \left[\frac{2.8 \times (4\pi Mr)}{F}\right]^2 \tag{2}$$

where $(4\pi Mr)$ is the remanence in gauss, and F is the frequency in megahertz. Similarly, the ratio κ/μ assumes the following simple form for the latched circulator:

$$\frac{\kappa}{\mu} = \frac{(4\pi\ Mr)}{F}. \tag{3}$$

In designing the circulator, it is important to consider the diameter of the ferrite post because circulation occurs primarily within that post and its direction is determined by the orientation of dc magnetization in the post [9]. Preliminary work [4], [5] on ring-and-post circulators indicated that latching circulators could be built with κ/μ values ranging

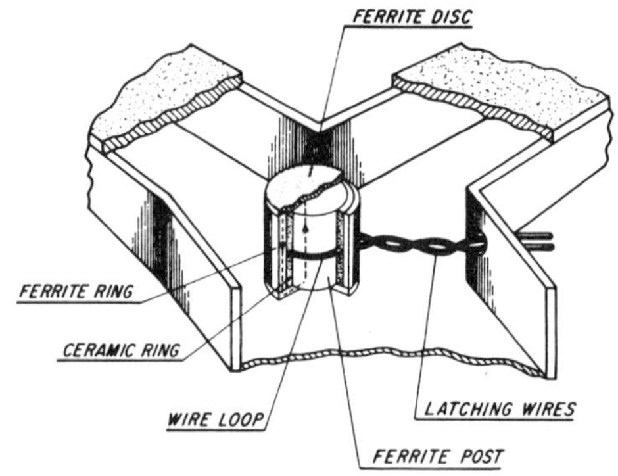

Fig. 1. Diagram of latching ring-and-post circulator in waveguide.

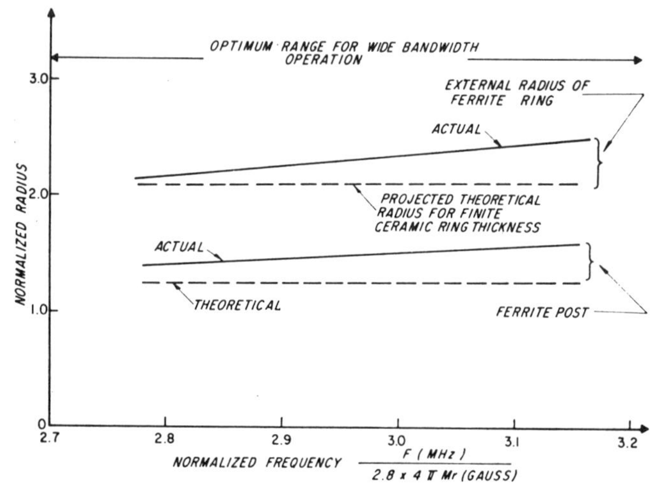

Fig. 2. Normalized radii of the ferrite ring and post as functions of normalized frequency in the optimum range of operation.

from 0.2 to 0.5. In the center of this range, the theoretical normalized external radius kr of the ferrite ring for negligible thickness of ceramic rings is 1.75. The normalized radius of the ferrite post, obtained by dividing the external radius of the ring[1] by $\sqrt{2}$ for preservation of equal cross-sectional areas, is 1.24. For this value of normalized post radius and finite practical thickness (typically 0.030 inch) of ceramic rings, the requirement for equal cross-sectional areas of ferrite ring and post leads to a normalized external radius of approximately 2.1 for the ferrite ring.

The experimental study described later indicates that it is necessary to combine proper ferrite diameter, remanence, and waveguide impedance to achieve maximum bandwidth and low insertion loss. Maximum bandwidth was obtained for the following parameters near the center frequencies: the normalized post and external ferrite ring radii were 1.5 and 2.3, respectively, the κ/μ ratio was approximately 0.34, and the waveguide impedance was approximately 590 ohms. The preceding radii are, respectively, 21 and 10 percent greater than the theoretical values of 1.24 and 2.1 presented earlier. Fig. 2 shows a comparison of the actual and theo-

[1] The theoretical normalized external radius of the ferrite ring, for negligible thickness of ceramic ring, is shown in [5] by the solid curve in the center of Fig. 2

etical normalized radii for the ferrite post and ring as functions of κ/μ in the optimum range of operation.

Experimental Results

The experimental evaluation of the ring-and-post circulator was carried out at C-band. The objective of the evaluation was to determine the optimum design for maximum bandwidth, minimum insertion loss, good temperature characteristics, and reasonably high power. The parameters investigated included remanence, ferrite diameter and height, and waveguide impedance.

Optimum results were obtained for a gadolinium-doped yttrium-iron garnet having a remanence of 670 gauss, a coercive force of approximately 1 oersted, and a squareness ratio of 0.8 for a magnetizing force of 5 oersted. The magnetic and temperature characteristics for the ferrite material are shown in Figs. 3 and 4, respectively.

Satisfactory RF impedance matches were obtained by use of only a single tuning post in each waveguide arm. Depending on the penetration and location of the tuning posts, the isolation characteristics could be adjusted for very high values across a narrow bandwidth, or for lower values across a wider bandwidth.

Fig. 5 shows the maximum bandwidth characteristics obtained by use of a WR-137 (0.622×1.372 inch) waveguide fixture and a ferrite assembly having a height essentially equal to that of the waveguide (0.622 inch). The solid curve shows the isolation as a function of frequency when the RF matches were adjusted for relatively wide bandwidth. The bandwidth for 20-dB isolation was 14 percent, and the insertion loss was 0.35 dB. The minimum isolation in the center of the band was 24 dB. When this isolation was reduced to 21 dB, the bandwidth increased to 17 percent. This bandwidth compares with approximately six percent for a 35-GHz latching circulator employing a triangular ferrite element [13] and from 15.5 to 19.4 percent for X-band switchable circulators employing Y-shaped ferrite elements [1], [8].

The upper dashed curve in Fig. 5 shows the isolation as a function of frequency for RF matches adjusted for high isolation across a narrower bandwidth. For this condition, the 20- and 30-dB isolation bandwidths were 10 and 5.5 percent, respectively.

The normalized design parameters for the ferrite assembly that provided the preceding characteristics are summarized in Fig. 2, which shows the normalized radii for the ferrite post and ring plotted as functions of the normalized frequency in the range of optimum performance. In the center of this range, the normalized external radii for the post and ferrite ring were 1.5 and 2.34, respectively. The corresponding ratio $\kappa/\mu[=2.8\,(4\pi Mr)/F]$ was 0.34, and the waveguide impedance was 590 ohms (WR-137 waveguide). The actual radius of the post was 21 percent greater than the theoretical value.

A full-waveguide-height ring-and-post assembly having an external radius 11 percent greater than that used for

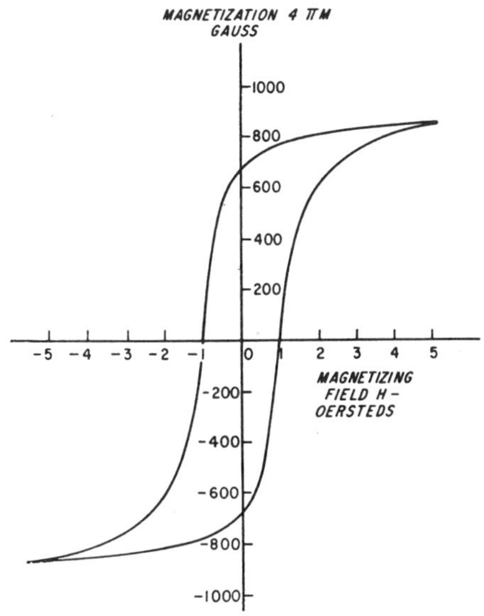

Fig. 3. Hysteresis loop for the ferrite material employed for construction of the latching ring-and-post circulator.

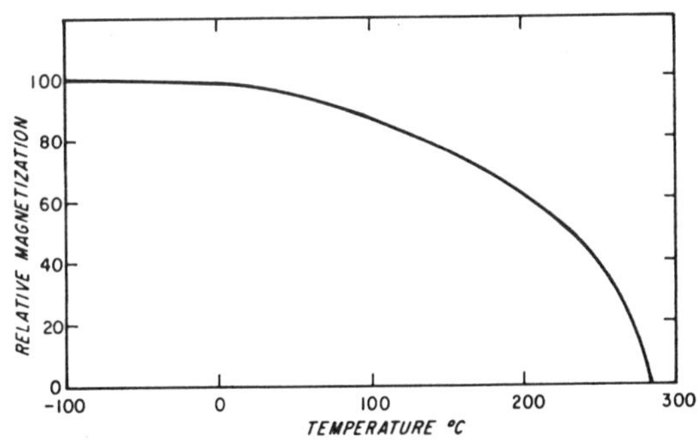

Fig. 4. Relative magnetization as a function of temperature for the ferrite material.

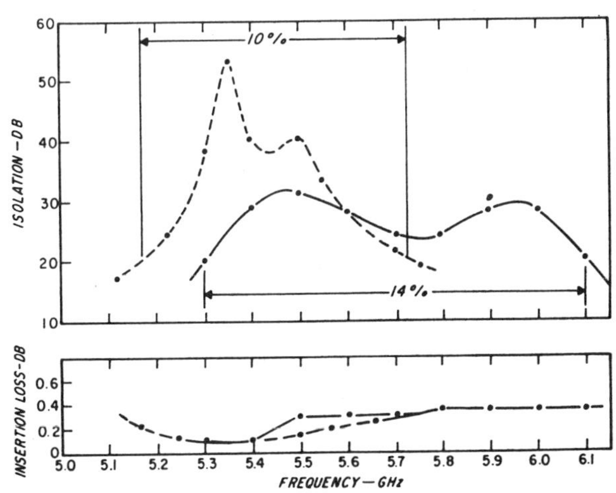

Fig. 5. Isolations for narrower (dashed) and wider (solid) bandwidth conditions and corresponding insertion losses as functions of frequency.

185

achievement of the characteristics shown in Fig. 5 and evaluated in a WR-137 waveguide fixture provided a 20-dB isolation bandwidth of ten percent at the center frequency of 4.9 GHz. The corresponding values for the post and external ferrite ring radii were 1.45 and 2.27. These numbers are very close to the optimum values (1.5 and 2.34) given previously for the smaller ferrite diameter.

A full-waveguide-height assembly having an external radius eight percent smaller than that used for achievement of the optimum results shown in Fig. 5, and evaluated in a WR-137 waveguide fixture, provided a bandwidth of only two percent. This experiment indicated that the impedance of the smaller ferrite was too large for achievement of wide bandwidth.

An experiment performed by use of the optimum-diameter but reduced-height (by 27 percent) ferrite assembly, supported by a teflon button in the WR-137 fixture, provided a maximum bandwidth of ten percent, i.e., smaller than that (17 percent) obtained by use of the full-height assembly.

The maximum bandwidth measured in the WR-187 (0.872×1.872 inch) waveguide fixture, by use of a full-waveguide-height ferrite assembly having a diameter that provided maximum (17 percent) bandwidth in the WR-137 fixture, was only three percent. Tests on different materials having remanences ranging from 400 to 900 gauss did not increase that bandwidth.

The following conclusions can be drawn from the above experiments for ferrite assemblies extending across the full height of the waveguide.

1) The ferrite radius is relatively critical. Maximum bandwidth was obtained for normalized post-and-ring radii in the vicinity of 1.5 and 2.34, respectively.
2) The optimum normalized remanence was approximately 0.12 gauss per MHz.
3) A relatively high waveguide impedance, in the vicinity of 600 ohms, is recommended. A significant reduction of this impedance results in a drastic reduction of bandwidth.

The following sections describe additional experiments performed by use of the full-height optimum-diameter ferrite assembly in the WR-137 waveguide fixture. Fig. 6 shows the temperature characteristics when the latched circulator was tuned for the higher isolation and narrower bandwidth condition. At room temperature (+25°C), the 20-dB isolation bandwidth was ten percent centered at 5.45 GHz. The decrease of center frequency with increasing temperature results from the corresponding decrease of magnetization, as shown in Fig. 4. The change of center frequency results from two reinforcing effects. The first is a shift of frequency for the optimum normalized radius of the circulator, and the second is a change of frequency at which optimum RF matching is achieved.

Equations (1) and (2) indicate that the change of frequency resulting from the first effect is proportional to the change of remanence, as shown by the following relation:

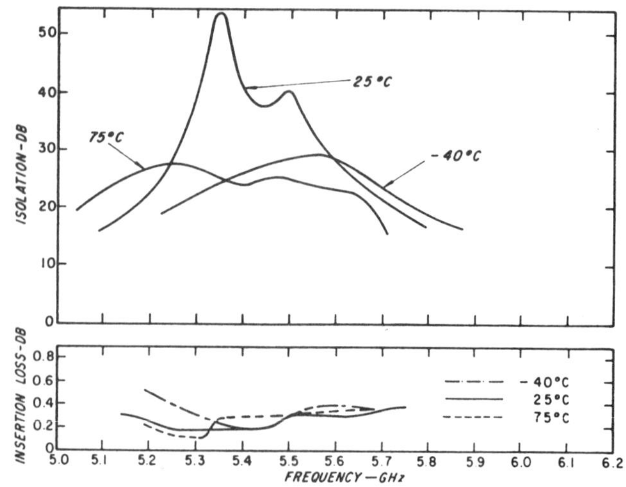

Fig. 6. Isolation and insertion loss as functions of frequency at temperatures of −40, +25, and +75°C.

$$\frac{dF}{F} = \left[\frac{2.8 \times (4\pi Mr)}{F}\right]^2 \frac{d(4\pi Mr)}{(4\pi Mr)}. \tag{4}$$

The preceding relation shows that the center frequency will decrease with decreasing remanence.

The second effect contributing to a decrease of frequency with increasing temperature is an increase of the ferrite wave impedance with decreasing remanence and, as a result, better RF impedance matches at lower frequencies, where the waveguide impedance is also higher.

Fig. 6 shows that at +75°C the 20-dB isolation bandwidth was 11.6 percent centered at 5.36 GHz. At that temperature the remanence of the ferrite was approximately seven percent lower than that at room temperature (as shown in Fig. 4). Substitution of this value in (4) indicates achievement of the optimum normalized radius at 5.41 GHz. The additional reduction to 5.36 GHz is attributed to improvement of the RF impedance matches, as discussed previously. At −40°C, the 20-dB isolation bandwidth was 9.5 percent centered at 5.51 GHz. The slight increase of the center frequency results from a corresponding increase of remanence with decreasing temperature (Fig. 4). The 20-dB isolation bandwidth in the −40°C to +75°C temperature range was 7.8 percent centered at 5.46 GHz.

Fig. 7 shows the isolation and insertion loss as functions of RF input power for the latched circulator.

The insertion loss remained unchanged up to a power of 7.5 kW and began to increase above that level as a result of nonlinear effects in the ferrite. The isolation remained essentially constant up to a power of 15 kW, and began to decrease beyond that power level. However, a relatively high value of 21 dB of isolation was measured up to the maximum available power of 100 kW.

Achievement of 21 dB of isolation at a power level of 100 kW permits the use of switchable circulators for transmitting-receiving (TR) functions in radars. A schematic diagram

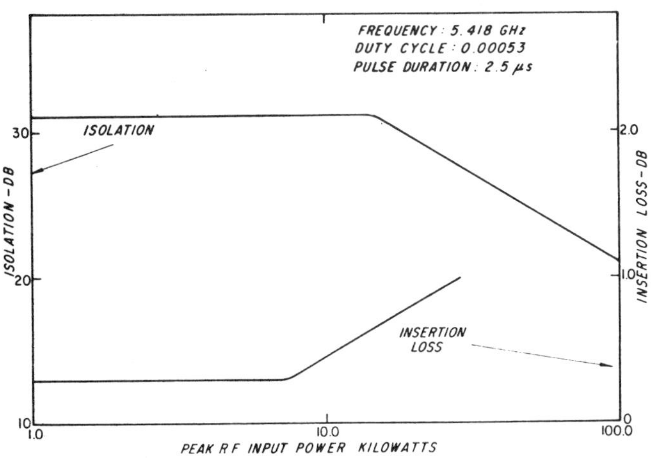

Fig. 7. Isolation and insertion loss as functions of pulsed RF power.

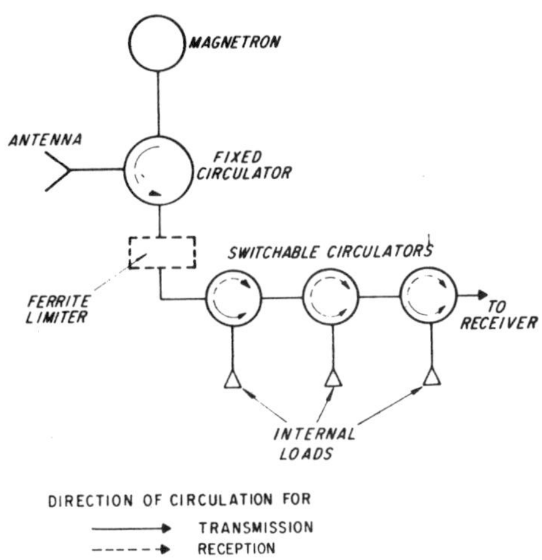

Fig. 8. Schematic diagram illustrating the use of switchable circulators for transmitting–receiving function in a radar.

for a TR box employing three switchable circulators in conjunction with a fixed circulator is shown in Fig. 8. During transmission of the high-power RF pulse, the circulators are switched in a direction that provides isolation of the receiver. RF power reflected at the antenna is absorbed by the internal loads. The three circulators are capable of providing approximately 65 dB of isolation in the −40° to +75°C temperature range. This amount of isolation is sufficient for receiver protection, even in case of complete RF breakdown at the antenna, for power levels of the order of 100 kW. Additional protection of approximately 20 dB can be obtained in case of RF breakdown by addition of a low-loss (0.1 dB), 10-dB ferrite limiter between the fixed and switchable circulators. The additional isolation of 10 dB is provided by the first switchable circulator as a result of power reduction from 100 to 10 kW and a corresponding increase of isolation from 21 to 31 dB (Fig. 7). For reception, the direction of circulation is reversed to permit flow of RF signals from the antenna directly to the receiver.

The switching energy for the circulator, not including driver losses, is approximately 30 μJ. The circulator can be switched in a fraction of a microsecond.

Conclusions

This paper has described the performance and design parameters for a latching ring-and-post waveguide ferrite circulator. A *C*-band circulator has provided 20-dB isolation bandwidths from 10 to 17 percent, an insertion loss of 0.35 dB, low-loss operation up to a pulsed power of 7.5 kW, and breakdown-free operation up to a pulsed power of at least 100 kW, in conjunction with an isolation of 21 dB. For the narrower bandwidth (10 percent) and higher isolation (50 dB) condition at room temperature, the 20-dB isolation bandwidth was 7.8 percent across the −40°C to +75°C temperature range.

Maximum bandwidth was obtained when the normalized radii for the post and ring were approximately 1.5 and 2.34, respectively, the ratio of remanence-to-frequency was approximately 0.12 gauss per megahertz, and the waveguide impedance was about 600 ohms at the center frequency. The switchable circulator is suitable for transmitting–receiving applications in high-reliability modern radars.

References

[1] P. C. Goodman, "A latching ferrite junction circulator for phased array switching applications," *1965 G-MTT Symp. Dig.*, pp. 123–126.

[2] W. C. Passaro and J. W. McManus, "A 35 GHz latching switch," *1966 G-MTT Symp. Dig.*, pp. 270–273.

[3] L. E. Davis, "Computed phase shift and performance of a latching 3-port waveguide circulator," *NEREM Rec.*, p. 96, 1966.

[4] W. W. Siekanowicz and W. A. Schilling, "A new type of latching switchable ferrite junction circulator," *Proc. IEEE* (Letters), vol 55, pp. 407–408, March 1967.

[5] ——, "A new type of latching switchable ferrite junction circulator," *IEEE Trans. Microwave Theory and Techniques*, vol. MTT-16, pp. 177–183, March 1968.

[6] F. Betts, D. H. Temme, and J. A. Weiss, "A switching circulator: *S*-band; stripline; remanent; 10 kW; 10 microsecond; temperature-stable," *1966 G-MTT Symp. Dig.*, pp. 275–279.

[7] J. W. Simon, W. K. Alverson, and J. E. Pippin, "Broadband latching waveguide circulator," *1967 G-MTT Symp. Dig.*, pp. 81–84.

[8] J. E. Pippin, "Microwave ferrite devices—1968," *Microwave J.*, vol. 11, pp. 29–55, April 1968.

[9] L. E. Davis and J. R. Bacon, "Latched 3-port waveguide circulators," *Electron. Letters*, vol. 4, pp. 490–492, November 1968.

[10] B. Lax and K. J. Button, *Microwave Ferrites and Ferrimagnetics.* New York: McGraw-Hill, 1962, pp. 150–151.

[11] C. E. Fay and R. L. Comstock, "Operation of the ferrite junction circulator," *IEEE Trans. Microwave Theory and Techniques*, vol. MTT-13, pp. 15–17, January 1965. (See p. 18, (2)–(4).)

[12] B. Lax and K. J. Button, see [10], p. 304, (7)–(20).

[13] W. C. Passaro and J. W. McManus, "A 35-GHz latching switch," *1966 G-MTT Internatl. Symp. Dig.*, pp. 270–273; also *IEEE Trans. Microwave Theory and Techniques*, vol. MTT-14, pp. 669–672, December 1966.

Microstrip Junction Circulator for Microwave Integrated Circuits

This author has reported recently on two forms of microstrip junction circulators.[1],[2] In the first type, a garnet disk was cemented into a ceramic substrate to form a composite ceramic substrate. Copper ground plane, quarterwave microstrip transmission lines, and disk were evaporated onto this substrate following a chrome flash. In the second form, the microstrip junction circulator was fabricated from a single garnet subtrate. The lines, ground plane, and disk were evaporated exactly as in the first case. The results were published as preliminary feasibility test data.[2] This all-garnet substrate circulator was identical in size with the composite ceramic substrate. The results were sufficiently promising to warrant a new smaller design size for the all-garnet microstrip junction circulator because the dielectric constant of the garnet substrate was larger than that of the ceramic.

Fig. 1 is a photograph of the revised x-band all-garnet microstrip junction circulator. The substrate is 0.023 inch thick with a copper disk center 0.21 inch in diameter. A small platinum–cobalt magnet, 0.1 inch long and 0.3 inch diameter, is located on the underside of the substrate ground plane. OSM connectors (and transition to type N) are used to permit testing of the device. The insertion loss data presented in this correspondence include the loss in the microstrip to OSM transition. The microstrip transmission lines are approximately a quarter wavelength long at 9.0 GHz, as calculated from design curves presented in a recent report.[3] The cylindrical inner conductor of the OSM connector is taken down to a rectangular cross section of approximately 0.010 inch thickness and 0.010 to 0.025 inch width (depending on the transmission linewidth). The inner conductor of the OSM connector extends 0.010 inch from the connector and makes a press contact with the copper line on the substrate.

Fig. 1. Photograph of an X-band ferrimagnetic substrate microstrip junction circulator (garnet).

Figs. 2 and 3 are graphs of early data obtained with the smaller size all-garnet junction circulator. Fig. 2 represents data taken with an electromagnet. The 20 dB isolation bandwidth is almost 13 percent with an insertion loss less than, or equal to, 0.45 dB over this entire band. The minimum insertion loss of 0.15 dB occurs at the frequency for maximum isolation. In Fig. 3 a broadbanding effect with the small platinum–cobalt magnet is evident. The 20 dB isolation bandwidth is about 20 percent while the insertion loss over this band is less than, or equal to, 0.45 dB with the exception of the lower frequency band edge where it rises to 0.6 dB and then drops. The minimum insertion loss is 0.15 dB and occurs at about the same frequency as with the electromagnet biasing field. However, the peak isolation frequency is shifted downward. The nonuniform dc magnetic biasing field from the platinum–cobalt magnet appears to stagger-tune the junction region, and hence, provides the improved bandwidth. This effect was observed with the early larger all-garnet circulator.[2] Some or all of this tuning effect must take place in the quarterwave transformer section on the magnetic substrate because the broadbanding phenomenon was not evident in the data on the composite ceramic substrate.[1] The increase in insertion loss at the low-frequency end was removed when the inner tab of the OSM connector was carefully machined flat. In fact, under these conditions 20 dB isolation bandwidths of 28 percent were obtained with an insertion loss less than 0.5 dB over this band.

Manuscript received June 7, 1967; revised July 31, 1967.

The previous data on the circulator were taken with about 10 mW of CW power at the circulator's input port. Following these results, some high-power tests were run to determine the power level at which nonlinear behavior occurs in the ferrimagnetic material. A 2-watt CW source was connected to the circulator and no nonlinear behavior was observed. A pulsed magnetron source with a 0.001 duty cycle and 2.5 μs pulsewidth was connected to the circulator. At a peak power of 956 watts no nonlinearities were observed. Additional tests at higher power levels are necessary to determine the onset of nonlinear behavior.

These promising results encouraged additional work as to the feasibility of a latched microstrip junction circulator fabricated completely from a ferrimagnetic material. A latched circulator does not require an external dc magnetic field. Operation of the circulator is at the remanent magnetization of the ferrimagnetic material. A loop of wire positioned in the material is pulsed so that the enclosed magnetic material is magnetically saturated. The saturation magnetization falls to the remanent magnetization value (of the hysteresis loop for the material) when the pulse is turned off, if a closed magnetic flux path is provided, i.e., energy is not required for the formation of magnetic poles at surfaces with magnetic field discontinuities. If the pulse of opposite sign is applied, the remanent magnetization of the material is oppositely directed. The two remanent states of the material permit switching the direction of circulation in a circulator, thereby providing a method by which the device can be used as a switch or a digital nonreciprocal phase shifter.

Preliminary experiments on a latched two-piece microstrip junction circulator were initiated. Fig. 4 is a photograph of the configuration employed. The two ferrimagnetic pieces are shown on the left-hand side of the photograph, while the component on the right-hand side is the assembled unit. The white ring visible in the part at the far left is an annulus 0.025 inch thick which is filled with a low-loss dielectric material whose dielectric constant

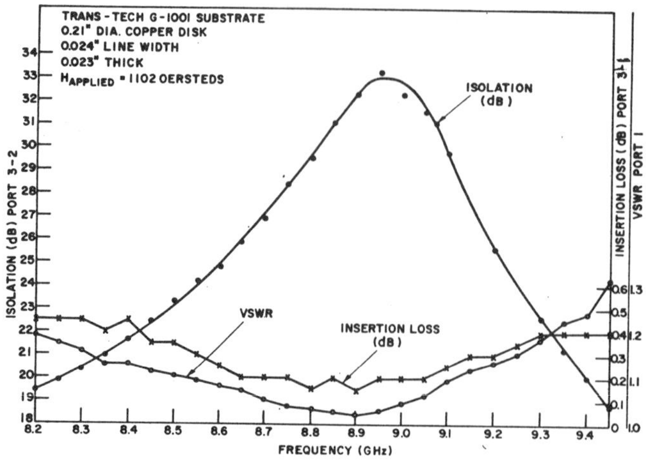

Fig. 2. Ferrimagnetic substrate microstrip junction circulator characteristics (electromagnet).

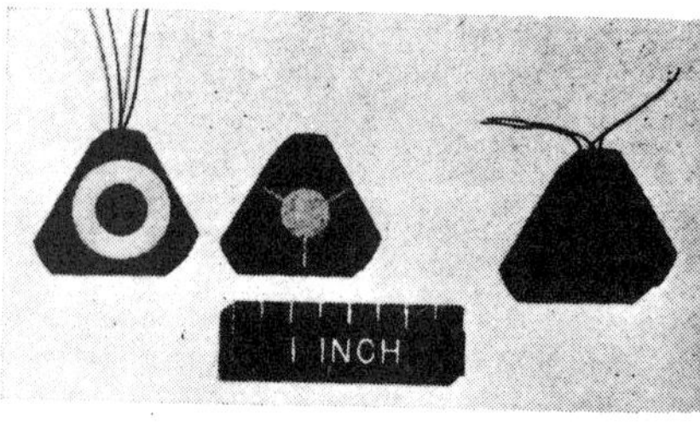

Fig. 4. Photograph of a two-piece latched microstrip junction circulator. (Right side of photograph is assembled circulator.)

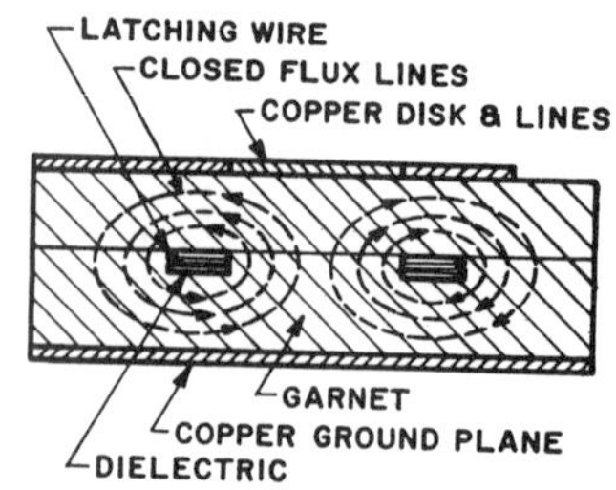

Fig. 5. Schematic cross-sectional view of latched circulator.

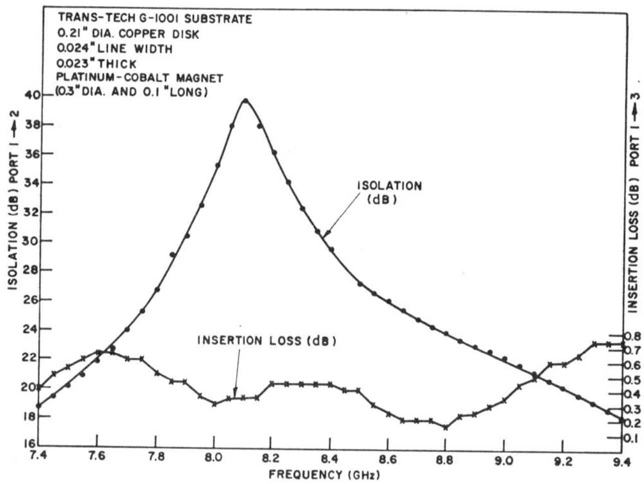

Fig. 3. Ferrimagnetic substrate microstrip junction circulator characteristics (permanent magnet).

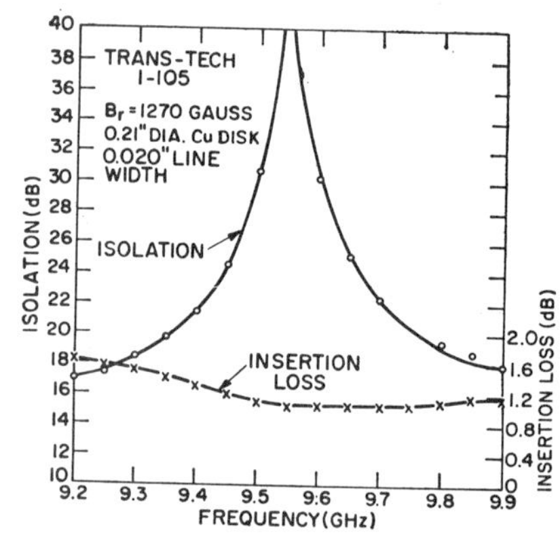

Fig. 6. Two-piece latched microstrip junction circulator characteristics.

matches the ferrimagnetic material. The dielectric is strictly for convenience because it eliminates the need to modify the quarter-wave linewidths for a proper impedance match in the annular region. A small lip on the underside of the dielectric holds the loop of wires used for the current pulse. Fig. 5 is a schematic cross-sectional view of the latched circulator showing the closed flux paths in the medium. Measurements of the latching energy required gave 0.8 μJ. This value is of the order of the experimental error. However, it is safe to assume that only a "few" microjoules of latching energy are required.

The preliminary experimental results established the feasibility of a latched microstrip junction circulator. Fig. 6 is a graph of the data obtained. A 350 MHz bandwidth for 20 dB isolation was measured, with an insertion loss varying from 1.15 to 1.3 dB over the band. This loss is considerably higher than that obtained with the thin ferrimagnetic substrates and small permanent magnets. Some of the difficulty lies in the thickness of the latched circulator. Its height was greater than a quarter of a guide wavelength and radiation was detected. A reduction in the circulator thickness reduced the insertion loss to 0.8 dB. There was still some radiation loss but this could only account for about 0.1 dB of the loss. The additional loss does not arise from operation at remanence because the loss was unchanged when the ferrimagnetic circulator was magnetically saturated in a uniform magnetic field. It is not obvious that the present geometry can give rise to low k-value (inverse wavelength) spin-wave excitation in the spin-wave manifold. However, this does remain as a possible explanation for the loss. The nature of this increased loss is under investigation at the present time.

These preliminary results were sufficiently encouraging to warrant an investigation as to the prospects for a one-piece latched circulator. The approach is almost identical with the RCA laminated memory sheet process[4] which uses the "doctor blading" technique to prepare sheets of sintered ferrimagnetic material which are then pressed together in a die. The latching wire is embedded in the center of the ferrimagnetic material layers by using a die to emboss a loop in the center of the laminated sheets. A platinum–sugar mixture adjusted for the shrinkage of the ferrimagnetic material is placed in the embossed region. The remaining sheets are placed on top, pressed, and fired. The resultant ferrimagnetic material is uniform and isotropic with no evidence of laminations and contains the platinum loop of wire for latching. Sheets 12 by 4 inches can be fabricated with wire loops in place. Evaporation of copper disks, lines, and ground plane then provide a sheet of latched circulators which can be cut off into individual circulators or hooked together via evaporation of connecting lines to form a large number of switchable circulators. It is possible to modify the copper disk shape[5] so as to construct a multibit digital nonreciprocal phase shifter for phased array systems.

The first one-piece latched microstrip junction circulator was identical in appearance with the photograph of the composite two-piece latched circulators shown on the far right side of Fig. 4. The platinum–sugar mixture was not used because an embossed die of the desired shape was not available. A 0.010 inch loop of platinum wire was placed on the sheets in the die and the remaining sheets were placed on top and pressed. After the final firing, evidence of shrinkage effects were observed as cracks in the vicinity of the platinum wire emerging from the specimen. Nevertheless, the experimental results established feasibility and, while far from outstanding, did substantiate the fact that the approach warrants further effort. The maximum isolation was 12 dB with a 10 dB isolation bandwidth of 350 MHz. The insertion loss was high, ranging from 2.0 to 5.0 dB over the band. However, a number of factors can account for this loss. The specimen was thicker than a quarter of a guide wavelength and the associated radiation loss was evident. The cracks in the material also contribute to the insertion loss. Furthermore, a thin specimen of the laminated ferrimagnetic material was prepared with size and shape identical with that shown in Fig. 1. This circulator was tested in an electromagnet where it showed a high insertion loss. The ferrimagnetic material was yttrium–iron–garnet, and it was a poor microwave material when compared to the commercially available materials. In all fairness it should be noted that this was the first attempt at RCA to prepare YIG material in laminated sheets since all previous experience has been with ferrite materials for laminated memory sheets. It is a reasonable assumption that a materials study would lead to materials comparable to commercially available material. However, before this is done, the insertion loss versus thickness problem previously discussed must be resolved. Following this, a laminated garnet materials study can be launched, or ferrite laminates could be used. The laminated technique is of significant importance because it provides an inexpensive means for batch fabrication of latched circulators. These circulators would find wide use as individual elements, or as inexpensive digital nonreciprocal phase shifters for phased array systems.[5]

This author's present work, and that of others at Syracuse University[6] where ferrimagnetic substrates have been used to build latched nonreciprocal devices, has established the role of these devices in microwave integrated circuits. Recent work at RCA Laboratories on new magnetic materials, chalcogenides,[7] has expanded this prospective role. Ferromagnetic insulators which can be locally doped to yield semiconductor materials with high mobilities and high carrier densities have been developed. It may be possible to prepare substrates of these materials, dope them locally to provide active devices such as amplifiers and oscillators, and then use the undoped portion of the substrate for passive nonreciprocal (and reciprocal) devices. Operation latched or with a permanent magnet should be feasible. These advances with both materials and devices make the prospect of microwave systems on magnetic substrates a distinct possibility.

ACKNOWLEDGMENT

The author wishes to acknowledge the help of H. L. Davis who assisted with the testing and was responsible for expediting and fabricating many of the components; and to C. Wentworth for his fabrication of the laminated circulator, and C. Horak, Sr., and R. E. Harwood who were invaluable in providing both precision work and useful advice for the design of various component parts.

B. HERSHENOV
RCA Labs.
Princeton, N. J.

REFERENCES

[1] B. Hershenov, "*X*-band microstrip circulator," *Proc. IEEE* (*Letters*), vol. 54, pp. 2022–2023, December 1966.

[2] ——, "All-garnet-substrate microstrip circulators," *Proc. IEEE* (*Letters*), vol. 55, pp. 696–697, May 1967.

[3] M. Caulton, J. J. Hughes, and H. Sobol, "Measurements on the properties of microstrip transmission lines for microwave integrated circuits," *RCA Rev.*, vol. 27, no. 3, September 1966.

[4] R. Shahbender, C. Wentworth, K. Li, S. Hotchkiss, and J. A. Rajchman, "Laminated ferrite memory," AFIPS, vol. 24, Fall Joint Computer Conference, 1963.

[5] "Development of a fast switching, high power digital phase-shifter at *X*-band," Western Microwave Labs., Santa Clara, Calif. Air R&D Com., Rome Air Develop. Ctr., Final Rept. Contract AF30(602)-3496, June 1965.

[6] G. T. Roome and H. A. Hair, "Thin ferrites for integrated microwave device," Special Projects Lab., Syracuse University Research Corp., Syracuse, N. Y., 4th Quart. Tech. Rept., Contract AF33(615)-3332, February 1967.

[7] H. W. Lehman and G. Harbeke, "Semiconducting and optical properties of ferromagnetic $CdCr_2S_4$ and $CdCr_2Se_4$," presented at 12th Annual Conf. on Magnetism and Magnetic Materials, November 1966.

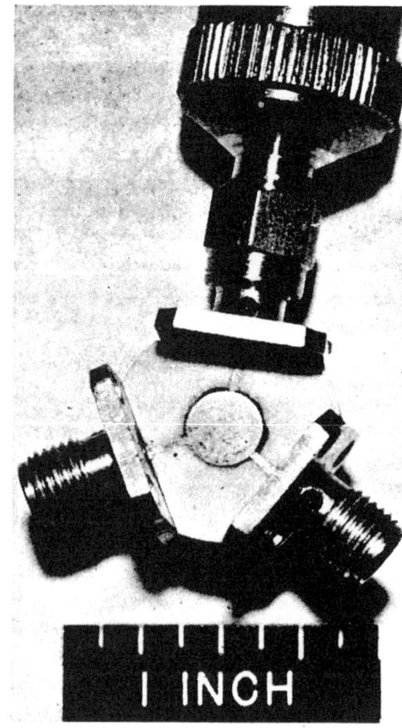

Fig. 1. Photograph of *X*-band microstrip circulator.

X-Band Microstrip Circulator

Recent progress in the fabrication of integrated microwave circuits has made the assembly of fairly sophisticated circuits possible. Some of these circuits require circulators. A ferrimagnetic junction microstrip circulator would provide an ideal compact circuit. At the present time, the author is unaware of any published data on microstrip circulators, although the literature is filled with results on Triplate and waveguide circulators. A feasibility study as to the practicality of such a device was initiated and preliminary results indicate that the device works extremely well and can be incorporated into microwave integrated circuits.

Figure 1 illustrates one such circulator. The OSM connectors (and transitions to type N) are used only to permit testing of the device. In practice the circulator would be joined directly to other components with microstrip line connections. The insertion loss presented in the data to follow does not include the loss of the OSM-to-type-N connectors. The size of the circulator, shown in Fig. 1, is made larger to provide mounting space for the connector. The inner conductor of the OSM connector extends 0.010 inch from the connector and is taken down to a 0.010 inch thickness. This portion of the conductor makes a press-contact with the evaporated copper lines on the substrate. The substrate itself is a 0.023 inch thick, high-purity alumina with a hole in the center into which is epoxied a Trans-Tech G-1001 garnet disc, 0.23 inch in diameter and 0.023 inch thick. The microstrip lines and copper-coating on the garnet are produced by copper evaporation following a chrome flash. The underside of the garnet and substrate which rest on a brass ground plane are completely coated with copper.

Figure 2 is a plot of the data obtained using an electromagnet. The isolation is greater than 20 dB over a 9.7 percent bandwidth, while the insertion loss is about 0.4 dB or less over most of this band. No attempt has been made at this time to maximize the bandwidth by optimizing the microstrip line dimensions or the contact transitions to the OSM connectors.

Another test was made with an Indox VI permanent magnet, 0.3

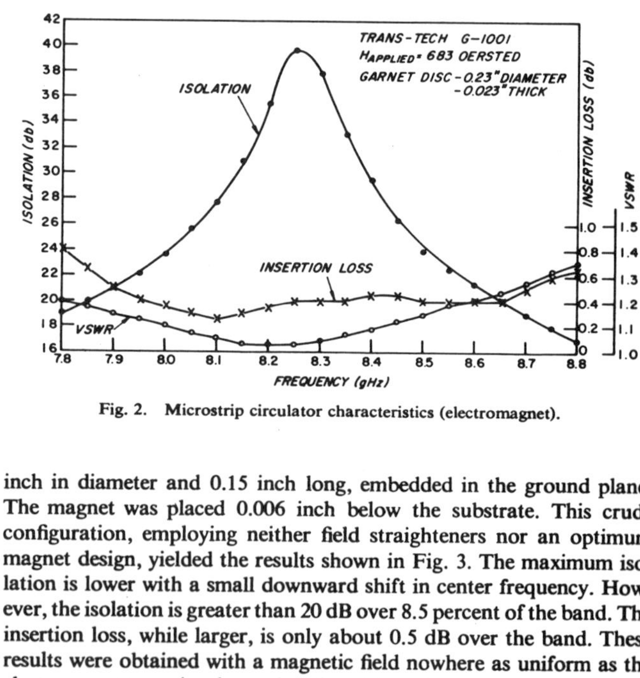

Fig. 2. Microstrip circulator characteristics (electromagnet).

inch in diameter and 0.15 inch long, embedded in the ground plane. The magnet was placed 0.006 inch below the substrate. This crude configuration, employing neither field straighteners nor an optimum magnet design, yielded the results shown in Fig. 3. The maximum isolation is lower with a small downward shift in center frequency. However, the isolation is greater than 20 dB over 8.5 percent of the band. The insertion loss, while larger, is only about 0.5 dB over the band. These results were obtained with a magnetic field nowhere as uniform as the electromagnet previously used. It is evident that an improved magnet design, with the magnet located at the substrate surface, would provide better field uniformity and, hence, lower insertion loss and higher isolation. The present package, magnet and substrate, is only 0.179 inch high. Different permanent magnets would permit a further decrease in the package height.

These preliminary results do show that a symmetrical three-part microstrip circulator is compatible with microwave integrated circuits. Improvements similar to those applied to Triplate circulators should further increase the bandwidth and reduce the insertion loss. The present device, just as is, is significantly smaller and lighter than present day x-band circulators.

Manuscript received October 3, 1966.

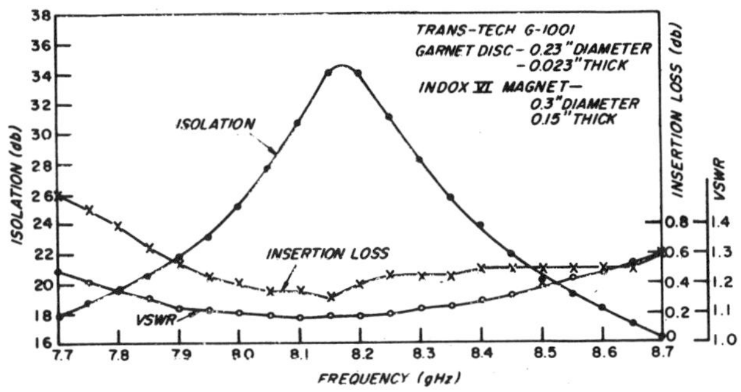

Fig. 3. Microstrip circulator characteristics (permanent magnet).

B. Hershenov
RCA Labs.
Princeton, N. J.

Reprinted from the PROCEEDINGS OF THE IEEE
VOL. 54, NO. 12, DECEMBER, 1966
pp. 2022-2023

V-3. NEW DESIGN TECHNIQUES FOR MINIATURE VHF CIRCULATORS

V. E. Dunn and R. W. Roberts

Melabs, Palo Alto, California

Since the feasibility of Y junction circulators was first reported in 1959, such circulators have been built to operate in frequency bands from 100 mc to the millimeter wavelength range. All of these circulators appear to operate on essentially the same principle. Using this approach the ferrite disc diameter is proportional to the wavelength. While junction circulators have been useful at microwave frequencies because of their simplicity and small size, such circulators at UHF and VHF frequencies have been expensive and inconveniently large because of the large ferrite volume required. This paper reports the development of a new type of circulator for the VHF and UHF bands. The approach utilizes the lumped element techniques which are natural and convenient for this frequency range and as a result the size of the circulator, and ferrite volume, do not increase with the wavelength. Lumped element circulators have also been developed recently in Japan by Konishi (Reference 1). Figure 1 illustrates the size reduction which has been obtained with the new technique.

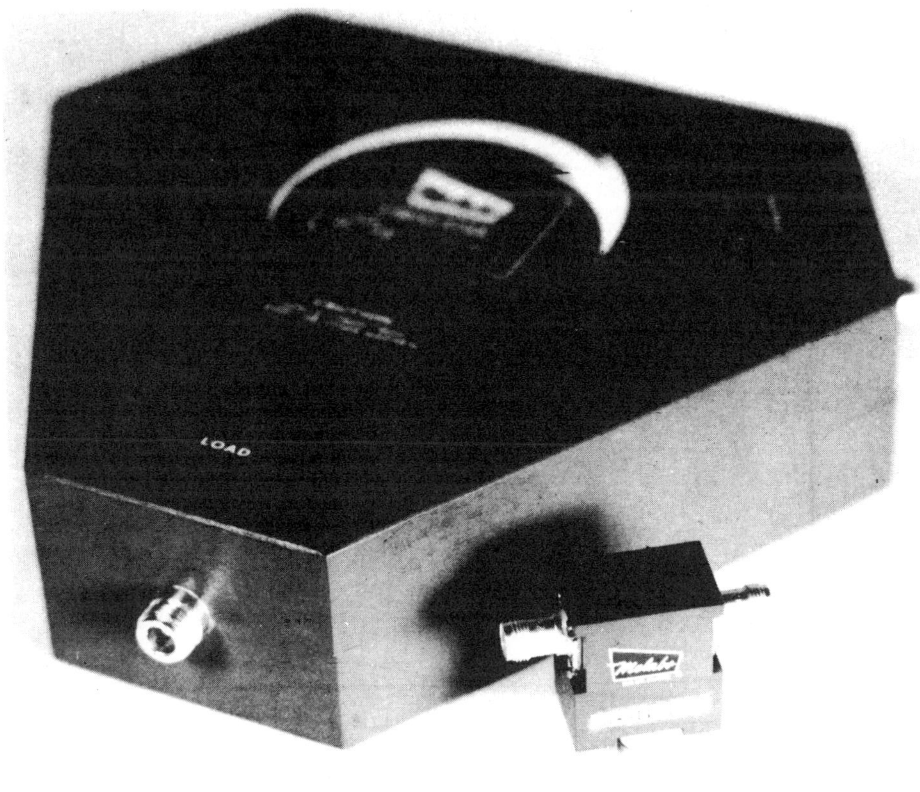

Figure 1. Conventional and Lumped Element 150 mc Circulators

The lumped, nonreciprocal element in the circulator consists of a ferrite disc with three coils wound on it so that the magnetic fields of the coils are oriented at 120 degrees with respect to each other. A d-c field sufficient in strength to bias the ferrite above resonance (i.e., ferromagnetic resonance frequency greater than the operating frequency of the circulator) is applied normal to the plane of the disc. This symmetrical but nonreciprocal element can be used to form a circulator by connecting capacitors either in series or shunt with the load and source impedances as shown in Figure 2. The analysis presented here will consider only the series case. The shunt case can be treated in an analogous manner.

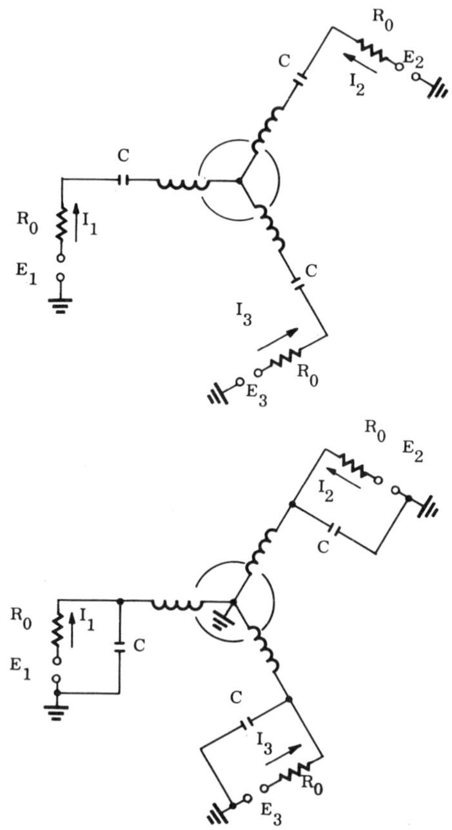

Figure 2. Schematic of Series and Shunt Tuned Circulators

The circulator can be analyzed by expressing E_1, E_2, and E_3 and I_1, I_2, and I_3 in terms of the symmetrical components, e.g.:

$$E_1 = E^o + E^+ + E^- \qquad I_1 = I^o + I^+ + I^-$$

$$E_2 = E^o + a^2E^+ + aE^- \qquad I_2 = I^o + a^2I^+ + aI^- \tag{1}$$

$$E_3 = E^o + aE^+ + a^2E^- \qquad I_3 = I^o + aI^+ + a^2I^-$$

$$a = e^{j2\pi/3}$$

For this case I^o is equal to zero.

To a good approximation the magnetic fields in the ferrite can be idealized by considering them to be circularly polarized for the plus and minus sequence voltages and currents. Then, for the case of the series capacitance:

$$\frac{E\pm}{I\pm} = R_o + j\left(\omega L\pm - \frac{1}{\omega C}\right). \tag{2}$$

$L\pm$ is proportional to the scalar permeabilities of the ferrite:

$$L\pm = L_o \mu\pm = L_o \left(1 + \frac{\omega_m}{\omega_o \mp \omega}\right) \tag{3}$$

$$\omega_m = \gamma 4\pi M_s$$

$$\omega_o = \gamma H_{int}$$

ω = angular signal frequency.

Application of the requirements for an ideal circulator at frequency ω, (i.e.: $E_2 = E_3 = I_3 = 0$) leads to the following conditions for circulation:

$$\frac{\omega}{2}\left(L^+ - L^-\right) = \frac{R_o}{\sqrt{3}} \tag{4}$$

$$\frac{\omega}{2}\left(L^+ + L^-\right) = \frac{1}{\omega C}. \tag{5}$$

Note that the inductance rather than the ferrite volume as such is the significant parameter. Equivalent performance can be obtained with a small ferrite disc with many turns in the coil around it and with a large disc with only one turn. Equations (4) and (5) do not specify the design uniquely. For example, with a given frequency, impedance level, and $4\pi M_s$ a large value of L_o can be used with the ferrite biased far from resonance ($\omega_o >> \omega$); or at the other extreme, a small value of L_o can be used with the ferrite biased close to resonance (ω_o only slightly greater than ω). This choice is not arbitrary, however, since the proximity to resonance determines the bandwidth and insertion loss of the circulator.

The input impedance and bandwidth of the circulator can be expressed simply if the frequency variation of μ^+ and μ^- is neglected. The input impedance is then that of a series resonant circuit in series with R_o. The bandwidth for 20 db isolation is given by:

$$\frac{\Delta\omega_{20\,db}}{\omega} = 0.2\sqrt{3}\,\frac{\mu^+ - \mu^-}{\mu^+ + \mu^-} = 0.2\sqrt{3}\,\frac{\omega_m \omega}{\omega_o^2 - \omega^2 + \omega_o \omega_m} \tag{6}$$

As a result the bandwidth increases as the bias approaches resonance. A more accurate analysis includes the frequency variation of the permeability, the effect of which is to decrease slightly the theoretical bandwidth, particularly close to resonance.

The insertion loss due to ferrite losses can be estimated by assuming the idealized Lorentzian resonance curve for the ferrite. The result in terms of the 3 db linewidth is:

$$L = 20 \log_{10}\left[\frac{4\left(\frac{\omega_o^2}{\omega^2} - 1\right) + \frac{3\gamma\Delta H}{\sqrt{3}\omega}\left(\frac{\omega_o^2}{\omega^2} + 1\right)}{4\left(\frac{\omega_o^2}{\omega^2} - 1\right) + \frac{\gamma\Delta H}{\sqrt{3}\omega}\left(\frac{\omega_o^2}{\omega^2} + 1\right)}\right] \tag{7}$$

As would be expected the loss increases as the linewidth increases; as the ferrite becomes magnetized closer to resonance; and, if $\frac{\omega o}{\omega}$ is held constant, as ω decreases.

It is seen that for this basic single tuned circulator, wide bandwidth can be obtained only at the expense of increased insertion loss, by biasing close to resonance. An alternative is the use of additional resonant circuits on each port to partially cancel the frequency variation of the single tuned circulator as shown in Figure 3.

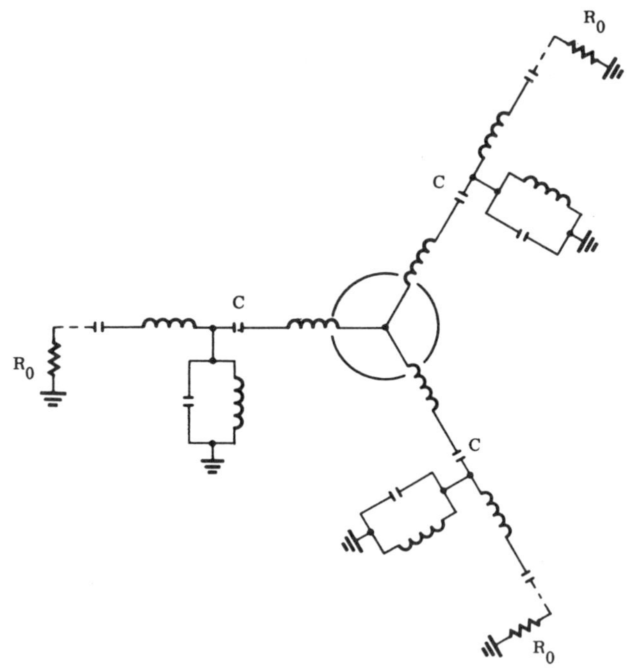

Figure 3. Schematic of a Multiple Tuned Circulator

198

Lumped element circulators have been operated over the 35 to 1000 mc range. For a given percentage bandwidth the insertion loss increases as the frequency decreases. With a 4 percent 20 db bandwidth the minimum insertion loss varies from about 0.3 db at 400 mc to 0.8 db at 100 mc. Figure 4 is a photograph of a single tuned circulator. Figure 5 compares the performance of a double tuned and single tuned circulator. The minimum loss is degraded slightly by the losses of the additional resonant circuits.

REFERENCE

1. Konishi, Y., "Lumped Element Y Circulator," International Conference on Microwaves, Circuit Theory, and Information Theory, Tokyo, Japan, September, 1964.

Figure 4. Photograph of a Single Tuned Lumped Element Circulator

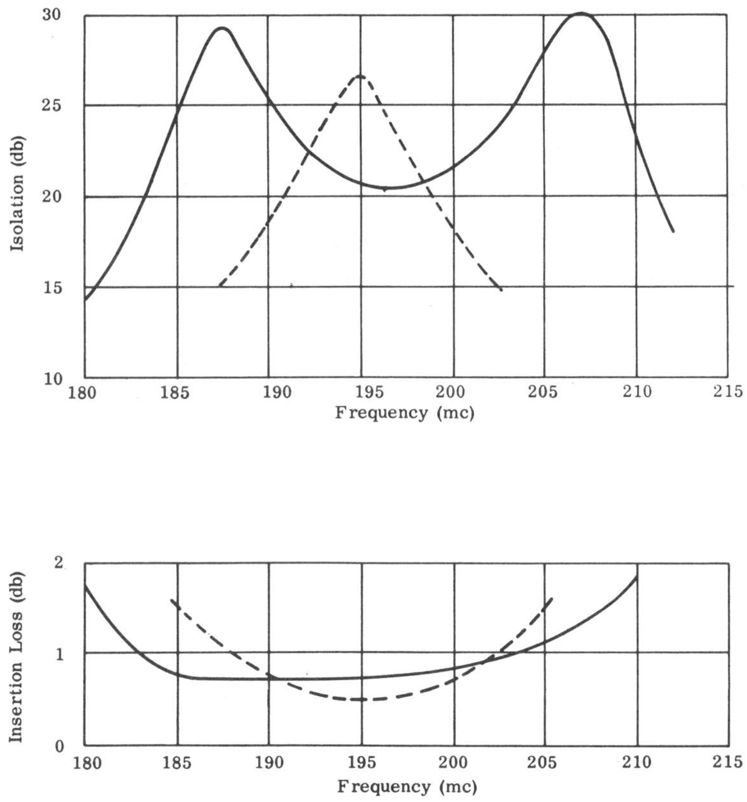

Figure 5. Performance of a Double Tuned and a Single Tuned Circulator

Lumped Element *Y* Circulator

YOSHIHIRO KONISHI, SENIOR MEMBER, IEEE

Abstract—A lumped element *Y* circulator using ferrite, having a mesh mechanism in place of the center conductor of an ordinary *Y* strip line circulator, is proposed. A theory is developed relating to bandwidth, insertion loss, and temperature dependence of the reactive energy and the tensor permeability of ferrite, and the bandwidth enlargement is discussed on the basis of the equivalent network.

The size of the circulator is approximately 1∼2 cm in diameter for VHF and UHF bands and the characteristics are, for example, about 50 percent bandwidth for 20-dB isolation, 2-dB insertion loss at 600-Mc/s center frequency. It is also shown that the insertion loss has its minimum value at a definite bandwidth and, as a practical example, a circulator is described with 0.25 dB of insertion loss and 6.5 percent bandwidth at 700 Mc/s.

I. Introduction

The ordinary *Y* strip-line circulator [1]–[3] requires a ferrite with suitable diameter to keep its wide bandwidth and lossless characteristics, and recently, a miniaturized circulator available to the receiver has been required for the purpose of reducing the radiation of the local oscillator.

This paper presents the theory to synthesize the lumped element *Y* circulator having the proposed mechanism [4] and also the experimental results to confirm the theory.

This circulator is approximately 2∼3 cm in diameter as shown on the photograph and the ferrite used is about 1∼1.5 cm in diameter and 1∼2 mm thick.

The center frequency of the circulator is adjustable by changing only the values of capacitances added at each terminal without changing the applied dc field and also internal mechanism around the ferrite. The experiment is made in the frequency range from about 10 Mc/s∼1000 Mc/s.

The theories are given especially for the relationship between a bandwidth, forward loss, and reactive energy corresponding to the eigenvector excitations.

These quantities are also connected to the complex tensor permeability of the ferrite to make it possible for the characteristics of the circulator to be estimated by the material used in it. Furthermore, the equivalent network of this circulator is obtained and the wide-band circulator with Chebyshev and Wagner characteristics is synthesized in the same way as the band-pass filter design, together with its previously mentioned equivalent network. As a practical example, the circulator with about 50 percent in bandwidth for 20 dB isolation at a center frequency 600 Mc/s is designed, and a small insertion loss circulator, one with 0.25 dB insertion loss at a center frequency 700 Mc/s is also shown.

The lumped element circulator has also been developed in the U.S.A. by V. Dunn and R. W. Robert.[1]

II. Construction and its Eigenvalues

Figure 1 shows the construction of the inside of the circulator.

Three conductors connected to corresponding terminal of the circulator are insulated from each other and knitted together to form a mesh and are connected to earth at the other end. The mesh mechanism contributes to maintain circularly rotating *rf* field over the entire disk for the rotating field excitation, and since most of *rf* magnetic field passes through two ferrite disks as shown in Fig. 2, the leakage inductance, which prevents the circulator's bandwidth from becoming broad, can be avoided. The symmetrical structure about the mesh plane can keep the radiation loss small.

This rotationally symmetrical circuit has the eigenvector of (1) [5], and its eigenvalues Z_i' of Z matrix can be expressed in terms of the magnetic and electric energy for the unit current excitation corresponding to the ith eigenvector. (See Appendix I.)

$$u_1 = \begin{bmatrix} 1 \\ 1 \\ 1 \end{bmatrix}, \qquad u_2 = \begin{bmatrix} 1 \\ \alpha^2 \\ \alpha \end{bmatrix}, \qquad u_3 = \begin{bmatrix} 1 \\ \alpha \\ \alpha^2 \end{bmatrix}. \tag{1}$$

However, when equal current flows only in conductors $1a$ and $1b$ in Fig. 2, the *rf* magnetic field surrounds $1a$ and $1b$ through the mesh between $2a$, $2b$, $3a$, and $3b$ as in Fig. 2(b), and is distributed almost uniformly on the ferrite surface as shown in Fig. 2(a). Simplifying the analysis, we take the simple assumption that the distribution of the field is completely parallel with the direction perpendicular to $1a$ and $1b$, although there exist the other higher modes to satisfy the boundary condition. We also assume the diameter is sufficiently small to be able to neglect the variation of the field intensity along $1a$ and $1b$, so that we can neglect the electric energy inside the ferrite compared with the magnetic energy. Under the assumption mentioned before, Z_i' in the part of ferrite without C added at each terminal can be expressed by (2) from the result in Appendix I.

Manuscript received February 11, 1965; revised August 9, 1965.
The author is with Technical Research Labs., Japan Broadcasting Co., Tokyo, Japan.

[1] See their article, "Miniature VHF-UHF circulators use lumped-element design," *Microwave J.*, pp. 46–47, December 1963.

$$Z_i' = j\frac{\omega}{3}\iiint_\tau H_i^* \boldsymbol{\mu} H_i d\tau \qquad (2)$$

where

τ: the integrated region of ferrite

$\boldsymbol{\mu} = \begin{bmatrix} \mu & -jk & 0 \\ jk & \mu & 0 \\ 0 & 0 & 1 \end{bmatrix}$: tensor permeability of ferrite with the dc magnetic field in the Z direction

E_i, H_i: high frequency electric and magnetic fields in the ferrite when it is excited by constant current of ith eigenvector

ω: angular frequency.

The frequency variation of Z_i', $\delta Z_i'$ is (see Appendix II)

$$\delta Z_i' = j\frac{\delta\omega}{3}\iiint_\tau H_i^*\left(\boldsymbol{\mu} + \frac{\partial \boldsymbol{\mu}}{\partial \omega}\omega\right) H_i d\tau. \qquad (3)$$

In VHF and UHF circulators, the applied dc magnetic field should be much stronger than the resonance field in order to make it of low forward loss. In this case, we can apply the relation $(\partial\boldsymbol{\mu}/\partial\omega)\omega \ll \boldsymbol{\mu}$ in (3). Taking account of the assumption of the field uniformity, the 2nd and 3rd eigenvector currents give the complete negative and positive rotating field inside ferrite. So, we get

$$Z_i' = j\omega L_i, \qquad L_i = \mu_i \xi,$$

$$\xi = \frac{1}{3}\iiint_\tau |H_i|^2 d\tau, \qquad (\mu_2 = \mu_-, \mu_3 = \mu_+)$$

from (2) and (3) where ξ is independent of frequency.

The 1st eigenvector currents give rise to no magnetic field inside the ferrite. This results in $Z_1 = 0$, which has been confirmed by experiments, as in Fig. 4, to be satisfactory in the VHF-UHF region. To examine the value of μ_i, the values of Z_2' and Z_3' are measured for dc magnetic field of several strengths, and the measured values $Z_2'/j\omega\xi = \mu_-$ and $Z_3'/j\omega\xi = \mu_+$ are plotted with a sign of $\times$ in Fig. 3, where $j\omega\xi$ are the values of Z_2', Z_3' under the dc magnetic field strong enough to saturate the $\mu_\pm$ values to one. The measuring technique is shown in Appendix III. The line in Fig. 3 shows the calculated values of μ_+ and μ_- by Polder's equation (15), and it shows good coincidence with measured values. This makes it possible not only to substitute $\boldsymbol{\mu}$ by μ_+ and μ_-, at least in the discussion of their relative values, but also to use (15) for analysis in introducing the design formula.

Therefore, considering the capacities at each terminal, the eigenvalues are expressed in the equivalent admittances in

$$y_1 = \infty,\; y_2 = j\omega c + \frac{1}{j\omega\mu_-\xi},\; y_3 = j\omega c + \frac{1}{j\omega\mu_+\xi}. \qquad (4)$$

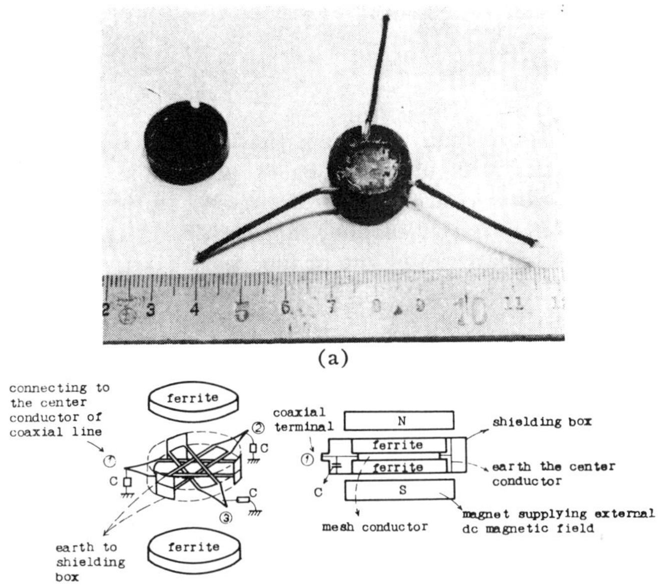

Fig. 1. (a). Photograph of lumped element Y circulator (scale indicates dimensions in cm). (b) Construction of proposed lumped element Y circulator. Left side—construction inside shielding box. Right side—sectional view in the plane including terminal ① and center axis.

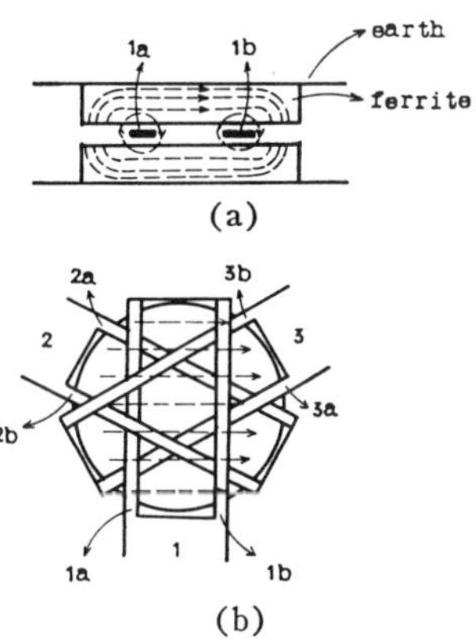

Fig. 2. Field distribution inside ferrite by currents on 1(a) and (b).

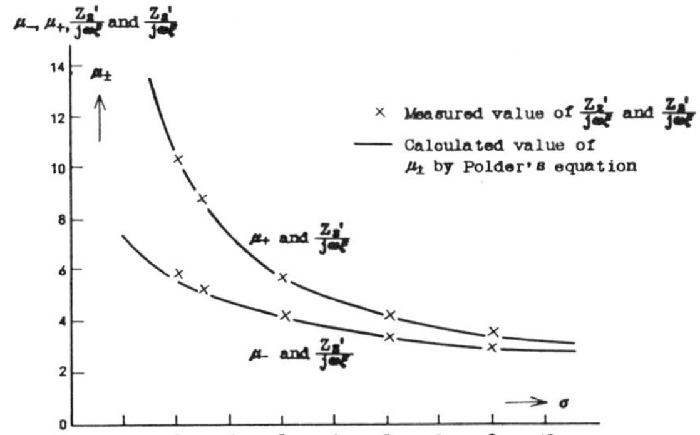

Fig. 3. Measured values of $Z_2'/j\omega\xi$ and $Z_3'/j\omega\xi$ at 150 Mc/s plotted by $\times$ signs, and calculated values of positive and negative polarized permeabilities $\mu_\pm$ shown by lines.

When

$$y_2 = -j\frac{1}{\sqrt{3}}, \qquad y_3 = j\frac{1}{\sqrt{3}},$$

a circulator is realized, where the terminal impedance is normalized to one ohm as simplification, and the normalization is also made throughout Section III.

The measured eigenvalues are shown on the Smith chart with an arrowed line in Fig. 4.

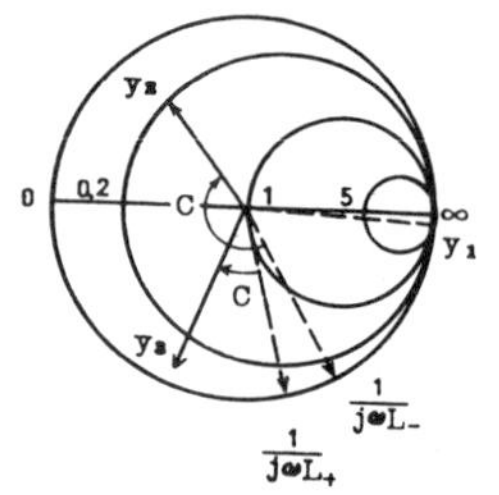

Fig. 4. Measured values of eigenvalues.
←— — before connecting C.
← after connecting C.

202

III. Electrical Characteristics

A. Bandwidth and Insertion Loss

The variation of eigenvalues δs_i takes place by the variation of frequency $\delta\omega$ from the center frequency, and the signal is transmitted also backward in the amount of $|S''|$ for the unit amplitude of the incident signal. The relation between $|S''|$ and $|\delta s_i|$ can be obtained as shown in

$$|S''| = \frac{1}{3}\sqrt{\sum_{i=1}^{3} |\delta s_i|^2 - \sum_{\substack{i,j=1 \\ i\neq j}}^{3} |\delta s_i|\cdot|\delta s_j|}\,. \qquad (5)$$

Nevertheless, there are relations of (6), between s_i, y_i and their frequency variations δs_i, δy_i;

$$s_i = \frac{1-y_i}{1+y_i}, \qquad \delta s_i = \frac{-\delta y_i}{(1+y_i)^2}. \qquad (6)$$

It is easily understood in Fig. 4 that (7) must be satisfied to realize a circulator.

$$\omega c = \frac{1}{2}\left(\frac{1}{\omega\xi\mu_-} + \frac{1}{\omega\xi\mu_+}\right), \qquad \frac{1}{\omega\xi\mu_-} - \frac{1}{\omega\xi\mu_-} = \frac{2}{\sqrt{3}}. \qquad (7)$$

Substituting (4) and (6) into (5), and applying the relation of (7) to it, $|S''|$ can be expressed in terms of C and $\mu_\pm$. It can be also connected to the time average total reactive energy $\tilde{W}_t$ stored inside the circulator in the case of constant voltage eigenvector u_2 and u_3 with unit amplitude.

The previously mentioned results, are shown in (8). (See Appendix IV.)

$$w = \frac{2|S''|}{\omega C\sqrt{1+\dfrac{1}{(2\omega C)^2}}} = \frac{2\sqrt{3}|S''|\eta}{\sqrt{1+\dfrac{3}{4}\eta^2}} = \frac{6|S''|}{\omega\tilde{W}_t\sqrt{1+\left(\dfrac{3}{2\omega\tilde{W}_t}\right)^2}}$$

$$\eta = \frac{\mu_+ - \mu_-}{\mu_+ + \mu_-}. \qquad (8)$$

In (8) w is the bandwidth normalized to the center frequency, at which the backward transmission is kept in the amount of desired values $|S''|$. Equation (8) shows that the bandwidth becomes narrower for the circulator with large C, when the required dc magnetic field should be stronger and ferrite operate at small η. In this operation, the corresponding reactive energy becomes larger.

Next, considering the quality factor of the permeability of the ferrite and the permittivity of C, the added capacitors at each terminal, we put

$$\mu_\pm = \mu_\pm'\left(1 - j\frac{1}{Q_\pm}\right), \qquad C = C'\left(1 - j\frac{1}{Q_c}\right). \qquad (9)$$

Substituting (9) into (4), we get the amount of forward loss as in (10)

$$L(\text{dB}) = \frac{4.96}{\eta Q_c} + \frac{2.48}{Q_+}\left(\frac{1}{\eta} - 1\right) + \frac{2.48}{Q_-}\left(\frac{1}{\eta} + 1\right). \qquad (10)$$

However, we have the relation between $\mu_\pm$ and μ_{eff} and the loss term of μ_{eff} is expressed by its quality factor Q_{eff} as in

$$\mu_{\text{eff}} = \mu_{\text{eff}}'\left(1 - \frac{1}{Q_{\text{eff}}}\right) = \frac{\mu^2 - \kappa^2}{\mu} = \frac{2}{\dfrac{1}{\mu_+} + \dfrac{1}{\mu_-}}. \qquad (11)$$

Substituting (9) into (11) one gets

$$\frac{1}{Q_{\text{eff}}} = \frac{\eta}{2}\left[\left(\frac{1}{\eta} - 1\right)\frac{1}{Q_+} + \left(\frac{1}{\eta} + 1\right)\frac{1}{Q_-}\right]. \qquad (12)$$

From (10) and (12), the insertion loss L (dB) can be expressed by the simpler form with Q_{eff} as in

$$L(\text{dB}) = \frac{4.96}{\eta}\left(\frac{1}{Q_c} + \frac{1}{Q_{\text{eff}}}\right). \qquad (13)$$

In (13), when dc magnetic field becomes stronger, Q_{eff} becomes larger and η smaller. So, L (dB) takes its minimum value at the proper strength of dc magnetic field, and this is determined by applying the measured values of Q_{eff} and η into (13) together with the value of Q_c, where η is obtainable from the measured values of μ_{eff} by the relation of

$$\eta = \frac{P}{\sigma(P+\sigma)-1}, \qquad \mu_{eff} = \frac{(\sigma+P)^2-1}{\sigma(\sigma+P)-1}. \tag{14}$$

The methods of the measurement of Q_{eff} and μ_{eff}, and the determination techniques are described in Appendix V. A practical example is shown in Section VII.

B. Adjustment of Center Frequency by Changing C

When the formula of (15) for μ_i is used,

$$\mu_i = 1 + \frac{P}{\sigma + (-1)^i}, \qquad (\mu_2 = \mu_-, \mu_3 = \mu_+)$$

$$\sigma = \frac{|\gamma| H_0}{\omega}, \qquad P = \frac{4\pi M_s |\gamma|}{\omega}, \tag{15}$$

where γ is gyro magnetic ratio and $4\pi M_s$ is saturation magnetization. The difference between both admittances of L_+ and L_- becomes independent of ω for $\sigma^2 \gg 1$, or $P^2 \gg 1$, where the former is satisfied in the case of strong dc field and the latter in the case of the low frequency as VHF band (Appendix VI).

Therefore, the circulator is always realized only by adjusting C proportional to ω^2 as in (16)

$$C = \frac{|\gamma| H_0}{\sqrt{3}\,\omega^2}\left(1 + \frac{H_0}{4\pi M_s}\right). \tag{16}$$

Substituting (16) into (10), the bandwidth w of this circulator is obtained in (17). It is understood that w becomes wide proportional to the center frequency and inversely proportional to σ.

$$w = \frac{2\sqrt{3}\,|S''|\,\omega}{\left(1 + \frac{H_0}{4\pi M_s}\right)|\gamma| H_0} = \frac{2\sqrt{3}\,|S''|}{\sigma\left(1 + \frac{\sigma}{P}\right)}. \tag{17}$$

Substituting (15) into (9), the value of ξ is obtained in

$$\omega\xi = \frac{\sqrt{3}}{2}\left(\frac{1}{\mu_-} - \frac{1}{\mu_+}\right) = \frac{\sqrt{3}\,P}{(\sigma+P)^2-1},$$

$$\sigma = \frac{|\gamma| H_0}{\omega}. \tag{18}$$

Therefore, the result of the measurement, in advance of ξ for several sizes of the mesh part, is available to decide the size of the mesh part and also ferrite disks required from (18). The measured values of ξ are indicated in Fig. 5, where t is the thickness of the ferrite disk.

While the previous description has been concerned with constant applied dc magnetic field, the situation with stronger dc magnetic field will be considered. In this case the larger value of C and the smaller value of ξ are required even at the same frequency and the bandwidth of w becomes smaller.

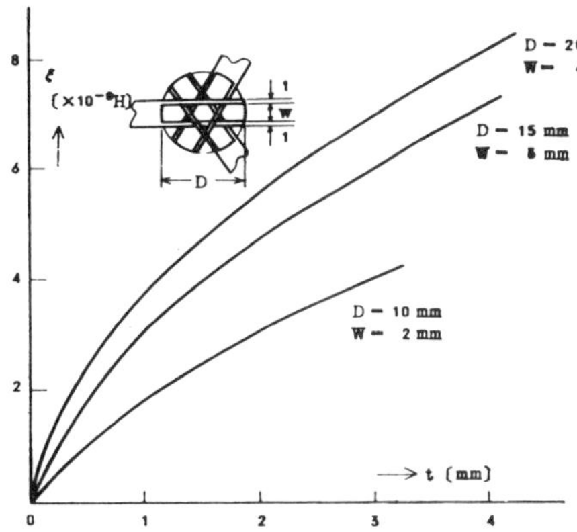

Fig. 5. Measured values of ξ of mesh part with several sizes, where ξ is the value $Z_i/j\omega$ when the strong dc magnetic field is applied to saturate $\mu_\pm$ to one. t is width of the part for the insertion of ferrite disks.

IV. The Synthesis of Wide-Band Lumped Element Y Circulator

A. Equivalent Circuit of Lumped Element Y Circulator

Taking the variation of y_i in (6) and applying (15) and its variation $\delta\mu_i$ we get the relation of

$$\delta y_i = j\delta\omega\left[C + \frac{1}{\omega^2\mu_i\xi} \cdot \left\{1 - \frac{(-1)^i P}{(\sigma+(-1)^i)(\sigma+P+(-1)^i)}\right\}\right].$$

For P and $\sigma \gg 1$, δy_2 and δy_3 take the same value of (19).

$$\delta y_2 = \delta y_3 = j\left\{\omega C + \frac{1}{\omega\xi\left(1 - \frac{P}{\sigma}\right)}\right\}\frac{\delta\omega}{\omega}. \tag{19}$$

However, y_1 takes always infinite value at any frequency. Under these conditions, that is,

$$\delta y_2 = \delta y_3, \qquad y_1 = \infty, \qquad \delta s_1 = 0, \tag{20}$$

the equivalent network has the construction consisting of one ideal circulator available to any frequency and three parallel tuned circuits connected at each terminal of the ideal circulator which is illustrated in the part surrounded by the dotted line in Fig. 6.

The frequency variation of the 2nd and the 3rd eigenvalues of Y matrix coincides with frequency variation of the parallel tuned circuits connected at each terminal, and $\delta y_2 = \delta y_3$ because they are reciprocal circuits. On the other hand, the 1st eigenvalues are not affected by the tuned circuits because they are connected at the zero impedance point of the 1st eigenvalue. So it satisfies the condition of $y_1 = \infty$, $\delta s_1 = 0$. From this reason, this equivalent network satisfies all the conditions described in (20).

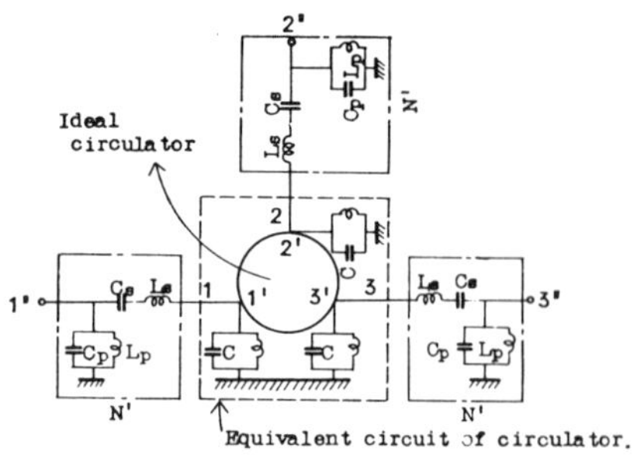

Fig. 6. Equivalent network of circulator and added network for wideband. N' is the network for wideband operation where cp, Lp are not necessary for $N=2$.

B. Bandwidth Enlargement with Chebyshev Characteristics

Let the terminals of the ideal circulator be $1'$, $2'$, and $3'$ as shown in Fig. 6. We consider the purely reactive four-terminal networks N' connected to each terminal 1, 2, 3 of the circulator and name the outside terminals of N', $1''$, $2''$, $3''$ respectively. Let the network, which contains N' and the parallel tuned circuits at terminals 1, 2, 3 be named N. When the terminals $2''$ and $3''$ are terminated with matched loads and an incident signal with unit amplitude is applied to terminal $1''$, it comes out of terminal $3'$. If there is a reflection in the amount of Γ at terminal $3'$, the reflected wave caused by Γ is transmitted again toward terminal 2 through the ideal circulator. Hence $|S''|=\Gamma$. Between $|\Gamma|$ and power-loss ratio of N, P_0/P_L, we have the relation of

204

$$\frac{P_0}{P_L}=\frac{1}{1-|\Gamma|^2}=1+\frac{1}{\frac{1}{|\Gamma|^2}-1}.$$

Therefore, to keep $|S''|$ below the value $|S''_{\max}|$ throughout the frequency band between ω_1' and ω_2', the network N should be designed with its power-loss ratio satisfying the requirements of a band-pass network with proper characteristics. In this section, the Chebyshev characteristics of (21) are discussed.

$$\left.\begin{aligned}\frac{P_0}{P_L}&=1+h^2T_n^2(\omega')\\ \omega'&=\frac{\omega_c}{K}\left(\frac{\omega}{\omega_0}-\frac{\omega_0}{\omega}\right)\\ \omega_1&=\frac{1}{2}\left\{-K\omega_c+\sqrt{(K\omega_c)^2+4\omega_0^2}\right\}\\ \omega_2&=\frac{1}{2}\left\{K\omega_c+\sqrt{(K\omega_c)^2+4\omega_0^2}\right\}\end{aligned}\right\}\tag{21}$$

where h is a constant determining the maximum insertion loss in a pass band, K is a constant relating the bandwidth of a band-pass filter to that of a low-pass filter, and T_n is a first kind nth order Chebyshev polynomial. In (21) h is related to minimum backward loss A dB by

$$h=\frac{1}{\sqrt{10^{A/10}-1}},\qquad A=20\log_{10}\frac{1}{|S_{\max}''|}.\tag{22}$$

These characteristics are synthesized from ordinary low-pass filter theory [7] and realized by ladder networks in which the series and parallel elements are, respectively, the series- and parallel-resonance circuits tuned to center frequency ω_0, where ω_0 is obtained from $\omega_0^2=\omega_1'\omega_2'$. When the total number n of all resonance circuits is odd, the input and output impedances of N take the same value, while they take different values when n is even. The network must always be synthesized in such a way that the 1st element is a parallel element, because the parallel tuned circuit of the equivalent network of a circulator is always put as the first element of network N.

The bandwidth is increased more with the increasing of n, and the results of the calculations are shown in Table I for the cases of backward loss A equal to 20 dB and 30 dB. It is seen that the improvement in bandwidth is much less remarkable for n larger than 2 or 3. Moreover, they are the most useful numbers of elements in practical components for VHF and UHF bands. So, only the results of $n=2$ and $n=3$ are shown in this paper. (For $n>3$, refer to (8).)

TABLE I

RELATION BETWEEN n AND BANDWIDTH INCREASING RATIO w_n/w_1

n	$A=20$ dB	$A=30$ dB
1	1	1
2	3.55	5.7
3	4.25	8.42
4	4.47	9.63
5	4.633	10.6
∞	6.16	14.3

A is the minimum backward loss in the specified bandwidth.

1) In the case of $n=2$

The network for the wide bandwidth consists of only the series-resonance circuit connected in series at each terminal, of which inductance L_s and capacitance C_s take the values shown in (23).

$$\begin{aligned}L_s&=1+\frac{4}{\zeta^2}R^2C,\qquad C_s=\frac{1}{\omega_0^2L_s}\\ \zeta&=\left(\frac{2}{a}\right)^{1/4}-\left(\frac{2}{a}\right)^{-1/4},\\ a&=\frac{1}{2(10^{A/10}-1)}=\frac{h^2}{2}\end{aligned}\tag{23}$$

where R is the terminal impedance of the wide-band circulator.

In this circulator, the terminal impedance, before increasing the bandwidth by L_s and C_s, must be R_e, which takes a slightly different value from R as shown in

$$R_e=\frac{1+\sqrt{2a}}{1-\sqrt{2a}}R.\tag{24}$$

This has the physical meaning that the circulator does not have infinite backward loss at a center frequency but at frequencies deviating slightly from the center.

The bandwidth is increased in the amount of 3.55 times for $A=20$ dB and 5.7 times for $A=30$ dB as shown in Table I. This is in good agreement with the experiment which will be described later.

2) In the case of $n=3$

The network constitutes one series-resonance circuit connected first in series, and one parallel-resonance circuit connected next in parallel as indicated in Fig. 6. The values in (25) denote the inductance and capacitance of series resonance, L_s and C_s, and those of parallel resonance, L_p and C_p,

$$L_s = \frac{2\zeta^2}{\zeta^2 + 3/4} R^2 C, \qquad C_s = \frac{1}{\omega_0^2 L_s}$$
$$L_p = \frac{1}{\omega_0^2 C}, \qquad C_p = C \tag{25}$$

where ζ takes the same value as in (23). In this circulator, the terminal impedance, before increasing the bandwidth, is equal to that of the increased bandwidth. This means that the backward loss must be adjusted to infinite value at the center frequency. The bandwidth is theoretically increased 4.25 times for $A=20$ dB and 8.42 times for $A=30$ dB as indicated in Table I.

C. Bandwidth Enlargement with Wagner Characteristics

The power-loss ratio of network N takes the form of

$$\frac{P_0}{P_L} = 1 + \omega'^{2n} \tag{26}$$

ω' is the same one used in (21) and shows the angular freqeency corresponding to that translated into the low-pass filter. The network construction is similar to the Chebyshev type with only the exception of their constants. The terminal impedance before enlarging the bandwidth is the same as that for the enlarged bandwidth. The circuit constants and the bandwidth increasing ratio for $n=2, 3$ are shown in (27).

1) $n = 2$, series elements $L_s = CR^2$, $C_s = \dfrac{1}{\omega_0^2 L_s}$

the bandwidth increasing ratio $= \sqrt{5}$

2) $n = 3$, series elements $L_s = 2CR^2$, $C_s = \dfrac{1}{\omega_0^2 L_s}$

parallel elements $L_p = \dfrac{1}{\omega_0^2 C_p}$, $C_p = C$

the bandwidth increasing ratio $= 2.33$. (27)

V. Temperature Dependence

The temperature dependence of a circulator is caused mainly by the deviation of the external dc magnetic field H_{ex} and the saturation magnetization Ms of the ferrite following the fluctuation of temperature.

The dc magnetic field H_0 inside the ferrite is also decreased from H_{ex} by the demagnetizing field in the amount of (28) for the ellipsoidal ferrite.

$$H_0 = H_{ex} - N_z \cdot M_s, \tag{28}$$

N_z is the demagnetizing factor in the Z direction. Equation (28) shows H_0 is affected by the change of H_{ex} and also M_s. Then $\mu_\pm$ vary their values by the change of M_s and H_0 following (15). These variations of permeability result in the deviation of the 2nd and the 3rd eigenvalues. Therefore, the center frequency of the circulator deviates, causing the backward loss at the operating center frequency to be changed by a variation of temperature. These temperature dependencies are quantitatively described in the following section.

A. Deviation of Center Frequency

The variation of μ, $\delta\mu$, is obtained from (15) as

$$\left.\begin{aligned} \delta\mu_- &= \frac{\pm P}{(\sigma \mp 1)^2} \cdot \frac{\delta\omega}{\omega} + \frac{P}{\sigma \mp 1}\left(1 + \frac{PN_z}{\sigma \mp 1}\right)\frac{\delta(4\pi M_s)}{4\pi M_s} \\ P &= \frac{4\pi M_s |\gamma|}{\omega}, \qquad \sigma = \frac{|\gamma| (H_{ex} - N_z M_s)}{\omega} \end{aligned}\right\} \tag{29}$$

$\delta(4\pi M_s)$ shows the deviation of $4\pi M_s$ caused by the change of temperature. As y_2 and y_3 are functions of $\mu_\pm$ and ω, their variation becomes

$$\delta y_i = j\frac{\delta\omega}{\omega}\left(\omega C + \frac{1}{\omega\xi\mu_i}\right) + \frac{j}{\omega\xi\mu_i} \cdot \frac{\delta\mu_i}{\mu_i},$$
$$(\mu_2 = \mu_-, \mu_3 = \mu_+). \tag{30}$$

However, as $\delta s_1 = 0$ as mentioned before, we have to apply the conditions $\delta s_2 = \delta s_3 = 0$ and $\delta y_2 = \delta y_3 = 0$ for satisfying the condition of the circulator. Substituting this relation into (30) and (29), one can get the deviation of the center frequency $\delta\omega$. Taking the approximations of $N_Z = 4\pi$ for thin ferrite disks, $\sigma \gg 1$ and $\omega C \fallingdotseq 1/\omega\xi_\mp\mu$, we get

$$\frac{\delta\omega}{\omega} = -\frac{\delta(4\pi M_s) - \dfrac{P}{\sigma + P}\delta H_{ex}}{2H_0}. \tag{31}$$

From (31), it is understandable that $\delta\omega$ is smaller for the circulator operating under the stronger H_{ex}. This means the effect of the demagnetizing field is small in this case.

B. Change of Backward Loss at Center Frequency Depending upon Temperature

As s_1 is independent of permeability, the variation for temperature δs_1 is also zero. Substituting $\delta s_1 = 0$ into (5) and considering the conditions $y_1 = \infty$, $y_2 = -j/\sqrt{3}$, $y_3 = j/\sqrt{3}$, and the relation of (6), one gets

$$|S'''| = \frac{1}{2}\sqrt{|\delta y_2|^2 - |\delta y_3|^2 - |\delta y_2| \cdot |\delta y_3|}. \tag{32}$$

205

As we consider the state at the same frequency, δy_i in (32) means the variation of y_i for only the variation of temperature and coincides with one substituted by $\delta\omega=0$ into the 1st term of (30). Also $\delta\mu_i$ is obtained by putting $\delta\omega=0$ into (29). That is,

$$\delta y_i = \frac{j}{\omega\xi\mu_i}\cdot\frac{\delta\mu_i}{\mu_i} \tag{33}$$

$$\delta\mu_i = \frac{P(P+\sigma)}{\sigma^2}\cdot\frac{\delta(4\pi M_s)}{4\pi M_s}. \tag{34}$$

Substituting (34) and the value of $\omega\xi$ of (18) into (33), and putting these results into (32), one gets the value of $|S''|$ in

$$|S''| = \frac{1}{2\sqrt{3}}\left(1+\frac{\sigma}{P}\right)\frac{\delta(4\pi M_s)}{H_{res}}. \tag{35}$$

It shows the circulator for the smaller frequency is more sensitive for the variation of temperature because H_{res} is proportional to ω in (35).

206

VI. Design Procedure and Its Practical Example

On the basis of the previous results, the design can be approached in the several ways. As an example, one can start with a specified bandwidth, or one can also design with a minimum insertion loss by operating under the optimum σ as described in 3A. The process for the former case is shown in Table II, where design formulas available to terminal impedance of R_e, are shown, though it has been normalized to one ohm in Section II and III.

As a practical problem, there exist some capacitances between conductors in the mesh part, and they work as a part of C at each terminal by 3 times its value. Therefore, it must be subtracted from the value of C calculated by Table II in advance for practical design, which takes different values depending mainly on the gap between conductors, actually being about 3–6 pF as determined by experiment.

Practical Example

We design a circulator which has a backward loss of 20 dB in the frequency range from 170 Mc/s to 230 Mc/s. A YIG polycrystalline material is used which has a $4\pi Ms$ of 1,000 Gs. The bandwidth ratio $w=0.3$ (30 percent). Following Table II,

1) $w_1=0.3/3.55=0.0845$ for $n=2$,
2) $\eta=0.25$,
3) $C=31.4$ pF
4) $L_S=0.113\ \mu H$, $C_S=5.6\ pF$, $R_e=60\Omega$,
5) $P=10$ (measured value of $|\gamma|/2\pi=2$ Mc/Oer is used), so $\sigma=3.13$
6) $\xi=3.9\ m\mu H$
7) $D=1.5$ mm $t=1.5$ mm from Fig. 5, t is thickness of ferrite disk.

Required capacitance at each terminal C is 25 pF, considering the floating capacitance mentioned before. The results of experiment, shown in Fig. 7(a), show good agreement with the specified characteristics.

VII. Result of Experiment

A. Experimental Results of VHF and UHF Wide-Band Circulator

Figure 7(a) is an example designed from results obtained in Section VI. The same experimental result in UHF band is illustrated in Fig. 7(b).

B. Variation of Center Frequency by Changing the Value of C

The experimental results of the adjustment of the center frequency by changing only the value of C, are shown in Fig. 8. It is understandable that the relationship between the values of C and the frequency, and also between w and the frequency are well satisfied by (16) and (17). The experiment was also performed on the circulator with the insertion loss of 2 dB and 1 per cent bandwidth at a center frequency of 12 Mc/s.

C. Determination of Inertion Loss and Bandwidth by Measurement of Q_{eff} and μ_{eff} of Ferrite

Q_{eff} and μ_{eff} are measured at 700 Mc/s under several strengths of dc magnetic field, and the estimated values of L(dB) and w are indicated in Fig. 9(a). This shows that the minimum insertion loss is 0.15 dB and the corresponding bandwidth is 6.5 percent. The experimental result shows 0.25 dB of insertion loss and 6.5 percent bandwidth as shown in Fig. 8(b).

The same measurements were made at several frequencies; the minimum insertion loss and corresponding bandwidth w are indicated in Fig. 9(b). This figure shows that the insertion loss becomes larger with the decreasing of the frequency.

D. Experiment on Temperature Dependence

Experiments were made on the temperature dependence of the circulator using MnMgAl ferrite disks and Ba-ferrite for H_{ex}. Its center frequency is 200 Mc/s, and the bandwidth is 17 Mc/s. The $4\pi\ Ms$ of ferrite is 1200 Gs and its variation $\delta(4\pi Ms)$ iz 81 Gs for the temperature variation from 20°C to 50°C. Following the design formula of Table II for $\sigma=3.35$, $P=16.7$ and $H_{res}=72.5$ Oer, we get $H_{ex}=N_z\cdot M_s+H_{res}=1435$ Oer. Ba-ferrite biasing magnet has a temperature coefficient such a 0.052 decrease in H_{ex} from 20°C to 50°C. Putting these values into (31), we get $\delta\omega/\omega=-4.1$ percent. The experimental results are shown in Fig. 10(a) and (b), making good coincidence with theoretical values.

TABLE II

DESIGN PROCEDURE AND FORMULAS

	Evaluated Values and Formulas Used	Known Values	
1)	$w_1 = w_n$/Bandwidth Increasing Ratio, the ratio is shown in Table I.	w_n(specified bandwidth)	
2)	$\eta = \frac{w_1}{2\sqrt{3}\|S''\|\sqrt{1-\left(\frac{1}{4\|S''\|}\right)^2}} \doteqdot \frac{w_1}{2\sqrt{3}\|S''\|}$	w_1, $\|S''\|$	Induced from (8)
3)	$C = \frac{2\|S''\|}{R_0\omega w_1}\sqrt{1-\left(\frac{w_1}{4\|S''\|}\right)^2} \doteqdot \frac{2\|S''\|}{\omega w_1 R_e}$	ω, w_1, $\|S''\|$	(8) is changed to that available to terminal impedance R_e.
4)	Network Constants for Wide Band. Cheb. Char. $n = 2$: $L_s = \left(1+\frac{4}{\zeta^2}\right)R^2C$, $C_s = \frac{1}{\omega_0^2 L_s}$, $R_e = \frac{1+\sqrt{2a}}{1-\sqrt{2a}}R$ $n = 3$: $L_s = \frac{2\zeta^2R^2C}{\zeta^2+\frac{3}{4}}$, $C_s = \frac{1}{\omega_0^2 L_s}$, $L_p = \frac{1}{\omega_0^2 C_p}$ $C_p = C$, $R_e = R$ $\zeta = \left(\frac{2}{a}\right)^{1/4} - \left(\frac{2}{a}\right)^{-1/4}$ $a = \frac{1}{2(10^{A/10}-1)}$ Wag. Char. $n = 2$: $L_s = CR^2$, $C_s = \frac{1}{\omega_0^2 L_s}$ $n = 3$: $L_s = 2CR^2$, $C_s = \frac{1}{\omega_0^2 L_s}$ $L_p = \frac{1}{\omega_0^2 C_p}$, $C_p = C$	A (Backward Loss) n: order of Cheb. or Wag. char.	L_s, C_s, L_p, C_p are the element constants in Fig. 7
5)	$\sigma = \frac{P}{2}\left(\sqrt{1+\frac{4(P+\eta)}{P^2\eta}}-1\right) \doteqdot \frac{P}{2}\left(\sqrt{1+\frac{4}{P\eta}}-1\right)$	$P = \frac{4\pi M_s\|\gamma\|}{\omega}$ $\eta = \frac{\mu_+ - \mu_-}{\mu_+ + \mu_-}$	Induced from (15) and the definition of η.
6)	$\xi = \frac{\sqrt{3}\,PR_e}{\omega\{(\sigma+P)^2-1\}} \doteqdot \frac{\sqrt{3}\,R_e}{\omega P\left(1+\frac{\sigma}{P}\right)^2}$	P, σ	(18) is changed to that available to terminal impedance R_e.
7)	Diameter $D(mm)$ and Thickness $t(mm)$ of ferrite are determined by the value of ξ and Fig. 5.	ξ	
8)	$H_{ex} = \frac{\omega}{\|\gamma\|}\sigma + N_z\cdot M_s$	ω, $4\pi M_s$ N_z, $\|\gamma\|$, σ	

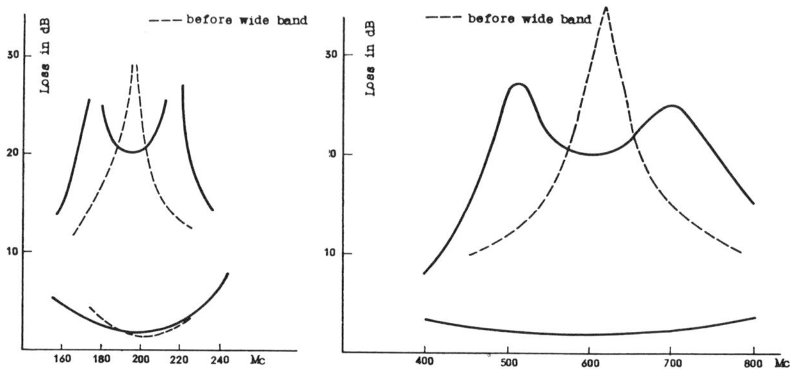

(a) (b)

Fig. 7. Measured values of characteristics of wideband circulator with 2nd Chebyshev characteristic. (a) Experiment of VHF circulator. (b) Experiment of UHF circulator.

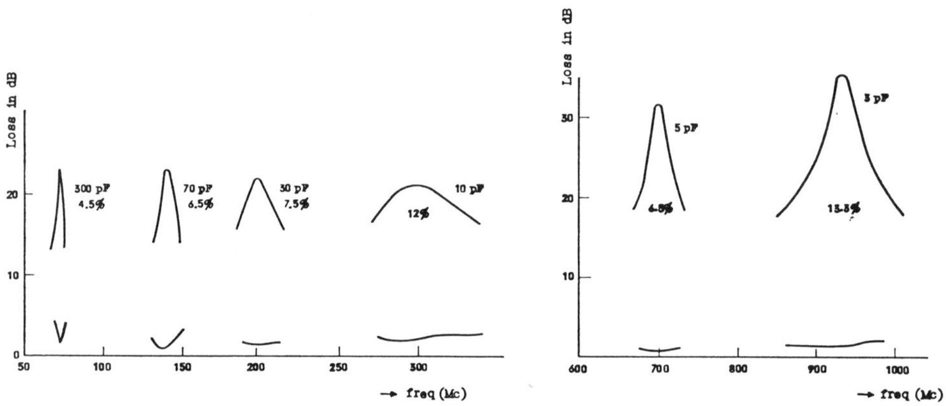

(a) (b)

Fig. 8. Experimental results of VHF, UHF lumped element Y circulator. (a) Adjustment of center frequency by only changing C. (b) Experiment of UHF circulator.

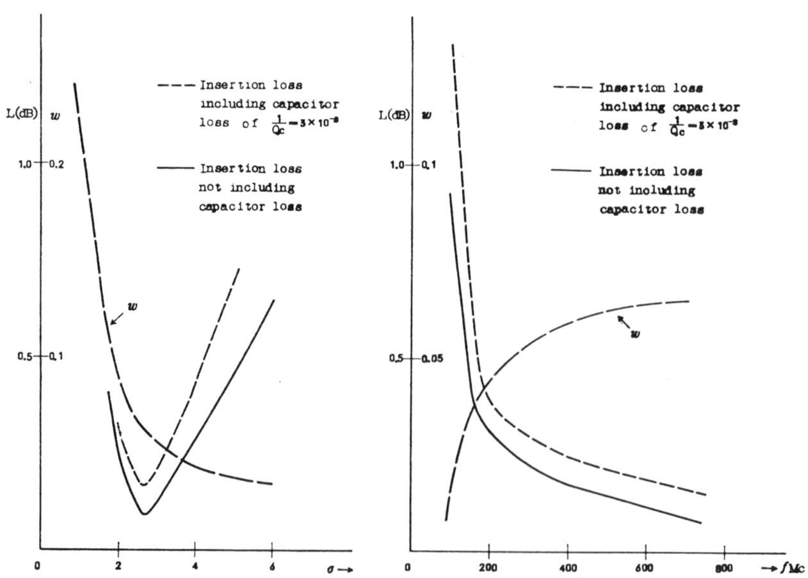

(a) (b)

Fig. 9. Determination of characteristics of circulator by measurements of permeability and its quality factor of ferrite. (a) Determination of insertion loss and bandwidth of circulator at 700 Mc/s for several operations of dc magnetic field. (b) Determination of minimum insertion loss and bandwidth of circulator for several frequencies.

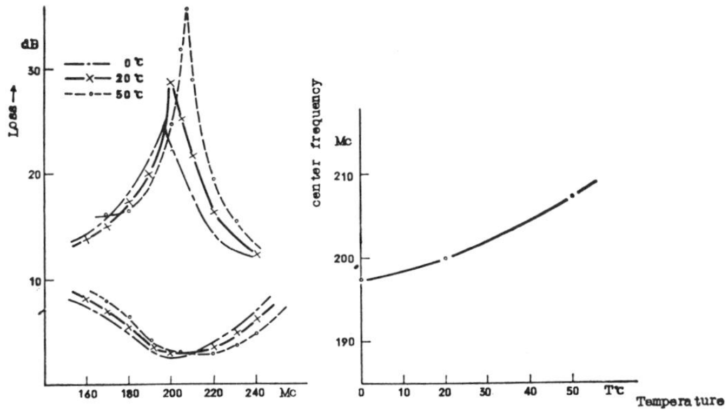

(a) (b)

Fig. 10. Measurement of temperature dependence of circulator. (a) Deviation of backward and forward loss with temperature variation. (b) Deviation of center frequency with temperature variation.

VIII. Conclusion

The miniaturized circulator available to VHF and UHF bands has been developed and we get good agreement between the theories and the experimental results.

The relationships between the electrical characteristics of the circulator and the characteristics of ferrite have been clarified, thus, one is able to estimate the characteristics or the device from the ferrite. The further development of ferrite material to improve the low field loss and also linewidth in the future, will contribute to making a circulator with smaller insertion loss and more wide-band performance.

Appendix I

Eigenvalues of Impedance and Admittance Matrix of Rotationally Symmetrical Circuit

Maxwell's equations in source free, are

$$\nabla \times H = j\omega \boldsymbol{\varepsilon} E \tag{36}$$

$$\nabla \times E = -j\omega \boldsymbol{\mu} H \tag{37}$$

Scalarly multiplying (37) by H^* and the conjugate of (36) by E, we obtain for the difference of the two resultant equations

$$\nabla \cdot H^* \times E = j\omega(H^* \boldsymbol{\mu} H - E\boldsymbol{\varepsilon}^* E^*). \tag{38}$$

Integrating (38) on the closed surface with n ports S_p ($p=1, 2, 3, \cdots, n$)

$$\sum_{p=1}^{n} \iint_{S_p} H_{tp}^* \times E_{tp} \cdot n dS = j\omega \iiint_{\tau} (H^* \boldsymbol{\mu} H - E\boldsymbol{\varepsilon}^* E^*) d\tau. \tag{39}$$

In (39) subscript p denotes pth port and t denotes transverse field at each port. The fields on surface S_p are expressed with mode voltages v_p, mode currents i_p and normalized transverse mode vector e_t, h_t. These relations are shown in the following.

$$E_{tp} = v_p e_t, \qquad H_{tp} = i_p h_t \tag{40}$$

$$\iint_{S_p} |e_t|^2 dS = \iint_{S_p} |h_t|^2 dS = -\iint_{S_p} n \cdot e_t \times h_t^* dS = 1 \tag{41}$$

From 40 and 39,

$$\sum_{p=1}^{n} i_p^* v_p \iint_{S_p} (e_t \times h_t^*) \cdot (-n) \cdot dS = j\omega \iiint_{\tau} (H^* \boldsymbol{\mu} H - E\boldsymbol{\varepsilon}^* E^*) d\tau. \tag{42}$$

When the ports are excited by eigenvector, we have this relation at each port,

$$v_p = z j_p. \tag{43}$$

Since the eigenvector has the same amplitude at each port for the rotationally symmetry, for $|i_p| = 1$, from (42) and (43), we get

$$z = \frac{j\omega}{n} \iiint_{\tau} (H^* \boldsymbol{\mu} H - E\boldsymbol{\varepsilon}^* E) d\tau. \tag{44}$$

When each port is excited by TEM mode, the ith eigenvalue of impedance matrix Z_i can be obtained from (44) as in

$$Z_i = \frac{j\omega}{n} \iiint_{\tau} (H_i^* \boldsymbol{\mu} H_i - E_i \boldsymbol{\varepsilon}^* E_i^*) d\tau. \tag{45}$$

In (45), H_i and E_i are the fields corresponding to the ith eigenvector with unit current. Similarly we get the eigenvalue of admittance matrix, as in

$$Y_i = \frac{j\omega}{n} \iiint_{\tau} (E_i \boldsymbol{\varepsilon}^* E_i^* - H^* \boldsymbol{\mu} H) d\tau \tag{46}$$

where E_i and H_i are the field corresponding to the eigenvector with unit voltage.

Appendix II

Frequency Variation of Eigenvalues of Pure Reactance Network

Taking the variation of (36) and (37), we get

$$\nabla \times \delta H - j\boldsymbol{\varepsilon}(\omega \cdot \delta E + E \cdot \delta\omega) - j\omega \cdot \delta\boldsymbol{\varepsilon} \cdot E = 0 \quad (47)$$

$$\nabla \times \delta E + j\boldsymbol{\mu}(\omega \cdot \delta H + H \cdot \delta\omega) + j\omega \cdot \delta\boldsymbol{\mu} H = 0. \quad (48)$$

Taking the difference between (47) and (48), and integrating on the closed surface with S_p, we get

$$\begin{aligned}\sum_{p=1}^{n} \iint_{S_p} (E \times \delta H^* - \delta E \times H^*) \cdot (-n) dS &= j\delta\omega \iiint_\tau \left\{ E\left(\boldsymbol{\varepsilon}^* + \frac{\partial \boldsymbol{\varepsilon}^*}{\partial\omega}\omega\right) E^* + H^*\left(\boldsymbol{\mu} + \frac{\partial\boldsymbol{\mu}}{\partial\omega}\omega\right) H \right\} d\tau \\ &+ j\omega \iiint_\tau (H^*\boldsymbol{\mu}\delta H - \delta H^*\boldsymbol{\mu} H) d\tau \\ &+ j\omega \iiint_\tau (E\boldsymbol{\varepsilon}^*\delta E^* - \delta E\boldsymbol{\varepsilon}^* E^*) d\tau. \quad (49)\end{aligned}$$

When the field is that of the eigenvector, v_p and i_p of (40) have the relation

$$y v_p = i_p. \quad (50)$$

Taking variation of (50), one gets

$$\delta y \cdot v_p + y \cdot \delta v_p = \delta i_p. \quad (51)$$

Substituting (40) and (51) into (49), one gets

$$\begin{aligned}\delta y \sum_{p=1}^{n} |v_p|^2 + y \sum_{p=1}^{n} (v_p \delta v_p^* - v_p^* \delta v_p) &= j\delta\omega \iiint_\tau \left\{ E\left(\boldsymbol{\varepsilon}^* + \frac{\partial \boldsymbol{\varepsilon}^*}{\partial\omega}\omega\right) E^* + \left(\boldsymbol{\mu} + \frac{\partial\boldsymbol{\mu}}{\partial\omega}\omega\right) H \right\} d\tau \\ &+ j\omega \iiint_\tau (H^*\boldsymbol{\mu}\delta H - \delta H^*\boldsymbol{\mu} H) d\tau \\ &+ j\omega \iiint_\tau (E\boldsymbol{\varepsilon}^*\delta E^* - \delta E\boldsymbol{\varepsilon}^* E^*) d\tau. \quad (52)\end{aligned}$$

The 1st term of the left and right hand side of (52) are both imaginary, and the 2nd and the 3rd term of the left and right hand side are real values. Therefore, when each port is excited by an eigenvector with unit voltage, from (52) one gets

$$\delta y = j\frac{\delta\omega}{n} \iiint_\tau \left\{ E\left(\boldsymbol{\varepsilon}^* + \frac{\partial \boldsymbol{\varepsilon}^*}{\partial\omega}\omega\right) E^* + H^*\left(\boldsymbol{\mu} + \frac{\partial\boldsymbol{\mu}}{\partial\omega}\omega\right) H \right\} d\tau. \quad (53)$$

In the same procedure, the variation of eigenvalues of impedance matrix can be expressed by

$$\delta z = j\frac{\delta\omega}{n} \iiint_\tau \left\{ E\left(\boldsymbol{\varepsilon}^* + \frac{\partial \boldsymbol{\varepsilon}^*}{\partial\omega}\omega\right) E^* + H^*\left(\boldsymbol{\mu} + \frac{\partial\boldsymbol{\mu}}{\partial\omega}\omega\right) H \right\} d\tau. \quad (54)$$

In the case

$$\frac{\partial\boldsymbol{\mu}}{\partial\omega}\omega \ll \mu, \qquad \frac{\partial\boldsymbol{\varepsilon}^*}{\partial\omega}\omega \ll \boldsymbol{\varepsilon}^*,$$

we get

$$\delta y = j\frac{\delta\omega}{n} \iiint_\tau (E\boldsymbol{\varepsilon}^* E^* + H^*\boldsymbol{\mu} H) d\tau \quad (55)$$

$$\delta z = j\frac{\delta\omega}{n} \iiint_\tau (E\boldsymbol{\varepsilon}^* E^* + H^*\boldsymbol{\mu} H) d\tau. \quad (56)$$

The E, H in (53) and (55) are the fields corresponding to unit current excitation, while those in (54) and (56) show the field for unit voltage excitation. The approximation of $(\partial\boldsymbol{\mu}/\partial\omega)\omega \ll \boldsymbol{\mu}$ is assured in the case of the operation under the strong dc magnetic field, and the approximation is easily verified from (15) assuming $\sigma^2 \gg 1$. In the case of ferrite, $\omega(\partial\boldsymbol{\varepsilon}^*/\partial\omega)$ can be also neglected from $\boldsymbol{\varepsilon}^*$.

Appendix III

Measurement of $\mu_\pm$

Three ports of the mesh part containing ferrite disks are excited by the eigenvector of (1), where a port of them is connected through a slotted line with standing-wave detector by which the impedance $Z_i' = j\omega\mu_i\xi$ is measured. The measured values μ_i are indicated with $\times$ signs in Fig. 3. Since under a dc magnetic field sufficiently strong so that $\mu_i \doteqdot 1$, the mesh part becomes reciprocal, ξ can be also determined by the following technique. The variable reactance or adjustable line is connected to port 2, and it is adjusted so that the signal applied to port 1 does not appear to port 3. The terminal impedance at port 1 in this state is the eigenvalue $j\omega\xi$. This is easily understood by the fact that the eigenvalues of $\boldsymbol{u}_2$ and $\boldsymbol{u}_3$ take the same value for a reciprocal symmetry circuit and the linear combination of $\boldsymbol{u}_2$ and $\boldsymbol{u}_3$ can satisfy the previously mentioned state.

Appendix IV

Relation Between Bandwidth and Total Reactive Energy in Lumped Element Circulator

From (46) eigenvalues y_2 and y_3 are expressed by

$$y_2 = \frac{2}{3} j\omega(\tilde{W}_e - \tilde{W}_{m_2}) = -j\frac{1}{\sqrt{3}},$$

$$y_3 = \frac{2}{3} j\omega(\tilde{W}_e - \tilde{W}_{m_3}) = j\frac{1}{\sqrt{3}}, \quad (57)$$

where $\tilde{W}_e$ is a time average of electric energy contained in C and $\tilde{W}_{m_2}$, $\tilde{W}_{m_3}$ are magnetic energies in ferrite. Taking the expression of

$$\frac{\tilde{W}_{m_2} + \tilde{W}_{m_3}}{2} = \tilde{W}_m, \quad \tilde{W}_{t2} = \tilde{W}_e + \tilde{W}_{m_2},$$

$$\tilde{W}_{t_3} = \tilde{W}_e + \tilde{W}_{m_3}, \tag{58}$$

the time average of total reactive energies are expressed by

$$\tilde{W}_{t_2} = 2\tilde{W}_m + \frac{\sqrt{3}}{2\omega}, \qquad \tilde{W}_{t_3} = 2\tilde{W}_m - \frac{\sqrt{3}}{2\omega} \cdot \tag{59}$$

From (6) and (55), we get

$$|\delta s_2| = \frac{3}{2}|\delta y_2| = \tilde{W}_{t_2}|\delta\omega|,$$

$$|\delta s_3| = \frac{3}{2}|\delta y_3| = \tilde{W}_{t3}|\delta\omega|. \tag{60}$$

However, since we have the relation $|\delta s_1| = 0$ as shown in (20), substituting the relation and (60) and (59) into (5), one gets

$$|S''| = \frac{\delta\omega}{3} \cdot \tilde{W}_t \sqrt{1 + \left(\frac{3}{2\omega\tilde{W}_t}\right)^2},$$

$$\tilde{W}_t = \frac{\tilde{W}_{t2} + \tilde{W}_{t3}}{2} = 2\tilde{W}_m. \tag{61}$$

From (57), we can also get the relation

$$\tilde{W}_{m_2} + \tilde{W}_{m_3} = \tilde{W}_t = 2\tilde{W}_e = 3C. \tag{62}$$

Substituting (62) into (61), one gets

$$w = \frac{2\delta\omega}{\omega} = \frac{6|S''|}{\omega\tilde{W}_t\sqrt{1 + \left(\frac{3}{2\omega\tilde{W}_t}\right)^2}}$$

$$= \frac{2|S''|}{\omega C\sqrt{1 + \left(\frac{1}{2\omega C}\right)^2}} \cdot \tag{63}$$

Since we can get the relation $\omega C = 1/\sqrt{3}\eta$ from (7), applying it to (63), we get (8).

Appendix V

Determination of Characteristics of Circulator by Measurement of μ_{eff}' and Q_{eff}

The means of measurement and the technique of the determination are as follows.

1) Measurements of μ_{eff}' and Q_{eff}

As μ_{eff}' is an equivalent permeability when dc magnetic field is applied along the wave propagation of a coaxial line filled with ferrite, μ_{eff}' is obtainable by measuring the admittances of coaxial line with and without ferrite at shortened end. The measurement is made for dc fields of several strengths.

The value of Q_{eff} is obtained by measuring the Q of resonator in both cases with and without ferrite by the following technique.

When signal is applied at SS' in Fig. 11(a), the amount of the signal at LL' takes the minimum value m as shown in Fig. 11(b) for the resonant state of the branched line with short end. The unloaded Q of the resonator with ferrite Q_0 is obtained by the relation of

$$Q_0 = \frac{Q_L}{m}, \qquad Q_L = \frac{f}{\Delta f}, \qquad \text{using Fig. 11(b).}$$

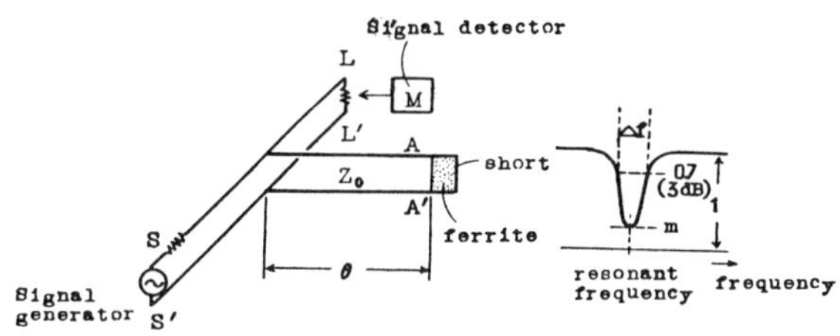

Fig. 11. Measurement of μ'_{eff} and Q_{eff}.

The unloaded Q of the resonator without ferrite, Q_{cav}, is also obtained by the same way. So, one can get the value of Q_{eff} by

$$\left.\begin{aligned} \frac{1}{Q_{eff}} &= \left(\frac{1}{Q_0} - \frac{1}{Q_{cav}}\right)\frac{1}{k_f} \\ k_f &= \frac{\tilde{W}_f}{\tilde{W}_t} = \text{filling factor,} \end{aligned}\right\} \tag{64}$$

where $\tilde{W}_f$ is a magnetic energy in the part of ferrite, and $\tilde{W}_t$ is a total reactive energy inside the resonator.

However, k_f is obtainable by

$$\left.\begin{aligned} k_f &= \frac{\tilde{W}_f}{\tilde{W}_t} = \frac{X_f}{Z_0\dfrac{\theta}{\cos^2\theta} + X_f} \\ \theta &= n\pi - \tan^{-1}\frac{X_f}{Z_0}, \end{aligned}\right\} \tag{65}$$

where $\frac{1}{X_f}$ is a imaginary part of the admittance at AA' with ferrite, n is the number of standing waves in resonator, and Z_0 is a surge impedance of the coaxial line in the resonator.

2) σ is obtained from

$$\sigma = \frac{P}{\mu'_{eff} - 1} \cdot \tag{66}$$

3) Insertion loss L(dB) is obtained from

$$L(\text{dB}) = 4.98 \frac{\sigma(\sigma + P)}{P} \cdot \frac{1}{Q_{eff}} \cdot \quad (67)$$

4) Bandwidth ratio w is obtained from

$$w = 2\sqrt{3}\,|S''| \frac{P}{\sigma(\sigma + P)} \cdot \quad (68)$$

5) We get Fig. 10(a) from (61) and (68).

Appendix VI

Reduction of $y_3 - y_2$ by Polder's Equation

$$\mu_{\pm} = 1 + \frac{4\pi M_s |\gamma|}{\omega(\sigma \mp 1)}, \qquad \sigma = \frac{|\gamma| H_0}{\omega} \cdot \quad (69)$$

From (69)

$$\frac{1}{\mu_-} - \frac{1}{\mu_+} = \frac{4\pi M_s |\gamma|}{\omega} \cdot \frac{\dfrac{2}{\sigma^2 - 1}}{1 + \dfrac{4\pi M_s |\gamma|}{\omega} \cdot \dfrac{2\sigma}{\sigma^2 - 1} + \dfrac{(4\pi M_s |\gamma|)^2}{\omega^2} \cdot \dfrac{1}{\sigma^2 - 1}} \cdot \quad (70)$$

Under the condition of $4\pi Ms|\gamma|^2/\omega \gg 1$ or $\sigma^2 \gg 1$,

$$\frac{1}{\mu_-} - \frac{1}{\mu_+} = \frac{4\pi M_s \omega}{|\gamma| H_0^2} \cdot \frac{2}{\left(1 + \dfrac{4\pi M_s}{H_0}\right)^2} \cdot \quad (71)$$

However, we have the relation

$$y_3 - y_2 = 2j\omega(\tilde{W}_{m_2}' - \tilde{W}_{m_3}')$$
$$= 2j\omega \frac{1}{2\omega^2 \xi}\left(\frac{1}{\mu_-} - \frac{1}{\mu_+}\right) = j \frac{1}{\omega \xi}\left(\frac{1}{\mu_-} - \frac{1}{\mu_+}\right). \quad (72)$$

Substituting (71) into (72),

$$y_3 - y_2 = j \frac{2}{\xi 4\pi M_s |\gamma| \left(1 + \dfrac{H_0}{4\pi M_s}\right)^2} \cdot \quad (73)$$

The value of $y_3 - y_2$ is independent of the frequency.

Acknowledgment

The author wishes to express his appreciation to the Director of the Laboratories, Mr. Tatsuji Nomura, for his encouragement and advice in connection with this study. Thanks are also due to technical assistants of this Laboratory for taking the experimental data.

References

[1] U. Milano, J. H. Saunders, and L. Davis, Jr., "A Y-junction strip lines circulator," *IRE Trans. on Microwave Theory and Techniques*, vol. MTT-8, pp. 346–351, May 1960.
[2] C. V. Beuhler A. F. Eikenberg, "A VHF high-power Y-circulator," *IRE Trans. on Microwave Theory and Techniques (Correspondence)*, vol. MTT-9, pp. 569–570, November 1961.
[3] Y. Konishi, I. Mochizuki, "Theory and design of Y-junction strip-line circulator," vol 15, no. 5, *Tech. J. Japan Broadcasting Corporation*, pp. 1–12, May 1963.
[4] Y. Konishi, "Lumped element Y-circulator," *I.C.M.C.I. (Tokyo*, pt. 1, M4–9, p. 61, September 1964.
[5] Montgomery, Dicke, and Purcell, *Principles of Microwave Circuits*, New York: McGraw-Hill, 1948, p. 413.
[6] B. A. Auld, "The synthesis of symmetrical waveguide circulators," *IRE Trans. on Microwave Theory and Techniques*, vol. MTT-7, pp. 238–246, April 1959.
[7] H. Takahashi, "On the Ladder-Type Filter Network with Tchebycheff Response," *J. Institute of Electrical Communication Engineering of Japan*, pp. 65–76, February 1951.
[8] Y. Konishi, "Wide band lumped element Y-circulator," *Institute of Electrical Communication Engineering of Japan Con. Rec.*, 4– 1964.

212

Broadband, Lumped Element UHF Circulator

A triple-tuned, broadband circulator covering the 400–700 MHz range has been developed utilizing lumped element techniques.[1]

The basic design is a shunt-tuned, lumped constant circulator biased above ferromagnetic resonance and triple tuned by means of two additional resonant circuits at each port. The equivalent circuit for this unit is shown in Fig. 1.

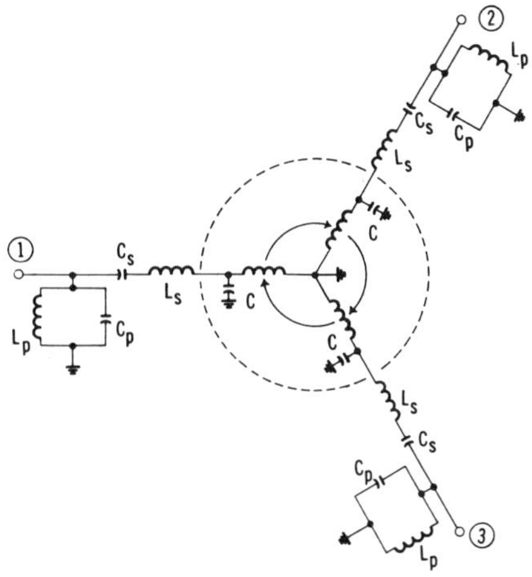

Fig. 1. Equivalent circuit of triple-tuned circulator.

The single-tuned circulator was designed and constructed using the techniques described by Dunn and Roberts[2] and Konishi.[3] According to the lumped constant circulator theory presented in these papers, the Q, and hence the bandwidth, of a basic single-tuned circulator is related to the bias field and saturation magnetization of the ferrite by

$$Q = \frac{\mu^+ + \mu^-}{\sqrt{3}\,(\mu^+ - \mu^-)}$$

where

$$\mu^{\pm} = 1 + \frac{\omega_m}{\omega_0 \mp \omega}$$

ω is the center frequency of the circulator, ω_0 is the frequency of ferromagnetic resonance, and ω_m is $\gamma 4\pi M_s$. This expression indicates that the bandwidth increases as the frequency of resonance approaches the operating frequency. Unfortunately, magnetic losses also increase as resonance is approached setting a limit to the bandwidth that can be achieved with tolerable losses. The basic single-tuned circulator was designed to operate as close to resonance as possible without introducing objectionable losses in the desired operating frequency band. The ferrite material was a polycrystalline garnet material with $4\pi M_s = 1000$ gauss. The impedance response of the single-tuned circulator is plotted in Fig. 2. It is seen that the response approximates very closely that of an ideal parallel resonant circuit, also shown in Fig. 2.

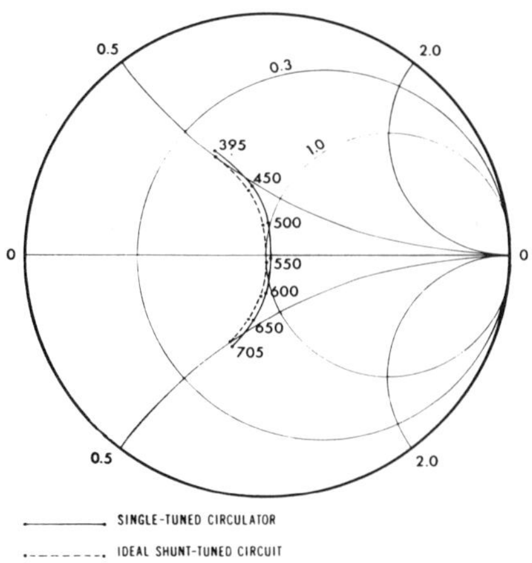

Fig. 2. Impedance response of the single-tuned circulator.

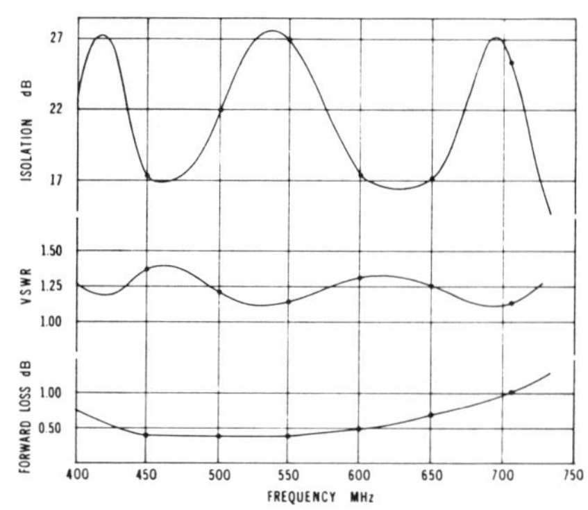

Fig. 3. Performance of triple-tuned circulator.

To realize the full bandwidth for the given single-tuned plot it was necessary to triple tune by adding the appropriate series and shunt-tuned circuits. The parameters of these circuits can be determined graphically, or from filter theory by using the circulator as a shunt resonator in a three-resonator bandpass filter with the desired Tchebyscheff response. Results for the triple-tuned circulator are plotted in Fig. 3. These performance curves show an equal ripple characteristic with a minimum isolation of 17 dB. The insertion loss reaches a maximum of 1.0 dB at the high end of the band and is typically 0.6 dB or less over most of the band. The corresponding impedance plot is indicated in Fig. 4. The packaged unit is $3\frac{11}{16}$ inches in diameter by $1\frac{1}{4}$ inches in height.

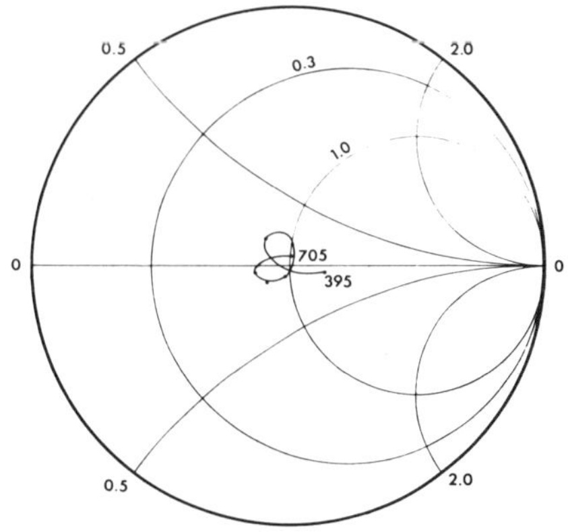

Fig. 4. Impedance plot of triple-tuned circulator.

In comparison with previously reported conventional junction circulators, this circulator has greater bandwidth for comparable isolation and insertion loss and has a possible cost advantage in that the ferrite volume required is considerably less.

G. McChesney
V. Dunn
Melabs
Palo Alto, Calif.

Manuscript received October 13, 1966.

[1] Patent pending.

[2] V. E. Dunn and R. W. Roberts, "New design techniques for miniature VHF circulators," presented at the G-MTT Symposium, Clearwater, Fla., 1965.

[3] Y. Konishi, "Lumped element Y-circulators," *IEEE Trans. on Microwave Theory and Techniques*, vol. MTT-13, pp. 852–864, November 1965.

A Compact Broad-Band Thin-Film Lumped-Element *L*-Band Circulator

REINHARD H. KNERR, CLARE E. BARNES, MEMBER, IEEE, AND FRIDOLIN BOSCH

***Abstract*—Impedance matrices including magnetic losses are developed for a number of lumped-element ferrite-loaded symmetrical three-port junctions. The scattering matrix eigenvalues corresponding to these matrices are determined as functions of frequency and circuit parameters and are used to analyze these three-ports with emphasis on their properties as circulators. A very compact broad-band thin-film lumped-element circulator is derived from the idealized equivalent circuit. An experimental model approximately represented by this circuit is shown to have a 20-dB isolation bandwidth of greater than 30 percent with an insertion loss of less than 0.6 decibel.**

A switchable circulator which requires no magnetic-field switching is treated using this same analytical approach.

It is suggested that this type of analysis together with additional experimental refinement of equivalent circuits will lead eventually to a computerized design of lumped-element circulators.

Manuscript received May 18, 1970; revised July 31, 1970. This paper was presented at the 1970 International Microwave Symposium, Newport Beach, Calif., May 11–14.

The authors are with Bell Telephone Laboratories, Inc., Allentown, Pa.

I. Introduction

THE constantly progressing integration of microwave circuits demands small circulators comparable in size and cost with other circuit elements. Distributed element circulators become excessively large at lower microwave frequencies. The development of the lumped-element circulator by Konishi [3] and others [9] was an important step in the miniaturization of circulators at and below *L*-band. The thin-film lumped-element circulator [4] was developed in the hope that the technique could be applied at higher microwave frequencies. This device was narrow band,

and conventional broad-banding techniques using lumped-element series resonance circuits were not very attractive because of the lack of good thin-film inductors at microwave frequencies. An existing need for an L-band circulator with 20-dB bandwidth in excess of 30 percent required a new approach to the broad-banding and tuning problem. This paper contains a complete analysis of several possible lumped-element circulators, permitting their eventual computerized design. The insight into the device behavior derived from this analysis leads to a new type of very broad-band low-loss easily tunable compact lumped-element circulator that is suited for batch processing and inexpensive mass production.

II. Analysis of the Thin-Film Lumped-Element Circulator

Assuming a structure of the type shown in Fig. 1 with, for the moment, negligible capacitance at the insulated crossovers, the currents and voltages were shown by Deutsch and Wieser [1] to be

$$[V] = [Z_1][i].^{1} \tag{1}$$

Where

$$[Z_1] = \begin{bmatrix} \alpha_1 & \beta_1 & \gamma_1 \\ \gamma_1 & \alpha_1 & \beta_1 \\ \beta_1 & \gamma_1 & \alpha_1 \end{bmatrix}$$

$$\alpha_1 = \omega G j\mu$$

$$\beta_1 = \omega G\left(-j\frac{\mu}{2} - \frac{k}{2}\sqrt{3}\right)$$

$$\gamma_1 = -\beta_1^* = \omega G\left(-j\frac{\mu}{2} + \frac{k}{2}\sqrt{3}\right) \tag{2}$$

with

- β^* complex conjugate of β
- ω angular frequency
- G geometry factor [1].
- μ, k complex elements of the "Polder" tensor:

$$[\mu_{ik}] = \mu_0 \begin{bmatrix} \mu & -jk & 0 \\ jk & \mu & 0 \\ 0 & 0 & 1 \end{bmatrix}.$$

The behavior of μ and k as functions of frequency and field have been extensively described in the literature [2]. Since the μ and k used are complex, the analysis includes the effect of magnetic losses.

In what follows, Deutsch and Wieser's concept is refined and applied to various structures leading up to a new broad-band low-loss compact lumped-element circulator. The first such structure is that of Konishi [3] as shown in Fig. 2. This structure is not completely

[1] The rectangular bracket indicates that the enclosed symbol is a matrix.

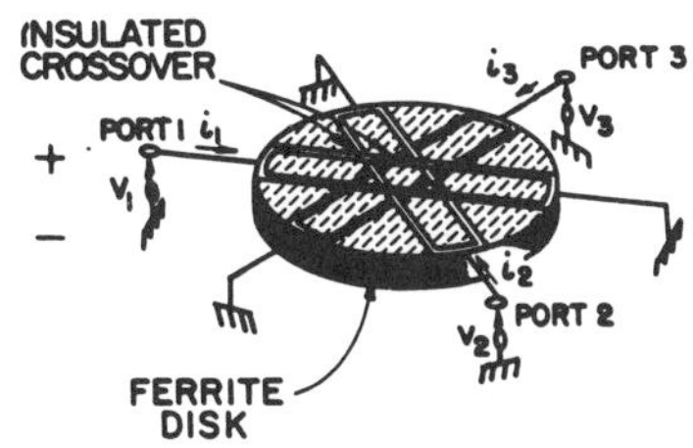

Fig. 1. Lumped-element circulator junction.

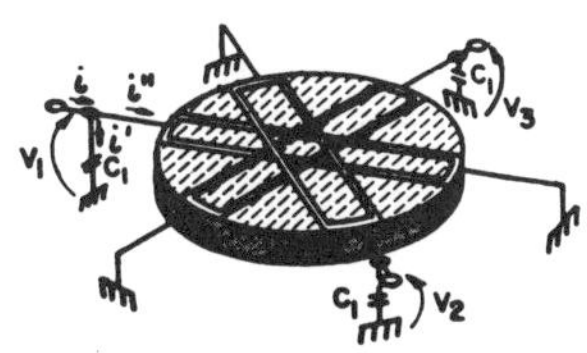

Fig. 2. Konishi-type single-tuned lumped-element circulator.

symmetrical since the conductors are not interwoven as in Fig. 1. Thin-film techniques permit interweaving by the appropriate design of the masks, and completely symmetrical structures have been obtained [4]. For the analysis a complete threefold symmetry is assumed. Applying (1) to Fig. 2,

$$[V] = [Z_1][i''] \tag{3}$$

and from Kirchhoff's law

$$[i] = [i''] + [i']$$

or

$$[i''] = [i] - [i'].$$

Substituting into (3),

$$[V] = [Z_1]\{[i] - [i']\}.$$

Since $[i'] = j\omega C_1[V]$,

$$[V] = [Z_1][i] - [Z_1]j\omega C_1[V]$$

or

$$[V] + [Z_1]j\omega C_1[V] = [Z_1][i]$$

and

$$[V] = \{[I] + j\omega C_1[Z_1]\}^{-1}[Z_1][i]$$

where $[I]$ equals the unity matrix; or

$$[V] = [Z_2][i] \tag{4a}$$

where

$$[Z_2] = \{[I] + j\omega C_1[Z_1]\}^{-1}[Z_1] = \begin{bmatrix} \alpha_2 & \beta_2 & \gamma_2 \\ \gamma_2 & \alpha_2 & \beta_2 \\ \beta_2 & \gamma_2 & \alpha_2 \end{bmatrix}. \tag{4b}$$

Equation (4a) is of the same form as (1). In (4b), α_2, β_2, and γ_2 are functions of C_1, α_1, β_1 and γ_1 and are easily obtainable by carrying out the necessary mathematical operations. In the case of the thin-film lumped-element circulator (Fig. 3), the capacitance due to the insulated crossover (Fig. 1) cannot be neglected and can be made large enough to assume the role of the capacitance C_1 in Fig. 2. An approximate equivalent circuit for this case is shown in Fig. 4(a). The crossover capacitors are replaced by equivalent capacitors C_1 and C_2. Using the Δ–Y transformation in conjunction with this specific coil configuration, the circuit in Fig. 4(a) can be easily transformed into that of Fig. 4(b) which is equivalent to the circuit of Fig. 2.

Another important basic circulator configuration is shown in Fig. 5. Keeping in mind that the sum of the input currents flows through the capacitor C, (1) becomes

$$[V] = [V'] + [V''] = [Z_1][i] + \frac{1}{j\omega C}\begin{bmatrix} 1 & 1 & 1 \\ 1 & 1 & 1 \\ 1 & 1 & 1 \end{bmatrix}[i]$$

or

$$\begin{bmatrix} V_1 \\ V_2 \\ V_3 \end{bmatrix} = \begin{bmatrix} \alpha_1 + X_c & \beta_1 + X_c & \gamma_1 + X_c \\ \gamma_1 + X_c & \alpha_1 + X_c & \beta_1 + X_c \\ \beta_1 + X_c & \gamma_1 + X_c & \alpha_1 + X_c \end{bmatrix}\begin{bmatrix} i_1 \\ i_2 \\ i_3 \end{bmatrix}$$

for

$$X_c = \frac{1}{j\omega C}$$

or

$$[V] = [Z_3][i] \tag{5a}$$

where

$$Z_3 = \begin{bmatrix} \alpha_3 & \beta_3 & \gamma_3 \\ \gamma_3 & \alpha_3 & \beta_3 \\ \beta_3 & \gamma_3 & \alpha_3 \end{bmatrix} \tag{5b}$$

and

$$\alpha_3 = \alpha_1 + \frac{1}{j\omega C}; \qquad \beta_3 = \beta_1 + \frac{1}{j\omega C}; \qquad \gamma_3 = \gamma_1 + \frac{1}{j\omega C}.$$

Finally, the combination of the circuits of Figs. 2 and 5 results in the circuit of Fig. 6 which is realized as indicated in Fig. 7. This structure will be shown to provide broad-band circulation without using the conventional series resonant broad-banding circuits shown in Fig. 8. Replacement of the relatively low-quality series resonant circuit by a high-quality capacitor (C of Fig. 6) results in a significantly smaller and lower loss device.

It should be noted that the circuits of Figs. 5 and 6 differ from the lower circuit forms of Figs. 1 and 2, respectively, by the addition of a capacitor C to ground. Thus the equations for Fig. 6 can be derived from (4) for Fig. 2 just as (5) for Fig. 5 was derived from (1) for

Fig. 3. Broad-band lumped-element circulator substrate.

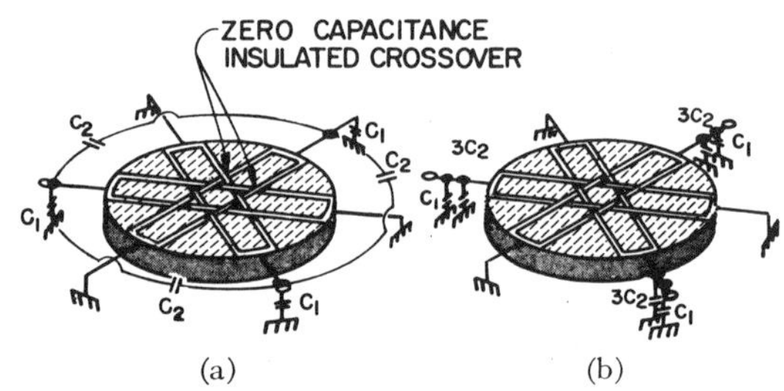

Fig. 4. Approximate equivalent structure of lumped-element circulator with nonnegligible crossover capacitance.

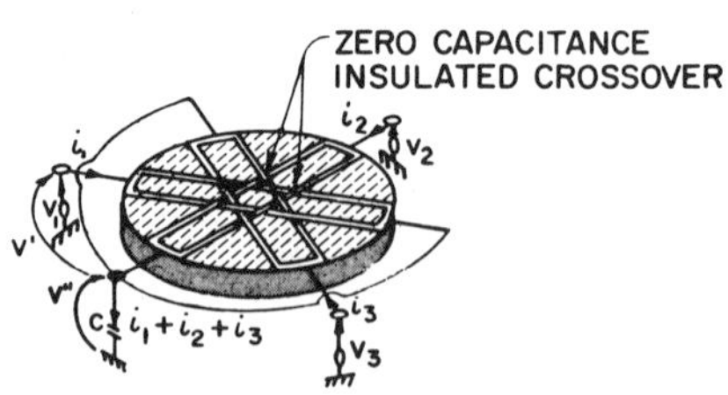

Fig. 5. Single-tuned lumped-element circulator.

Fig. 1. This leads to

$$\begin{bmatrix} V_1 \\ V_2 \\ V_3 \end{bmatrix} = \begin{bmatrix} \alpha_2 & \beta_2 & \gamma_2 \\ \gamma_2 & \alpha_2 & \beta_2 \\ \beta_2 & \gamma_2 & \alpha_2 \end{bmatrix}\begin{bmatrix} i_1 \\ i_2 \\ i_3 \end{bmatrix} + \frac{1}{j\omega C}\begin{bmatrix} 1 & 1 & 1 \\ 1 & 1 & 1 \\ 1 & 1 & 1 \end{bmatrix}[i]$$

and, therefore,

$$\begin{bmatrix} V_1 \\ V_2 \\ V_3 \end{bmatrix} = \begin{bmatrix} \alpha_4 & \beta_4 & \gamma_4 \\ \gamma_4 & \alpha_4 & \beta_4 \\ \beta_4 & \gamma_4 & \alpha_4 \end{bmatrix}\begin{bmatrix} i_1 \\ i_2 \\ i_3 \end{bmatrix} \tag{6a}$$

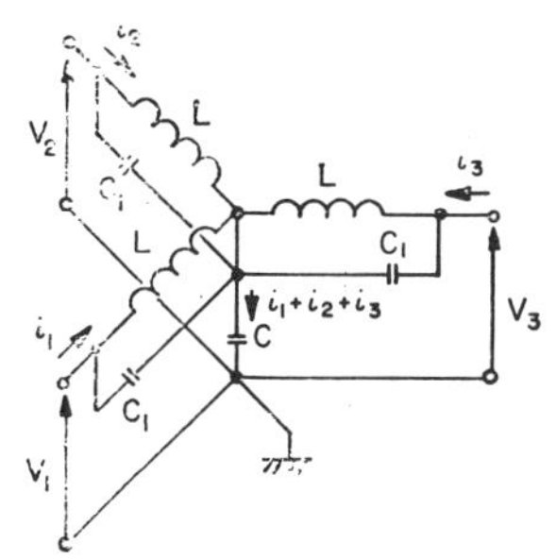

Fig. 6. Approximate equivalent circuit of broad-band lumped-element circulator. $L = L(\omega, G, 4\pi M, \mu, k, [a]_n)$.

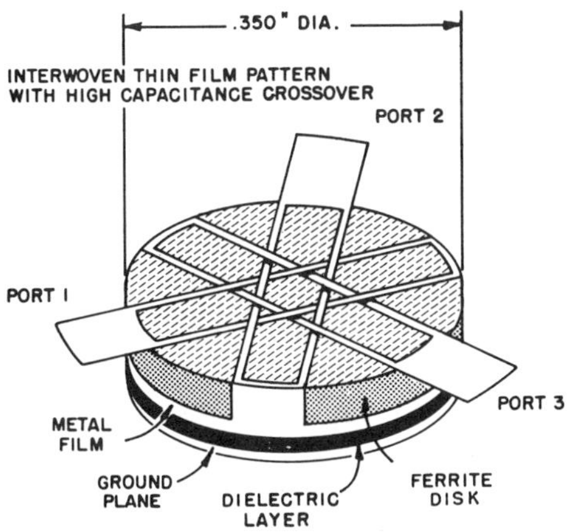

Fig. 7. Thin-film broad-band lumped-element circulator.

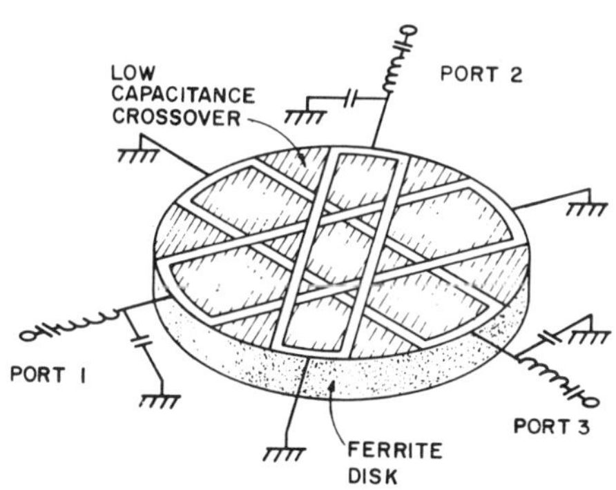

Fig. 8. Conventional broad-band lumped-element circulator.

where

$$\alpha_4 = \alpha_2 + X_c$$
$$\beta_4 = \beta_2 + X_c$$
$$\gamma_4 = \gamma_2 + X_c$$

or

$$[V] = [Z_4][i]. \tag{6b}$$

Equations (6a) and (6b) represent the most complicated configuration to be considered here.

In this section the voltage–current relations for different lumped-element lossless three-ports were derived. Those relations were all shown to be of the same form. Although the equations are now formally very simple, it should be noted that they are matrix equations whose elements are frequency- and field-dependent complex numbers. Their solution requires a digital computer. The results obtained here will be used in the next section to find numerical solutions for the eigenvalues of the equations, which in turn permit predictions on the broad-band behavior of the device.

III. Eigenvectors and Eigenvalues Applied to the Lumped-Element Circulator

Eigenvalue equations are classical mathematical problems and are easily solved for symmetrical boundary conditions. The mathematical theory of eigenvectors and eigenvalues is extensively treated in the literature [5]. The fact that the Maxwell's equations are invariant under several transformations, such as rotation or reflection, was used early to apply the eigenvalue technique to symmetrical waveguide junctions [6]. Since Maxwell's equations are linear in $\boldsymbol{E}$ and $\boldsymbol{H}$, the electric and magnetic field, respectively, their superposition is permitted. This permits a very simple physical interpretation in the case of n-port junctions. The eigenvectors represent the set of n special excitations which are applied to the n-port and the sum of which are equivalent to excitation of a single port. The response or the output quantities of the n-port will be determined by the physical structure of the junction as such. This structure is then taken into account by proportionality factors (which may be matrices) that relate input and output for the different eigenvector matrices. If, for example, the eigenvectors are given in the form of input voltages and the output in the form of currents, then the eigenvalues will have the dimensions of impedances and actually represent the impedance "seen" by the different eigenvector excitations. The same argument is valid for scattering parameters and wave-amplitude excitations.

In practice it is also true that the eigenvectors are related to actual sets of field modes in the junction. For a three port there exist three eigenvectors and at least three field modes that, because of the linearity of Maxwell's equations, can be superimposed to give the complete field pattern. As will be demonstrated subsequently, for some applications it is useful to consider the behavior of these modes or their eigenvalues separately.

In the literature most derivations are done in terms of the dimensionless scattering parameters. Since in the case of the lumped-element circulator, the impedance matrix of the ferrite-loaded nonreciprocally coupled inductors is known (1) and the impedance matrices for more complicated structures [(4b), (5b), and (6b)] have been derived, most computations will be done in terms of impedance eigenvalues, and at the appropriate point results will be transformed into scattering-parameter terminology to permit easy comparison with already existing basic results.

The eigenvectors for a symmetrical three-port are [6]

$$[a]_1 = \begin{bmatrix} 1 \\ 1 \\ 1 \end{bmatrix}; \qquad [a]_2 = \begin{bmatrix} 1 \\ b_2 \\ b_1 \end{bmatrix}; \qquad [a]_3 = \begin{bmatrix} 1 \\ b_1 \\ b_2 \end{bmatrix} \tag{7}$$

where in rectangular form

$$b_1 = -\frac{1}{2} + \frac{j}{2}\sqrt{3}; \qquad b_2 = -\frac{1}{2} - \frac{i}{2}\sqrt{3}. \tag{8}$$

If, for example, the eigenexcitation $[a]_2$ is applied to the circuit in Fig. 1 in the form of voltages, they would be of the form

$$V_1 = 1$$

$$V_2 = -\frac{1}{2} - \frac{j}{2}\sqrt{3} = 1 \exp(-j120^\circ)$$

$$V_3 = -\frac{1}{2} + \frac{j}{2}\sqrt{3} = 1 \exp(+j120^\circ).$$

The impedance matrix connecting these voltages to the corresponding currents is the matrix of the second eigenvalue. The eigenvalue λ_n can be found from [6]

$$\{[Z] - \lambda_n[I]\}[a]_n = 0, \qquad n = 1, 2, 3 \tag{9}$$

or

$$\det([Z] - \lambda_n[I]) = 0$$

where $[I]$ is the unity matrix and the λ_n are the impedance eigenvalues corresponding to the three eigenexcitations.

It is well known from the literature [7], [8] that the condition for circulation is equivalent to an angular separation of 120° (electrical degrees) of the scattering coefficients λ_n'.

The scattering-coefficient eigenvalues λ_n' are related to the impedance eigenvalues λ_n by

$$\lambda_n' = \frac{1 - \dfrac{\lambda_n}{Z_0}}{1 + \dfrac{\lambda_n}{Z_0}} \tag{10}$$

Z_0 being the characteristic impedance of the transmission line connected to the three-port.

Recall that (1), (4a), (5a), and (6b) for the four basic three-ports were of the general form

$$[V] = [Z_m][i], \qquad m = 1, 2, 3, 4$$

where

$$[Z_m] = \begin{bmatrix} \alpha_m & \beta_m & \gamma_m \\ \gamma_m & \alpha_m & \beta_m \\ \beta_m & \gamma_m & \alpha_m \end{bmatrix}. \tag{11}$$

This covers the four basic circuit configurations discussed in Section II. Using this general notation and (9), the impedance eigenvalues for the lumped-element three-port can be found:

$$\lambda_1 = (\alpha_m + \beta_m + \gamma_m)$$

$$\lambda_2 = \left[\alpha_m + \beta_m\left(-\frac{1}{2} + \frac{j}{2}\sqrt{3}\right) + \gamma_m\left(-\frac{1}{2} - \frac{j}{2}\sqrt{3}\right)\right]$$

$$\lambda_3 = \left[\alpha_m + \beta_m\left(-\frac{1}{2} - \frac{j}{2}\sqrt{3}\right) + \gamma_m\left(-\frac{1}{2} + \frac{j}{2}\sqrt{3}\right)\right] \tag{12}$$

and α_m, β_m, γ_m for $m = 1, 2, 3, 4$ are defined in (2), (4b), (5b), and (6a).

Case of $m = 1$

For the case of $m = 1$,

$$\lambda_1 = 0$$

$$\lambda_2 = j\omega G \frac{3}{2}(\mu + k) = j\omega G \frac{3}{2}\mu_- = \lambda_-$$

$$\lambda_3 = j\omega G \frac{3}{2}(\mu - k) = j\omega G \frac{3}{2}\mu_+ = \lambda_+. \tag{13}$$

The behavior of μ_+ and μ_- as functions of frequency is well known [2] and needs no further discussion. It is interesting to note that λ_1 is zero and that λ_2 and λ_3, which correspond to the counterrotating modes, are influenced by the product of the geometry factor G and ω, which is directly related to the inductive reactance of the conductors, and by the magnetic biasing field upon which μ_+ and μ_- are functionally dependent. That λ_1 should be zero is quite apparent if it is kept in mind that for equal phase excitation the magnetic-field components are 120° geometrically displaced and therefore cancel out, making the inductor look like a short circuit. It should be obvious from (10) that the λ' will be functions of the same variables as the corresponding λ.

Case of $m = 2$

For $m = 2$, the addition of C_1 (Fig. 2) affects only λ_2 and λ_3 leaving $\lambda_1 = 0$. The eigenvalue λ_1 is still zero since, as in the case of $m = 1$, the inductors still represent a short circuit that now shorts out C_1.

Fig. 9 shows a plot of the phases of λ_1', λ_2', and λ_3' using (9), (10), and (12) for different values of the capacitance C_1, the applied field being constant. $\angle\lambda_1'$ is independent of C_1 as expected and is $-180°$ corresponding to reflection from a short circuit. λ_2' and λ_3' show resonant responses since C_1 now can resonate with L. In Fig. 9 and all subsequent computer-generated plots, a ferrite linewidth $\Delta H = 35$ Oe has been assumed in order to include the effects of magnetic losses.

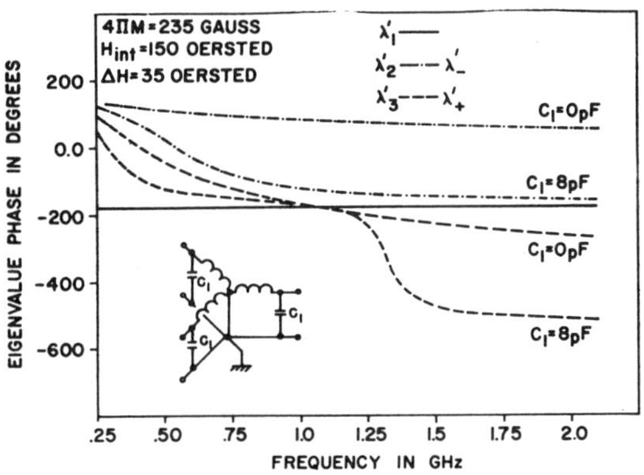

Fig. 9. *S*-parameter eigenvalue phase as function of frequency for structure in Fig. 2.

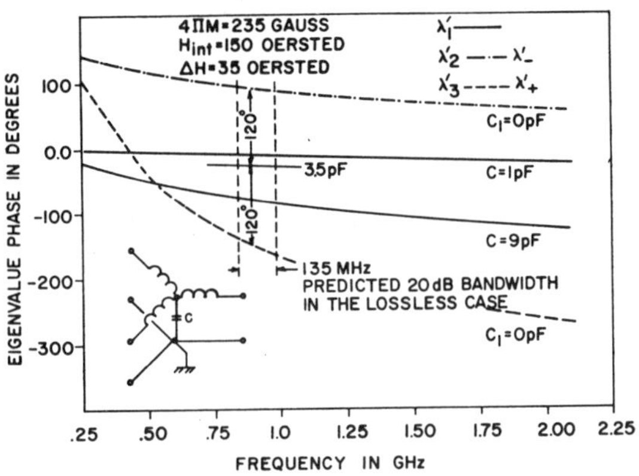

(a)

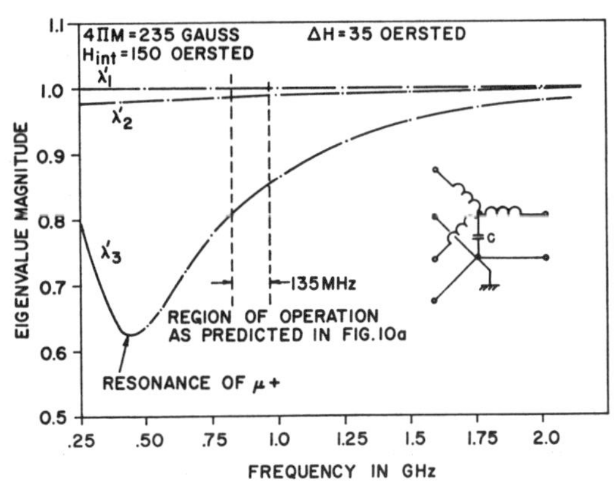

(b)

Fig. 10. *S*-parameter eigenvalue phase and magnitude for structure in Fig. 5.

Case of $m=3$

Using (9) in connection with Fig. 5 results in

$$\lambda_1 = \frac{3}{j\omega C}$$

$$\lambda_2 = j\omega G \frac{3}{2}(\mu + k) = j\omega G \frac{3}{2}(\mu_-) = \lambda_-$$

$$\lambda_3 = j\omega G \frac{3}{2}(\mu - k) = j\omega G \frac{3}{2}(\mu_+) = \lambda_+. \tag{14}$$

Equation (14) demonstrates clearly that the capacitor C influences λ_1 solely and not λ_2 or λ_3, which of course can be influenced by the applied biasing field. That C should not affect λ_2 and λ_3 is to be expected, since for these two eigenvalues i_1, i_2, and i_3 are 120° out of phase in C and therefore cancel. Fig. 10(a) shows the phases of the eigenvalues λ' for different values of C as in the structure of Fig. 5. Although there is a possible operation point for $C=3.5$ pF as indicated in Fig. 10(a), the 20-dB bandwidth is relatively small due to the rapid divergence of the eigenvalue phases, which is typical for this type of circulator structure. In addition magnetic losses can invalidate the 120° requirement for circulation. Fig. 10(b), shows the magnitude of the scattering-coefficient eigenvalues for the case of Figs. 5 and 10(a). The magnitude decreases with increasing losses. Conductor, dielectric, or radiation losses have not been taken into account. The region of high losses around 0.5 GHz coincides, as can be easily verified in (14), with the resonance of μ_+. It would be desirable to have a structure that combines the features of $m=2$ and $m=3$, i.e., a circuit that permits the adjustment of λ_1, λ_2, and λ_3 independently for constant H_{int}. This would serve two purposes: first, it permits high-loss regions to be avoided; and second, by proper choice of the geometry and material parameters, there exists a possibility of keeping the phase separation required for circulation approximately constant over a relatively broad band.

The structure which permits the independent adjustments of λ_1, λ_2, and λ_3 is presented in Figs. 6 and 7. The eigenvalue phases and magnitudes for this structure ($m=4$) are plotted in Fig. 11(a) and (b), respectively. In Fig. 11(a) the capacitor C influences only λ_1' while C_1 affects only λ_2' and λ_3'. There is a single circuit resonance indicated in λ_2' and two circuit resonances for λ_3'. These are resonances of C_1 and $L(\mu_\pm)$ and can be predicted from a knowledge of $\mu_\pm$ [2]. From Fig. 11(b) it can be seen that there is no loss for the in-phase mode, i.e., $|\lambda_1'|=1$. This is not surprising since the only loss mechanism accounted for is magnetic loss, and the model assumes a net zero magnetic field in the ferrite for the in-phase mode. The magnetic loss is most pronounced for λ_3' as might be expected since λ_3' is a function of μ_+ [(14)], which undergoes gyromagnetic resonance and has a relatively large imaginary part in the vicinity of this resonance. There is a single loss peak in λ_2' and two loss peaks in λ_3'. These loss peaks occur at the circuit resonances and for λ_3' increase as the circuit resonances approach the resonance of μ_+. This increase in the loss peak will continue until there is a match between the resonance losses and the characteristic impedance of the line, at which time the magnitude of λ_3' will be zero. Although a closer approach to gyromagnetic resonance will result in decreased peak losses, this area is not of practical interest for circulation because of the very rapid changes in the phases of the eigenvalues. One can conclude from Fig. 11(a) and (b) that the

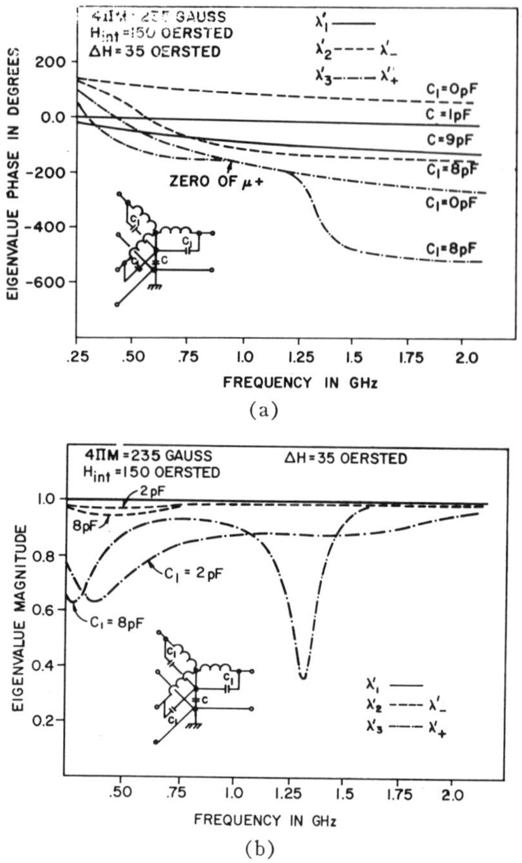

Fig. 11. S-parameter eigenvalue phase and magnitude variation for broad-band unit of Figs. 6 and 7.

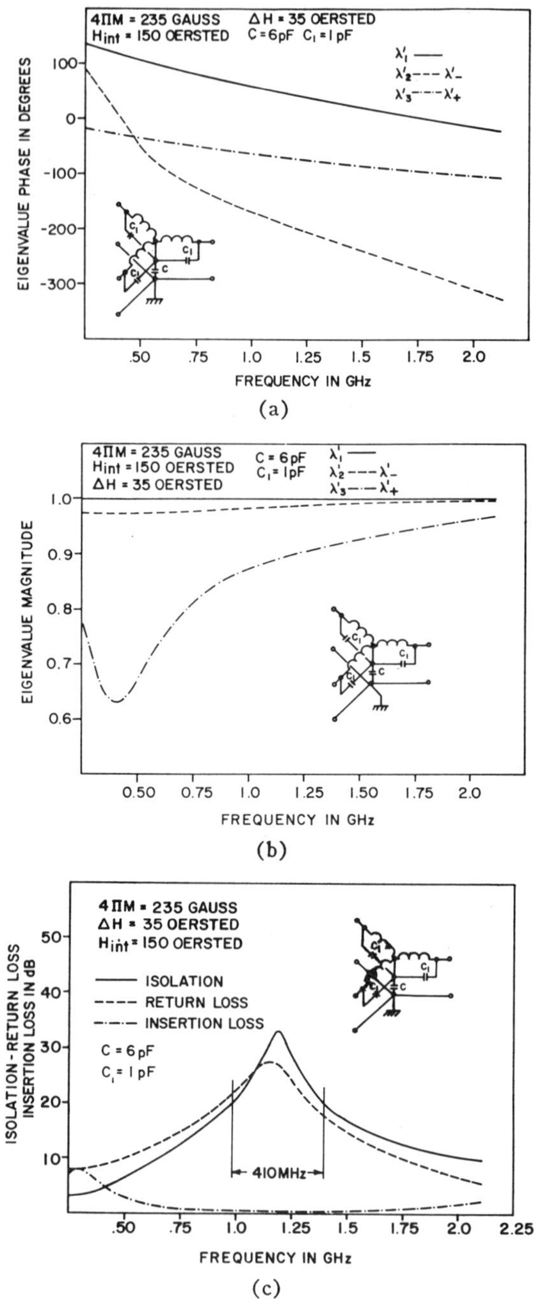

Fig. 12. S-parameter eigenvalue phase, magnitude, and computed performance for structure in Fig. 7.

eigenvalue-phase requirements for circulation can be satisfied in the vicinity of the circuit resonances and that minimum loss requires that these resonances be removed as far as possible from the gyromagnetic resonance.

Fig. 12(a)–(c) shows the results of selecting (but not necessarily optimizing) C and C_1 for circulation. All other parameters were kept the same as for Fig. 11(a) and (b). Fig. 12(a) and (b) gives the resultant eigenvalue phases and magnitudes. The importance of including magnetic losses can be indicated by calculating the circulator transmission and return losses, first from Fig. 12(a) with the eigenvalue magnitudes assumed to be unity, and second from the phases of 12(a) with the magnitudes shown in 12(b). In the first case the calculated 20-dB isolation bandwidth is approximately 410 MHz centered at 1.16 GHz; the return loss and isolation peak are at the same frequency at over 40 dB, and the insertion loss at this peak is essentially zero. The calculated circulator characteristics including the eigenvalue-magnitude variations are plotted in Fig. 12(c). The 20-dB isolation bandwidth is again approximately 410 MHz but is shifted to a center frequency of 1.19 GHz; the isolation and return loss peak are at slightly different frequencies and are significantly different in level; and the insertion loss is approximately .3 dB. Once again it is not safe to generalize from these characteristics, but they illustrate the kinds of effects resulting from losses in the circulator. It is interesting to note in passing that a second circulator is indicated at the low-frequency circuit resonance of λ_3' (field above gyromagnetic resonance) and that the sense of circulation is reversed. These two regions of circulation and the reversal in circulation are well known in the art.

The calculated curves for Fig. 12(c) were obtained as follows. The scattering matrix $[S]$ for a three-port is

$$[S] = \begin{bmatrix} S_{11} & S_{12} & S_{13} \\ S_{21} & S_{22} & S_{23} \\ S_{31} & S_{32} & S_{33} \end{bmatrix}.$$

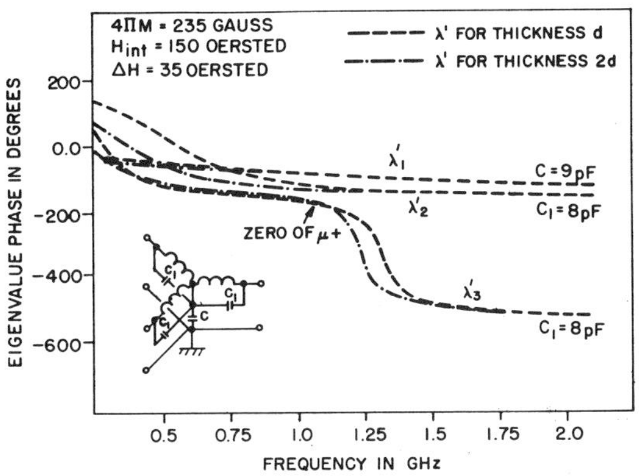

Fig. 13. *S*-parameter eigenvalue phase for different substrate thicknesses.

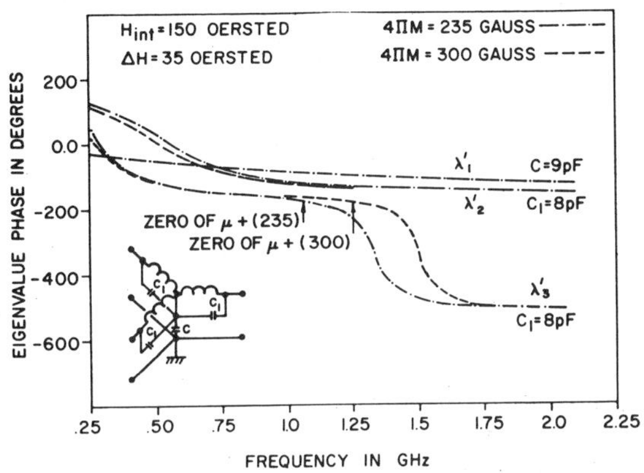

Fig. 14. *S*-parameter eigenvalue phase for different $4\pi M$.

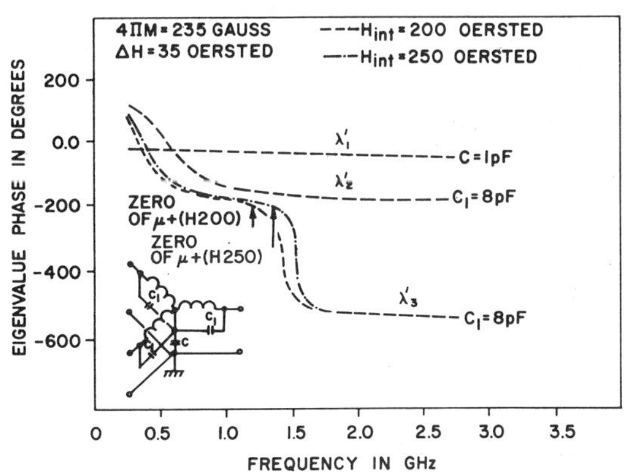

Fig. 15. *S*-parameter eigenvalue phase for different magnetic biasing fields.

From symmetry

$$S_{11} = S_{22} = S_{33}, \qquad S_{12} = S_{23} = S_{31}$$

and

$$S_{13} = S_{21} = S_{32}.$$

The S_{ij} expressed in terms of the λ_n' are

$$S_{11} = \tfrac{1}{3}(\lambda_1' + \lambda_2' + \lambda_3')$$

$$S_{12} = \tfrac{1}{3}(\lambda_1' + e^{j(2\pi/3)}\lambda_2' + e^{-j(2\pi/3)}\lambda_3')$$

$$S_{13} = \tfrac{1}{3}(\lambda_1' + e^{-j(2\pi/3)}\lambda_2' + e^{j(2\pi/3)}\lambda_3')$$

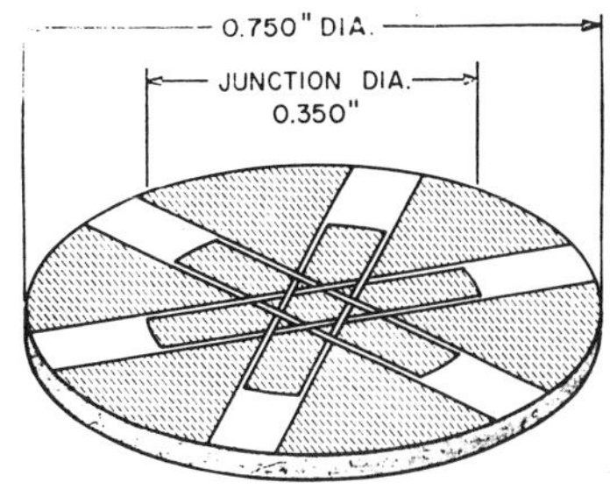

Fig. 16. Substrate dimensions used for measurements in Fig. 17.

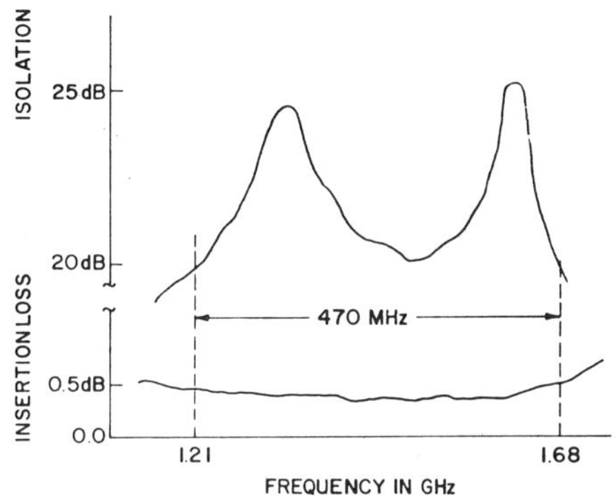

Fig. 17. Performance.

and the junction losses L are

$$A_1 = -20 \log |S_{21}|, \qquad \text{port 1 to 2}$$

$$A_2 = -20 \log |S_{31}|, \qquad \text{port 1 to 3}$$

$$20 \log |S_{11}|, \qquad \text{return loss}$$

where $A_{1,2}$ is either insertion loss or isolation depending on the sense of circulation.

In order to maximize the bandwidth it is necessary to vary the remaining parameters. Fig. 13 shows the variation of the eigenvalue phases for different thicknesses of the garnet substrate. To keep the plot simple, the value of the capacitance is held constant. The curves show that an increase of substrate thickness lowers the frequency of the circuit resonances. This is a useful adjustment parameter operating on the geometry factor G which in turn is related to the inductance L.

The λ_3' eigenvalue phase reacts to a change in the $4\pi M$ in a way related to the resultant change in the permeability μ_+ of the material; i.e., the circuit resonances shift with the resonance of μ_+. This is shown in Fig. 14 for two different $4\pi M$. Fig. 15 shows the effect of the applied magnetic biasing field on the eigenvalue phases. Once again the response is predictable from the known variation of μ_+ and μ_-.

Measurements were taken using 0.750-inch-diameter substrates (Fig. 16) in a housing with n-type connectors. The best results obtained in preliminary tests are shown in Fig. 17. Insertion loss less than 0.6 dB and isolation greater than 20 dB over a bandwidth greater than 30 per-

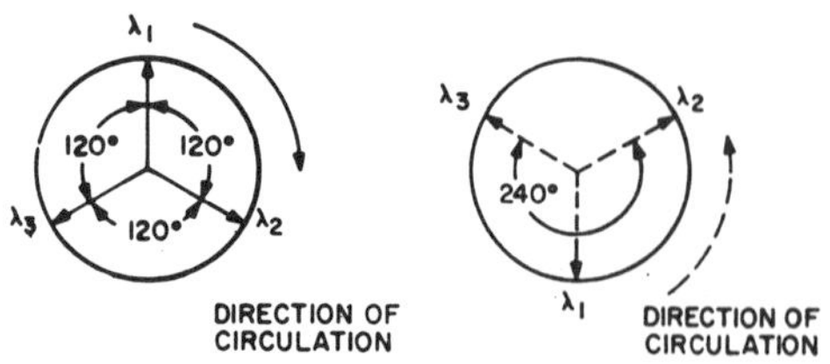

(a) (b)

Fig. 18. *S*-parameter eigenvalues for switching circulator.

cent have been obtained routinely. The double-peaked isolation of Fig. 13 was obviously not predicted by the calculations that yielded Fig. 12(b). The equivalent circuit for the in-phase eigenexcitation is incomplete in that it should include some series inductance arising from the fact that the magnetic fields due to this excitation do not completely cancel. The importance of this inductance is expected to arise from the continuing analysis and experimentation, which will yield more quantitative verification of the curves and a refined equivalent circuit of the device.

IV. Switching Without Changing the Applied Magnetic Biasing Field

Assume the circulator is adjusted so that the phase of the eigenvalues of (10) are

$$\angle\lambda_3' - \angle\lambda_2' = \angle\lambda_2' - \angle\lambda_1' = 120°$$

as indicated in Fig. 18(a). A change of the capacitance C_1 (Fig. 2) will leave λ_1' unaffected while increasing the separation of λ_2' and λ_3' so that

$$\angle\lambda_3' - \angle\lambda_2' = 240°.$$

$\angle\lambda_1'$ can then be adjusted by a change in C (Fig. 5) so that $\angle\lambda_3' - \angle\lambda_1' = 120°$ as in Fig. 18(b). Obviously the order of the subscript and hence the sense of circulation have been reversed [8]. This reversal has been observed experimentally, but no attempt has been made to build a switching circulator. Conceptually, at least, the variation in capacitance could be accomplished by means of semiconductor devices, an accomplishment which implies possibilities for a new generation of very fast switching circulators.

V. Conclusion

An analysis of the lumped-element circulator in terms of its impedance matrix and eigenvalues has been presented. The analysis is applicable to any lumped-element circulator. Besides giving a deep insight into the functional mechanisms and the influence of the different parameters, the method can be adapted to a complete computerized design. The particular broad-band circulator design reported here is compact, and although its performance compares favorably with that of other lumped-element circulators we are aware of, we are certain that further optimization is possible.

Acknowledgment

The authors wish to thank D. J. Thibault who initially observed the broad-banding effect of the capacitor C and carried out most of the experiments related to this work; and O. J. Duff, J. A. Hare, III, C. C. Shiflett, W. Worobey and all those who willingly assisted when consulted by the authors in the course of establishing crossover processes.

The experimental models reported here were processed in the Allentown Thin Films Models Laboratory using standard techniques [10].

References

[1] J. Deutsch and B. Wieser, "Resonance isolator and Y circulator with lumped elements at VHF," *IEEE Trans. Magn.*, vol. Mag-2, September 1966, pp. 278–282.

[2] B. Lax and K. J. Button, *Microwave Ferrites and Ferrimagnetics*. New York: McGraw-Hill, 1962.

[3] Y. Konishi, "Lumped element *Y* circulator," *IEEE Trans. Microwave Theory Tech.*, vol. MTT-13, November 1965, pp. 852–864.

[4] R. H. Knerr, "A thin film lumped element circulator," Int. Microwave Symp., Dallas, Tex., May 1969.

[5] Birkhoff and MacLane, *A Survey of Modern Algebra*. New York: Macmillan, 1941.

[6] Montgomery, Dicke and Purcell, *Principles of Microwave Circuits*. New York: Dover, 1965.

[7] J. B. Davies and P. Cohen, "Theoretical design of symmetrical junction stripline circulators," *IEEE Trans. Microwave Theory Tech.*, vol. MTT-11, November 1963, pp. 506–512.

[8] U. Milano, J. H. Saunders, and L. Davis, Jr., "A *Y*-junction strip-line circulator," *IRE Trans. Microwave Theory Tech.*, vol. MTT-8, May 1960, pp. 346–351.

[9] Von Aulock and Fay, *Linear Ferrite Devices for Microwave Applications*. New York: Academic Press, 1968.
V. E. Dunn and R. W. Roberts, "New design techniques for miniature VHF circulators," presented at the 1965 Int. Microwave Symp., Clearwater, Fla.
E. Pivit, "Zirkulatoren aus konzentrierten Schaltelementen," *Telefunken J.*, vol. 38, 1965, p. 206.

[10] H. Basseches and A. Pfahnl, "Crossovers for interconnections on substrates," *1969 Proc. Electronics Components Conf.*, Washington, D. C., pp. 78–82.

An Improved Equivalent Circuit for the Thin-Film Lumped-Element Circulator

REINHARD H. KNERR, MEMBER, IEEE

Abstract—A program written for the Hewlett-Packard automatic network analyzer permits the S-parameter eigenvalue phases and magnitudes to be displayed. The thus measured eigenvalues of a lumped-element circulator lead to an improved equivalent circuit which explains the observed "double hump" characteristic. The influence of different circuit parameters on the eigenvalues is measured and found in good agreement with the author's previously published theory. It is concluded from this theory and the measurements that, for the lossy circulator in general, maximum isolation, return loss, and minimum forward loss do not occur at the same frequency.

I. Introduction

IN A PREVIOUS PAPER [1] the author developed a theory on the thin-film lumped-element circulator using the eigenvalue analysis [2]. The computer results which were obtained from this theory improved our understanding of circulators in general and were applied to a new broad-banding principle [1], as well as to a new switching principle for circulators [3].

The availability of a computerized network analyzer has since made it practical to measure eigenvalue phases and magnitudes as a function of frequency using the proper computer program. These measurements have permitted the principal results of the theory to be verified and have revealed additional information that has led to an improved equivalent circuit for the thin-film lumped-element circulator. Furthermore, the study has advanced our understanding of the losses in the lumped-element circulator and has permitted us to draw conclusions as to the behavior of lossy circulators in general.

Manuscript received September 15, 1972; revised January 21, 1972.

The author is with Bell Telephone Laboratories, Inc., Allentown, Pa. 18103.

II. Comparison of the Experimental Lumped-Element Circulator with the Calculation for the Earlier Equivalent Circuit

The broad-band thin-film lumped-element circulator (described in a previous paper by the author [1]) is shown schematically in Fig. 1. It uses the crossover capacitances as the sole means of resonating the circulator junction. Broad-band behavior is achieved with a capacitor formed by the dielectric layer between the metal film and ground plane of Fig. 1. This capacitor is designated C_0 in Fig. 2, which shows the approximate equivalent circuit of the structure in Fig. 1. The crossover capacitances are represented by the discrete capacitors C_1, while the split conductors of Fig. 1 are delineated by the inductors L, which are functions of the angular frequency ω, the geometry factor [1] G, the saturation magnetization, and the applied magnetic biasing field H_{dc}.

From this structure and its analysis [1] it was deduced that the capacitor C_0 influences only the eigenvalue of the in-phase excitation, and the capacitors C_1, i.e., the crossover capacitances, influence only the eigenvalues due to the rotating excitations or eigenvectors. (For details on eigenvalues and eigenvectors see the

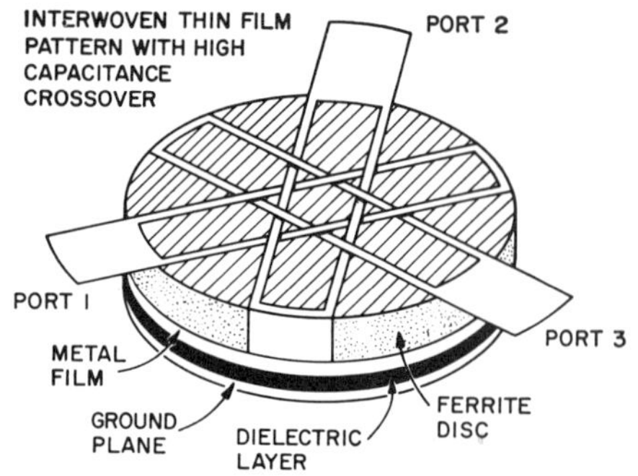

Fig. 1. Thin-film broad-band lumped-element circulator.

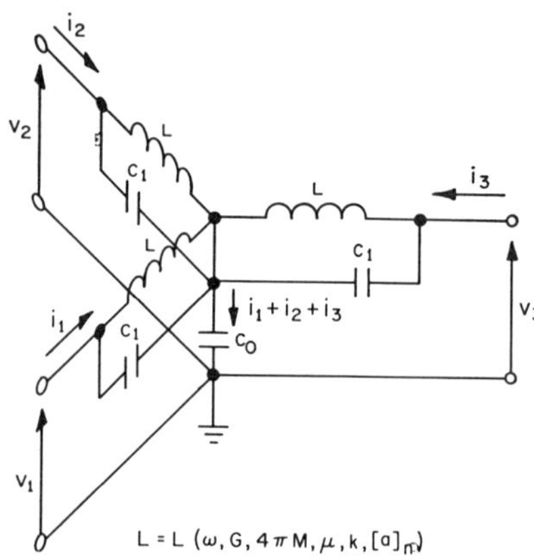

Fig. 2. Previous equivalent circuit used in [1].

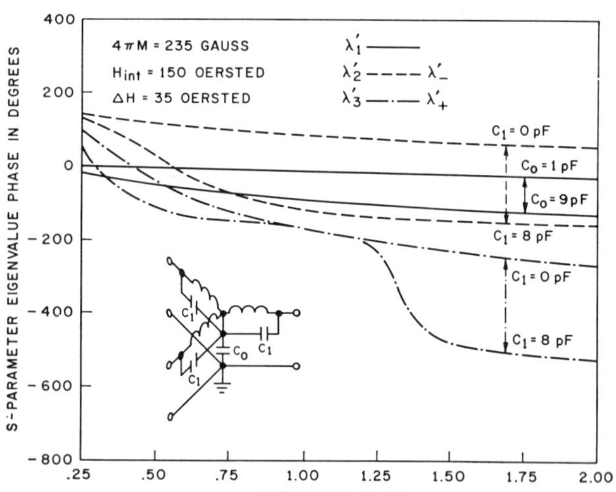

Fig. 3. Computed influence of capacitance on the S-parameter eigenvalue phases using the approximate equivalent circuit in Fig. 2.

Appendix.) This implies the possibility of adjusting the eigenvalue phases so that they trace with the proper phase separation over a wide frequency range; this represents a simple broad-banding mechanism. The experimental verification will be documented in a later paragraph. The theoretical curves from [1] are shown in Fig. 3, which indicates the movement of the eigenvalue phases for different values of the capacitors C_0

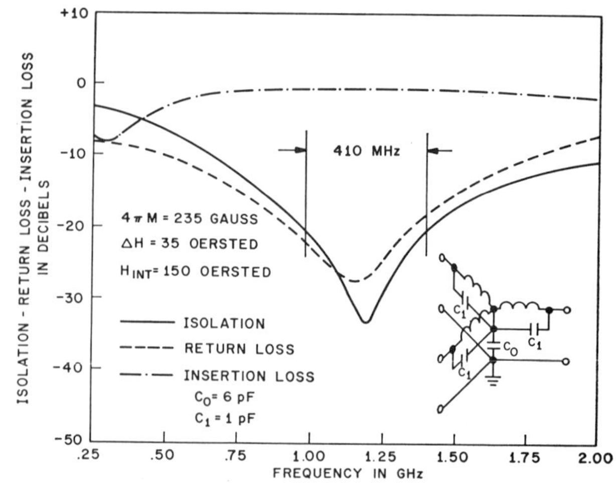

Fig. 4. Computed performance of the lumped-element circulator using the approximate equivalent circuit in Fig. 2.

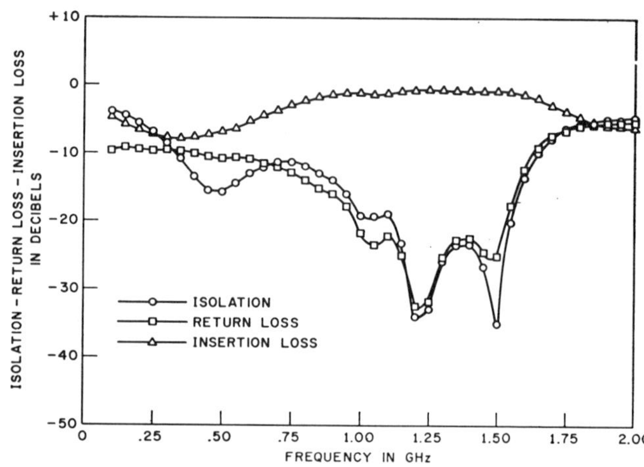

Fig. 5. Measured performance of circulator in Fig. 1.

and C_1. Fig. 4 shows the computed performance of a lumped-element circulator. The main difference between computed and measured performance (Fig. 5) is a very distinct "double hump" characteristic in the measured isolation as compared to the "single hump" in the computed curve. This deviation could not be fully explained with the previous equivalent circuit (Fig. 2) and will be analyzed in the following section using the measured eigenvalues of the device.

III. Eigenvalue Measurements with the Hewlett-Packard Automatic Network Analyzer

The eigenvalues of the S-parameter matrix for a symmetrical 3-port are given by [1]

$$\lambda_0' = S_{11} + S_{12} + S_{31}$$

$$\lambda_+' = S_{11} + S_{12} \exp(+j120°) + S_{13} \exp(-j120°)$$

$$\lambda_-' = S_{11} + S_{12} \exp(-120°) + S_{13} \exp(+j120°). \quad (1)$$

As the subscripts of the S parameters in these equations indicate, measurements at all three ports are required. This is inconvenient for fast high-precision measurements, since the device has to be connected and disconnected in order to take measurements between the corresponding 2-ports. It turns out, however, that for a symmetrical 3-port [2]

$$S_{13} = S_{21} = S_{32}. \tag{2}$$

This then reduces the problem to a 2-port measurement with the third port terminated and the device connected during all measurements. Measurements between ports 1 and 2, port 3 being terminated, yield S_{11}, S_{12}, and S_{21}, which according to (2) and (1) fully determine the eigenvalues.

The Hewlett-Packard (HP) computerized network analyzer, which very accurately measures the S parameters of 2-ports, was programmed to compute the eigenvalue phases and magnitudes. Either of these computed parameters can be displayed on an oscilloscope or an $X-Y$ recorder.

IV. Refinement of the Lumped-Element Circulator Equivalent Circuit from the Eigenvalue Measurements

Fig. 6 shows the measured S-parameter eigenvalue phases as a function of frequency for the device whose performance is pictured in Fig. 5. The angular phase separation is pointed out for some frequencies within the band of interest. The phase separation is not 120° as in the lossless case [2]. This is in agreement with the theory [1] in which magnetic losses have been taken into account by assuming the elements of the polder tensor to be complex [4]. As expected, the deviation from 120° can be quite significant. Even at 35-dB isolation there is a deviation of ±7°, and for 20 dB it may deviate as much as 23° in our special example. Comparing the measured phase curves with the computed ones in Fig. 3 shows a good qualitative agreement for eigenvalue phase λ_+'. The phase of λ_-' corresponds to the theoretical values in the range of circulator operation. At lower frequencies there are two irregularities R_3 and R_6 which were not predicted by the theory and have to be explained. The eigenvalue phase λ_0' changes from 0° to 360° passing through 180° in the center of the circulator band. This typical series resonant behavior was not predicted by our previous theory, which assumed unity coupling between the nonreciprocally coupled areas; i.e., it assumed the inductors to be short circuits for the in-phase excitation [1]. This assumption is in practice not always warranted; there is actually a part of the inductance which does not participate in the nonreciprocal coupling. This inductance is seen by a linearly polarized wave and is mainly a function of μ_{eff}, the permeability for a plane linearly polarized wave, which is defined as

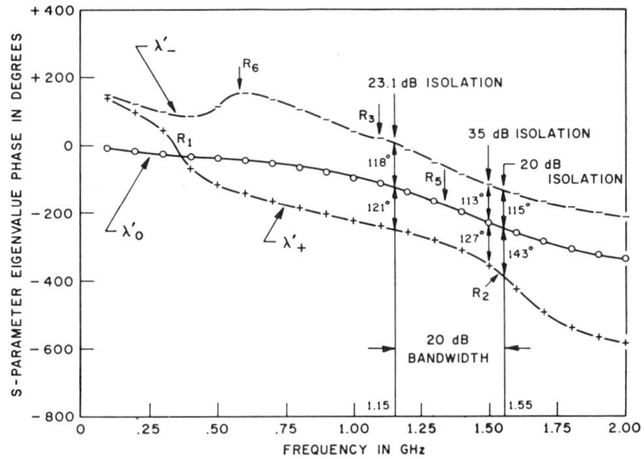

Fig. 6. Measured eigenvalue phase of circulator in Fig. 1.

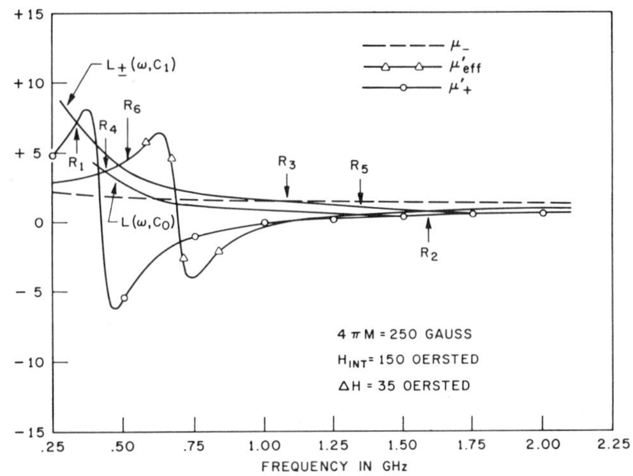

Fig. 7. Permeabilities for linearly and circularly polarized waves and possible circuit resonances.

$$\mu_{\text{eff}} = \frac{\mu^2 - k^2}{\mu} \tag{3}$$

μ and k being the elements of the polder tensor [6], [4]. The rotating modes "see" μ_+ and μ_-, respectively, where

$$\mu_+ = \mu + k$$
$$\mu_- = \mu - k. \tag{4}$$

These permeabilities for one specific material with a fixed magnetic biasing field are shown in Fig. 7 as a function of frequency. They will help to explain the resonances of the different eigenvalues. The inductance formed by the split conductor pattern is loaded with μ_+ or μ_- for the rotating excitations and resonates with the capacitors C_1. The inductance not participating in the nonreciprocal action is μ_{eff} loaded and "seen" by all three modes of excitation and is in series resonance with the capacitor C_0 for the in-phase excitation. These inductances, divided by their respective geometry factors,

225

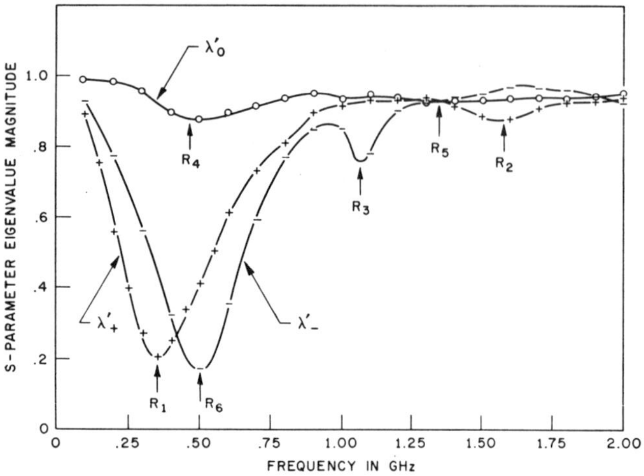

Fig. 8. Measured eigenvalue magnitude of circulator in Fig. 1.

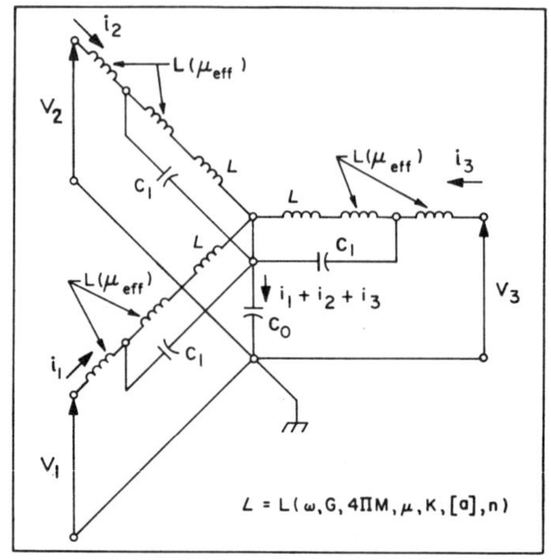

Fig. 9. Improved equivalent circuit of circulator in Fig. 1.

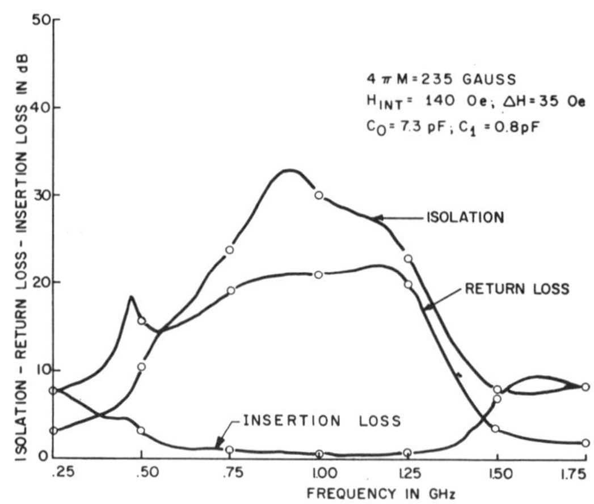

Fig. 10. Computer generated performance using improved equivalent circuit of Fig. 9.

intersect the permeability curves in Fig. 7 at several points (R_1–R_5). These intersection points of $L_\pm(\omega, C_1)$ with μ_+ and μ_- correspond to possible circuit resonances for the rotating excitations, and the intersection of $L(\omega, C_0)$ with μ_{eff} to possible resonances of the in-phase excitation. These resonances are expressed by $(0° + n\ 360°)$ or $(180° + n\ 360°)$ phase angles of the eigenvalue phases ($R_1 - R_5$ in Fig. 6), depending on whether it is a series or parallel resonance. These resonances and the increased losses associated with them can also be deduced from the magnitude of the eigenvalues, which are shown as a function of frequency in Fig. 8. If mismatch losses are not taken into account the magnitudes decrease with increasing losses. The loss peaks $R_1 - R_5$ then can be easily identified with the resonances in Figs. 6 and 7. There is one additional loss peak (R_6) in the magnitude of the negative rotating mode. This is probably due to the polarization of the wave tending to change from circular to linear from the center to the edge [5], which is substantiated by the presence of this peak at the same frequency as the peak for the in-phase excitation (R_4), i.e., near the μ_{eff} resonance. This polarization change is probably also influencing the position of R_1.

As a result of these measurements, an improved equivalent circuit which takes into account the effect of the μ_{eff} loaded inductance at each port can be drawn (Fig. 9). Using this improved equivalent circuit, the computed performance as shown in Fig. 10 does exhibit the experimentally observed double hump.

V. Experimental Control of the Eigenvalues

In order to obtain the broad-band lumped-element circulator characteristics also described in [1], it is essential to be able to adjust the in-phase eigenvalue independently of the rotating eigenvalues or vice versa. Since the refined equivalent circuit now adequately describes the eigenvalues, the influence of different parameters on the eigenvalues can be measured and better understood. Fig. 11(a) shows the measured phases of the eigenvalues as a function of frequency (1 GHz–2 GHz) for two different values of crossover capacitance, all other parameters held constant. The change of the crossover capacitance was achieved by encapsulating the air crossover with the Dow Corning RTV 3140. The results clearly show that the change in crossover capacitance produced the predicted changes in the phases of the rotating eigenvalues. This phase change corresponds to a downshift in frequency of the minima of the magnitudes of the rotating eigenvalues [Fig. 11(b)], since an increase in crossover capacitance increases the LC product, and therefore decreases the resonance frequency in Fig. 7.

A change of the capacitance C_0 in Fig. 2 has been predicted to influence the in-phase excitation only. Fig. 12 verifies this effect. The magnitudes of the eigenvalues stay essentially unchanged for this limited change in C_0.

An analysis of the improved equivalent circuit of Fig. 9 shows that a change of the magnetic biasing field

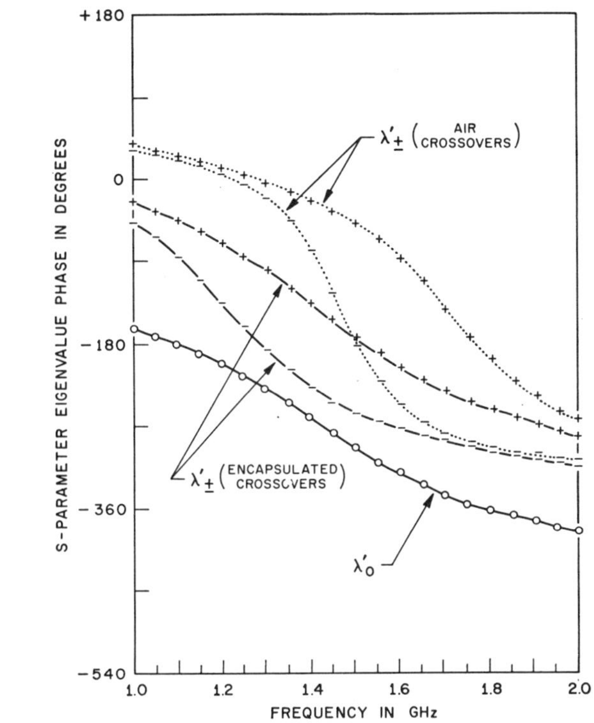

(a)

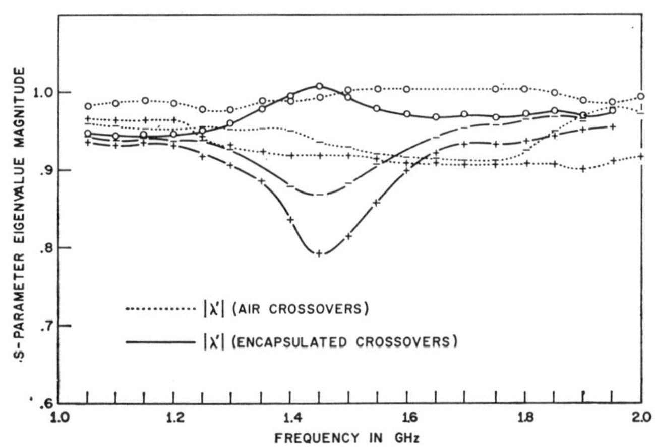

(b)

Fig. 11. (a) Influence of the crossover capacitance on the S-parameter eigenvalue phase. (b) Influence of the crossover capacitance on the S-parameter eigenvalue magnitude.

should influence all three eigenvalues. The large inductance L, which is a function of $\mu_\pm$ and is seen only by the rotating excitations, will be more influenced than the small inductance $L(\mu_{eff})$. Thus, for small changes in the applied magnetic field, only the rotating eigenvalues are affected in phase [Fig. 13(a)] as well as in magnitude [Fig. 13(b)].[1] This is especially valid if the circulator is operated below resonance. As expected for an in-

[1] The data of Fig. 13 were taken using the circulator with the encapsulated crossover (Fig. 11).

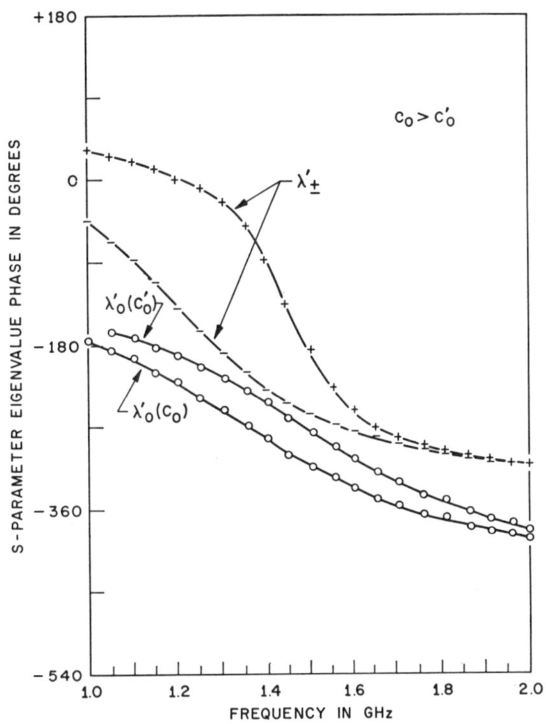

Fig. 12. Influence of the capacitance C_0 of Fig. 9 on the S-parameter eigenvalues.

creased field, the resonance of the eigenvalue following μ_+ shifts up in frequency since μ_+, and hence L, decrease for increasing field H.

These measurements, which verify the theory [1], and the insight gained therefrom permit us to make some more general statements about the lossy circulator applicable to any type of circulator.

VI. The "Lossy" Circulator

Carlin [7] defines the ideal 3-port circulator to be identical to a lossless 3-port, matched at all three ports. This implies that 1) the eigenvalue magnitudes of this 3-port are unity, and 2) the S-parameter eigenvalue phase separation is 120°. Since by definition losses from port 2 to 1:

$$A_1 = 20 \log_{10} |S_{12}| \tag{5a}$$

losses from port 3 to 1:

$$A_2 = 20 \log_{10} |S_{13}| \tag{5b}$$

and

$$\text{return loss} = 20 \log_{10} |S_{11}| \tag{5c}$$

where

$$S_{11} = \tfrac{1}{3}(\lambda_1' + \lambda_2' + \lambda_3') \tag{6a}$$

$$S_{12} = \tfrac{1}{3}(\lambda_1' + \lambda_2' \exp(j120°) + \lambda_3' \exp(-j120°)) \tag{6b}$$

$$S_{13} = \tfrac{1}{3}(\lambda_1' + \lambda_2' \exp(-j120°) + \lambda_3' \exp(j120°)) \tag{6c}$$

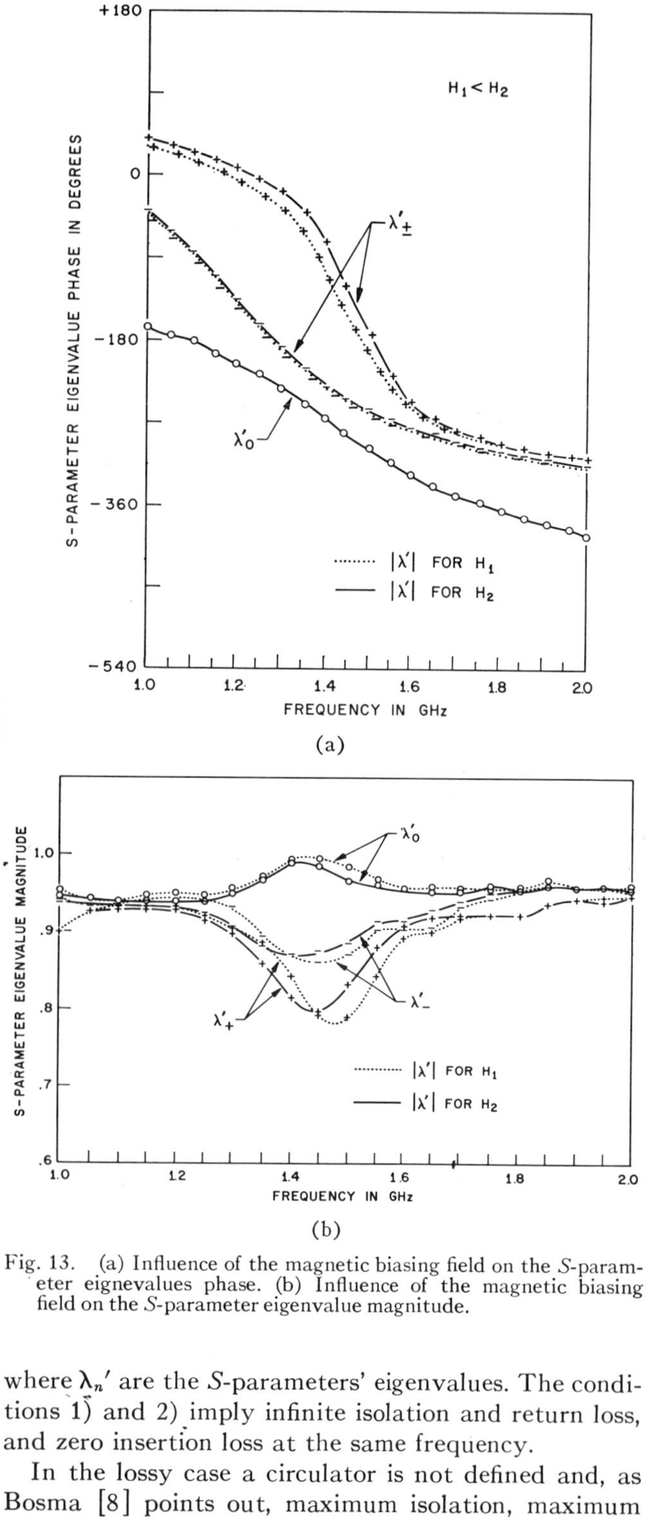

Fig. 13. (a) Influence of the magnetic biasing field on the S-parameter eignevalues phase. (b) Influence of the magnetic biasing field on the S-parameter eigenvalue magnitude.

where λ_n' are the S-parameters' eigenvalues. The conditions 1) and 2) imply infinite isolation and return loss, and zero insertion loss at the same frequency.

In the lossy case a circulator is not defined and, as Bosma [8] points out, maximum isolation, maximum return loss, and minimum insertion loss may not occur at the same frequency. This can be easily explained. It has become already obvious from our measurements and theory [1] that the different modes of excitation "see" different μ's (μ_+, μ_-, μ_{eff}), which have different resonant frequencies and therefore different losses at any

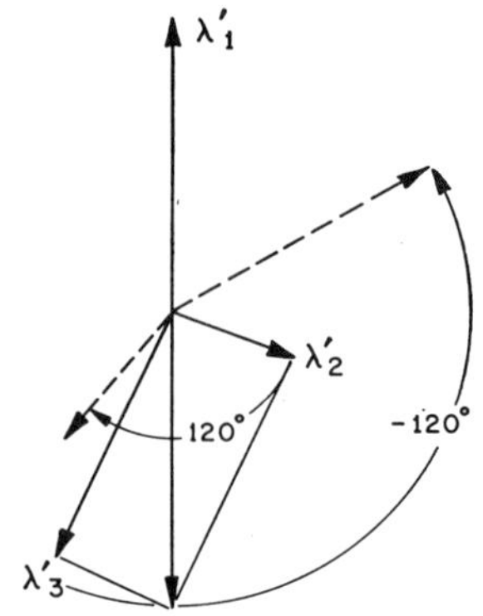

Fig. 14. Phase diagram of S-parameter eigenvalues at a fixed frequency for the "lossy" circulator.

one frequency. The forward loss of a circulator is therefore the combination of the losses due to these different modes of excitation.

If one assumes that in the example of Fig. 14 the eigenvalues[2] λ_n' have been adjusted at one fixed frequency (indicated by the solid arrows), their geometrical sum equals zero. This makes S_{11} in (6a) equal to zero, and causes an infinite return loss according to (5c). If one then satisfies (6b) by the corresponding rotation of the eigenvalues as indicated by the dashed arrows, their geometrical sum is no longer zero, i.e., the isolation is not infinite at this frequency. Certainly, (6b) may have a minimum at some other frequency, since the phase separations as well as the magnitudes of the vectors in Fig. 14 are frequency dependent. A similar argument is valid if one maximizes (6a) for minimum forward loss. Consequently, it can be concluded that, for the lossy circulator in general, maximum isolation, return loss, and minimum forward loss do not occur at the same frequency. Theoretically, there is one exception possible; if all three magnitudes are equal, then the three functions of (6) go through an extremum at the same frequency. As far as the magnetic losses are concerned, this is only true very far from ferromagnetic resonance and, therefore, is of no interest for circulators. If dielectric losses are taken into account, additional external elements such as capacitors, placed into the circuit in such a way as to be recognized by the individual modes of excitation, could conceivably make the magnitudes equal, but there is no general rule as to the amount of electric or magnetic energy to be dissipated to achieve this. For any practical application, the losses are small enough to make the displacement negligibly small.

VII. Conclusion

Eigenvalue measurements of the thin-film lumped-element circulators using the HP computerized network analyzer have verified to a large extent the author's

[2] Magnitudes different from unity indicate losses.

earlier theoretical results. New insight into the device behavior has been gained, and an improved equivalent circuit has been derived. The results are applicable to circulators in general and give valuable information on the loss mechanisms in circulators.

Appendix

Derivation of the S-Parameter Eigenvalues

The impedance matrix of a symmetrical 3-port is of the form

$$[Z] = \begin{bmatrix} \alpha & \beta & \gamma \\ \gamma & \alpha & \beta \\ \beta & \gamma & \alpha \end{bmatrix}.$$

This 3-port is lossless if [9] Re $(\alpha) = 0$ and $\beta = -\gamma^*$. Any symmetrical 3-port can be completely described by three eigenvectors or eigenexcitations which in this case are voltages:

$$[V]_1 = \begin{bmatrix} 1 \\ 1 \\ 1 \end{bmatrix}$$

$$[V]_2 = \begin{bmatrix} 1 \\ 1 \exp(+j120°) \\ 1 \exp(-j120°) \end{bmatrix}$$

$$[V]_3 = \begin{bmatrix} 1 \\ 1 \exp(-j120°) \\ 1 \exp(+j120°) \end{bmatrix}.$$

The impedances seen by these eigenexcitations are the impedance eigenvalues

$$\lambda_1 = (\alpha + \beta + \gamma)$$

$$\lambda_2 = \left(\alpha + \beta\left(-\frac{1}{2} + \frac{j}{2}\sqrt{3}\right) + \gamma\left(-\frac{1}{2} - \frac{j}{2}\sqrt{3}\right)\right)$$

$$\lambda_3 = \left(\alpha + \beta\left(-\frac{1}{2} - \frac{j}{2}\sqrt{3}\right) + \gamma\left(-\frac{1}{2} + \frac{j}{2}\sqrt{3}\right)\right).$$

The scattering coefficient eigenvalues can be computed from the impedance eigenvalues by

$$\lambda_n' = \frac{1 - \dfrac{\lambda_n}{Zo}}{1 + \dfrac{\lambda_n}{Zo}}, \qquad n = 1, 2, 3$$

Zo being the characteristic impedance of the connecting transmission line. The matrix elements α, β, γ for the lumped-element circulator are given in [1].

Acknowledgment

The author wishes to thank D. J. Thibault, who made most of the measurements.

References

[1] R. H. Knerr, C. E. Barnes, and F. Bosch, "A compact broad-band thin-film lumped-element L-band circulator," *IEEE Trans. Microwave Theory Tech. (1970 Symposium Issue)*, vol. MTT-18, pp. 1100–1108, Dec. 1970.
[2] B. A. Auld, "Synthesis of symmetrical waveguide circulators," *IRE Trans. Microwave Theory Tech.*, vol. MTT-7, pp. 238–246, Apr. 1959.
[3] R. H. Knerr, "A proposed lumped-element switching circulator principle," *IEEE Trans. Microwave Theory Tech.*, vol. MTT-20, pp. 396–401, June 1972.
[4] R. F. Soohoo, *Theory and Applications of Ferrites*. Englewood Cliffs, N. J.: Prentice-Hall, 1960, pp. 64–65.
[5] C. E. Fay and R. L. Comstock, "Operation on the ferrite junction circulator," *IEEE Trans. Microwave Theory Tech.*, vol. MTT-13, pp. 15–27, Jan. 1965.
[6] D. Polder, "On the theory of ferromagnetic resonance," *Phil. Mag.*, vol. 40, p. 99, 1949.
[7] H. J. Carlin, "Principles of gyrator networks," in *Proc. Symp. Modern Advances in Microwave Tech.*, Polytech. Inst. of Brooklyn, Brooklyn, N. Y., Nov. 1954.
[8] H. Bosma, "Performance of lossy H-plane Y circulators," *IEEE Trans. Magn.*, vol. MAG-2, pp. 273–277, Sept. 1966.
[9] E. Pivit, "Zirkulatoren aus konzentrierten Schaltelementen," *Telefunken J.*, vol. 38, p. 206, 1965.

Design of a New Broad-Band Isolator

YOSHIHIRO KONISHI, SENIOR MEMBER, IEEE, AND NORIO HOSHINO

Abstract—**The design theory and experimental results of a proposed broad-band technique of an isolator are presented. The isolator not only has a broad-band backward loss but also a low broad-band insertion loss. The network constants and the performance are obtained in connection with the magnetic loss of ferrite, and they are confirmed by experimental data. The trial ultrawide-band isolator shows the performance of 20-dB isolation over the frequency band from 70 MHz to 800 MHz.**

I. Introduction

IN THE wide-band technique of circulators and isolators, the resonant networks were used in the past [1] as shown in Fig. 1 (a). The insertion loss of this wide-band isolator, however, rapidly increases in the frequency region outside the center frequency range as shown in Fig. 1 (b), so that it is impossible to connect several isolators in cascade. This paper, therefore, proposes a method that not only simplifies the wide-band network but also makes it possible to maintain the low insertion loss even in the frequency range outside the center frequency region. The design theory of this proposed technique for a lumped-element circulator and an extension to a distributed circulator are herein presented. We also show that the network constant is largely controlled by the magnetic loss of the ferrite. Therefore, the design theory is related to the insertion loss and wide-band performance. Since the design used for a circulator is different from that for a cascade connection of several isolators, designs are shown for each of the above cases in this paper. Finally, the performance of a trial isolator with the extremely wide band covering the frequency range from 70 MHz to 800 MHz is shown.

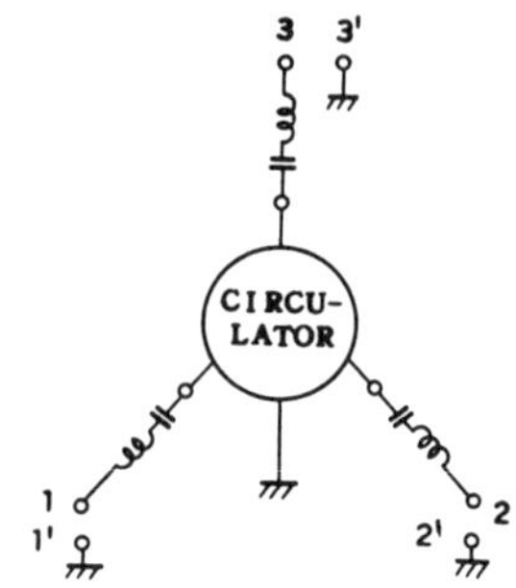

(a)

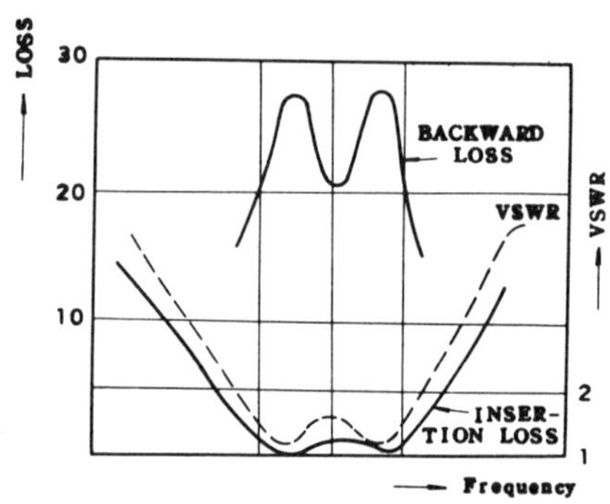

(b)

Fig. 1. Construction and performance of an ordinary wide-band circulator. (a) Construction. (b) Performance.

II. Construction and Physical Explanation

Fig. 2 (a) shows a proposed construction where a series resonant network is connected between a ground plate and the center of the outer conductor of a circulator insulated from a ground plate. This construction is realized by a lumped-element network on the principle [2] that an effective method to broad band is to increase the reactive energy only for the first eigenvector [1] of a circulator. In this mechanism, the current flows only for the first eigenvector through the resonant network and it increases the reactive energy, whereas current does not flow through the network for the second and the third eigenvector [1]. In a circulator before wide band, however, the eigenimpedance for the first eigenvector is zero for any frequency so it has no frequency variation, whereas eigenimpedances for the second and the third eigenvectors have some frequency variations as shown in Fig. 3 [2]. Therefore, the properly designed wide-band network makes it possible for the eigenimpedances of the first eigenvector to take the same frequency variation as that of the second and third eigenvector and to create a wide-band performance.

In the frequency range far from the center frequency band operating as a nonreciprocal circuit, the additional resonant network presents a high impedance and the circulator takes a low impedance, so that the equivalent network becomes a T branch with terminals 1, 2, and 3 as shown in Fig. 2 (c).

In such a frequency range, therefore, the insertion loss and the standing wave ratio of a circulator or isolator approach 3 dB and 2, respectively, as shown in Fig. 4 (d), because the terminal resistance is connected between terminal 3 and a ground plate as shown in Fig. 4 (c). On the other hand, when the absorbing resistance is connected between terminal 3 and the outer conductor of an original circulator before wide band as shown in

Manuscript received March 31, 1970; revised September 1, 1970.
The authors are with Nippon Hoso Kyokai (Japan Broadcasting Corporation), Tokyo, Japan.

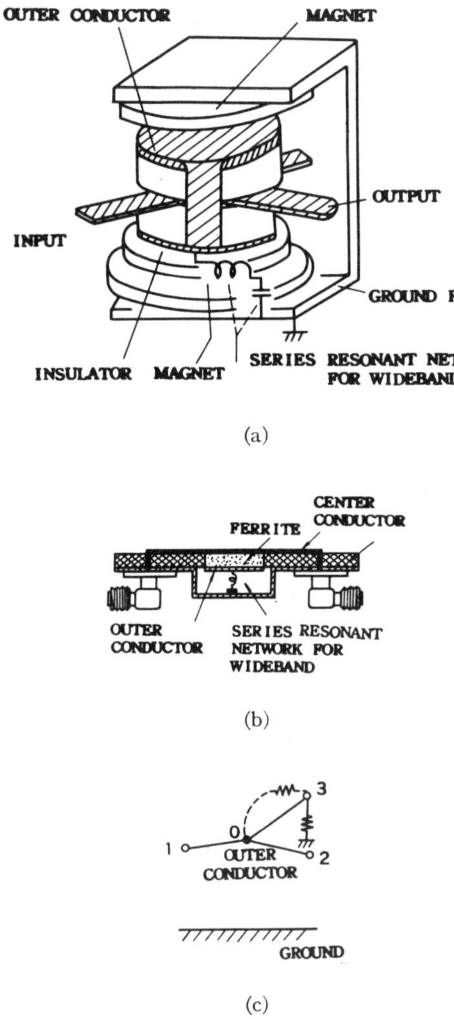

Fig. 2. Construction of proposed wide-band circulator and equivalent network at frequency range far from center frequency. (a) Wide-band lumped-element circulator. (b) Wide-band stripline circulator. (c) Equivalent network at frequency range far from center frequency.

Fig. 4 (a), it is understood from the equivalent network of Fig. 2 (c) that the insertion loss approaches 0 dB and the standing wave ratio approaches 1 in the frequency range far from the center frequency with the performance of Fig. 4 (b). This enables low insertion loss to be maintained over an extremely wide band and enables several isolators to be connected in cascade to obtain the extremely wide-band isolator shown later, or to get an arbitrary performance on a backward loss.

Furthermore, the bandwidth of the backward loss of an isolator is still more widened as shown in Fig. 4 (f) by connecting another series resonant network in series with the absorbing resistance as shown in Fig. 4 (e). In his case the low insertion loss is still maintained over an extremely wide band.

In the following, the design theory of a wide-band lumped-element isolator and circulator is described for each case mentioned before. Finally the theory is extended to a stripline Y circulator.

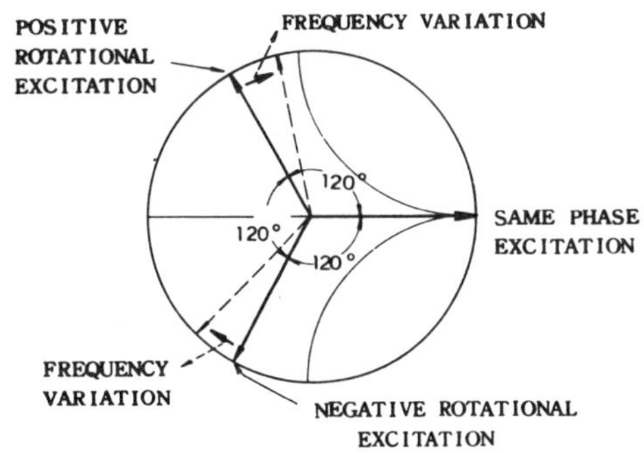

Fig. 3. Frequency variation of eigenvalues of circulator before wide band.

III. Design Theory

A. The Case of Connecting a Resistance Between a Terminal and the Outer Conductor of a Circulator

In Fig. 4 (a), 1, 2, and 3 show the terminal of a circulator and 1″, 2″, and 3″ show the outer conductor of the circulator before wide band. Eigenvalues of admittances of the first, the second, and the third eigenvectors, y_1, y_2, and y_3, are [1]

$$y_1 = \infty, \qquad y_2 = j\omega c + \frac{1}{j\omega\mu_-\xi}, \qquad y_3 = j\omega c + \frac{1}{j\omega\mu_+\xi} \tag{1}$$

where μ_+ and μ_- are negative and positive polarized relative permeability, C is the capacity at each terminal, and $\omega\xi$ is the eigenimpedance for the second and the third eigenvectors when the ferrite disks are removed. At the center angular frequency ω_0, y_2, and y_3 of a lossless circulator take the values of [1]

$$y_{\substack{2\\3}} = \mp j\frac{Y_0}{\sqrt{3}} \tag{2}$$

where Y_0 is a terminal admittance.

Therefore, we have a relation

$$\frac{1}{\omega_0\xi}\left(\frac{1}{\mu_{-0}} - \frac{1}{\mu_{+0}}\right) = \frac{2}{\sqrt{3}}Y_0 \tag{3}$$

where μ_{-0} and μ_{+0} are the values of μ_- and μ_+ at ω_0. If $\mu_\mp$ is assumed to take the values of Polder's equation,

$$\mu_\mp = 1 + \frac{P}{\sigma \pm 1} = 1 + \frac{P_0}{\sigma_0 \pm x}$$

$$P = \frac{|\gamma|\,Ms}{\omega}, \qquad \sigma = \frac{|\gamma|\,H_0}{\omega}, \qquad x = \frac{\omega}{\omega_0} \tag{4}$$

where

Ms	saturated magnetization
$\lvert\gamma\rvert$	gyromagnetic ratio
σ_0	value of σ at ω_0
μ_0	magnetic permeability at vacuum
P_0	value of P at ω_0
H_0	internal dc magnetic field.

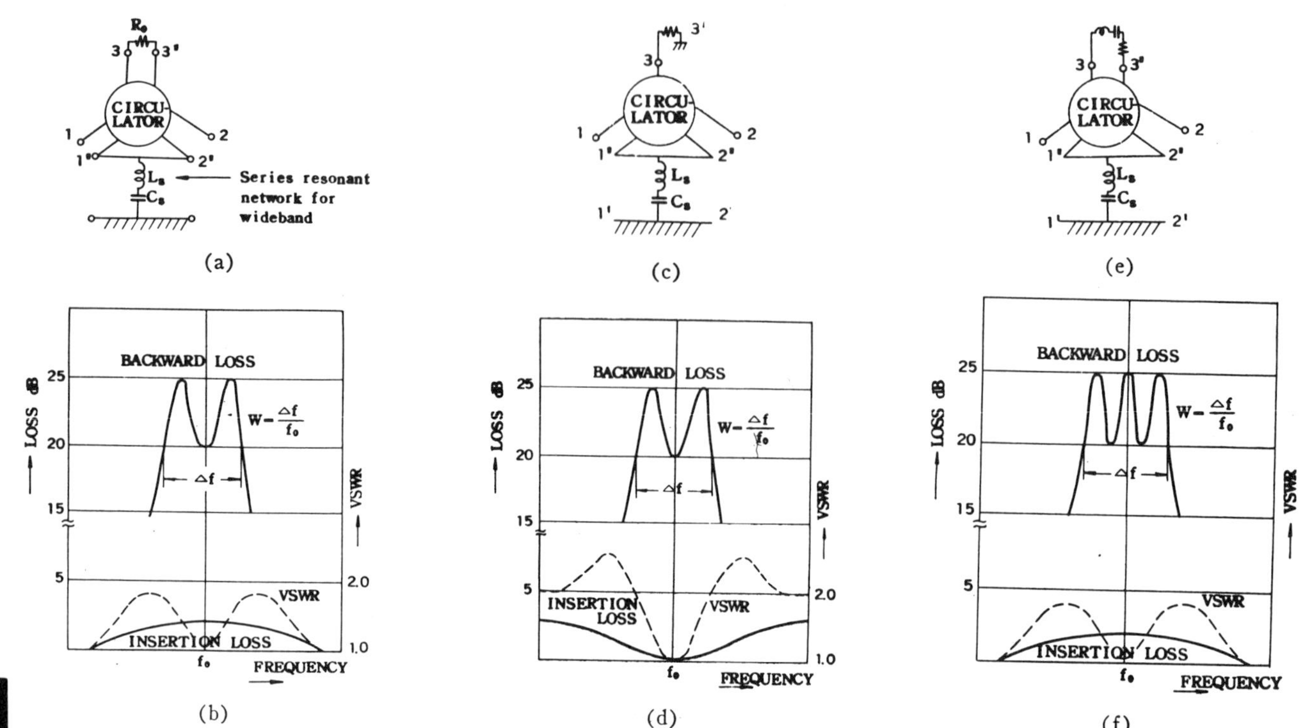

Fig. 4. Several kinds of wide-band methods and performances.

Substituting (3) and (4) into (1) and considering (2),

$$y_{\frac{2}{3}} = jQY_0\left[x - \frac{(\sigma_0 + P_0)^2 - 1}{\sqrt{3}P_0Qx}\cdot\frac{1}{1 + \dfrac{P_0}{\sigma_0 \pm x}}\right],$$

$$Q = \omega_0 C Z_0, \qquad Z_0 = \frac{1}{Y_0}. \qquad (5)$$

The values of $\mu_{\mp}$, however, take the loss components $\mu_{\mp}''$ and are related to their quality factor Q by

$$\mu_{\mp} = \mu_{\mp}' - j\mu_{\mp}'' = \mu_{\mp}'\left(1 - j\frac{1}{Q_{\mp}}\right) \qquad (6)$$

where $Q_{\mp}$ are the Q values of $\mu_{\mp}$.

The lossless components $\mu_{\mp}'$ are generally related to $\mu_{\mp}''$ by the Kramers–Kronig relation and therefore take different values from that shown by (4). However, we approximate the values of $\mu_{\mp}'$ by (4) to simplify our consideration. Concerning a lossy circulator, however, Bosma states that the loss term caused by electric and magnetic losses of three eigenvectors should take the same values to realize a circulator [3]. In our proposed circulator, the loss term of the first eigenvalues is controlled only by Q_w; that of the second eigenvalues by Q_- and Q_c; and that of the third eigenvalues by Q_+ and Q_c; where Q_c is a quality factor of capacitors and Q_w the quality factor of a series resonant network constituted of L_s and C_s. In practice, however, the lack of coincidence between the input matching frequency and the maximum backward loss frequency becomes important only in a circulator with large insertion loss. In such a case, we arrive at a reasonable approximation of $Q_+ \ll Q_-$, Q_c, and, Q_w through experimentation. We, therefore, consider only Q_+ as to the input match and backward loss. Q_+, however, can be related to Q_{eff}, Q values of effective permeability μ_{eff}, by [1]

$$Q_{\mathrm{eff}} = \frac{2\sqrt{3}Q}{\sqrt{3}Q - 1}Q_+. \qquad (7)$$

The values of y_3, therefore, deviate from the values of (2) by the increase of the loss term $\delta y_3''$ caused by Q_+ as shown by

$$y_3 = j\frac{Y_0}{\sqrt{3}} + \delta y_3''. \qquad (8)$$

The amount of $\delta y_3''$ can be related to Q_+ by substituting (6) into μ_+ of (1), and can also be related to Q_{eff} by using the relation of (7) as follows:

$$\frac{\delta y_3''}{Y_0} = \frac{2Q}{Q_{\mathrm{eff}}}. \qquad (9)$$

To satisfy the backward condition of a circulator under such a state, the amount of the imaginary part of y_2 at ω_0 should be slightly deviated from that of (2) where $\delta y_2'$ is obtained later, as shown by

$$y_2 = -j\frac{Y_0}{\sqrt{3}} + j\delta y_2'. \qquad (10)$$

However, it can be proved that the values of $\delta y_3''$ and $\delta y_2'$ at ω_0 are almost maintained even at the frequency range outside ω_0. Therefore, we add $\delta y_3''$ and $\delta y_2'$ at y_3 and y_2 of (5), respectively.

On the other hand, it is also shown later that an original circulator with the magnetic loss of ferrite should be previously designed under the slightly lower terminal admittance $aY_0(a<1)$ than Y_0 to satisfy the matching condition of an isolator realized by connecting an absorbing resistance R_0 at an arbitrary terminal as illustrated in Fig. 4 (a).

Considering the above mentioned matter, y_2 and y_3 should take the form of (11) in the case of the existence of a magnetic loss:

$$y_2 = jaY_0\left[Q\left\{x - \frac{(\sigma_0+P_0)^2-1}{\sqrt{3}P_0Qx}\cdot\frac{1}{1+\dfrac{P_0}{\sigma_0+x}}\right\} + \frac{\delta y_2'}{Y_0}\right]$$

$$y_3 = jaY_0Q\left\{x - \frac{(\sigma_0+P_0)^2-1}{\sqrt{3}P_0Qx}\cdot\frac{1}{1+\dfrac{P_0}{\sigma_0-x}}\right\} + \frac{\delta y_3''}{Y_0}aY_0. \tag{11}$$

At ω_0, they take the values of

$$y_2 = j\left(\frac{-1}{\sqrt{3}} + \frac{\delta y_2'}{Y_0}\right)aY_0$$

$$y_3 = j\frac{aY_0}{\sqrt{3}} + \frac{\delta y_3''}{Y_0}aY_0. \tag{12}$$

Now, we get the impedance matrix concerning the pair of terminals of 11″ and 22″ when the absorbing resistance R_0 is connected between terminals 3 and 2″ in Fig. 4 (a).

Cascade connecting the impedance matrix above obtained with that concerning a pair of terminals of 1′1″ and 2′2″, we can get the scattering matrix concerning a pair of terminals of 11′ and 22′ as shown by

$$[S] = \begin{bmatrix} S_{11} & S_{12} \\ S_{21} & S_{22}\end{bmatrix} = \frac{1}{z_{11}^2 + 2z_{11}Z_0 - z_{12}z_{21} - Z_0^2}\cdot\begin{bmatrix} z_{11}^2 - z_{12}z_{21} - Z_0^2 & 2z_{12}Z_0 \\ 2z_{21}Z_0 & z_{11}^2 - z_{12}z_{21} - Z_0^2\end{bmatrix}$$

$$z_{11} = A' - \frac{B'C'}{A'+R_0} + z,$$

$$z_{12} = B' - \frac{C'^2}{A'+R_0} + z,$$

$$z_{21} = C' - \frac{B'^2}{A'+R_0} + z$$

$$A' = \frac{y_2+y_3}{3y_2y_3},$$

$$B' = \frac{\alpha^2 y_2 + \alpha y_3}{3y_2y_3},$$

$$C' = \frac{\alpha y_2 + \alpha^2 y_3}{3y_2y_3},$$

$$\alpha = e^{j(2\pi/3)}, \qquad z = j\omega L_s - j\frac{1}{\omega C_s} \tag{13}$$

where L_s and C_s are inductance and capacitance of a series resonant network in Fig. 4 (a), and y_3^2 takes the values of (5). In the following, the several characteristics are induced from the scattering matrix of (13).

The backward loss L_b(dB) is obtained by S_{12} of (13) as

$$L_b = 20\log_{10}\frac{1}{|S_{12}|}. \tag{14}$$

It can be shown by an approximate analytical method that the backward loss of a lossless isolator can be widened in a double hump performance similar to second Chebyshev characteristics by using R_0 of

$$\frac{R_0}{Z_0} = \frac{1}{1-2H}, \qquad L_b = 20\log_{10}\frac{1}{|H|} \tag{15}$$

where H is the value of S_{12} at ω_0.

However, if we require a more accurate performance taking into account the ferrite loss and the dispersion of the permeability, we have to use (11) and (12) and a computer-aided method.

First, we obtain the values of R_0 to satisfy L_b dB of a backward loss at ω_0 before we get a frequency performance. Since, however, increasing or decreasing the amount of a in (11) corresponds to increasing or decreasing a terminal admittance, R_0 should be inversely proportional to a in the form

$$R_0 = \frac{b_0}{a}Z_0. \tag{16}$$

On the other hand, considering $z=0$ at ω_0 and substituting (16) and (12) into (13), S_{12} $(=H)$ at ω_0 is given by

$$H = \frac{-2\left(b_0 + 1 + \dfrac{\delta y_3''}{2Y_0}b_0 + \dfrac{\sqrt{3}}{2}\dfrac{\delta y_2'}{Y_0}b_0\right) - jb_0\left(\sqrt{3}\dfrac{\delta y_3''}{Y_0} + \dfrac{\delta y_2'}{Y_0}\right)}{\left(\sqrt{a} + \dfrac{1}{\sqrt{a}}\right)^2 b_0}. \tag{17}$$

233

Putting the imaginary part of (17) to zero to make the absolute value of H as small as possible, we get

$$\delta y_2' = -\sqrt{3}\delta y_3''. \tag{18}$$

Substituting (18) into (17), we get

$$b_0 = \frac{1}{1 - q - \dfrac{(\sqrt{a} + 1/\sqrt{a})^2}{2} H}. \tag{19}$$

On the other hand, the value of Γ, the input reflection coefficient at ω_0, is obtained by substituting (12) and (16) into S_{11} of (13) and using (19) as shown by

$$\Gamma = \frac{\left\{b_0(1 - a^2) - 3b_0(2a^2 - 1)\dfrac{\delta y_3''}{Y_0} - a^2\dfrac{\delta y_3''}{Y_0}\right\} + j\sqrt{3}a\dfrac{\delta y_3''}{Y_0}\{a(1 + 2b_0) - b_0\}}{4a^3}. \tag{20}$$

Setting the real part of Γ as zero, we get

$$b_0 = \frac{a^2\dfrac{\delta y_3''}{Y_0}}{1 - a^2 - 3(2a^2 - 1)\dfrac{\delta y_3''}{Y_0}}. \tag{21}$$

The imaginary part of Γ can be made zero by connecting an additional reactance at the input terminal. From (19) and (21), we get the relation

$$\frac{a^3}{2}Hq - a^2(1 + 7q - qH + q^2) + \frac{a}{2}qH + (1 + 3q) = 0$$

$$q = \frac{\delta y_3''}{Y_0} = \frac{2Q}{Q_{eff}} \tag{22}$$

and the value of a is obtained in parameters of L_b from (22) as shown in Fig. 5 (a). Using these values of a and b_0 obtained from (19), we get R_0 from (16) as functions of q and L_b, the result of which is shown in Fig. 5 (b).

The reactance to be connected between the input terminal and the outer conductor of the circulator is obtained by substituting a and b_0 into (20), and it becomes capacitive.

This capacity C' is shown in Fig. 5 (c), where C' is expressed in terms of Q' by

$$\omega C' Z_0 = Q'. \tag{23}$$

Since $S_{11} = S_{22}$ from (13), an output impedance can be also matched by connecting the same capacity C' between an output terminal and the outer conductor of a circulator.

Putting R_0, a, and C' obtained as functions of L_b and q into (13), we can compute the frequency performance of the backward loss, the standing wave ratio, and the insertion loss as follows.

B. Backward Loss

To obtain the optimum value of L_s of a wide-band network, the bandwidth ratio W to maintain 20-dB backward loss was computed for several values of m $(= Y_0^2(L_s/C))$ as parameters of Q_{eff} and Q. The result shows that the QW product varies with m and Q_{eff} in the case of $Q \gg 1$ as shown in Fig. 6 (a). Since we have the relation of $QW = 0.2$ [1] in the case of $L_s = 0$, $C_s = \infty$, that is, the condition that existed before wide band, the bandwidth increasing ratio can be obtained from Fig. 6 (a) as $QW/0.2$.

C. Standing Wave Ratio

Designing a circulator for the terminal admittance of aY_0 and connecting C' at input and output terminals, the standing wave ratio becomes 1 at ω_0. In the frequency range far from ω_0, it also approaches 1 as mentioned before. Therefore, it takes a maximum value of SWR_{max} at the joint frequency band, as indicated in Fig. 4 (b). As the result of the computation of SWR_{max} for several values of m as parameters of Q_{eff} and Q, SWR_{max} becomes a function only of Q_{eff} and m for $Q \gg 1$, the result of which is shown in Fig. 6 (b). This shows that $m = 0.4 \sim 0.5$ is a preferable condition for wide-band matching.

D. Insertion Loss

The insertion loss can be obtained from S_{21} of (13), and it generally takes the maximum values L_i at the center frequency as shown in Fig. 4 (b). The values of L_i are obtained in connection with Q_{eff}, Q_c, and Q_w, where Q_c is a quality factor of a capacity and Q_w is the noload Q of a series resonant network constituted of L_s and C_s, as shown by

$$L_i[\text{dB}] = \frac{4.96}{\eta Q_c} + \frac{4.96}{\eta Q_{eff}} + \frac{17.34\, mQ}{Q_w} = 8.67Q\left(\frac{1}{Q_c} + \frac{1}{Q_{eff}} + \frac{2m}{Q_w}\right). \tag{24}$$

The first and second terms are the same terms as those obtained for a circulator [1], and the third term is that caused by the additional network for wide band. This third term is easily obtained by considering that the input current flows through the pass of $1 \to 1'' \to 1'$ in the same phase with the output current through the pass of $2 \to 2'' \to 2'$.

The merit of the low insertion loss of this component outside its frequency band is that it becomes possible to connect several isolators at any frequency band in cascade.

E. Design Procedure and Practical Example

Based upon the theory mentioned earlier in Section III-A–C, the design procedure is summarized in Table I.

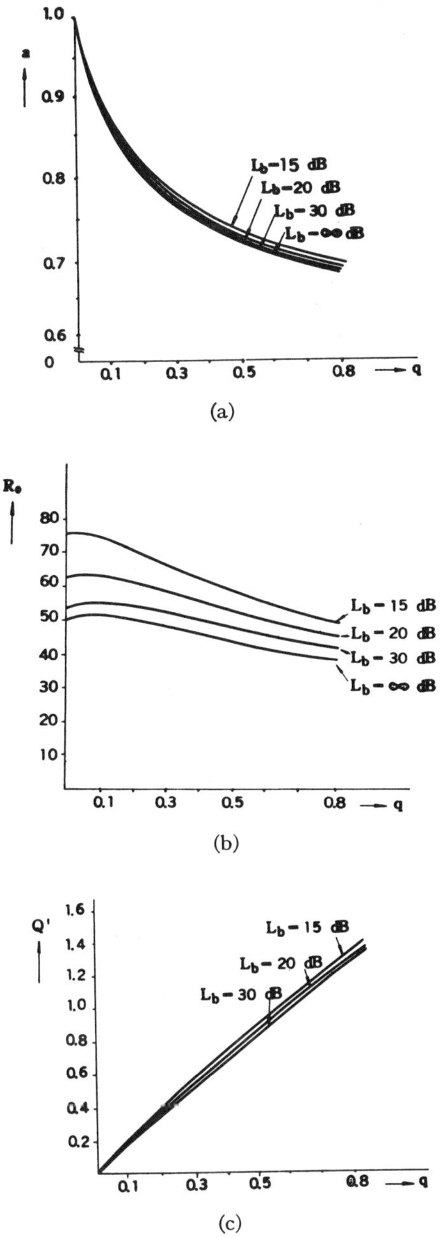

Fig. 5. Design monograph for construction of Fig. 4 (a). (a) Relation between a and q. (b) Relation between R_0 and q. (c) Relation between Q' and q.

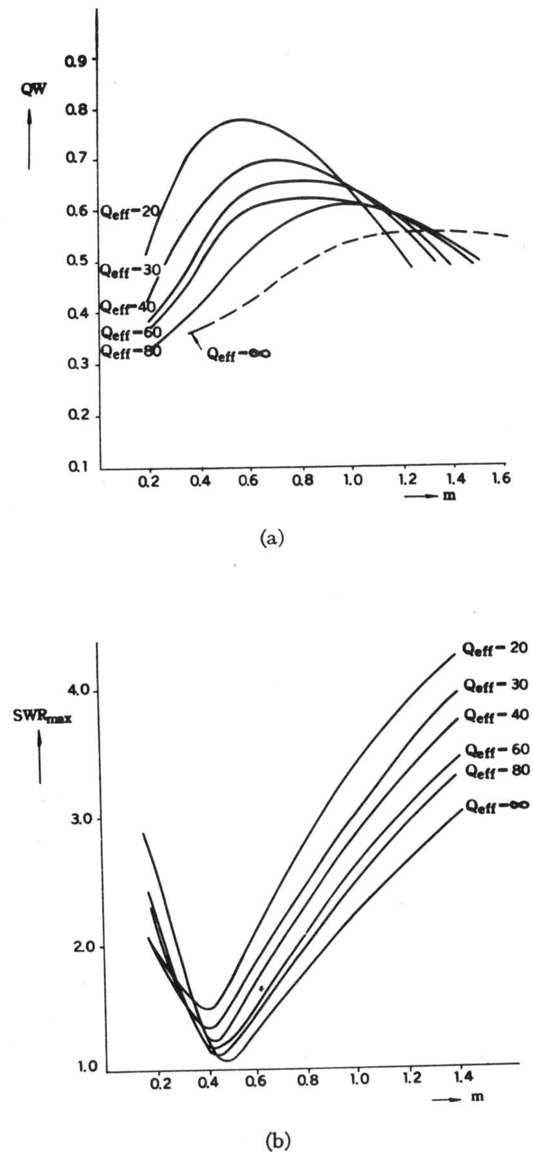

Fig. 6. Performance of wide-band isolator of Fig. 4 (a). (a) Relation between QW product and m for $L_b=20$ dB. (b) Relation between SWR_{max} and m for $L_b=20$ dB.

Referring to Table I, we will show a practical example of the design of an isolator using A1–YIG polycrystalline with Ms of 60 Oe and ΔH of 80 Oe.

The circulator before broadbanding has the bandwidth ratio W_0 of 0.062 at the center frequency of 200 MHz. From Table I-1, we get $C=51$ pF, therefore $Q=3.2$. By measuring, we get $Q_c=300$ and $Q_{eff}=25$. From Table I-3, we get $q=0.256$. Next, from Fig. 5 (a)–(c), we get $a=0.8$, $R_0=60\ \Omega$, $Q'=0.42$, and $C'=6.7$ pF.

Applying these values to Fig. 6 (a) and (b), the optimum value of m becomes 0.45, so that $L_s=5.7\times10^{-8}[H]C_s=11$ pF. The theoretical bandwidth ratio becomes 21 percent, and the SWR_{max} and insertion loss become 1.3

TABLE I
DESIGN PROCEDURE

1	$C=\dfrac{2H}{\omega_0 W_0 Z_0}$	[W_0: bandwidth ratio before wide band]
2	$Q=\omega_0 C Z_0$	
3	$q=\dfrac{2Q}{Q_{eff}}$	[Q_{eff}: measured values]
4	a, R_0, Q'	[from Fig. 5(a), (b), (c)]
5	$C'=\dfrac{Q'}{\omega_0 Z_0}$	
6	m, W, SWR_{max}	[from Fig. 4(a), (b)]
7	$L_s=Z_0^2 Cm,\quad C_s=\dfrac{1}{\omega_0^2 L_s}$	
8	$L_i=8.67Q\left(\dfrac{1}{Q_c}+\dfrac{1}{Q_{eff}}+\dfrac{2m}{Q_w}\right)$	[measured values]

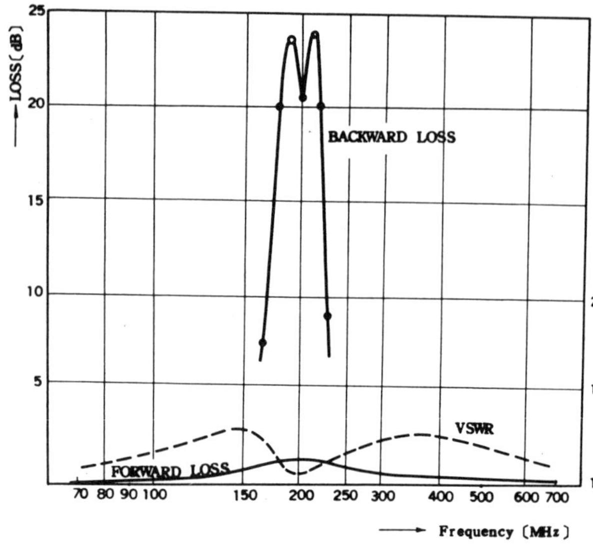

Fig. 7. Performance of trial isolator.

and 1.4 dB, respectively, all of which coincide with the experimental performance shown in Fig. 7 with excellent results.

IV. The Case of Connecting a Resistance and a Series Resonant Network Between a Terminal and the Outer Conductor of a Circulator

By connecting a series resonant network constituted of L_s' and C_s' in series with an absorbing resistance as shown in Fig. 4 (e), the bandwidth of the backward loss is widened further into a characteristic similar to the third Chebyshev characteristics as shown in Fig. 4 (f). This performance can be obtained by replacing R_0 of (13) with Z of

$$Z = R_0 + j\omega L_s' + \frac{1}{j\omega C_s'}, \qquad \omega_0^2 L_s' C_s' = 1. \quad (25)$$

In the same manner as in Section III, we obtain the values of R_0, a, and c' to satisfy the infinite backward loss and complete matching at ω_0, which can be easily known from the corresponding point to $L_b = \infty$ in Fig. 5.

As the result of computation with the values obtained above, QW product becomes maximum for $n(=Y_0^2(L_s'/C_s'))=1$, and it can be expressed by Q_{eff} and m under the condition $Q \gg 1$ and $n=1$, of which result is shown in Fig. 8 (a). Comparing Fig. 8 (a) with Fig. 6 (a), it is understood that the bandwidth ratio is widened still more by the additional $L_s'C_s'$ network. The frequency performance of SWR is almost similar to that described in Fig. 6 (b), and $SWR_{\max}$ is shown in Fig. 8 (b). The insertion loss also becomes maximum at ω_0 and takes the same values as those obtained from (24).

As a practical example, we obtained an extremely wide-band isolator with 20-dB isolation and 3-dB insertion loss over the frequency range 70 MHz to 800 MHz, by connecting in cascade six isolators with the construction shown in this section and each with a different center frequency. The performance is shown in Fig. 9.

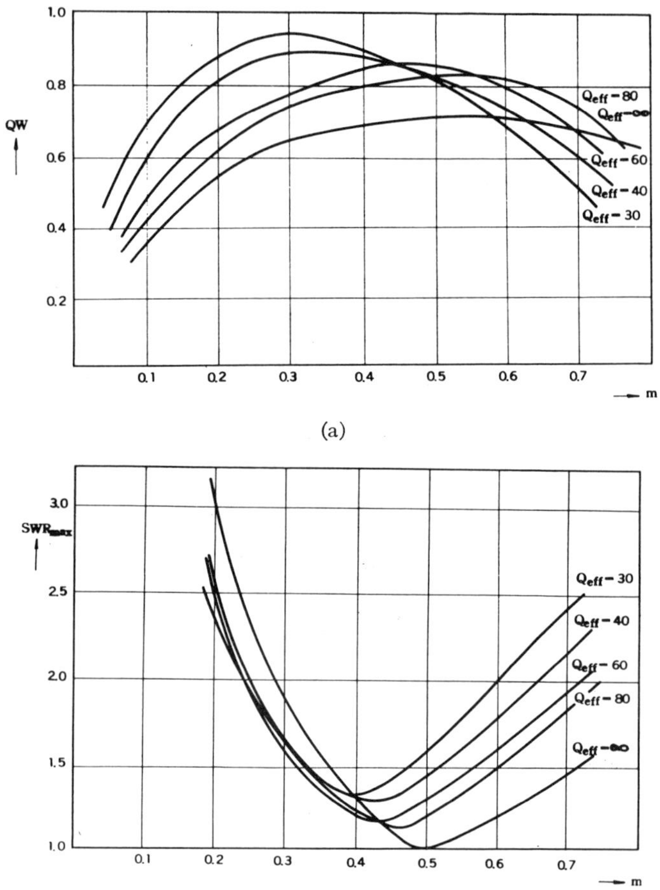

Fig. 8. Performance of isolator with the connection of Fig. 4 (e). (a) Relation between QW and m for $L_b=20$ dB. (b) Relation between $SWR_{\max}$ and m for 20 dB.

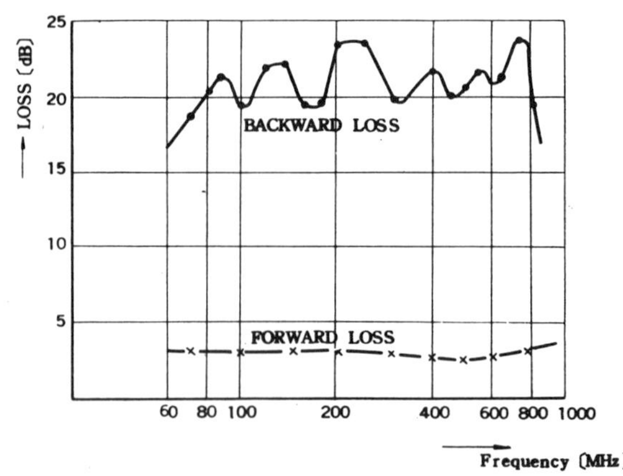

Fig. 9. Performance of trial ultrawide-band isolator.

V. Case of Connecting an Absorbing Resistance Between a Terminal and a Ground

Since the network of L_s and C_s is series resonant at the center frequency, the characteristics are identical to those described in Section III. At the frequency band far from ω_0, however, the insertion loss and the standing wave ratio approach 3 dB and 2, respectively, as mentioned in Section II. The bandwidth of the backward loss is widened for the reason mentioned in Section II. In this case, it is obtained through approximate analysis that the bandwidth is widened with a double hump performance by designing a circulator with a terminal impedance slightly different from Z_0. However, the accurate performance should be obtained from the following scattering matrix under the computer-aided method.

The scattering matrix $[S_c]$ of the circulator is obtained as follows.

$$[S_c] = \begin{bmatrix} A & B & C \\ C & A & B \\ B & C & A \end{bmatrix}. \tag{26}$$

$$A = \frac{1}{3}(S_1 + S_2 + S_3), \qquad B = \frac{1}{3}(S_1 + \alpha^2 S_2 + \alpha S_3),$$

$$C = \frac{1}{3}(S_1 + \alpha S_2 + \alpha^2 S_3)$$

$$S_1 = \frac{3z - Z_0}{3z + Z_0}, \qquad S_2 = \frac{Y_0 - Y_2}{Y_0 + Y_2}, \qquad S_3 = \frac{Y_0 - Y_3}{Y_0 + Y_3}$$

$$\alpha = e^{-j(2\pi/3)}, \qquad z = j\omega L_s + \frac{1}{j\omega C_s}, \qquad Z_0 = \frac{1}{Y_0}. \tag{27}$$

Since $y_1 = \infty$, $y_3^2 = \mp j(Y_0/\sqrt{3})$, $S_1 = -1$, $S_2 = -\alpha^2$, and $S_3 = -\alpha$ at ω_0 for a lossless circulator.

However, $|S_3|$ becomes slightly smaller than 1 due to ferrite loss. Now let us assume to adjust S_2 and S_3 to the values of (28) to obtain the isolation and matching of the circulator.

$$S_2 = -\alpha^2 e^{j\delta_2'}, \qquad S_3 = -\alpha e^{j\delta_3'} e^{-\delta_3''}. \tag{28}$$

Substituting those values of S_2 and S_3 into (27), we get

$$A = -\frac{1}{6}(\sqrt{3}\delta_2' - \sqrt{3}\delta_3' + \delta_3'') + \frac{j}{6}(\delta_2' + \delta_3' + \sqrt{3}\delta_3'')$$

$$B = -\frac{1}{6}(\sqrt{3}\delta_3' - \sqrt{3}\delta_2' + \delta_3'') + \frac{j}{6}(\delta_2' + \delta_3' + \sqrt{3}\delta_3'') \tag{29}$$

where $|\delta_2'|, |\delta_3'|, |\delta_3''| \ll 1$.

To get a complete match, A should be zero, so that $|B| = (2/3)\delta_3''$ from (29). On the other hand, B should be zero to get an infinite isolation, so that $A = (2/3)\delta_3''$ from (29). δ_3'', however, is related to $\delta y_3''$ by the relation

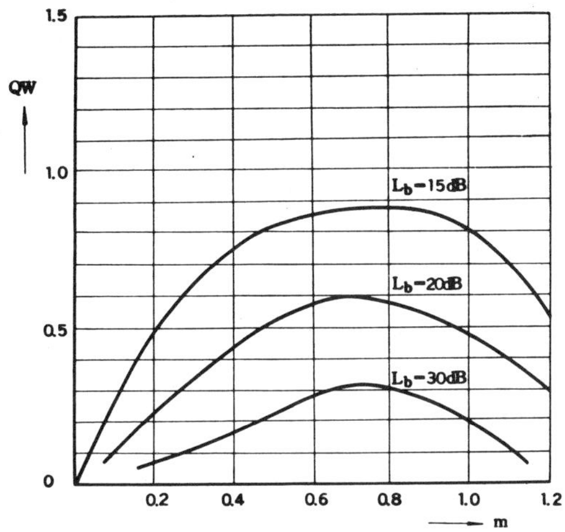

(a)

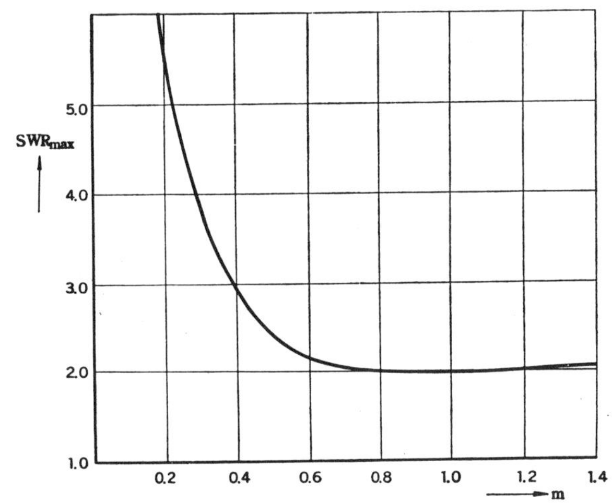

(b)

Fig. 10. Performance of lossless circulator of Fig. 4 (c). (a) Relation between QW product and m. (b) Relation between SWR_{max} and m.

of $(\delta y_3''/Y_0) = (4/3)\delta_3''$ and furthermore related to Q/Q_{eff} by (9). Therefore, in the case of Γ (input reflection coefficient) $= 0$, $L_b = 20 \log_{10} Q/Q_{eff}$, and in the case of

$$L \text{ (backward loss)} = \infty, \ \Gamma = Q/Q_{eff} \text{ or } SWR = 1 + 2\,Q/Q_{eff}. \tag{30}$$

From this standpoint, Q/Q_{eff} should be as small as possible to get the ideal operation of a circulator. Therefore, we obtained the performance of L_b, L_i, and the standing wave ratio from (27) only in the case of a lossless circulator, i.e., $Q_{eff} = \infty$. As a result, QW product becomes a function of m and L_b for $Q \gg 1$ as shown in Fig. 10 (a). The standing wave ratio has the performance shown in Fig. 4 (d) and SWR_{max} is obtained in terms of m and L_b as shown in Fig. 10 (b). The insertion loss becomes a minimum at ω_0 and the value is obtained from (24).

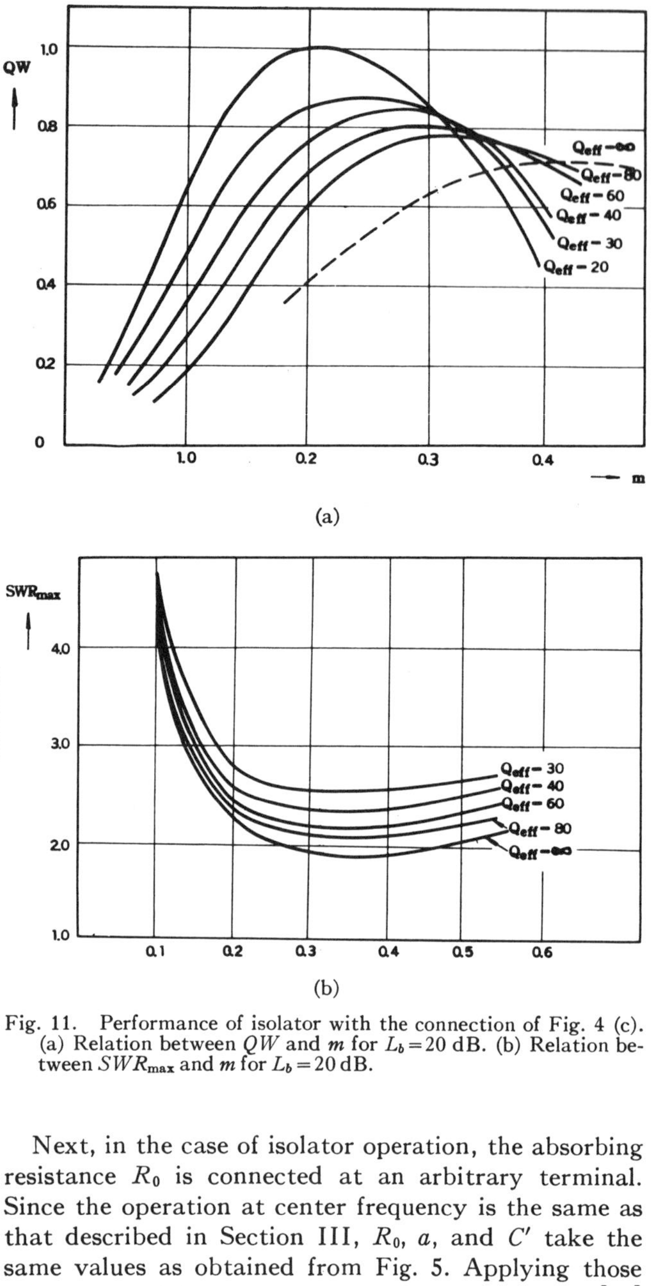

Fig. 11. Performance of isolator with the connection of Fig. 4 (c). (a) Relation between QW and m for $L_b = 20$ dB. (b) Relation between SWR_{max} and m for $L_b = 20$ dB.

Next, in the case of isolator operation, the absorbing resistance R_0 is connected at an arbitrary terminal. Since the operation at center frequency is the same as that described in Section III, R_0, a, and C' take the same values as obtained from Fig. 5. Applying those constants to (27), we can get the scattering matrix $[S_i]$ of an isolator as follows.

$$[S_i] = \begin{bmatrix} S_{11}^i & S_{12}^i \\ S_{21}^i & S_{22}^i \end{bmatrix}$$

$$S_{11}^i = S_{22}^i = A + \frac{BC\Gamma_W}{1 - \Gamma_W A} - \Gamma'$$

$$S_{12}^i = B + \frac{C^2\Gamma}{1 - \Gamma_W A}, \qquad S_{21}^i = C + \frac{B^2\Gamma_W}{1 - \Gamma_W A}$$

$$\Gamma_W = \frac{R_0 - Z_0}{R_0 + Z_0}, \qquad \Gamma' = \frac{Q'}{2 + Q'}. \tag{31}$$

The backward loss and the standing wave ratio can be computed from S_{12}^i and S_{11}^i, and the results are indicated in terms of Q_{eff} and m in Fig. 11.

This result shows that the QW product is larger than that of Section III. Therefore, this type is better than that of Section III when it is used alone without any other isolator in cascade. In fact, cascade connection with other isolators adjusted at other frequencies is undesirable because of the increase of the insertion loss outside of the center frequency.

As a practical example, applying this technique to the lumped-element circulator with ferrite disks of 10-cm diameter, we get the circulator at 96 MHz with 13-MHz bandwidth for 20-dB isolation, 0.2-dB insertion loss and the available power of CW 2 kW, which is used for a transmitter.

We described above the three kinds of wide-band techniques for a lumped-element isolator, but this design can be extended to a stripline Y circulator as will now be discussed.

VI. Extension to a Stripline Y Circulator

A circulator with a small frequency variation of the eigenvalue for the first eigenvector can be approximated by the equivalent network of an ideal circulator with a parallel tuned LC circuit at each terminal [1]; The frequency variation of the eigenadmittances for the second and the third eigenvectors is expressed by

$$\delta y \doteqdot 2jC\delta\omega. \tag{32}$$

On the other hand, the frequency variation of the eigenadmittance of a general rotationally symmetric 3-port circuit is expressed by the time average total reactive energy $\tilde{W}_t'$ as [1]

$$\delta y = j\,\frac{2}{3}\,\tilde{W}_t'\delta\omega. \tag{33}$$

Therefore, the capacity c equivalently inserted at each terminal is expressed by

$$C = \frac{\tilde{W}_t'}{3}. \tag{34}$$

Let us consider a stripline Y circulator with disks of radius r and width t which are inserted in a space h between the center conductor and the outer conductor of a circulator. The value of $\tilde{W}_t'$ for the second and third eigenvectors with unit voltage is [2]

$$\tilde{W}_t' = 0.0725\,\frac{r^2}{\omega h\lambda\left(1 - t' + \dfrac{t'}{\epsilon_{2s}}\right)} \tag{35}$$

where $t' = t/h$, ϵ_{2s} is a specific dielectric constant of ferrite, and λ is a free-space wavelength. Substituting (35) into (34), we get

$$C = 0.0242\,\frac{r^2}{\omega h\lambda\left(1 - t' + \dfrac{t'}{\epsilon_{2s}}\right)}. \tag{36}$$

Hence,

$$Q = 0.0242 \frac{r^2 Z_0}{h\lambda\left(1 - t' + \frac{t'}{\epsilon_{2s}}\right)}. \tag{37}$$

In a simple case of $t'=1$ and no matching capacity at each terminal, we have [2]

$$\frac{r}{\lambda} = \frac{0.293\left(1 - \frac{\eta^2}{2}\right)}{\sqrt{\epsilon_{2s}\,\mu_{\mathrm{eff},s}}} \tag{38}$$

$$\frac{r}{h} = \frac{1.63\,(\mu_s/\epsilon_{2s})^{1/2}}{\eta}, \qquad \eta = \frac{\mu_+ - \mu_-}{\mu_+ + \mu_-},$$

$$\mu_s = \frac{\mu_+ + \mu_-}{2\mu_0}. \tag{39}$$

Putting (38) and (39) into (37) yields

$$Q \doteq 0.00708 \frac{Z_0}{\eta}. \tag{40}$$

Applying the Q values of (37) or (40) to Fig. 6 (a), 10 (a), or 11 (a) together with a consideration of Q_{eff}, we can design the wide-band network.

In a triplate type stripline circulator, a wide-band network should be connected at upper and lower sides, when the amount of L_s at each side must be twice that obtained from Fig. 6 (a), 10 (a) or 11 (a). However, in an unbalanced type circulator, only one wide-band network is connected as shown in Fig. 2 (b).

VII. Conclusion

The new wide-band technique and the design theory were presented, and it was shown that the designs are largely controlled by the effect of ferrite loss in lumped-element and stripline circulators used for VHF and UHF bands. Furthermore, a method to make the matching frequency coincide with the maximum isolation frequency in the isolator even if the third eigenvalue takes the loss term, is described. These techniques will also apply to an isolator with different loss terms of each eigenvalue. It is also shown that the connecting points of the resistance should be changed according to the case either used for the cascade connection or used alone.

References

[1] Y. Konishi, "Lumped element *Y* circulator," *IEEE Trans. Microwave Theory Tech.*, vol. MMT-13, Nov. 1965, pp. 852–864.

[2] Y. Konishi, "A high-power UHF circulator," *IEEE Trans. Microwave Theory Tech.*, vol. MTT-15, Dec. 1967, pp. 700–708.

[3] H. Bosma, "Performance of lossy H-plane Y circulators," *IEEE Trans. Magn.*, vol. MAG-2, Sept. 1966, pp. 273–277.

Ferrite-Junction Circulator Bibliography

RAMESH C. KUMAR

***Abstract*—To date an enormous amount of work has been done in the field of ferrite-junction circulators, but as yet no up-to-date bibliography is available. The present bibliography has been compiled to cover the period from January 1956 to September 1969 in order to help researchers working in this field. The bibliography provides references on ferrite circulators including entries for theory, stripline and waveguide versions, low and high powers, low-temperature circulators, lumped-element circulators, switching circulators, slot circulators, microstrip circulators, etc. It also includes references for typical designs and constructions as well as for experimental results and applications.**

As in the compilation of any bibliography, some valuable and pertinent article may have been overlooked.

Manuscript received August 25, 1969; revised April 27, 1970.
The author is with the Central Electronics Engineering Research Institute, Pilani (Rajasthan), India.

1956

[1] Allen, P. J., "The turnstile circulator," *IRE Trans. Microwave Theory Tech.*, vol. MTT-4, pp. 223–227, October 1956.

[2] Treuhaft, M. A., "Network properties of circulators based on the scattering concept," *Proc. IRE*, vol. 44, pp. 1394–1402, October 1956.

1957

[1] Calvin, A., and E. A. Ohm, "A broadband microwave circulator," *IRE Trans. Microwave Theory Tech.* (Correspondence), vol. MTT-5, pp. 164–165, April 1957.

[2] Rizzi, P. A., "High-power ferrite circulators," *IRE Trans. Microwave Theory and Tech.*, vol. MTT-5, pp. 230–237, October 1957.

959

[1] Arams, F. R., and G. Krayer, "Low-loss L-band circulator," *Proc. IRE* (Correspondence), vol. 47, p. 442, March 1959.
[2] Auld, B. A., "The synthesis of symmetrical waveguide circulators," *IRE Trans. Microwave Theory Tech.*, vol. MTT-7, pp. 238–246, April 1959.
[3] Boyet, H., S. Weisbaum, and I. Gerst, "Design calculation for UHF ferrite circulators," *IRE Trans. Microwave Theory Tech.*, vol. MTT-7, pp. 475–476. October 1959.
[4] Chait, H. N., and T. R. Curry, "Y-circulator," *J. Appl. Phys.*, Suppl., vol. 30, pp. 152–153, April 1959.
[5] ——, "New microwave circulators," *Electronics*, vol. 32, p. 81, December 18, 1959.
[6] Yoshida, S., "X-circulator," *Proc. IRE* (Correspondence), vol. 47, p. 1150, June 1959.
[7] ——, "An *E*-type *T* circulator," *Proc. IRE* (Correspondence), vol. 47, p. 2018, November 1959.
[8] ——, "*E*-type *X* circulator," *Proc. IRE* (Correspondence), vol. 47, pp. 2017–2018, November 1959.

1960

[1] Davis, L., Jr., U. Milano, and J. Saunders, "A strip-line *L*-band compact circulator," *Proc. IRE* (Correspondence), vol. 48, pp. 115–116, January 1960.
[2] Freiberg, L., "Lightweight *Y*-junction strip-line circulator," *IRE Trans. Microwave Theory Tech.* (Correspondence), vol. MTT-8, p. 672, November 1960.
[3] Grace, M., and F. R. Arams, "Three-port ring circulators," *Proc. IRE* (Correspondence), vol. 48, pp. 1497–1498, August 1960.
[4] Lax, B., and K. Button, "Ferrimagnetics and applications from UHF to millimeter waves," *Microwave J.*, vol. 3, pp. 49–56, November 1960.
[5] Levey, L., and L. M. Silber, "A fast-switching X-band circulator utilizing ferrite toroids," *IRE WESCON Conv. Rec.*, vol. 4, pt. 1, pp. 11–20, 1960.
[6] Masse, D., "A high power S-band ferrite circulator" (in French), *Onde Elec.*, vol. 44, pp. 150–154, February 1960.
[7] Milano, U., J. H. Saunders, and L. Davis, Jr., "A *Y*-junction strip-line circulator," *IRE Trans. Microwave Theory Tech.*, vol. MTT-8, pp. 346–351, May 1960.
[8] Privit, E., and W. Stösser, "The microwave circulator" (in German), *Frequenz*, vol. 14, pp. 77–84, March 1960.
[9] Thaxter, J. B., and G. S. Heller, "Circulators at 70 and 140 KMC," *Proc. IRE* (Correspondence), vol. 48, pp. 110–111, January 1960.
[10] Yoshida, S., "Strip-line *Y*-circulator," *Proc. IRE* (Correspondence), vol. 48, pp. 1337–1338, July 1960.
[11] ——, "*J*-band strip-line *Y* circulator," *Proc. IRE* (Correspondence), vol. 48, p. 1664, September 1960.

1961

[1] Arams, F., B. Kaplan, and B. Peyton, "Octave-bandwidth UHF/*L*-band circulator," *IRE Trans. Microwave Theory Tech.*, vol. MTT-9, pp. 212–216, May 1961.
[2] Buehler, G. V., and A. F. Eikenberg, "Stripline *Y*-circulators for the 100 to 400 Mc region," *Proc. IRE*, vol. 49, pp. 518–519, February 1961.
[3] ——, "A VHF high-power *Y*-circulator," *IRE Trans. Microwave Theory Tech.* (Correspondence), vol. MTT-9, pp. 569–570, November 1961.
[4] Clark, J., "Perturbation techniques for miniaturised coaxial Y-junction circulators," *J. Appl. Phys.*, Suppl., vol. 32, pp. 323–324, March 1961.
[5] Clark, J., and J. Brown, "Miniaturized temperature stable coaxial *Y*-junction circulators," *IRE Trans. Microwave Theory Tech.* (Correspondence), vol. MTT-9, pp. 267–269, May 1961.
[6] Clavin, A., "Higher-order mode resonances in strip-line *Y*-junction circulators," *IRE Trans. Microwave Theory Tech.* (Correspondence), vol. MTT-9, p. 575, November 1961.
[7] Freiberg, L., "Pulse-operated circulator switch," *IRE Trans. Microwave Theory Tech.* (Correspondence), vol. MTT-9, p. 266, May 1961.
[8] Kavka, W., "Circulator properties of a T-junction containing a ferrite" (in Polish), *Prace PIT*, no. 34, pp. 57–77, 1961.
[9] Koop, W. S., and A. K. Jordan, "Design note on an *L*-band strip-line circulator," *IRE Trans. Microwave Theory Tech.* (Correspondence), vol. MTT-9, pp. 571–572, November 1961.
[10] Schuegraf, E., "On microwave circulators using a Faraday rotator" (in German), *Frequenz*, vol. 15, pp. 266–271, August 1961.

1962

[1] Benzie, R. J., P. A. Kerrigan, and J. A. Ponny, "Some ferrite components for isolation and controlled coupling," *Proc. Inst. Elec. Eng.*, vol. 109B, pp. 277–283, May 1962.
[2] Bosma, H., "On the principle of stripline circulation," *Proc. Inst. Elec. Eng.*, vol. 109B, Suppl. 21, pp. 137–146, January 1962.
[3] Clark, J., and G. R. Harrison, "Miniaturized coaxial ferrite devices," *Microwave J.*, vol. 5, p. 108, June 1962.

[4] Davies, J. B., "An analysis of the *m*-port symmetrical *H*-plane waveguide junction with central ferrite post," *IRE Trans. Microwave Theory Tech.*, vol. MTT-10, pp. 596–604, November 1962.

[5] Feoktistov, V. G., "Diffraction model of a Y-circulator," *Radio Eng. Electron. Phys.* (USSR), vol. 7, pp. 1632–1637, October 1962.

[6] Germ, A. I., "Certain problems in theory of a symmetrical ring circulator," *Radio Eng. Electron. Phys.* (USSR), vol. 7, pp. 250–259, February 1962.

[7] Goller, J., "The hybrid cross or turnstile junction, theoretical basis and application of a new form of microwave seven port device" (in German), *Hochfreq. Elektroakust.*, vol. 71, pp. 205–211, December 1962.

[8] Humphreys, B. L., and J. B. Davies, "The synthesis of *N*-port circulators," *IRE Trans. Microwave Theory Tech.*, vol. MTT-10, pp. 551–554, November 1962.

[9] Krzycki, Z., "Ferrite circulators for 3 cm wavelength band" (in Polish), *Arch. Elektrotech* (Warsaw), vol. 11, no. 1, pp. 172–174, 1962.

242

[10] Lewandowski, S. J., "On some design problems of a star type ferrite circulator," *J. Inst. Telecommun. Eng. New Delhi*, vol. 8, pp. 103–106, March 1962.

[11] ——, "Ferrite circulator type Y for 3 cm band" (in Polish), *Przegl. Telekomun.*, no. 1, p. 18, 1962.

[12] Mita, S., T. Takcya, S. Mayashi, K. Kakizaki, and T. Yoshida, "Experimental results of pulse radars employing a circulator and travelling wave tubes," *J. Inst. Electron. Commun. Eng. Jap.*, vol. 45, pp. 902–910, July 1962.

1963

[1] Aitken, F. M., and R. McLean, "Some properties of waveguide circulators," *Proc. Inst. Elec. Eng.*, vol. 110, pp. 256–260, February 1963.

[2] Butterweck, H. J., "The Y-circulator" (in German), *Arch. Elek. Übertragung*, vol. 17, p. 163, April 1963.

[3] Clavin, A., "Reciprocal and nonreciprocal switches utilizing ferrite junction circulators," *IEEE Trans. Microwave Theory Tech.* (Correspondence), vol. MTT-11, pp. 217–218, May 1963.

[4] Collins, J. H., and A. R. Watters, "A field map for the waveguide Y-junction circulator," *Electron. Eng.* (London), vol. 35, pp. 540–543, August 1963.

[5] Comstock, R. L., "A cryogenic four-port circulator using single crystal Yttrium Iron Garnet," *Proc. IEEE* (Correspondence), vol. 51, pp. 1768–1769, December 1963.

[6] Davies, J. B., and P. Cohen, "Theoretical design of symmetrical junction stripline circulators," *IEEE Trans. Microwave Theory Tech.*, vol. MTT-11, pp. 506–512, November 1963.

[7] Davis, L. E., and S. R. Longley, "*E*-plane 3-port *X*-band waveguide circulators," *IEEE Trans. Microwave Theory Tech.* (Correspondence), vol. MTT-11, pp. 443–445, September 1963.

[8] De Gruyl, J. A., W. W. Heinz, and S. Okwit, "Helium-cooled Y-junction ferrite circulator switch," *Proc. IEEE* (Correspondence), vol. 51, p. 947, June 1963.

[9] Krzycki, Z., "A broad-band ferrite circulator for 10 cm band" (in Polish), *Arch. Elektrotech.* (Warsaw), vol. 12, no. 4, pp. 767–768, 1963.

[10] Mason, P. R., "Analysis of four port circulators using non-reciprocal phase shifters and directional couplers with an application to a design for the 5.925–6.425 Gc/s band," *Proc. Inst. Elec. Eng.*, vol. 110, pp. 869–880, May 1963.

[11] Skomal, E. N., "Theory of operation of a 3-port *Y*-junction ferrite circulator," *IEEE Trans. Microwave Theory Tech.*, vol. MTT-11, pp. 117–122, March 1963.

1964

[1] Bosma, H., "On stripline *Y*-circulation at UHF," *IEEE Trans. Microwave Theory Tech.*, vol. MTT-12, pp. 61–72, January 1964.

[2] ——, "The theory of the Y-junction circulator" (in French), *Acta Electron.*, vol. 8, p. 155, April 1964.

[3] Davis, L. E., M. D. Coleman, and J. J. Cotter, "Four-port crossed-waveguide junction circulators," *IEEE Trans. Microwave Theory Tech.*, vol. MTT-12, pp. 43–47, January 1964.

[4] Dessert, E., "The study and production of Y-junction circulators in the frequency range 450–1000 Mc/s," *Acta Electron.*, vol. 8, p. 175, April 1964.

[5] Gandhi, O. P., and C. Dattatreyan, "A L-band broad bandwidth Y-circulator," *Proc. Inst. Elec. Eng.*, vol. 111, pp. 1285–1286, July 1964.

[6] Nakahara, S., and H. Kurebayashi, "Behaviour of waveguide circulators and improvements of their performances," Mitsubishi Denki, Japan, Lab. Rep., vol. 5, pp. 207–222, April 1964.

[7] Rowley, J., and E. Sheehan, "Design consideration for ferrite junction circulators," *Proc. IEEE* (Correspondence), vol. 52, p. 221, February 1964.

[8] Silber, L. M., and A. Weis, "A fast switching high power C-band ferrite circulator," *IEEE Int. Conv. Rec.*, vol. 12, pt. 2, pp. 32–36, 1964.

[9] Sokolov, J., "Below resonance operation of the strip line *Y*-circulator," *IEEE Trans. Microwave Theory Tech.* (Correspondence), vol. MTT-12, pp. 568–569, September 1964.

[10] Volman, V. I., "Principle of operation and design of waveguide Y-circulators," *Telecommun. Radio Eng.* (USSR), vol. 19, pp. 78–82, March 1964.

1965

[1] Chandra, K., R. Parshad, N. Mohanty, and V. K. Aggarwal, "An experimental 3-port Y-circulator for X-band," *J. Inst. Telecommun. Eng. New Delhi*, vol. 11, pp. 531–534, December 1965.

[2] Comstock, R. L., and C. E. Fay, "Performance and ferrimagnetic material considerations in cryogenic microwave devices," *J. Appl. Phys.*, vol. 36, pt. 2, pp. 1253–1258, March 1965.

[3] Damiano, R., and J. Kliphuis, "Cryogenically cooled *Y*-junction circulators for parametric amplifier applications," *Proc. IEEE* (Correspondence), vol. 53, pp. 198–199, February 1965.

[4] Davies, J. B., "Theoretical design of wideband waveguide circulators," *Electron. Lett.*, vol. 1, pp. 60–61, May 1965.

[5] Emmrich, P., H. Junker, and E. Pivit, "Ferrite components for telecommunication engineering" (in German), *Telefunken*, vol. 38, no. 2, p. 175, 1965.

[6] Fay, C. E., and R. L. Comstock, "Operation of the ferrite junction circulator," *IEEE Trans. Microwave Theory Tech.*, vol. MTT-13, pp. 15–27, January 1965.

[7] Gandhi, O. P., and G. S. Sidhu, "Design and development of X-band ferrite components," *J. Inst. Telecommun. Eng. New Delhi*, vol. 11, pp. 181–187, June 1965.

[8] Heinz, W. W., and S. Okwit, "Cryogenic to ambient temperature operation of circulators for ultra-low-noise applications," *Proc. IEEE* (Correspondence), vol. 53, p. 408, April 1965.

[9] Kraszewski, A., "Some experimental results on the X-band junction circulator," *IEEE Trans. Microwave Theory Tech.* (Correspondence), vol. 13, pp. 382–383, May 1965.

[10] Krzycki, Z., "Recent developments of microwave ferrite devices in Poland" (in Polish), *Przegl. Elektron.*, vol. 6, no. 8, p. 365, 1965.

[11] Konishi, Y., "Lumped element *Y* circulator," *IEEE Trans. Microwave Theory Tech.*, vol. MTT-13, pp. 852–864, November 1965.

[12] Motz, H., and H. W. Wrede, "Ferrites for isolators and circulators" (in German), *Telefunken*, vol. 38, no. 2, p. 187, 1965.

[13] Mukherjee, A., S. Sundaram, and G. S. Mani, "Low power ferrite devices," *J. Inst. Telecommun. Eng. New Delhi*, vol. 11, pp. 493–497, December 1965.

[14] Pivit, E., "Circulators from lumped elements" (in German), *Telefunken*, vol. 38, no. 2, pp. 206–213, 1965.

[15] Roberts, R. W., and J. E. Lahey, "Ferrites for VHF frequencies," *IEEE Trans. Veh. Commun.*, vol. VC-14, pp. 117–121, March 1965.

[16] Schlöman, E., J. J. Green, and J. H. Saunders, "Ultimate performance limitations of high-power ferrite circulators and phase shifters," *IEEE Trans. Magn.*, vol. MAG-1, pp. 168–174, September 1965.

[17] Sidhu, G. S., and O. P. Gandhi, "A four-port waveguide junction circulator and effects of dielectric loading on its performance," *IEEE Trans. Microwave Theory Tech.* (Correspondence), vol. 13, pp. 388–389, May 1965.

[18] Simon, J. W., "Broadband strip-transmission line *Y*-junction circulators," *IEEE Trans. Microwave Theory Tech.*, vol. MTT-13, pp. 335–345, May 1965.

[19] Towpik, R., "3-way S-band ferrite circulator" (in Polish), *Przegl. Elektron.*, vol. 6, no. 12, pp. 609–615, 1965.

[20] Veselkov, G. I., "Autostabilization of the angle of rotation of the plane of polarisation of a wave in a circulator," *Sov. Radio Eng.*, vol. 8, pp. 265–266, May–June 1965.

[21] Volman, V. I., "Design of a waveguide Y-circulator," *Telecommun. Radio Eng.* (USSR), vol. 19, pp. 85–92, March 1965.

[22] Weiss, J. A., "Circulator synthesis," *IEEE Trans. Microwave Theory Tech.*, vol. MTT-13, pp. 38–44, January 1965.

[23] Yoshida, S., K. Kurihara, and S. Takahashi, "Strip-line *Y*-circulator at 4.2°K in the UHF region," *Proc. IEEE* (Correspondence), vol. 53, pp. 1805–1806, November 1965.

1966

[1] Antsiferov, V. A., "Slot circulator," *Telecommun. Radio Eng.* (USSR), vol. 21, pp. 79–82, November 1966.

[2] Benda, O., S. Novak, and L. Uher, "Some experimental results on a switching Y-circulator in the X-band" (in Czech.), *Elektrotech. Cas.*, vol. 17, no. 8, pp. 579–591, 1966.

[3] Betts, F., D. H. Temme, and J. A. Weiss, "A switching circulator: S-band; stripline remanent 15 kilowatts; 10 microseconds; temperature stable," *IEEE Trans. Microwave Theory Tech.*, vol. MTT-14, pp. 665–669, December 1966.

[4] Bonfeld, M. D., D. F. Linn, and M. Omori, "A novel strip-line circulator," *IEEE Trans. Microwave Theory Tech.* (Correspondence), vol. MTT-14, pp. 98–99, February 1966.

[5] Bosh, F., and G. Schneider, "Wide band tuning of Y-circulators by means of variable line transformers" (in German), *Arch. Elek. Übertragung*, vol. 20, pp. 526–528, September 1966.

[6] Bosma, H., "Performance of lossy H-plane Y-circulators," *IEEE Trans. Magn.*, vol. MAG-2, pp. 273–277, September 1966.

[7] Buchta, G., "Miniaturized broadband *E*-tee circulator at *X*-band," *Proc. IEEE* (Letters), vol. 54, pp. 1607–1608, November 1966.

[8] Deutsch, J., and B. Wieser, "Resonance isolator and Y-circulator with lumped elements at VHF," *IEEE Trans. Magn.*, vol. MAG-2, pp. 278–282, September 1966.

[9] Emmrich, P., "Stripline circulator for low temperatures" (in German), *Arch. Elek. Übertragung*, vol. 20, p. 237, April 1966.

[10] Hagelin, S., "A flow graph analysis of 3- and 4-port junction circulators," *IEEE Trans. Microwave Theory Tech.*, vol. MTT-14, pp. 243–249, May 1966.

[11] Helszajn, J., "Analysis of a coupled waveguide circulator using non-reciprocal attenuation," *Radio Electron. Eng.*, vol. 31, pp. 319–323, May 1966.

[12] ——, "A ferrite ring stripline junction circulator," *Radio Electron. Eng.*, vol. 32, pp. 55–60, July 1966.

[13] Hershenov, B., "*X*-band microstrip circulator," *Proc. IEEE* (Letters), vol. 54 pp. 2022–2023, December 1966.

244

[14] Knop, C. M., and G. Lichtenwalter, "Radar echo reduction using circulators," *IEEE Trans. Antenna Propagat.* (Communications), vol. AP-14, pp. 789–790, November 1966.

[15] Kraszewski, A., "Waveguide junction circulators," (in Polish), *Rozpr. Elektrotech.*, vol. 12, no. 2, pp. 379–401, 1966.

[16] Krynetskii, B. B., G. P. Kuzmin, and A. V. Shirkov, "A cooled circulator in the 3 cm band," *Radio Eng. Electron. Phys.* (USSR), vol. 11, p. 1988, December 1966.

[17] Maher, B., D. Mitchell, and A. Reeves, "A 150 Mc/s circulator," *IEEE Trans. Microwave Theory Tech.* (Correspondence), vol. MTT-14, pp. 42–43, January 1966.

[18] Okean, H. C., "Synthesis of negative resistance reflection amplifiers employing band-limited circulators," *IEEE Trans. Microwave Theory Tech.*, vol. MTT-14, pp. 323–337, July 1966.

[19] Passaro, W. C., and J. W. McManus, "A 35-GHz latching switch," *IEEE Trans. Microwave Theory Tech.*, vol. MTT-14, pp. 669–672, December 1966.

[20] Schwartz, E., "Empirical synthesis of lossless symmetrical circulator" (in German), *Arch. Elek. Übertragung*, vol. 20, pp. 623–625, November 1966.

[21] Smoczynski, L., "Some properties of the scattering matrix coefficients of Y-circulator" (in Polish), *Przegl. Elektron.*, vol. 7, no. 11, pp. 541–545, 1966.

[22] Volman, V. I., "Waveguide Y-circulator with ferrite sphere," *Telecommun. Radio Eng.* (USSR), vol. 21, pp. 85–88, February 1966.

[23] ——, "A method for controlling the parameters of ferrite cylinders," *Telecommun. Radio Eng.*, vol. 21, pp. 121–123, July 1966.

[24] Wantuch, E., and D. Lepore, "An extremely high power X-band three port circulator," *J. Appl. Phys.*, vol. 37, pp. 1079–1080, March 1966.

1967

[1] Ackers, J. K., "Contributions to the theory of stripline junction circulator," *Microwave J.*, vol. 10, pp. 57–59, July 1967.

[2] Anderson, L. K., "An analysis of broadband circulators with external tuning elements," *IEEE Trans. Microwave Theory Tech.*, vol. MTT-15, pp. 42–47, January 1967.

[3] Antsiferov, V. A., "Slotted circulators," *Electron Express*, vol. 8, no. 11, p. 11, 1967.

[4] Barsony, P., "Strip transmission line Y-circulator," *Hiradastechnika*, vol. 18, p. 138, May 1967.

[5] Bogdanov, A. G., I. V. Iretskaya, and V. B. Kartazhov, "Experimental research on the field structure in a waveguide X-circulator," *Radio Eng. Electron. Phys.* (USSR), vol. 12, pp. 144–146, January 1967.

[6] Chandra, K., R. Parshad, and V. K. Agarwal, "Broadbanding of waveguide junction circulators," *J. Inst. Telecommun. Eng. New Delhi*, vol. 13, pp. 145–148, April 1967.

[7] Darwent, A. P., "Symmetry in a high power circulator for 35 GHz," *IEEE Trans. Microwave Theory Tech.* (Correspondence), vol. MTT-15, p. 120, February 1967.

[8] Hagler, T. A., "Microwave N-port ferrite circulator," Sperry Microwave Electronic Div., Clearwater, Fla., Tech. Rep. ECOM-01992-F, Final Rep., Contract DA-28-043-AMC-01992(E), November 1967.

[9] Hancock, K. E., "An introduction to microwave ferrite devices—Circulators and isolators," *Wireless World*, vol. 73, pp. 184–188, April 1967.

[10] Helszajn, J., "An H-plane high power TEM ferrite junction circulator," *Radio Electron. Eng.*, vol. 38, pp. 257–262, April 1967.

[11] Hershenov, B., "All-garnet-substrate microstrip circulator," *Proc. IEEE* (Letters), vol. 55, pp. 696–697, May 1967.

[12] Kobierzycki, J., "The 8 mm ferrite circulator" (in Polish), *Prace PIT*, no. 56, pp. 69–76, 1967.

[13] Konishi, Y., "A high-power UHF circulator," *IEEE Trans. Microwave Theory Tech.*, vol. MTT-15, pp. 700–708, December 1967.

[14] Kumar, R. C., "S-band three port stripline circulator," *J. Inst. Telecommun. Eng. New Delhi*, vol. 13, pp. 354–358, September 1967.

[15] Longley, S. R., "Experimental 4-port E-plane junction circulators," *IEEE Trans. Microwave Theory Tech.* (Correspondence), vol. MTT-15, pp. 378–380, June 1967.

[16] McChesney, G., and V. Dunn, "Broadband,

lumped element UHF circulator," *IEEE Trans. Microwave Theory Tech.* (Correspondence), vol. MTT-15, p. 198, March 1967.

[17] Medoks, A. G., "Design of a waveguide Y-circulator," *Telecommun. Radio Eng.* (USSR), vol. 21, pt. 2, pp. 90–94, September 1967.

[18] Mikaelyan, A. L., and A. K. Stolyarov, "Development and applications of microwave ferrites," *Radio Eng. Electron. Phys.* (USSR), vol. 12, pp. 1860–1872, November 1967.

[19] Ogasawara, N., "A latching millimeter-wave *T*-circulator utilizing a single ferrite located inside the waveguide structure," *Proc. IEEE* (Letters), vol. 55, pp. 1640–1641, September 1967.

[20] Penney, J. A., "Non-reciprocal-ferrite devices at microwave frequencies," *Point-to-Point Telecommun.*, vol. 11, pp. 121–138, April 1967.

[21] Siekanowicz, W. W., and W. A. Schilling, "A new type of latching switchable ferrite junction circulator," *Proc. IEEE* (Letters), vol. 55, pp. 407–408, March 1967.

[22] Smoczynski, L., "Ferrite circulator for the 22 cm band" (in Polish), *Arch. Elektrotech.* (Warsaw), vol. 16, p. 1043, 1967.

[23] Stolyarov, A. K., and I. P. Tyukov, "On the analysis of waveguide Y-circulators in terms of their extrinsic characteristics," *Radio Eng. Electron. Phys.* (USSR), vol. 12, pp. 346–348, February 1967.

[24] ——, "Electrical parameters of waveguide Y-circulator," *Radio Eng. Electron. Phys.* (USSR), vol. 12, pp. 1954–1962, December 1967.

[25] Towpik, R., "Wideband Y- and T-type ferrite circulators for the 10 cm band" (in Polish), *Arch. Elektrotech.*, vol. 16, no. 4, p. 1047, 1967.

[26] Tyukov, I. P. "Theory of symmetric waveguide circulators," *Radio Eng. Electron. Phys.* (USSR), vol. 12, pp. 244–253, February 1967.

[27] ——, "Analysis of symmetrical y-junctions filled with ferrite dielectrics," *Radio Eng. Electron. Phys.* (USSR), vol. 12, pp. 341–345, February 1967.

[28] Whiting, K., "Design data for UHF circulators," *IEEE Trans. Microwave Theory Tech.* (Correspondence), vol. MTT-15, pp. 195–198, March 1967.

968

[1] Allen, J. L., and D. R. Taft, "Ferrite elements for hybrid microwave integrated systems," *IEEE Trans. Microwave Theory Tech.*, vol. MTT-16, pp. 405–411, July 1968.

[2] Bergman, J. O., and C. Christensen, "Equivalent circuit for a lumped-element Y circulator," *IEEE Trans. Microwave Theory Tech.* (Correspondence), vol. MTT-16, pp. 308–310, May 1968.

[3] Bosch, F., and Z. Moursi, "On the determination of low field working point of stripline circulators" (in German), *Arch. Elek. Übertragung*, vol. 22, pp. 605–607, December 1968.

[4] Bosma, H., "A general model for junction circulators; choice of magnetization and bias field," *IEEE Trans. Magn.*, vol. MAG-4, pp. 587–596, September 1968.

[5] Davis, L. E., "Central pin tolerances in broadband 3-port waveguide circulators," *Electron. Lett.*, vol. 4, pp. 307–308, July 1968.

[6] Davis, L. E., and J. R. Bacon, "Latched 3-port waveguide circulators," *Electron. Lett.*, vol. 4, pp. 490–492, November 1968.

[7] Dunn, V. E., and A. J. Domenico, "Recent advances in microstrip circulators," *IEEE Trans. Microwave Theory Tech.* (Correspondence), vol. MTT-16, pp. 1060–1061, December 1968.

[8] Feoktisov, V. G., "Design of a stripline Y-circulator," *Radio Eng. Electron. Phys.* (USSR), vol. 13, pp. 1111–1115, July 1968.

[9] Gardiol, F. E., "Quick answers on high power circulators," *Microwaves*, vol. 7, pp. 106–107, June 1968.

[10] Helszajn, J., and C. R. Buffler, "Adjustment of the 4-port single junction circulator," *Radio Electron. Eng.*, vol. 35, pp. 357–360, June 1968.

[11] Konishi, Y., "General theory of coaxial Y-junction circulator and design consideration on its UHF high power application," *Tech. J. Japan Broadcast. Corp.*, vol. 20, no. 2, p. 19, 1968.

[12] Konishi, Y., T. Murayama, and N. Hoshino, "VHF power circulator with slow-wave circuits," *Electron. Commun. Japan*, vol. 51A, August 1968.

[13] Krzycki, Z., and L. Smoczynski, "A 10 cm broadband stripline ferrite circulator" (in Polish), *Przegl. Elektron.*, vol. 9, no. 2, p. 87, 1968.

[14] Masse, D., "Broadband microstrip junction circulators," *Proc. IEEE* (Letters), vol. 56, pp. 352–353, March 1968.

[15] Medoks, A. G., "Theory of symmetric waveguide Y-junctions," *Radio Eng. Electron. Phys.* (USSR), vol. 13, pp. 106–109, January 1968.

[16] ——, "Design algorithms of a waveguide Y-circulator," *Telecommun. Radio Eng.* (USSR), vol. 23, pp. 84–87, September 1968.

[17] Parsonson, C. G., S. R. Longley, and J. B. Davies, "The theoretical design of broadband 3-port waveguide circulators," *IEEE Trans. Microwave Theory Tech.* (Correspondence), vol. MTT-16, pp. 256–258, April 1968.

[18] Pippin, J. E., "Microwave ferrite devices—1968," *Microwave J.*, vol. 11, pp. 29–39, April 1968.

[19] Roome, G. T., and H. A. Hair, "Thin ferrite devices for microwave integrated circuits," *IEEE Trans. Microwave Theory Tech.*, vol. MTT-16, pp. 411–420, July 1968.

[20] Schwartz, E., "Broadband matching of resonant circuits and circulators," *IEEE Trans. Microwave*

Theory Tech., vol. MTT-16, pp. 158–165, March 1968.

[21] Shipley, C. A., "Effect of 'ring resonance' in circulator time-delay circuit," *IEEE Trans. Microwave Theory Tech.* (Correspondence), vol. MTT-16, pp. 202–203, March 1968.

[22] Siekanowicz, W. W., and W. A. Schilling, "A new type of latching switchable ferrite junction circulator," *IEEE Trans. Microwave Theory Tech.*, vol. MTT-16, pp. 177–183, March 1968.

[23] Smoczynski, L., "Design of broadband stripline ferrite circulators," *Rozpr. Elektrotech.*, vol. 14, no. 3, pp. 539–554, 1968.

[24] ——, "Wideband ferrite circulator for 15 cm band" (in Polish), *Arch. Elektrotech.* (Warsaw), vol. 17, no. 1, p. 203, 1968.

[25] Stern, R. A., and J. P. Agrios, "A 500 kW *X*-band air-cooled ferrite latching switch," *IEEE Trans. Microwave Theory Tech.*, vol. MTT-16, pp. 1034–1037, December 1968.

[26] Vamberskiy, M. V., and V. I. Kazantsev, "Design of H-plane waveguide Y-circulator," *Telecommun. Radio Eng* (USSR), vol. 23, pp. 76–81, October 1968.

[27] ——, "Design of decimeter wavelength stripline Y-circulator," *Telecommun. Radio Eng.* (USSR), vol. 23, pp. 77–81, November 1968.

[28] Volman, V. I., N. D. Kozyrev, and V. I. Redechkin, "Five arm waveguide circulator," *Telecommun. Radio Eng.* (USSR), vol. 23, pp. 141–142, October 1968.

[29] Wright, W. H., Jr., and J. W. McGowan, "High-power *Y*-junction *E*-plane circulator," *IEEE Trans. Microwave Theory Tech.* (Correspondence), vol. MTT-16, pp. 557–559, August 1968.

1969

[1] Bogdanov, A. G., "Design of waveguide X-circulators," *Radio Eng. Electron. Phys.* (USSR), vol. 14, pp. 528–536, April 1969.

[2] Fliegler, E., "A 350-MHz broad-band lumped element circulator as a protective isolator," *IEEE Trans. Microwave Theory Tech.* (Correspondence), vol. MTT-17, pp. 275–277, May 1969.

[3] Freibergs, E., "Half-height waveguide *Y* and *T* circulators," *IEEE Trans. Microwave Theory Tech.* (Correspondence), vol. MTT-17, pp. 729–730, September 1969.

[4] Gertsenshtein, M. E., V. R. Magnushevskiy, and L. G. Solovey, "Matching error when measuring small losses in circulators and reciprocal four terminal network," *Telecomnun. Radio Eng.* (USSR), vol. 24, pt. 2, July 1969.

[5] Hagelin, S., "Analysis of lossy symmetrical three-port networks with circulator properties," *IEEE Trans. Microwave Theory Tech.*, vol. MTT-17, pp. 328–333, June 1969.

[6] Hershenov, B., and R. L. Ernst, "Miniature microstrip circulators using high-dielectric-constant substrates," *RCA Rev.*, vol. 30, pp. 541–543, September 1969.

[7] Kumar, R. C., "S-band Y-junction stripline circulator with triangular ferrite," *Electron. Eng.* (London), vol. 41, pp. 214–215, February 1969.

[8] ——, "A 6.3 GHz stripline circulator," *J. AEU* (Japan), vol. 11, no. 3, pp. 23–24, 1969.

In this section, six papers describing the theory and operation of YIG filters are reprinted. The first paper, an early paper by P.S. Carter (1) provides a good introduction to the behavior of such components. The second paper by Comstock (2) presents a synthesis procedure for bandpass filters with Chebyshev response using orthogonal circuit resonators coupled by the YIG resonator. The other papers are concerned generally with the design of various filters. The paper by Fjerstad (3) describes YIG filter structures which operate at frequencies as high as 40 GHz.

YIG FILTERS

III

References

1. P.S. Carter, Jr., "Magnetically Tunable Microwave Filters Using Single Crystal Yttrium Iron Garnet Resonators," MTT-9, May 1961, pp. 252-260.
2. R.L. Comstock, "Synthesis of Filter-Limiters Using Ferrimagnetic Resonators," MTT-12, Nov. 1964. pp. 599-607
3. R.L. Fjerstad, "Some Design Consideration and Realizations on Iris Coupled YIG Tuned Filters in the 12-40 GHz Region" MTT-18, Apr. 1970, pp. 205-212.

Magnetically-Tunable Microwave Filters Using Single-Crystal Yttrium-Iron-Garnet Resonators*

PHILIP S. CARTER, Jr.†, MEMBER, IRE

Summary—**A new type of magnetically-tunable band-pass microwave filter that makes use of ferrimagnetic resonance in single-crystal yttrium iron garnet is presented. The 3-db bandwidth can be adjusted from about 6 Mc to 100 Mc at *X* band, and the center frequency can be tuned over a wide range of frequencies, by means of a varying dc field. A theoretical analysis of the operation and behavior of this type of filter is presented. Descriptions of single-resonator and two-resonator filters which can be tuned over the *X*-band frequency range are given and experimental data are presented showing their tuning range, insertion loss, and bandwidth.**

I. Introduction

WITH the increased use of continuously tunable microwave receivers, there has been a demand for a nonmechanically tuned, narrow-band filter which is tunable over a broad band of frequencies in the microwave frequency range. Previous attempts to solve this problem have included the use of cavities containtaining ferroelectrics,[1] "Varicaps" or voltage-tuned back-biased diodes,[2,3] and ferrites.[4–6] In general, tuning ranges of a few per cent are possible using these techniques while maintaining a high unloaded resonator $Q(Q_u)$.

The main difficulty with previous tuning techniques is high losses, which are greatly reduced by the use of the ferrimagnetic material, yttrium-iron-garnet (YIG). Single crystals of this material (now available commercially) show extremely low losses when used as the resonant elements in a band-pass (or band-reject) filter. The Q_u of this material considered as a resonator is typically between 2000 and 4000 at X-band frequencies and therefore compares favorably with transmission-line and hollow-cavity resonators.

This new approach to the problem of nonmechanical tuning makes use of the equivalence between a resonant circuit, such as an inductively coupled cavity or lumped-element series-resonant circuit, and an inductively coupled magnetic resonator biased with a dc magnetic field.

This equivalent circuit was first worked out analytically by Bloembergen and Pound[7] and was first applied by DeGrasse[8] to the design of an S-band limiter, using a single-crystal YIG resonator.

The tunability feature results from the fact that the resonant frequency is nearly a linear function of dc magnetic biasing field. This feature allows a very broad tuning range, limited mainly by the bandwidth of the microwave coupling structure employed.

The first part of this paper consists of a mathematical analysis of the equivalent circuit of a ferrimagnetic resonator coupled to external loads by means of loops, transmission lines, and waveguides. Analytical formulas for the external $Q(Q_e)$ of the ferrimagnetic resonator are given for loops, strip-transmission-lines and TE_{10} mode rectangular waveguide. A formula is given which relates the unloaded $Q(Q_u)$ of the ferrite resonator to the properties of the ferrite (damping time and gyromagnetic ratio).

The second part of this paper consists of descriptions of, and performance data taken on, two types of magnetically tunable filters operating in the X-band region and above. These are: 1) a single-resonator filter which has a tuning range from 7.0 kMc to 11.2 kMc, and 2) a two-resonator filter whose tuning range extends from 8.2 to 14.0 kMc.

II. Basic Principle of Operation

The basic idea of the magnetic resonance filter is illustrated in Fig. 1. Two coils have their axes at right angles to each other, and a small ferrite sample is placed at the intersection of the coil axes. When the sample is not magnetized, no power is transferred between the coils because the loop axes are perpendicular to each

* Received by the PGMTT, July 15, 1960; revised manuscript received, March 6, 1961. Sponsored by the U. S. Army Signal Research and Development Lab., Fort Monmouth, N. J., under Contract No. DA 36-039-74862. A shorter version of this paper was presented at the 1960 IRE International Convention (1960 IRE International Convention Record, pt. 3, pp. 130–135).

† Stanford Research Inst., Menlo Park, Calif.

[1] W. J. Gemulla and R. D. Hall, "Ferroelectrics at microwave frequencies," *Microwave J.*, vol. 3, pp. 47–51; February, 1960.

[2] E. M. T. Jones, G. L. Matthaei, S. B. Cohn, and B. M. Schiffman, "Design Criteria for Microwave Filters and Coupling Structures," Stanford Research Inst., Menlo Park, Calif. Tech. Rept. 5, SRI Project 2326, Contract No. DA 36-039 SC-74862; March, 1959.

[3] A. Uhlir, Jr., "The potential of semiconductor diodes in high-frequency communications," Proc. IRE, vol. 46, pp. 1099–1115; June, 1958.

[4] G. R. Jones, J. C. Cacheris, and C. A. Morrison, "Magnetic tuning of resonant cavities and wideband frequency modulation of klystrons," Proc. IRE, vol. 44, pp. 1431–1438; October, 1956.

[5] C. E. Fay, "Ferrite-tuned resonant cavities," Proc. IRE, vol. 44, pp. 1446–1449; October, 1956.

[6] C. E. Nelson, "Ferrite-tunable microwave cavities and the introduction of a new reflectionless tunable microwave filter," Proc. IRE, vol. 44, pp. 1449–1455; October, 1956.

[7] N. Bloembergen and R. V. Pound, "Radiation damping in magnetic resonance experiments," *Phys. Rev.*, vol. 95, pp. 8–12; July 1, 1954.

[8] R. W. DeGrasse, "Low-loss gyromagnetic coupling through single crystal garnets," *J. Appl. Phys.*, Suppl. to vol. 30, pp. 155S–156S; April, 1959.

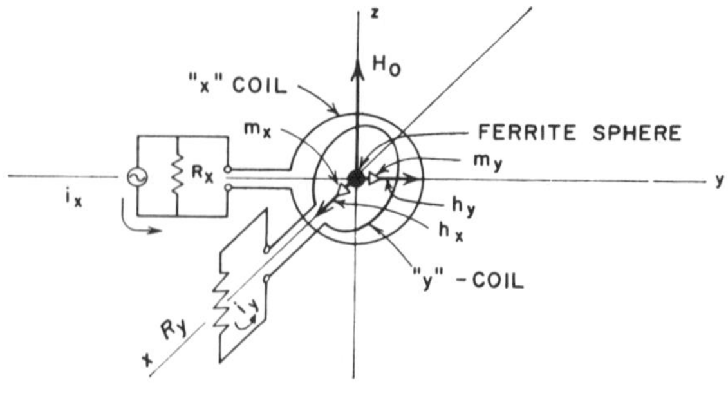

Fig. 1—Magnetic resonance filter.

other and there is no interaction with the ferrite. When a dc field H_0 is applied along the z axis, and an RF driving current $i_x e^{j\omega t}$ is applied to the terminals of the x coil, the magnetic moments of the electrons in the ferrite precess around the x axis, producing an RF magnetic moment along the y axis and inducing a voltage in the y circuit. The precession angle is largest and thus the induced voltage in the y-coil is maximum, at ferromagnetic resonance given by $\omega_0 = \mu_0 \gamma H_0$ for a spherical ferrite,[9] where γ is the gyromagnetic ratio and is very close to 1.759×10^{11} for electrons in most ferrites. The drop-off in response away from resonance is determined by the degree of coupling of the external loads R_x and R_y and the internal losses in the ferrite. It is shown in Section III that the equivalent circuit of this device is a resonant circuit, inductively coupled to the output
250
lines or loads. This circuit is also a gyrator; the phase shift in one direction through the circuit differs from the phase shift in the other direction by 180°.

One requirement on the resonator in a low-insertion-loss, narrow-band-pass filter is that Q_u be high. Although it is necessary to specify the application in more detail in order to specify what exactly is meant by "high-Q_u," it may be stated in general that Q_u's of several hundred or higher are useful for filter design. Q_u's of this order are obtained at microwave frequencies with single-crystal-yttrium-iron-garnet.

The second requirement to obtain low insertion loss in a narrow-band, band-pass filter is that the external Q, Q_e of the resonators be considerably lower than Q_u. It was not apparent from previous work by DeGrasse[8] that sufficiently low Q_e could be obtained over a wide tuning range. It is shown below that sufficient coupling can actually be obtained without the use of the narrow-band impedance matching employed by DeGrasse.[8]

Two other characteristics of single crystal ferrimagnetic resonators must be considered in designing a magnetically tunable filter: 1) magnetocrystalline anisotropy, and 2) the higher-order magnetostatic modes. Their effects on filter characteristics and design are discussed below.

[9] These quantities are defined in Section III.

III. Analysis and Measurement of Coupling Characteristics and Equivalent Circuits of Ferrite Resonators in Lumped-Element Circuits, Transmission Lines, and Waveguides

In the following analysis, the ferrite resonator is assumed to be small enough that the RF field is substantially uniform throughout the volume of the sample and therefore that higher-order modes are not excited. Also magnetocrystalline anisotropy is neglected. In this case with the dc magnetic field H_0 applied along the z-axis the RF magnetizations in the x and y directions are given by the following relations:[10]

$$m_x = \chi_{xx}{}^e h_x + \chi_{xy}{}^e h_y, \tag{1a}$$

$$m_y = \chi_{yx}{}^e h_x + \chi_{yy}{}^e h_y, \tag{1b}$$

where

$m_x = x$ component of RF magnetization within ferrite,
$m_y = y$ component of RF magnetization within ferrite,
$h_x = x$ component of applied RF field,
$h_y = y$ component of applied RF field,
$\chi_{xx}{}^e = \chi_{yy}{}^e = x-x(y-y)$ component of effective RF tensor, susceptibility,[11]
$\chi_{xy}{}^e = -\chi_{yx}{}^e = x-y(y-x)$ component of effective RF tensor susceptibility.

Flammer[12] has developed analytical formulas for the effective susceptibility that include the loss in the ferrite. These expressions were derived using the Bloch-Bloembergen[13] formulation of the equations of motion of magnetization, including the effect of the shape-dependent demagnetizing factors N_x, N_y, and N_z for a

[10] MKS units are used here and throughout this paper.

[11] The effective or external susceptibility is the ratio of the RF magnetic moment to the *applied* RF magnetic field (not the RF magnetic field inside the ferrite).

[12] C. Flammer, "Resonance Phenomena in Ferrites," Stanford Research Inst., Menlo Park, Calif. unpublished memorandum; 1956. Stanford plans to publish this memorandum in the near future.

[13] N. Bloembergen, "Magnetic resonance in ferrites," Proc. IRE, vol. 44, pp. 1259–1269; October, 1956. It has been pointed out by Flammer that (5) in this reference contains an error. The term $-M_0/\gamma$ should be deleted.

general ellipsoidal sample.[14] These formulas for the effective tensor susceptibility are

$$\chi_{xx}^e = \frac{\omega_m\left[\omega_0 + (N_y - N_z)\omega_m + \dfrac{j\omega/\tau}{\omega_0 - N_z\omega_m}\right] + \dfrac{\omega_m}{(\omega_0 - N_z\omega_m)^2}\left[\omega_0 + (N_y - N_z)\omega_m\right]\cdot\dfrac{1}{\tau^2}}{\left\{\begin{array}{l}\left[\omega_0 + (N_y - N_z)\omega_m + \dfrac{j\omega/\tau}{\omega_0 - N_z\omega_m}\right]\left[\omega_0 + (N_x - N_z)\omega_m + \dfrac{j\omega/\tau}{\omega_0 - N_z\omega_m}\right] - \omega^2 \\ \quad + \dfrac{1}{(\omega_0 - N_z\omega_m)^2\tau^2}\cdot\left[\omega_0 + (N_x - N_z)\omega_m\right]\left[\omega_0 + (N_y - N_z)\omega_m\right]\end{array}\right\}}, \tag{2a}$$

$$\chi_{yy}^e = \frac{\omega_m\omega_0 + (N_x - N_z)\omega_m + \dfrac{j\omega/\tau}{(\omega_0 - N_z\omega_m)^2} + \dfrac{\omega_m}{(\omega_0 - N_z\omega_m)^2}\left[\omega_0 + (N_x - N_z)\omega_m\right]\dfrac{1}{\tau^2}}{\left\{\begin{array}{l}\left[\omega_0 + (N_x - N_z)\omega_m + \dfrac{j\omega/\tau}{\omega_0 - N_z\omega_m}\right]\left[\omega_0 + (N_y - N_z)\omega_m + j\dfrac{\omega/\tau}{\omega_0 - N_z\omega_m}\right] - \omega^2 \\ \quad + \dfrac{1}{(\omega_0 - N_z\omega_m)^2\tau^2}\left[\omega_0 + (N_x - N_z)\omega_m\right]\left[\omega_0 + (N_y - N_z)\omega_m\right]\end{array}\right\}}, \tag{2b}$$

$$\chi_{xy}^e = -\chi_{yx}^e = \frac{-j\omega_m\omega}{\left\{\begin{array}{l}\left[\omega_0 + (N_x - N_z)\omega_m + \dfrac{j\omega/\tau}{\omega_0 - N_z\omega_m}\right]\left[\omega_0 + (N_y - N_z)\omega_m + \dfrac{j\omega/\tau}{\omega_0 - N_z\omega_m}\right] - \omega^2 \\ \quad + \dfrac{1}{(\omega_0 - N_z\omega_m)^2\tau^2}\left[\omega_0 + (N_x - N_z)\omega_m\right]\left[\omega_0 + (N_y - N_z)\omega_m\right]\end{array}\right\}}, \tag{2c}$$

in which

$\omega_m = \mu_0\gamma M_0 = \gamma_0 M_0$,
$\omega_0 = \mu_0\gamma H_0$,
γ = gyromagnetic ratio $= g(e/2m)$,
N_x, N_y, N_z = effective demagnetizing factors in x, y, and z directions,
τ = Bloch-Bloembergen phenomenological relaxation time,
g = Landé "g" factor $\cong 2.00$ for electrons in most ferrites,
e/m = ratio of charge e to mass m of electron, $= 1.759 \times 10^{+11}$ coulombs/kg,
M_0 = saturation magnetization in amperes per meter,
μ_0 = intrinsic permeability of free space $= 1.256 \times 10^{-6}$ henries per meter,
H_0 = applied dc field in amperes per meter.

For a spherical geometry, $N_x = N_y = N_z = \frac{1}{3}$; and (2) becomes after some algebraic reduction and elimination of terms of order $1/\tau^2$ or smaller,

$$\chi_{xx}^e = \chi_{yy}^e = \frac{\omega_0\omega_m}{\omega_0^2 - \omega^2 + \dfrac{2j\omega}{\tau}\left(1 + \dfrac{\omega_m/3}{\omega_0 - \omega_m/3}\right)} \tag{3a}$$

$$\chi_{xy}^e = -\chi_{yx}^e = \frac{-j\omega_m\omega}{\omega_0^2 - \omega^2 + \dfrac{2j\omega}{\tau}\left(1 + \dfrac{\omega_m/3}{\omega_0 - \omega_m/3}\right)}. \tag{3b}$$

[14] C. Kittel, "On the theory of ferromagnetic resonance absorption," *Phys. Rev.*, vol. 73, pp. 155–161; January 15, 1948.

The complete detailed equivalent circuit[15] of the ferrimagnetic resonator located at the intersection of the two loops is shown in Fig. 2. It consists of a resonant circuit coupled inductively to the signal source i_x, to the source impedance R_x, and to the load impedance R_y. In addition to the series-resonant circuit, there is a nonreciprocal phase shifter which produces, with the coil arrangement shown in Fig. 1, an additional *negative* 90° phase shift in the output voltage across R_y. With the signal generator connected to the y terminals an additional *positive* 90° phase shift is produced across R_x.

The complete analytical formulation of the circuit representation of the ferrimagnetic resonator has been carried out for the following cases:

1) Ferrimagnetic resonator located at the intersection of the axes of the two orthogonal loops shown in Fig. 1.

2) Resonator located at a position of an equivalent short circuit along a short-circuited two-wire TEM-mode transmission line, shown in Fig. 3(b).

3) Same as case 2 in a strip-center-conductor line.

4) Resonator located at a position of an equivalent short circuit at the center of a short-circuited TE_{10}-mode waveguide.

[15] P. S. Carter, Jr. and G. L. Matthaei, "Design Criteria for Microwave Filters and Coupling Structures," Stanford Research Inst., Menlo Park, Calif. Tech. Rept. 8, SRI Project 2326, Contract No. DA 36-039 SC-74862; September, 1959.

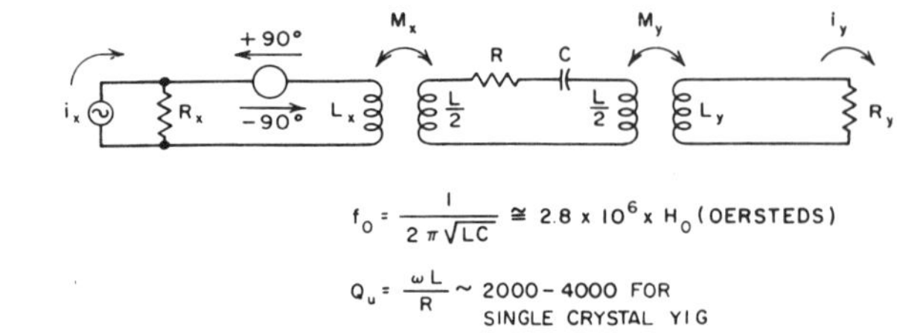

Fig. 2—Equivalent circuit of magnetic resonance filter.

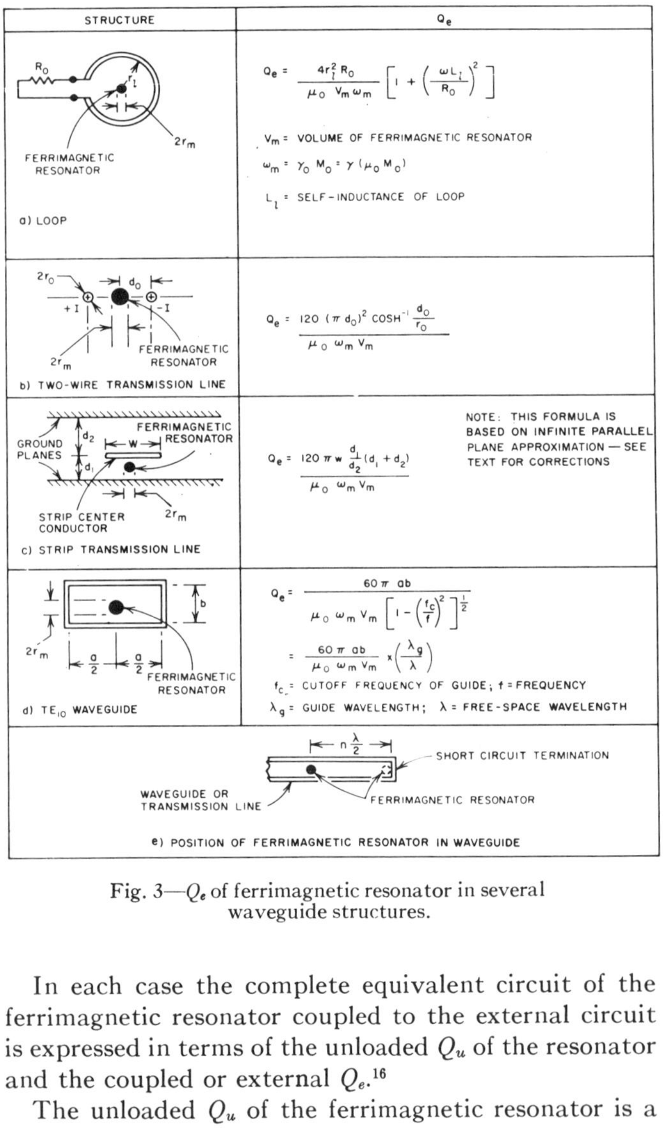

Fig. 3—Q_e of ferrimagnetic resonator in several waveguide structures.

In each case the complete equivalent circuit of the ferrimagnetic resonator coupled to the external circuit is expressed in terms of the unloaded Q_u of the resonator and the coupled or external Q_e.[16]

The unloaded Q_u of the ferrimagnetic resonator is a property of the ferrimagnetic material and is independent of the external circuit.[17] An analytical formula for Q_u derived from the equivalent circuit is given by[15]

$$Q_u = (\omega_0 - N_z\omega_m)\tau/2 = \frac{H_0 - N_z M_0}{\Delta H^i}, \qquad (4)$$

where ΔH^i is the line width given by

$$\Delta H^i = \frac{2}{\gamma_0 \tau}.$$

Eq. (4) predicts low values of unloaded Q_u for low resonant frequencies, and also predicts, in particular, that Q_u will reach zero at value of dc field and RF frequency given by $\omega_0 = N_z\omega_m$. For a spherical YIG resonator, this low cutoff frequency occurs at $f_0 = \omega_m/3\times 2\pi = 1670$ Mc. Measurements made of Q_u in the frequency range of this cutoff support this theory.[18]

Eq. (4) predicts that Q_u depends on $N_z\omega_m$. In particular, a long-rod geometry with the dc magnetic field along the axis of the rod (so that $N_z \approx 0$) should yield a maximum value of Q_u. This effect of the shape on Q_u has not yet been measured. Finally, (4) shows that, for a constant relaxation time τ, Q_u is higher in the case of a lower saturation magnetization material. For frequencies below around 2000 Mc, a low-M_0, narrow-linewidth material is desirable. This requirement is discussed in Section VII.

The external or coupled Q_e depends upon the volume of the ferrimagnetic resonator V_m, its saturation magnetization M_0, and the type and geometry of the coupling structure. Formulas for Q_e have been derived for the four cases listed. The resulting formulas, tabulated in Fig. 3, may be applied to the calculation of the Q_e of the uniform precessional mode of any ellipsoidally shaped magnetic resonator (rods, disks, etc).

The formulas for Q_e given in Fig. 3 were derived using two different methods: The first, which employs the reciprocity theorem,[15] is valid for lumped-parameter coupling devices such as loops, and for TEM-mode transmission lines. The equation for Q_e of the magnetic resonator in the TE_{10}-mode rectangular waveguide [Fig. 3(d)] was derived using a more general coupling formula discussed by Slater.[19] For the waveguide case, a *linearly-polarized* magnetic moment was assumed. The RF magnetic moment of a magnetic resonator is in general *elliptically* polarized, therefore, both perpendicular components of RF magnetic moment can interact with an elliptically polarized waveguide field. The formula for Q_e given above is thus valid only when the magnetic sample is located in the linearly-polarized field at the center of the guide.[20]

[16] C. G. Montgomery, R. H. Dicke, and E. M. Purcell, "Principles of Microwave Circuits," M.I.T., Rad. Lab. Ser. No. 8, McGraw-Hill Book Co., Inc., New York, N. Y., pp. 228–239; 1948.

[17] This is true for the essentially unperturbed resonator located a sufficient distance from the boundaries of the waveguide. For close wall-to-resonator spacings, resistive losses due to image currents in the waveguide walls and the detuning effects of these images are significant to the resonance characteristics.

[18] P. S. Carter, Jr. and C. Flammer, "Unloaded Q of single crystal garnet resonator as a function of frequency," IRE Trans. on Microwave Theory and Techniques, vol. MTT-8 (Correspondence), pp. 570–571; September, 1960.

[19] J. C. Slater, "Microwave Electronics," D. Van Nostrand Co., Inc., New York, N. Y., ch. 7, p. 150; 1950.

[20] Recently while this paper was being revised, the author was made aware of a more general scattering formula which is applicable in the case where the ellipsoidal sample is at an arbitrary position in the guide. See H. J. Shaw and L. K. Anderson, "Ferrimagnetic resonance in ferrites," in "Interaction of Microwaves with Matter," Microwave Laboratory, Stanford University, Stanford, Calif., Tech. Status Rept. No. 9, Section 1, Contract AF 49(638)-415; May 1–July 31, 1960.

Measurements were made of the Q_e of single-crystal YIG resonators in strip transmission line and TE_{10}-mode rectangular waveguide, and the results of these measurements are shown in Figs. 4 and 5. Fig. 4 shows measurements made on spherical single-crystal YIG resonators of various diameters mounted on a wafer of polyfoam in a section of one-quarter height, standard width, X-band, rectangular waveguide. The waveguide was short-circuited and the YIG sphere was mounted one-half wavelength from the short circuit. The measurements were made in this configuration at 10.0 kMc, using the impedance method described by Ginzton.[21] The solid curve represents the theoretical formula for Q_e. Close agreement between the experimental and theoretical values is evident. When resonator diameters greater than about 0.070 inch were used, measurement accuracy was diminished by the high degree of coupling between the external circuit and the higher-order magnetostatic modes. The effect of the higher-order modes on filter performance is discussed in Section IV.

Fig. 5 shows measurements of Q_e of a strip-transmission-line coupling circuit. The strip-transmission-line configuration (Fig. 5) has a 0.500-inch-wide 0.020-inch-thick center conductor spaced 0.135 inch from the outer conductor on both sides of the center conductor. The inside width of the outer conductor is 1.00 inch. Q_e was measured at a resonant frequency, $f_0 = 3000$ Mc. In these measurements the YIG resonator was mounted close to the short-circuited end of the strip transmission line—the position occupied by the resonator in a magnetically tunable filter where a broad tuning range is desired—since this remains a position of nearly maximum magnetic field as the frequency is varied. The minimum distance between the sphere and the short-circuited end is prescribed by the maximum allowable coupling to higher-order modes. With the resonator touching the short-circuit, appreciable coupling to higher-order modes occurs. In the case shown in Fig. 5 with 0.125-inch spacing between the resonator and the end wall, experiment showed that the coupling to higher-order modes was very loose, *i.e.*, $Q_e > 15{,}000$, for the most strongly coupled higher-order mode. The coupling to the uniform precession resonance at 3000 Mc was not measurably decreased by this 0.125-inch spacing.

The theoretical values for Q_e of the strip transmission line are shown in Fig. 5 as the solid-line curve. This curve was calculated using the formula given in Fig. 3(c), which is very accurate where $w \gg d_1$ and d_2, but not quite so accurate where $w = 0.500$ inch and $d_1 = d_2 = 0.135$ inch (as in Fig. 5). An empirical correction was applied to improve the accuracy of this formula by replacing w in Fig. 3(c) by a slightly larger value, w', which takes into account the fringing of the field at the edge of the

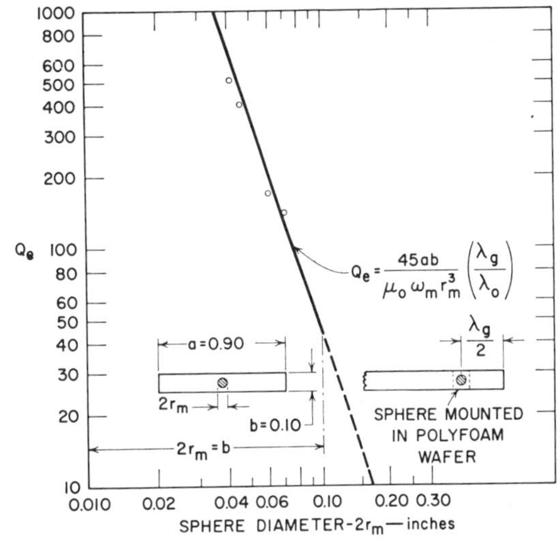

Fig. 4—Theoretical and experimental Q_e of YIG resonators in reduced-height (one-quarter), standard-width X-band waveguide at 10,000 Mc.

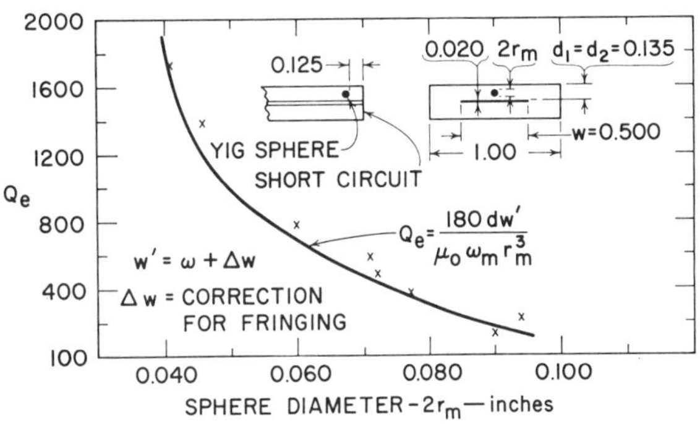

Fig. 5—Theoretical and experimental values of Q_e of YIG resonator using stripline coupling at 3000 Mc.

strip center conductor. This corrected width w' was obtained as follows: the characteristic impedance Z_0' of the actual configuration shown in Fig. 5 was obtained from an accurate formula[22] which includes the effect of the finite thickness of the center conductor and the fringing of the field at the two edges. This value of characteristic impedance Z_0' was then used to compute an effective strip width w', based on the parallel-plane formula, *i.e.*,

$$w' = \frac{\eta_0}{Z_0'} \frac{d_1 d_2}{d_1 + d_2},$$

again using the actual conductor spacings d_1 and d_2, where $\eta_0 = 377$ ohms.

The agreement between the measured and theoretical values of Q_e was quite good (Fig. 5). The measured values of Q_e differ appreciably from the calculated values only for the largest values of YIG sphere diameter

[21] E. L. Ginzton, "Microwave Measurements," McGraw-Hill Book Co., Inc., New York, N. Y., ch. 9, pp. 405–417; 1957.

[22] S. B. Cohn, "Problems in strip transmission lines," IRE Trans. on Microwave Theory and Techniques, vol. MTT-3, pp. 119–126; March, 1955.

253

(around 0.090 inch). For these largest-sphere diameters the measured value of Q_e is also sensitive to the exact location of the sphere with respect to the conducting walls. This shows that the boundary conditions imposed on the motion of the magnetization by the cônducting surfaces of the strip transmission line affect the coupling appreciably.

The strip transmission line offers more design flexibility than the waveguide. Examination of the formula in Fig. 3(c) for the Q_e of a strip transmission line shows that it is possible to adjust both the characteristic impedance of the line and Q_e independently, by adjusting the strip outer conductor width w and the ground-plane spacings d_1 and d_2.

IV. Development of a Single-Resonator Tunable Filter

A single-resonator tunable filter using waveguide coupling was constructed which is tunable over the X-band frequency range from 7.00 kMc to 11.00 kMc. The coupling arrangement (Fig. 6) is analogous to the crossed loops in Fig. 1. The single-crystal YIG sphere is mounted on a styrofoam support in the open iris region common to both the input and output waveguides. The depth of this region of interpenetration, as well as the guide height and resonator size, can be varied to control the coupling between the guides and the
254 resonator. The bias field is applied along the common axis of the guides.

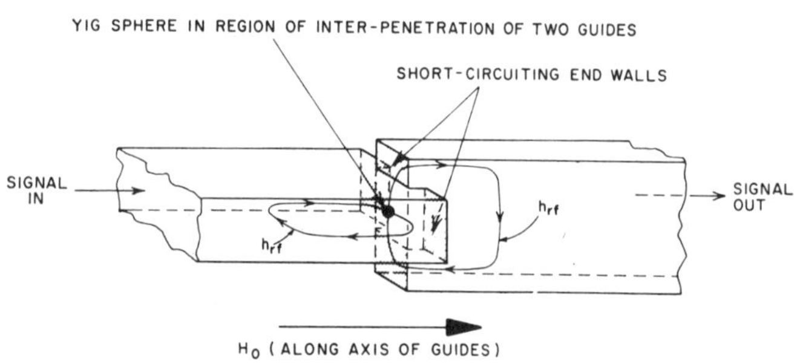

Fig. 6—Coupling principle of single-resonator waveguide filter.

Figs. 7 and 8 show the construction of the experimental single-resonator filter employing a standard-width X-band waveguide. The half-height ($b=0.200$ inch) coupling sections used in this experimental filter are followed by E-plane circular bends (mean radius $=0.300$ inch). These E-plane bends permit the use of a "c" type electromagnet for biasing along the axis of transmission of the guide. Immediately following the E-plane bends are 2.20-inch-long straight tapers joining the half-height sections to the full-height X-band input and output guides. The interpenetration of the two guides was accomplished by milling out 0.035-inch-deep slots in the face of the opposite section, as shown in Figs. 7 and 8.

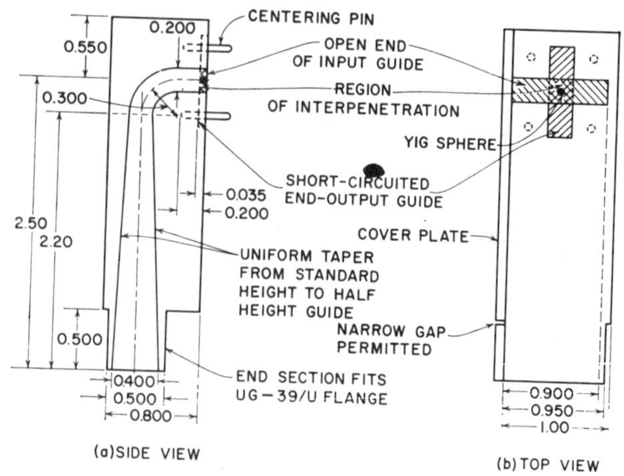

Fig. 7—Construction of half-section of single-resonator waveguide filter.

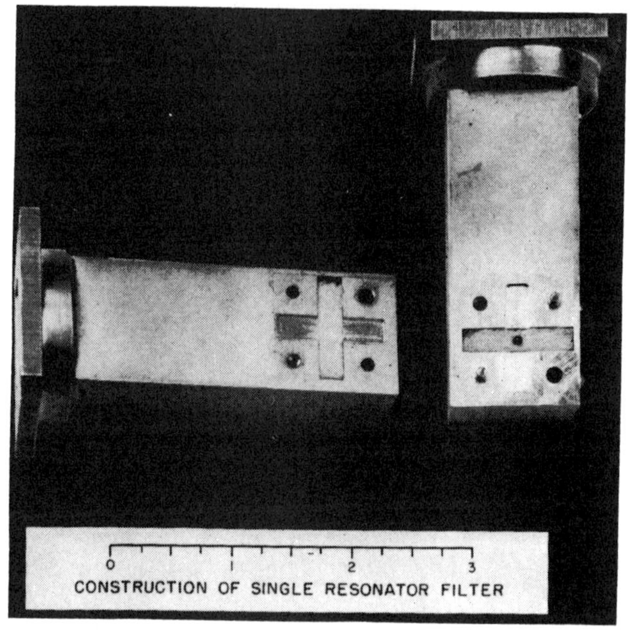

Fig. 8—Experimental single-resonator filter.

The measured responses when this single-resonator filter was tuned to resonate successively at 8.20, 9.00, 10.00, and 11.00 kMc are shown in Fig. 9. Table I shows the measured insertion losses at these frequencies and also at 7.00 kMc, 3-db bandwidths, and the peak response of the strongest subsidiary mode, and the values of Q_u calculated from the insertion loss and bandwidth data (using equations given by Ginzton[23]).

A prominent feature of this single-resonator filter response is the presence of many subsidiary responses which are due to higher-order magnetostatic resonances within the ferrimagnetic sample. These resonances, which were first analyzed by Walker,[24] result from the

[23] Ginzton, *op. cit.*, pp. 403–405.
[24] L. R. Walker, "Magnetostatic modes in ferromagnetic resonance," *Phys. Rev.*, vol. 105; January 15, 1957.

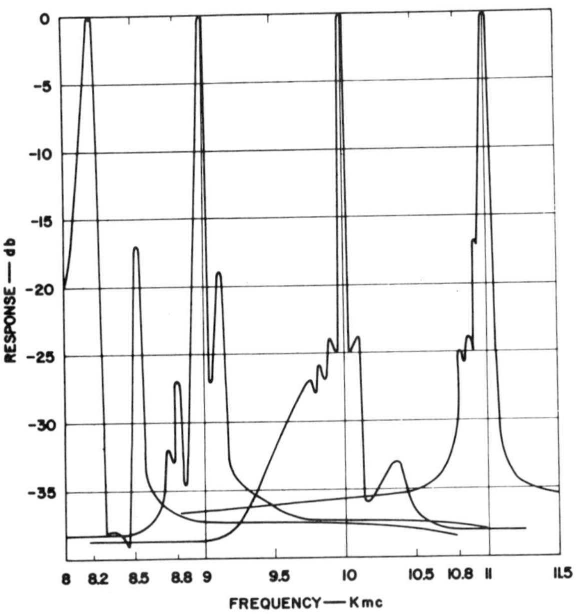

Fig. 9—Single-resonator filter response curves.

TABLE I
PERFORMANCE OF SINGLE-RESONATOR FILTER

f (kMc)	Insertion Loss at Resonance (db)	3-db Bandwidth (Mc)	Q_u	Strongest Magnetostatic Mode Response (db Below Transmission at Main Resonance)
7.00	2.0	not measured	—	—
8.20	1.5	14	3600	−17
9.00	1.2	20	3400	−18
10.00	1.3	25	3200	−23
11.00	1.1	28	3200	−17

nonuniform motion of the magnetization within the ferrimagnetic sample and the resulting dipolar interactions between the magnetic moments. Walker showed that the resonances are characterized by: 1) independence of the size of the sample, *i.e.*, they depend on the shape only, and 2) occurrence only at frequencies given by

$$\omega_0 + \gamma_0 \frac{M_0}{2} \geq \omega \geq \omega_0,$$

where

$$\omega_0 = \gamma_0 H_0.$$

Thus the resonant frequencies of the nonuniform resonances lie somewhere within a spectrum which is $\gamma_0 M_0/2$ cycles wide. For yttrium-iron-garnet in which the saturation magnetization M_0 is approximately 1395×10^2 amperes per meter, the width of this frequency band is 2450 Mc. Coupling to these nonuniform modes occurs when the RF coupling field is nonuniform over the dimensions of the sample. The maximum allowable response to higher-order modes thus dictates the largest-ferrimagnetic resonator that can be used in a specific coupling circuit. In practice, this restriction limits the diameter of the spherical resonator used in the present X-band single-resonator filter to less than 0.100 inch.

The spurious responses due to the higher-order magnetostatic modes are greatly reduced by the use of two or more ferrimagnetic resonators coupled directly together by means of their external RF fields. This is described in detail in Section VI.

A characteristic of this single-resonator filter response using TE_{10}-mode waveguide coupling is the increase of the bandwidth and the decrease of the insertion loss as the center frequency is raised. This increase in bandwidth, and decrease in insertion loss, is due to the decrease of the external Q_e of the waveguide coupling circuit as the frequency increases. Variation of the bandwidth over the tuning range is minimized by the use of a constant characteristic impedance coupling circuit, such as the TEM-mode strip transmission line shown in Fig. 3(c).

Another important aspect of the behavior of a YIG resonator is the saturation occurring at high power levels. This effect is explained in terms of the nonlinear coupling between the uniform mode and the short-wavelength "spin waves," and has been investigated and explained by Suhl.[25,26] This saturation effect was found by experiment to limit the peak power of the X-band single-resonator filter to less than about 15 watts. This effect has been discussed further and experimental results have been presented by Carter and Matthaei.[27] Considerably lower saturation thresholds occur at lower frequencies in the S-band range, both in theory[26] and in experiment.[8]

It has been shown[28] that Q_u is critically dependent on the achievement of a high surface polish on a yttrium-iron-garnet sphere. The spheres used here were ground and polished by a simple tumbling procedure which has been described by Bond.[29] The spheres were ground to shape from the raw crystals by tumbling, using No. 120 grade silicon carbide abrasive paper. They were then ground and polished using this same tumbling technique with successively finer grades of abrasive grinding paper and powder, as follows:

silicon carbide, No. 360, No. 600, No. 900,
emery polishing paper, No. 2/0 and No. 4/0,
Linde fine abrasive, No. B-5125.

[25] H. Suhl, "The non-linear behavior of ferrites at high signal levels," PROC. IRE, vol. 44, pp. 1270–1284; October, 1956.
[26] H. Suhl, "The theory of ferromagnetic resonance at high signal powers," *J. Phys. Chem. Solids*, vol. 1, pp. 209–217; January, 1957.
[27] P. S. Carter, G. L. Matthaei, "Design Criteria for Microwave Filters and Coupling Structures," Stanford Research Inst., Menlo Park, Calif., Final Rept. SRI Project 2326, Contract No. DA 36-039 SC-74862; January, 1961.
[28] R. C. LeCraw, E. G. Spencer, and C. S. Porter, "Ferromagnetic resonance line in yttrium-iron-garnet single crystals," *Phys. Rev.*, vol. 110, pp. 1311–1313; June 15, 1958.
[29] W. L. Bond, "Making small spheres," *Rev. Sci. Instr.*, vol. 22, pp. 344–345; May, 1951.

V. Development of a Two-Resonator Filter

Fig. 10 shows the application of two garnet spheres to form a two-resonator filter. Coupling between the resonators is provided by the long-slot iris in a 0.010-inch-thick conducting wall separating the input and output waveguides.[30] The long vertical coupling slot permits the vertical components of RF magnetic moment to be coupled together through their external RF fields, and prevents the horizontal magnetic fields of the waveguides and horizontal RF magnetic moments from being coupled together appreciably.

The two spheres are synchronously tuned by mounting them in dielectric capsules so that they can turn in any direction. Application of a dc biasing magnetic field results in a torque that causes one of the "easy axes" of magnetization to align itself along the dc magnetic field. Thus, detuning that would be caused by magnetocrystalline anisotropy[31] is easily eliminated.

Fig. 11 shows the disassembled two-resonator filter with the coupling iris, and the garnet spheres in 0.080-inch-diameter dielectric capsules. The capsules mounted on polystyrene rods are adjusted to vary the spacing of the spheres. Two 90° *E*-plane bends make it possible to use a "*c*"-type electromagnet.

In this experimental model, the input and output guides are standard-height *X*-band guides. Here, as for the single-resonator filter, the guide height or the sphere diameter can be varied to obtain the desired filter re-
256 sponse. The two YIG spheres used in this model were 0.064 inch and 0.060 inch in diameter.

Fig. 12 shows the measured responses of the two-resonator filter tuned to 8.23 kMc, 9.49 kMc, 10.99 kMc, 12.00 kMc, 13.00 kMc, and 14.25 kMc. The initial adjustment of the filter was made at 9.49 kMc by varying the coupling slot dimensions and the spacing of the resonators to give the slightly overcoupled response with greater than 35-db insertion loss at frequencies far removed from the 9.49-kMc center frequency. This last requirement, *i.e.*, on rejection loss at frequencies far removed from resonance, limits the maximum width of the coupling slot, since the coupling between the guides in the absence of the garnet resonators is proportional to the width of the slot. The isolation obtained in this way can be calculated from the slot dimensions.[32,33] The slot dimensions used in this filter were 0.070 inch by 0.300 inch and were determined experimentally to give greater than 35-db off-channel rejection. The spacing between the centers of the YIG spheres was 0.164 inch. It was found that almost any desired degree of overcoupling could be obtained by closer spacing of the gar-

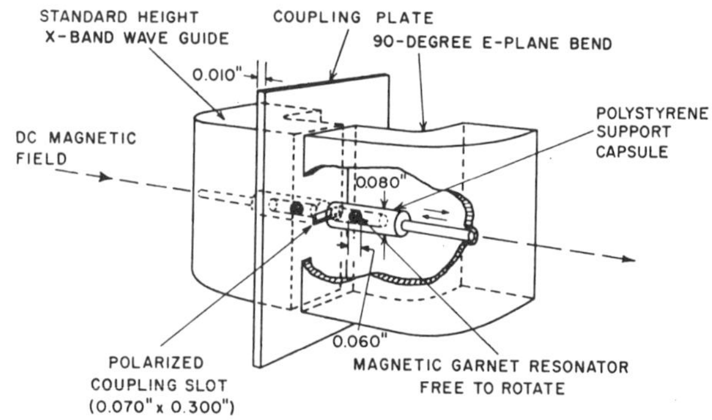

Fig. 10—Internal construction of two-resonator tunable filter.

Fig. 11—Disassembled two-resonator filter.

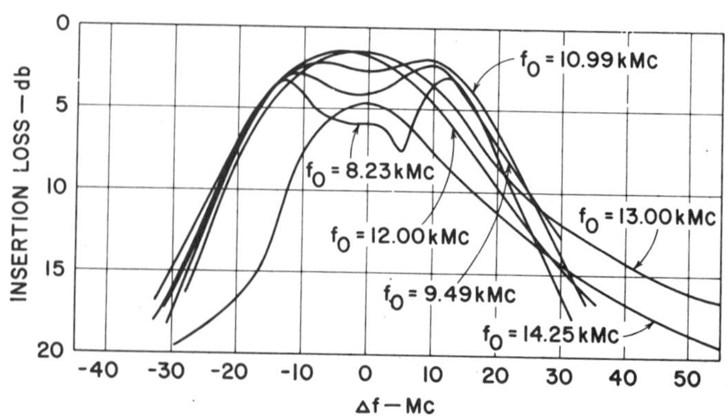

Fig. 12—Measured frequency responses of two-resonator filter.

[30] The use of polarized slot coupling was suggested by E. M. T. Jones of Stanford Research Inst., Menlo Park, Calif.

[31] C. Kittel, "On the theory of ferromagnetic resonance absorption," *Phys. Rev.*, vol. 73, pp. 155–161; January 15, 1948.

[32] H. A. Bethe, "Lumped Constants for Small Irises," M.I.T. Rad. Lab., Cambridge, Mass., Rept. No. 43-22; March 24, 1943.

[33] S. B. Cohn, "Microwave coupling by large apertures," Proc. IRE, vol. 40, pp. 696–699; June, 1952.

net spheres. The 3-db and 10-db bandwidths are nearly constant at about 32 Mc and 56 Mc, respectively, as the center frequency is changed with varying frequency.

The trends observed in Fig. 12 are: 1) a decrease of the degree of over-coupling with increasing frequency, 2) an approximately constant bandwidth throughout the tuning range, and 3) a decrease of insertion loss with increasing frequency. These trends are attributed to the same effect which caused the increase in bandwidth of the single-resonator filter, *i.e.*, the increase, with increasing frequency, of the coupling between the resonators and the waveguides.

The maximum tuning range of this experimental two-resonator filter appears to be limited by the leakage at 15.5 kMc through the coupling slot. This is shown in Fig. 13, which is a plot over a broadband of the response of the filter tuned to 12 kMc. This leakage is due to resonance of the slot at 15.5 kMc; it is then one-half wavelength long, and it couples energy between the two guides since it is not exactly parallel to the electric field of the waveguide. Evidently this slot response might be reduced by more accurate construction—this slot was cut by hand. Also, a shorter, wider slot might give nearly the same response and insertion loss, while moving the half-wave slot resonance to a higher frequency.

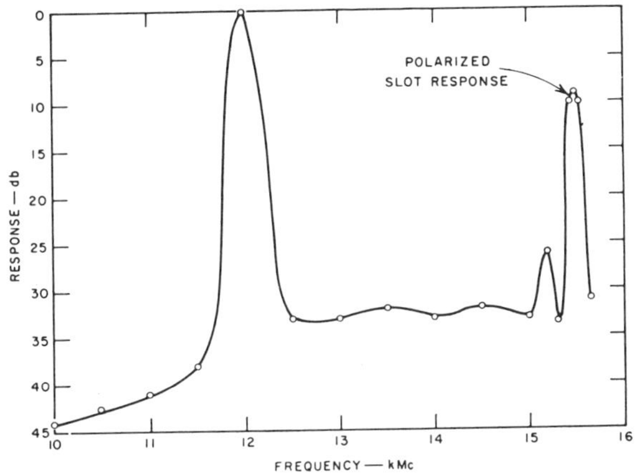

Fig. 13—Response of two-resonator filter tuned to 12 kMc showing polarized slot leakage at 15 kMc.

Fig. 14 shows the response of the two-resonator filter down to −30 db below the peak response, with the center frequency at 11.0 kMc. No spurious responses are evident here, as they were with the single resonator filter. This greater rejection of the higher-order modes is explained by the fact that the RF fields of the higher-order modes decrease very rapidly with distance away from the spheres, more rapidly than do the RF fields of the uniform mode. The higher-order modes of the two resonators are therefore coupled together very weakly through their external RF fields, making the insertion loss of these modes very high compared to that of the uniform mode. The absence of spurious responses in the two-resonator filter is an important advantage of this type of filter over the single-resonator type.

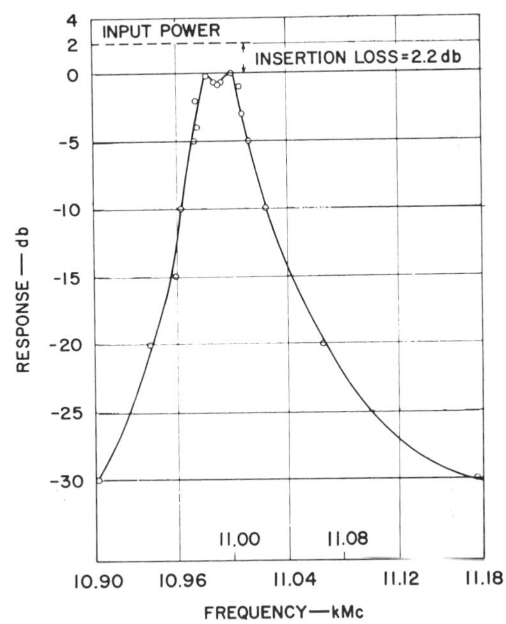

Fig. 14—Response of two-resonator filter tuned to 11 kMc.

VI. Future Developments

A number of useful extensions can be made of this type of tunable filter. It appears that filters can be constructed at frequencies lower and higher than *X*-band frequencies, and employing strip-transmission-line and coaxial-line coupling structures. Filters operating at lower frequencies (below *S*-band) will require a narrow-line-width magnetic material with a lower saturation magnetization than the yttrium-iron-garnet employed here. One material which appears to satisfy these requirements is gallium-substituted yttrium-iron-garnet.[34]

It appears possible to extend the tuning range of these filters beyond that reported here. This can presumably be accomplished by the use of ridged-waveguide, strip-transmission-line, and other inherently-wide-band coupling structures.

VII. Acknowledgment

The author is indebted to a number of his associates at Stanford Research Institute, Menlo Park, Calif., for important contributions. Special acknowledgment is made to Dr. E. M. T. Jones for some of the original ideas presented here. The support and general assistance of Dr. George Matthaei and Dr. Seymour B. Cohn are gratefully acknowledged. The general cooperation and assistance in the over-all supervision of this program by Nathan Lipetz of the U. S. Army Signal Research and Development Laboratory, Fort Monmouth, N. J., is also acknowledged.

[34] E. G. Spencer and R. C. LeGraw, "Line width narrowing in gallium substituted yttrium iron garnet," *Bull. Am. Phys. Soc.*, Ser. 2, vol. 5, pt. 1, p. 58; January 27, 1960.

Reprinted from IEEE TRANSACTIONS
ON MICROWAVE THEORY AND TECHNIQUES
Volume MTT-12, Number 6, November, 1964

PRINTED IN THE U.S.A.

Synthesis of Filter-Limiters Using Ferrimagnetic Resonators

R. L. COMSTOCK, MEMBER, IEEE

***Summary*—A synthesis procedure is developed for microwave band-pass filters with the Chebyshev response using orthogonal circuit resonators coupled by a ferrimagnetic resonator. A stripline ferrimagnetic resonator filter is analyzed in detail. Equations and graphs are given which allow the selection of ferrimagnetic material and size of the ferrimagnetic sample necessary to achieve a desired bandwidth and insertion loss for a given pass-band response. The theoretical behavior of these circuits as microwave power limiters is discussed and it is shown that the ratio of the limiting threshold to the filter bandwidth is a constant depending only on the pass band response shape. Experimental confirmation of the design information is discussed as well as some practical methods of varying the limiting threshold.**

Manuscript received June 6, 1964; revised July 24, 1964.
The author is with the Lockheed Missiles and Space Company, Research Laboratories, Palo Alto, Calif. He was formerly with the Bell Telephone Laboratories, Inc., Murray Hill, N. J.

I. Introduction

THE COUPLING of small ferrimagnetic ellipsoids to microwave transmission lines has been discussed by a number of authors [1]–[3]. It is found that when the ferrimagnet is excited in the uniform precession it behaves like a lumped constant resonator with, in certain cases, a nonreciprocal phase behavior. The resonant frequencies of such a resonator can be tuned with a dc magnetic field. Its unloaded $Q(Q_u)$ depends on the surface polish of the sample and doping with impurity ions. For example, with highly polished yttrium iron garnet (YIG) spheres Q_u can be as great as 10,000 at C band. It is further observed that the coupling to such a resonator can be tight enough to allow a large amount of energy transfer through the resonator, and with the large unloaded Q's available, with low insertion loss. This paper will extend these results by discussing the coupling of ferrimagnetic resonators not to transmission lines but rather to a class of resonant microwave circuits. It is shown that ferrimagnetic resonators can be fitted into the scheme developed by Dishal [4] and Taub and Bogner [5] for optimizing direct coupled resonator filters using lumped elements. Also, the use of ferrimagnetic filter structures as microwave power limiters will be discussed; in particular, a detailed analysis of the resonant stripline circuit first described by De Grasse [1] will be presented.

In order to use the results given by Dishal, and Taub and Bogner, it is necessary to describe the microwave resonators by lumped-constant equivalent circuits. This always can be done providing the results are not extended to large (>10 per cent) bandwidths. The lumped-element equivalent circuit of the ferrimagnetic resonator is also only valid over a small bandwidth since, in general, the element values are functions of frequency. In case it is desired to develop devices having even larger bandwidths, the design procedure developed by Cohn [6] for wideband direct-coupled resonator filters can be extended to include ferrimagnetic resonators in a similar manner.

II. The Ferrimagnetic Resonator and Dipolar Coupling

In this section some properties of the ferrimagnetic resonator along with some information on microwave circuit coupling to be used in the next section on band-pass filter synthesis will be discussed.

The important parameters for describing a resonator are its unloaded $Q(Q_u)$ and degree of coupling. For a generalized ellipsoidal ferrimagnetic sample excited in the uniform precession magnetostatic mode the unloaded Q is,

$$Q_u = \frac{\omega_0}{2\eta_0}, \tag{1}$$

where η_0 is a loss parameter which in the case of a sphere is given by $\mu_0\gamma\Delta H/2$ where ΔH is the resonance linewidth, and ω_0 is the uniform precession resonance frequency $=\mu_0\gamma H_0$ for spheres, where H_0 is the external dc magnetic field.

In general there are other microwave resonances that can exist in these samples, *e.g.*, magnetostatic modes and short wavelength spin waves. While magnetostatic modes (other than the uniform precession) may be important in certain applications of ferrimagnetic resonators, they will not be discussed further in the theoretical part of this paper since they have not proven to be of significance in the experiments to be described (Section VI). The effects of spin waves on the properties of these resonators will be discussed in the section on limiting.

It is desired to extend the theory of lumped-element band-pass transmission type filters consisting of direct-coupled resonators to include ferrimagnetic resonators. It is necessary to have input and output ports for each resonator which are uncoupled except by energy transfer through the resonator. As shown by De Grasse [1], this can be achieved by coupling normally orthogonal microwave circuits through transverse components of the dipolar fields of a ferrimagnetic sphere excited in the uniform precession mode [*e.g.*, see Fig. 1(a)]. These orthogonal circuits can be crossed wires, loops, striplines, waveguides, cavity modes, etc.

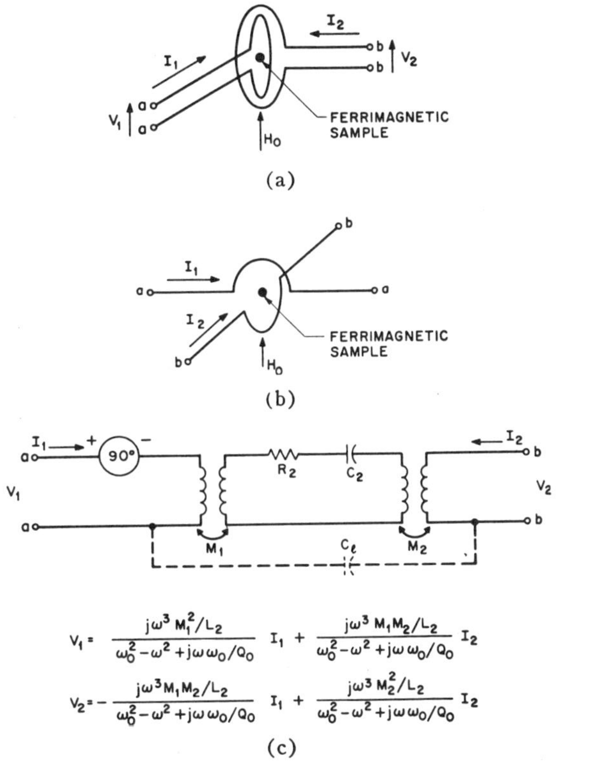

Fig. 1—Loop (a) and semi-loop coupling (b) by a ferrimagnetic sphere. The equivalent circuit representation (c) for these circuits derived from (10) and (11) also is shown along with the circuit equations.

The dipolar coupling between a ferrimagnetic sample and a pair of orthogonal circuits can be calculated by assuming currents to exist in the two circuits which create magnetic fields at the position of the dipole. A vector operator can then be defined,

$$I = \begin{pmatrix} I_1 \\ I_2 \end{pmatrix}, \tag{2}$$

where I_1 and I_2 are the currents in the two orthogonal circuits. The resulting external magnetic field at the sample is then,

$$h^e = \begin{pmatrix} h_x \\ h_y \end{pmatrix} = \mathbf{C}\cdot I, \tag{3}$$

where,

$$\mathbf{C} = \begin{pmatrix} C_{12} & 0 \\ 0 & C_{23} \end{pmatrix} \tag{4}$$

259

is a coupling matrix whose elements depend only on the geometry of the circuit and the position of the sample [2]. The rectangular coordinates x, y are centered at the center of symmetry of the sample. The transverse magnetization in the sample is

$$m = \chi^e \cdot h^e = \chi^e C \cdot I, \tag{5}$$

where χ^e is the external susceptibility tensor for the ferrimagnetic sample, the components of which are, in the case of a sphere,

$$\chi_{xx}{}^e = \chi_{yy}{}^e = \frac{\omega_M \omega_0}{\omega_0{}^2 - \omega^2 + 2j\eta_0\omega_0} \qquad \chi_{xy}{}^e = \chi_{yx}{}^e = \frac{\omega_M \omega}{\omega_0{}^2 - \omega^2 + 2j\eta_0\omega_0}, \tag{6}$$

where $\omega_M = \gamma\mu_0 M$. The voltages induced in the two orthogonal circuits are given by

$$V_i = j\omega \int_{\text{circuit } i} b \cdot da, \qquad i = (1, 2), \tag{7}$$

where b is the magnetic induction arising from the magnetic dipolar fields and $V_{1,2}$ is identified with the $i=1$ and $i=2$ integrals, respectively. The maximum dipolar induction is given by

$$b = \mu_0 h = \mu_0 \frac{m v_s}{2\pi r^3}, \tag{8}$$

where r is the radial distance from the magnetic dipole and $v_s = (\pi/6)(D_s)^3$ is the volume of the sample. After combining the above equations it is found that

$$V_i = \frac{j\omega\mu_0 v_s}{2\pi} \int_{\text{circuit } i} \frac{\chi^e \cdot C \cdot I \cdot da}{r^3}. \tag{9}$$

As an example of the use of this result consider the coupling to two identical orthogonal loops (radius r_0) by a spherical sample in the geometrical center as shown in Fig. 1(a). In this case

$$h^e = \frac{1}{2r_0} I$$

and

$$C_{12} = C_{23} = \frac{1}{2r_0}.$$

When integrated over the area of the loops, (3) results in

$$V_{\text{loop}} = \begin{bmatrix} V_1 \\ V_2 \end{bmatrix}_{\text{loop}} = j\omega\mu_0 \frac{v_s}{2r_0{}^2} \begin{bmatrix} \chi_{xx}{}^e I_1 + \chi_{xy}{}^e I_2 \\ \chi_{yx}{}^e I_1 + \chi_{yy}{}^e I_2 \end{bmatrix}, \tag{10}$$

where in this case I_1 and I_2 are currents flowing in the loops and V_1 and V_2 are voltages developed around the loops. Eq. (9) has been previously derived by De Grasse [1] and Carter [2]. In the remainder of this paper the analysis will be restricted to the coupling between ferrimagnetic spheres and "necked-down" striplines whose shapes in the vicinity of the sample are in the form of semi-loops as shown in Fig. 1(v). For this case the coupling is approximately half that for the full loop, resulting in

$$V_{\frac{1}{2}\text{ loop}} = \begin{bmatrix} V_1 \\ V_2 \end{bmatrix}_{\frac{1}{2}\text{ loop}} = \frac{j\omega\mu_0 v_s}{8r_0{}^2} \chi^e \cdot I. \tag{11}$$

An equivalent lumped electrical circuit can be found to represent (11), and using (6) the result is shown in Fig. 1(c). The circuit equations given in the figure are of the same form as (10) and (11). The nonreciprocal phase characteristic of the gyromagnetic coupler is indicated by the fact that $\chi_{xy}{}^e = -\chi_{yx}{}^e$ and is shown in the equivalent circuit by the nonreciprocal 90° phase shift. The nonreciprocal phase shift is demonstrated experimentally in Section VI.

The ferrimagnetic sample appears as a resonator inductively coupled to the wire semi-loops. The lumped constant equivalent circuit for the semi-loop coupled ferrimagnetic resonator can now be used in any of the available synthesis procedures for band-pass filters. As an example, the synthesis procedure developed by Dishal [4] and extended by Taub and Bogner [5] will be generalized to include these resonators. Dishal's synthesis procedure provides values of the coupling coefficients and loaded Q's of lumped constant direct coupled resonators which are necessary to achieve a specified pass-band response. In the case of inductively coupled resonators the coupling coefficients provided by Dishal's procedure are

$$K_{12,23} = \frac{M_{1,2}}{\sqrt{L_{1,3}L_2}},$$

where $L_{1,3}$ is the series inductance connected across terminals aa and bb of the circuit in Fig. 1(c), respectively. By comparing the original parameters for the ferrimagnetic resonator from (11) and the equivalent circuit parameters from Fig. 1(c), the coupling coefficient for a semiloop results,

$$K^2{}_{12,23} = \frac{\mu_0 v_s}{L_{1,3}} \frac{\omega_M}{\omega_0} \frac{1}{8r_0{}^2}. \tag{12}$$

The inductances L_1 and L_3 are the inductances of the series LC equivalent circuits for the microwave resonators coupled to the ferrimagnetic resonator. The inductances are calculated for TEM resonators in Section IV by making the resonant frequencies and reactance slopes equal for the actual and lumped equivalent circuits. In the following analysis only symmetrical networks will be considered for which $K_{12} = K_{23} = K$.

SECTION III

III. Band-Pass Filter Synthesis

Dishal has analyzed a class of lumped-constant band-pass circuits which have specified pass-band characteristics as shown in Fig. 2. This response shape is the well-known Chebyshev response which reduces to the maximally flat response in the limit of vanishing pass-band ripple amplitude.[1] Dishal's analysis allows the specification of response shape for a filter but does not allow for a specification of the minimum insertion loss (l_{min}). Taub and Bogner have analyzed *three* resonator band-pass filters of the type considered by Dishal and using his equations have imposed the additional constraint of minimum insertion loss in the pass band. Their results will be used in what follows and the analysis is restricted to three resonator structures.

The circuits considered by Dishal consist of shunt (series) resonators coupled by series (shunt) lumped elements. Schematically the three resonator filter appears as shown in Fig. 3. The results of the synthesis procedure developed in [4] and [5] for band-pass filters are indicated briefly in this section with the circuit parameters defined as follows:

$d = 1/Q$ = total resonant circuit decrement.
$\delta = d/(\Delta f/f_0)$ = normalized decrement.
$\Delta f/f_0$ = total 3-db fractional bandwidth.
f_0 = resonant frequency of each resonant circuit.
$\omega_0 = 2\pi f_0$ = angular resonant frequency.
K_{ij} = resultant coefficient of coupling between resonant circuits.
$k_{ij} = K_{ij}/\Delta f/f_0$ = normalized coefficient of coupling $= K_{ij}Q_L$.
l = midband insertion loss in db.
P_a = available input power.
P_{out} = output power.
$Q_1 = Q$ of first resonant circuit loaded only by the input.
$Q_2 = Q_u$ = unloaded Q of resonators (assumed equal).
$Q_3 = Q$ of third resonant circuit loaded only by the output.
$Q_L = f_0/\Delta f$.

Dishal has shown that for a specified over-all bandwidth the required minimum unloaded Q (or maximum value of δ_2, $\delta_{2\,max}$) for the resonators (assumed to be the same[2]) is specified completely by the ripple level (peak to valley ratio of the transmission coefficient) in the pass band; for example, with 0 db and 0.5 db pass-band ripple the value of $\delta_{2\,max} = Q_L/Q_{0\,min}$ is 0.5 and 0.271 respectively [5]. Taub and Bogner have shown that the minimum insertion loss in the pass band which, along with the ripple level, describes the pass band can be specified approximately for any ripple level by a

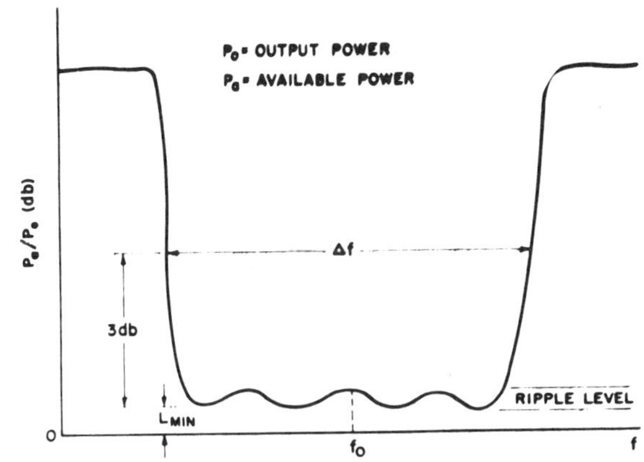

Fig. 2—Specification of band-pass filter characteristic.

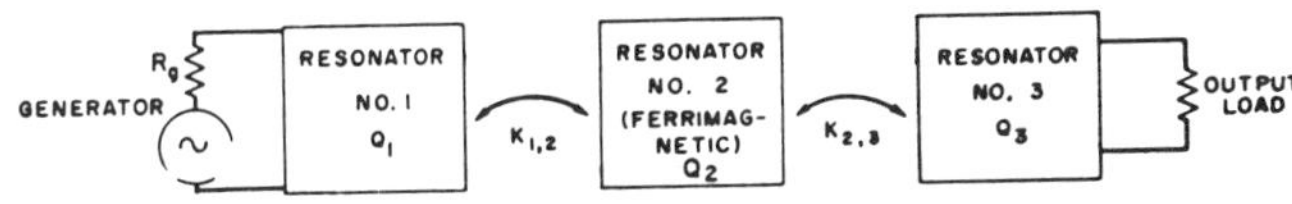

Fig. 3—Block diagram of three resonator filter.

specification of $\delta_2/\delta_{2\,max}$. For $l_{min} < 2$ db, their results show

$$l_{min} \approx 10 \frac{\delta_2}{\delta_{2\,max}} = \frac{10 f_0}{\Delta f Q_u \delta_{2\,max}}, \tag{13}$$

where l_{min} is expressed in db. In the limit of vanishing insertion loss the coupling coefficient $k_{12} = k_{23} = k$ and $\delta_1 = \delta_3$ depend only on the pass-band ripple; for example, with 0 db and 0.5 db pass-band ripple the normalized coupling coefficients are $k = 0.7$ and 0.65 and the normalized resonator decrements are $\delta_1 = 1.0$ and 0.55, respectively. Extensive curves of k and δ_1 for various values of $\delta_{2\,max}$ vs $\delta_2/\delta_{2\,max}$ and hence ripple level are given in [5].

No consideration has been given so far to the shape of the filter characteristic outside the pass band. For an n resonator filter the skirt selectivity (sharpness of the filter cutoff) increases rapidly as n increases.[3] In case the selectivity resulting from a 3 resonator (3 pole) filter is inadequate for the requirements at hand it will be necessary to extend these results for $n > 3$.

IV. Crossed Stripline Filter

The information given in Section II on ferrimagnetic resonators and in Section III on band-pass filters will now be combined to show how a 3-resonator filter using a ferrimagnetic resonator as the center resonator can be synthesized. First, a choice must be made as to what type of resonator will be used as the first and last resonator. The resonators which will be discussed here are resonant sections of strip transmission line with a semi-loop located in the center (Fig. 4) as first described by

[1] These pass-band responses are optimum in the sense of having the sharpest cutoff in the stop band for a given pass-band ripple.
[2] As shown in [4] this is not a serious limitation if all resonators exceed the minimum Q_u specified in [4].
[3] See, for example, "Reference Data for Radio Engineers," 4th Ed., International Telephone and Telegraph Corp.; 1956.

De Grasse [1]. The resonators can have $l_r/2 = n(\lambda/2)$ $(n = 1, 2, \cdots)$ in which case the ternimals must all be short circuited and coupling to external transmission lines supplied by, say, loops placed in the vicinity of the short circuits. It is also possible for $l_r/2 = n(\lambda/4)$ (n odd) in which case the terminals would need to be open circuited and coupling supplied by, say, series capacitance or quarter wavelength transmission lines. Note that coupling to external transmission lines is only necessary at the terminals marked "1" and "2," as shown in Fig. 4. The lumped constant equivalent circuit for one strip-line resonator coupled to a ferrimagnetic resonator (located in the semi-loop as shown in Fig. 4) is shown in Fig. 5. The end couplings have been represented by coupled coils with mutual inductance $M_{0,3}$.[4] This representation of the end couplings can be shown to be quite general and will be used exclusively in what follows. The lumped element equivalent circuit for the stripline resonator can be found in a number of ways [7], one of which is to equate the slopes of the two reactance functions at resonance, *i.e.*,

$$\left.\frac{dX(\omega_0)}{d\omega}\right|_{\text{equivalent lumped constant resonator}} = \left.\frac{dX(\omega_0)}{d\omega}\right|_{\text{stripline resonator}}$$

In case $l_r/2 = n(\lambda/2)$ (short circuit ends) this condition results in [7]

$$L_1 = Z_{01}\frac{n\pi}{\omega_0}$$

$$\omega_0^2 = \frac{1}{L_1C_1},$$

where Z_{01} is the characteristic impedance of the strip-line resonators. If $l_r/2 = n\lambda/4$ (n odd for open circuit ends) the lumped elements are [7]

$$L_1 = \frac{Z_{01}n\pi}{4\omega_0}$$

$$\omega_0^2 = \frac{1}{L_1C_1}.$$

In arriving at the lumped constant equivalent circuit the resonator losses have been neglected. The losses will be represented by the unloaded resonator Q's. The results of Section II and the lumped element equivalent for the stripline resonators can be used to evaluate the coupling coefficient, $K = K_{1,2} = K_{2,3}$

[4] The coupling parameters $M_{0,3}$, can be calculated for a given microwave structure as shown in vol. 8 of the radiation Laboratory Series (Principles of Microwave Circuits). It has not been found necessary to carry out such a calculation since the end couplings are invariably adjusted experimentally. One convenient way to adjust the external Q's ($Q_{1,3}$) experimentally is to observe the signal reflected from the filter when the dc magnetic field for the ferrimagnetic resonator is adjusted far from resonance.

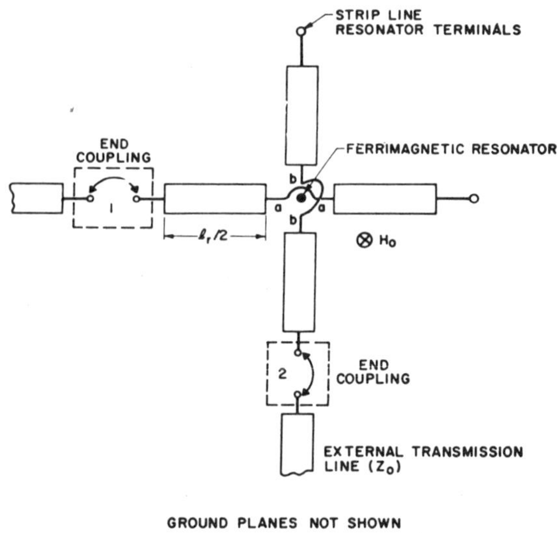

Fig. 4—Crossed strip-line filter with ferrimagnetic resonator.

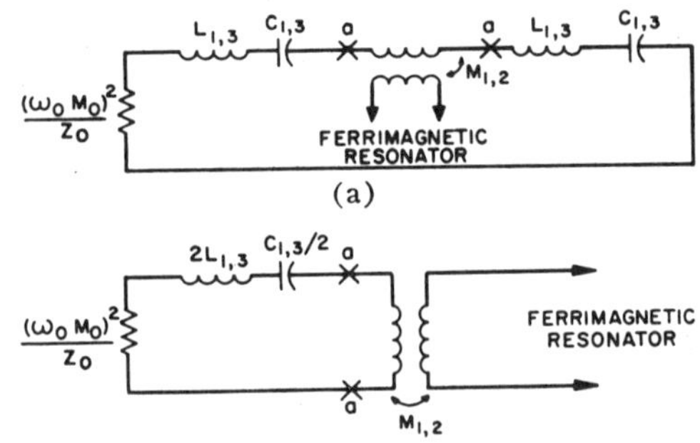

Fig. 5—Lumped equivalent circuit of strip-line resonator coupled to ferrimagnetic resonator.

$$K^2\Big|_{\text{short circuit ends}} = \frac{\mu_0 v_s \omega_M}{Z_{01}} \frac{1}{n\pi 8 r_0^2}\left(\frac{l_r}{2} = \frac{n\lambda}{2}\right) \quad (14a)$$

$$K^2\Big|_{\text{open circuit ends}} = \frac{\mu_0 v_s \omega_M}{Z_{01}} \frac{1}{n\pi 2 r_0^2}$$
$$\left(\frac{l_r}{2} = n\frac{\lambda}{4},\ n \text{ odd}\right). \quad (14b)$$

The complete equivalent circuit for the crossed stripline ferrimagnetically coupled filter is shown in Fig. 6.

The experiments to be described in Section IV have been restricted to open circuit resonators and the design procedure will also be restricted to this case. Using the value for K given in (8) it is now possible to combine the results of this section with those of Section III resulting in a design procedure for these filters as follows:

1) Choose bandwidth, ripple level and maximum pass-band insertion loss.
2) From curves of $\delta_{2\,\max}$ vs pass-band ripple given in [5] and (7), evaluate the required minimum value for Q_u.
3) Evaluate the coupling coefficient and loaded resonator decrements from curves in [5].

SECTION III

4) From (8), evaluate the ratio $(\omega_m/Z_0)D_S^3/r_0^2$. These parameters can then be used in a trade-off in any convenient way.

To illustrate the above design considerations the following case is considered

$l_{max} = 0.5$ db
ripple $= 0.5$ db
$Z_{01} = 100\ \Omega$
$r_0 = 0.150''$
$4\pi M = 1750$, 3640 gauss.

The first value of $4\pi M$ is appropriate to yttrium iron garnet, and the second to lithium ferrite. From [5] (see Section III), the value of k is 0.65, and from (7), the minimum value for Q_u is found to be $7.38/(\Delta f/f_0)$. Using the normalized value of k given along with (8) it is possible to solve for the value of the required sphere diameter as a function of the 3-db bandwidth. The results of this calculation are shown in Fig. 7 for the two values of $4\omega M$. It is seen from the value of Q_u given above that extremely narrow lindwidths are not needed for low insertion loss filters with moderate bandwidths; for example, for 10 per cent bandwidth at S-band (3 Gc), $Q_u > 74$ or $\Delta H < 14.5$ oe. The loaded resonator Q's are then adjusted by adjusting the external coupling.

From the equivalent circuit given in Fig. 6, it is possible to relate the coupling coefficient to the easily measured mormal mode resonances of the three element filter under conditions that the input and the output of the filter are loosely coupled ($Q_1 \gg 1$). That is, K

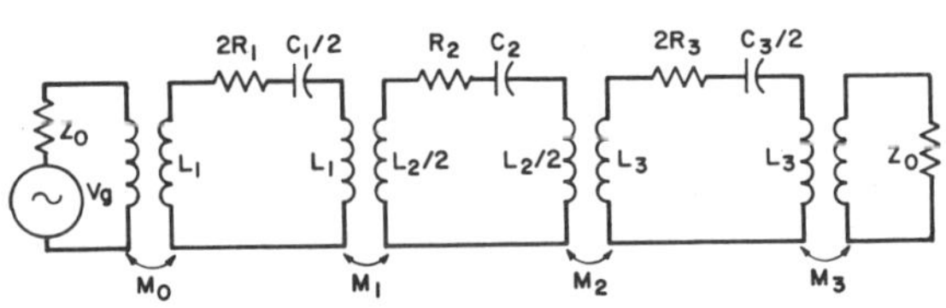

Fig. 6—Lumped constant equivalent circuit for strip-line ferrimagnetic resonator filter.

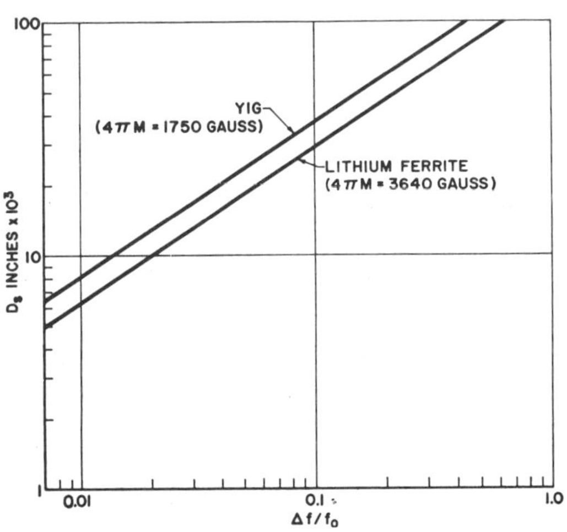

Fig. 7—Required diameter of YIG or lithium ferrite sphere for specified filter bandwidth for 0.5-db pass-band ripple.

can be related to the frequencies ($\omega^{\pm}$) for which the input impedance vanishes when the output is open circuited. In the lossless case, this condition results in

$$(\omega^{\pm})^2 = \omega_0^2(1 \pm \sqrt{2}\,K), \quad \text{or}$$

$$K = \frac{(\omega^+)^2 - (\omega^-)^2}{2\sqrt{2}\,\omega_0^2}. \tag{15}$$

In Section VI measurements will be described of $\omega^{\pm}$ for various sphere diameters using yttrium iron garnet in a representative crossed stripline filter. These measurements can then be used to determine experimentally the coupling constant.

V. Limiting

As has been discussed by Shul [8] the resonance susceptibility of a ferrimagnet can be expected to decrease above some critical value of transverse magnetization due to the parametric excitation of spin waves. In what follows only the first order spin wave instability [8] will be considered since it has resulted [9]–[12] in limiting characteristics which are superior to second order process limiters [13].

In the first order process, the resonance susceptibility has a constant value χ_0'' for RF fields less than a critical value, h_{crit} but declines as

$$\chi'' = \frac{\chi_0''}{h/h_{crit}} \tag{16}$$

for $h > h_{crit}$. The threshold field depends on the uniform precession and spin wave linewidths as discussed in [8].

The decline in the susceptibility leads to power limiting in filters of the type under consideration. This can be seen by rewriting (4) in terms of RF fields, with the assumption that all susceptibility components decline with magnetic field as given by (16),

$$V_1 = (V_1)_{crit} = \frac{j\omega\mu_0 v_s}{2\pi}\int_{\text{circuit 1}} \left[(\chi_{xx0}'')^e (h_x)_{crit} + (\chi_{xy0}'')^e (h_y)_{crit}\right]\frac{da}{r^3}$$

$$V_2 = (V_2)_{crit} = \frac{j\omega\mu_0 v_s}{2\pi}\int_{\text{circuit 2}} \left[(\chi_{yx0}'')^e (h_x)_{crit} + (\chi_{yy0}'')^e (h_y)_{crit}\right]\frac{da}{r^3}. \tag{17}$$

The voltages are limited since they are constants and no longer depend on the circuit currents.

The threshold output power for a limiter can be calculated once h_{crit} is known and such a calculation for the crossed stripline circuit described in Section IV is given below.

From Fig. 6 the impedance coupled into the output resonator is

$$Z_L = \frac{(\omega_0 M_3)^2}{Z_0},$$

and if the internal damping of the resonator is neglected, which is a satisfactory approximation for wide-band operation, the loaded resonator Q is

$$Q_3 = \frac{\pi Z_{03} Z_0}{4(\omega_0 M_3)^2}.$$

The output power from the limiter at the threshold is, again neglecting internal resonator losses,

$$(P_0)_{\text{crit}} = \frac{I^2_{\text{crit}}}{2} \frac{(\omega_0 M_3)^2}{Z_0},$$

and with

$$h_{\text{crit}} = I_{\text{crit}}/4r_0,$$

$$(P_0)_{\text{crit}} = \frac{h^2_{\text{crit}} Z_{03} 4 r_0^2 \pi}{Q_3}.$$

The loaded resonators Q can be related to the over-all bandwidth of the filter by

$$\delta_3 = \frac{f_0/\Delta f}{Q_3},$$

which δ_3 is determined from information on the insertion loss and ripple level (see [5]). Thus

$$(P_0)_{\text{crit}} f_0/\Delta f = h^2_{\text{crit}} Z_{03} 4 r_0^2 \pi \delta_3. \tag{18}$$

From this expression it is seen that for a given h_{crit}, fixed circuit geometry (r_0, Z_{01}), pass-band ripple, and insertion loss the ratio of threshold power to fractional bandwith is a constant. This result is reminiscent of the familiar (gain)×(bandwidth) limitation of amplifiers and indicates that an increase in the bandwidth of the filter (*e.g.*, by using a larger sample) will raise the threshold of the limiter. Limiting thresholds as low as −35 dbm have been observed in narrow-band limiters using narrow linewidth single crystal yttrium iron garnet [9] while wide-band limiters ($\Delta f/f_0 > 10$ per cent) have been made which have thresholds in the milliwatt range as described in Section VI. The "dynamic range"[5] of these limiters is typically 25–30 db and is prevented from obtaining larger values by leakage represented in Fig. 1 by the capacitor C_l. The leakage capacitance can be minimized by making the mutual area of the crossed strip lines as small as possible commensurate with maintaining the minimum value of the unloaded resonator Q's. Capacitive shields between the strip lines have been used also.

VI. Experimental Results

First of all it is necessary to establish the validity of the equation for dipolar coupling given in Section IV. As mentioned in that section, it is possible to relate the coupling constant K to the normal mode resonances of the three-element filter. This relationship was checked experimentally by measuring these frequencies for various diameters of YIG spheres in an air dielectric stripline structure at 5 Gc.

The results of this investigation are shown in Fig. 8 which is a plot of the measured values of K^2 [calculated from (15)] vs the diameter of the spheres used. Using (14b) with the relevant parameter values ($r_0 = 0.15$ inch, $4\pi M = 1750$ gauss, $Z_{01} = 100\ \Omega$) the theoretical curve is also plotted in Fig. 8. Agreement is seen to be good except at the largest values of coupling where the observed coupling is reduced from the theoretical value. The disagreement is believed to be caused by the failure of dipolar fields to adequately describe the coupling when the sample diameter approaches the loop diameter.

Using the design procedure outlined in Section IV, a filter was designed using 0.003 inch copper stripline in irradiated polyethylene dielectric[6] with quarter wave sections of high impedance transmission line for the end sections. The impedance of the quarter wavelength lines was varied in order to vary the end coupling and hence the external $Q(Q_{1,3})$. The ferrimagnetic material was yttrium iron garnet with a line width of 0.5 oe which corresponds to an unloaded Q of 2800 at C-band. The sphere diameter was 0.105 inch. The response characteristics are shown in Fig. 9. With an external Q of 17.6 it is seen that the system has two well defined normal modes (the microwave circuit resonators had the same resonant frequency which reduces the number of normal modes from three to two) as discussed in [12]. The value of $K(K = 0.54)$ calculated from these two frequencies using (15) agrees closely with the value given in Fig. 8 (measured curve). For tighter coupling the loaded Q's of the normal modes are lowered resulting in the band-pass filter characteristic shown for $Q_e = 9.80$ and 3.52. In the latter case, the bandwidth was observed to be over 1 Gc. The observed bandwidth with $Q_e = 3.52$ (25 per cent) is smaller than predicted by the curve in Fig. 7 (45 per cent) since K is observed to be reduced from the theoretical result for large samples as shown in Fig. 8. For larger input microwave powers the response of the filter is limited as shown in Fig. 10 for the case $Q_e = 9.8$. The threshold power for limiting is in the milliwatt range in contrast with previously reported experiments using narrow bandwidth filters [9]. The theoretical limiting threshold calculated from (18) and Suhl's theory [8] for h_{crit} agrees in order of magnitude with the experimental values. The uncertainty in the theoretical result is partly due to uncertainties in the value of the spin wave linewidth for the particular spin waves excited in the first order instability. The limiting curves were taken with a fixed dc magnetic field. The dynamic limiting range of this limiter was typically 25 db. For frequencies outside of the band shown in Fig. 10 the dynamic range becomes significantly degraded. The bandwidth over which limiting is observed is indicated

[5] Dynamic range denotes the range of input power above the threshold over which the output power remains constant, say to within one db.

[6] Manufactured by Tellite Corp., Orange, N. J.

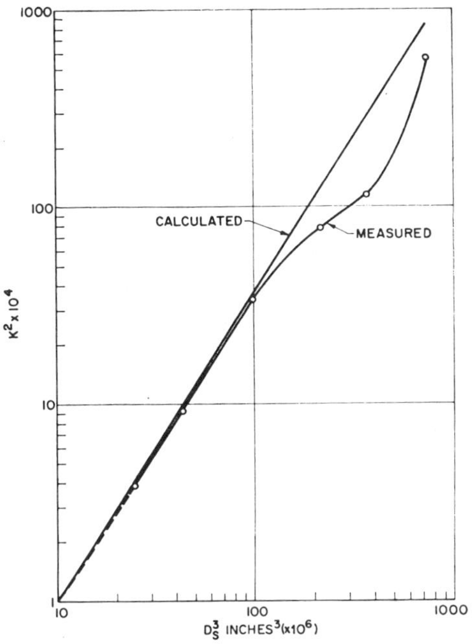

Fig. 8—Measured value of coupling coefficient vs diameter of yttrium iron garnet sphere. The calculated value corresponds to Z_{01} (resonator) $= 100\ \Omega$, $r_0 = 0.150$ inch.

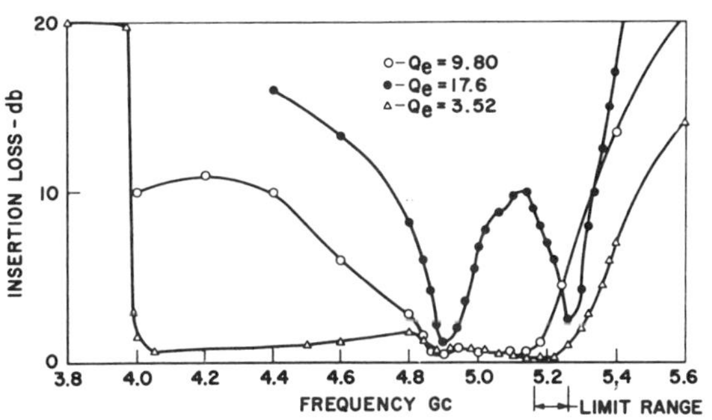

Fig. 9—(After Comstock and Dean [11]) Measured response of C-band ferrimagnetic filter for 3 values of filter end coupling. The yttrium iron garnet sphere diameter was 0.105 inch.

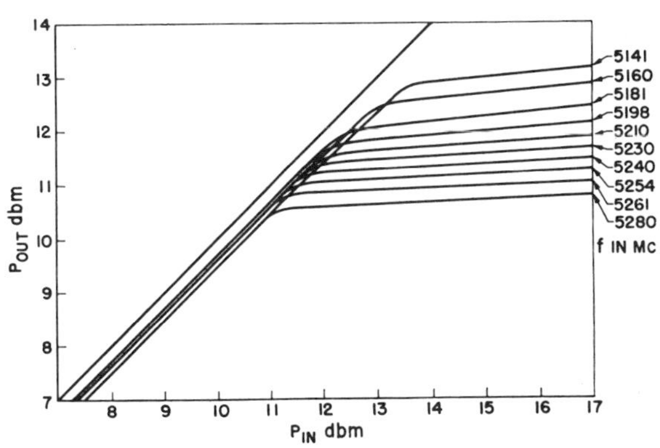

Fig. 10—(After Comstock and Dean [11]) Limiting characteristic taken with filter having characteristics given in Fig. 9. The external Q was 9.8. For frequencies outside the band shown the dynamic range is degraded.

on Fig. 9 which illustrates the conclusion made in [11] that first order process limiting for $\omega > (2/3)\omega_n$ should be observed only in the vicinity of the higher frequency normal mode.

In order to raise the threshold power level with a fixed bandwidth and dynamic range for this class of limiters, it has been found feasible to use ferrimagnetic materials with wider linewidths. As shown in [8] the threshold field is directly proportional to the ferromagnetic resonance linewidth for the first order process. For example, a limiter using terbium doped yttrium iron garnet [14] ($\Delta H \approx 2.3$ oe), has been constructed with the characterstics shown in Table I. The characteristics obtained with undoped YIG are also shown in this table. With a 0.5 db increase in insertion loss the threshold is seen to increase by 9 db using the wider linewidth material with a 3 db decrease in dynamic range. Another approach to increasing the threshold is to use samples with unpolished surfaces since then the contribution to the linewidth from surface scattering can dominate the loss mechanism [15]. Wide-band crossed stripline filter-limiters have also been made with polycrystal YIG ($\Delta H \approx 30$ oe) in which case the limiting thresholds are typically a watt or more.

TABLE I

Operating Characteristics of Strip-Line Filter-Limiter

	YIG ($\Delta H = 0.5$ oe)	0.02 per cent Tb doped YIG ($\Delta H = 2.3$ oe)
Operating Frequency	2.43 Gc	2.88 Gc
Bandwidth	90 Mc	60 Mc
Threshold (P_{crit})	−6 dbm	+ 3 dbm
Dynamic Range	25 db	22 db
Insertion Loss ($P < P_{crit}$)	0.5 db	1 db

Recently first order process power limiting was reported at X-band using lithium ferrite [12]. A photograph of the X-band filter-limiter is shown in Fig. 11. The stripline resonators are shown crossed over in the center of the ground plane (the top ground plane is removed). The high impedance lines are quarter wavelength coupling sections. The spherical sample is placed between the two resonators and biased with a dc magnetic field normal to the ground planes. The limiting characteristics are shown in Fig. 12. The filter bandwidth of this limiter was 320 Mc and limiting with a reasonably flat slope was observed over approximately 100 Mc. The dynamic range using low average power to avoid heating effects exceeded 25 db. The preliminary design of this limiter based on the results shown in Fig. 7 for a diameter of 0.024 inch predicted a bandwidth of approximately 10 per cent while the observed was closer to 4 per cent. However, the coupling constant was difficult to estimate since the coupling loops were far from circular in this case.

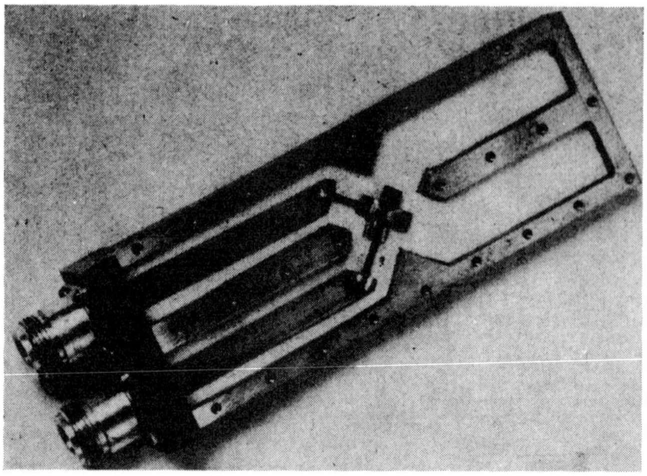

Fig. 11—Photograph of X-band crossed strip line filter-limiter. The lithium ferrite sample is placed between the two half-wave resonators (shown necked down in the center).

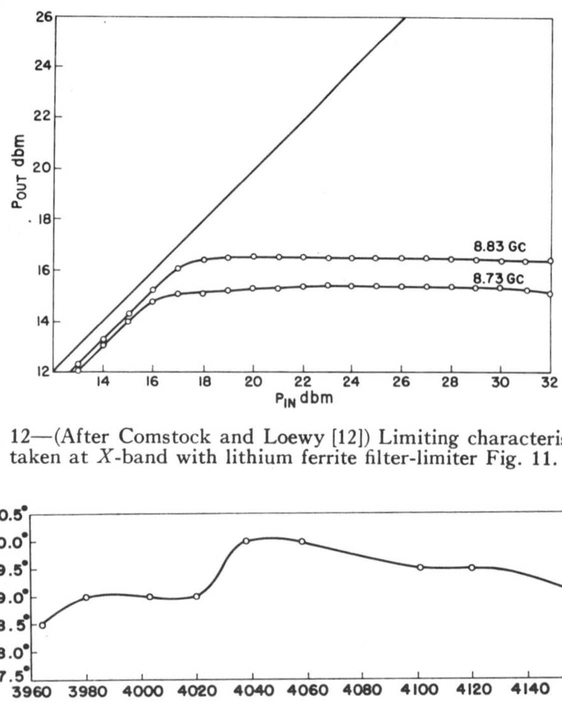

Fig. 12—(After Comstock and Loewy [12]) Limiting characteristics taken at X-band with lithium ferrite filter-limiter Fig. 11.

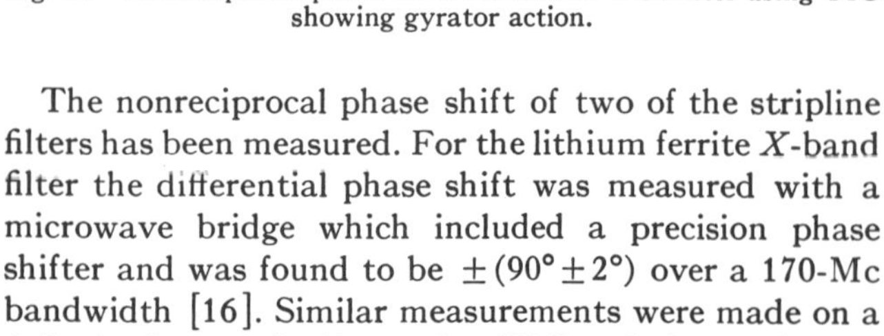

Fig. 13—Nonreciprocal phase characteristic of 4-Gc filter using YIG showing gyrator action.

266

The nonreciprocal phase shift of two of the stripline filters has been measured. For the lithium ferrite X-band filter the differential phase shift was measured with a microwave bridge which included a precision phase shifter and was found to be $\pm(90° \pm 2°)$ over a 170-Mc bandwidth [16]. Similar measurements were made on a 4-Gc ferrimagnetic filter using YIG and the results are shown in Fig. 13, which is a plot of one-half the measured differential phase shift vs frequency. The gyrator element formed from the ferrimagnetic filter has been found useful in constructing hybrid-gyrator four-port circulators [16].

VII. Conclusions

The theory of dipolar coupling presented in this paper has been found useful in preliminary design of ferrimagnetically coupled filter-limiters. The ferrimagnetic sample, when excited in the uniform precession mode, behaves in much the same way as a microwave cavity and can thus be incorporated into microwave filter structures. In particular, ferrimagnetic resonators have been found to be useful in conjunction with stripline circuitry in building wide- or narrow-band microwave filters. The first order spin wave instability has been found to result in excellent limiting characteristics in these filters. Limiting occurs over the entire filter bandwidth only if $\omega < (2/3)\omega_m$. For frequencies not satisfying this condition limiting occurs near the upper band edge of the filter response. By varying the linewidth of the uniform precession, by either varying the sample polish or by adding impurities such as the rare earths to narrow linewidth yttrium iron garnet, the limiting threshold can be changed with only a small variation in insertion loss. The nonreciprocal phase shift of these filters has been demonstrated and shown to be useful in constructing gyrators.

References

[1] R. W. De Grasse, "Low-loss gyromagnetic coupling through single crystal garnets," *J. Appl. Phys.*, Supp. to vol. 30, pp. 1558–1568; 1959.

[2] P. S. Carter, Jr., "Magnetically-tunable microwave filters using single-crystal yttrium iron garnet resonators," IRE Trans. on Microwave Theory and Techniques, vol. MTT-9, pp. 252–260; May, 9161.

[3] L. K. Anderson and H. J. Shaw, "The representation of waveguides containing small ferrimagnetic ellipsoids," presented at the PGMTT National Symposium, Boulder, Colo.; May, 1962.

[4] M. Dishal, "Design of dissipative band pass filters producing exact amplitude-frequency characteristics," Proc. IRE, vol. 37, pp. 1050–1069; September, 1949.

[5] J. J. Taub and B. F. Bogner, "Design of three-resonator dissipative band-pass filters having minimum insertion loss," Proc. IRE, vol. 45, pp. 681–687; May, 1957.

[6] S. B. Cohn, "Direct-coupled resonator filters, Proc. IRE, vol. 45, pp. 187–196, February, 1957; and "Dissipation loss in multiple-coupled resonator filters," Proc. IRE, vol. 47, pp. 1342–1348, August, 1959.

[7] G. L. Ragan, "Microwave Transmission Circuits," Radiation Laboratory Series, vol. 9, McGraw-Hill Book Co., Inc., New York, N. Y., 1948.

[8] H. Suhl, "The nonlinear behavior of ferrites at high signal levels," Proc. IRE, vol. 44, pp. 1270–1284; October, 1956.

[9] F. J. Sansalone and E. G. Spencer, "Low-level microwave power limiter," IRE Trans. on Microwave Theory and Techniques, vol. MTT-9, pp. 272–273; May, 1961.

[10] F. C. Rossol, "Power limiting in the 4-kMc to 7-kMc frequency range using lithium ferrite," Proc. IRE, vol. 49, p. 1574; October, 1961.

[11] R. L. Comstock and W. A. Dean, "Extension of coincidence limiting frequency range in ferrimagnets," *J. Appl. Phys.*, vol. 34, pp. 1275–1277; April, 1963.

[12] R. L. Comstock and M. Loewy, "An X-band coincidence ferrimagnetic limiter using lithium ferrite," Proc. IEEE (*Correspondence*), vol. 51, pp. 1769; December, 1963.

[13] J. Clark and J. Brown, "A miniaturized ferrimagnetic high power coaxial duplexer-limiter," *J. Appl. Phys.*, vol. 33, pp. 1270–1271S; March, 1962.

[14] J. F. Dillon, Jr. and J. W. Nielsen, "Effects of rare earth impurities on ferrimagnetic resonance in yttrium iron garnet," *Phys. Rev. Letters*, vol. 3, July, 1959.

[15] R. C. LeCraw, E. G. Spencer, and C. S. Porter, "Ferrimagnetic line width in yttrium iron garnet single crystals," *J. Appl. Phys.*, vol. 29, pp. 429–430; March, 1958.

[16] R. L. Comstock, "A cryogenic four-port circulator using single crystal yttrium iron garnet," Proc. IEEE (*Correspondence*), vol. 51, p. 1769; December, 1963.

Side-Wall-Coupled, Strip-Transmission-Line Magnetically Tunable Filters Employing Ferrimagnetic YIG Resonators

PHILIP S. CARTER, MEMBER, IEEE

Abstract—This paper describes a new type of band-pass filter configuration for multiple-coupled-resonator magnetically tunable microwave filters. For two or more resonators, this configuration, which employs a coupling slot in the common side wall between two strip-transmission lines, results in the smallest possible air gap and, therefore, the least amount of leakage or fringing flux, and the smallest ampere-turns requirement on the bias magnet.

Response curves including pass band insertion-loss, bandwidth, stop-band rejection, spurious response levels and bandwidths, VSWR, and the effect of temperature on these characteristics are presented for a two-resonator band-pass filter of the type employing YIG resonators and making use of a ferrite core electromagnet to obtain the bias field. Performance data are also given for an experimental side-wall-coupled three-resonator filter.

Manuscript received March 2, 1964; revised January 18, 1965. Portions of this paper are based on work which was started while the author was employed at Stanford Research Institute and continued at Physical Electronics Labs. That part of the work which was performed at Stanford Research Institute was performed under Contract DA 36-039-SC-87398 for the US Army Signal R&D Lab., Ft. Monmouth, N. J. A shorter version of the contents of this paper was delivered at the 1963 IRE WESCON, San Francisco, Calif.

The author is with the Physical Electronics Labs., Menlo Park, Calif.

Introduction

PREVIOUSLY developed types [1]–[5] of magnetically tunable filters have employed one or more single-crystal YIG (yttrium-iron-garnet) resonators in overlapping transmission-line configurations similar to those shown in Fig. 1. Filters have been built using strip-transmission line, coaxial transmission line, and waveguide. The lines or guides may be oriented at right angles as in the single resonator filter shown in Fig. 1(a), or in parallel, as in the two-resonator filter shown in Fig. 1(b).

The band-pass filters shown in Fig. 1 have in common the geometrical arrangement of resonator centers in a line along the direction of the magnetic bias field H_0. The magnet air gap length required for this arrangement of resonators increases as the number of resonators is increased. The ampere turns required to furnish this bias field increases in nearly direct proportion to this air gap. Also, the percentage of leakage flux increases as the ratio of the gap to the pole face diameter is increased.

The magnetic flux energy can be reduced only to a certain extent by changing the lengths and proportions of the magnet core and the size and distribution of the winding on the core. The only way to further reduce the flux energy is to reduce the height of the filter itself.

An approach to the reduction of the height of the filter is to use the configuration of resonators shown in Fig. 2, with a coupling slot in the common side wall between the two transmission lines or waveguides. In the strip-transmission-line configuration shown in Fig. 2(a), a slot cut in the common side wall between the resonators produces magnetic coupling between the resonators due to the RF magnetic moments of the ferrimagnetic resonators which lie along the direction of the slot. At the same time, the slot causes very little coupling between the transmission lines since it is perpendicular to the magnetic fields produced by these lines.

The slot is made as narrow and as thick as possible, consistent with the amount of magnetic coupling that is needed, in order to minimize the direct coupling between the transmission lines.

In the three-resonator filter, shown in Fig. 2(b), the center resonator is mounted between the input and output resonators in the coupling iris or slot. One of the advantages of the three-resonator filter, aside from its increased selectivity, is its increased stop-band attenuation compared with that of the two-resonator filter. This increased stop-band attenuation is due to the thicker slot between the transmission lines.

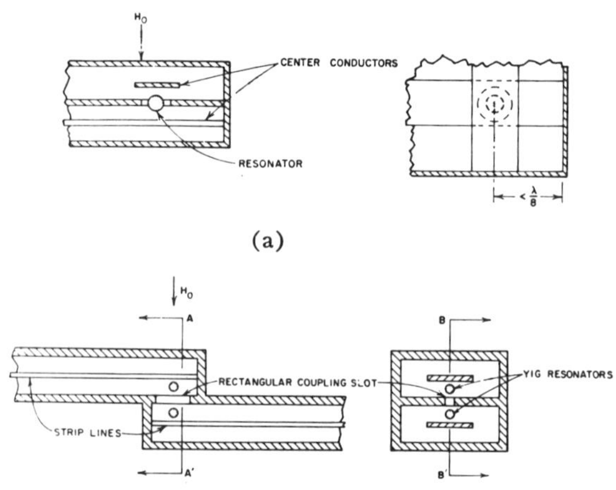

Fig. 1. Top- and bottom-wall-coupled magnetically tunable filter configurations.

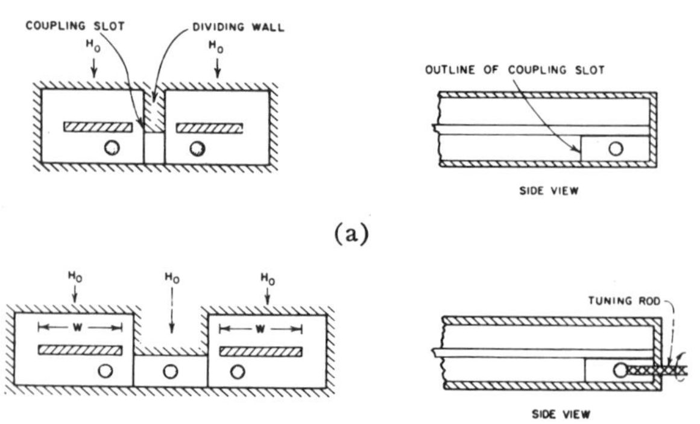

Fig. 2. Side-wall-coupled filter configurations.

Calculation of Filter Design Parameters

Figure 3 gives an approximation to the lumped-element circuit representation of the reciprocal multiple-ferrimagnetic-resonator band-pass filter. From this representation a systematic design procedure is carried out as follows, using previously developed formulas [6]–[9].

$$(Q_e)_A = \frac{\omega_1' g_1 g_0}{w} \tag{1}$$

$$(Q_e)_B = \frac{\omega_1' g_n g_{n+1}}{w} \tag{2}$$

$$k_{i,i+1} = \frac{w}{\omega_1' \sqrt{g_i g_{i+1}}} \tag{3}$$

where

$(Q_e)_A$, $(Q_e)_B$ = External Q's of input and output resonators.

$k_{i,i+1}$ = Coupling coefficient between ith and $(i+1)$th resonators.

$g_0, g_1, \cdots, g_n, g_{n+1}$ = Normalized low-pass prototype element values for desired response shape [9].

ω_1' = Cutoff frequency (radians/sec) of normalized low-pass prototype defined in Fig. 4.
= 1 for normalization chosen in Matthaei, et al. [9].

$w = (f_2 - f_1)/f_0$ = Fractional bandwidth, where f_1 and f_2 are defined in Fig. 4 and $f_0 = \sqrt{f_1 f_2}$.

n = Number of resonators.

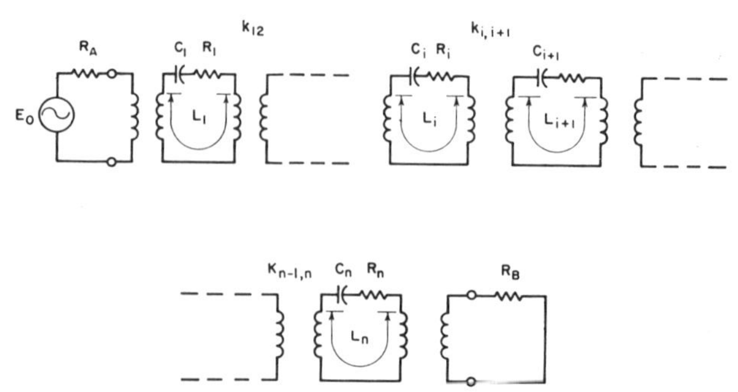

Fig. 3. Equivalent circuit of reciprocal multiple-coupled-resonator ferrimagnetic resonance filter.

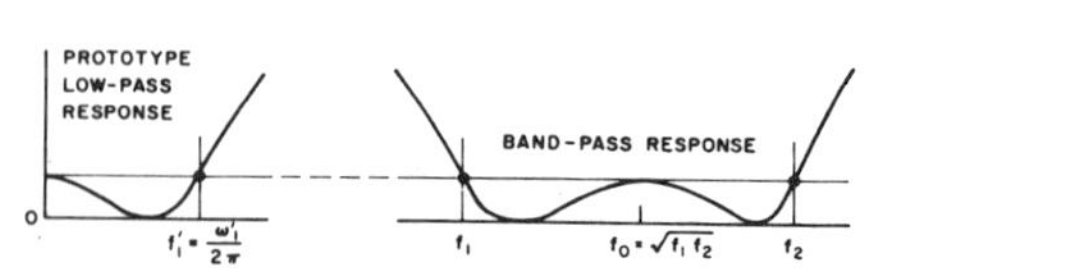

Fig. 4. Definition of low-pass-prototype and band-pass response nomenclature.

The center frequency dissipation loss L_d (not including reflection loss) is given approximately by [10]

$$L_d = 4.343 \frac{\omega_1'}{w} \sum_{i=1}^{n} \left(\frac{g_i}{Q_{ui}}\right) \text{dB.} \tag{4}$$

Here the prototype element values are assumed to be normalized so that $g_0=1$ and the Q_{ui} are the unloaded Q's of the resonators.

Dimensions of Coupling Structures

The dimensions of the coupling circuits for the input and output resonators are computed using previously developed theoretical formulas for the external Q's [2]. Figures 5 and 6 give curves of theoretical Q_e vs. resonator diameter for various combinations of line dimensions for 1) YIG resonators at the center of the short-circuited standard-height TE_{10} rectangular waveguide and 2) short-circuited symmetrical strip-transmission line. The curves in Fig. 5 can be used to obtain Q_e when TE_{10}-mode rectangular guides of other than standard height are used, by applying the following formula

$$Q_e' = Q_e \frac{b'}{b} \tag{5}$$

where

Q_e' = External Q in working guide.
Q_e = External Q of standard guide from Fig. 5.
b' = Height of working guide.
b = Height of standard guide.

The curves can also be used for all different shapes of ellipsoidal resonators by replacing D_m with D_m^e where

$$D_m^e = 2(abc)^{1/3} \tag{6}$$

in which a, b, and c are the semi-axes of the ellipsoid.

The curves of Figs. 5 and 6 can be used for materials other than YIG, e.g., Ga-YIG, by means of the following formula

$$Q_e^{\text{OM}} = \frac{M_s^{\text{YIG}}}{M_s^{\text{OM}}} Q_e^{\text{YIG}} \tag{7}$$

where

Q_e^{YIG} = External Q for YIG, from Fig. 5.
Q_e^{OM} = External Q for material other than YIG.
M_s^{YIG} = Saturation magnetization of YIG (1750 gauss at room temperature).
M_s^{OM} = Saturation magnetization of other material.

The dimensions of the coupling slot and the spacing between resonators are determined largely by experiment. Generally it is desirable to space the resonators as far apart as possible in order to minimize spurious responses due to magnetostatic modes [3]. This procedure, however, leads to a large coupling slot and consequently to a lower stop-band rejection. A trade-off, determined largely by experimental procedures, between high stop-band rejection and spurious response level is made.

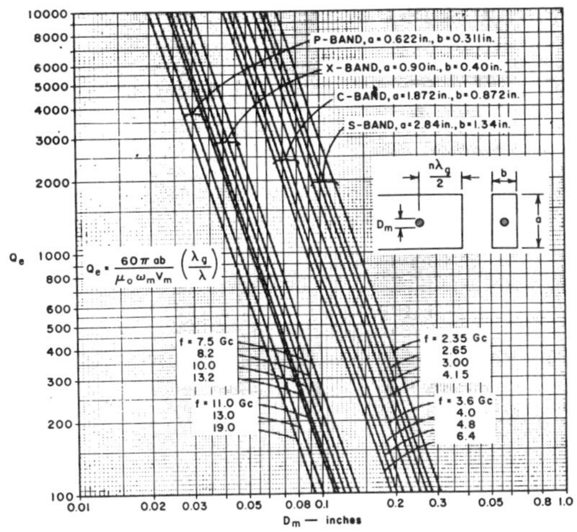

Fig. 5. Theoretical Q_e of YIG resonator in short-circuited TE_{10} rectangular waveguide.

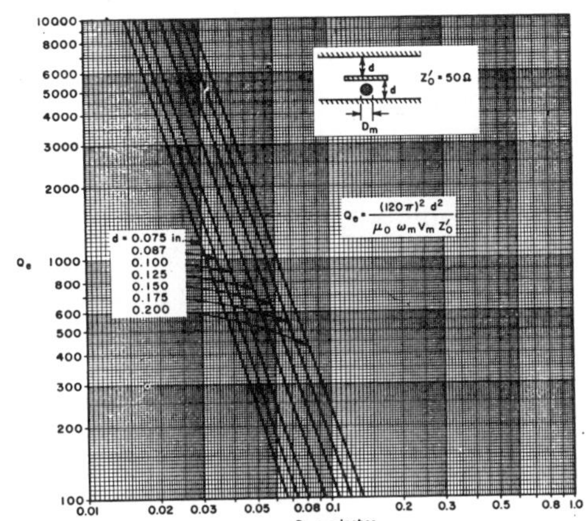

Fig. 6. Theoretical Q_e of YIG resonator in short-circuited strip-transmission line.

Two side effects of the presence of the coupling iris in the common wall between the two resonators are: 1) to cause the values of Q_e to be higher than those which are shown in Figs. 5 and 6, and 2) to cause Q_u to be higher than that which is measured when the coupling iris is not present. The practical consequence of the existence of these two side effects of the coupling slots is to require some additional experimental adjustment of the waveguide and resonator dimensions. Experience has shown that an approximate 25 per cent increase in the value of Q_e is obtained with typical coupling-slot configurations.

Properties of Ferrimagnetic Resonator Materials

Several different single-crystal materials are known to have possible application in magnetically tunable filter design including yttrium-iron-garnet (YIG), gallium-substituted yttrium-iron-garnet (Ga-YIG), lithium ferrite, and barium ferrite. The properties, including the unloaded $Q(Q_u)$ of these materials have been presented in previous publications [11]–[16].

TABLE I
PROPERTIES OF SINGLE-CRYSTAL FERRIMAGNETIC MATERIALS FOR MAGNETICALLY TUNABLE FILTERS AT ROOM TEMPERATURE

Material	$4\pi M_s$ (Gauss)	K_1/M_s (Oersteds)	ΔH* (Oersteds)	T_c (°C)
Yttrium-Iron-Garnet** YIG	1750	−43	0.22 (4Gc, [17])	292
Gallium-Substituted** Yttrium-Iron-Garnet	50–1750	—	—	—
	600	−55.8	—	160
	950±50	−41.7	0.7–2.0 (at 4.4 Gc)	206
Lithium Ferrite	3550±40 [13]	—	3*** (5 Gc)	—
"Planar" Ferrite "Zn_2Y" ($Ba_{12}Zn_2Fe_{12}O_{22}$)	2850 ([14])	4950 ([14])	10(*X*-band, ([14])	—

* These values of ΔH were measured in cavities. They may be considerably larger when measured in a filter structure.
** These materials were supplied by Microwave Chemicals Lab., 282 Seventh Ave., New York City, N. Y.
*** Private communication, J. W. Nielsen, Airtron Div. of Litton Industries, Morris Plains, N. J.

The physical constants of ferrimagnetic materials that affect the filter design are the saturation magnetization M_s, the anisotropy field constant K_1/M_s, the Curie temperature T_c, and the resonance linewidth ΔH.

Table I lists values of these physical constants measured at room temperature. The values of K_1/M_s, given in the table, were measured using two samples of Ga-YIG of different degrees of substitution of gallium. Values of the constants of the other materials in the table are taken from the given references.

The saturation magnetization M_s and the shape of the resonator determine the lowest resonant frequency f_0^{min}, which can be reached due to desaturation. Figure 7 shows f_0^{min} of spheroidal resonators as a function of $4\pi M_s$ in gauss with the ratio of principal axes as a parameter. The most commonly employed resonator shape is spherical, due to the ease with which it can be fabricated and polished to give resonators with very high values of Q_u. In principle it is possible to achieve lower resonant frequencies by the use of a thin disk-shaped resonator. Polished disks having low resonant frequencies and narrow linewidths are now available.[1]

The unloaded Q, Q_u, is usually given by the manufacturer in terms of an equivalent quantity, the linewidth ΔH.[2] To obtain Q_u from the linewidth, the following formula is used:

$$Q_u = \frac{f_0^{Mc}}{2.8\Delta H} \tag{8}$$

f_0^{Mc} = Resonant frequency, megacycles per second.
ΔH = Linewidth, oersteds.

The Q_u of YIG increases about linearly with increasing frequency in the 2-to-10 Gc frequency range. The values of Q_u obtained in a practical filter structure where the coupling is relatively tight, and consequently the conducting boundaries are near the resonator, are usually lower than the values specified for the material by the manufacturer. Figure 8 shows the values of Q_u that have been obtained with a 0.074-inch diameter YIG resonator in different coupling configurations, including a full-height *S*-band waveguide.

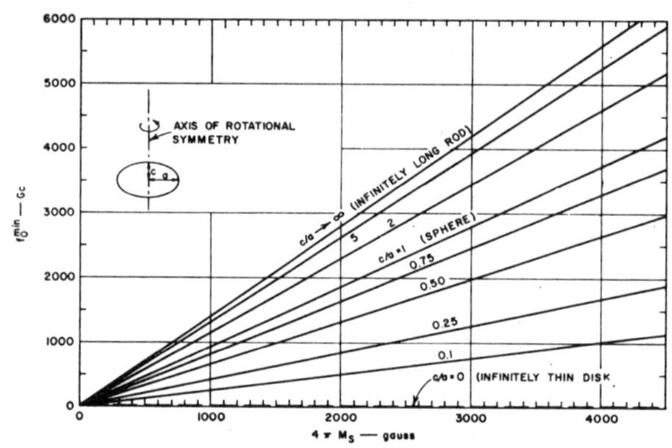

Fig. 7. Theoretical minimum resonant frequencies of spheroids.

The effect of magneto-crystalline anisotropy is that the biasing magnetic field required to produce resonance varies according to the orientation of the crystal axis with respect to the biasing magnetic field. Figure 9 shows the biasing field required to resonate a YIG sphere at room temperature. The resonator is rotated around a (110) crystal axis which is perpendicular to the biasing field. For crystals with cubic symmetry such as single-crystal YIG or gallium-substituted YIG [18],

$$H_0 = H_{eff} - \left(2 - \frac{5}{2}\sin^2\theta - \frac{15}{8}\sin^2 2\theta\right)\frac{K_1}{M_s}, \text{ oersted} \tag{9}$$

where

$$H_{eff} = \frac{f_0^{Mc}}{2.8}$$

θ = angle between H_0 and a [100] axis.

Formulas for the resonant frequency have been given for several other crystal structure types by Yager, et al. [18], Kittel [19], and Artman [20].

[1] Disks fabricated to specifications can be obtained from Litton Industries Airtron Div., Morris Plains, N. J., and from Microwave Chemicals Lab., New York City, N. Y.

[2] Linewidth is here defined as the difference between the two values of biasing field for which the imaginary part of the intrinsic susceptibility equals the real part, while frequency is held constant.

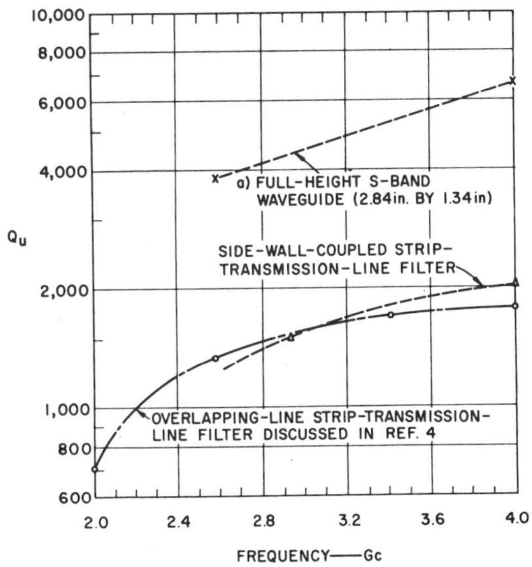

Fig. 8. Q_u of YIG resonator vs. frequency in waveguide and strip-transmission-line filter configurations.

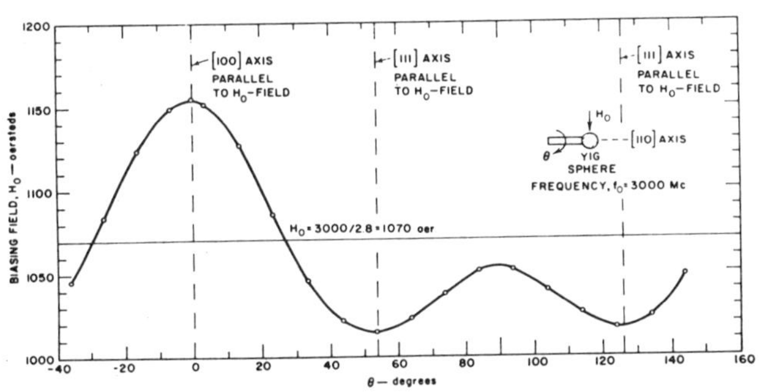

Fig. 9. Field strength for YIG sphere resonance at 3000 Mc as resonator is rotated about a [110] axis.

Design of Two-Resonator Side-Wall-Coupled Filter

The foregoing methods were used to design a two-resonator side-wall-coupled filter with the following characteristics:

Tuning Range	2–4 Gc
Bandwidth between 3-dB rejection frequencies	20 Mc

Type of response—maximally flat at center of tuning range.

From (1) and (2):

$$(Q_e)_A = (Q_e)_B = \frac{\omega_1' g_1 g_0}{w} = \frac{1 \times 1.414 \times 1}{0.0071} = 202$$

where

$$w = \frac{f_2 - f_1}{f_0} = \frac{20}{2820} = 0.0071.$$

$f_0 = 2820$ Mc is the geometric mean of the limits of the tuning range, i.e.,

$f_0 = \sqrt{2000 \times 4000}$. From (3)

$$k_{12} = \frac{w}{\omega_1' \sqrt{g_1 g_2}} = \frac{0.0071}{1 \times \sqrt{(1.414)^2}} = 0.00495.$$

k_{12} is realized experimentally by adjustment of the dimensions of the coupling iris.

The resonator material is chosen using the requirement that the minimum center frequency is 2 Gc. From Fig. 7 a maximum $4\pi M_s$ of 2150 gauss for a spherical resonator is seen to be allowable. Yttrium-iron-garnet ($4\pi M_s = 1750$ gauss) satisfies this requirement; furthermore, its unloaded Q, Q_u, is still quite high at frequencies as low as 2 Gc (see Fig. 8). From Fig. 6 we then obtain the strip-transmission-line center-conductor-to-ground-plane spacing using $Q_e = 150$, rather than the value of 202 calculated above, in order to allow for a 33 per cent increase due to the decoupling effect of the iris. The resonator diameter $D_m = 0.074$ inch and center-conductor-to-ground-plane spacing $d = 0.110$ inch were chosen. The strip-transmission-line dimensions for the 50-ohm characteristic impedance line given in Fig. 10 were then obtained by conventional procedures.

Construction and Alignment of Filter

The body of the filter, which is shown in Fig. 11, was machined from LEXAN.[3] A thin shell, about 0.0005 inch thick, of copper was plated onto the LEXAN body. The purpose of this procedure was to avoid eddy currents in the filter body when using a rapidly swept bias field. The YIG resonators were mounted on Rexolite rods which could be rotated on an axis perpendicular to the direction of the biasing field. The YIG resonators were aligned with their [110] axes along the rod axis. This orientation was obtained using the ferromagnetic crystal orienter previously described in the literature [23]. The strip-transmission-line center conductor was supported by means of Rexolite slabs in the regions between the type N input and output connectors and the short section of transmission-line coupling to the YIG resonator. Split-block construction was used in fabricating the body of the filter.

The first step in aligning the filter was to adjust the spacing of the resonators to give the desired maximally flat response shape at 2.83 Gc. The next step was to orient the YIG crystal axes so that the variation of the resonant frequency with temperature was minimized. Examination of Fig. 9 or (9) shows that there is an angle θ at which the effect of the anisotropy field K_1/M_s is reduced to zero. With the sphere in this orientation, which is 29°45′ from a [100] axis, the variation of the resonant frequency due to a variation of K_1/M_s with

[3] Trade name, manufactured by General Electric Co. This material is a high-strength plastic dielectric material which retains its mechanical and electrical characteristics at high temperature.

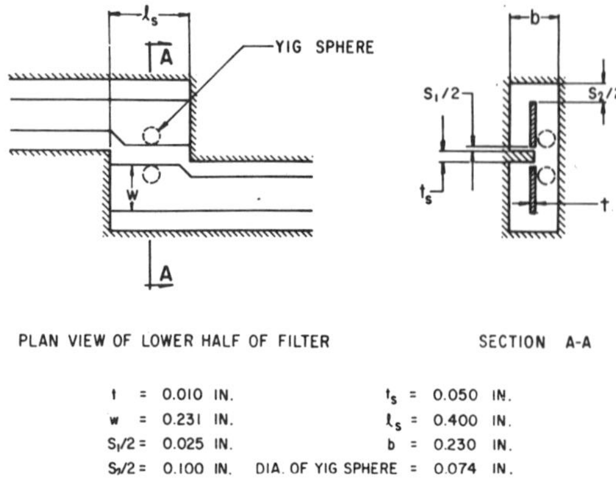

Fig. 10. Dimensions of two-resonator filter.

272

Fig. 11. Two-resonator filter with shield lid removed.

temperature is eliminated [23]. The resonator was mounted in the filter and rotated so that the desired zero-temperature-coefficient axis was obtained. This operation is conveniently carried out by operating the device with only one resonator in place, and measuring the small coupling to the output which is the result of leakage through the coupling slot in the sidewall. For YIG the zero-temperature-coefficient resonant frequency is 241 Mc above the minimum resonant frequency or 161 Mc below the maximum resonant frequency. The second resonator was then rotated to give the same resonant frequency.

In taking measurements on this type of filter using YIG resonators, it is important to operate the device below the saturation levels, which are as low as -10 dBm over part of the 2-to-4 Gc tuning range [4].

A highly uniform magnetic bias field is required for this filter since a nonuniform field causes the resonators to be tuned to different frequencies. Several references are available in which various aspects of electromagnet design are discussed [24], [25], and [26]. The magnet shown in Fig. 11 used two ferrite E-cores, the center legs of which had been shortened to provide the air gap. The pole-face dimensions were 0.685 inch by 0.685 inch and the air gap was 0.300 inch.

Filter Performance

Figure 13 summarizes the operating characteristics of the filter as it performed at room temperature. The quantities plotted in Fig. 13 refer to the filter response characteristics which are defined in Fig. 12.

The insertion loss at band center varies between a maximum value of $L_0 = 2.4$ dB at $f_0 = 2.0$ Gc and a minimum value of $L_0 = 1.2$ dB at $f_0 = 3.0$ Gc.

The center frequency loss could be reduced somewhat by using resonators with higher unloaded Q's. YIG resonators with unloaded Q's 25 to 50 per cent higher than those used in this filter can be obtained. The resonators used here are typical of those made of high-grade YIG material now available.[4]

The 3-dB and 30-dB bandwidths of the filter vary only slightly as the center frequency is tuned over the 2-to-4 Gc range. This constancy of the bandwidth is one of the features of the ferrimagnetic-resonance magnetically tunable filter which cannot easily be obtained with other types of mechanically and electronically tunable filters. This constant bandwidth characteristic results from the fact that the coupling coefficient k_{12} between magnetic resonators is inversely proportional to frequency while Q_e remains approximately constant [6]. An examination of the following expression for the insertion loss L shows why this results in a constant bandwidth.

$$L = 10 \log \left\{ \left[\frac{\left(1 + \frac{Q_e}{Q_u}\right)^2}{2kQ_e} + \frac{kQ_e}{2} \right]^2 + 2\left[\frac{\left(1 + \frac{Q_e}{Q_u}\right)^2}{k^2} - Q_e^2 \right] \left(\frac{f - f_0}{f_0} \right)^2 + \frac{4Q_e^2}{k^4} \left(\frac{f - f_0}{f_0} \right)^4 \right\}. \tag{10}$$

For a maximally-flat response $Q_e = k^{-1}$. Thus the variation of L with frequency is predominantly due to the third term on the right in (10). If $k \propto (f_0)^{-1}$ and $Q_e =$ constant, then this term is proportional to $(f - f_0)^4$, resulting in an approximately constant pass-band bandwidth.

A minimum stop-band rejection $L_R^{\min} = 40$ dB is obtained in this filter. It is probable that a somewhat greater value of stop-band rejection could be obtained with an elliptical coupling slot geometry.

The maximum amplitude of the spurious response at 730 Mc below the main response of this filter was at least 30 dB below the main response, as shown in Fig. 13. The three spurious responses shown here are at

[4] These YIG resonators were supplied to us by Microwave Chemicals Lab., Inc., New York, N. Y.

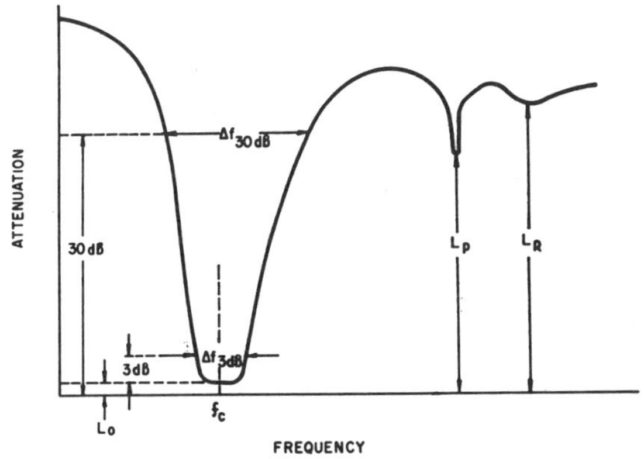

Fig. 12. Filter response nomenclature.

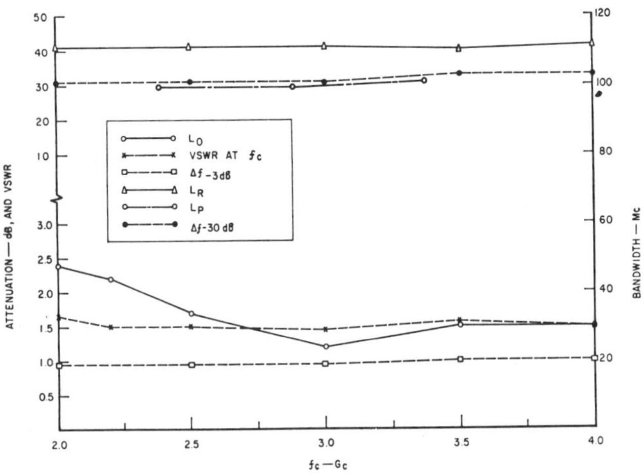

Fig. 13. Characteristics of two-resonator filter at room temperature.

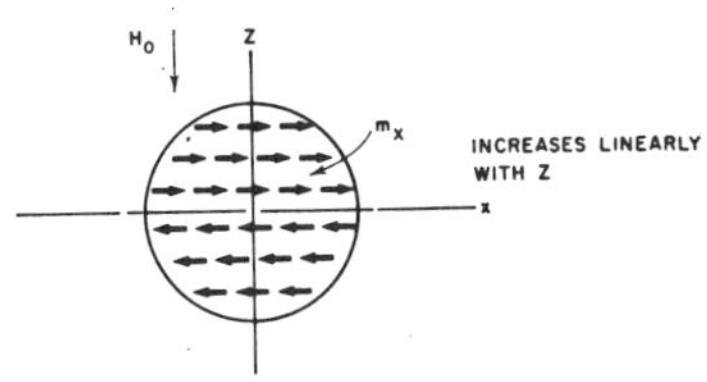

Fig. 14. Distribution of the RF magnetization in the (210) magnetostatic mode.

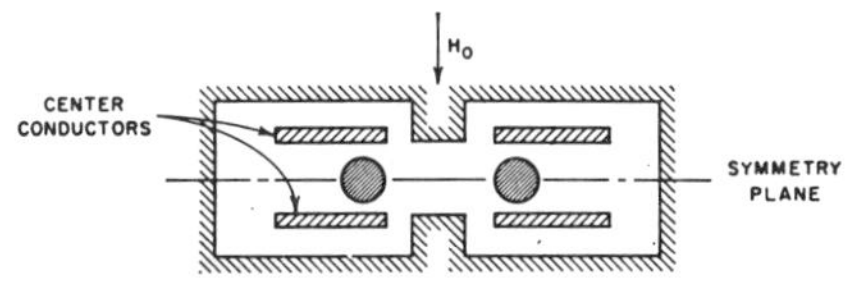

Fig. 15. A possible structure to eliminate the (210) magnetostatic mode spurious response.

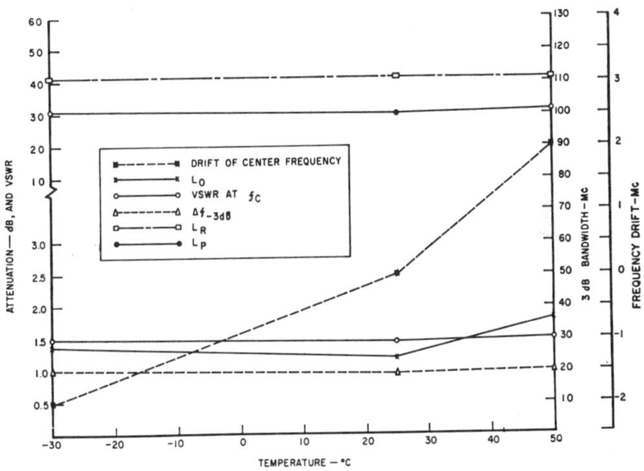

Fig. 16. Characteristics of two-resonator filter as a function of temperature at 3 Gc.

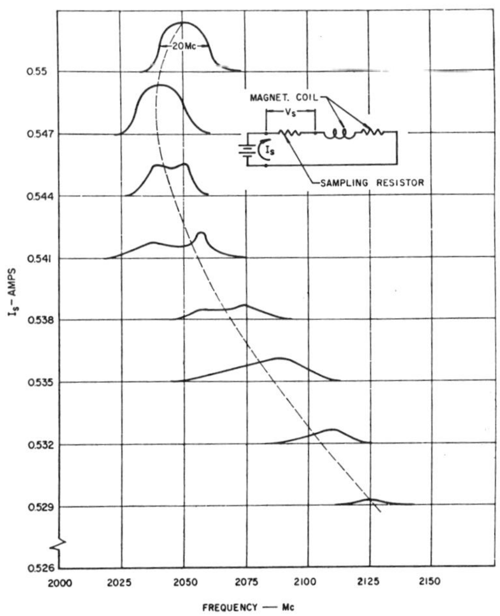

Fig. 17. Pass band response around 2 Gc vs. magnet current at −40°C.

2.270, 2.770, and 3.270 Gc, corresponding to main responses at 3.000, 3.500, and 4.000 Gc. The level of the spurious response depends on the uniformity of the biasing magnetic fields as well as the uniformity of the RF magnetic fields. A uniform bias field is needed to minimize the level of the spurious responses as well as to maintain synchronous tuning of the resonators.

The spurious response which occurs at a frequency approximately 730 Mc below the main response is caused by the excitation of the (210) magnetostatic mode. The distribution of the m_x component of RF magnetization of this mode is shown in Fig. 14. It is seen from this figure that this mode may be excited by the component of the RF magnetic field which has odd symmetry around the equator of the sphere. This odd-symmetrical component of RF field could be minimized by the use of a balanced symmetrical strip-transmission-line coupling structure such as the one shown in Fig. 15.

Figure 16 shows the effect of varying the temperature of the filter, which was tuned to a center frequency of 3.0 Gc at room temperature. The ambient temperature was varied between −30°C and +50°C. It is seen that the operating characteristics, including the center frequency, remain very nearly constant over this temperature range. The center frequency remained nearly constant—a total drift of 4 Mc was measured. Similar results were obtained when the filter was tuned to 2 Gc and to 4 Gc.

Above about 50°C the filter response shape deteriorated, broadening considerably as the temperature was increased. This behavior was believed to be due to saturation of the ferrite magnet core material. Filters using silicon steel magnet cores have been operated at ambient temperatures of up to 100°C with very little deterioration of performance.

A change of the response shape and the tuning characteristic occurs at the low end of the tuning range at low temperatures.[5] Figure 17 shows the responses which were obtained by varying the coil current at an ambient temperature of −40°C. As can be seen from Fig. 17, there is a "turnaround" point after which, when the magnet coil current is decreased, the center frequency begins to increase. Below this point the insertion loss also increases.

Three-Resonator Filter

A prototype three-resonator, side-wall-coupled filter was also constructed using the principles outlined previously. One of the main purposes for building this filter was to ascertain whether or not three resonators could be synchronously tuned over a wide range of frequencies. Before the filter was built, this point was somewhat in doubt, since the center resonator of a three-resonator filter is exposed to different field conditions than the two end resonators. As it turned out, differences in the resonant frequency of the center resonator could be compensated by rotating the crystal axis of the center resonator with respect to the bias field.

The dimensions and configuration of the input and output coupling lines and resonators were made exactly the same as those which were used in the two-resonator filter previously described. The only part of the two-resonator filter structure which was changed was the coupling iris, which was made wider, i.e., thicker, to accommodate the center resonator. A 0.230-inch-thick iris was used. The assembled experimental filter is shown in Fig. 18, without the biasing magnet.

The measured performance of the three-resonator filter is shown in Figs. 19 and 20. A laboratory-type electromagnet was used to supply the magnetic field. The output of the filter is shown as the magnetic field was varied through resonance while keeping the frequency f_c into the filter constant. These plots are closely related to the frequency-response properties of the filter since the RF impedance of the ferrite is a function of the quantity $\omega-\gamma H_0$. A change in H_0 thus is equivalent to an equal but opposite change in ω. Thus, the high field side of the field-response curve resembles the low-frequency side of the frequency-response curve and vice versa. The only approximation is that the frequency variations of the coupling circuit are not simulated. However, over the narrow pass band of the filter, this variation of the filter structure properties should be small.

Figure 19 shows the pass band responses of the filter when it was tuned to various center frequencies between 2.0 Gc and 7.0 Gc. The synchronism of the resonators is maintained over a very large tuning range as was hoped. Figure 19 shows that synchronous tuning is still obtained at a center frequency of 7.0 Gc. It is evident that at this highest frequency the response has become somewhat undercoupled as would be expected from the theory of coupled magnetic resonators [6].

Figure 20 shows the stop band and spurious mode responses of the filter as it was tuned to different frequencies between 2.0 and 5.0 Gc. As can be seen from these curves the rejection in the stop band is considerably better than that of the two-resonator filter. It is likely that the actual stop-band rejection (excluding spurious mode responses) is better than the −66 dB, shown for $f_c=5.0$ Gc, which was the limit of detection of the measurement set up.

The spurious, (210) mode, response which occurs at about 740 Mc below the center frequency is completely absent in Fig. 20. Its position, if it were present in these response patterns, would be at about 265 oersteds higher than the center of the main response. This increased insertion loss of the (210) mode response is due to the very weak coupling between (210) modes in the three resonators.

A spurious response appears at $f_c=2.0$ Gc as the field is decreased below the value required for the main response. This spurious response appears about 100 oersteds below the main resonance, a value of field at which no spurious response would be expected with an undoped YIG resonator. It is probable that this spurious response is actually the "turnaround" effect which was shown in Fig. 17 for the two-resonator filter at a low temperature. In application of this filter, this response would be avoided by keeping the bias field above the value at which this secondary response appears.

Table II lists some of the properties of this filter. Included here are frequency bandwidths, which were measured directly. The VSWR was low except at $f_c=2.0$ Gc where, it was suspected, the resonators were slightly detuned. The 30-dB bandwidth stays nearly constant over the tuning range. The ratio of the 30-dB bandwidth to the 3-dB bandwidth runs around 2.8 for the three-resonator filter design. Although the cutoff rate was considerably sharper for the three-resonator design, the performance of the three-resonator filter probably was not optimum because of a small amount of mistuning and because no effort was made to optimize the shape of the

[5] This phenomenon was first pointed out to us by Thomas Miles, Applied Electronics Lab., Stanford University, Calif.

Fig. 18. Three-resonator filter shown assembled, without magnet.

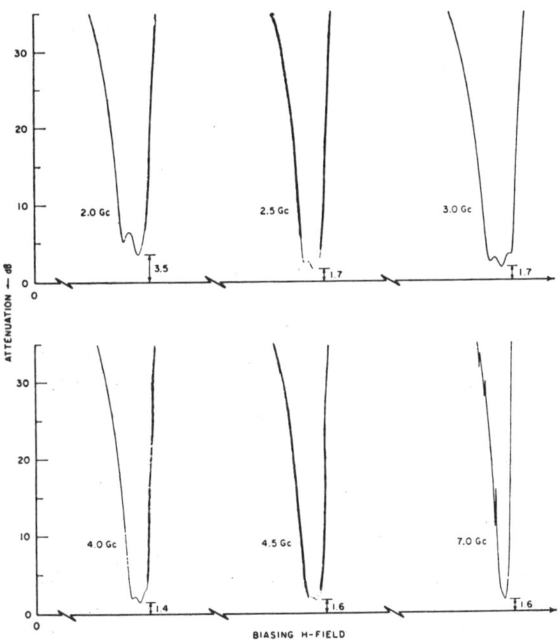

Fig. 19. Pass band responses of three-resonator filter.

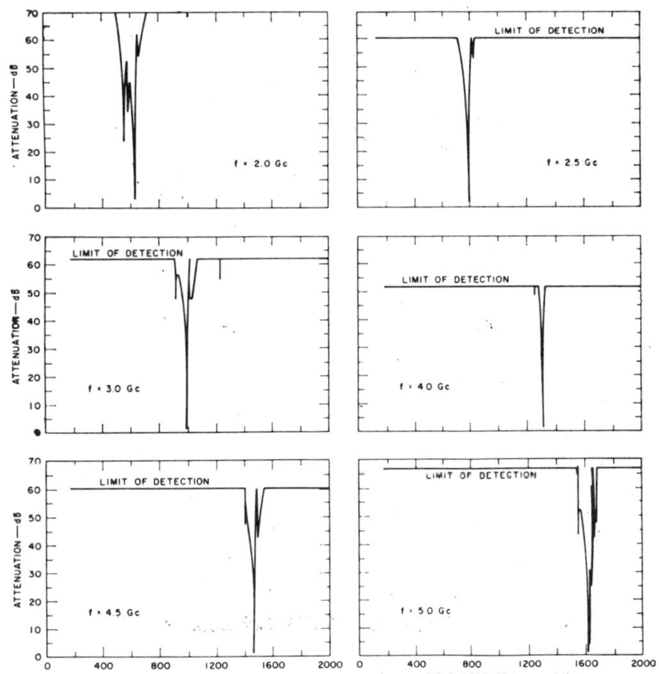

Fig. 20. Stop band responses of three-resonator filter.

TABLE II
MEASURED CHARACTERISTICS OF THREE-RESONATOR YIG FILTER

f_c (Gc)	L_0 (dB)	Δf^*-3dB (Mc)	$\Delta f-30$dB (Mc)	VSWR	H_0 (oersteds)
2.0	3.5	32.9*	91.4	2.50	610.0
2.5	1.7	34.2	90.1	—	867.1
3.0	1.7	33.6	90.6	1.23	990.0
4.0	1.4	30.9	87.2	1.10	1310.0
4.5	1.6	24.0	89.1	—	1586.8
7.0	1.6	29.6	—	—	—

* The bandwidths given are measured between points which are 3 dB and 30 dB below L_0.

pass band response. Filters designed to correspond to equal-element prototypes would probably be desirable for most tunable filter applications. Such filters would result in minimum midband dissipation loss for a given specified 30 dB (or other specified level) stop-band bandwidth [28].

Signal Limiting with the Magnetically Tunable Filter

An important property of filters employing ferrimagnetic resonators is their ability to suppress undesired high-level signals without affecting the response to the desired low-level signal. This limiting property occurs as a result of the parametric coupling between the main (uniform precession) resonance and the higher order "spin-waves," and causes signals in the pass band of the filter to be limited to output levels of the order of 0 dBm to −10 dBm in certain frequency ranges and +20 to +30 dBm in others. Several ferrimagnetic limiters having low limiting levels in the 2-to-10 Gc frequency range have been built [29]–[31].

Summary and Conclusions

A new type of coupling structure has been demonstrated which is applicable to YIG band-pass filters having two or more resonators, which minimizes the air gap between the pole faces of the bias magnet. It has been demonstrated, by constructing two- and three-resonator models of this filter, that at least an octave tuning range can be obtained while maintaining useful insertion loss and selectivity characteristics. It has also been shown that the two-resonator filter can be temperature stabilized to a degree which allows it to be operated over approximately a −30°C to 50°C range. Finally, a technique using a copper-plated dielectric filter body was demonstrated which allows the filter to be swept at high speeds.

Acknowledgment

Dr. G. Matthaei of Stanford Research Institute made many suggestions during the course of the work. Y. Sato conducted the testing with great skill at Stanford Research Institute.

Acknowledgment is made to the entire staff of Physical Electronics Laboratories for suggestions and assistance in completing the development of the two-resonator filter. E. Sheehan's efforts in the assembly and testing of this unit are also gratefully acknowledged.

References

[1] Carter, P. S., Jr., G. L. Matthaei, and W. J. Getsinger, Design criteria for microwave filters and coupling structures, Tech Rept 8, Project 2326, Contract DA 36-039 SC-74862, Stanford Research Inst., Menlo Park, Calif., Sep 1959.[6]

[6] Report submitted under referenced contract. Copies of this and other reports listed are available to qualified requestors from the Armed Services Technical Information Agency, or by writing to the U. S. Army Electronics Research and Development Lab., Ft. Monmouth, N. J., Attention: W. Dattilo.

[2] Carter, P. S., Jr., Magnetically-tunable microwave filters using single-crystal yttrium-iron-garnet resonators, *IRE Trans. on Microwave Theory and Techniques*, vol MTT-9, May 1961, pp 252–260.

[3] Matthaei, G. L., Carter, P. S., Jr., et al., Design criteria for microwave filters and coupling structures, final rept, SRI Project 2326, Contract DA 36-039 SC-74862, Stanford Research Inst., Menlo Park, Calif., Jan 1961, ch 28, pp 475–512.

[4] Carter, P. S., Jr., et al., Microwave filters and coupling structures, Rept 3, SRI Project 2326, Contract DA 36-039 SC-87398, Stanford Research Inst., Menlo Park, Calif., Oct 1961, pp 3–35.

[5] Kotzebue, K. L., Broadband electronically-tuned microwave filters, *1960 WESCON Conv. Rec.*, vol 4 pt 1, pp 21–27.

[6] Matthaei, G. L., L. Young, and E. M. T. Jones, Design of microwave filters, impedance-matching networks, and coupling structures, Stanford Research Inst., Menlo Park, Calif., vol 2, Jan 1963, ch 17, pp 993–1077. (To be published by McGraw-Hill.)

[7] Carter, P. S., Jr., et al., Microwave filters and coupling structures, Rept 7, SRI Project 3527, Contract DA 36-039 SC-87298, Stanford Research Inst., Menlo Park, Calif., Oct 1961, pp 31–70.

[8] Cohn, S. B., et al., Research on design criteria for microwave filters, final rept, SRI Project 1331, Contract DA 36-039 SC-74862, Stanford Research Inst., Menlo Park, Calif., Jan 1961.

[9] Matthaei, G. L., L. Young, and E. M. T. Jones, Design of microwave filters, impedance-matching networks, and coupling structures, SRI Project 3527, Contract DA 36-039 SC 87398, Stanford Research Inst., Menlo Park, Calif., vol I, Jan 1963, ch 4, pp 83–157.

[10] Cohn, S. B., Dissipation loss in multiple-coupled-resonator filters, *Proc. IRE*, vol 47, Aug 1959, pp 1342–1348.

[11] Carter, P. S., Jr., and C. Flammer, Unloaded Q of single crystal yttrium-iron-garnet resonator as a function of frequency, *IRE Trans. on Microwave Theory and Techniques* (*Correspondence*), vol MTT-8, Sep 1960, pp 570–571.

[12] Spencer, E. G., and R. C. LeGraw, Line width narrowing in gallium substituted yttrium-iron-garnet, *Bull. Am. Phys. Soc.*, vol 5, pt 1, Jan 1960, p 58.

[13] Nielsen, J. W., D. A. Lepore, J. Zneimer, and G. B. Townsend, Effect of mechanical, thermal, and chemical treatment of the ferrimagnetic resonance linewidth on lithium ferrite crystals, *J. Appl. Phys.*, vol 33S, Mar 1962, pp 1379–1380.

[14] Buffler, C. P., Resonance properties of single-crystal hexagonal ferrites, *J. Appl. Phys.*, vol 33, Mar 1962, pp 1360–1362.

[15] Dixon, S., High-power characteristics of single-crystal ferrites with planar anisotropy. *J. Appl. Phys.*, vol 33S, Mar 1962, pp 1368–1369.

[16] Tauber, A., R. O. Savage, R. J. Gambino, and C. G. Whimfrey, Growth of single-crystal hexagonal ferrites containing Zn, *J. Appl. Phys.*, vol 33S, Mar 1962, pp 1381–1382.

[17] Matthaei, G. L., P. S. Carter, Jr., et al., *op. cit.*, pp 490–491.

[18] Yager, W. A., J. K. Galt, F. R. Merritt, and E. A. Wood, Ferromagnetic resonance in nickel ferrite, *Phys. Rev.*, vol 80, Nov 1950.

[19] Kittel, C., On the theory of ferromagnetic resonance absorption, *Phys. Rev.*, vol 73, 1948, pp 155–161.

[20] Artman, J. O., Microwave resonance relations in anisotropic single-crystal ferrites, *Proc. IRE*, vol 44, Oct 1956, pp 1284–1293.

[21] Matthaei, G. L., L. Young, and P. S. Carter, Jr., Microwave filters and coupling structures, Rept 5, SRI Project 3527, Contract DA 35-039 SC-87398, Stanford Research Inst., Menlo Park, Calif., May 1962.

[22] Reference Data for Radio Engineers, 4th ed. Internat'l Telephone and Telegraph Corp., New York, N. Y., 1959, pp 190–198.

[23] Sato, Y., and P. S. Carter, A device for rapidly aligning and mounting ferromagnetic single crystals along any desired axis, *IRE Trans. on Microwave Theory and Techniques* (*Correspondence*), vol MTT-10, Nov 1962, pp 611–612.

[24] Brailsford, F., *Magnetic Materials*, 3rd ed. London: Methuen, 1960.

[25] Rotors, H. C., *Electromagnetic Devices*, 1st ed. New York: Wiley, 1941.

[26] Livingston, M. S., and J. P. Blewett, *Particle Accelerators*, 1st ed. New York: McGraw-Hill, 1962, ch 8.

[27] Lax, B., and K. S. Button, *Microwave Ferrites and Ferrimagnetics*, sec 5-5, 1st ed. New York: McGraw-Hill, 1962, pp 227–231.

[28] Taub, J. J., Design of minimum loss band-pass filters, *Microwave J.*, vol 6, Nov 1963, pp 67–76.

[29] Lax, B., and K. J. Button, *op. cit.*, sec 12-13.

[30] Brown, J., Ferrimagnetic limiters, *Microwave J.*, vol 4, Nov 1963, pp 74–79.

[31] Kotzebue, K. L., Frequency-selective limiting in YIG filters, *J. Appl. Phys.*, vol 33, Feb 1962, p 747.

276

Nonreciprocal YIG Filters

R. E. TOKHEIM, MEMBER, IEEE, C. K. GREENE, SENIOR MEMBER, IEEE, J. C. HOOVER, MEMBER, IEEE, AND R. W. PETER, FELLOW, IEEE

Abstract—**Several techniques for achieving nonreciprocity in YIG filters employing stripline or miniature coaxial line construction are presented. Nonreciprocal filters can be designed with either single-ended or double-ended nonreciprocity. In either case, the circuit element possesses isolator and circulator properties as well as that of a ferrimagnetic filter. Experimental results have indicated that nonreciprocity greater than 10 dB over octave bandwidths is possible with passband characteristics similar to those of conventional YIG filters. The insertion loss may be somewhat higher than the latter, depending on the technique of nonreciprocity utilized. Nonreciprocal YIG filters are useful with TWT microwave receiver systems as preselectors and post-selectors and obviate the need for isolators. These filters are also useful as tunable microwave circulators in conjunction with tunnel diode amplifiers and parametric amplifiers. An individual filter may be used as a diplexer, while several filters will provide multiplexer operation.**

Introduction

A YIG FILTER has generally come to mean a low-insertion loss narrowband microwave ferrimagnetic bandpass filter, which is typically tunable over at least one octave frequency range. The ideas on low-loss gyromagnetic coupling for ferrimagnetic resonators, which are used extensively in these filters, were presented by DeGrasse in 1959.[1] These ideas were extended by Carter, Jr., Matthaei, and Getsinger,[2],[3] Comstock,[4] and Kotzebue,[5] who enhanced the capability of filter-limiters with ferrimagnetic resonators.

Since the original work, reciprocal bandpass and band reject ferrimagnetic filters and limiters utilizing gyromagnetic coupling have been developed at UHF and microwave frequencies. The optimum filter or limiter performance from the standpoint of low-insertion loss, minimum spurious magnetostatic mode excitation, and limiting properties has generally been best achieved by utilizing YIG or gallium-doped YIG spheres. Hence, a ferrimagnetic filter is often referred to as a YIG filter. Typically the unloaded Qs of microwave ferrimagnetic resonators are comparable to those of conventional cavities, while being about two orders of magnitude smaller in size. The size of the filter itself, however, is determined chiefly by the size of the RF coupling and dc magnetic circuits.

Nonreciprocity in ferrimagnetic devices is not new. The basic requirements of nonreciprocity in microwave ferrimagnetic devices were reviewed by Hogan in 1956.[6] Nonreciprocity is made possible by the dominantly right-hand, circularly polarized motion of the magnetization in the uniform precessional mode. The usual method of utilizing this nonreciprocity is to couple to the ferrimagnet with an RF transmission line which will excite a circularly polarized RF magnetic field. Energy propagating in the reverse direction will not couple to the resonator due to the wrong sense of circular polarization. The rectangular waveguide propagating in the dominant mode has been the most frequently used transmission line possessing the aforementioned conditions. The work of Duncan, Swern, Tomiyasu, and Hannwacker[7] followed by that of Fleri and Hanley[8] showed that these conditions were equally possible with dielectric-loaded TEM mode transmission lines which they used to demonstrate low-loss, ferrimagnetic isolators. It was not until 1961, however, that a nonreciprocal, tunable YIG filter using waveguide transmission lines and a single ferrimagnetic resonator was reported by Patel.[9] In 1962, Shaw utilized Fleri and Hanley's results to achieve nonreciprocity in tunable YIG filters employing two resonators.[10] Disadvantages of the latter approach, however, were indicated as inflexibility of design, large variation in nonreciprocity with frequency, and constructional difficulties.

Another approach to nonreciprocity has also been shown, namely, that of using the gyrator or 180°-phase nonreciprocity property of the orthogonal gyromagnetic coupler. Comstock showed the feasibility of this approach with his cryogenic 4-port circulator.[11] Although the phase shift through the gyrator is nonreciprocal, RF coupling to the gyrator is reciprocal.

Discussion

For the approaches considered in this paper, nonreciprocity was achieved by coupling the ferrimagnetic resonator with a circularly polarized RF magnetic field. This field is generated by utilizing stripline or miniature coaxial line construction. A dielectric loaded waveguide approach is also included. The process of generating the circular polarization obtains through 1) quadrature phase orthogonal RF loop coupling, or 2) TEM mode conversion. An elliptically polarized RF magnetic field is set up by a combination of linearly polarized fields in the former case and by a quasi-TEM mode in the latter case.

It is well known that an elliptically polarized RF field may be separated into a right-hand circularly polarized

Manuscript received May 17, 1967. The work reported in this paper was supported in part by the U. S. Department of Defense.
C. K. Greene is with the Air Force Avionics Laboratory, Electronic Technology Division, Wright-Patterson Air Force Base, Dayton, Ohio.
R. E. Tokheim, J. C. Hoover, and R. W. Peter are with the Solid-State Division, Watkins-Johnson Company, Palo Alto, Calif.

component which couples to the resonator, and a left-hand circularly polarized component which does not couple. The result given by Patel[12] for the external Q_e due to coupling of two orthogonal RF magnetic fields is

$$Q_{e_1} = \frac{Zo_1}{\gamma 4\pi M_s \mu_0 V(k^2_{x_1} + k^2_{y_1})},$$

where the sub-ones refer to the input coupling. A similar expression is obtained for the output coupling. The quantity Zo_1 is the impedance of the input line; $4\pi M_s$ is the saturation magnetization; V is the sample volume; and k_{x_1}, k_{y_1} are RF coupling constants defined by

$$k_{x_1} = \frac{h_{x_1}}{I}$$

and

$$k_{y_1} = \frac{h_{y_1}}{I},$$

where I is the total input current amplitude, and h_{x_1}, h_{y_1} are orthogonal RF magnetic field amplitudes.

The unloaded resonator Q_u for a typical filter design is about two orders of magnitude larger than the external Q_e. Thus, the approximate bandwidth B_d is obtained for a 2-stage double-tuned filter, assuming a factor of $\sqrt{2}$ for bandwidth broadening. Hence

$$B_d \approx \frac{2\sqrt{2}f_0}{Q_{e_1}} = \frac{2\sqrt{2}f_0 4\pi M_s \mu_0 V(k^2_{x_1} + k^2_{y_1})}{Zo_1}.$$

This relation is useful in comparing bandwidths of the different types of nonreciprocal coupling.

A 2-stage nonreciprocal ferrimagnetic filter with double-ended nonreciprocity is shown schematically in Fig. 1(a) as comprising two nonreciprocal coupling transmission line structures (NR), each coupled to a ferrimagnetic resonator (F). The coupling from one resonator to another is reciprocal and indicates the practical design approach. The equivalent representation for each coupling structure and resonator is shown in Fig. 1(b) as a resonance circulator (R). The equivalent, a 4-port circulator, is shown in Fig. 1(c).

Assuming symmetrical coupling with respect to the point O, where the resonators couple reciprocally to each other, the on-resonance asymmetrical insertion loss matrix shown in Table I is obtained.

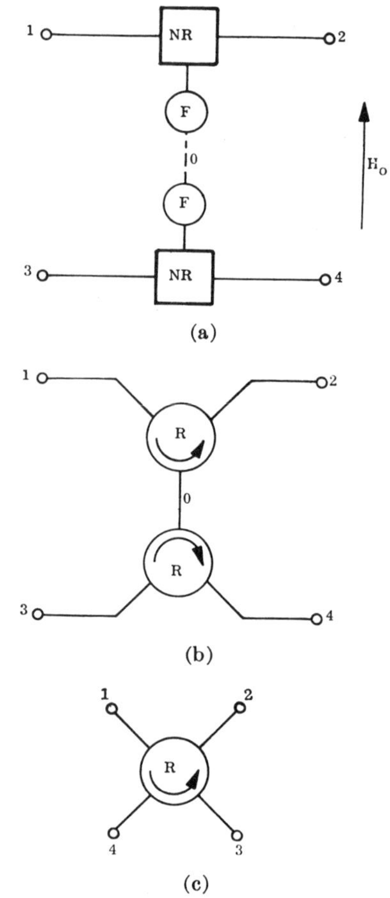

Fig. 1. Nonreciprocal filter equivalent representations.

The diagonal components of the matrix are zero, assuming that all ports are perfectly matched. The quantity L represents the insertion loss for each stage

$$L = L_{10} = L_{02} = L_{30} = L_{04}$$

and I represents the corresponding reverse isolation

$$I = I_{01} = I_{20} = I_{40} = I_{30}.$$

The values L' and I' are the insertion loss and isolation for each of the straight-through ports 2-1, or 4-3. The insertion loss L' of the straight-through ports is usually only a few tenths of a dB since energy does not couple to the resonator in the reverse direction.

TABLE I

Incident Power Port	On-Resonance Reflected Power Port 1	2	3	4	Incident Power Port	Off-Resonance Reflected Power Port 1	2	3	4
1	—	I'	$L+I$	$2L$	1	—	L'	I''	I''
2	L'	—	$2I$	$I+L$	2	L'	—	I''	I''
3	$L+I$	$2L$	—	I'	3	I''	I''	—	L'
4	$2I$	$L+I$	L'	—	4	I''	I''	L'	—

The nonreciprocal filter parameters of isolation $\mathcal{I}$, insertion loss $\mathcal{L}$, and nonreciprocity N in terms of the quantities defined above, are given by

$$\mathcal{I} = 2I$$

$$\mathcal{L} = 2L$$

$$N = 2I - 2L.$$

The off-resonance properties of the device are also indicated in Table I. The quantity I'' is the cross-coupling isolation, and is typically greater than 40 dB. Typical values for a nonreciprocal 2-stage filter are $L = 2$ dB, $I = 14$ dB, $I' = 15$ dB, and $L' = 0.3$ dB, thereby giving an isolation $\mathcal{I}$ of 28 dB, an insertion loss $\mathcal{L}$ of 4 dB, and a nonreciprocity N of 24 dB.

In rectangular waveguide the orthogonal RF magnetic field components are always in quadrature, but their relative amplitudes are frequency-dependent. In the TEM structures considered here, the phase is frequency-dependent and the relative field amplitudes are nearly constant.

It can be shown by considering the positive and negative circularly-polarized component amplitudes h^+ and h^- of an elliptically polarized RF magnetic field that the nonreciprocity N and the resulting insertion loss $\mathcal{L}$ for a double-ended nonreciprocal device are given by

$$N = 20 \log_{10} \left|\frac{h^+}{h^-}\right|^2$$

$$= 20 \log_{10} \frac{1 + e^2 + 2e \sin \phi}{1 + e^2 - 2e \sin \phi}$$

$$\mathcal{L} = 20 \log_{10} \left[1 + \left|\frac{h^-}{h^+}\right|^2\right]$$

$$= 20 \log_{10} \frac{2(1 + e^2)}{1 + e^2 + 2e \sin \phi}$$

where the ellipticity e is defined as the ratio of the orthogonal field amplitudes

$$e = \frac{h_{x_1}}{h_{y_1}} = \frac{k_{x_1}}{k_{y_1}},$$

and ϕ is the phase angle between the two field components. Figure 2 shows a plot of these expressions. Maximum directivity is obtained with equal amplitude components ($e = 1$) and quadrature phase ($\phi = \pi/2$).

Assuming that ϕ is proportional to frequency, maximum nonreciprocity over an octave requires $\phi = 60°$, $120°$ at extremes of the frequency range. Infinite nonreciprocity is never achieved in practice due to the finite cross-coupling isolation I''. Also, the ellipticity is, practically speaking, not unity on the average over the sphere due to unsymmetrical loop coupling.

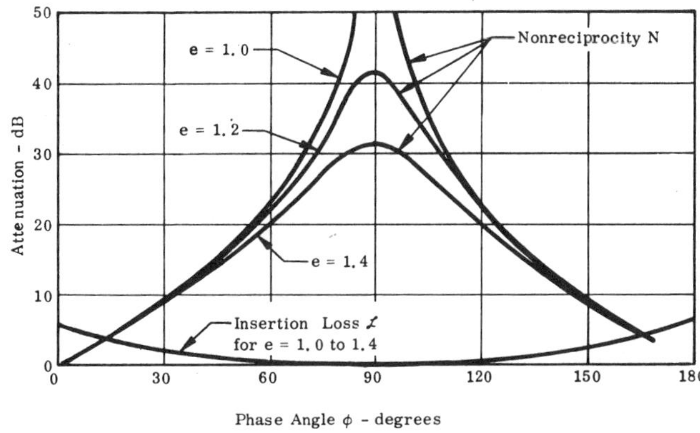

Fig. 2. Nonreciprocity and resulting insertion loss of double-ended nonreciprocal filter as a function of phase angle between orthogonal RF magnetic field components, with ellipticity as a parameter.

Quarter Wave Nonreciprocal Filter

An external view of the quarter wave nonreciprocal filter is shown in Fig. 3. In this design, double ground plane stripline was employed and coupling from one sphere to the other takes place through an iris coupling hole. In such a design, the top and bottom RF circuits are identical. Utilizing the equivalent circuit for each gyromagnetic coupler, the equivalent circuit for the quarter wave nonreciprocal filter will be as indicated in Fig. 4. The phase angle $\phi = 2\pi/\lambda$ is proportional to frequency and affects the degree of nonreciprocity and insertion loss of the filter. By considering gyromagnetic coupling coefficients and impedances, it can be shown that the effective coupling and hence bandwidth of the quarter wave approach is one-half that of the reciprocal filter. Data taken of nonreciprocity is plotted in Fig. 5. The design center frequency for this filter was 2.5 GHz. The nonreciprocity data closely follows the calculated curve shown in Fig. 2 for $e = 1.4$.

The off-resonance input VSWR and bandwidth are shown in Fig. 6. Fine structure variation in VSWR is due to linelength effects. The limiting level curve is shown in Fig. 7 for the coincidence mode limiting region. It may be noted that the quarter wavelength approach is similar to the resonant ring in that energy is circulated in the delay line by transferred energy through the resonator.

Dual Delay Nonreciprocal Filter

This RF filter housing design is the same as described previously for the quarter wave approach. The equivalent circuit for the dual delay approach is shown in Fig. 8. Quarter wavelength lines are inserted to provide the required quadrature phase relationship for nonreciprocity. The output quarter wavelength line functions primarily to provide a matched transmission line from ports 1 and 2 and 3 to 4 for off-resonance signals. Stripline power dividers are designed with 50-ohm input and 100-ohm output lines. It can be shown that the effective coupling in this case is one-fourth that of the reciprocal filter.

Measured data for nonreciprocity is shown in Fig. 9. The nonreciprocity is approximately as predicted in the previous section with $e = 1.3$. The on-resonance VSWR and bandwidth are shown in Fig. 10.

Fig. 3. Quarter wave nonreciprocal filter assembly.

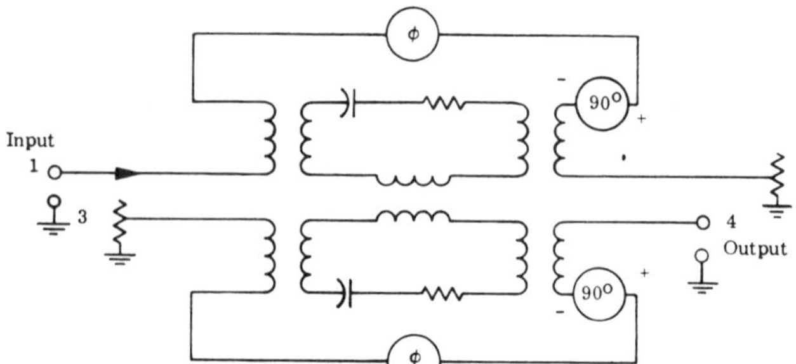

Fig. 4. Equivalent circuit of quarter wave nonreciprocal filter.

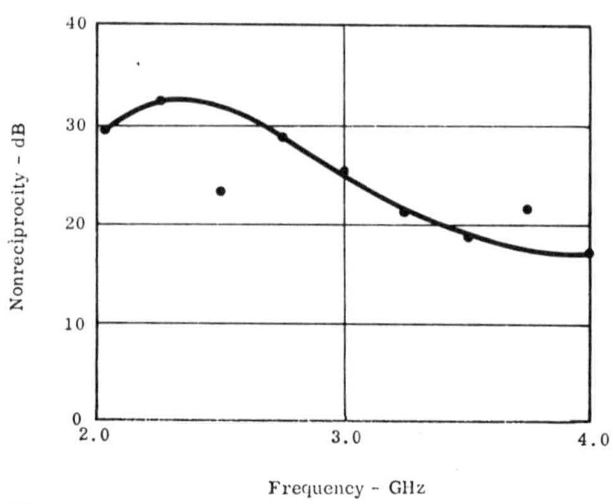

Fig. 5. Nonreciprocity and insertion loss of quarter wave nonreciprocal filter.

Three-dB Hybrid Nonreciprocal Filter

The 3-dB hybrid nonreciprocal filter assembly is shown in Fig. 11. As in the previous approaches, this filter also has double-ended nonreciprocity; that is, nonreciprocal coupling structures are employed in both input and output stages. An enlarged photograph of the RF structure is shown in Fig. 12, which indicates the miniature coaxial line construction. The equivalent circuit for the 3-dB hybrid nonreciprocal filter is portrayed in Fig. 13. It can be shown that the bandwidth of the hybrid approach is equal to that of the reciprocal filter.

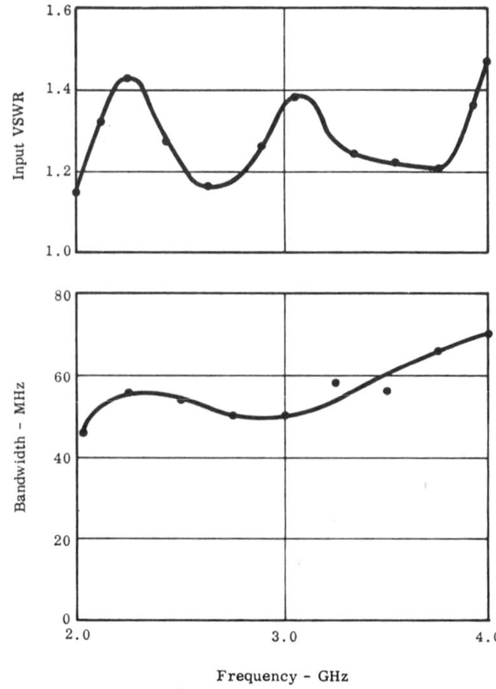

Fig. 6. The VSWR and bandwidth of quarter wave nonreciprocal filter.

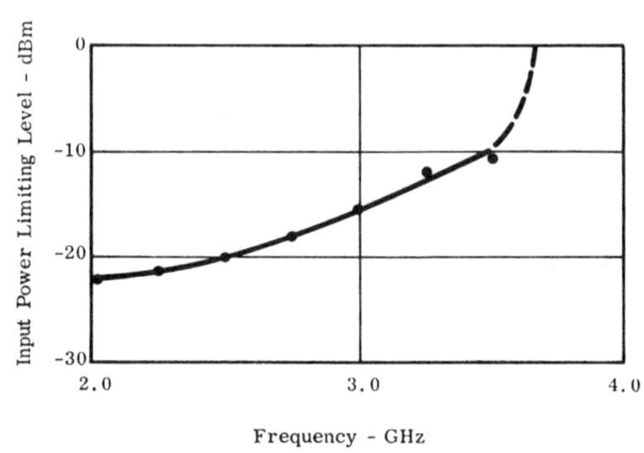

Fig. 7. Input power limiting level for quarter wave nonreciprocal filter.

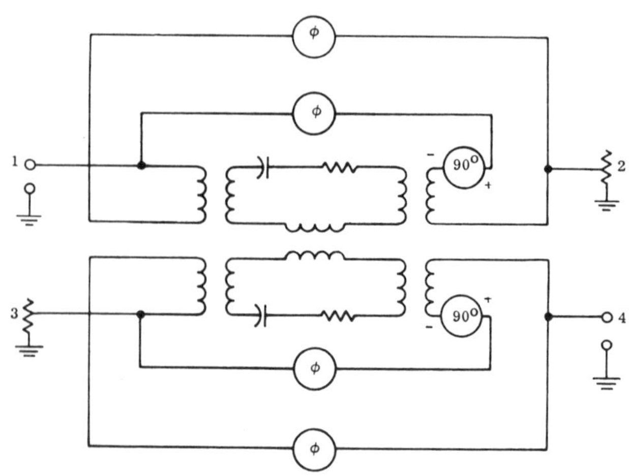

Fig. 8. Equivalent circuit of dual delay nonreciprocal filter.

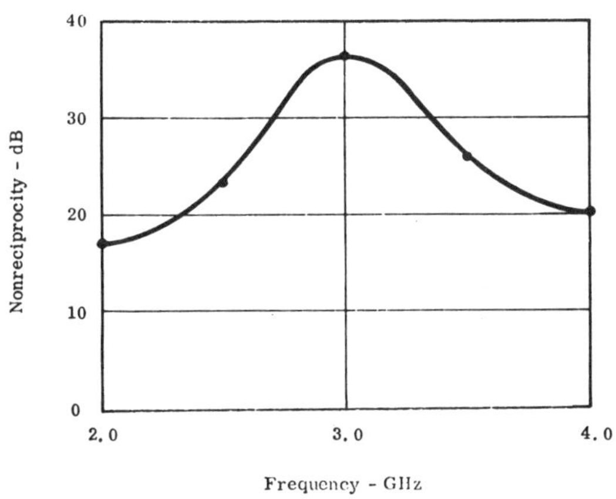

Fig. 9. Nonreciprocity and insertion loss of dual delay nonreciprocal filter.

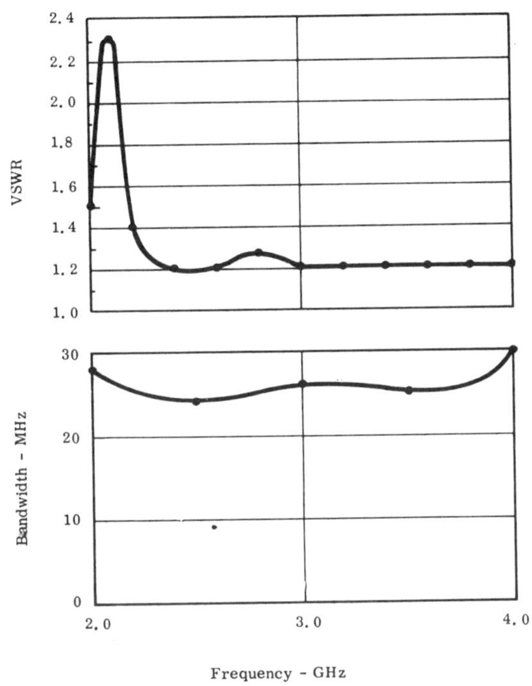

Fig. 10. The VSWR and bandwidth of dual delay nonreciprocal filter.

The curve exhibiting nonreciprocity shown in Fig. 14 indicates satisfactory performance over greater than one octave with nonreciprocity typically 30 dB. Such a filter will perform well over the useful range of the hybrid.

The bandwidth is shown in Fig. 15. Variation in bandwidth is due to VSWR mismatch and phase degradation effects in the 3-dB hybrids, as well as mismatch effects resulting from the RF construction. Such effects show up even more significantly above 4.0 GHz, which is the upper end of the specified hybrid range of operation. The effect on insertion loss is also evident in this range. Notice the significantly higher input power limiting level than was displayed for the quarter wave structure; the higher limiting level of the former case is due chiefly to the

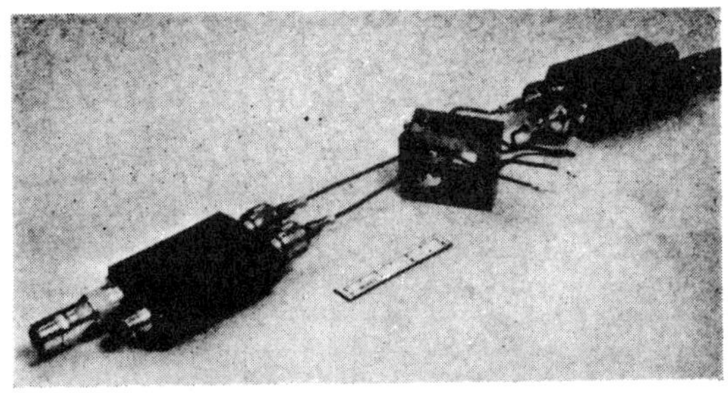

Fig. 11. A 3-dB hybrid nonreciprocal filter assembly.

Fig. 12. Enlarged view of RF structure of 3-dB hybrid nonreciprocal filter.

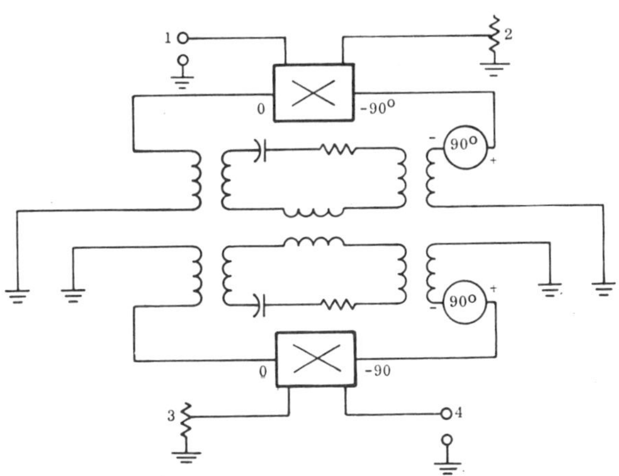

Fig. 13. Equivalent circuit of 3-dB hybrid nonreciprocal filter.

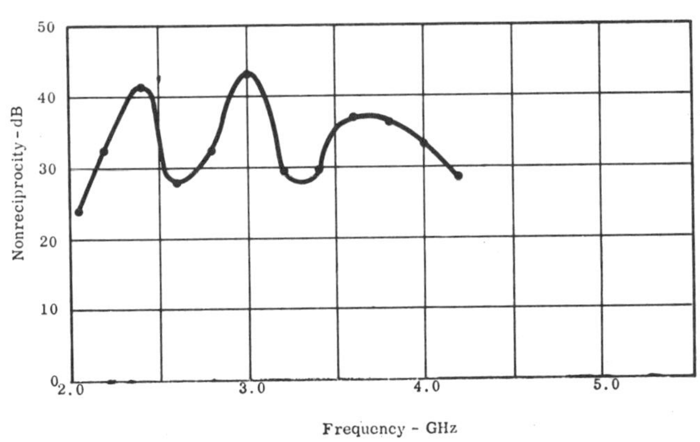

Fig. 14. Nonreciprocity of 3-dB hybrid nonreciprocal filter employing external 3-dB hybrids.

281

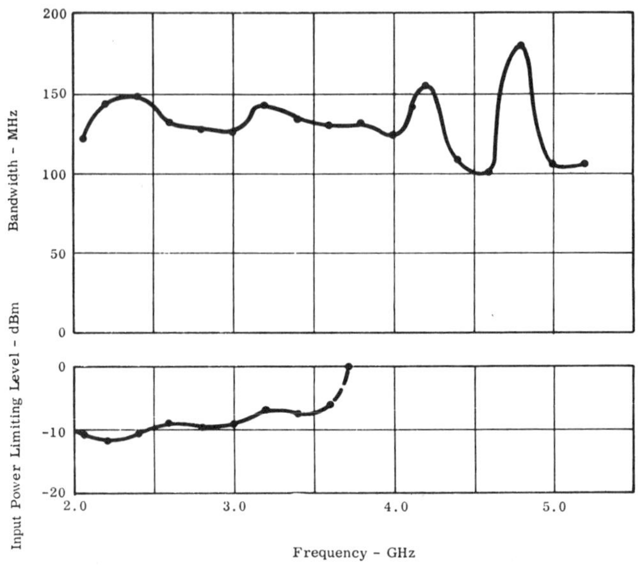

Fig. 15. Bandwidth and input power limiting level of broadband nonreciprocal filter employing external 3-dB hybrids.

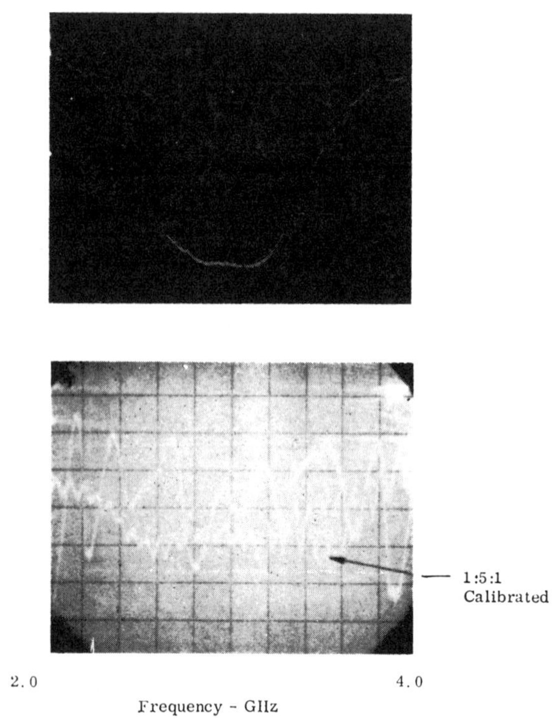

Fig. 16. Typical passband and VSWR of 3-dB hybrid nonreciprocal filter.

appreciably greater bandwidth. A typical passband and input VSWR are shown in Fig. 16.

Dielectric-Loaded Stripline Nonreciprocal Filter

The dielectric-loaded stripline nonreciprocal filter is sketched in Fig. 17. This approach utilizes Stycast Hi-K dielectric in a top-wall coupled, double ground plane stripline configuration. A 90° bend in the dielectric and stripline enhances the quasi-TEM mode conversion obtainable by approaches previously described by Fleri and Hanley[8] and Shaw.[10] As shown in Fig. 18, significant nonreciprocity was obtained over a wide frequency range and insertion loss was low over nearly two octaves. The bandwidth increased slightly over the range as shown in Fig. 19.

Nonreciprocal Band Reject Filter

The RF circuit configuration for the nonreciprocal band reject filter stripline design is shown in Fig. 20. The TEM mode conversion used here is effectively the same type of phenomenon as was achieved in the quarter wave nonreciprocal filter. However, the construction is simpler and lends itself particularly well to higher microwave frequency designs. The actual mode conversion takes place at the junction of the two orthogonal striplines. Simply stated, however, it is apparent that the length of the stub must be an integral multiple of 1/8 of a wavelength, thereby making the total phase delay a multiple of 90° at the design frequency.

Figure 21 shows the typical band reject response and the disassembled RF structure of the band reject filter.

Data for the 1/8 wavelength stub is given in Fig. 22. Greater than 14 dB of isolation was obtained over X-band. The low-input VSWR on resonance was quite good for this design; the off-resonance VSWR was about 1.3. The insertion loss was less than 1 dB over nearly half the range. This is somewhat larger than that obtainable with the previously outlined stripline approaches with the exception of the dielectric-loaded design. The additional losses incurred are due to the TEM mode conversion which takes place at the junction.

Additional data was taken, as shown in Fig. 23, for the 3/8 wavelength stub. In this case, the effect of increasing the stub length was to broaden the overall response and to increase the rate of isolation cutoff near the band edges.

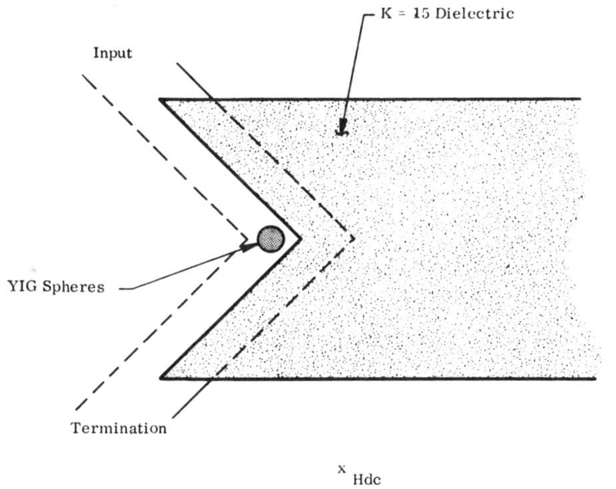

Fig. 17. Sketch showing top view of RF structure of 2-stage dielectric-loaded nonreciprocal filter.

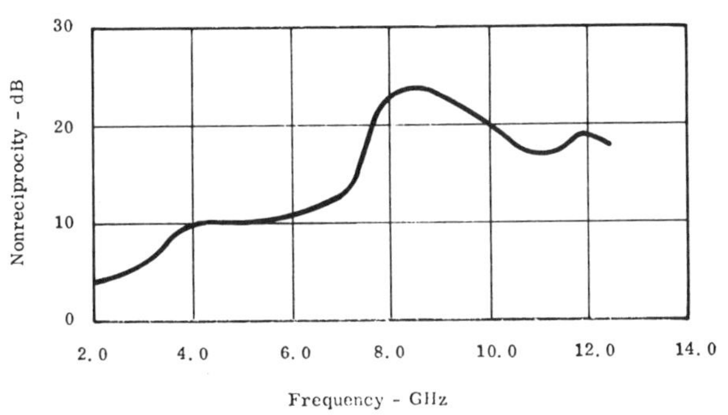

Fig. 18. Nonreciprocity and insertion loss of dielectric-loaded nonreciprocal filter.

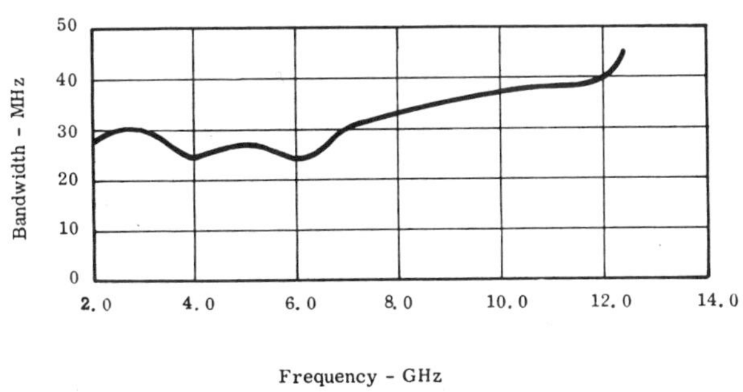

Fig. 19. Bandwidth of dielectric-loaded nonreciprocal filter.

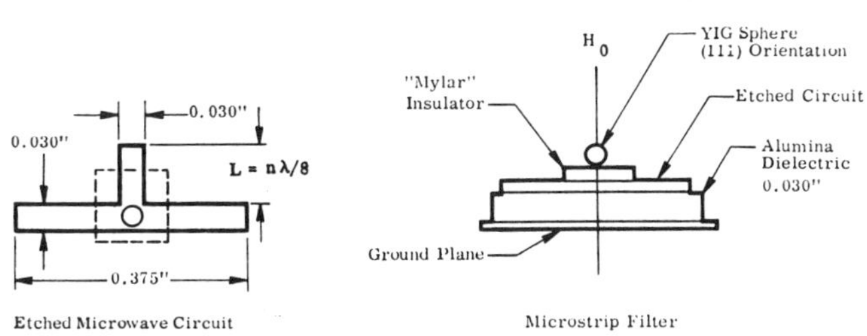

Fig. 20. An RF circuit configuration of nonreciprocal band reject filter.

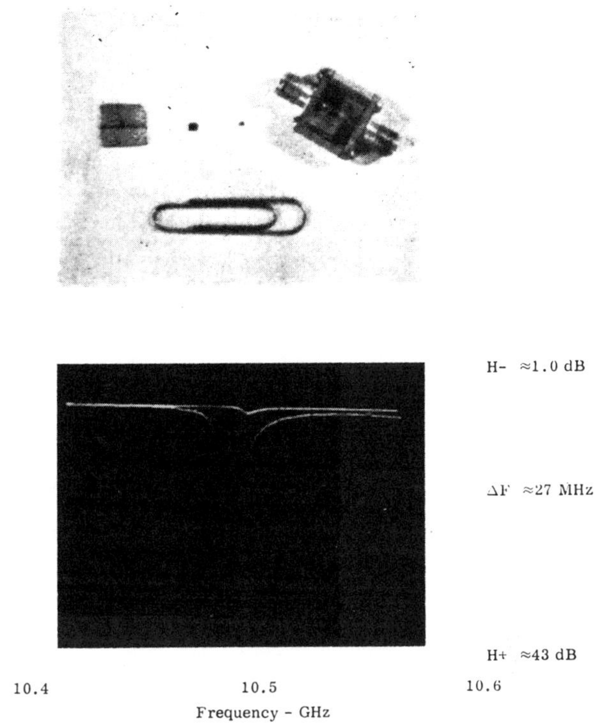

Fig. 21. Typical passband and assembly and disassembled RF structure of nonreciprocal band reject filter.

However, the tunable frequency range was reduced by a factor of four, over which a minimum of 10 dB of isolation was obtainable. Insertion loss, however, improved slightly.

283

Dielectric-Loaded Waveguide Nonreciprocal Filters

Dielectric-loaded waveguide filters have been made nonreciprocal for 2-stage filtering by employing top wall coupling, with YIG spheres placed at the position of maximum circular polarization at the design center frequency. By employing a dielectric having a constant of 15, it is possible to design an S-band bandpass filter with waveguide width less than one inch with a reduced height of 0.070-inch. The bandwidth for such a device with loss of 3 dB would be about 25 MHz. Such a design can be made quite small by utilizing a recent development for a miniature coaxial line to dielectric-loaded waveguide transition.[13] Such a transition is shown in Fig. 24. The design entails a three-step quarter wavelength transformer as a stripline metallized area on top of the waveguide. For S-band, such a transition is no longer than 0.75-inch due to the high K dielectric material. For octave bandwidth coverage it has been found possible to utilize a similar construction by employing ridge waveguide.

The chief advantage of using a dielectric-loaded waveguide is that very low insertion loss of the straight-through transmission line can be obtained. The insertion loss of the waveguide may be as low as 0.1 dB/in. As a

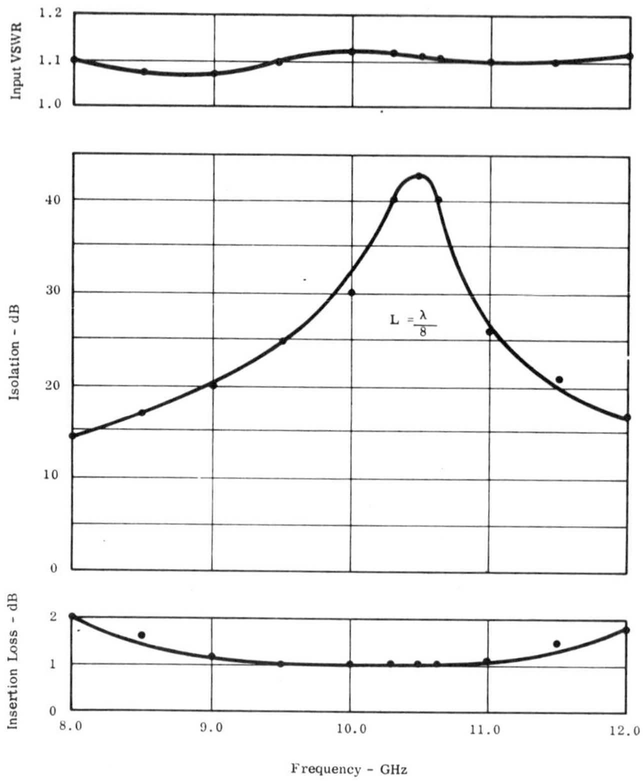

. 22. The VSWR, isolation, and insertion loss of nonreciprocal band reject filter.

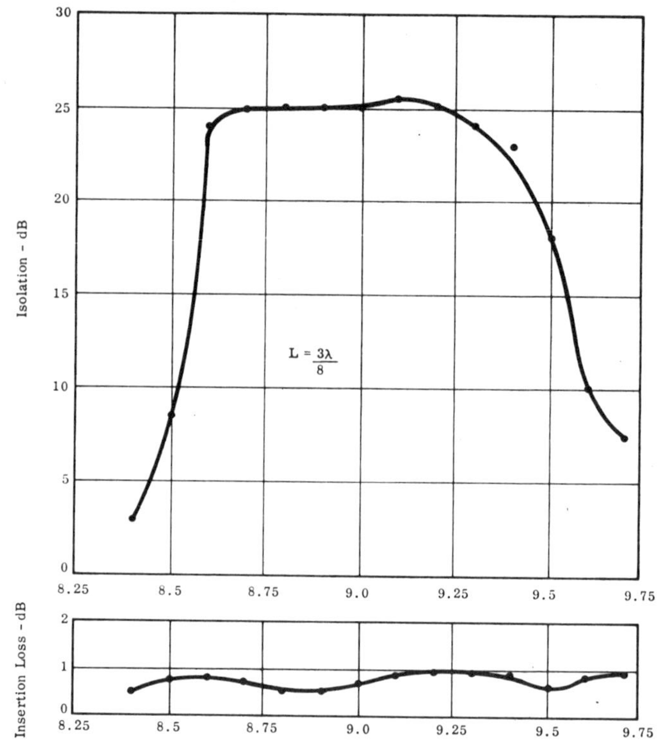

Fig. 23. Isolation and insertion loss of nonreciprocal band reject filter having $L = 3\lambda/8$.

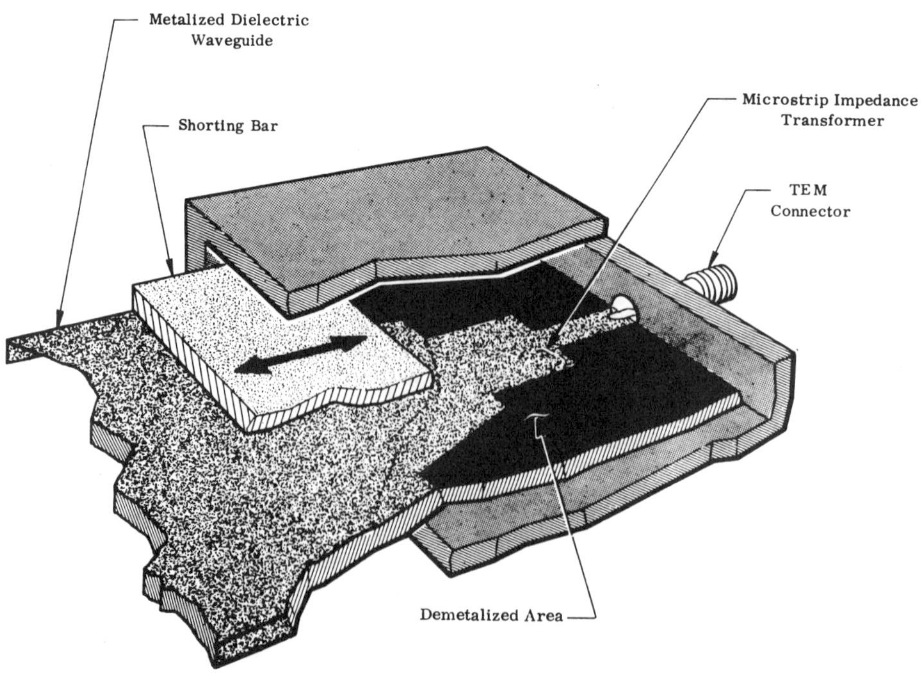

Fig. 24. A TEM adapter cross section.

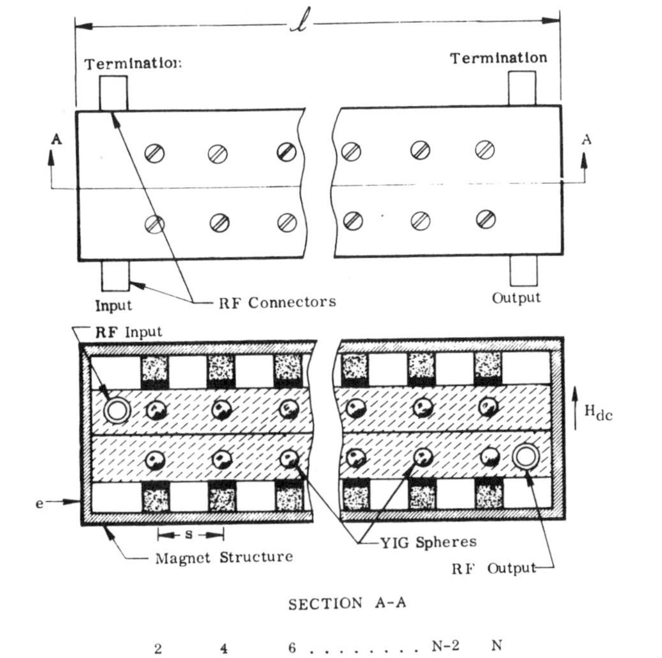

Fig. 25. Multiple channel dielectric-loaded waveguide filter-limiter.

result the waveguide approach is generally most useful for multiplexer and multiple channel filter devices.

Such a multiple channel dielectric-loaded waveguide filter-limiter was designed as sketched briefly in Fig. 25. Permanent magnet biasing was employed for fixed-tuned broad instantaneous bandwidth.

This filter-limiter had a 400-MHz instantaneous bandwidth in S-band with output limiting level less than −25 dBm. With a higher limiting level, much broader bandwidths would be possible since individual channel bandwidths would not be so narrow. Nonreciprocity was greater than 15 dB and insertion loss varied from 5.5 to 8.3 dB. A photograph of the swept frequency insertion loss is shown in Fig. 26. The fine structure is due to narrow bandwidth channels and associated pass-band overlap effects.

A typical limiting level curve for this device is shown in Fig. 27. Due to the large number of channels, off-resonance isolation reduced the effectiveness of the normally flat limiting level obtainable.

Such broadband limiters are useful in preventing saturation of a low-noise traveling-wave tube while also providing frequency-selective limiting. With a higher limiting level specified, fewer channels would make possible the scheme shown in Fig. 28, utilizing 3-dB hybrid nonreciprocal filters in cascade. Multiplexers, of course, could be similarly made with output lines separately terminated.

Applications

The most immediate application of the tunable 4-port nonreciprocal filter per se is as a preselector filter which does not require the usual isolators.

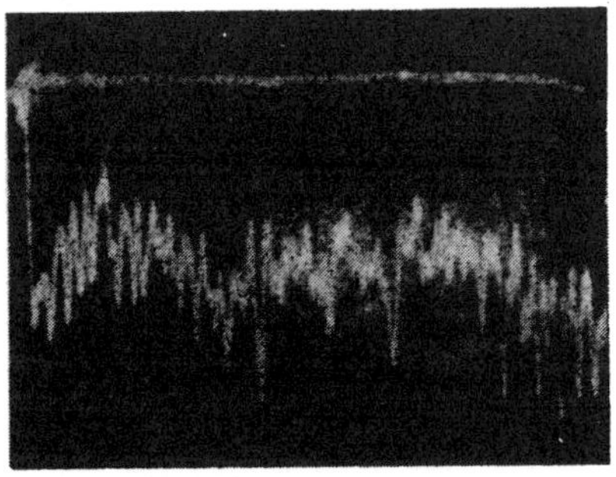

Fig. 26. Insertion loss versus frequency of multiple channel dielectric-loaded waveguide filter limiter.

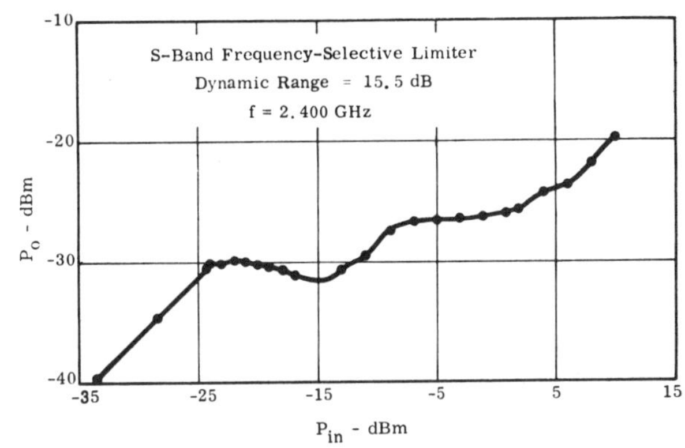

Fig. 27. Typical limiting curve of multiple channel dielectric-loaded waveguide limiter.

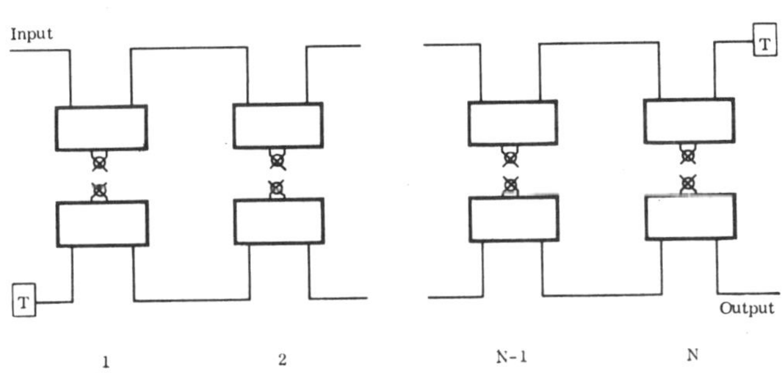

Fig. 28. Sketch of broadband limiter scheme indicating N cascaded wide bandwidth nonreciprocal hybrid filter channels.

As a tunable circulator and preselector, the nonreciprocal device is ideal for applications with a tunnel diode or other negative resistance amplifiers, parametric amplifiers, or phase-locked oscillators. The configuration suggested is shown in Fig. 29. Recall that preselection takes place between ports 1 and 4. Insertion loss between ports 4 and 3 is merely the straight-through loss L', a few tenths of a dB.

Also shown in Fig. 29 is a duplexer system. The 4-port nonreciprocal filter provides all functions of the system by acting simultaneously as a preselector, circulator, diplexer and even as a frequency-selective limiter when desired.

285

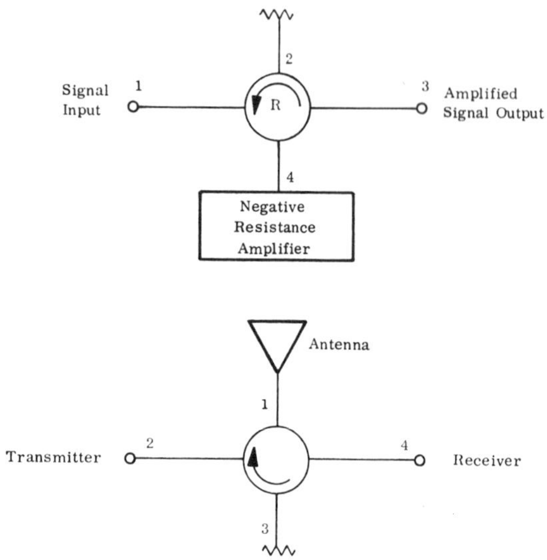

Fig. 29. Applications of 4-port nonreciprocal YIG filter-circulator.

Acknowledgment

The authors would like to thank G. F. Johnson for his assistance on the experiments, W. Capogeannis who assisted on the dielectric-loaded nonreciprocal filter, and C. Hughes and D. Brown who assisted on the dual-delay nonreciprocal filter.

References

[1] R. W. DeGrasse, "Low-loss gyromagnetic coupling through single crystal garnets," *J. Appl. Phys.*, vol. 30, pp. 155S-156S, April 1959.

[2] P. S. Carter, Jr., G. L. Matthaei, and W. J. Getsinger, "Design criteria for microwave filters and coupling structures," Stanford Research Institute, Menlo Park, Calif., Tech. Rept. 8, Contract DA-36-039-SC-74862, p. 3, October 1959.

[3] P. S. Carter, Jr., "Magnetically-tunable microwave filters using single-crystal yttrium-iron-garnet resonators," *IRE Trans. Microwave Theory and Techniques*, vol. MTT-9, pp. 252–260, May 1961.

[4] R. L. Comstock, "Synthesis of filter-limiters using ferrimagnetic resonators," *IEEE Trans. Microwave Theory and Techniques*, vol. MTT-12, pp. 599–607, November 1964.

[5] K. L. Kotzebue, "Frequency-selective limiting," *IRE Trans. Microwave Theory and Techniques*, vol. MTT-10, pp. 516–520, November 1962.

[6] C. L. Hogan, "The elements of nonreciprocal microwave devices," *Proc. IRE*, vol. 44, pp. 1345–1368, October 1956.

[7] B. J. Duncan, L. Swern, K. Tomiyasu, and J. Hannwacker, "Design considerations for broad-band ferrite coaxial line isolators," *Proc. IRE*, vol. 45, pp. 483–490, April 1957.

[8] D. Fleri and G. Hanley, "Nonreciprocity in dielectric loaded TEM mode transmission lines," *IRE Trans. Microwave Theory and Techniques*, vol. MTT-7 pp. 23–27, January 1959.

[9] C. N. Patel, "Investigations of magnetically tunable narrow bandpass nonreciprocal filters using ferrimagnetic resonators," Stanford Electronics Laboratory, Stanford University, Stanford, Calif., Tech. Rept. 411-1, Contract Nonr 255(24), NR 373 360, p. 93, April 24, 1961.

[10] A. W. Shaw, "Wide band electronically tunable filters," Rome Air Development Center, Griffiss AFB, N. Y., Contract AF 30-(602)-2625, RADC-TDR-64-170, May 1964.

[11] R. L. Comstock, "A cryogenic four-port circulator using single crystal yttrium iron garnet," *Proc. IEEE (Correspondence)*, vol. 51, pp. 1768–1769, December 1963.

[12] C. N. Patel, [9] p. 46.

[13] J. C. Hoover and R. E. Tokheim, "Microstrip transmission-line transitions to dielectric-filled waveguide," *IEEE Trans. Microwave Theory and Techniques (Correspondence)*, vol. MTT-15, pp. 273–274, April 1967.

Reprinted from IEEE TRANSACTIONS ON MAGNETICS
Vol. MAG-3, No. 3, September 1967
pp. 383–392

Equivalent Circuit of Orthogonal-Loop-Coupled Magnetic Resonance Filters and Bandwidth Narrowing Due to Coupling Inductance

PHILIP S. CARTER, MEMBER, IEEE

Abstract—**The equivalent circuit of an orthogonal-loop-coupled magnetic resonance filter is shown to consist of a gyrator, two ferrite-induced inductances, and two coupling loop inductances. The effects of the coupling inductances on the passband and stopband responses are shown to be significant by means of calculations based on this equivalent circuit.**

It is proved that the maximum passband bandwidth $\Delta f_{-3dB}=f_0 (L_f/L_c)$, where f_0 is the center frequency, and L_f and L_c the ferrite-induced and the coupling-loop inductance, respectively. Other unusual insertion-loss characteristics of this filter which differ from those of a conventional reciprocal-element bandpass filter are shown. Finally, a test circuit for determining experimentally the coupling inductance ratio L_c/L_f and the external Q, Q_f of a ferrite resonator is presented.

Manuscript received April 21, 1969; revised June 9, 1969.
The author is with Physical Electronics Laboratories, Menlo Park, Calif. 94025.

The Orthogonal-Loop-Coupled Ferrite Filter and its Equivalent Circuit

THE PURPOSE of this paper is to summarize and evaluate some numerical results recently obtained from a theoretical analysis of a fundamental type of magnetically tunable ferrimagnetic resonance filter as shown in Fig. 1. The circuit consists of two concentric coupling loops arranged at right angles to each other, with a small ferrite ellipsoid at the center of the loops. A dc magnetic field H_0 is applied along the axis of intersection of the two loops. A filter circuit which made use of this coupling principle was first proposed and demonstrated by DeGrasse.[1]

[1] R. W. DeGrasse, "Low-loss gyromagnetic coupling through single crystal garnets," *J. Appl. Phys.*, vol. 30S, April 1959.

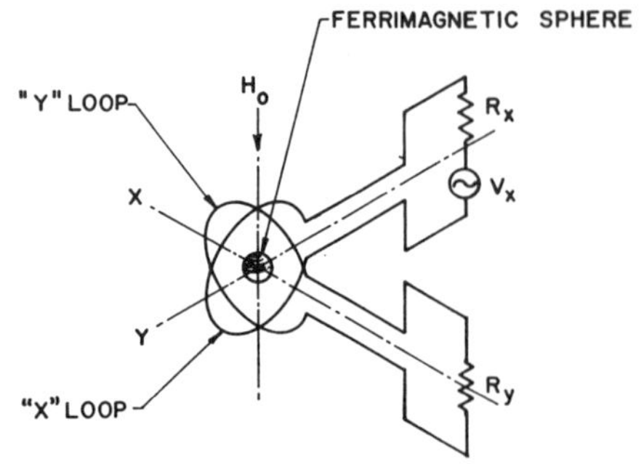

Fig. 1. Loop-coupled magnetic resonance filter.

It was later analyzed by the author.[2] According to this analysis, the open-circuit impedances Z_{xx}, Z_{yy}, Z_{xy}, and Z_{yx} are given by the following formulas.

$$Z_{xx} = j\omega L_{cx} + j\omega\mu V K_x^2 \chi_{xx}^e \tag{1a}$$

$$Z_{yy} = j\omega L_{cy} + j\omega\mu V K_y^2 \chi_{yy}^e \tag{1b}$$

$$Z_{xy} = j\omega\mu V K_x K_y \chi_{xy}^e \tag{1c}$$

$$Z_{yx} = j\omega\mu V K_y K_x \chi_{yx}^e \tag{1d}$$

where Z_{xx} and Z_{yy} are the open-circuit input impedances, and Z_{xy} and Z_{yx} are the mutual impedances. Also,

L_{cx}, L_{cy} = self inductance of loops in the absence of the ferrite resonators (henries)

ω = radian frequency of signal source, $2\pi f$ (radians/second)

μ = permeability of medium in which entire coupling unit is immersed (henries/meter)

K_x, K_y = ratio of field produced at center of ferrite ellipsoid to current flowing in either loop in the absence of the ferrite, and with a current generator connected to the loop in question[2] (meters^{-1})

χ_{xx}^e, χ_{yy}^e = diagonal components of RF external susceptibility of the ferrite ellipsoid (dimensionless)

χ_{xy}^e, χ_{yx}^e = off-diagonal components of RF external magnetic susceptibility of the ferrite ellipsoid (dimensionless)

V = volume of ferrite ellipsoid (meters^3).

MKS units are used exclusively in the formulas presented in this paper.

Let us consider the case, often encountered in practice, in which the effect on the filter response shape of the RF losses in the ferrite is negligible. Also, since a sphere is the resonator shape used almost exclusively in practical realizations of this type of filter, we shall narrow our considerations to that shape. For spherical resonators, the components of the external tensor susceptibility, χ_{xx}^e, χ_{yy}^e, χ_{xy}^e, χ_{yx}^e, are

[2] P. S. Carter, "Microwave filters and coupling structures," Stanford Res. Inst., Menlo Park, Calif., Rept. 7, SRI Proj. 2327, Contract DA36-039 SC87298, October 1960.

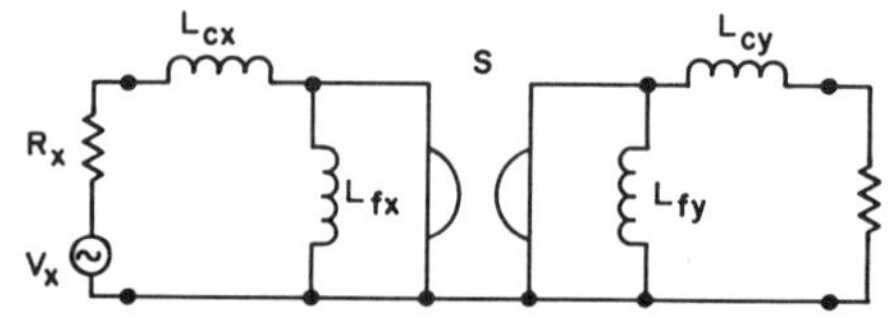

Fig. 2. Equivalent circuit of loop-coupled magnetic resonance filter.

given by the following relationships, in which the RF loss term has been omitted:

$$\chi_{xx}^e = \chi_{yy}^e = \frac{\omega_m \omega_0}{\omega_0^2 - \omega^2} \tag{2a}$$

$$\chi_{xy}^e = -\chi_{yx}^e = \frac{j\omega_m \omega}{\omega_0^2 - \omega^2} \tag{2b}$$

where

$\omega_m = \mu_0 \gamma M_0$

$\omega_0 = \mu_0 \gamma H_0$

μ_0 = permeability of free space $= 1.257 \times 10^{-6}$ henries/meter

γ = gyromagnetic ratio $= 1.761 \times 10^{11}$ radians/second/weber/meter^2

M_0 = saturation magnetization of ferrite, amperes/meter

H_0 = external, i.e., applied dc bias field, amperes/meter.

An exact equivalent circuit of this mathematical formulation is shown in Fig. 2. The open-circuit impedances of this circuit are given in terms of the circuit element values by the following formulas.

$$Z_{xx} = j\omega L_{cx} + \frac{j\omega L_{fx} S^2}{S^2 - \omega^2 L_{fx} L_{fy}} = j\omega L_{cx} + \frac{j\omega\omega_r^2 L_{fx}}{\omega_r^2 - \omega^2} \tag{3a}$$

$$Z_{yy} = j\omega L_{cy} + \frac{j\omega L_{fy} S^2}{S^2 - \omega^2 L_{fx} L_{fy}} = j\omega L_{cy} + \frac{j\omega\omega_r^2 L_{fy}}{\omega_r^2 - \omega^2} \tag{3b}$$

$$Z_{yx} = -Z_{xy} = \frac{S L_{fx} L_{fy} \omega^2}{\omega^2 L_{fx} L_{fy} - S^2} = \frac{\sqrt{L_{fx} L_{fy}}\,\omega_r \omega^2}{\omega^2 - \omega_r^2} \tag{3c}$$

where S, the gyration resistance of the gyrator, is related to the resonant frequency ω_r by

$$S^2 = \omega_r^2 L_{fx} L_{fy}. \tag{4}$$

The general form of (3) is the same as that of (1), after substituting (2). The magnetic resonance filter may therefore be represented by the circuit of Fig. 2, provided that

$$\omega_r = \omega_0 \tag{5a}$$

$$L_{fx} = \mu V K_x^2 \omega_m / \omega_r \tag{5b}$$

$$L_{fy} = \mu V K_y^2 \omega_m / \omega_r \tag{5c}$$

$$S = \mu V K_x K_y \omega_m. \tag{5d}$$

Different values of L_{fx} and L_{fy} are required to represent the filter at different center frequencies ω_r.

Effect of Coupling Inductances on Passband Characteristics

Some of the characteristics of this filter which are created due to the presence of the coupling loop inductances L_{cx} and L_{cy} will now be examined. An important consequence of their presence, which is demonstrated and discussed below, is the narrowing of the passband. Furthermore, as is also shown below, a limitation on the maximum bandwidth is created by the presence of the coupling inductance.

Analysis of the circuit for the commonly employed case where $L_{cx}=L_{cy}=L_c$, $L_{fx}=L_{fy}=L_f$, and $R_x=R_y=R_0$, yields the following formula for the power insertion-loss ratio (ILR), i.e., ratio of power output to the maximum (matched load) power output.

$$\text{ILR} = \left[4\left\{\frac{L_c}{L_f}\times\left(\frac{\omega}{\omega_r}-\frac{\omega_r}{\omega}\right)-\frac{\omega_r}{\omega}\right\}^2 + \left\{\frac{\omega_r L_f}{R_0}\left(1+\frac{2L_c}{L_f}\right)-\left(\frac{\omega_r L_c}{R_0}\right)\left(\frac{\omega}{\omega_r}\right)\cdot\left(\frac{L_c}{L_f}\right)\left(\frac{\omega}{\omega_r}-\frac{\omega_r}{\omega}\right) + \left(\frac{R_0}{\omega_r L_f}\right)\left(\frac{\omega_r}{\omega}\right)\left(\frac{\omega}{\omega_r}-\frac{\omega_r}{\omega}\right)\right\}^2\right]\Big/4. \quad (6)$$

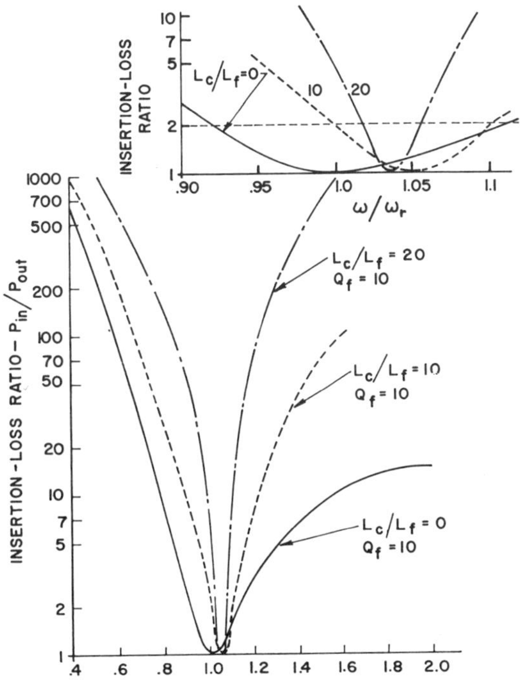

Fig. 3. Effect of coupling inductance on filter response.

The effect of the coupling inductance on the passband is shown in Fig. 3 where the ILR has been plotted for the case $Q_f=R_0/\omega_r L_f=10$. This parameter $R_0/\omega_r L_f$ corresponds to the external Q, Q_e of a passive parallel-resonant-circuit filter having an inductance L_f, a resonant frequency ω_r, and an external load R_0. The effect of the coupling inductance on the passband shape is seen to be considerable for the values of the ratio L_c/L_f chosen here. Note also the appreciable detuning. This detuning goes through a maximum at a value of $L_c/L_f=10$, after which it decreases with increasing L_c/L_f.

A curious property of this filter occurs in the case $L_c/L_f=0$. For this case, the ILR approaches a maximum value as $\omega\to\infty$ given by

$$\text{ILR}\big|_{\omega\to\infty} = \frac{\left(\frac{1}{Q_f}+Q_f\right)^2}{4}. \quad (7)$$

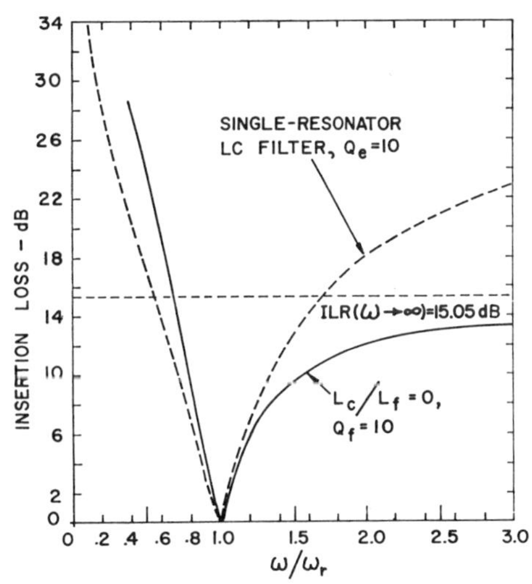

Fig. 4. Comparison between magnetic resonance filter and parallel-resonant-circuit filter.

This characteristic is illustrated in Fig. 4 for the case $Q_f=10$, for which case $10\log_{10}\text{ILR}_{\omega\to\infty}=15.05$ dB. For comparison the response of a parallel-resonant-circuit filter having the same external Q, $Q_e=10$ is also shown here.[3] Substantial differences between the two responses are evident.

That the insertion loss of a gyrator filter should approach a constant at very high frequencies can be made obvious from an inspection of the equivalent circuit of Fig. 2. Consider the case where $R_x=R_y=R_0$, $L_{cx}=L_{cy}=0$, for very high frequencies at which $\omega L_{fx}=\omega L_{fy}\gg R_0$. In this case, the insertion loss is a mismatch loss due to the gyrator. The impedance R_{in} looking into either terminal pair[4] is equal to $S^2R_0=R_0/Q_f^2$. For the case, $Q_f=10$, $R_{in}=R_0/100$. This mismatch loss is calculated to be equal to 15.05 dB.

A rule which is useful for determining whether or not the passband is narrowed appreciably by the coupling inductance can be stated as follows. The effect of L_c on the re-

[3] This is the single-ended external Q due to either source or load resistance. The loaded Q of the filter due to both source and load resistance would be half this value, or 5.

[4] R_{in} is the impedance presented at the "X" terminal pair with R_x and the generator disconnected, or the impedance presented at the "Y" terminal with R_y disconnected.

289

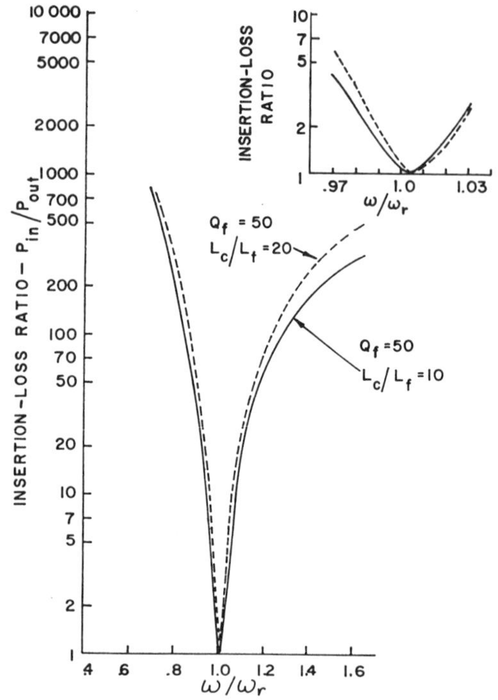

Fig. 5. Effect of coupling inductances when bandwidth is relatively narrow.

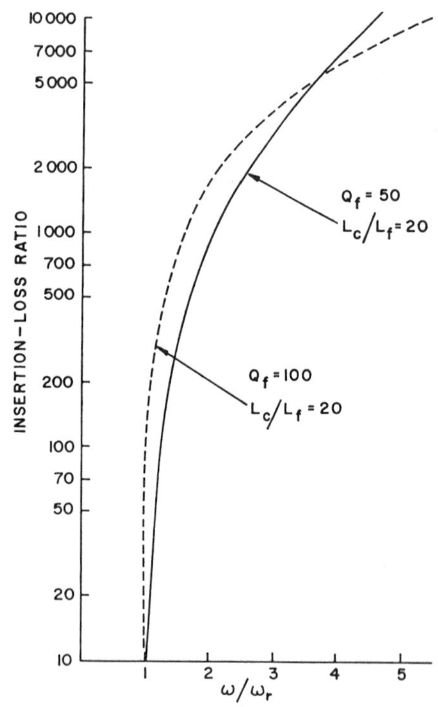

Fig. 6. Effect of coupling inductances on rejection band.

sponse shape is important in those cases for which the ratio L_c/L_f is of the same order as, or greater than, the external Q, Q_f. Fig. 5 shows response calculations for two cases in which $Q_f=50$, and $L_c/L_f=10$ and 20. It is seen from these calculations that the response is affected very slightly in both of these cases.

Fig. 6 shows an interesting effect which occurs in the rejection band on the high side of the center frequency. The attenuation characteristics of two filters having the same ratio $L_c/L_f=20$, but possessing different external Q's, $Q_f=50$ and 100, are seen to intersect. This effect is explained as follows. The insertion loss at frequencies far above resonance is due to the combined effects of the gyrator and the inductances L_f and L_c. As we observed previously, the effect on the insertion loss of the inductances L_f decreases with increasing frequency, until at very high frequencies, far above resonance, their effect is very small. At these frequencies the slope of the insertion loss versus frequency characteristic is determined mainly by the series coupling inductance L_c. A little analysis shows that L_c is larger for the case $Q_f=50$ than for $Q_f=100$, and therefore the slope of the insertion-loss curve for the $Q_f=50$ case will be larger. However, at frequencies close to resonance, the case $Q_f=100$ has the larger slope, the slope in this frequency range being determined almost completely by the resonator external Q.

It is useful to know to what extent the bandwidth may be increased by lowering the external load resistance R_0. It can be shown that a bandwidth maximum occurs as R_0 is decreased. An approximate analysis, thought to be quite accurate for practical purposes, has been made by noting from Fig. 4 that, near ω_r, the single-resonator parallel-resonant-circuit filter has approximately the same response as the ferrite-resonator filter. Suppose that coupling inductances L_c are placed in series in the input and output circuits of the parallel-resonant-circuit filter as shown in Fig. 6, analogous to the ferrite-resonator filter. It can be shown that the external Q, Q_e, of the parallel-resonant-circuit filter reaches a minimum, and consequently the bandwidth a maximum, when $Q_e=L_c/L_f$. The maximum fractional 3-dB bandwidth of the filter Δ_{3dB}^{max}/f is, correspondingly, equal to L_f/L_c.

Experimental Determination of Q_f and L_c/L_f

Numerical values of the external Q, Q_f, of the ferrimagnetic resonator, as well as the inductance ratio L_c/L_f, are required in order to predict the filter response. L_c/L_f can be measured as follows in the simple easy-to-construct test circuit shown in Fig. 7(a), along with the equivalent circuit of the device in (b), and a plot showing the insertion loss versus frequency in (c). The transmission response has two resonances, due to the parallel resonant circuit at ω_r, and a series resonance at ω_s. It can be shown that

$$\omega_s = \omega_r\left(1 + \frac{L_f}{L_c}\right)^{1/2}. \tag{8}$$

This test circuit is useful for examining the coupling constants of various loop and resonator configurations. From (8) we get

$$\frac{L_f}{L_c} = \left(\frac{\omega_s}{\omega_r}\right)^2 - 1 \tag{9}$$

so that L_f/L_c can be calculated from accurate measurements of ω_s and ω_r. In order to find Q_f, it is necessary to measure

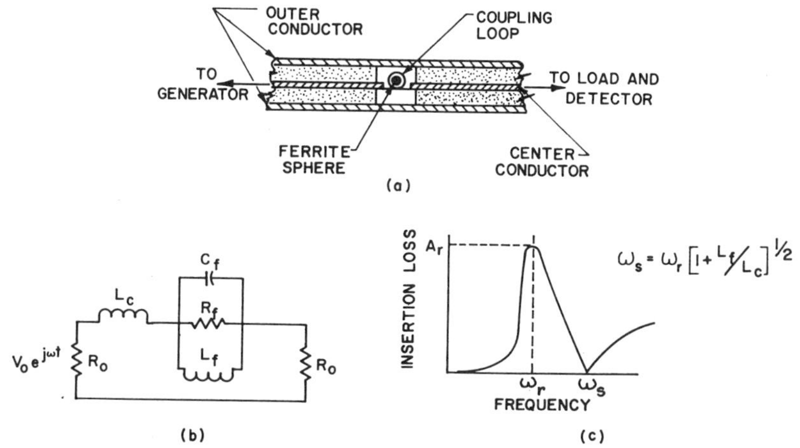

Fig. 7. Magnetic resonator test circuit. (a) Test circuit configuration. (b) Equivalent circuit of test device. (c) Insertion loss versus frequency of test circuit.

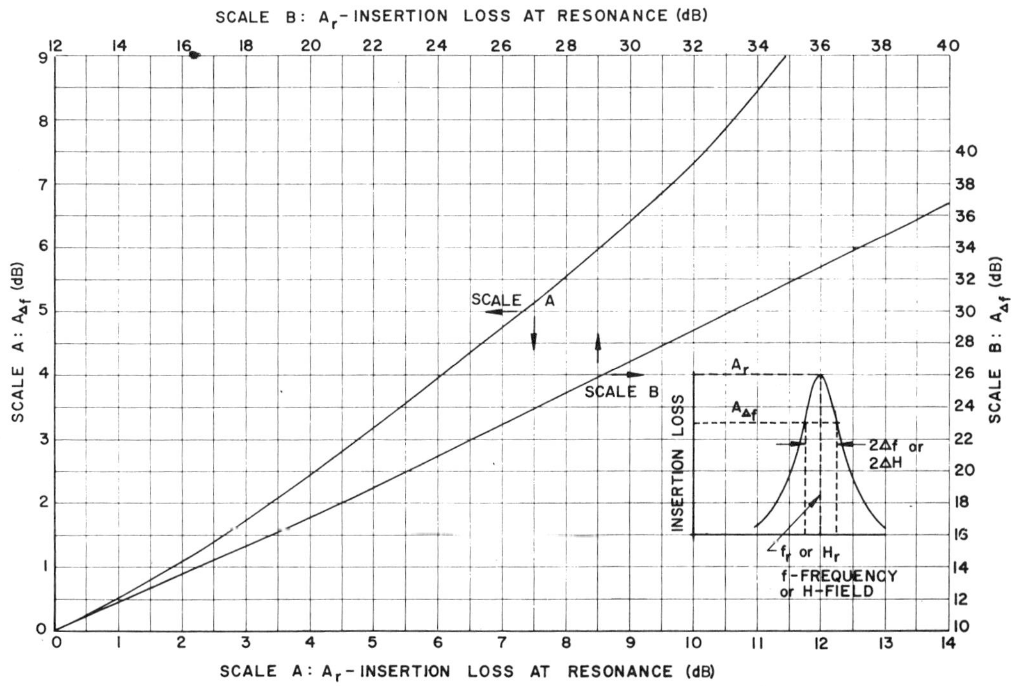

Fig. 8. Chart for use in measuring Q's of magnetic resonators.

the insertion loss A_r at resonance and at frequencies displaced from resonance. Fig. 8 gives a chart from which both Q_u and Q_f may be determined by the following procedure.

1) Measure the insertion loss A_r at resonance.
2) Determine $A_{\Delta f}$ from Fig. 8.
3) Detune the resonator by varying either the field H_0 or the frequency so that the insertion loss is decreased to the value $A_{\Delta f}$. Measure the frequency difference $2\Delta f$ (or the field difference $2\Delta H$) between the two frequencies (or fields) at which the insertion loss is equal to $A_{\Delta f}$.
4) Calculate the unloaded Q from

$$Q_u = \frac{f_r}{2\Delta f} \tag{10}$$

or

$$Q_u = \frac{H_r}{2\Delta H}. \tag{11}$$

5) Q_f can be calculated from

291

$$Q_f = \frac{Q_u}{10^{(A_r+6)/20} - 2}. \qquad (12)$$

The circuit analysis presented above can be extended to cases of multiple-resonator filters employing ferrite resonators. Phase-shift and impedance calculations, as well as insertion-loss calculations, are being carried out and will be presented in a paper which is now in preparation.

Summary and Conclusions

A circuit containing a gyrator and inductances only has been used to analyze the performance of a single-resonator ferrite resonator filter. The effect of the coupling inductance has been analyzed and a maximum bandwidth limitation criterion devised as follows. The maximum fractional 3-dB bandwidth that can be achieved by lowering the source and load resistances to achieve the lowest circuit Q is equal to the ratio L_f/L_c of the ferrite equivalent inductance L_f to the coupling inductance L_c.

An experimental technique by means of which the resonator Qs and the inductance ratio L_c/L_f can easily be measured has been presented.

Acknowledgment

The author appreciates the help of Dr. R. Craig and Dr. L. Young who made suggestions regarding this paper, and Miss M. Tabellija who typed the manuscript.

292

Some Design Considerations and Realizations of Iris-Coupled YIG-Tuned Filters in the 12-40 GHz Region

ROGER L. FJERSTAD, MEMBER, IEEE

Abstract—YIG filters covering the frequency ranges 12 to 18, 18 to 26, and 26 to 40 GHz have been designed and developed. The filters in the 18- to 40-GHz range were of three sections and achieved 80-MHz bandwidth and greater than 40 dB of higher order mode rejection. The filter in the 12- to 18-GHz region was of four sections utilizing staggered spheres in order to achieve a larger bandwidth filter with a minimum magnet gap. The bandwidth of this filter was 50 to 60 MHz with greater than 55 dB of higher order mode rejection. Useful information for the design of these filters was obtained with a special test fixture built to measure mutual coupling between spheres as a function of the angle between the line joining the sphere centers and the dc magnetic field. This data is reported for various iris sizes, shapes, and sphere spacings and indicates large differences in mutual coupling as the angle is varied from 0 to 90°. The effect of various types of discontinuities and iris sizes and shapes on higher order mode excitation and coupling are discussed as well as constructional problems and the design of high field magnets from vanadium permendur.

Introduction

DESIGNING bandpass filters utilizing yttrium iron garnet (YIG) involves calculations of the coupling of energy to the first YIG resonator and of coupling of energy between resonators. There are also various other considerations of importance such as temperature properties, unwanted mode suppression, signal handling capability, etc. Many of these design requirements have been considered by others. For example, Carter [1] has provided design relationships for coupling to the resonator, and Walker [2] has provided a means of identifying the higher order modes. Information on factors influencing temperature behavior and signal handling capability are also available [4], [7], [10], [11]. What is not available is adequate information on coupling between YIG resonators and means of designing filters for the minimum excitation and coupling of unwanted modes. One of the goals of this work was the determination of those physical parameters which affect the coupling of the fundamental mode and those which influence the excitation of spurious modes.

The major objective was the development of filters which cover the three waveguide bands from 12 to 40 GHz. A minimum bandwidth of 50 MHz was required in the 18- to 26.5- and 26.5- to 40-GHz ranges. In these two ranges the band center frequency had to be accurately selected with a linear selection voltage which had to be directly proportional to frequency. As a result of this requirement, a bandwidth somewhat broader than 50 MHz was desirable in order to allow for linearizer errors. A linearizer would be necessary due to large magnet hysteresis and nonlinearity in these two frequency ranges.

Manuscript received June 24, 1969; revised October 27, 1969. This work was supported in part by U. S. Army Electronics Command, Ft. Monmouth, N. J., under Contract DAAB07-67-0528.

The author is with Sylvania Electronics Systems, Mountain View, Calif.

One of the principal problems in designing YIG filters with broad bandwidths is the difficulty of achieving the required intersphere coupling while at the same time maintaining a sphere environment which does not excite higher order modes to an undesirable extent. With this problem in mind, efforts in the early development stages were aimed at learning as much as possible about the properties of direct sphere-to-sphere coupling of the fundamental mode and factors influencing the excitation and coupling of higher order modes. The parameters of interest are sphere-to-sphere spacing, iris size and shape, intersphere medium, and any possible relationships between filter properties and the direction of the intersphere axis relative to the dc magnetic field. A stripline test fixture was constructed in which a few of these parameters could be varied in order to determine an optimum coupling arrangement. The information obtained from these measurements proved useful in designing filters with maximum bandwidth and minimum higher order mode excitation and coupling.

The following discussion of 12- to 40-GHz filters includes the development of an 80-MHz bandwidth three-section YIG filter in the 18- to 26- and 26- to 40-GHz bands and a novel four-section filter in the 12- to 18-GHz range utilizing staggered spheres to achieve a larger bandwidth while minimizing the size of the magnet gap. Also discussed are construction problems and the design of high-field magnets from vanadium permendur.

Coupling Between Spheres

The data to follow include mutual coupling between spheres as a function of the angle between the line joining the sphere centers and the dc magnetic field. These data were taken for two circular and two slot irises and two sphere spacings. The primary purpose of the investigation was mutual coupling of the fundamental mode. Some useful information was also obtained on coupling of unwanted modes.

The first observation of coupling versus angle was made in a TE_{10} waveguide filter. Fig. 1 shows schematic side views of the two-section TE_{10} waveguide filter which was constructed

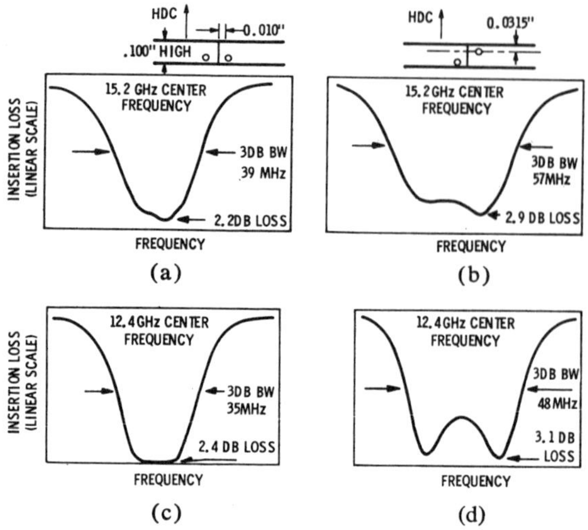

Fig. 1. 12- to 18-GHz two-section filter.

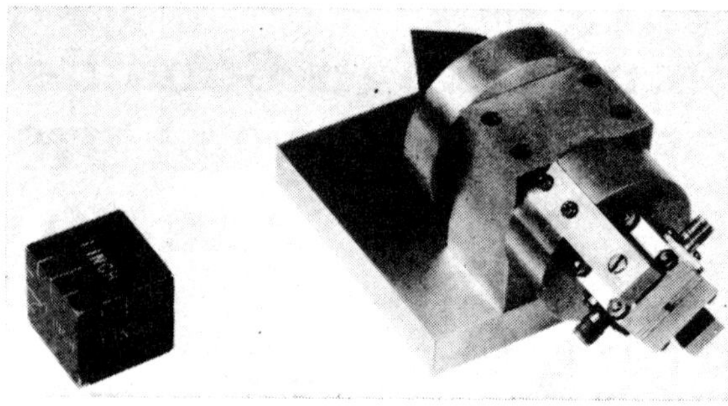

Fig. 2. Device for measuring angle dependence of sphere-to-sphere coupling.

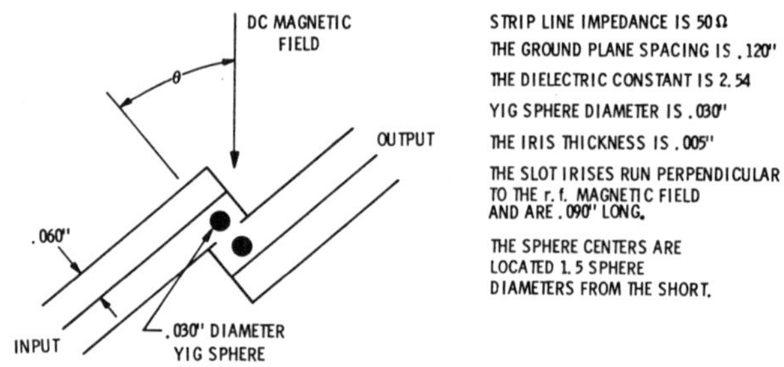

Fig. 3. Schematic of stripline filter used for measuring angle dependence.

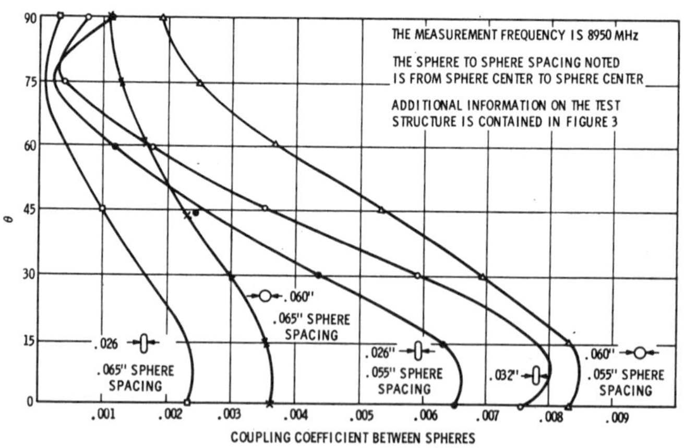

Fig. 4. Coupling coefficient between spheres as function of θ, iris shape and size, and distance between spheres.

for the 12- to 18-GHz range. The direction of the dc magnetic field is perpendicular to the broad waveguide wall. The iris was a narrow slot iris of 0.025-inch width, the length running the full waveguide height, and the spheres (0.030-inch diameter) were located halfway across the broad waveguide dimension. In the narrow waveguide dimension the spheres were located one third of the waveguide height from the bottom, as shown in the upper left of Fig. 1. Further description of this filter can be obtained from the description of the four-section filter as shown in Figs. 7 and 8. The first two sections of the four-section filter are essentially the same as the two-section filter shown in Fig. 1. Fig. 1(a) and (b) shows the passband. At this stage of development the filter was examined in the arrangement shown in the upper right of Fig. 1 with the second sphere located one third of the waveguide height from the top. The passband was again measured as shown in Fig. 1(c) and (d). It should be noted that in spite of the fact that the spheres were farther apart in the staggered position, the mutual coupling was much stronger.

In order to further investigate this phenomenon of directional coupling between spheres, the fixture in Fig. 2 was constructed. This consists of a two-section stripline filter (shown schematically in Fig. 3) and a device for mounting and rotating the filter in the dc magnetic field of a large laboratory electromagnet. With this large magnet the applied magnetic field was uniform over the entire volume of space occupied by the filter. Fig. 4 shows plots of the measured coupling coefficient as a function of angle θ, that is, the angle between the dc magnetic field and the line joining the sphere centers. The data are recorded for various iris shapes and sizes and intersphere spacing. The ratio of the mutual coupling at θ equal 0° to the coupling at θ equal 90° is seen to be a number greater than unity, with a minimum value of this ratio of 3.3 for the circular iris with 0.065-inch sphere spacing and a maximum of 9.7 for the slot iris with 0.055-inch sphere spacing. Errors in the determination of the coupling, especially near 0°, would possibly be the cause of some of this range of maximum to minimum coupling. Another cause may be related to the various iris sizes and shapes, with the slot iris having somewhat larger variations than the circular iris. The variation in coupling between spheres as a function of angle (independent of iris configuration) is believed to be due to a very basic property of a system which can be described to some approximation by a pair of hertzian dipoles as will now be explained.

After discovering this dependence of coupling on angle empirically, an article in the *Radiotekhnica i Elektronika* [5] (*Radio Engineering and Electronic Physics*) was discovered wherein the authors solved for the coupling between two magnetic dipoles, spaced far apart compared to their dimensions, for two configurations corresponding to θ equal to 0°

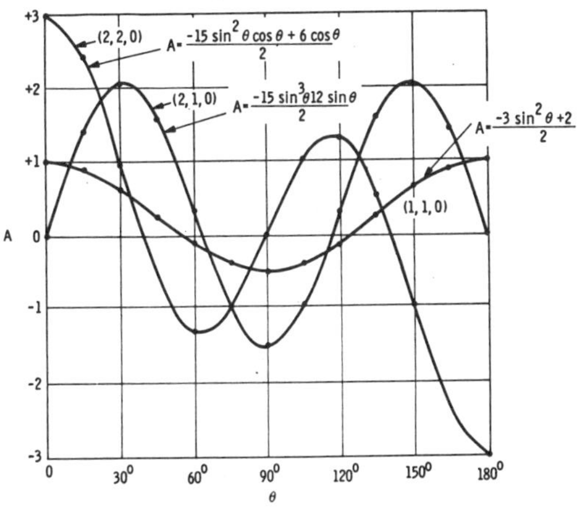

Fig. 5. Quantity A (proportional to circularly polarized component of field h) versus θ.

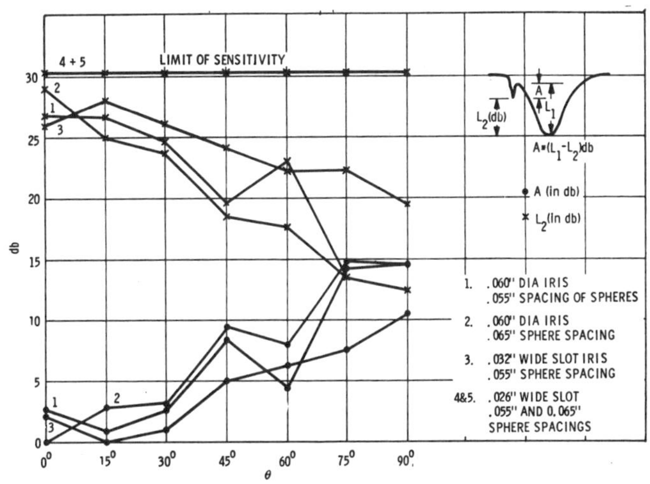

Fig. 6. Level ($L2$) and amplitude (A) of (2, 2, 0) mode versus θ and various irises and sphere spacings.

and 90°. The difference in coupling between the two configurations, as derived from their equations using the 0.055-inch sphere spacing of the filter of Fig. 3, was found to be 1.7 to 1, indicating that some variation of coupling exists even without the effect of an iris.

This angular dependence of the coupling of the various modes, including the fundamental or (1, 1, 0) mode, is also somewhat understandable from an examination of the magnetic field equations as derived from the magnetostatic approximation, that is, such that $\nabla \times H = 0$ implying that $H = \nabla \psi$ can be written. Fletcher and Bell [3] have derived the scaler potential function ψ for the various modes for a sphere. Their scaler potential function equations can be taken a step further by taking their gradient, which gives the magnetic field, and then solving for the correctly polarized component to couple to the other sphere. This has essentially been done by Egan [17] for the (1, 1, 0) mode. If this is done, the magnetic field components which couple for the three modes (1, 1, 0), (2, 2, 0), and (2, 1, 0) are, respectively,

$$h_1{}^1 = \frac{D_1}{r^3}\left(\frac{-3\sin^2\theta + 2}{2}\right)e^{j(\phi+\omega t)}$$

$$h_2{}^2 = \frac{D_2{}^2}{4}\left(\frac{-15\sin^3\theta\ 12\sin\theta}{2}\right)e^{j(2\phi+\omega t)}$$

$$h_2{}^1 = \frac{D_2{}^1}{r^4}\left(\frac{-15\sin^2\theta\cos\theta + 6\cos\theta}{2}\right)e^{j(\phi+\omega t)}.$$

The quantities in parentheses are plotted in Fig. 5. These quantities cannot be used alone to describe the phenomena in Figs. 4 and 6 which are the result of many influencing factors; nevertheless, they may be useful in helping to understand the more complicated "practical" case. Of particular interest is the 2:1 ratio of field strength indicated between $\theta = 0$ and 90° for the (1, 1, 0) mode, where a 3.3:1 ratio of coupling between spheres was noted in Fig. 4 for the case of larger sphere spacing and iris diameter. Also of interest are the variations in the (2, 1, 0) and (2, 2, 0) modes. These will be discussed in some detail in the section on higher order modes.

The Coupling Equation

The coupling coefficient between two YIG spheres in space is [5], [12]

$$K = \frac{C 4\pi M_s V}{f_0 r^3}$$

where

V = volume of a sphere,
r = distance from center of one sphere to center of the other,
$4\pi M_s$ = saturation magnetization of resonator,
C = constant,
f_0 = resonant frequency of resonator.

The constant C is determined by Mikaelyan and Ya'Anton'yants [5] for the two angles θ equal to 0° and 90°. For the configurations shown in Fig. 4 the constant C can be determined for any angle using the above equation and the data of Fig. 4. A couple of simple calculations using the coupling equation and Fig. 4 will also show that the power of r may be more than 3 for small sphere spacings.

Higher Order Modes

Some useful information concerning the (2, 2, 0) and (2, 1, 0) modes (defined in [2]) was also obtained from the experiment. The (2, 2, 0) mode level was found to vary in a manner opposite to that of the fundamental mode coupling with changes in θ. The (2, 2, 0) mode level was also found to be sharply reduced in the case of a slot iris which is slightly smaller in width than the sphere diameter. These facts are described in Fig. 6 which shows the amplitude A and level L_2 of the mode as a function of θ, sphere spacing, and iris. The reduction of (2, 2, 0) mode amplitude as θ approaches 0° is quite clear. Again, $\theta = 0°$ is a position of large fundamental mode coupling. The amplitude of the (2, 2, 0) mode is shown to be unmeasurable for all angles with the 0.026-inch wide slot iris (0.030-inch diameter sphere) for both sphere spacings. Although the attenuation level of the (2, 2, 0) mode was rapidly increased as the slot width was changed from 0.002-inch larger to 0.004-inch smaller than the 0.03-inch sphere diameter, the coupling of the fundamental and isolation outside the passband were left relatively unchanged. Measurements of the excitation or single sphere external Q of the (2, 2, 0) mode showed that the external Q was not significantly different for the various iris shapes. For example, a (2, 2, 0) mode reflection loss of approximately 0.8 dB was observed at $\theta=90°$ and 45° for all the configurations tested. This indicates that the reduction in (2, 2, 0) mode amplitude as θ is changed from 90° to 45° may be due to a change in intersphere coupling of the mode.

Since the coupling between spheres is proportional to the field strength, reference to Fig. 5 may be helpful in understanding one contributing factor to the observed data. The 0 and 90° point for the (2, 2, 0) mode is in agreement as a contributing factor to the data of Fig. 6 in that the field is weakest at the point corresponding to $\theta=0°$. A similar agreement was observed in the case of the (2, 1, 0) mode which came through strongly for $\theta=0°$ and weakly for $\theta=90°$. In the case of the (2, 1, 0) mode, however, strong differences in apparent mode excitation were noticed between $\theta=0$ and 90° showing the mode most strongly excited at $\theta=0°$. In the case of the (2, 2, 0) mode, as θ approached 0° the reflection loss of the (2, 2, 0) mode reduced to about 0.1 dB. This again appears to be a separate phenomena and may possibly be due to the fact that the RF exciting magnetic field, determined by the transmission line shape, is not perfectly linear over the sphere dimensions due to the finite stripline width, and as a result the configuration perpendicular to the dc magnetic field changes somewhat as the device is rotated. Other modes were problems, but their cause was usually linked with structural imperfection. For example, sphere placement and iris size were critical on the 26- to 40-GHz units to ± 0.0002 inch, and misplacements of ± 0.0005 inch caused the excitation of various modes. Poor electrical contact between the iris plate and the TE_{10} waveguide walls invariably generated crossing modes. More will be said on the mechanical problems and critical sphere placement in the following discussion of 12- to 40-GHz filters.

12- to 18-GHz Staggered In-Line Filter

The greater coupling in the over-under position was utilized to advantage in filter designs in the 12- to 40-GHz range. Figs. 7 and 8 show a compact four-section 12- to 18-GHz filter using a staggered sphere arrangement which utilizes some of the larger over-under coupling, yet still allows a fairly small magnet gap. This reduces the package size considerably (at least by a factor of 7) from a strictly four-section over-under arrangement. The top and side views of Fig. 8 show that narrow slot irises were used between the first and second, and between the third and fourth spheres. The very large and thick iris is used in the center in an attempt to make boundary conditions for all spheres equal, which seems to be of some importance in building YIG filters which minimally excite and couple unwanted modes. The actual filter construction shown in Fig. 7 is of plated plastic. The structure in the foreground is an unplated filter. The irises were formed by sawing slots in from the side and then filling the slots with plating. The structure in the background provides input and output waveguide and support. This filter achieved a bandwidth of 50 to 60 MHz with greater than 55 dB higher order (unwanted) mode rejection as shown in filter response (Fig. 9). The only higher order mode to present a problem in the development of this filter was the (2, 2, 0) mode lying about 320 MHz on the high side of the principal response. Again, the iris slot width proved to be of critical importance in reducing the magnitude of this mode. An iris slot width of 0.030 inch or greater

Fig. 7. 12.4- to 18-GHz four-section TE_{10} waveguide filter using staggered spheres.

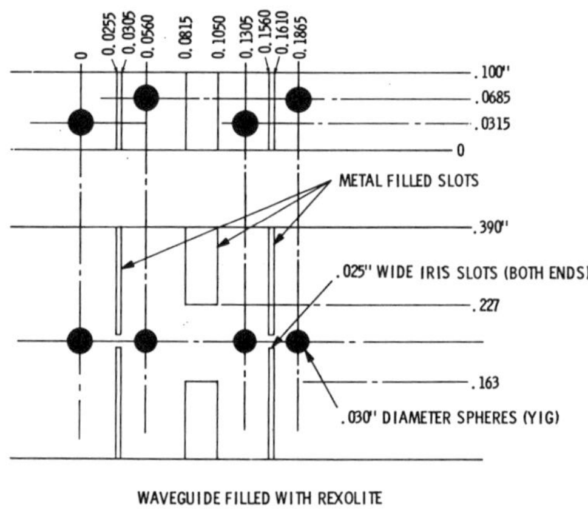

Fig. 8. 12- to 18-GHz four-section TE_{10} waveguide filter using staggered spheres.

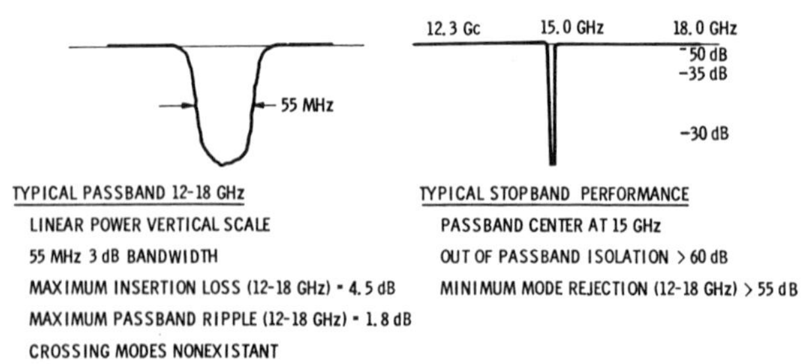

Fig. 9. Four-section TE_{10} staggered sphere filter performance.

attenuated the mode by less than 40 dB. The actual iris used, which attenuated the (2, 2, 0) mode by more than 55 dB, was 0.025 inch, i.e, only 0.005 inch smaller than the 0.030-inch iris!

Higher order modes which cross the fundamental as it is tuned were noticed in all of the filters that were constructed but were found in all the 12- to 40-GHz filters to be caused by inadequate contact between the iris and the waveguide. The use of soft annealed copper iris plates is most helpful in eliminating this problem. More will be said on the (2, 2, 0) mode as well as the (5, 4, 0) and (2, 1, 0) modes in the following discussion on 18- to 40-GHz three-section filters.

18- TO 40-GHz FILTERS

A set of filters was designed for the 18- to 40-GHz range. Due to the large bandwidth requirement which was 80 MHz, the spheres were arranged in the over-under configuration for maximum coupling between spheres. Fig. 10 shows the 26- to 40-GHz three-section over-under filter. The iris plates are not shown. The figure shows the input tapering from standard-height to reduced-height guide in Rexolite. The spheres are located halfway across the guide, and the iris is a narrow slot iris running perpendicular to the direction of the microwave magnetic field. The 18- to 26-GHz filter is similar with the exception of the input taper to reduced-height guide being replaced with a step transformer. Fig. 11 shows the dimensions of the filter coupling regions for 18- to 26-GHz and 26- to 40-GHz. The filter housings were made from a 70–30 copper nickel alloy which has a resistivity of about 36.3 Ω·cm compared with 2.8 Ω·cm for aluminum. This high-resistivity material was chosen in order to reduce the eddy current losses in the housing, which it does in direct relation to the resistivity when the magnetic field is swept. It is also very nonmagnetic and fairly easy to machine. For comparison, it is much easier to machine and far less magnetic than the nonmagnetic stainless steels. Its disadvantage is that it is a comparatively soft metal and care must be taken when machining and handling in order to avoid warpage. Even so, it is much easier to hold tolerances in this material than it is in the "platable plastics."

Fig. 12 shows a typical 18- to 26- or 26- to 40-GHz passband shape. The maximum insertion loss of either filter over their frequency ranges was 3.8 dB. The maximum passband ripple was 1.0 dB in the 26- to 40-GHz unit and 1.8 dB (just at the low end of the band) in the 18- to 26-GHz unit.

Temperature control of the center sphere of the three-section filter is required because the center sphere sees a different dc magnetic field than the two outside ones as a result of the dc magnetization of the outside spheres. The microwave field interaction between the spheres is also enough to cause a significant pulling of the center sphere relative to the outside ones. The total shift of the center sphere relative to the outside ones is about 90 MHz, of which approximately half was shown to be due to dc interaction and half microwave pulling. This total shift is far too great for proper temperature compensation of all three spheres. In contrast to this filter the four-section staggered filter of Fig. 7 exhibited good temperature properties due to the fact that each sphere sees approximately the same environment.

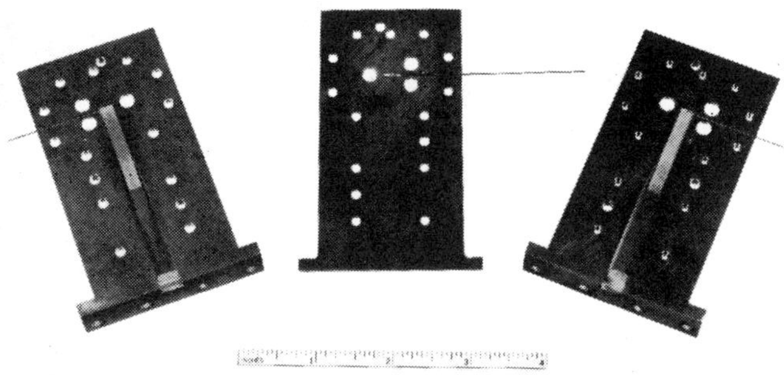

Fig. 10. 26.5- to 40.0-GHz YIG filter.

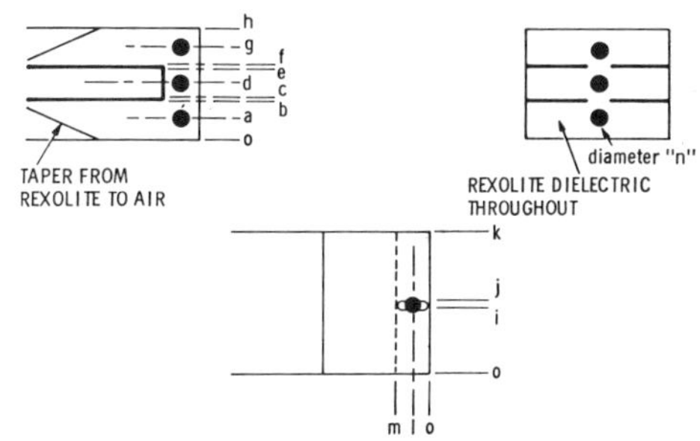

	a	b	c	d	e	f	g
18-26 GHz	.0208	.0380	.0400	.0572	.0744	.0764	.0936
26-40 GHz	.0173	.0300	.0320	.0447	.0574	.0594	.0721

	h	i	j	k	l	m	n
18-26 GHz	.1144	.1225	.1405	.263	.0300	.0600	.0202
26-40 GHz	.0894	.0816	.0943	.176	.0225	.0450	.0150

Fig. 11. Coupling structure details for 18- to 26-GHz and 26- to 40-GHz filters.

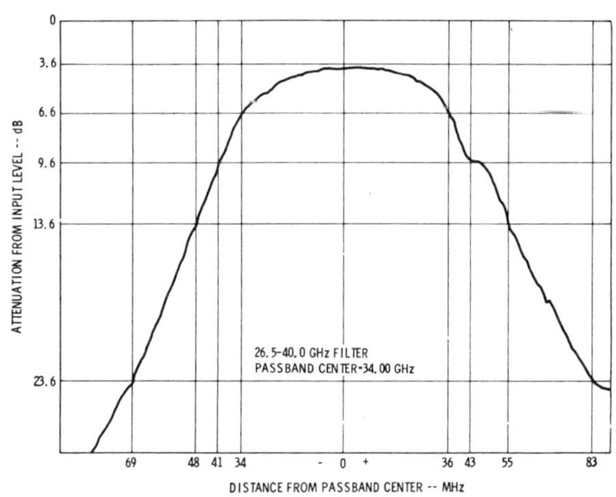

Fig. 12. Passband and skirt characteristics.

Spurious modes in the filters can be described in three categories: those that lie in the passband, those that lie in the range defined as filter skirts, and those that lie in the range defined as filter stopband. The filter stopband in this case will be defined as the frequencies outside the 60-dB bandwidth. The skirts are therefore those frequencies between the 3- and the 60-dB rejection points. The passband modes were 1 dB maximum. The modes on the high-fre-

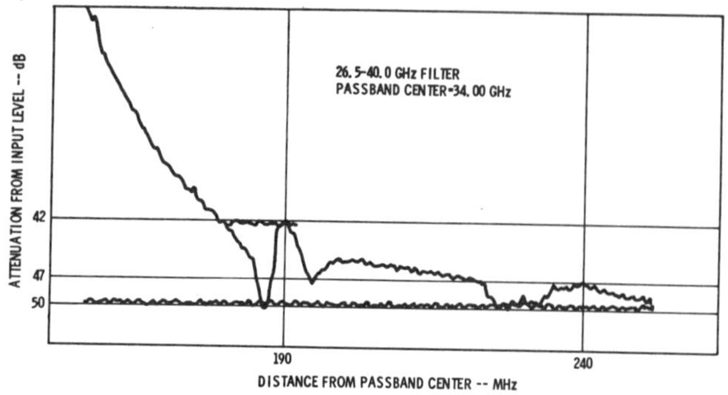

Fig. 13. Filter skirts showing amplitude and level of skirt modes.

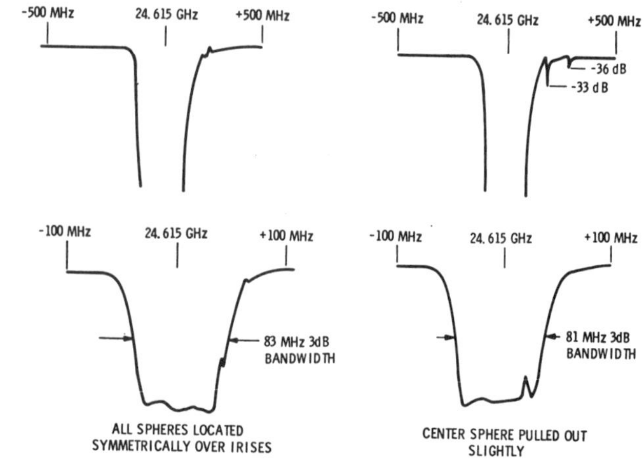

Fig. 14. Demonstration of mode excitation due to slightly misplaced sphere in 18- to 26-GHz three-section filter.

quency side of the filter were all within the normal 60-dB skirt rejection level. No skirt modes were observed from the passband center to 150 MHz on the high side. Fig. 13 shows the extent of the skirt modes from 150 to 250 MHz on the high side of the passband center. Spectrum analyzer data showed two more skirt modes on the high side, one located 370 MHz from passband center and attenuated 40 dB or more, and one located about 400 MHz from passband center and attenuated 50 dB or more. No skirt modes existed on the low-frequency side. One stopband mode was observed at about 650 MHz below the passband center and attenuated 40 to 50 dB. This data was part of extensive data taken on the 18- to 26- and 26- to 40-GHz filters after extensive handling and modification for temperature control devices, which severely warped the housings causing some poor contact between the iris and the waveguide edges, the effect being the enhanced excitation of modes and VSWR's. It is unfortunate that extensive recorded data were not taken on the units prior to these housing modifications. For example, prior to the modifications, the mode lying 650 MHz below the passband was not observable with 50 dB of sensitivity, and skirt modes on the high side were only noted as 1 or 2 dB fluctuations in skirt response in the worst case, again with sensitivity of only 50 dB. Passband shape was also somewhat better before the modifications. In addition to the above mentioned warpage problem, higher order modes were a problem in the three-section over-under filter as a result of the very tight tolerances required on sphere location. The (2, 2, 0) and the (5, 4, 0) modes, lying approximately 280 and 150 MHz, respectively, on the high side of the fundamental were found to be greatly excited when the spheres were slightly displaced from a perfectly central location with respect to the iris. Fig. 14 shows two conditions of sphere location in the 18- to 26-GHz three-section over-under filter. In the case of the centrally located spheres the two modes were only slightly noticeable. In the case of the spheres slightly displaced, the modes were highly excited and coupled. It should be noted here that the displacement was small enough that fundamental coupling was not changed appreciably. Both sphere proximity to the iris edge as well as the unsymmetrical location may be contributors to the mode excitation. Very accurate hole depths were required in order to locate the spheres in order to eliminate complicated tuning adjustments involving essentially simultaneous observation of passband and stopband performance. Sphere depth location accuracy to ± 0.0002 inch was achieved by measuring the depth of holes with a round-ended rod (the same diameter as a sphere).

The distance between spheres is extremely critical. For example, in the 26.5- to 40-GHz unit, a ± 0.0002-inch error in sphere location is predicted to cause about a 10 percent change in sphere-to-sphere coupling. An error of 0.001 inch is observed to move the spheres close enough together, or to an iris, to cause large excitation of undesired modes. In order to achieve adequate sphere location, very close tolerances were put on the fit of the Rexolite insert in the waveguide (Fig. 11). The Rexolite part ideally must be 0.0001 to 0.0002 inch below the waveguide height in order to allow the waveguide edges to bite into the iris which forms the corners and one broad wall of the waveguide. Good contact between iris and waveguide is essential in order to minimize the excitation of higher order modes, especially those types which cross through the passband. The spheres are mounted[1] on Rexolite rods which are 0.0005 inch smaller in diameter than the rod holes, thereby allowing a maximum of ± 0.0002-inch error in sphere location in the hole. Normally, a clearance of only 0.0005 inch is an interference fit on the 0.015-inch diameter rod in the hole, because of machining irregularities and bends in the rod. This interference is overcome by spraying the rods with a thin film of *fluorocarbon*, dry lubricant, and release agent.[2]

THE MAGNET

Fig. 15 shows the magnet used to tune to 40 GHz. It is a laminated core, made from vanadium permendur. The laminations are 0.014 inch thick and are coated with magnezium metholate. Pillars of zero temperature-coefficient glass are placed in the gap, and a pressure mechanism forces the magnet against them, thereby eliminating variations in the gap with temperature. The magnet pole taper was found to be essential in order to reach 40 GHz (14.3 kG) without encountering severe saturation effects. Also, cold working of the gap face (even surface lapping after the final anneal) was

[1] Spheres are glued on with Resiweld, Compound 7004.
[2] MS-122, made by Miller Stephenson Chemical Co.

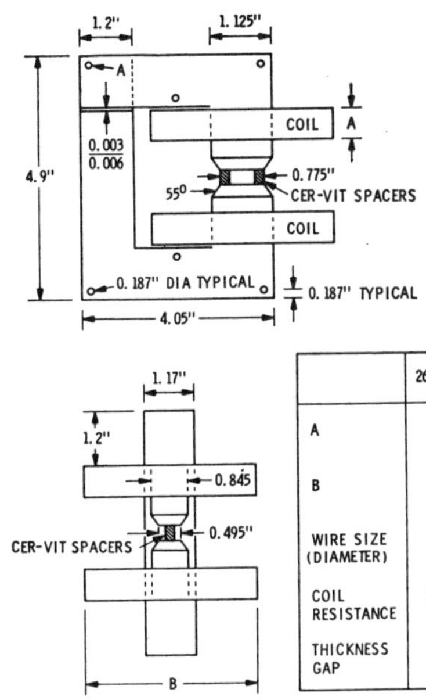

	26-40 GHz	18-26 GHz
A	1.030	0.880
B	4.140	4.014
WIRE SIZE (DIAMETER)	0.040"	0.036"
COIL RESISTANCE	12.0Ω	15.5Ω
THICKNESS GAP	0.115"	0.139"

Fig. 15. Magnet with coils.

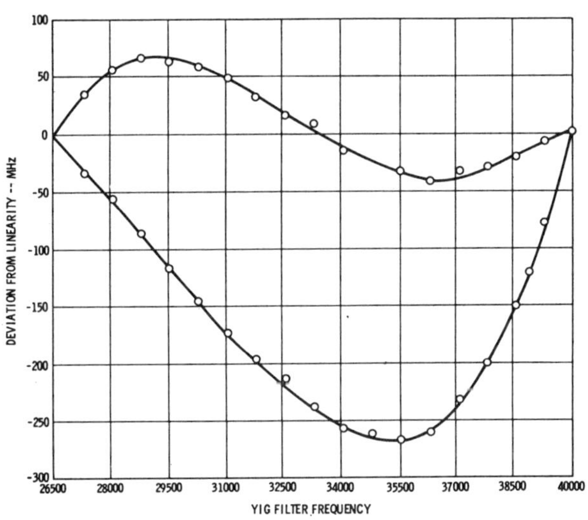

Fig. 16. 26.5- to 40-GHz magnet hysteresis.

found to significantly deteriorate the magnet performance. A typical 26- to 40-GHz magnet hysteresis loop is shown in Fig. 16. This plot was obtained by computing a straight line between the end points of the actual hysteresis curve and then calculating the deviation of the actual hysteresis curve from the straight line. The design of magnets has been well covered by others [13]–[15]. Much of the work on magnets has been done on the large laboratory type, but the information on these is certainly applicable to the much smaller scaled down versions used for tuning YIG filters.

This is a good place to discuss total filter size since the magnet is the bulk of the YIG filter assembly. For a given magnetic field to be generated and for a given allowable temperature rise, the volume of the magnet is nearly proportional to the cube of the gap spacing. The size problem is therefore one of reducing the gap thickness. It was previously indicated that it is easier to get good wide bandwidth filter performance with the spheres in the over-under arrangement. Unfortunately, this is also the arrangement which yields by far the largest package. Due to the smaller gap provided by the staggered sphere filter, its use could reduce the package volume by a factor of ten. Larger bandwidths from the staggered sphere filter may be obtainable through the use of resonator materials with a higher $4\pi M_s$ such as lithium ferrite. A few more comments on this will be made in the following conclusion.

Conclusions

Information on sphere-to-sphere coupling and details on the 18- to 40-GHz filters have been presented which should be of help to those who are building YIG filters in any frequency range for the first time regardless of the bandwidth being the same size, slightly larger, or smaller. It should be noted that the coupling coefficient between spheres required for a given bandwidth is inversely proportional to the frequency. It should then be noted that the actual coupling coefficient achieved from a given sphere-to-sphere arrangement is also inversely proportional to frequency. The apparently obvious deduction from these two statements is that, for example, the arrangement of spheres used in the 18-to 40-GHz range to achieve an 80-MHz bandwidth coefficient of mutual coupling will also produce an 80-MHz bandwidth coefficient of coupling in any other frequency range where pure YIG can be used. For the same bandwidth it would be necessary to provide the required coupling to the sphere which would change as the frequency changes. This may not always be a trivial matter but can be accomplished by various well-known means [1], [12], [18].

The ratios of sphere diameter to sphere spacing and sphere diameter to iris size that were given are also useful for smaller bandwidths, in that further significant improvement in out-of-band modes was observed when the 0.018 inch wide slot iris on the 18- to 26-GHz filter was reduced to 0.015 inch in width, although the bandwidth was reduced from 80 to about 65 MHz. For very small bandwidths it would be desirable to move the spheres away from the iris in order to loosen the tolerance on sphere location. The iris width can then be used to adjust the coupling so long as the width is kept somewhat smaller than the sphere diameter. Much larger bandwidths may be obtainable using this direct sphere-to-sphere approach through the use of a resonator material such as lithium ferrite. This material has a saturation magnetization of about 3900 gauss. For example, if this material were directly substituted for YIG, a bandwidth enlargement ratio of 3900/1750 would be realized, increasing the 80-MHz bandwidth to 178 MHz.

Acknowledgment

The author would like to thank J. P. Agrios and W. P. Dattilo, both of the U. S. Army Electronics Research Laboratory, who did much to further this work through their interest and direction. R. Rempel ably served as mechanical engineer in the design of the 18- to 40-GHz filters, reported in more detail in the USAECOM report. Much credit is due J. A. Sine, who assembled and tested the devices with great skill. The author is also grateful to Dr. E. W. Matthews, D. B.

Weller, and D. Shipley for their encouragement, handling of administrative duties, and reviewing the manuscript. Thanks are also due many others who reviewed the manuscript, made helpful suggestions, and in many other ways did much to further this work.

References

[1] P. S. Carter, Jr., "Magnetically-tunable microwave filters using single-crystal yttrium-iron-garnet resonators," *IRE Trans. Microwave Theory and Techniques*, vol. MTT-9, pp. 252–260, May 1961.

[2] L. R. Walker, "Magnetostatic modes in ferromagnetic resonance," *Phys. Rev.*, vol. 105, pp. 390–399, January 15, 1957.

[3] P. C. Fletcher and R. O. Bell, "Ferromagnetic resonance modes in spheres," *J. Appl. Phys.*, vol. 30, pp. 687–698, May 1959.

[4] G. H. Thiess, "Theory and design of tunable YIG filters," *Microwaves*, September 1964.

[5] A. L. Mikaelyan and V. Ya'Anton'yants, "Phenomena of mutual coupling in a system of magnetized ferrite specimens," *Radiotekhnika i Elektronika*, p. 586, April 1962.

[6] J. F. Dillion, Jr., "Magnetostatic modes in ferrimagnetic spheres," *Phys. Rev.*, vol. 112, pp. 59–63, October 1958.

[7] B. M. Lebed and K. V. Shevlyagin, "Investigations of the temperature and frequency dependencies of the ferromagnetic resonance line width in ferrite single crystals near the Curie temperature," *Soviet Phys.—Solid State*, vol. 7, pp. 1220–1222, November 1965.

[8] L. Young and D. B. Weller, "A 500–1000 MHz magnetically tunable bandpass filter using two YIG-disk resonators," *IEEE Trans. Microwave Theory and Techniques*, vol. MTT-15, pp. 72–86, February 1967.

[9] P. S. Carter, Jr. and C. Flammer, "Unloaded Q of single crystal yttrium-iron-garnet resonator as a function of frequency," *IRE Trans. Microwave Theory and Techniques* (Correspondence), vol. MTT-8, pp. 570–571, September 1960.

[10] H. Suhl, "The nonlinear behavior of ferrites at high microwave signal levels," *Proc. IRE*, vol. 44, pp. 1270–1284, October 1956.

[11] H. Suhl, "The theory of ferromagnetic resonance at high signal powers," *J. Phys. Chem. Solids*, vol. 1, pp. 209–217, January 1957.

[12] G. L. Matthaei, L. Young, and E. M. T. Jones, *Microwave Filters, Impedance-Matching Networks, and Coupling Structures.* New York: McGraw-Hill, 1964.

[13] Y. Ishikawa and S. Chikazumi, "Design of high power electromagnets," *Japan. J. Appl. Phys.*, vol. 1, September 1962.

[14] L. R. Blake, "The design of electromagnets," *Electronic Engrg.*, part 1, September 1953, part 2, October 1953.

[15] H. Roters, *Electro-Magnetic Devices*, New York: Wiley, 1941.

[16] E. Jahnke and F. Emde, *Tables of Publications.* New York: Dover, 1945.

[17] W. F. Egan, "Microwave propagation in periodic ferrite structures," Hansen Labs. of Physics, Stanford University, Stanford Calif. Microwave Lab. Rept. 1535, May 1967.

[18] L. Young, C. L. Matthaei, E. G. Cristal, D. B. Weller, L. A. Robinson, B. M. Shiffman, and D. K. Adams, "Novel microwave filter design techniques," Stanford Research Institute, SRI Project 4344, February 1965.

Two features of nonlinear phenomenon in ferrites may be used in the realization of ferrite limiters. These are the collapse of susceptibility at resonance at a threshold rf power level and the rise of a subsidiary absorption peak in the normally low loss region between zero dc field and the resonance field (1). Both types of limiters have received attention. In 1959, De Grasse achieved a coincidence limiter by coupling crossed strip transmission lines with a small YIG sphere (2).

In this section, a selection of letters and short papers describing various limiter developments are reprinted. The first paper by Arams, Grace, and Okwit describes some good early efforts in limiter work (3).

MICROWAVE POWER LIMITERS

IV

References

1. H. Suhl, "Nonlinear Behavior of Ferrites at High Microwave Signal Levels," Proc. IRE, Vol. 44, Oct. 1956, pp. 1270-1284.
2. R.W. DeGrasse, "Low-Loss Gyromagnetic Coupling Through Single Crystal Garnets," J. Appl. Phys. Vol. 30, Apr. 1959, pp. 155-156S.
3. F.R. Arams, M. Grace, S. Okwit, "Low-Level Garnet Limiters," Proc. IRE, Vol. 49, Aug. 1961, pp. 1308-1313.

Low-Level Garnet Limiters*

F. R. ARAMS†, SENIOR MEMBER, IRE, M. GRACE‡, MEMBER, IRE, AND S. OKWIT†, SENIOR MEMBER, IRE

Summary—**S- and L-band low-level passive limiters preselectors using polished single crystals of narrow line-width yttrium-iron-garnet have been developed. The S-band limiter employed unsubstituted YIG and has an insertion loss of 0.6 db, a limiting threshold of −26 dbm, and a dynamic range greater than 30 db.**

The L-band limiter used specially grown gallium-substituted YIG crystals, which, when ground into spheres and polished to a high finish, exhibited extremely narrow line widths. The L-band limiter had an insertion loss of 1.1 db, a limiting threshold of −21 dbm, and a dynamic range greater than 25 db. A tuning range from 1120 to 1305 Mc was obtained. Data on frequency sensitivity and SWR are given and indicate that such limiters have a high off-band rejection and a fair power-handling capability. A leading-edge spike of 0.1-μsec duration was observed.

The decline of the various components of the magnetic susceptibility tensor can be used to construct other novel types of limiters. The operation of three such devices—electronically-tunable preselectors, and cavity-type and comb-type limiters—will be described and experimental data given.

I. Introduction

FOR a long time there has existed a need for a microwave limiter which would protect microwave receivers against overload and burnout. In addition, requirements exist for microwave power levelers which eliminate power output variations in microwave power sources and in systems. Such limiters should preferably be passive and phase-distortionless. This paper discusses such limiter devices employing single crystals of unsubstituted and gallium-substituted yttrium-iron garnet as nonlinear elements. Limiters have been developed for L and S bands which have exceedingly low-limiting levels.

II. Nonlinearity in Ferrites

Experiments had shown that nonlinear effects appeared in ferrites at power levels only about 1 per cent of the level predicted by the simple saturation theory.[1,2] The main resonance line decreases in amplitude and broadens as the power level is increased. An additional rather broad absorption peak appears—usually at a value of applied static magnetic field, (H_z) lower than that required for the main resonance. This latter absorption has been called the subsidiary resonance.

The above effects have been explained by Suhl.[3,4] The decline of the main (uniform precessional) resonance is due to the excitation within the ferrite medium of z-directed spin waves at the same frequency as the driving-signal frequency. The appearance of the subsidiary absorption peak involves the parametric excitation of non-z-directed spin waved near one half the driving-signal frequency. These spin waves can be thought of as spatial disturbances in the normally uniform ferrite magnetization that propagate as plane waves through the medium. As the RF magnetic field is increased beyond an often sharply-defined critical level, the parametric coupling to these half-frequency spin waves increases, so that a subsidiary absorption peak will occur that increases with increasing signal level.

Thus, a limiter can be constructed in which the ferrite is placed in a coaxial line or waveguide, and the magnetic field is biased to an optimum value below resonance. Such subsidiary-resonance limiters have been investigated at X-band by Soohoo[5] and Uebele.[6]

For certain combinations of signal frequencies, ferrite geometry, and saturation magnetization, the range of magnetic field in which the subsidiary instability can occur extends through the main resonance. The coincidence of the subsidiary and main resonances is a condition particularly favorable to *low-level* limiting because at resonance maximum energy transfer occurs between the input signal and the spin-magnetization vector. This is shown by a calculation of the critical RF magnetic field (h_{crit}) at which nonlinearity sets in:[3]

$$h_{crit} = \frac{\Delta H \Delta H_k}{4\pi M_s} g(\theta). \tag{1}$$

where

ΔH = line width of uniform precessional resonance,
ΔH_k = line width of spin wave having wave-number k,
$4\pi M_s$ = saturation magnetization,
θ = angle between H_z and the direction of propagation of the half-frequency spin waves.

* Received by the IRE, November 30, 1960; revised manuscript received, May 23, 1961. This work was supported in part by the Dept. of Defense. It was presented at the 1961 Internatl. Solid-State Conf., Philadelphia, Pa., February 16, 1961.

† Airborne Instruments Lab., Div. of Cutler-Hammer, Inc., Melville, L. I., N. Y.
‡ Airtron, Inc., Morris Plains, N. J. Formerly with Airborne Instruments Lab., Div. of Cutler-Hammer, Inc., Melville, L. I., N. Y.

[1] R. W. Damon, "Relaxation effects in ferrimagnetic resonance," *Rev. Modern Phys.*, vol. 25, pp. 239–245; January, 1953.
[2] N. Bloembergen and S. Wang, "Relaxation effects in para- and ferrimagnetic resonance," *Phys. Rev.*, vol. 93, pp. 72–83; January, 1954.
[3] H. Suhl, "Nonlinear behavior of ferrites at high microwave signal levels," Proc. IRE, vol. 44, pp. 1270–1284; October, 1956.
[4] H. Suhl, "Theory of ferromagnetic resonance at high signal powers," *J. Phys. Chem. Solids*, vol. 1, pp. 209–227; January, 1957.
[5] R. F. Soohoo, "Power limiting using ferrites," 1958 IRE National Convention Record, pt. 1, pp. 36–47.
[6] G. S. Uebele, "Characteristics of ferrite microwave limiters," IRE Trans. on Microwave Theory and Techniques, vol. MTT-7, pp. 18–23; January, 1959.

For regions of limiting described in this paper, the function $g(\theta)$ approaches unity,[3] which is its minimum value.

Single-crystal yttrium-iron-garnet (YIG) is particularly suited to low-level limiting, since it has a very narrow resonance line width, a narrow ΔH_k, and a saturation magnetization of about 1800 gauss.

LeCraw, Spencer, and Porter[7] have demonstrated that for a single-crystal YIG sphere, the line width improves directly with the surface polish. LeCraw and his co-workers obtained resonance line widths as low as 0.52 oersted at 9200 Mc by polishing the sphere to a surface finish of better than 0.1 μ.

Experimental verification of (1) for the limiting threshold under coincidence of the main and subsidiary absorption peaks was also obtained.[8] This coincidence occurs when

$$2N_t 4\pi M_s \geq \frac{f}{\gamma} \geq N_t 4\pi M_s, \tag{2}$$

where

f = signal frequency in Mc,
γ = 2.8 Mc per oersted
(The definition for γ used here differs by 2π from the γ used in the References, where it is defined in terms of the angular frequency.)
N_t = transverse demagnetizing factors for spheroid ($=N_x=N_y$).

For a YIG sphere ($N_t=N_z=\frac{1}{3}$), having $4\pi M_s=1800$ gauss, (2) yields lower- and upper-frequency limits of 1680 to 3360 Mc, respectively. Spencer, *et al.*,[8] observed nonlinear effects in the frequency range from 1600 to 3200 Mc with a limiting level at 2200 Mc in reasonable agreement with (1). These measurements were recently extended by Rossel.[9]

A general plot of (2) for various spheroidal geometries is shown in Fig. 1. For example, for low-level limiting (coincidence of subsidiary and main resonance) at *L* band, a YIG disk with $N_z=0.670$ and $N_t=0.154$ will have optimum operation near 1200 Mc, and a theoretical operating range from 860 to 1720 Mc.

The practical use of single-crystal YIG as a low-level and low-loss passive limiter was first demonstrated by DeGrasse.[10] He used two crossed-strip-transmission-line half-wave resonators oriented at right angles to one another to obtain maximum decoupling between resonators. We have based some of our work on this same circuit arrangement.

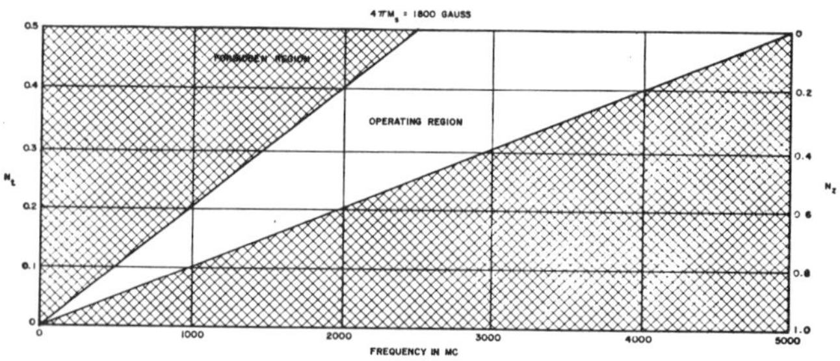

Fig. 1—Frequency region for low-level limiting as a function of ferrite geometry for YIG.

Fig. 2—*S*-band low-level limiter.

III. *S*-Band Limiter

An *S*-band limiter (with the top ground plane removed) is shown in Fig. 2. The strip-transmission-line resonators are open circuited at each end, so that there is an RF magnetic-field maximum at their centers, where the resonators cross. A 0.020-inch-diameter single-crystal YIG sphere is placed between the strips at their crossover point and biased to ferrimagnetic resonance (uniform precession of the spins). The resonators will then be coupled through the off-diagonal component of the susceptibility tensor. This process can be visualized in the following manner.

The driving RF magnetic field in the input resonator will fan out the spin magnetization vector. The vector will precess about the dc magnetization and in turn will induce an RF magnetic field in the output resonator. Thus, essentially all the RF energy is coupled through the YIG sphere, which can be considered a miniature microwave resonator with a very high unloaded Q. For a bandwidth ($\gamma\Delta H$) of 1 Mc, yields an unloaded Q of the order of 3000 at *S*-band. Such a miniature high Q resonator has many other potential applications in microwave circuits: electronically tunable preselector, filter, and panoramic receiver.

The single-crystal YIG resonator is coupled to and, therefore, loaded down by the strip-transmission-line

[7] R. C. LeCraw, E. G. Spencer, and C. S. Porter, "Ferrimagnetic line width in yttrium iron garnet single crystals," *Phys. Rev.*, vol. 110, pp. 1311–1313; June, 1959.

[8] E. G. Spencer, R. C. LeCraw, and C. S. Porter, "Ferromagnetic resonance in yttrium iron garnet at low frequencies," *J. Appl. Phys.*, vol. 29, pp. 429–430; March, 1958.

[9] F. C. Rossel, "Subsidiary resonance in the coincidence region in yttrium iron garnet," *J. Appl. Phys.*. vol. 31, pp. 2273–2275; December, 1960.

[10] R. W. DeGrasse, "Low-loss gyromagnetic coupling through single-crystal garnets," *J. Appl. Phys.*, vol. 30, pp. 155–156; Suppl. April, 1959.

resonators. This explains why the device has a 3-db bandwidth of approximately 70 Mc instead of a bandwidth of about 1 Mc. In fact, our theoretical investigation of the device has been based on an analysis of a three-resonator filter for minimum insertion loss.[11] This analysis shows that for minimum insertion loss, the input and output couplings should be tight, while the coupling to the center resonator (the YIG sphere) should be relatively loose. The resonators have been substantially reduced in width in the region where the YIG sphere is located in order to increase the RF magnetic field.

The transmission characteristic for the S-band limiter is shown in Fig. 3. The insertion loss is 0.6 db and the limiting threshold is -26 dbm (3 μw). The dynamic range is greater than 30 db. The power output actually drops several db from its threshold limiting value. (This characteristic can be changed to a slightly rising output-power slope, if desired, by changing the coupling coefficients.) The limiting threshold level, where the spin precession angle "sticks," is quite pronounced. This is seen more clearly when the YIG sphere is placed in a microwave transmission cavity and insertion loss is measured as a function of input power. The dissipative part of the magnetic susceptibility χ'' can be computed from this measurement. Fig. 4 is a plot of the decline in χ'' as a function of power level. The correlation of the measured values to the theoretical curve is very good.

The YIG sphere used for the above measurements had a line width $\Delta H=0.37$ oersted measured at 5200 Mc. The dc magnetic field is adjusted to uniform precessional resonance, that is, a signal frequency of 2000 Mc would correspond to 715 gauss. Care must be taken in adjusting the magnetic field to ensure that it is not biased to one of a number of magnetostatic modes[12] in the YIG sphere that have higher insertion loss and, apparently, higher limiting threshold.

Although the limiting characteristic (Fig. 3) was measured at 2000 Mc, we have obtained similar performance at frequencies between 1850 and 3400 Mc by substituting other resonators and readjusting the coupling parameters and the magnetic field.

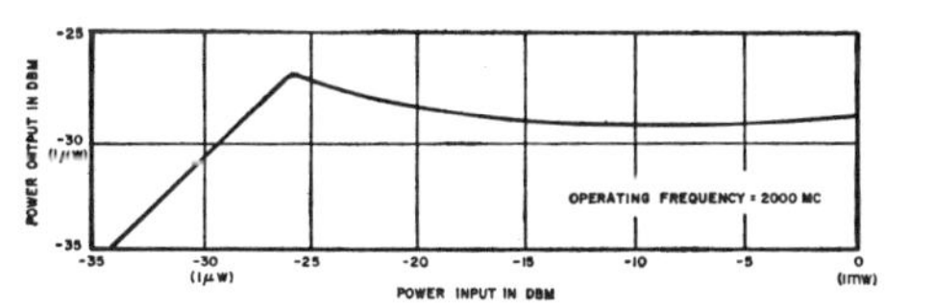

Fig. 3—Transmission characteristic of S-band limiter.

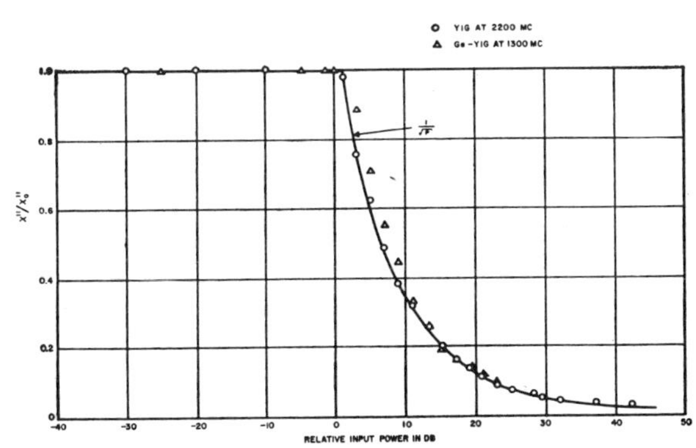

Fig. 4—Magnetic susceptibility of YIG and Ga-YIG spheres as a function of input-power level.

IV. L-Band Limiter

As Fig. 1 indicates, for low-level limiting at 1200 Mc using pure YIG, an oblate ellipsoid of revolution is required, having $N_z=0.670$ and $N_t=0.165$. However, to fabricate such an ellipsoid with a diameter of about 0.030 inch with an optical finish is difficult. A disk was therefore chosen to approximate the ellipsoid.

To achieve the lowest line width possible, YIG disks were X-ray oriented, cut with the hard axis [100] perpendicular to the plane of the disk, and then polished.

Line width and decline of magnetic susceptibility as a function of frequency and power level were measured for three disks of various thickness-to-diameter ratios. These measurements showed that the line width increases rapidly as the frequency is decreased, being 40 oersteds at 1200 Mc. The increase is caused by the YIG disks not being completely saturated at these low frequencies, as well as edge effects.[13] Limiting at low power levels was not obtained.

Another approach is the use of a sphere (easy to polish) having a reduced saturation magnetization, whose value can be determined from (2). Thus, using a narrow line-width material with a saturation magnetization of 1000 gauss, a limiter with a theoretical operating range from 950 to 1900 Mc can be constructed.

Experiments had been reported on decreasing the saturation magnetization of YIG by gallium substitution.[14] The line width remained narrow, increasing linearly with decreasing saturation magnetization in a 1:1 ratio.

Single crystals of Ga-YIG, having a saturation magnetization of 980 gauss and a Curie temperature of 206°C, were obtained from Microwave Chemicals Laboratory. Several spheres of this material were ground and polished to a Linde A finish. Measurements of the polished spheres indicated a very narrow line width (0.72 oersted measured at 5200 Mc), and low-level limiting was successfully obtained.

[11] J. Taub and B. Bogner, "Design of three-resonator dissipative band-pass filters having minimum insertion loss," Proc. IRE, vol. 45, pp. 681–687; May, 1957.

[12] R. L. Walker, "Magnetostatic modes in ferromagnetic resonance," *Phys. Rev.*, vol. 105, pp. 390–399; January, 1957.

[13] R. C. LeCraw, private communication.

[14] E. G. Spencer and R. C. LeCraw, "Line width narrowing in gallium substituted yttrium iron garnet," *Bull. Am. Phys. Soc.*, Ser. II, vol. 5, p. 58; January 27, 1960.

A plot of the magnetic susceptibility as a function of power level that has the same sharp break as pure YIG is shown in Fig. 4. The decline has the $P^{-1/2}$ dependence predicted by Suhl.

An *L*-band limiter with an adjustable permanent magnet is shown in Fig. 5. The measured value for the limiting-threshold power level was -21 dbm (8 μw), which was lower than expected. The operating band over which limiting was measured extended from 1000 to 1500 Mc. The results in an *L*-band crossed-resonator limiter are shown in Fig. 6. An insertion loss of 1.1 db was obtained at the center frequency (1305 Mc). The 0.2-db and 3-db bandwidths for fixed magnetic field are 8 and 43 Mc, respectively; SWR is 1.06 at center frequency.

Fig. 5—*L*-band limiter (shown with adjustable permanent magnet).

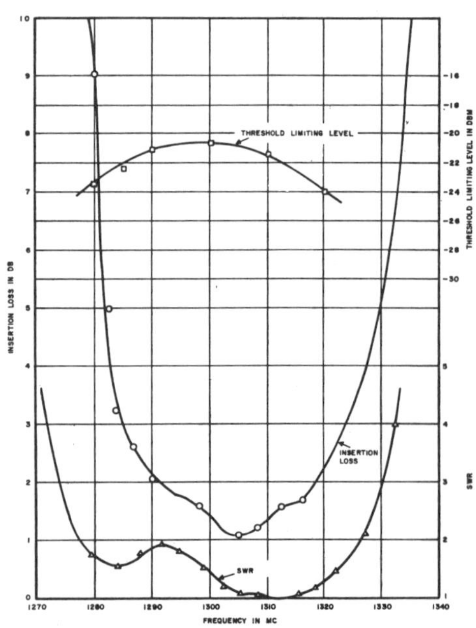

Fig. 6—Operating characteristics of *L*-band limiter

The output-power vs input-power characteristic is shown in Fig. 7, and indicates a dynamic range greater than 25 db. The SWR is also plotted as a function of the input-power level. The reflected portion of the incident power increases sharply as the input power is raised. The remainder of the input power is dissipated in the spin waves within the garnet.

Above limiting threshold, instabilities within the ferrite give rise to oscillations believed to be in the kilocycle region. In response to fast-rising pulses, the limiter exhibits a leading-edge spike about 0.1 μsec long, caused by the finite build-up time of the spin waves that result in limiting action.

A simple method of tuning the limiter was investigated, that is, capacitively loading the ends of the half-wave resonators with shorting plungers and by optimizing the magnetic field. Measurements showed a linear increase in limiting threshold and insertion loss with decreasing frequency (Fig. 8); a tuning range of 190 Mc was obtained. It is possible to devise other tuning methods in which the insertion loss is essentially independent of frequency.

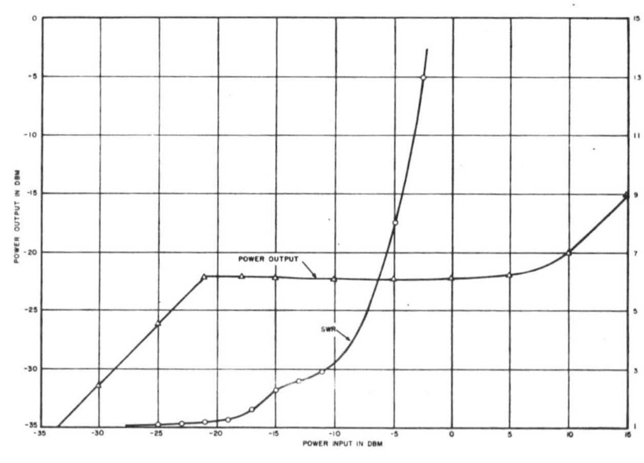

Fig. 7—Transmission characteristics of *L*-band limiter.

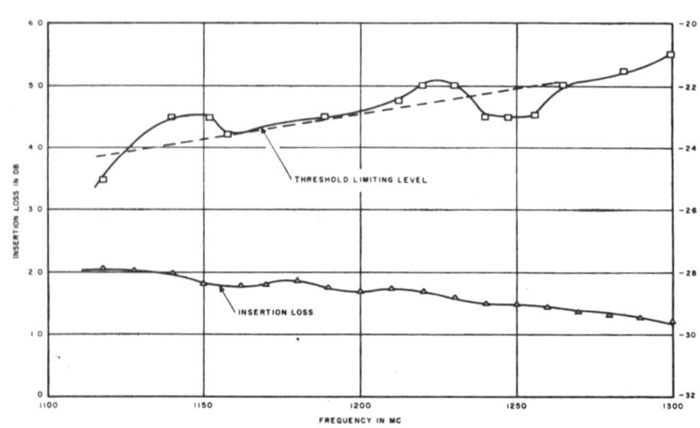

Fig. 8—Tuning characteristics of *L*-band limiter.

V. Electronically Tunable Limiter-Preselector

Another interesting type of limiter has been operated in which the resonators were replaced by nonresonant transmission lines terminated in open circuits a quarter-wave from the YIG crystal. Thus, the principal frequency-determining parameter is the magnetic resonant frequency of the sphere, and is determined by the applied magnetic field.

By varying the magnetic field, an electronically tunable limiter-preselector was obtained. The test results (using an inferior sphere) were:

Frequency tuning range	2000 to 3400 Mc
3-db bandwidth	20 Mc
Threshold limiting level	−20 to −25 dbm
Insertion loss	4.4 db

Insertion loss can be reduced by sacrificing threshold limiting level.

VI. Other Limiting Configurations

In the limiters described, the κ' component of the off-diagonal element of the ferrite susceptibility tensor[15] is used to couple two normally-decoupled orthogonal resonators. However, all of the χ and κ elements in the susceptibility tensor are dependent on the RF power level and can be used for the design of other limiters. Two methods for obtaining limiting that use other elements of the susceptibility tensor in appropriate microwave structures are the cavity limiter and the comb limiter.

A. Cavity Limiter

A cavity limiter has been operated that uses the selectivity of a high-Q single-tuned resonator and the χ' component of the diagonal element of the susceptibility tensor. The operation of this limiter is as follows.

A quarter-wave resonator is (Fig. 9) tuned to a frequency f_0. A ferrite sample is placed on the center conductor near the shorted end of the resonator where the RF magnetic field is maximum. The dc magnetic field is oriented at right angles to the RF magnetic field and adjusted to a field strength slightly less, or greater, than that required for uniform precession at cavity frequency f_0. A reactive component will be introduced into the cavity circuit by the ferrite that will retune the cavity to a new resonant frequency f_1. The cavity, with RF power levels below P_{crit}, will pass a band of frequencies centered about f_1 with a minimum of insertion loss. However, when the RF signal is increased above the critical threshold, the decline in the reactive component χ' will detune the cavity and thereby increase its insertion loss at frequency f_1.

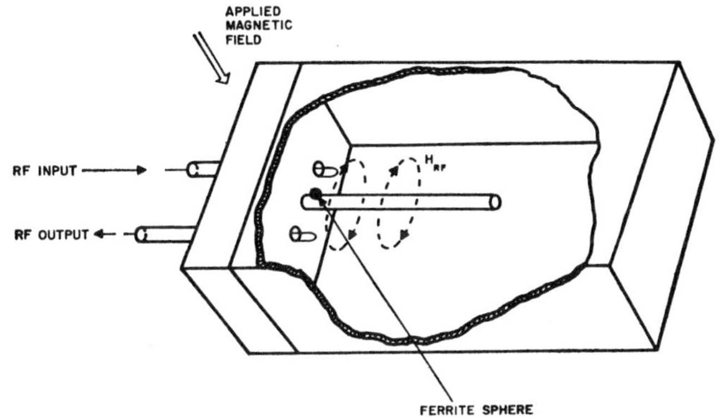

Fig. 9—Slab-line single-tuned cavity power limiter.

An experimental cavity limiter using this technique was constructed in slab-line (Fig. 9). This limiter incorporates a single-crystal YIG sphere with a resonance line width of about 1 oersted. The limiting characteristic is a plateau similar to that in Fig. 6. The measured characteristics are:

Center frequency	2590 Mc
Threshold limiting level	−20 dbm
Insertion loss (low-level)	2 db
Dynamic range	20 db

The insertion loss of the limiter can be minimized by using high-Q microwave circuits, and operating the ferrite at a compromise dc magnetic field. In this way the magnetic losses are small but the reactive component is high enough to sufficiently detune the cavity from its unperturbed resonance.

The power level at which limiting occurs (once a ferrite material has been selected) can be varied by moving the ferrite axially along the center conductor to regions of lower or higher RF magnetic field strength.

B. Comb Limiter

A limiter has been operated that uses the inherent decoupling between two adjacent quarter-wave TEM-mode resonators. The garnet spin system is then used as an RF power-level-dependent coupling element between the two resonators (Fig. 10). This configuration has the advantages of compactness and ease in cascading. Furthermore, it is of special interest when used in conjunction with a traveling-wave maser amplifier, where the comb-type slow-wave structure has found wide application.[16,17]

The measured characteristics of a comb limiter are:

Center frequency	2638 Mc
Threshold limiting level	−15 dbm
Dynamic range	25 db
Insertion loss	10 db

[15] The susceptibility tensor is given by

$$\begin{bmatrix} \chi & -j\kappa & 0 \\ j\kappa & \chi & 0 \\ 0 & 0 & 1 \end{bmatrix}$$

where $\chi = \chi' - j\chi''$ and $\kappa = \kappa' - j\kappa''$.

[16] R. W. DeGrasse, E. O. Schulz-DuBois, and H. E. D. Scovil, "Three level solid-state traveling wave maser," *Bell Sys. Tech. J.*, vol. 38, pp. 305–334; March, 1959.

[17] S. Okwit, F. Arams, and J. G. Smith, "Electronically tunable traveling-wave masers at *L* and *S* Bands," Proc. IRE, vol. 48, pp. 2025–2026; December, 1960.

The comb limiter that was fabricated is an experimental model for testing feasibility, and it is believed that better performance can be obtained. The comb limiter uses the dissipative part of the diagonal or off-diagonal elements of the susceptibility tensor, or a combination of these elements depending upon the positioning of the garnet sphere in plane A with respect to the comb fingers (Fig. 10). The garnet sphere used for this experiment was the same sphere used for the cavity limiter.

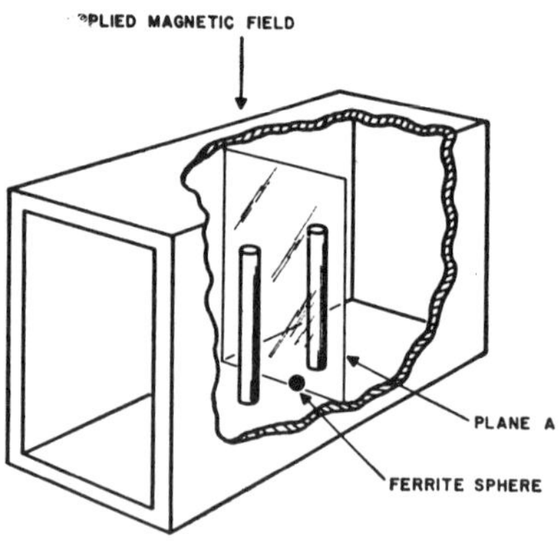

Fig. 10—Comb limiter (coupling not shown).

VII. Conclusions

The use of a polished single-crystal YIG of narrow line width in S-band and L-band low-level-limiter preselectors has been investigated.

The S-band limiter uses unsubstituted YIG and has an insertion loss of 0.6 db, a limiting threshold of −26 dbm, and a dynamic range greater than 30 db.

The L-band limiter uses specially grown gallium-substituted YIG crystals. When ground into spheres and polished to a high finish, these crystals exhibit extremely narrow line widths. The L-band limiter has an insertion loss of 1.1 db, a limiting threshold of −21 dbm, a dynamic range greater than 25 db, and a tuning range from 1120 to 1305 Mc. Detailed data on frequency sensitivity and SWR indicates that such limiters have high off-band rejection and fair power-handling capability. A leading-edge spike of about 0.1 μsec duration was observed.

The decline of the various components of the magnetic-susceptibility tensor can be used to construct other types of limiters. The operation of three of these —the electronically-tunable preselector, the cavity limiter, and the comb limiter—has been described and data on experimental models given.

308

A Temperature-Stabilized Microwave Power Limiter for Communication Satellite Use*

A ferrimagnetic microwave power limiter has been developed for use in an experimental active communication satellite repeater. At a given frequency the limiting level can be varied with the magnetic bias field in contrast to coincidence limiters.[1–4] Temperature stability of the device characteristics was achieved over a temperature range $55° \leq T \leq 120°F$ required for satellite operation. This limiter uses as its nonlinear element a large single crystal sphere of yttrium iron garnet (YIG) biased to the subsidiary absorption,[5,6] and placed in a transmission microwave cavity. Limiting occurs when the power level exceeds the threshold for the subsidiary absorption (parametric excitation of half-frequency spin waves) in the garnet. The losses associated with this absorption increase with power above the threshold, lowering the unloaded cavity Q and increasing the cavity insertion loss.

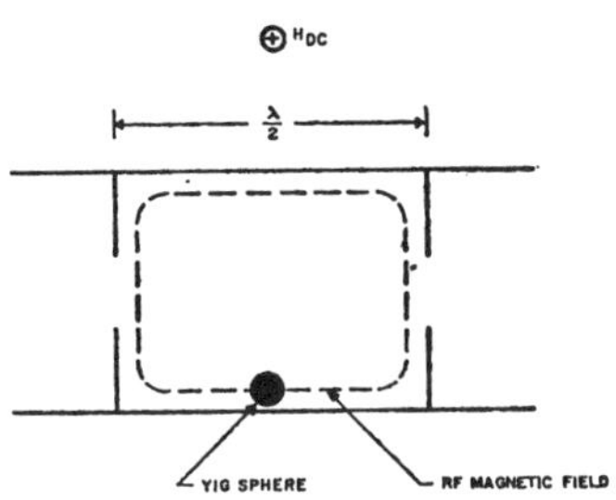

Fig. 1—Schematic diagram of subsidiary absorption limiter.

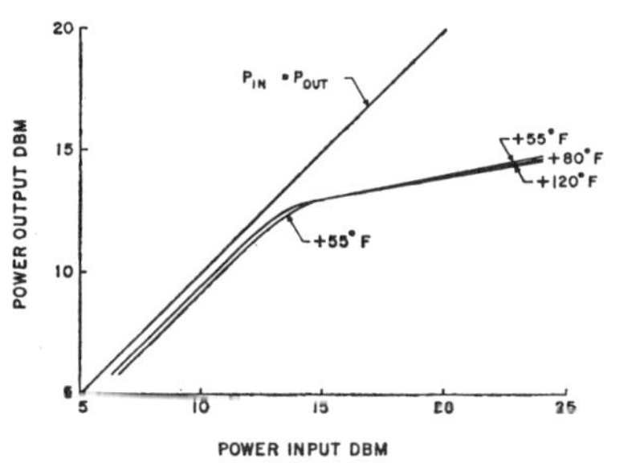

Fig. 2—Measured characteristics of the 4-kMc limiter at 3 temperatures. The dc magnetic field was 1100 oersteds.

A limiter using this principle can be constructed at any frequency, depending only on the availability of an adequate dc magnetic field, whose amplitude is somewhat smaller than the ferrimagnetic resonance value. Theoretical considerations show that the filling factor, which is proportional to the sample volume divided by the cavity volume, must be large in order to achieve limiting over a significant range of power. The availability of large single crystals of yttrium iron garnet now makes it possible to obtain these large filling factors along with narrow linewidths.

Fig. 1 is a schematic of a 4 kMc model of this limiter which used a polished sphere of single crystal YIG, 0.260 inches in diameter. The desired limiting value was 14 dbm. This level can be adjusted to as low as 3 dbm with the dc magnetic field. A reduced-height (0.400 in) transmission cavity with a loaded Q of 200 was operated in the TE_{101} mode with the sample placed in the maximum RF magnetic field close to the side wall of the cavity. At low power levels the sample has very little effect on cavity transmission properties. Above a well-defined threshold level sample losses increase rapidly and limit the transmission power level.

Temperature changes can have large effects on the characteristics of this device The limiting power level changes with $4\pi M$, ΔH_k, H_{dc} and H_a, the anisotropy field.[7] In addition, there are low-level insertion loss variations due to thermal cavity detuning effects. These include cavity expansion and RF susceptibility variations dependent on $4\pi M$, ΔH_k, H_{dc}, and H_a. A major source of thermal changes in limiter performance can be avoided by the use of a spherical sample, for which the ferromagnetic resonance frequency is independent of the saturation magnetization and its associated temperature variation. It can also be shown that orientation of the anisotropy field can be used to minimize further the temperature sensitivity. The samples used here were of a purity such that ΔH_k had only a slightly negative temperature coefficient at room temperature. For these samples orientation of the magnetically hard [100] crystallographic axis along the biasing field gave optimum temperature characteristics.

The limiter performance over the design temperature range from 55° to 120°F is shown in Fig. 2. The sharp break in characteristic at the threshold is important for efficient power leveling. It is dependent on the use of single-crystal narrow-linewidth material and the surface polish. In this case ΔH_k was approximately 0.15 oe at 4 kMc. Spike leakage was observed but not investigated; however, this limiter should have characteristics similar to the coincidence limiter[2] in this respect.

Because of weight limitations in the satellite application, the cavity was fabricated from silver-plated magnesium resulting in a total weight, including magnet, of 11 ounces. While the limiter was designed at a particular frequency with particular requirements, it should prove useful in other applications at different frequencies requiring precise control of power level.

The authors wish to acknowledge the contributions of R. C. LeCraw, E. G. Spencer, F. C. Rossol, R. Kordos, J. Degan, and V. Czarniewski to this development.

R. L. Comstock
W. A. Dean
L. J. Varnerin
Bell Telephone Labs.
Murray Hill, N. J.

* Received by the IRE, January 12, 1962.

[1] R. W. DeGrasse, "Low-loss gyromagnetic coupling through single crystal garnets," *J. Appl. Phys.*, vol. 30, pp. 155–156S; April, 1959.

[2] F. J. Sansalone and E. G. Spencer, "Low-temperature microwave power limiter," IRE Trans. on Microwave Theory and Techniques, vol. MTT-9, pp. 272–273; May, 1961.

[3] F. C. Rossol, "Power limiting in the 4 kMc to 7 kMc frequency range using lithium ferrite," Proc. IRE, vol. 49, p. 1574; October, 1961.

[4] F. R. Arams, M. Grace and S. Ikurit, "Low-level garnet limiters," Proc. IRE, vol. 49, pp. 1308–1313; August, 1961.

[5] H. Suhl, "Nonlinear behavior of ferrites at high microwave signal levels," Proc. IRE, vol. 44, pp. 1270–1284; October, 1956.

[6] R. L. Martin, "High-power effects in ferrite slabs at X-band," *J. Appl. Phys.*, vol. 30, pp. 159S–160S; April, 1959.

[7] J. F. Dillon, Jr., "Ferrimagnetic resonance in yttrium iron garnet," *Phys. Rev.*, vol. 105, pp. 759–760; January, 1957.

Reprinted from the PROCEEDINGS OF THE IRE
VOL. 50, NO. 4, APRIL, 1962

PRINTED IN THE U.S.A.

Reprinted from JOURNAL OF APPLIED PHYSICS, Vol. 34, No. 4 (Part 2), 1275–1277, April 1963

Spin-Wave Instabilities and Magnetoelastic Coupling

E. G. SPENCER, *Chairman*

Extension of Coincidence Limiting Frequency Range in Ferrimagnets

R. L. COMSTOCK AND W. A. DEAN
Bell Telephone Laboratories, Inc., Murray Hill, New Jersey

Power limiting using the first-order spin-wave instability has been observed in a ferrimagnetically coupled filter-limiter at higher frequencies than had previously been believed possible. For example, a filter-limiter, using orthogonal strip-line resonators coupled by the dipolar fields of a YIG sphere, was constructed at 4000 Mc. Output power was limited with the sharp break at the threshold and flat limiting curve characteristic of coincidence limiters. This result is in contrast with the operation of similar limiters at this frequency which use the second-order spin-wave instability in which the limiting characteristic has a much more gradual slope transition above the threshold. Coincidence limiters have previously been believed to have a maximum operating frequency $\omega_c=2\omega_M N_T$ (3.27 kMc for a YIG sphere) since, with an isolated sample, coincidence of the uniform precession resonance and the subsidiary absorption occurs only for $\omega<\omega_c$.

These results are observed for a coupled mode resonance in these ferrimagnetically coupled circuits which occurs for dc magnetic field values in the subsidiary absorption region below the uncoupled uniform precession value (ω/γ). It is found that a minimum value of the circuit-sample coupling is required to observe this extension of coincidence type limiting. Limiting thresholds as low as −10 dBm have been observed with the strip-line circuit and lower levels are believed to be possible. Wide low-level bandwidths (>10%) are also obtained with these circuits, but with an increase in the limiting threshold.

The above results are explained by computing the normal modes of the microwave-ferrimagnetic resonator circuit. It is found for tight enough coupling that one of the branches of the normal mode spectrum is in coincidence with the subsidiary absorption at a higher frequency than the uniform precession resonance. The usual coincidence limiter operation $(\omega<\omega_c)$ is thus the small coupling case of the more general type of operation.

INTRODUCTION

THIS paper describes some theoretical and experimental results concerning a method of extending the frequency range of coincidence type ferrimagnetic limiters[1] which were previously believed to have a maximum operating microwave frequency.[2,3] Coincident type limiters, which use the first-order spin-wave instability,[4] are distinguished by their sharp slope transition at the threshold and flat limiting characteristic above the threshold. The operation of these limiters is in contrast to similar limiters which use the saturation of the uniform precession resonance (second-order spin-wave instability).[5] With second-order process limiters, the slope transition above the threshold is not abrupt, the limiting characteristic is not flat, and the thresholds are much higher than with coincidence limiters. It is, therefore, important to extend the range of coincident type limiting to higher frequencies.

These limiters, typically, consist of orthogonal microwave resonators coupled by the dipolar fields of a biased ferrimagnetic resonator, thus forming a three-resonator, bandpass filter network.[6] The limiters used in the present investigation consisted of two crossed half-wave strip-line resonators, which were tapered down to form half-loops at the crossover, with a YIG sphere mounted in the center. This strip-line circuit is similar in form to one which was previously reported.[2] Limiting in these circuits is observed when the rf magnetic field at the sample (h_{rf}) exceeds the threshold for the subsidiary absorption, which depends on the applied dc magnetic bias field. However, it was previously believed that the dc magnetic bias field must have the value ω_r/γ, where ω_r is the unperturbed resonant frequency of the uniform precession mode, i.e., it was believed that the uniform precession resonance should be "coincident" with the subsidiary absorption. Under this condition, it can be shown that the frequency of operation is restricted to the range $\omega<\omega_c=2\omega_M N_t$, where $\omega_M=\gamma 4\pi M_s$ ($4\pi M_s$ is the saturation magnetization) and N_t is the transverse demagnetizing factor for an ellipsoid ($\frac{1}{3}$ for a sphere). In the case of a YIG sphere f_c is 3.27 kMc. In contrast, we have observed coincidence type limiting as high as 5.3 kMc using a YIG sphere. This result arises since the ferrimagnetic resonator is tightly coupled to the strip-line resonators and thus the combined circuit-sample system can violate the above frequency restriction. The manner in which

[1] An extensive review of microwave ferrimagnetic power limiters is given by B. Lax and K. Button in *Microwave Ferrites and Ferrimagnetics* (McGraw-Hill Book Company, Inc., New York, 1962).

[2] F. J. Sansalone and E. G. Spencer, IRE Trans. on Microwave Theory Tech. **9**, (1961); and F. C. Rossol, Proc. IRE **49**, 1574 (1961).

[3] E. G. Spencer, R. C. LeCraw, and C. S. Porter, J. Appl. Phys. **29**, 429 (1958).

[4] H. Suhl, J. Phys. Chem. Solids **1**, 209 (1957).

[5] J. Clark and J. Brown, J. Appl. Phys. **33**, 1270 (1962).

[6] R. W. DeGrasse, J. Appl. Phys. **30**, 151S (1959).

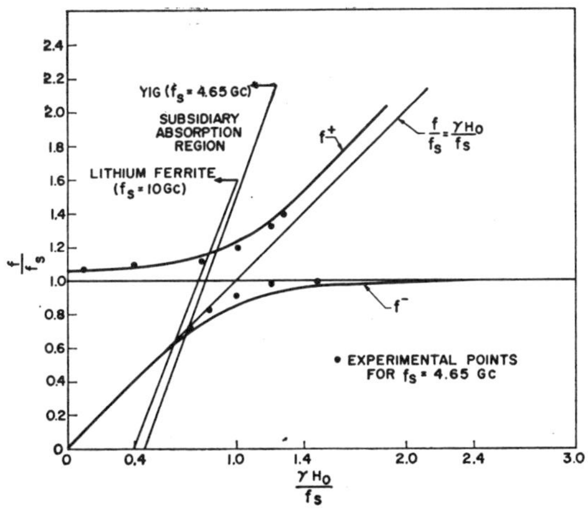

FIG. 1. Normal mode resonant frequencies normalized to strip-line resonator frequency (f_s) vs normalized, uncoupled, uniform precession frequency. Also shown are the locus of the minimum threshold for the subsidiary absorption for the case of YIG at 4.65 kMc and lithium ferrite at 10 kMc. The intersection of these curves with the upper branch of the normal mode spectrum (f^+) can be used to observe coincidence type, power limiting.

this is accomplished can be explained by considering the normal modes of the combined three-resonator, bandpass filter network. These normal modes can be calculated by the use of a lumped element equivalent circuit and an illustrative diagram of the normal mode frequencies is shown in Fig. 1. This diagram shows the resulting two branches of the normal mode spectrum labeled $f\pm$ and normalized with respect to the strip-line resonator resonant frequency (f_s), plotted against the normalized, uncoupled, uniform precession resonant frequency. The dipolar coupling was calculated from the lumped constant equivalent circuit resulting in a coupling coefficient[7]

$$K^2 = (\mu_0\omega_M/3Z_{0r})(r_s^3/r_0^2),$$

in the case of strip-line resonators (Z_{0r}) with centrally located half-loops (radius r_0) coupled to a ferrimagnetic sphere (radius r_s). The parameters for an experimental model of a strip-line limiter were used in the calculation and the resulting experimental points are also shown on the diagram. The locus of the conditions for the minimum threshold for the subsidiary absorption are also shown for YIG in the case f_s is 4.65 Gc and for lithium ferrite ($4\pi M = 3550$ Oe) at 10.0 Gc. Thus, there is a coincident condition between the subsidiary absorption and the upper branch of the normal mode spectrum for frequencies greater than f_s. Since the combined circuit-sample system is resonant at this frequency the insertion loss of the circuit below the threshold can be made small by tightly coupling the external loads to the three resonator filter. The intersection of the subsidiary absorption with the uniform precession is not a normal mode frequency so that in general at this point the insertion loss would be expected to be high. If it is desired to operate at even higher frequencies it is only necessary to increase f_s. However, as this is done the dc magnetic field may drop below the magnetic saturation value ($4\pi M/3$), resulting in excessive losses and an increase[3] in h_{crit}. If this results it is only necessary to increase the dipolar coupling to increase the coincidence frequency. Therefore, the upper limit for the coincidence of the subsidiary absorption with the upper branch of the normal mode spectrum is only limited by the magnitude of the dipolar coupling. Note, with a higher saturation material like lithium ferrite this coincidence condition can be obtained at frequencies greater than 10 kMc for the same value of K which resulted in coincidence with YIG near 4.65 kMc.

[7] Milton Dishal, Proc. IRE **39**, 1448 (1951).

For microwave powers greater than the threshold for the subsidiary absorption the precession angle of the uniform mode sticks at its threshold value decoupling the microwave resonators and causing power limiting. The bandwidth of the limiter below the threshold can be large, although it can be shown that the ratio of the limiting threshold to the bandwidth is a constant for a given material and strip line configuration.

EXPERIMENTAL RESULTS

Figure 2 shows a limiting curve (power output vs incident power) obtained with a 0.075 in. single-crystal YIG sphere in a strip line limiter at 4000 Mc. The dynamic range of this limiter exceeds 34 dB and the 3-dB bandwidth below the threshold (−4 dBm) is 180 Mc. By narrowing the filter bandwidth it was possible to reduce the threshold to −10 dBm. By tighter coupling (larger sample radius or smaller loop radius) the bandwidth can be increased. For example, at 5300 Mc a limiter was constructed which had a threshold of 10 dBm, a low level 1-dB bandwidth of 430 Mc, and a 0.5-dB midband insertion loss. However, limiting was observed only over the upper 100 Mc of the low level bandpass response curve.

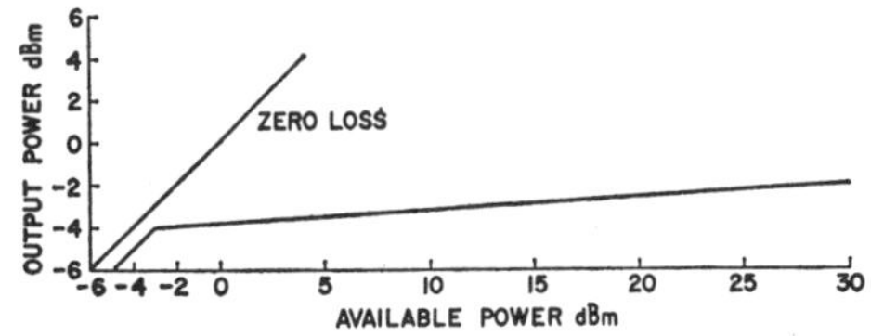

FIG. 2. Limiting characteristic obtained at 4000 Mc with a 0.075-in. YIG sphere. The dc field was 1200 Oe corresponding to the subsidiary absorption region. The dynamic limiting range was not measured but exceeds 34 dB.

CONCLUSIONS

It has been shown that it is possible to build coincident type limiters at any microwave frequency if a coupled mode resonance of the ferrimagnetic sample microwave circuit system can be realized in the subsidiary absorption region. Limiters which have a coincidence of the uniform precession and the subsidiary absorption are thus the small coupling case of this more general system. The limiting threshold obtained with this system are somewhat higher than those obtained in these circuits for f less than f_c. This is believed to be the result of a smaller precession angle per unit power under the present circumstances since here a larger part of the stored energy is in the form of electromagnetic energy in the more tightly coupled circuit resonances.

It can be shown that for sufficient coupling the normal modes described here may lie outside of the magnetostatic mode spectrum.[8] In this case they may be useful in other applications, e.g., harmonic generators and ferrite amplifiers, where it is desired to have circuit-sample resonances which are separated from other sample modes.

[8] B. A. Auld and H. J. Shaw (private communication).

An X-Band Coincidence Ferrimagnetic Limiter Using Lithium Ferrite*

Ferrimagnetic power limiters which use the uniform precession resonance to couple two orthogonal microwave circuits have been extensively investigated using both yttrium iron garnet[1-4] and lithium ferrite.[5] In these circuits the limiting threshold is particularly low if the bias field and frequency are chosen so that the subsidiary and uniform precession resonances *coincide.* It can be shown that this coincidence condition is equivalent to the condition that at the uniform precession resonance the operating frequency is less than $f_c=(\frac{2}{3})\gamma 4\pi M_S$ (for yttrium iron garnet and lithium ferrite f_c equals 3260 and 7280 Mc, respectively). However, it was reported recently that this coincidence condition can be extended by tightly coupling the uniform precession to the orthogonal microwave resonators.[6] For low insertion loss in the tightly coupled case the system must be adjusted to resonance for a normal mode of the combined circuit-sample system, not to the uniform precession resonance. In this case, coincidence type limiting with YIG was obtained as high as 5300 Mc by using the upper branch of the normal mode spectrum.[6]

A criterion for the existence of the extended coincidence condition can be established by noting that, for a three resonator filter consisting of identical end resonators, the two branches of the normal mode spectrum are given by

$$(\omega^{\pm})^2 = \omega_0^2(1 \pm \sqrt{2}K) \tag{1}$$

where ω_0 is the midband resonant frequency (for the uniform precession $\omega_0=\gamma H$) and K is the coefficient of coupling between the center and either end resonator. In the case of a ferrimagnetic sample excited in the uniform precession, the coupling constant can be written $K^2=CM_SV_S$ where the constant C depends on the geometry of the coupling arrangement and V_S is the sample volume. However, the subsidiary absorption has a small threshold only for[7]

$$\omega > 2(\omega_0 - \omega_m N_t) \tag{2}$$

where $\omega_m=\gamma 4\pi M_S$ and for a sphere the demagnetizing factor N_t equals $\frac{1}{3}$. Using (2) in (1), eliminating ω_0 and expanding the square root for small K, we find

$$f_c/f^+ > 1 - \sqrt{2}K. \tag{3}$$

In order to satisfy this inequality for large f^+ we see that large values of M_S are required to make f_c and K as large as possible. Typically, in our strip-line circuits, values of K of order 0.1 have been achieved.

In this communication we report the extension of coincidence limiting to X band using lithium ferrite. The circuit used was similar to those previously reported[2,5] and consisted of two orthogonal half-wave strip-line resonators coupled by the uniform precession mode in a lithium ferrite sample[8] 0.0318 inches in diameter and having a small signal linewidth at 6.2 Gc of 4.25 oersteds. By tightly coupling the sample to the strip-line resonators, a 3-db bandwidth of 410 Mc was achieved below the limiting threshold. The limiting characteristic observed with this limiter with a fixed bias field ($H=2750$ oersteds) is shown in Fig. 1 for two frequencies in X band, 100 Mc apart. The insertion loss was less than 1 db and the threshold was approximately 16 dbm. For narrower low level bandwidths it is expected that this limiting threshold can be reduced. Outside of this 100 Mc range the limiting slope increases.

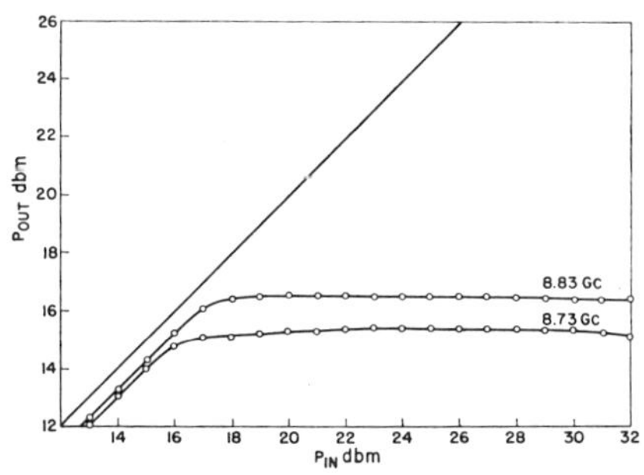

Fig. 1—Typical limiting characteristics obtained with a lithium ferrite extended coincidence limiter.

The dynamic range exceeded 25 db. However, heating effects caused the limiting action to disappear for average powers above 1 watt. We believe that the major effect is the change in the anisotropy field (the sample was oriented with H along [111]) resulting from the temperature sensitivity of the magnetization. This has the effect of detuning the filter and increasing the threshold.[7]

R. L. Comstock
M. Loewy
Bell Telephone Labs., Inc.
Murray Hill, N. J.

* Received September 20, 1963.

[1] R. W. DeGrasse, "Low-loss gyromagnetic coupling through single crystal garnets," *J. Appl. Phys.*, vol. 30, pp. 155S–156S; April, 1959.
[2] F. J. Sansalone and E. G. Spencer, "Low-level microwave power limiter," IRE Trans. on Microwave Theory and Techniques (*Correspondence*), vol. MTT-9, pp. 272–273; May, 1961.
[3] F. R. Arams, M. Grace, and S. Okwit, "Low-level garnet limiters," Proc. IRE, vol. 49, pp. 1308–1313; August, 1961.
[4] J. Clark and J. Brown, "A miniaturized ferrimagnetic high power coaxial duplexer-limiter," *J. Appl. Phys.*, vol. 33, pp. 1270–1271; March, 1962.
[5] F. C. Rossol, "Power limiting in the 4-kMc to 7-kMc frequency range using lithium ferrite," Proc. IRE (*Correspondence*), vol. 49, p. 1574; October, 1961.
[6] R. L. Comstock and W. A. Dean, "Extension of coincidence limiting frequency range in ferrimagnets," *J. Appl. Phys.*, vol. 34, pp. 1275–1277; April, 1963.
[7] H. Suhl, "Theory of ferromagnetic resonance at high signal powers," *J. Phys. Chem. Solids*, vol. 1 pp. 209–227; April, 1957.
[8] F. C. Rossol, "Microwave radiation hazards," Proc. IRE (*Correspondence*), vol. 49, p. 1574; October, 1961.

313

A High Power *X*-Band Limiter

The narrow ferrimagnetic linewidth exhibited by single crystal yttrium iron garnet forms the basis for several novel microwave devices. These may be divided into two main categories, namely, magnetically tunable filters and microwave limiters. Single crystal YIG spheres may be associated with additional microwave resonators to enhance one of these desired characteristics and to suppress the other.

In the present program, our task was to design a high power *X*-band limiter with the following characteristics:

Center frequency: 9600 Mc/s
Instantaneous bandwidth: 100 Mc/s
Insertion loss: 0.5 ± 0.2 dB for low-level signals
Power handling capability: 50-kW peak; no catastrophic failure up to 250-kW peak at a 0.0013 duty cycle
Power output: Shall be sufficiently low so that a diode switch can provide the additional protection required in a typical radar system (less than 100-watt peak).

A three-resonator configuration was chosen. The configuration originally proposed by DeGrasse, in a strip-line version, seemed the best starting point. The two outer cavities consist of conventional microwave resonators which are made a full wavelength long so that the YIG sphere may be placed at a convenient position of maximum RF magnetic field. The two cavities are oriented at right angles to each other, with the YIG sphere located in a coupling hole; it thereby provides the required mutual coupling when it is biased to gyromagnetic resonance.

The filter may be considered a three-cavity direct coupled configuration. In order to obtain a maximumly flat design, all the external Q's may be computed. In the case of the outer two resonators, a loaded Q of approximately 75 is required. For the middle resonator, which consists of the YIG sphere, the loaded Q was determined empirically by observing the overall band-pass filter characteristics on a microwave sweeper.

314

The relatively wide bandwidth requires a rather large YIG sphere. The problem of avoiding spurious responses is rather serious. Some improvement may be obtained by reducing the waveguide height of the auxiliary resonators to obtain a larger filling factor for a given size sphere. In our initial experiments, the auxiliary resonators were constructed of 0.100- by 0.900-inch ID waveguide. Subsequently, comparisons were made with 0.150- by 0.900-inch ID resonators. A 0.3-dB reduction in overall insertion loss was noted with the higher waveguide resonators.

The common wall between the crossed resonators was 0.010-inch. The diameter of the coupling hole was chosen about 1.5 times the sphere diameter.

A second important consideration con-

Manuscript received May 6, 1965; revised May 19, 1965. The work reported here was partially supported under Contract DA-28-043-AMC-00280(E) with the U. S. Army Electronics Labs., Fort Monmouth, N. J.

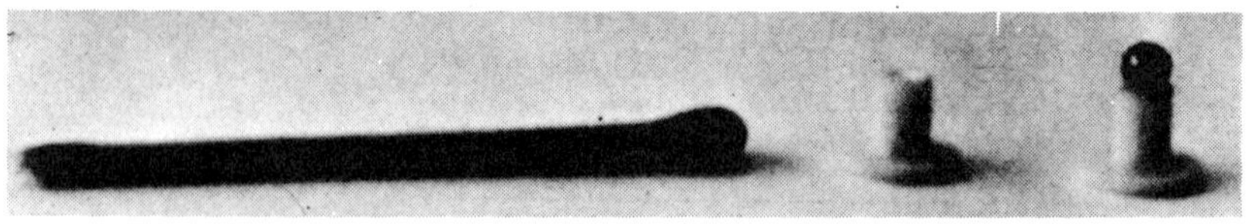

Fig. 1. Beryllium oxide supports for YIG sphere.

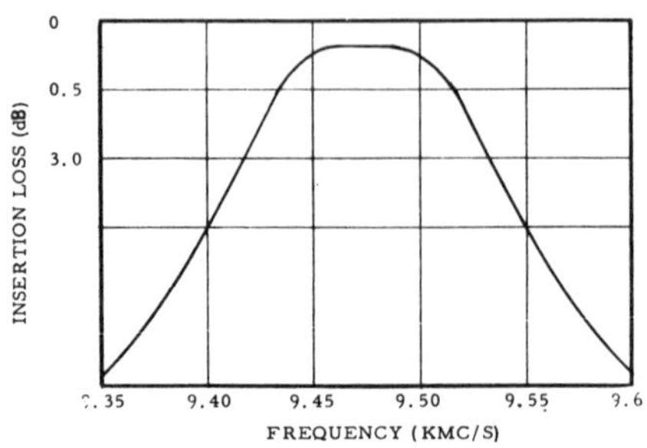

Fig. 2. Low power response curve.

Fig. 3. Prototype YIG limiter.

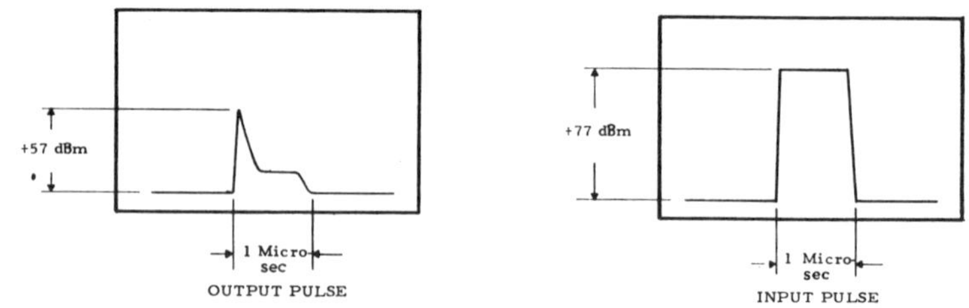

Fig. 4. Detected RF envelope (20-Mc/s bandwidth oscilloscope).

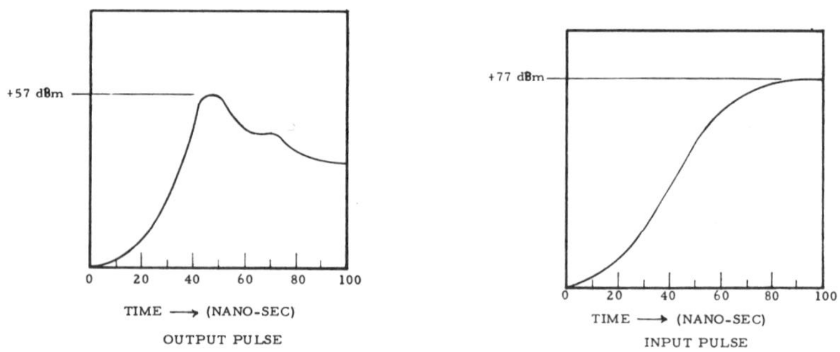

Fig. 5. Detected RF envelope (1000-Mc/s bandwidth oscilloscope).

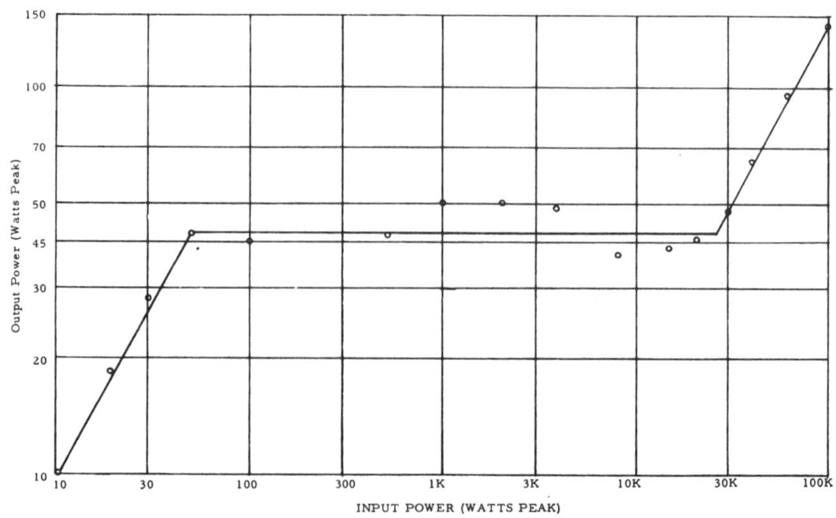

Fig. 6. YIG limiter performance.

cerns the heat transfer from the YIG sphere to the waveguide wall. The YIG sphere was supported on beryllium oxide posts equal to the full height of both resonators. These components are shown in Fig. 1.

A third important consideration concerns the quality of the yttrium iron crystal available. The resonance linewidth of typical crystals used in this program is less than 0.25 oersted when measured at 5000 Mc/s. This linewidth may be related to the unloaded Q of the YIG sphere and it assumes a value of approximately 6000. These measurements are based on a configuration where the sphere represents a slight perturbation of a microwave cavity, whereas in a limiter, a tightly coupled configuration is required. The effective Q of the YIG sphere is thereby reduced by a factor of 2. With a loaded Q of approximately 50 it contributes approximately 0.2-dB insertion loss.

A narrow linewidth lithium ferrite sphere was substituted for the YIG sphere It raised the overall device insertion loss to 3 dB, in agreement with its relatively broader resonance linewidth.

The overall insertion loss, with a YIG sphere 0.090-inch diameter, is shown in Fig. 2. It was 0.3 dB at the band center, and 0.5 dB at the edges of a 75-Mc/s band.

The assembled device, including the biasing magnet, is shown in Fig. 3.

Figure 4 shows a typical high power response curve. The lower trace represents a 50-kW peak pulse with one microsecond duration. The upper trace is taken under 100 times higher sensitivity. One may observe a short duration spike at the leading edge of the RF pulse with flat leakage following during the RF pulse interval.

Figure 5 shows the spike leakage measured on a sampling oscilloscope of 1000-Mc/s bandwidth. Due to the short duration of this spike, wider bandwidth is required to obtain a true picture of the spike leakage. The lower trace represents the leading edge of a 50-kW input pulse. The upper trace, displayed at 100 times higher sensitivity, represents the spike leakage through the device. Total horizontal sweep time is 100 ns. The peak amplitude of the spike is approximately 500 watts.

Figure 6 shows a graph of output peak power vs. input peak power. It may be seen that limiting occurs at approximately 50-watts peak. The slope of the output power curve is less than 2 dB for a 30-dB range of input power.

The leakage by the device has been reduced sufficiently to be handled by a semiconductor or switch to provide the total receiver protection required in a typical radar system.

This work therefore represents a substantial advancement in the state of the art of solid-state receiver protectors.

E. Wantuch[1]
Fairleigh Dickinson University
Teaneck, N. J.
M. King
Airtron Div.
Litton Precision Products, Inc.
Morris Plains, N. J.

[1] Formerly with Airtron Div., Litton Precision Products, Inc., Morris Plains, N. J.

Reprinted from IEEE TRANSACTIONS
ON *MICROWAVE THEORY AND TECHNIQUES*
Volume MTT-17, Number 4, April, 1969
pp. 231-232

X-Band Ferrite-Varactor Limiter

***Abstract*—A ferrite-varactor limiter combination operating at *X* band has been investigated. The ferrite limiter employs a combination of polycrystalline YIG for dynamic range and single-crystal YIG for reduced threshold. Various design techniques employing ridge waveguide and dielectric loaded ferrite limiters were studied.**

The ferrite-varactor package has a threshold of about 30 mW and a dynamic range up to 10 kW. The overall insertion loss is less than 1.0 dB in the frequency range of 8.9 to 9.5 GHz. The flat leakage is about 30 mW and the spike leakage energy is less than 0.3 erg at 10 kW of peak power input.

INTRODUCTION

The conventional method of achieving reliable receiver protection for sensitive microwave receivers is to use a limiter consisting of a plasma stage and a varactor stage. The plasma limiter is used to reduce the incident power to a level that can be safely limited by the varactor device. The major power incident on the varactor is the spike leakage pulse caused by the slow response time of the plasma device. The fast response time of the varactor makes it ideal for reducing the spike energy. The varactor cannot be used alone because of its limited power-handling capacity and a catastrophic failure mechanism.

A ferrite-varactor combination has several advantages over the plasma-varactor package: 1) there is no problem with gas cleanup; and 2) the recovery time is reduced from microseconds to nanoseconds at a peak power input of 10 kW.

A ferrite-varactor limiter having a low-level insertion loss less than 1.0 dB over the frequency range of 8.9 to 9.5 GHz and spike leakage less than 0.3 erg over a peak power range of 0 to 10 kW has been successfully designed. The recovery time is less than 20 ns over the same peak power range. The limiter package can be combined with a ferrite circulator to provide reliable solid-state *X*-band duplexers at peak power levels in excess of 500 kW.

DESIGN CONSIDERATIONS

The major effort was to develop a ferrite limiter compatible with the characteristics of existing varactor limiters. Ferrimagnetic limiters can be designed using nonlinear effects, which appear at high signal levels in ferrimagnetic materials. These nonlinear effects are subdivided into three categories:

1) subsidiary resonance;
2) premature decline of main resonance;
3) coincidence of subsidiary resonance with the main resonance.

Subsidiary resonance was selected as the most desirable for the following reasons:

Manuscript received September 18, 1968; revised November 25, 1968.

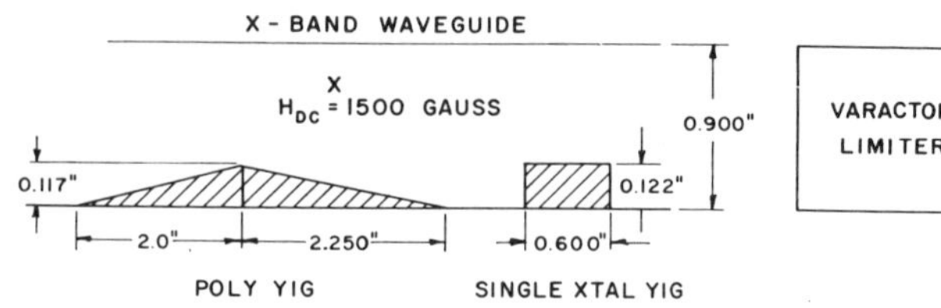

Fig. 1. Composite ferrite structure.

1) the subsidiary resonance effect does not have the frequency limitations associated with the other two processes; 2) the frequency bandwidth is wider for the subsidiary resonance process; and 3) although the other two effects have lower critical fields, the critical field for subsidiary resonance is still low enough to be compatible with the power-handling capacity of the second-stage varactor limiter.

Subsidiary resonance limiting manifests itself in the form of power absorption above a certain critical level of microwave field intensity with a magnetic biasing field substantially less than that required for main gyromagnetic resonance. The expression of interest for the RF magnetic field intensity in a subsidiary resonance limiter is given by the following equation [1]:

$$h_{\text{crit}} = \frac{2\omega(\Delta H_k)[(\omega - \omega_{\text{res}})^2 + (\gamma \Delta H)^2]^{1/2}}{\omega_m[(\omega/2) + \omega_H - \omega_m N_Z]} \qquad (1)$$

where

ω = signal frequency,
ΔH_k = spin-wave linewidth,
ω_{res} = resonant frequency,
γ = gyromagnetic ratio,
ΔH = uniform precessional linewidth,
$\omega_m = \gamma 4\pi M_s$: $4\pi M_s$—saturation magnetization,
$\omega_H = \gamma H$; H—dc magnetic field,

and

N_Z = demagnetizing factor.

To construct a ferrimagnetic limiter at a given frequency, the following have to be considered: a) selection of a ferrimagnetic material or materials that will result in a low threshold, e.g., minimize (1) for h_{crit}; and b) reduction of the threshold power by concentrating the microwave magnetic field in the ferrimagnetic material.

Step a) can be accomplished in a number of ways. One method used to reduce h_{crit} is to use a ferrimagnetic material that has a small value of ΔH_k and ΔH. Single-crystal YIG (yttrium iron garnet) material was chosen as part of the limiter, since it has the lowest ΔH_k and ΔH values and hence reduces the limiting threshold h_{crit}.

Uebele [2] investigated several techniques for accomplishing step b). He used ridge waveguide since ridge waveguide can be designed with much lower characteristic impedance than conventional rectangular waveguide. The RF magnetic field is concentrated in the volume of the ridge gap. Thus, the interaction between the ferrite material and the RF magnetic field can be enhanced by placing the ferrite in the ridge gap. A dielectric ferrite sandwich was also utilized to concentrate the RF magnetic field.

It was not possible to obtain satisfactory dynamic range and insertion loss utilizing these techniques in a single-stage limiter. A limiter design utilizing two different ferrimagnetic materials was also considered.

A single-crystal ferrite material was used to provide the lowest possible threshold power, and polycrystal ferrite material was then utilized to provide limiting at power levels exceeding the dynamic range of the single-crystal material. A two-stage structure that combined polycrystalline YIG and single-crystal YIG material, shown in Fig. 1, was investigated. The structure consists of tapered polycrystal ferrite YIG [3] followed by a rectangular slab of single-crystal YIG. The spacing between the two samples is not critical but is greater than a guide wavelength to prevent direct coupling between the samples.

EXPERIMENTAL RESULTS

The low-level insertion loss for the composite structure, varactor limiter and complete limiter, is plotted as a function of frequency in Fig. 2. The value of the dc biasing field for the ferrite structure was 1500 Oe. The high-level characteristics for the composite structure are shown in Fig. 3. Fig. 3 also contains a drawing of a typical limiter pulse for the ferrite limiter. The leakage pulse has two distinct sections. The first section, known as the spike, is a result of the finite time required to generate the spin waves that perform the limiting function. The flat section of the pulse is due to the attenuation of the pulse amplitude by the spin-wave absorption. The threshold power for the structure is 28 watts and the amplitude of the flat leakage does not exceed this value at a peak input power level of 10 kW. This is a dynamic range of at least 25 dB. Some limiting was also observed on the spike-leakage pulse, which reached a value of 2900 watts at a peak input power of 10 kW. These values of spike-leakage power and flat-leakage power were considered compatible with the varactor limiter to be used as the second stage of the package.

The varactor junction diode exhibits a voltage-dependent capacitance that may vary from a very small value below 10 pF to infinity, depending on the incident voltage. A power-dependent attenuator or limiter is achieved by employing this variable capacitance in a suitable microwave circuit. The

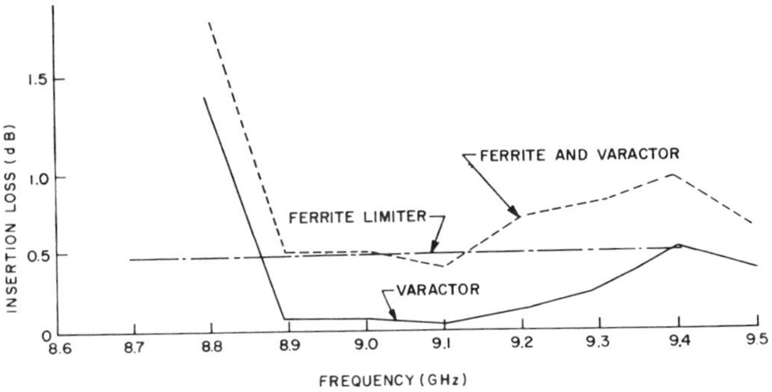

Fig. 2. Low-level performance.

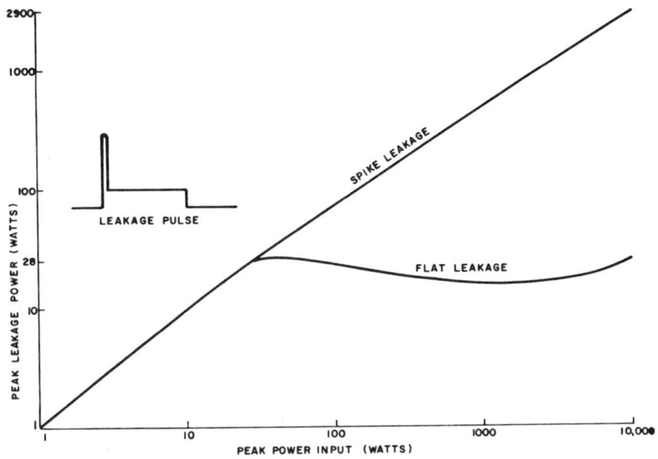

Fig. 3. High-level characteristics of composite ferrite structure.

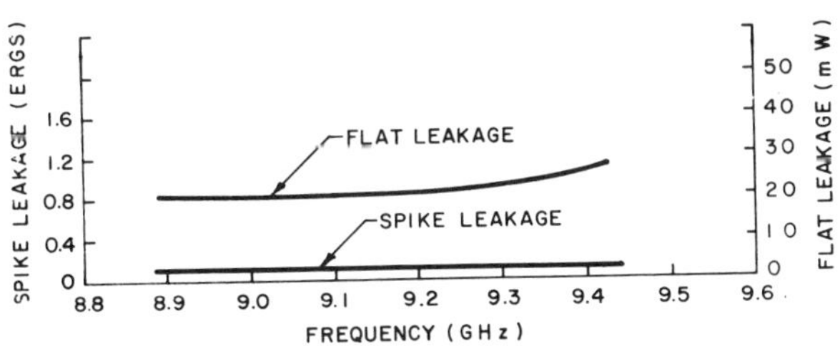

Fig. 4. Limiting characteristics of varactor-limiter. Power input—500 watts.

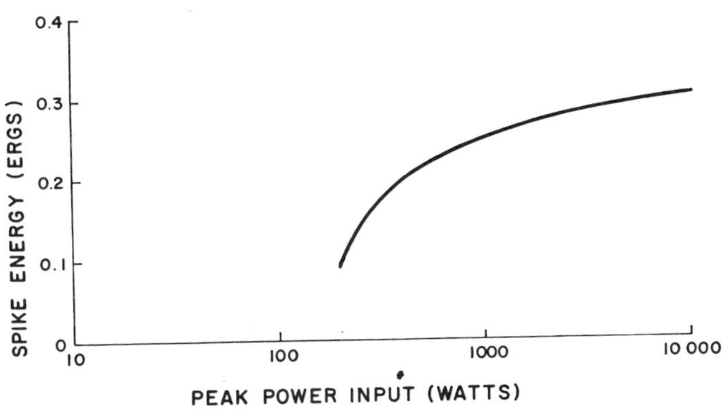

Fig. 5. Spike leakage of ferrite-varactor limiter as a function of peak input power.

varactor has a very fast reaction time; therefore, the spike-leakage energy is very small (<0.05 erg) and the recovery time is less than 10 ns.

It should be noted from Fig. 2 that the useful frequency band for the varactor limiter is 200 MHz less than that of the ferrite limiter. The limiting characteristics of the varactor are presented in Fig. 4, where the flat-leakage energy and the spike-leakage energy are plotted as a function of frequency for an incident power of 500 watts. This limiter had a threshold power of 30 mW and a maximum attenuation of 30 dB at a peak power of 300 watts.

The ferrite limiter and varactor limiter were then combined, as shown in Fig. 1. The ferrite limiter attenuates the 10 kW incident pulse to a spike value of 2.9 kW and a flat-leakage value of approximately 30 watts. The spike and flat portions of the leakage pulse through the ferrite limiter are attenuated 30 dB more by the varactor limiter.

The spike-leakage energy for the ferrite-varactor limiter is plotted in Fig. 5 for input power levels from 100 watts to 10 kW. The maximum spike-leakage energy is 0.3 erg for an incident power of 10 kW. The flat leakage power was not plotted, since it remained less than 1 mW at input power levels up to 10 kW. The recovery time of the limiter was less than 0.2 μs at input power levels of 10 kW.

Conclusions

The results of this investigation indicate that it is feasible to utilize ferrimagnetic and varactor limiter technologies to design a complete solid-state limiter for application in radar systems. The device that resulted from this investigation is satisfactory for incident peak powers of 10 kW and 100 watts average in the 9.0 GHz frequency range. This limiter can be combined with an appropriate ferrite circulator to achieve a complete duplexer for peak-power levels as high as 500 kW and average powers of 500 watts. The ferrite-varactor limiter, with none of the problems of a gas TR tube, has a life expectancy that approaches infinity. Shelf life is increased and handling problems are reduced. These advantages are not achieved without some compromise. The length of the prototype ferrite-varactor limiter is about twice the length of the TR limiter. The weight is also greater because of the biasing magnet required. It is believed the size and weight of the ferrite-varactor limiter can be reduced by improved packaging.

John L. Carter
Joseph W. McGowan
Electronic Components Lab.
U. S. Army Electronics Command
Fort Monmouth, N. J. 07703

References

[1] B. Lax and R. J. Button, *Microwave Ferrites and Ferrimagnetics.* New York: McGraw-Hill, 1961.
[2] G. S. Uebele, "Characteristics of ferrite microwave limiters," *IEEE Trans. Microwave Theory and Techniques,* vol. MTT-7, pp. 18–23, January 1959.
[3] W. F. Krupke, T. S. Hartwick, and M. T. Weiss, "Solid-state X-band power limiter," *IEEE Trans. Microwave Theory and Techniques,* vol. MTT-9, pp. 472–480, November 1961.

A Compact Subsidiary-Resonance Limiter

JAMES L. ALLEN

***Abstract*—A compact subsidiary-resonance limiter with a dynamic range of greater than 200 kW is described. The structure consists of a multilayer ferrite–dielectric "sandwich" placed on the narrow wall of a reduced-height reduced-width rectangular waveguide. Below-the-threshold insertion loss is less than 0.5 dB over the 5.4–5.9-GHz band, and the threshold level is about 20 W.**

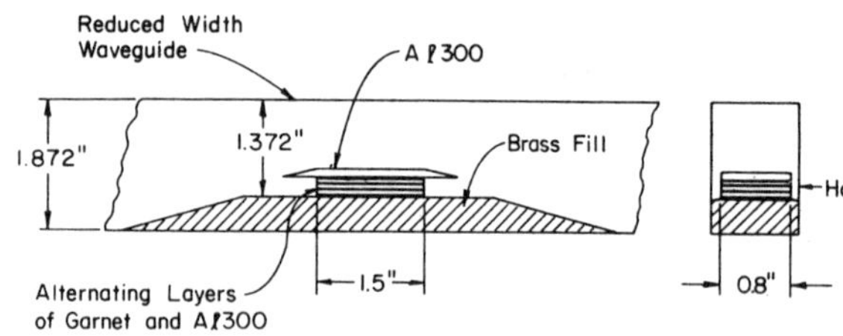

Fig. 1. Cross section of a multilayer limiter. The thicknesses of the layers, beginning with the garnet layer adjacent to the brass fill, are 0.030, 0.050, 0.030, 0.030, 0.070, 0.030, 0.030, and 0.125 in, respectively.

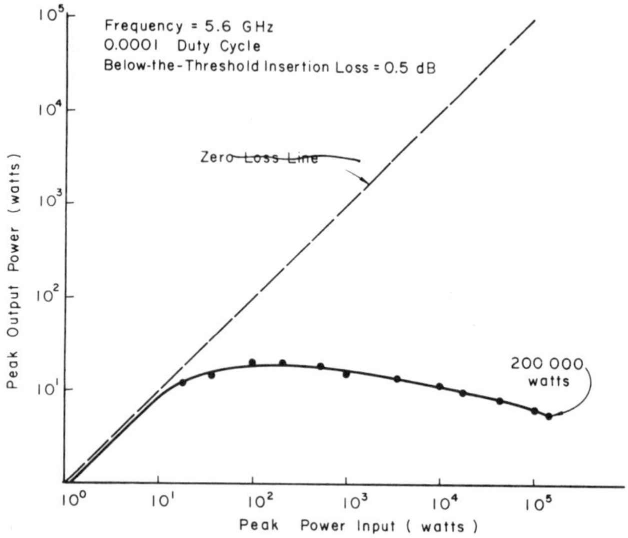

Fig. 2. High-power characteristics for the multilayer subsidiary-resonance limiter.

Subsidiary-resonance limiters have been studied extensively [1]–[4] for application in duplexing circuits. A recent paper by Carter *et al.* [5] describes a laminar limiter structure consisting of a cascade of rectangular rods of ferrite and dielectric along the narrow wall of a rectangular waveguide. The laminar structure has the distinct advantage that its threshold power level can be very low (2.8 W). The dynamic range of the laminar limiter is controlled by the number (and size) of the ferrite rods. Dynamic range of a conventional single-slab type of limiter is controlled by the length of the ferrite slab. In both cases obtaining wide dynamic range necessarily requires a relatively large structure with a long magnet. The purpose of this correspondence is to describe a multilayer loading technique that permits a wide dynamic range and a low threshold to be obtained with a physically small structure and a short magnet.

Cross-sectional views of a typical multilayer limiter structure are given in Fig. 1. This structure consists of four slabs of YIG interleaved with four slabs of Al-300 dielectric placed on the narrow wall of a reduced-height reduced-width rectangular waveguide. The ferrite and dielectric slabs are of various thicknesses selected experimentally to obtain maximum dynamic range. The ferrite slabs are only $1\frac{1}{2}$ in long. Thicknesses and other pertinent data are provided in the diagram. This multilayer limiter has a below-the-threshold insertion loss of less than 0.5 dB in the 5.4–5.9-GHz frequency band. The threshold power level is about 20 W and the dynamic range is greater than 200 kW. Characteristics of the limiter were essentially constant over the frequency range of the magnetron, 5.4–5.9 GHz. A plot of output versus input power is given in Fig. 2. Notice that the limit of the dynamic range has not been reached at a peak power input of 200 kW, which was the maximum power available from the magnetron. The finite excitation time of the nonlinear effect permits the leading edge of the input pulse to "leak through" essentially unattenuated. This is referred to as the spike leakage and its amplitude very nearly follows the zero loss line shown in Fig. 2. The width of the leakage spike

Manuscript received May 17, 1971; revised August 16, 1971. This work was supported by Sperry Microwave Electronics Division, Sperry Rand Corporation, Clearwater, Fla.

The author is with the Department of Electrical Engineering, Colorado State University, Fort Collins, Colo. 80521.

Fig. 3. Comparison of a multilayer limiter with a single-slab limiter of comparable dynamic range.

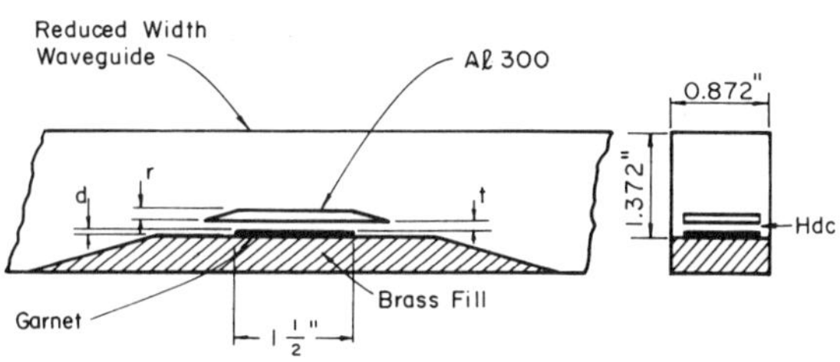

Fig. 4. Single-slab limiter.

was about 30 ns. A single-slab limiter with roughly the same peak power characteristic (40-W threshold, greater than 200-kW dynamic range) is compared with the multilayer limiter in Fig. 3. The length of the ferrite slab in the single-slab limiter is 7 in.

The multilayer limiter as described above is suitable for low average power operation. Higher average power capability can be achieved by using high thermal conductivity dielectrics, such as beryllium oxide or boron nitride, as the dielectric layers in the ferrite–dielectric "sandwich." Fins and liquid cooling channels may also be utilized to boost the average power capabilities.

For comparison with the multilayer configuration, the characteristics of the single-slab structure shown in Fig. 4 were measured for a large number of combinations of dielectric and ferrite thickness and interslab spacings. The length of the ferrite was the same as that in the multilayer structure. Some of the combinations had very low threshold levels (about 10 W) but none of the single-slab configurations exhibited a wide dynamic range. Best results were obtained with a YIG, Al-300 structure having $t=0.170$ in, $r=0.125$ in, and $d=0.40$ in. This combination had a threshold of 150 W and a dynamic range of 10 kW.

Although conclusive evidence is not available, the wide dynamic range of the multilayer structure is apparently obtained through control of the RF field intensity profile. The ferrite slabs nearest the center of the waveguide begin to limit for a lower input power level than those near the waveguide wall. As the input power is increased, the ferrite slabs nearest the waveguide axis reach a saturation value of limiting but the slabs further from the axis begin to limit maintaining the output power level approximately constant. This process continues until all slabs approach their saturation limiting value, at which time the output power begins to increase with additional increase in input power and the structure is considered to have reached the limits of its dynamic range. To maintain a constant output level the slab arrangement must be such that each additional slab begins to limit at just the right input power level to provide smooth transition.

In conclusion, a multilayer loading technique has been described for subsidiary-resonance limiters which permits a wide dynamic range and a low threshold power level to be obtained with a physically small structure and a short magnet.

References

[1] G. S. Uebele, "Characteristics of ferrite microwave limiters," *IRE Trans. Microwave Theory Tech.*, vol. MTT-7, pp. 18–23, Jan. 1959.
[2] J. Brown, "Ferromagnetic limiters," *Microwave J.*, vol. 4, pp. 74–97, Nov. 1961.
[3] W. F. Krupke, T. S. Hartwick, and M. T. Weiss, "Solid-state *X*-band power limiter," *IRE Trans. Microwave Theory Tech.*, vol. MTT-9, pp. 472–480, Nov. 1961.
[4] B. Lax and K. J. Button, *Microwave Ferrites and Ferrimagnetics*. New York: McGraw-Hill, 1962.
[5] J. L. Carter and J. W. McGowan, "A laminar subsidiary-resonance limiter," *IEEE Trans. Microwave Theory Tech.* (Corresp.), vol. MTT-18, pp. 652–654, Sept. 1970.